# THE DŽARYLGAČ SURVEY PROJECT

VOLUME 1

BLACK SEA STUDIES

14

THE DANISH NATIONAL RESEARCH FOUNDATION'S
CENTRE FOR BLACK SEA STUDIES
&
GRONINGEN INSTITUTE OF ARCHAEOLOGY,
UNIVERSITY OF GRONINGEN

# THE DŽARYLGAČ SURVEY PROJECT

## VOLUME I: Text

*Edited by*
*Pia Guldager Bilde, Peter Attema & Kristina Winther-Jacobsen*

Aarhus University Press |

THE DŽARYLGAČ SURVEY PROJECT, VOL. 1


Cover design Grafisk SIGNS
Photos by Pia Guldager Bilde and Kristina Winther-Jacobsen
Typeset and printed by Narayana Press, Gylling

ISBN 978 87 7934 521 8

Printed in Denmark 2012

Aarhus University Press
Langelandsgade 177
DK-8200 Aarhus N
www.unipress.dk

White Cross Mills
Hightown, Lancaster, LA1 4XS
United Kingdom
www.gazellebooks.co.uk

ISD
70 Enterprise Drive, Suite 2
Bristol, CT 06010
USA
www.isdistribution.com

The publication of this volume has been made possible by a generous grant from The Danish National Research Foundation, Aarhus University's Research Foundation, and the Augustinus Foundation.

# Contents

## CHAPTER 4

## CHAPTER 5

## CHAPTER 6



## CHAPTER 7

# Foreword

*P. Attema & P. Guldager Bilde*

The Džarylgač Survey Project (DSP) is a multidisciplinary archaeological survey project, which aimed at investigating the rural landscape in the hinterland of the ancient settlement of Panskoe I at both sides of the Džarylgač Lake, the Tarchankut Peninsula, Northwestern Crimea (Ukraine; Figs. 0.1-2).[1] The purpose of the study is to reach a better understanding of man's interaction with the Northwest Crimean landscape in a diachronic perspective from antiquity until early modern times. The project is a co-operation between the Danish National Research Foundation's the Centre for Black Sea Studies (DNRFCBSS) in Aarhus (Denmark), the Ukrainian National Academy for Scientific Research (Ukraine) in Simferopol' and the Groningen Institute of Archaeology (GIA) of the University of Groningen (the Netherlands). The methodologies applied are systematic field walking and collecting of surface finds supplemented with extensive site reconaissance, as well as geomagnetic and other surface measurements, GPS recording, and trial excavations at selected sites.

A first visit to the area with preliminary field-walking undertaken in 2006 (Attema, Guldager Bilde, Stolba) fully demonstrated the potential for systematic fieldwork in the region. In May 2007, the first campaign with large-scale survey was undertaken. This was followed by a second campaign in May 2008.

The Džarylgač Survey Project was initially planned as a three-year pilot project that might have led to a larger project on the Tarchankut Peninsula. However, due to disagreement with the Crimean Branch of the Institute of Archaeology, on the initiative of our Ukrainian collaborators, the project was regrettably discontinued after its second campaign. At the same time, the Ukrainian government planned to construct a large-scale wind power plant with more than 100 very large windmills on the inland slopes.[2] This work was scheduled to start in the autumn of 2008. According to the homepage of Nova-Eco, the company could only carry out the construction provided that "In compliance with the Ukrainian legislation, geological, archaeological and botanical researches were conducted on the selected routes, and also necessary agreements are executed".[3] The mentioned archaeological investigation was carried out by the Institute of Archaeology of the Simferopol' branch of the Ukrainian Academy of Sciences led by S.D. Koltuchov. Mainly kurgans were discovered during this work and the construction of the power plant on the hills S of Černomorskoe was therefore not problematized (Fig. 0.3). However, already in 2007 a number of well-preserved settlements were found by the DSP and many more were found in 2008 (see Chapter 4.4 and Appendix 1). These findings added urgency to the Džarylgač Survey Project, and resources were made available from our side that would ensure a thorough investigation of the area threatened by the wind-mill plant (Fig. 0.4). However, our work was much restricted by the Crimean authorities as represented by the Institute of Archaeology, Simferopol' and as mentioned above, on the Institute's initiative, the collaboration contract was discontinued, which implied a cancellation of first a major geological campaign in the fall of 2008 as well as the third scheduled survey campaign.

On 17 September 2008, the project of Nova-Eco LLC passed the evaluation by state experts and obtained a positive resolution of the Ukrainian State Construction Expertise (http://www.nova-eco.kiev.ua/static/14/). The current state of the project is unclear. Rumours have it that the project has been suspended. Some parts of Nova's homepage seem to suggest that the project has been moved to the Donuzlav Lake area. In the light of the immanent threat of a unique cultural landscape as testified by the results of the first two campaigns of the DSP, which calls for the need for a large-scale revision of the settlement history of Northwestern Crimea, we deplore the fact that international collaboration over this project was discontinued before it had reached its completion.

---

1 Guldager Bilde et al. 2007; Attema 2008; Attema, P. & T. de Haas 2008; Attema, P. et al. 2009; Attema, Guldager Bilde & Williamson 2010; Handberg 2011.

2 Http://www.Nova-eco.Kiev.Ua/static/20/ accessed 11 august 2009.

3 http://www.nova-eco.kiev.ua/static/13/ accessed 7 January 2009.

*Permanent staff*
Directors: Peter Attema (GIA); Pia Guldager Bilde (DNRFCBSS)
Ukrainian partner: Sergej B. Lancov
Tymon de Haas (GIS; GIA), Søren Handberg (pottery specialist; DNRFCBSS), Wieke de Neef (team leader; GIA), Tatiana Smekalova (geophysicist; DNRFCBSS), Vladimir Stolba (liason; DNRFCBSS), Christina Williamson (team leader; GIA), Kristina Winther-Jacobsen (liason, team leader Acting Director 2008; DNRFCBSS).

*Further participants*
Mariëlle Bannink (2008), Signe Barfoed (2007), Sophie Bengtsson (2008), Cecilie Brøns (2007-2008), Andrej Čudin (2007-2008), Tycho Derks (2007), Jan Delvigne (2008), Sanne Hoffmann (2008), Jakob Munk Højte (2008), Ann Sølvia Jacobsen (2008), Alexander Karjaka (2008), Tijm Lanjouw (2007), Marlies van Kruining (2007-2008), Maria Mawla (2008), Aisha Mellah (2008), Karina Bechmann Olesen (2008), Alexei Pasumanski (2008), Lars Rønje (2008), Sander Tiebackx (2007-2008), Marina Unger (2008), Kees van der Veer (2007-2008), Elly Weistra (2007).

In 2008, the Berlin-based company Eastern Atlas, directed by Cornelius Meyer and Burkart Ullrich and with participation by Lilli Keller, Rudolf Kniess, Dana Pilz, Ekkehard Schönherr, Henning Zöllner carried out geophysical prospection for the project. This part of the project was most generously financed by the "Excellenz Cluster" *Topoi* (Prof. Dr. Friederike Fless, Institut für klassische Archäologie, Freie Universität, Berlin and Prof. Dr. Brigitta Schütt, Institut für Geographische Wissenschaften, Physische Geographie, Freie Universität, Berlin). We are most grateful for this contribution, which made it possible to survey in detail part of the area destined to the construction of the before-mentioned wind mill plant.

The editors would like to thank all contributors as well as to acknowledge the effort of friends and colleagues, who assisted us in producing this volume. The inking of all artefact drawings was done by Martin Uildriks (Groningen Institute of Archaeology). Finally, it is a pleasure to record our gratitude to the three foundations, which have supported the publication of the present volume: The Danish National Research Foundation, Aarhus University's Research Foundation, and the Augustinus Foundation.

CHAPTER I

# Introduction

*Compiled by P. Guldager Bilde*
*with contributions by P. Attema & V. Stolba*

## 1.1 OVERVIEW AND RESEARCH GOALS | P. GULDAGER BILDE & P. ATTEMA

The Džarylgač Survey Project (DSP) is a multidisciplinary project, which aims at investigating the rural landscape at both sides of the Džarylgač Lake in the hinterland of the ancient settlement of Panskoe I. In antiquity as well as today, this region was located in the periphery of regional powers. Because of the much varied ethnic composition of the peoples settled or otherwise frequenting the area, past research has foremost been concerned with the question of the relation between settled and nomadic groups, and in line with much Soviet research, strong focus has been the economic capacity of the landscape.

DSP targeted the same questions, but in contrast to previous research, which is fractioned along chronological lines, the project has a holistic approach to the study of the region. This implies that we study the landscape in a long-term, diachronic perspective, because we believe that settlement patterns and land use prior to the Greeks settling the region as well as in post antique and (early) modern periods may help throwing light on antiquity too. In addition, the project combines the study of the productive and the funerary landscape alongside with that of the settlements themselves. Our study has resulted in a better understanding of the use and function of the Northwestern Crimean landscape as well as pastoralist and agricultural strategies enacted by the peoples using the land from antiquity until the present period. More specifically, we hope that the DSP data add to our understanding of the interaction between Greek settlers and indigenous peoples and more in general, to the dynamics of colonization.

## 1.2 ENVIRONMENTAL CONTEXT | P. GULDAGER BILDE & P. ATTEMA

The Džarylgač Survey zone comprises five distinct landscape zones that differ according to relief, soils, land use, archaeological visibility and known archaeological features (Fig. 3.4). These consist of (a) the coastal Lowland Ridge, where the majority of the previously known sites is located; (b) the Pediment S of Lake Džarylgač, where a few scatters of pottery were recorded in the past; (c) the Hillsides, and (d) the Plateaus/Uplands, where known archaeological features consist of kurgan burial mounds, cists, wells and deserted villages, and (e) the Vodopojnoe Ravine, where a number of deserted Tatarian villages are located and where several Bronze Age sites have previously been recorded. The region is dominated by the two lakes flanking on each side the small promontory upon which is located the settlement of Panskoe I, to the W Lake Sasyk (or Lake Panskoe), to the E Lake Džarylgač. Neither was a lake in antiquity but the outer end of a *balka* (ravine). Around 4,700 BP a sand barrier was formed cutting off the outflow of the *balkas* from the sea.

## 1.3 RESEARCH HISTORY | P. GULDAGER BILDE & V. STOLBA

The archaeological investigation of the Tarchankut Peninsula prior to the DSP has a long history, which goes back to the first half of the 19th century (Fig. 1.1).[4] Already in the 1830s, a small-scale excavation in the area of Ak-Mečet (post-WW

[4] Recent overviews in English: Vnukov 2001; Ščeglov 2002a, 11-25; Lantsov & Uzhentzev 2007.

II Černomorskoe) was conducted by A.I. Šmakov, who has also left a description of the ancient settlement associated in the later scholarship with Kalos Limen (Šmakov 1844). In an attempt to localize the Greek cities mentioned by ancient Greek sources, P.O. Buračkov (1815-1894) conducted regional investigation of the area in the 1870s, having produced a valuable description of the coast from the Bakal' Spit to the bay of Karadža and further E towards the Lake of Donuzlav and Eupatoria (Buračkov 1875). Yet, it is only from the first half of the 20th century with the investigations by L.A. Moiseev and P.N. Schulz that the West Crimean antiquities were addressed systematically and from a truly scientific point of view.

While Moiseev's works in 1916-1917 and in the late 1920s, later on taken over by M.A. Nalivkina, concerned mainly Kerkinitis and Kalos Limen (Ak-Mečet/Černomorskoe Settlement), Schulz, between 1933 and 1934, investigated the W coast of the Crimea from Karadža to the S end of Lake Sasyk-Šivaš (Šul'c 1937, 252-254; 1941, 265-277) and again in 1948, together with G.D. Smirnov, the W coast of the Crimea from the River Alma to Perekop as well as the S coast of Lake Donuzlav. Furthermore, in the beginning of the 1960s he investigated archaeological monuments of the N coast of Lake Donuzlav (Lancov 2004, fig. 2). These works remain by and large unpublished.

A new period in the investigations of the Northwestern Crimea is associated with the works of the Tarchankut Expedition directed in 1959-1986 by A.N. Ščeglov and since 1987 by V.F. Stolba (for a summary, see Ščeglov 1985; 2002a, 11-17, pl. 4).[5] The expedition was launched from the Chersonesos Museum, Sevastopol', in 1959 with the aim of investigating the Chersonesean chora and verifying Schulz's concept about the simultaneous appearance of the Greek and Scythian fortified settlements. In 1969 it was transferred to the Leningrad Branch of the Institute of Archaeology, Academy of Sciences of USSR (LOIA AN SSSR; now The Institute for the History of Material Culture, Russian Academy of Sciences, St Petersburg [IHMC RAS]). During the first decade of its functioning (1959-1968) the coast line of Western Crimea from the Kača River to the S to the Bakal' Spit to the N, more than 250 km along the coast and up to 10 km inland, was investigated by means of *razvedki*-type of survey, that is of a non-systematic, exploratory kind (Ščeglov 1985, 3). During the same years excavations took place at the multi-period fortified settlement of Tarpanči (1960-1966; Ščeglov 2002a, pl. 4.1.8-10) and the Early Hellenistic farmhouse at Vetrenaja Bay (1963-1966; Ščeglov 2002a, pl. 4.1.22). In addition, aerial photos were investigated as sources for the reconstruction of the ancient agricultural landscape of Western Crimea, first of all its regular divisions into plots (Ščeglov 1989).

From 1969 attention was turned to the settlement of Panskoe I (Ščeglov 2002a, pl. 4.1.25). Between 1969 and 1977 the large building U [=*usadba*, farm house] 6 was excavated (Panskoye I 2002) as well as U13 and U14. From 1979 the largest complex, U7, in the central part of the settlement was the subject of study. Two more building complexes, so-called U2 and U10, have been partially investigated at the E and SE periphery of the site in 1987-1992 and 1994 respectively. Alongside the works at the settlement, in 1969-1986 and 1994 a large part of its necropolis was excavated (Rogov 2011). A small scale investigation carried out in 1993 located the 'settlement' Panskoe V between Panskoe I and II (Randsborg 1994). In 1978 the remains of an isolated Early Hellenistic farmstead at the locality of Panskoe III was completely unearthed (Ščeglov 1987, 240, fig. 2.2; Ščeglov 2002a, pl. 4.1.23).

Between 1981 and 1987, a multi-period fortified farmhouse at Bol'šoj Kastel' was excavated (unpublished; Ščeglov 2002a, pl. 4.1.19) and between 1983-1984, remains located *ex situ* of a rural sanctuary were investigated at the cliffs of Džangul' (unpublished; Ščeglov 1988; Ščeglov 2002a, pl. 4.1.17).

Ščeglov's long-term study of Western Crimea has resulted in a number of written contributions. Most important are the two synthetic monographs *Polis i chora (Archeologičeskie pamjatniki Kryma)* from 1976 (a revised French edition appeared in 1992: *Polis et chora: Cité et territoire dans le Pont-Euxin*) and *Severo-Zapadnyj Krym v antičnuju epochu* from 1978. In addition is to be mentioned the well-documented publication of *Panskoye I. The Monumental Building U6*, which saw the light in 2002. Even though it is concerned with a single building complex, it gives a good overall introduction to the Tarchankut Peninsula in antiquity in English.

The 1960s and the early 1970s are characterized by a number of systematic excavations primarily carried out by the North Crimean Expedition of the Crimean branch of the Institute of Archaeology, Simferopol. This expedition was launched in 1961 as an outcome of the plans for the construction of the Severo-Krymskij Kanal (North Crimean Canal; Ščepinskij & Čerepanova 1969). This canal, completed in 1971, is the main waterway of Northern Crimea supplying Crimea with water from the Dnieper River. The first years (to and including 1964) were concentrated in the Perekop area far N of

[5] Please note that the site names attributed to the numbers 27-29 are not correct. They should be: 27: Mežvodnoe; 28: Vodopojnoe; 29: Burun Eli or Groty; 30: Vladimirovka/Masliny.

and thus not of interest for the DSP zone. Even though the main aim of the expedition was the investigation of kurgans in the inland area, since 1966 the *razvedki*-type of surveys were also focusing on the coastal area of the Razdol'noe and Černomorskoe districts including the area of the DSP, where a string of Hellenistic settlements was detected. The expedition was directed successively by P.N. Schulz (1961, 1964), E.V. Vejmarn (1962-1963), A.A. Ščepinskij (1965-1976), V.N. Korpusova (1977-1978), V.A. Kolotuchin (1981-1990) and S.G. Koltuchov (1992-1995).

In 1971 five kurgans near Vodopojnoe were excavated. They are dated to the second half of the 5th-beginning of the 4th century BC (unpublished), and in 1972 further four contemporary kurgans over Snežnoe (Ščepinskij 1973, 355) and four kurgans near Mežvodnoe were investigated. In 1987 a *razvedki*-type of survey along the coast between Bakal' and Mežvodnoe was carried out (unpublished; reference in Lancov 2004, 62; Lantsov & Uzhentzev 2007, 681). On that occasion 10 'settlements' situated at the coast and up to 5 km inland were located. They date to the 4th-3rd century BC, but in most cases material found was slight. Furthermore, in 1991-1992 nine barrows, which might have represented two different burial grounds, were excavated near the village of Dalekoe. Out of 29 attributed burials excavated in these kurgans, 25 date to the Early through Late Bronze Age (Pit Culture [2], Catacomb Culture [13], *Kultura Mnogovalikovoj Keramiki* [5], Timber Grave Culture [4]), the Scythian and the Late Medieval periods being represented only by three and one graves respectively (Koltuchov, Kislyj & Toščev 1994, 101-102). In the 1990s S.G. Koltuchov carried out investigations in the ravine E of Vodopojnoe with the aim of studying the prehistoric landscape of the region.

In addition to the work carried out by the North Crimean Expedition is to be mentioned the contributions of the Donuzlav Expedition of the Moscow Institute of Archaeology and the Eupatoria Museum headed by O.D. Daševskaja (1964 till present) with investigations of the Zapadno-Donuzlavskoe, Južno-Donuzlavskoe and Beljaus settlements on the S coast of Tarchankut and the excavations of the expedition of the Leningrad branch of the Institute of Archaeology and the Moscow State University under the directorship of A.N. Karasev in Čajka.

As far as the DSP area and the closely adjacent territory is concerned, attention must be drawn to the work of the Institute of Archaeology of the Ukrainian Academy of Sciences at the settlement of Mežvodnoe directed by T.N. Vysotskaja (1967-1968) and the Charkov University detachment of the North Crimean Expedition headed by V.A. Latyševa, which in 1972-1986 carried out excavations at the Hellenistic settlements of Masliny (Vladimirovka), Groty and Skalistoe.

Since the last campaign in 1994 at Panskoe I, archaeological work taking place in the Tarchankut has been primarily concentrated in the Černomorskoe Settlement (since 1988 carried out by the West-Crimean Expedition of the Simferopol Institute of Archaeology directed by V.A. Kutajsov), in Beljaus (Donuzlav Expedition of the Moscow Institute of Archaeology and Eupatoria Museum; director O.D. Daševskaja) and in Kulčuk (directed by A.S. Golencov; first as a detachment of the Moscow Institute of Archaeology expedition then as an independent expedition of the Institute for the History of Material Culture, St Petersburg; currently directed by S. Lancov), and in 2005-2006 excavations have been conducted in the Karadža Settlement. The only excavations taking place in the DSP study zone are the ongoing investigations of the Černomorskoe Settlement.

Three main themes have pervaded the scientific investigation and analysis of the region in previous research schemes: (i) the ethnicity of the inhabitants of the various settlements, (ii) the maximum extension of the Chersonesean chora, and (iii) the production capacity of the settled land.

## 1.4 DSP'S POSITION IN REGIONAL SURVEY | P. GULDAGER BILDE & P. ATTEMA

### 1.4.1 Black Sea region

As a particular discipline of archaeological fieldwork in eastern as well as western scholarship, regional investigations (in Russian *razvedki*) were of a non-systematic, exploratory nature. Usually work was (and is) done on foot and assumed ancient locations were visited and sampled. This method was frequently combined with gathering information from local informers (shepherds, landowners, villagers with special knowledge of local affairs, such as bar- or restaurant owners), and the local intelligence was then tested in the field. The result was the coverage of large tracts of land, but because 'sites' were mostly visited and sampled based on the knowledge that there either were ancient remains or because the landscape type adduced the researchers to believe that based on analogy there would be ancient remains, this type of survey tended to confirm already existing knowledge.

In the West, over the past 40 years, survey methodology has developed into a highly specialized field method in its own right (Barker and Mattingly (eds.) 1999-2000; David & Thomas 2008; see below, Section 1.4.2). Over the years, in the Mediterranean landscapes, of especially Greece and Italy, the need was felt to map regional landscapes in ever more detail. This was given by a combination of heritage concerns as the archeological record appeared to be rapidly destroyed through increasing urbanization and mechanical farming and of academic interest in the potential of the surface record as a means *in se* for detailed reconstruction of past settlement and land use histories.

In Soviet archaeology surface investigations have remained above all focused on highly visible features in the landscape (e.g. kurgans, remains of buildings, or land divisions) and the tradition of the *razvedki*-type of survey has continued somewhat longer in the Black Sea region than in the West. There are two main reasons for this. One is that, as an effect of the Iron Curtain, there has been little scholarly exchange concerning the development of the discipline, wherefore eastern and western scholarship has developed in two different directions during most of the 20th century. Another reason is the notorious lack of (detailed) topographical maps, which until recently were kept as military secrets. This has made topographical studies and not least their publication exceedingly difficult (Guldager Bilde & Stolba 2006, 8).

The *razvedki*-type of survey in the Northern Pontic region has produced a number of interpretative historical maps featuring dots marking 'sites'. In the Tarchankut region we may mention Kolotuchin 1996, fig. 1; 2003, fig. 1 concerning the Bronze Age; Ščeglov 2002a, pl. 4 concerning the Hellenistic period (Fig. 1.1); Lancov 2004, fig. 1 on sites in the Chersonesean chora of the 4th-3rd century BC. None of the mentioned maps contain any detailed topographic information. In fact, in many cases it is impossible to determine the more precise location of the individual dot. This is problematic, because many of these 'sites' have never been published further than as the dot on the map. In fact, before the 1990s not a single 'site' found in a survey and not subjected to following excavation has been published with full documentation.

Moreover, dots on maps are commonly taken as persuasive testimonies of history, but they are foremost a mapping of research activity. And if this activity interacts with existing knowledge it tends to confirm established ideas or research stereotypes. Concerning the Tarchankut earlier regional investigations took as point of departure the shore line of the sea and of the lakes. This led to the discovery of ancient artifacts, especially Greek, at a number of localities and this occasioned the historical conclusion that the Greeks settled along the coast and the lakes, whereas inland areas were the arena of nomadic groups, who left their burial mounds as markers of their territory (e.g. Ščeglov 1978, 32 fig. 8; 2002 pl. 4 = Fig. 1.1). As we shall see, the results of the DSP systematic, intensive survey has proven that this historical interpretation needs serious reconsideration.

However, things are changing. Historical and contemporary maps are now publicly available along with aerial and space photos, and the early 1990s saw an intensification of East-West scholarly cooperation. Systematic, intensive field survey is now occasionally practiced in the Black Sea region mainly as part of collaborative projects. To be mentioned is the Ukrainian-Polish investigation of Nymphaion's *chora* made in 1993-1997 (Scholl & Zinko 1999; Zin'ko 2006), O. Doonan's survey around Sinope carried out between 1996 and 1999 (Doonan 2004; 2006), as well as the survey made by a French-Russian team on the Taman' Peninsula in 1997 through 1999 (Müller et al. 1998; 1999; 2000; Bolikhovskaja et al. 2002; Gorlov et al. 2002). In addition to the above-mentioned surveys with a primary focus on the Greek period, the British-Bulgarian survey around the Roman city Nikopolis ad Istrum carried out between 1999 and 2001 (Poulter 2000) as well as the German-Bulgarian investigation of the territory in the hinterland between the Roman forts in Iatrus and Novae which took place from 1997 to 2003 are to be mentioned too (Conrad 2006).

We strongly recommend that modern landscape archaeology become much more widely used in the Pontic region, because it is an invaluable method to recover prehistoric and historic data in a landscape that is currently threatened and rapidly being destroyed by natural and anthropogenic factors such as coastal erosion,[6] illicit digging as well as by urban, infrastructural and agricultural expansion. Further initiatives on the recording and preservation of the unique archaeological landscape of Northwestern Crimea are therefore badly needed.

### 1.4.2 The Mediterranean

In Mediterranean landscape archaeology the awareness of a rich but rapidly vanishing archaeological surface record has led to increasingly sophisticated field methods and data analysis ever since Sir Ward Perkins' famous South Etruria surveys

---

6 Alone in the DSP study zone, the sites of the Černomorskoe Settlement, Masliny, Vetrenaja Bay farm, and Skalistoe 5 are partly eroded away.

during the 1950s to the 70s.[7] Since about 1980, however, the initially still extensive surveys, focusing above all on obtrusive remains in the landscape such as necropoleis and Roman villa remains, have gradually given way to ever more intensive and systematic surveys aimed at recording all past activity in the landscape. Especially over the last decade significant advancements have been made in both field methodology and data analysis, which the DSP has sought to introduce in the archaeology of the Northern Black Sea coast.[8] Below, we mention some relevant aspects:

As to field methods we note:

(a) *The ever increasing intensity of systematic pedestrian artefact survey of ploughed agricultural terrain in contiguous gridded fields.*

In the DSP coverage of surveyed units was aimed at 20%. Such highly intensive surveys are especially useful to detect settlement patterning on the local scale and aim at recording the smallest of sites. This type of survey was deemed suitable for the cultivated zones within the DSP survey area consisting of large stretches of evenly ploughed fields (see Chapter 2.3.1).

(b) *A growing attention for zones with low ground visibility covered in vegetation and for the survey of which effective field strategies still have to be developed.*

In the DSP the Hillsides with their steppe vegetation posed this challenge. In Chapter 2.3.1.2 the methodology of the systematic survey of the Hillsides is described.

(c) *The implementation of handheld computers equipped with global positioning devices for fast and spatially accurate recording.*

In the DSP the efficiency of working with digital recording in the field for storage in a Geographical Information System was further improved (see Chapter 2.5.1).

(d) *The incorporation of remote sensing techniques as an integrated tool in the field survey.*

In the DSP the focus was on geomagnetic survey which was applied to known archaeological sites but was also successfully used to detect and map off-site features (e.g. animal pens), as discussed in Chapter 2.3.3.

(e) *The incorporation of landscape analysis.*

In the DSP this took the form of a landscape classification to be used as a tool to explain variations in settlement patterning and to comment on land use potential, as discussed in Chapter 2.3.4.1.

As to data analysis we note:

(a) *The attention for biases in the surface archaeological record caused by, for instance, surface visibility, post-depositional processes, and methods of collecting finds.*[9]

In the DSP surface visibility was systematically recorded, while a number of sites were augered to establish whether subsurface strata had been preserved. Surveyors were instructed to, in principle, collect all surface material found on their transects (see Chapter 2.3).

(b) *The establishing of criteria for the classification of surface sites as an interpretive tool.*[10]

In the DSP all sites were classified and related to the overall landscape classification (see Chapter 6.1).

(c) *An increasing interest in chrono-typological and functional analysis of survey pottery assemblages of all periods including coarsewares.*[11]

In the DSP these aspects are evident from the discussions in Chapter 5.3.

It is our feeling that with the DSP we have demonstrated the potential of the application to the Crimean landscape of survey methodologies developed in the Mediterranean. At the same time it offers a possibility for comparison between settlement development in landscapes around the Mediterranean and the Black Sea. So far, limited exchange of methodological insights between landscape archaeologists working in the West and the East has implied that sampling techniques

---

[7] Then Director of the British School at Rome, Ward Perkins undertook these surveys to record the archaeological landscape around Rome when deep ploughing destroyed the archaeological heritage of the countryside bringing vast quantities of fresh archaeological material to the surface (Barker & Loyd (eds.) 1991, Introduction).

[8] Over the years, an ample bibliography on survey methodology and data analysis in the Mediterranean has come into existence.

[9] Issues relating to biases in field survey and analysis are discussed in Van Leusen 2002, chapter 4.

[10] A recent example of site classification of a regional survey in Italy is found in Van Leusen et al. 2005.

[11] Winther-Jacobsen 2010a and b.

and the strategies behind them differ to an extent that comparison of data between Mediterranean and Black Sea landscapes is in the present situation virtually impossible.[12] The DSP is unique in the sense that the same field methodology was applied both in South and Central Italy and in Northwestern Crimea, allowing direct comparison of finds distributions and site densities and of trends in the development of settlement organization (Attema, Guldager Bilde & Williamson 2010). Conferences drawing together landscape archaeologists working in the Mediterranean and in the Black Sea area are therefore necessary fora in light of exchange of methodological insights (Guldager Bilde & Stolba 2006).

Finally, it needs to be stressed here that the application of the instrumentation of modern landscape archaeology to the Northwestern Crimean landscape has been exceptionally felicitous because the archaeological record of the Northwestern Crimea is much better preserved compared to many Mediterranean landscapes. Intensive field walking in combination with non-invasive investigations such as applied in the DSP have led to very rapid and exceptional results as shown in this publication. A thorough and rapid documentation of the unique archaeological landscape of Northwestern Crimea and subsequent initiatives for its preservation we highly recommend to anticipate future developments in the field of mechanized agriculture, coastal tourism, and industry. Also this is something to be learned from many Mediterranean landscapes. It is the project's fundamental belief that the best possible results are only gained through international collaboration, a multi- and interdisciplinary approach and in a constructive relationship with local political and administrative authorities.

The data of the DSP are permanently deposited on the web hosted by *Data Archiving and Networked Services* (DANS EASY), The Hague, The Netherlands. Permanent web identifier: http://persistent-identifier.nl/?identifier=urn:nbn:nl:ui:13-0mg-hmc. The access to the data requires a password. Instructions are available on the web page.

---

12 Comparison of survey results is at the heart of the current debate in Mediterranean survey archaeology. See for instance the contributions in Alcock and Cherry (eds.) 2004.

CHAPTER 2

# Field methodology, recording and analytical procedures

*Compiled by P. Attema*
*with contributions by P. Guldager Bilde, T. de Haas, S. Handberg, C. Meyer, W. de Neef, D. Pilz, T. Smekalova, V. Stolba, B. Ullrich, C. Williamson & K. Winther-Jacobsen*

## 2.1 INTRODUCTION | P. ATTEMA

In this chapter, the various methods are discussed that were employed in the course of the Džarylgač Survey Project. Field methods ranged from intensive artefact survey in ploughed fields, aimed at documenting site and off-site patterns, to the systematic survey of large tracts of overgrown steppe landscape, recording any visible cultural features, and from extensive survey with diagnostic pottery sampling at selected sites, building reference collections for the classification of the survey pottery, to trial trenching, in order to better understand the function and chronology of sites. Equally important in the field methodology were the geophysical surveys to investigate possible subsurface features in both known archaeological sites and outside these.

In addition, environmental surveys were aimed at establishing a landscape classification based on geomorphologic characteristics and at a description of the soils to provide the landscape context for the archaeological data. Besides the surveys themselves, the various recording and analytical procedures connected with the field methodology are also discussed in this chapter. These procedures concern, for example, the way in which field data and artefacts were processed and recorded in the database and in a GIS to create the various artefact distribution maps underlying our interpretations and analyses. Finally, the way in which sites were designated as such and were classified is discussed.

The base for determining the extent of the research area was a wide diversity of cartographic materials, from nineteenth century maps to aerial photography and satellite images of Google Earth. Furthermore, these were an invaluable source of information, not only for orientation in the field, but especially for spatial analyses and understanding the changes that have taken place in the landscape through time. This chapter therefore begins with a detailed discussion of these sources before continuing with the methodology employed in the project for fieldwork and data recording, processing and analysis.

The DSP area of study comprises ca. 450 km$^2$, in which a 16 km wide N-S transect of ca. 200 km$^2$ was delineated for more intensive research. The wider area of study encompasses part of the central Uplands of the Tarchankut Peninsula and part of the coastal zone N of it, between Černomorskoe and Vodopojnoe, while the transect centres on Lake Džarylgač (Fig. 2.1).[13] The transect cross-cuts various types of landscape that are characteristic of the wider area of study and as such may be taken as representative for the archaeology found in each land type. From S to N and from high (170 m a.s.l.) to low (immediately above sea level) these land types consist of Plateaus/Uplands, Hillsides, Pediment, the Coastal Lake and Lowland Ridge. Chapter 3.3 gives a detailed description of the land types in the transect.

Fundamental to the present discussion of the survey methodologies used by the DSP team are the two contrasting forms of land use within the transect and their related potential for the detection of surface archaeology. Whereas the lower parts of the landscape (the Pediment and Lowland Ridge) are regularly under the plough, the higher areas (Hillsides, Plateaus/Uplands) are for the larger part covered in steppe vegetation. Fig. 2.1 shows how the transect can be divided into two main 'visibility zones'. The lower area is parcelled out into vast agricultural fields with a very high overall visibility when ploughed; ploughed fields, once gridded into uniform units, are exceptionally fit for standard intensive field walking and artefact collection. The higher area, however, is largely grassland. This called for a different design of a field

[13] See Chapter 1 for the motivation behind the selection area of study.

walking method suited to deal with low ground visibility and the lack of agricultural plots as survey units. The methods employed in the two visibility zones are described in Sections 2.3.1.1 (the block survey in the Pediment and Lowland Ridge) and 2.3.1.2 (the slope survey) respectively.

In order to better comprehend the cultural landscape of the wider area of study, site-oriented, extensive investigations were carried out at a number of previously known sites (and their vicinity) and kurgans both within and outside the transect. The methods employed in the site-oriented surveys are described in Section 2.3.1.3.

A number of the kurgans were targeted for more intensive three-dimensional mapping using the Total Station (see Chapter 6.5.2 and Hill 19 in the site catalogue). The aim of this was to produce a digital terrain model which can be used for more detailed analyses within the context of wider research on the funerary landscape of the NW Crimea. This approach is discussed in Section 2.3.2.

Selected areas of archaeological interest were subjected to geomagnetic survey to obtain indications of possibly preserved subsurface archaeological features. This was conducted by two teams, whose methods are discussed in Section 2.3.3.1. Manual augering was carried out at specific points across the transect to establish the nature and thickness of the soil on top of the limestone bedrock, and at specific sites to discern the presence of cultural layers below the plough zone. The methods used for this are discussed in Section 2.3.4. Certain sites were targeted for more detailed subsurface investigation through the excavation of trial trenches, in order to gain better insight into the nature and chronology of specific features discovered through the other field methods. The method and approach used for this is discussed in Section 2.3.5.

Besides the surveys themselves, the various recording and analytical procedures connected with the field methodology are also discussed in this chapter. These procedures concern, for example, the way in which field data and archaeological finds were processed and recorded in the database; this is discussed in Sections 2.4 and 2.5. Underlying the various approaches to the survey was the use of a Geographic Information System (GIS). Besides mapping finds, GIS was used together with cartographic data to target areas suitable for survey, and to prepare the groundwork for field walking ahead of time. Using this data on handheld computers (or PDA, for Personal Digital Assistant) with Global Positioning Systems (GPS) receivers made systematic survey much more efficient in the seemingly endless fields. The spatial data was linked to a database containing detailed information from each unit (on finds, field conditions etc.) allowing for specific distribution maps and further spatial analyses. This is all discussed in Section 2.5.

Finally, the distribution maps and the density of archaeological material and features have together led to the identification of sites. How these were determined and further classified is discussed in Section 2.6.

## 2.2 CARTOGRAPHIC BASE OF THE DSP | P. GULDAGER BILDE & T. SMEKALOVA

Historical maps and aerial photos of the Crimea, which are preserved in Russian archives, have formed an invaluable source in the archaeological and environmental investigations of the DSP study area in the context of the wider region of Northwestern Crimea. Below, we will discuss the topographical maps that were used in the project. The use of Google Earth as a tool within the project will then be discussed.

### 2.2.1 Historical topographical maps | T. Smekalova

Fig. 2.2 shows a map dating from 1837 that is based on the topographic surveys of Colonel Betev as well as on triangulation measurements of the Crimea that were carried out in 1836-1938 by the Corps of Military Topographers under the direction of F.F. Schubert and D.D. Oberg. The map clearly shows the relief of various landscape features, major and minor roads and the numerous settlements present in the landscape. It also shows, for the first time, water sources and a substantial number of kurgans. The proportions of the peninsula practically coincide with those of modern maps.

On the basis of this map, a further reconnaissance of Crimea was conducted in 1865 as part of a large-scale mapping carried out between 1845-1870. During these years, Poland and 22 provinces of Central and Southern Russia, including the Tauric Province, were surveyed in the scale of one-verst (1 *verst* is 1.067 km). The resulting map depicts a larger number of roads, settlements, kurgans and wells than the previous map (emendations in red). Also the ethnic composition of the individual towns and villages are corrected.

In the early 1870s, ordnance surveys almost completely ceased in the internal provinces of Russia, but fortunately were continued in the border regions, including the Crimea. At this time they were conducted on the half-verst scale (1:21,000) (Fig. 2.3). In the half-verst military topographic map of the Crimea issued in the late 19th-early 20th century, a vast number of kurgans is marked by out-of-scale symbols showing height specifications for the largest mounds. This specific attention for the kurgans was due to their value as strategic military points. However, water sources and wells were also recorded in detail, as well as roads and paths, including those that were no longer in use. On this map the relief of landscape features is rendered in contour lines at an interval of 2 *sažens* (about 4.3 m).

The latter map of the Crimea was extensively used in the DSP as it is an exceptionally rich source for the study of the spatial patterning of Early Modern settlements and ancient kurgans over the landscape and it can be used to detect traces of ancient land division and to locate the presence of ancient ditches and ramparts (Fig. 2.4). As the proportions of this map correspond exactly to modern cartographic renderings of Crimea, the features identified in this map can be transferred with relative ease onto new maps or air and space photographs facilitating comparison of features appearing on either of these sources. Used was also a map made in the mid-20th century at a scale of 1:25,000 and with contour lines drawn at an interval of 5 m (Fig. 2.5) as well as a more recent map at the scale of 1:100,000 (Fig. 2.6). The former map was revised in the 1980s on the basis of air photos.

## 2.2.2 Aerial photography | T. Smekalova

Besides historical topographical maps, aerial photos were also valuable sources of information in the DSP for revealing particular classes of archaeological objects, especially for linear structures such as land divisions and traces left by land usage. Aerial photos taken in three different years were used for analyses: 1956, 1971, and 1978. The separate images were adjusted and combined using Adobe Photoshop. This made it possible to detect long features that continued across multiple photos, mostly traces of boundaries of old and ancient fields. Individual architectural features, such as ancient houses, turned out to be better visible on Google Earth than on the aerial photographs (see Section 2.2.3).

## 2.2.3 Google Earth | P. Guldager Bilde

In the course of the DSP, Google Earth satellite images have been used for various purposes:

- As landscape images
- To help orientate and to select fields for survey
- To detect or verify landscape features
- To detect man-made features (buildings, kurgans, roads) (Figs. 4.148, 160, 167, 234 and 6.16, 21-23, and others)

Google Earth is a proprietary virtual globe program, which renders the earth by series of superimposed images obtained from satellite imagery, aerial photography and GIS 3D globe. In mid-2006 it was released to the public as a free tool. Since its release the resolution has been gradually improving. Most land is now covered in at least 15 m resolution. The images are not all taken at the same time, but are generally current to within three years. The internal coordinate system of Google Earth is geographic coordinates (latitude/longitude) on the World Geodetic System of 1984 (WGS84) datum. The program is a forceful research tool for any research dealing with spatial data – but equally so for persons engaged in illicit excavation activities.

During the time of the fieldwork, the Tarchankut Peninsula was covered by images of much varying resolution and the full map is based on a patchwork of images made during a number of different passes primarily in summer time and early autumn. Until recently, the very tip of the peninsula including the town of Karadža as well as the strip between latitude 32.4742 and 32.5700 covering the central part of the DSP study zone was published in a rather low resolution, whereas the section between the two mentioned strips are covered by several high resolution maps (August 2004 and October 2004 respectively). Immediately E of Lake Džarylgač is a narrow strip photographed in June 2003 in high resolution. Further E of this is a broad sector in high resolution dated in August 2007. Over the winter 2009/2010, the maps of the DSP study zone were finally updated with high resolution images shot 8 April 2009.

For a close study of the area it is necessary to employ Google Earth's feature 'historical maps', because it enables one to study the same section of the landscape through maps taken at different times. Only thus does one catch the largest number of individual features.

## 2.3 FIELD METHODS | P. ATTEMA

### 2.3.1 Field walking | P. Attema, P. Guldager Bilde, T. de Haas, W. de Neef, C. Williamson & K. Winther-Jacobsen

2.3.1.1 Systematic survey of the Pediment and Lowland Ridge | T. de Haas & P. Attema
The land relief in the N part of the survey area (in Chapter 3 referred to as the Pediment on the S side of Lake Džarylgač) and the Lowland Ridge (on the N side of Lake Džarylgač) is gently undulating with relatively deep soils, making it suitable for cultivation. This area is divided into extensive arable plots that are ploughed in a rotating system (see Chapter 3.5 on land use). Except for a number of previously mapped kurgans, the partially excavated settlement of Panskoe, part of the Hellenistic cadastre system, and the large Bronze Age site Skalistoe 2 (see Chapter 1.3), few archaeological sites had hitherto been recorded in these extensive fields.

In May of 2006 a pilot survey was conducted (by Attema, Guldager Bilde and Stolba) around Lake Džarylgač to gain an indication of the nature of the presence and distribution of archaeological material and to determine the suitability of a standard survey approach in the cultivated areas of Tarchankut Peninsula. Four ploughed fields were extensively surveyed and were shown to yield finds in very discrete scatters (see Chapter 4.4.1 and 4.4.5). The good and regular surface visibility on the extensive ploughed fields around the lake was therefore shown to offer an excellent opportunity to map human activity in this landscape on the smallest of scales, supporting one of the primary aims of the DSP: to investigate the archaeology of the area around the lake and more specifically to discover traces of rural occupation during the Hellenistic period (see Chapter 1.1). To this end an off-site survey methodology known as the *block survey*, which has been devised by the Groningen Institute of Archaeology (GIA) for ploughed fields, was employed and further developed.

2.3.1.1.1 The block survey
The block survey was developed by the GIA from the late 1990s onwards for intensive field surveys in Central and South Italy.[14] In a block survey, agricultural fields – traditionally the most important survey unit – are further divided into a grid of regular units, in principle 50 by 50 m. Each of these units is traversed by five walker transects, 10 m apart, from which all archaeological artefacts are collected. Assuming a total pickup of artefacts from a 2 m wide swath, this would result in a 20% sampling coverage of each unit. Data recording on paper unit forms is supplemented by digital recording of spatial information using handheld computers (PDAs). Intensive block surveys thus provide systematic and detailed information on the distribution of archaeological artefacts over the ploughed areas of the landscape (e.g. Figs. 4.44-47).

The use of newly acquired GPS and PDA mapping software helped us to further optimize the survey efficiency with a pre-established grid of survey units (see Section 2.5.1 below on the application of GIS software). Each field team consisted of 5 to 7 people: 5 or 6 transect walkers, one of whom was the team leader responsible for the paper administration, and a PDA operator, who marked the grids in advance of the surveyors and measured in both natural and archaeological features (Fig. 2.7). The general lack of orientation in these vast fields made it important to flag the corners of each unit. This was done with small barbecue sticks with fluttering red-and-white plastic tape, which had been prepared in large numbers beforehand. Grids were set out per field in units of 50 or 60 m across to keep the coverage as close to 20% as possible depending on the expected number of field walkers, and were usually square.

To expedite the fieldwork, two teams of field walkers were deployed as long as conditions allowed. In the large fields, the teams typically began at opposite ends and worked towards each other, going back and forth across the shorter sides of the fields. Separate numbering ranges for the units were used by each team to facilitate record keeping. In 2007, one team used numbers beginning with 1000, while the other used 2000 numbers; in 2008 the series of 6000, 7000, and 8000 were used for the block survey units. The 3000 series was used for the samples taken outside the block surveys from the entire campaign in both years (see below).

14 This method was developed in the course of the *Regional Pathways to Complexity Project* of the Groningen Institute of Archaeology (GIA) and the Archaeological Centre of the Free University of Amsterdam (ACVU). This multi-disciplinary project started in late 1997 and was published as Attema, Burgers & van Leusen 2010. Articles focusing on the field surveys carried out in the RPC-project and which made use of the field methodology practiced in the DSP are Burgers, Attema & van Leusen 1998, Attema, van Joolen & van Leusen 2001, and Van Leusen & Attema 2003. See of recent also Attema & de Haas 2008.

Finds from each unit were collected in bags marked with the project code, the date, unit number and bag number with sample type, quantified and summarily described, and then recorded on the unit forms (Fig. 2.8). These forms were filled out by the team leaders during and/or after the transect walking of each unit. Besides finds data, the unit forms also contain a sketch of the unit showing any possible archaeological features, as well as sampled areas and concentrations, and notes on field properties and visibility conditions.[15] Additional notes were kept in each team's field notebook, including general information on survey conditions and observations of the field or situation.

In certain circumstances, e.g. with relatively high amounts of finds or conspicuous features such as kurgans, additional sampling took place. This typically involved the collection of artefacts from a limited area, the so-called *Total Sample*: in this case field walkers were 2 m apart, implying a 100% coverage. This close inspection was also more readily applied when handmade wares were found, since they are generally more difficult to discern in the field. *Grab Samples*, for finds randomly collected, were also taken in a few instances.[16] Finds from the incidental samples were kept separate from the standard samples so as not to distort the overall density maps, which are drawn from the regular unit surveys based on 20% coverage. All of the samples were registered on the unit forms and are recognizable by their unique Bag-ID, a combination of the unit number and an individual bag number. Concentrations and sampling areas were subsequently measured in with the GPS mapping software on the PDA.

#### 2.3.1.1.2 Data processing

The information gathered from the surveys and during material processing (see Section 2.5) was digitized on a daily basis. This consisted of data entry from the unit forms into an MS Access relational database and uploading the spatial features recorded with GPS and the handheld computers to the central desktop GIS environment (see Section 2.5.1). Doing this on a daily basis not only allowed for control and accuracy checks, but also for provisional analyses as work progressed. Furthermore, the daily reports from each team, drawn from observations made in the field notebooks, were recorded by the team leaders in text files (MS Word).

### 2.3.1.2 Systematic survey of the slopes | K. Winther-Jacobsen

The S part of the survey area consists of a series of Hillsides, incised by erosive ravines of varying depths, which gently descend from the Plateau down towards Lake Džarylgač and the Black Sea in the NW. In this area, a large number of kurgans had already been mapped (Chapter 6.5), but no other archaeological evidence had previously been recorded. The initial aim of the Hillside survey in 2007 was therefore to test the presumed dearth of material culture and to work out a method to investigate the archaeology of this specific landscape type and its role within the regional settlement system. More specifically, the survey aimed at tracing remains of Hellenistic settlements in relation to the groups of kurgans on the slopes.

Although there are numerous traces of previous agricultural episodes, the slopes are currently largely uncultivated and are covered by steppe vegetation. This accounts for the low visibility, making the slopes unsuitable for block survey aimed at mapping artefact scatters. Consequently, survey in this area in 2007 was largely experimental. However, although very few artefacts were found during the 2007 season, the survey did reveal the presence of several clearly visible stone structures, enough to warrant a full and systematic investigation of this uncharted area. Furthermore, by the end of the 2007 season it became apparent that the area would soon be inaccessible due to the construction of a large wind power plant scheduled to start in 2008 (see the Foreword). As a result, for the 2008 season the methodology was adapted to investigate as much of the area as possible within the time permitted.

#### 2.3.1.2.1 Methods

In 2007, a start was made with the survey of the Hillsides for which Hill 2 was targeted. This resulted in the observation of several interesting features (stone structures, surface pottery) that were recorded with GPS coordinates. The results of this investigation led to an intensive survey of four further Hillsides. The targeted hills had in common that all were located next to slopes where groups of kurgans had already been mapped, while not featuring kurgans themselves, this

[15] Factors that can affect visibility include the degree of shade, dust, vegetation, modern material, stoniness, and the level to which the field has been ploughed; these factors are discussed in detail in Van Leusen 2002.

[16] Grab samples within the unit boundaries followed the unit number; grab samples taken outside the unit boundaries were given a number starting with 3000, and the coordinates were taken.

in order to detect a possible functional differentiation between hills. For the intensive survey, walkers were spaced at an interval similar to that used in the block survey (10 m). However, for orientation and data recording, the survey relied entirely on the PDAs, using GPS and ArcPad mapping software (see Section 2.5), since cartographic features are scarce in the area. The outer (and sometimes central) field walkers were equipped with a PDA to guide the team in the designated direction and to record their tracks. All artefacts were recorded as spatial points in the PDA and collected as grab samples (numbered in the 3000 series, see above Section 2.3.1.1.1). In case of poor visibility, the immediate contexts of the finds were surveyed more intensively. Finds were most often recorded in relation to animal holes, where the burrowing exposed some of the soil. Features were also recorded with the PDA: small structures such as small concentrations of stones were recorded as points, while larger structures such as ditches and linear elevations and stone structures were recorded as lines in the GIS (see Section 2.5.1), giving a rough indication of their size and shape. In addition to the digital data recording, team leaders kept written notes in their field notebooks.

Based on the results of the 2007 season and the challenge of a tight schedule, some changes had to be made to the methodology for the 2008 season. A map showing the locations of the 102 mills of the wind power plant provided us with a basis for designating the trajectories of surveys in advance using GIS software (Section 2.5.1). The primary features on the map consisted of the foundations for the mills and the cable lines and roads connecting them. The secondary features consisted of archaeological sites identified from the investigations in 2007 (see below). We chose to lay out the survey units along the cable lines and roads, using the planned foundations as starting and ending points. The actual coordinates of these foundations were not made available, but were only roughly indicated on a low resolution map (Fig. 0.4). This also made these guiding survey lines especially instrumental in orientation.

The survey line formed the base for a trajectory which typically passed first from N to S along the W side of the line, and then returning from S to N again down the E side of the line: each pass was recorded as a separate 'unit' in a numbering system starting from 20000 (so as not to interfere with the block survey numbering system). Following this survey line, which could be up to two km long, was crucial to orientation. A consistent team of five members walked at 10 m intervals, with one team member following the line as indicated by the GPS position on the PDA, and guiding the remaining team who were spaced out to the right of the line. On the return pass, the member with the PDA remained on the line, both for the sake of orientation and to compensate for his or her having to follow the GPS signal while surveying. Thus a path of 45 m wide was covered in each direction; in this way a total path of 90 m wide was surveyed along the axis of each line.

In the field, the foundation areas were marked with sticks and the coordinates of these were registered wherever possible, since these were the base for survey trajectory. The sticks, however, were often up to 200 m away from where they were indicated on the map.

Features were recorded in the same manner as in 2007; sherds collected outside the line-units were recorded in the PDA and given grab sample numbers starting with 3200. All archaeological observations and observations concerning the conditions of the survey, such as weather and vegetation, were recorded in narrative form by the team leader in the field notebook, rather than on unit forms, which are tailored to the block survey. Consequently, the level of topographical recording is on a lower resolution compared to the ploughed zone (Table 2.1). Furthermore, the units in the individual

<table>
<tr><td colspan="6">Unit 20019</td></tr>
<tr><td>Administrator</td><td>Walkers</td><td>Date</td><td>Time</td><td>Weather</td><td>Average visibility</td></tr>
<tr><td>KWJ</td><td>5</td><td>16.5.</td><td>12:25</td><td>Sunny</td><td>4</td></tr>
<tr><td colspan="6">Line starts at foundation 47 and goes S towards foundation 46 on the W side, measuring in both foundations. Three cores taken at stick 47 each 10 m from the stick. I photograph them all: Photos 16.05.08.2-4. They prove that although the soil looks deeper it is in fact in this place no more than 0.2 m deep.<br>Topography: Areas of shallow and deeper soils alternating.<br>Features: 20019/01: flint (45.529050/32.90681).<br>According to the map there should be five kurgans W of the line, but none are visible; however, there is a cultivated field approximately 70 m W of the line.<br>Three cores have also been drilled at foundation 46, but only two of them were left on the surface. I photograph them: Photos 16.05.08.5-6.<br>Just past foundation 46 we reach an approximately 12 m long, slight elevation oriented NS, which is marked with a question mark on the map. We take a polyline along its top.<br>Line stops 95 m S of foundation 46.</td></tr>
</table>

Table 2.1 Example of a short unit record.

lower fields in the ploughed zone are all within the same geomorphologic type, whereas the unit-lines on the slopes, although within the same macro-geomorphologic type, exhibited a much higher degree of micro-variation in terms of changes in soil and vegetation (see also Chapter 3). It soon became obvious that some of these changes reflect different human activities, and it would have been useful to investigate this aspect more fully.

The 2008 Hillside survey worked alongside a team of geophysicists from Eastern Atlas, Berlin, surveying the same areas as the team of Tatiana Smekalova (see Section 2.3.3.1). Consequently, we have correlating archaeological and geophysical data sets from the Hillsides investigated in 2008, although the spatial accuracy of the data is affected by the low resolution of the wind power plant map. Because of this wind power plant project, the Crimean Branch of the Institute of Archaeology required that we produce a prioritized list of areas for survey (Table 2.2). This map also included preliminary identifications of archaeological features based on the 2007 season results and investigations continued by T. Smekalova, S.B. Lancov and S.D. Koltuchov in June, and further work done by Lancov and Koltuchov in the autumn of 2007 (Fig. 0.4).

Despite the urgency, our work on the slopes in 2008 was constrained by the agreement to only work in those areas with known archaeological features, and to survey larger coherent areas rather than focus on the mill foundations. We therefore prioritized coherent areas where several preliminarily identified sites would be affected by projected foundations and/or infrastructure. Therefore, much time was spent re-examining known phenomena such as kurgans, and the 'stone circles' which were natural rather than manmade (see Chapter 3.3.2.6). Although the strategy forced upon us was detrimental to finding new sites, we were, nevertheless, able to record two new large settlements and two diffused scatters of pottery, besides making several surveys of complete sites.

| Preliminary site no. | Preliminary identity | Windmill Park Project feature<br>(nos. refer to foundations) |
|---|---|---|
| 34, 36 and 43-44 | Bronze age and Tatarian settlement, kurgans | 31, 32, 34 and cables between them |
| 21-22 | Stone circles | 41, 40, 39 and cables between them |
| 47, 26, 27 and 28 | Settlement and stone circles | 42, 43, 44 and cables between them |
| No numbers | Kurgan group | 47, 46 and cables between them |
| No numbers | Kurgan group | 48, 49, road to and cables between them |
| 48, 30 and 59 | Settlements, stone circles, kurgan group | (49), 54, 59 cables and roads between them |
| | | (54), 55, cables and roads between them |
| No numbers and Hill 7.3 | Kurgans, settlement | 60, 62, 63, cables and roads between them |
| Hill 1 and no number | Two large settlements | 64, 65, 66, cables and roads between them |
| No numbers | Kurgan groups | 68, 69, 70, 71, cables and roads between them |
| 49 | Kurgan group, stone circles | 75, 76, 77, 78, cables and roads between them |
| 17-1 | Stone circles | 79, 80, 81, 82, cables and roads between them |
| 14 | Stone circles | 86, 87, cables and roads between them |
| 10 | Stone circles and kurgan group | 92, 93, 94 cables and roads between them |
| 15 (Hill 13) | Settlement | |
| Hill 12 | Settlement | 71 |
| A total of 36 foundations of 1000 m$^2$ and 17.3 km cables and roads between them | | |

Table 2.2 Prioritized list of areas surveyed.

2.3.1.3 Extensive site survey | P. Guldager Bilde

In 2007, as a supplement to the systematic surveys described above, brief visits to previously known sites in the DSP area were carried out in the company of S.B. Lancov and S.D. Koltuchov (Attema, Guldager Bilde, Smekalova, Stolba). The purpose of these visits was to become familiar with the region, to build an archaeological reference context for the systematic surveys within the transect, and to document any new observations of these sites.

The targeted sites for extensive survey range in period from Bronze Age to Early Modern, and include the areas (Fig. 2.9):

- Along the coast between Masliny and Černomorskoe
- Along part of the N shore of Lake Džarylgač
- In the ravine E of Vodopojnoe

These site visits led to non-systematic sampling (grab samples) of 11 known sites as well as to the localisation of 12 more sites hitherto unknown, which (with one exception) were also sampled by means of grab samples (see below Chapter 4.2).

GPS coordinates of the sites were taken with the PDAs and registered in the central GIS environment. Grab samples from all sites were processed, classified and entered in the database. Apart from serving as dating tools for the sites themselves, the pottery functioned as a reference collection for pottery from the intensive survey.

## 2.3.2 Total Station surveys | P. Attema

During the 2007 and 2008 campaign, part of the DSP team was involved in the mapping and documenting of kurgans within the DSP study area. Methods employed to document the kurgans consisted of recording their geographical position and exact dimensions by means of GPS and Total Station. This was done in order to obtain the necessary spatial data to create digital terrain models that could further be elaborated in AutoCAD software. As a rule, the digital terrain models comprised groups of kurgans rather than single specimens. Geomagnetic surveys were aimed at revealing details on the internal structure of individual kurgans and to detect possible subsurface features in between kurgans relating to cultic activities or *fossa* graves (see for the method Section 2.3.3.1). On the basis of the acquired data a catalogue was compiled containing standard descriptions of each kurgan, photographs, a schematic plan and cross-sections (see Appendix 1, DSP07-H19 and DSP08-H19 listings). Of kurgan groups where geomagnetic surveys were carried out, maps and interpretation of the results were added to the catalogue entries. In two campaigns a total of 31 kurgans were recorded using Total Station, of these nine were also surveyed by magnetometer. Besides the selected kurgans, the Total Station was also used to map architectural remains in the area of study especially of Panskoe I (see Chapter 4.3.2.1) and of a Hellenistic house on Hill 10 (DSP07-H10-01).

## 2.3.3 Geophysical surveys | C. Meyer, D. Pilz, T. Smekalova & B. Ullrich

### 2.3.3.1 Geomagnetic prospection

In the DSP, extensive use was made of geomagnetic prospection to determine the physical characteristics of the soils, and allow for the visualization of buried man-made structures. Below we will first discuss some general principles of this technology and will then focus on the specific methods that were applied by two teams that participated in the DSP. These were a Russian team headed by T. Smekalova of DNRFCBSS and a German team of the company Eastern Atlas Geophysikalische Prospektion (Berlin).

#### 2.3.3.1.1 General principles

Buried archaeological features can be detected as long as there is a significant contrast in one or more geophysical parameters. Each geophysical method varies in depth range, is sensitive to a specific physical attribute and is influenced by certain environmental and soil conditions. Geophysical methods for the investigation of buried features were initially developed for studying geological structures, but during recent years they have become more and more important for archaeology. Of all the geophysical methods, magnetic survey is one of the most effective and universal ones used in archaeology, because many archaeological objects have specific magnetic properties, which allow them to be detected through the surface of the site by the magnetic anomalies they create.

To carry out magnetic prospecting, one measures the Earth's magnetic field within a small area done very close to the surface, and at many points within a fixed grid. For archaeological features to show up on the magnetic map they must contain magnetic material that differs from those in their surroundings, thus producing 'anomalies' in the retrieved recordings. Geomagnetic anomalies are caused by changes in the complex magnetic properties of the soil that may be induced by human activities. The amplitude of the magnetic anomalies is determined by the contrast of magnetic susceptibility between archaeological structures and the surrounding unaffected soil. Iron oxides like haematite, magnetite or maghaemite are carriers of magnetization. Those minerals occur ubiquitously in the soil, forming microscopically small grains. Usually magnetic anomalies are very small. For that reason, highly sensitive probes are required.

2.3.3.1.2 The 2007 and 2008 campaigns

In the 2007 campaign, geomagnetic prospection was undertaken by T. Smekalova assisted by A. Čudin. Prospection was aimed at complementing the results of the field walking survey, especially on the Hillsides (Hill01, Hill02 and Hill06 to Hill11) where various settlement structures were detected on the surface, like walls, pits and ditches, and the 'stone circles', that were later identified as natural features (see Chapter 3.3.2.6).

In 2008 the geomagnetic prospection was extended to two teams. The Russian team continued to focus on sites with high archaeological potential, and a German team, Eastern Atlas from Berlin, was invited to conduct geomagnetic survey on the slopes S of Lake Džarylgač in the areas to be affected by the construction of the projected wind park (see also Section 2.3.1.2). The Eastern Atlas investigations were directed by B. Ullrich and C. Meyer assisted by L. Keller, R. Kniess, R. Meyer, D. Pilz, E. Schönherr and H. Zöllner.

In order to check the consistency of the data yielded by the two different teams, a control check was performed by using the Fluxgate array to resurvey site DSP07-H01-01, and comparing the data with that produced by T. Smekalova using the Overhauser gradiometer in 2007. Both systems yielded the same result.

2.3.3.1.3 Methodology of T. Smekalova

During the 2007 campaign the magnetic survey by the Russian team was carried out with an Overhauser gradiometer (Fig. 2.10). This is a magnetometer with two sensors produced by Gem systems (Ontario, Canada). The measurements were made along straight parallel lines with an intermediary space of 0.5 m. The magnetometer was operated in so-called 'walking mode' measuring every 0.2 seconds. The intermediary distance between measurements along the lines did not exceed 0.2-0.3 m. The height of the sensor above the surface of the ground was ca. 0.3 m. Two Gem magnetometers were used, one serving as a main instrument that was moved on the plot to measure the magnetic field along the lines of a coordinate system. The other was placed at a fixed point where it automatically measured the temporal daily variations of the Earth's magnetic field every 15 seconds. The combined data from the two magnetometers was used to obtain the measurements of the magnetic field on the plot. With the help of a special program, the necessary corrections were made to remove the temporal changes of the global magnetic field.

In the 2008 campaign, a four-sensor device was used; the sensors of the Overhauser magnetometer GSM-19 of Gem Systems were mounted on a cart with an intermediary distance of 0.5 m between each sensor. This increased the speed of the magnetic survey by several times.

| System | Overhauser GSM-19WG magnetometer with 4 sensors |
|---|---|
| Sensors | One, two or four |
| Measurement category | Total field in nT |
| Configuration | 1, 2 or 4 channels, mounted on cart |
| Sensor separation | 0.5 m |
| Sensor height | 0.25-0.35 m |
| Distance measurement | GPS coordinates of the corners of investigation area |
| Investigated area | 250,000 m$^2$ = 25 ha |
| Data format | Data export in standard X, Y, Z (line-oriented to facilitate software compatibility) |
| Grid | 0.5 x 0.3 m |
| Processing and filters | Corrections for daily variations of the Earth's magnetic field |

Table 2.3 Summary of geomagnetic prospection parameters of the Smekalova team using the Overhauser magnetometer.

To localize the surveyed areas, GPS coordinates were taken of the corner points of each plot using the same GPS hardware and software and coordinate system as the intensive survey (see Section 2.5.1). This data was later incorporated into the central GIS environment.

As a rule, plots were 50 m long (or less) and as wide as necessary for the magnetic survey to cover the investigated area. Two long measuring tapes were set out along the two opposite sides of the plot, with shorter (50 m or less) measuring tapes used to mark the intervals along which the parallel lines had to be walked.

As for the elaboration of the data, two different presentations of the magnetic data were prepared with the help of Surfer software (Golden, Colorado): coloured contour maps and grey-scale maps. On the contour maps, positive anomalies are shown in blue, and the negative ones in red. On the grey-scale maps the positive anomalies appear dark colour, and the negative ones light. The contour interval was usually 2 or 5 nT (Nanotesla).

2.3.3.1.4 Methodology of Eastern Atlas

The Eastern Atlas team used for their investigations an array of 5 and 6 Fluxgate Gradiometer probes respectively, mounted on a cart (see Fig. 2.11). The two sensors of each probe measured the vertical component of the earth's magnetic field with an accuracy of 0.1 nT. The measured gradient is insensitive to the typical large fluctuations of the earth's magnetic field and is determined only by the magnetization of the ground. The technical details concerning the investigation are summarized in Table 2.4. The grid images shown in the maps were produced by transforming the measured values to a scale of 256 grey levels. The geomagnetic maps in this case are displayed in dynamic scales of ±6 nT.

| System | Array of five or six Foerster fluxgate gradiometers |
|---|---|
| Sensors | 3 Foerster 4.021, 3 Foerster 4.032 |
| Data logger | heslog DLAD62001 |
| Measurement category | Vertical gradient in nT/m |
| Configuration | 6 channels, mounted on cart |
| Sensor separation | 0.5 m/0.8 m/1 m |
| Sensor height | 0.35 m |
| Distance measurement | Odometer (survey wheel), coordinates of investigation area by GPS Mobile Mapper |
| Investigated area | 1,300,000 m$^2$ = 130 ha |
| Data format | MRC, MDT (MARPLOT), images as PNG |
| Resampling (images) | 0.25 m x 0.25 m/0.2 m x 0.2 m |
| Processing and filters | – |

Table 2.4 Summary of geomagnetic prospection parameters of Eastern Atlas using the Foerster Fluxgate gradiometer.

The investigated sites were localized using a handheld GPS receiver (Thales Mobile Mapper) with an accuracy of within 0.5 m; additional GPS coordinates were taken when necessary to compare topographic features with geomagnetic anomalies. These coordinates were later incorporated into the central GIS environment.

On two selected archaeological sites, aerial photographs were also taken by the Eastern Atlas team using a kite which was set up with a digital camera PENTAX Optio A20 (see Fig. 2.12). To rectify the pictures, GPS points were defined and marked in the field and were used to set spatial reference points on the picture. Fig. 4.301 shows the aerial photography of Hill 12 which is a composition of several rectified and referenced pictures. The obtained image facilitates easy comparison of visible surface features and geomagnetic anomalies.

## 2.3.4 Soil surveys and augering | W. de Neef

In the 2008 campaign, manual augerings were conducted to gain more information on the landscape of the Lake Džarylgač research area (Fig. 2.13). Augerings were conducted for two reasons. The first was to gather information on soil properties, consistency and profiles in order to better understand the geology, hydrology and soils of the area of study (see Chapter 3). The second reason was to study the soil profile and to check the degree of disturbance of the soil at specific site locations, reflecting the degree to which the archaeological layers are still intact (see Chapter 4). The individual augering profiles are presented in Appendix 3, whereas the augering profiles made in DSP sites are included in Appendix 1.

2.3.4.1 Studying soil, sites and landscape

The study of soils in a landscape can reveal natural and human-induced changes in the landscape, such as erosion or deforestation, which may have influenced human occupation and activity as well as the deposition and preservation of archaeological remains. As the underlying factors and processes can differ greatly from area to area, it is necessary to observe the soils at different locations. Prior to the start of the project, little was known of the soils in the area of study apart from the 1967 soil chart of the Crimea (scale 1:250,000; Fig. 3.1). A greater level of detail was required to produce the necessary landscape description for the specific research area.

In 2008, systematic observations and manual augerings were conducted in the area to collect this detailed information. Although intended as the initial phase of observation, to acquire a first impression of the different land types in the area, these soil studies could not be continued due to the termination of the project after the 2008 campaign. Therefore, the collected data as presented in Chapter 3 is rather summary.

One method to obtain a first impression of the soils in a landscape is to describe their variations. In the DSP research area, a total of 29 soil augerings were carried out in a transect between the Black Sea coast to the N and the Hillsides to the S. This made it possible to draw a section of the landscape showing the relationship between topography and soil types. These samples were taken at approximately 100 m intervals.

To collect detailed information on the soil material, soil horizons, and the degree of disturbance of the soil profile at specific locations, an additional 31 augerings were conducted at seven sites throughout the area. The purpose of these site-oriented augerings was to gain information about the relationship between the archaeological remains on the surface, the soil underneath, and possible strata.

At each augering location data were collected to answer the following questions:

- What type of soil is present at the augering location?
- Is the soil profile undisturbed, or is it disturbed by, for instance, ploughing?
- Can we detect archaeological indicators, such as pottery fragments, bone, charcoal and flint, in the soil profile?
- How deep is the bedrock?

Augerings were conducted with an Edelman auger, a manually operated device with a drilling head of 8 cm in diameter. The drilling head is a hollow cylindrical tube with a length of 25 cm. The Edelman auger used in the 2008 season could be extended to a length of 2 m. At most augering locations, bedrock was reached before that, except for augerings 15 and 16 on the S and N shores of Lake Džarylgač. The drilling head is brought into the ground in a turning motion, the inserted bits at the bottom of the drilling head cutting the soil and filling the cylinder, whereby the soil structure is slightly disturbed. Cores are then brought up and laid out in the same sequence as the soil in the ground, after which the soil profile can be described.

Of each augering, the following information was recorded:

- Thickness of soil horizons in cm below the surface
- Texture and texture changes of the soil horizons
- Colour of the soil horizons (Munsell values)
- Mottles of different colour
- Humidity of the soil
- Effervescence of the soil material
- Types of natural inclusions (e.g. stones, plant remains)
- Types of archaeological inclusions (pottery fragments, charcoal, flint)
- Depth of bedrock below the surface

Where necessary, the amount of stones on the surface surrounding the augering location was recorded, as well as vegetation, land use, and humus content of the layers. The coordinates and elevation of each augering location were taken with the same PDA-GPS system used by the field survey teams, so they could easily be plotted into the central ArcGIS environment.

### 2.3.5 Trial trenches | V. Stolba

As a final step in the methodological sequel, trial trenches were made at selected sites discovered in the field walking and geomagnetic surveys and during the revisit of known coastal sites. All of the soundings were aimed at producing additional data for the cultural and chronological attribution of these sites, when this could not be established based on the exposed features and/or surface finds.

The geophysical prospection, which produced detailed geomagnetic maps of sites prior to excavation, proved very valuable for picking out an optimal location for the trenches. The orientation of the latter generally depended on the features detected on the geomagnetic map. In other cases it followed the four main cardinal points or the grid of the magnetic map. The standard trench size used was 2 x 2 m, but in two cases (the trial trenches dug in DSP07-H08-01 Feature 2 and DSP07-H10-01 Feature 5) this was exceeded in order to completely expose the objects (storage pits) or to provide an extent that allowed a better understanding of their basic purpose and outline. All of the trenches were excavated and recorded down to the bedrock, with layer-wise clearing of the fill; finds from each layer were recorded in separate find lists.

## 2.4 FINDS PROCESSING AND ANALYSIS | S. HANDBERG

Finds collected in the DSP were washed, dried and subsequently processed on a daily basis. The actual recording of the finds was limited to one person (S. Handberg) in order to secure consistency in the use of terminology as well as taxonomy. Basically, all of the finds were classified per bag into functional groups (Architecture, Handmade Ware, Transport Amphora, Tableware, Utilities, Bone, Lithics and Indeterminable) and then subdivided by shape, fabric, type and date (see Chapter 5.3 for more on the classification).

An MS Access database, which could be integrated with ArcGIS, was used for the recording. All of the finds were first recorded in a FINDS table. This table formed the basis for the creation of the distribution maps (see also Section 2.5.3). All diagnostic fragments were selected for numbering, more detailed description, drawing and photographing. They were then entered into a separate DIAGNOSTICS table. The information for the finds catalogue pertaining to the individual sites was later extracted from this table.

Finds were acquired in various ways; systematic collection during field walking of the ploughed zone and the slopes, diagnostic sampling during extensive surveys, random sampling during the geomagnetic and soils surveys and total collection from trial trenches. A more detailed treatment of artefact classification is given in the introductions to the site catalogue (Appendix 1) and the finds database (Appendix 2).

## 2.5 GIS AND DATABASE APPLICATION | T. DE HAAS

GIS and database applications were used in every part of the DSP. This section discusses first the use of mobile and desktop GIS applications in the preparation, execution and processing of the different types of fieldwork. Following this, the structure of the database that was used to store the non-spatial data of the surveys and pottery studies is given. The final part of this section describes how the database was integrated with GIS in order to create the distribution maps that form the base for site identification and further analyses, as given in Chapters 4 and 6.

### 2.5.1 GIS usage

GIS software, both desktop and mobile (combined with Global Positioning Systems [GPS] receivers), was a central component of the surveys. This section first discusses the actual hardware used in the field for the field walking surveys, as well as the mapping software employed to prepare, conduct, and register survey data. Following this is a description of the way in which this equipment was used in each of the methods of field research in the DSP.

2.5.1.1 Hardware and software

The central GIS environment used by the GIA team for this facet of the project is ESRI ArcGIS 9. In principle, the GIS environment was set up similar to that used in previous fieldwork in Italy, and will be briefly described below. As the choice for intensive survey methods both on the slopes and in the ploughed zone was expected to lead to large amounts of data, spatial information was digitally recorded in the field in order to limit data processing afterwards. Collection of specific geographical coordinates was done using handheld computers (PDA) with digital maps, connected to external GPS receivers.[17] The software used on the PDA was ESRI ArcPAD 7.0 package for mobile devices, since it is well integrated with ArcGIS. Each field team was equipped with at least one such PDA set. Although ArcPAD was a new program for the GIA team, a one-day trial in the field proved it to be sufficiently stable, efficient, and user-friendly to allow it to be further modified according to the specific needs of the DSP. All GIS data as well as the database, daily reports, Total Station and geophysical data will be made available online.[18]

2.5.1.2 GIS-use in the block surveys

The block survey depended on the use of desktop and mobile GIS applications (Section 2.3.1.1). As field boundaries could not reliably be copied from topographic maps, they were measured in before survey using the PDA. These field boundaries were then uploaded to the desktop GIS and used to create the survey grids that were subsequently loaded onto the PDAs again. Creating the grids beforehand proved very efficient, as experience from other surveys showed that doing this while surveying is time-consuming and prone to inaccuracies, especially in large fields.

The grid in the PDA was used to mark the individual unit boundaries in the field with small flags (Section 2.3.1.1.1). During the survey, any potentially relevant feature (points, lines and polygons) was measured. Such features include the extent of artefact scatters, soil discolouration or elevations, sample points and areas, and various other features.

Unit IDs as recorded in the field and on unit forms were later added to the shapefile (spatial file in ArcGIS) that contained the survey grids. The field data gathered on the PDAs was uploaded on a daily basis to the desktop GIS and subsequently edited to create the following separate files:[19]

- Fields and units (polygons)
- Features (points and polygons)
- Site extents (polygons)
- Site centroid (points)
- Sample areas (points and polygons)

2.5.1.3 GIS-use in the slope survey

The slope survey relied heavily on the use of the PDA, as topographic maps contain very few features of reference in this area (Section 2.3.1.2). In general, during the survey the PDAs were used to map both the transects, that were actually walked on the slopes, by means of a tracklog, and specific observations, both (individual) artefacts and structures. After the survey, the field data was uploaded and edited to create the following separate files:

- Hills and units (polygons)
- Features (points, lines and polygons)
- Site centroids (points)
- Samples (points)

2.5.1.4 GIS-use in the extensive survey

In 2006 the coordinates of sites and fields were registered using a Garmin GPS receiver; the data were later manually entered into the desktop GIS.

The PDAs were used in the extensive survey mainly to map sample and site locations (Section 2.3.1.3). After the survey, the field data was uploaded and edited to create the following separate files:

17 The devices are standard HP ipaq machines (of the h2200 an hx2490 series). The devices communicate via bluetooth connections with GPS receivers (Royaltek STAR3 and Socket D1598).

18 The archive is deposited at DANS, the Dutch Data Archiving and Networked Services. The data as well as the terms and conditions for use can be found via www.edna.nl or directly at http://persistent-identifier.nl/?identifier=urn:nbn:nl:ui:13-0mg-hmc.

19 All GIS files use the coordinate system WGS 1984 UTM zone 36N.

- Samples (points)
- Site centroids (points)

2.5.1.5 GIS-use in the Total Station survey
The PDAs were used in the Total Station survey in order to connect the Total Station measurements to a real world-coordinate system (Section 2.3.2). For this, the coordinates of various grid points used during this survey were measured with the PDA. After the survey, the coordinates of these control points were used to import the Autocad data into the desktop GIS.[20]

2.5.1.6 GIS-use in the magnetometer survey
The PDAs were also used in the magnetometer survey to connect the magnetometer measurements to a real world-coordinate system (Section 2.3.3.1). As with the Total Station method, the points of the grid used during this survey were measured with the PDA. After the survey, the coordinates of these control points were used to georeference the Tiff maps of the magnetometer survey data into the desktop GIS.[21]

2.5.1.7 GIS-use in the soil surveys and augerings
The PDA was used to map various observations during the soil surveys as well as the locations of the augerings (Section 2.3.4). After the survey, the field data was uploaded and edited to create the following separate files:

- Quarries (lines)
- Augering locations (points)

2.5.1.8 GIS-use in the test trench excavations
The coordinates of some of the test trenches were taken with the PDA; the remaining trenches were measured with a separate GPS device.[22] After fieldwork, the PDA data and the coordinates recorded with GPS devices were used to create a file with the locations of the test trenches (points).

2.5.1.9 GIS and cartography
For reference in the field, the 1:21,000 half-verst maps were used (see above Section 2.1). Unfortunately the digital versions of these maps that were spatially referenced proved to have some distortions, so that in certain cases there is a slight deviance between locations as measured in the field and their location on the map. This deviation can be up to ca. 100 m.[23]

The following topographical features were digitized from these maps:

- Roads (lines)
- Settlements, fishing stations, ruins and saltworks (polygons)

The location of kurgans and springs had previously been digitized by T. Smekalova in Mapinfo and were imported into the desktop GIS.

A shaded relief model is used as background for the visual representation of the area surveyed based on data provided in the ESRI ArcGIS 9 kit of data and maps. This model is based on the world elevation model of the US Geological Service, at a resolution of 90 m; this shaded relief model is used for the larger overview maps as background, while the more detailed maps use the half-verst map.

---

20 The extent of each kurgan was added to the polygon features file of the slope survey.

21 The interpretations from these maps (i.e. all subsurface features) were digitized and added to the (line and polygon) features files of the slope survey.

22 Three trenches were measured in detail with the Total Station.

23 The deviations may partly be explained by the scale of mapping, but could also be due to slight distortions in the map that arose during the process of scanning and-or copying of the original maps.

### 2.5.2 Database structure

All non-spatial data from the block survey as well as the results of all pottery studies were stored in an MS Access database. This database was modified after those used for the GIA's surveys in Italy.[24] For the DSP survey, the database initially consisted of four tables, related to each other by their unique IDs for units and samples. These four tables include:

- The UNIT table, that contains data on the individual units (i.e. info from the Unit forms);
- The BAGS table, containing data on the samples (info from Unit forms and from the finds processor;
- The FINDS table, that contains records for each find class per sample (see also Section 2.4 and Chapter 5.3);
- The DIAGNOSTICS table, recording data on individual selected finds (see also Section 2.4 and Chapter 5.3).

After the completion of the survey, two more tables were added to the database:

- A FIELDS table with summary data (number of units, size, unit size) on the surveyed fields;
- A SITES table, with summary data on the sites that were identified after the survey, based on the distribution maps (see below).

### 2.5.3 GIS and database integration for distribution maps

In order to come to an analysis of the research results, the data stored in both the GIS and the database were integrated into distribution maps. The following section describes how this was done for the block survey and the slope survey; the maps of the slopes include data from the magnetometer survey, the excavations, and the Total Station survey.

2.5.3.1 The block survey maps

The analysis of the survey data as presented in Chapter 4.4 presents data for each field and for each individual site. The distribution maps are all based on the finds collected during standard sampling and present densities per grid unit. The densities have been corrected for coverage and indicated by graded colour values.[25] It was decided not to correct these densities for visibility for two reasons: on the one hand, the level of visibility was roughly consistent (all the fields were ploughed, and almost all were surveyed under moderate to excellent conditions, excepting the weather at times); on the other hand, considering the very low sherd count we fear that applying additional correction factors may 'create' patterns based in reality only on single finds.[26]

Thus, for each field an overview map was made that displays all finds, as well as period maps for all of the relevant periods: Bronze Age, Late Classical to Early Hellenistic, Byzantine, Medieval, or Early Modern, and Modern. After a description of these distribution patterns, the areas with relevant artefact scatters are analyzed, forming the basis for the identification of (possible) sites (see Appendix 1). For these areas we present more detailed distribution maps for each relevant period that also show the absolute number of finds per unit (black, italics), the scatter/site extent as well as the location of additional samples and the number of finds from these additional samples (indicated by white circles).

2.5.3.2 The slope survey maps

The slope survey data are presented in maps for both entire hills/units and for individual sites.[27] Two overview maps are presented for entire hills or, where applied, survey units (Chapter 4.5): the one presents the outline of the hill/unit

---

24 Cf. Van Leusen 2002, chapter 8, paragraphs 3.1 and 3.2.

25 Units were normally surveyed with a 20% coverage; this means that all absolute artefact counts are multiplied by a factor 5 (raw count x 5 = 100/% coverage). These figures are subsequently normalised to densities per hectare.

26 See further Section 2.3.1.1 on the application of visibility factors.

27 We note here that in some cases the margin of error in positioning the topographic background map causes displacement of this background; in some cases hill outlines appear to run on across ravines. Similarly, the location of kurgans as indicated on the topographic maps sometimes differs from the location as measured with PDA and/or Total Station. The kurgans measured in during the Total Station survey on Hill 19 are up to 40 m W of their location on the topographical map.

together with the magnetometer survey results and the location of sites, the other displays the outline of the hill/unit together with the samples, the features recorded and again the site locations.

The site maps zoom in on site areas, displaying all relevant site data (magnetometer survey data, samples, features, excavation trenches).

2.5.3.3 The extensive survey maps

For the extensive survey maps, all sample locations and sites are displayed, including the digitized topographical features such as roads and villages (Chapter 4.2-3).

## 2.6 SITE DETERMINATION AND CLASSIFICATION | P. ATTEMA

### 2.6.1 Site and off-site definition

In the surveys of the DSP, the term site is used to denote all artefact scatters and structural features that are indicative of past human activity at circumscribed locations in the landscape, provided these remains are detected 'in situ'. Sites encountered in the DSP can be subdivided into various types: artefact scatters, stone structures, artificial elevations and other interventions in the landscape (such as pits, quarries, and kilns). Single lithics and potsherds as a rule are considered as off-site finds unless clearly related to an archaeological feature, e.g. a burial mound (kurgan). Off-site artefacts, especially pottery, are in the context of the DSP valued as indicators of past land use and were therefore consistently recorded and analysed.

Lithics and potsherds appear on the surface foremost in the cultivated zone (i.e. Pediment and Lowland Ridge in the land type classification presented in Chapter 3.3.1), due to the regular ploughing that takes place there. The definition of a site on the basis of artefact distribution in this study depends on a combination of a high finds density relative to its surroundings and/or spatial discreteness of the artefacts and chronological coherence (see Chapter 4.3). In the non-cultivated zone (i.e. the Hillside and Plateaus/Uplands in the land type classification presented in Chapter 3.3.1), artefact scatters are rare, and site definition here, as a rule, is dependent on the surfacing remains of stone structures, artificial elevations and other visible interventions in the landscape as well as subsurface features that were detected through geophysical techniques.

### 2.6.2 Site classification

In order to interpret the different kinds of sites mentioned above, several classes of sites are defined in Chapter 6.1, using the data provided in Chapters 4 and 5 and in the site catalogue (Appendix 1). These site classes are provisory functional categories, and are based on the presence of such indicators as individual artefacts or artefact scatters, structures in stone, or features such as elevations, and discussed in relation to the land types in which they occur.[28] Sites consisting of artefact scatters can substantially differ in scatter size, density, artefact assemblage and chronology, and these variables help in establishing hypotheses on site function. For instance, while some of the larger Bronze Age scatters are certainly the remains of permanent nucleated settlements, the often very small Hellenistic scatters may represent small settlements, or even non-settlements, such as temporary or seasonal facilities.

Depending on their shape, the main stone structures found in the DSP area of study are interpreted as encircling walls of animal pens or as house foundations of different sizes. So far these pens and foundations have been found exclusively on the hillside. Although isolated examples occur, they are typically found together and are thus interpreted as (complex) settlements. In the Lowland Ridge there is evidence of large fortified farmhouses[29], while the elevated parts of the landscape gave no indication of this class of farmstead, but rather of modest to primitive rural housing.

Most of the artificial elevations in the landscape are burial mounds (kurgans) from various periods from the Bronze

[28] See also Attema et al. 2009.

[29] As large fortified farmhouses were not encountered in the survey, they do not figure as a class in the site classification presented in Chapter 6.1 that was exclusively based on sites defined within the transect. Of course they are incorporated in the concluding analysis on settlement development in the wider study area.

Age onward, belonging thus to the site class of funerary area. They appear both in the elevated parts of the landscape and in the lower zones. Many have been looted, while in the cultivated zone a substantial number has been ploughed away. Other elevations in the landscape pertain to ash heaps or dams, mainly from the Early Modern period; land divisions were also noted in the field, however their origin was uncertain. Other past interventions in the landscape are pits dug into the soil that in most cases can be related to houses, and kilns and quarries, the latter of course related to specific geological situations (see Chapter 3 passim).

The site classes and their correspondence with land types will then be used for the wider interpretation of the settlement history of the DSP area of study that is presented in Chapter 6.

CHAPTER 3

# Environmental aspects of the DSP landscape

*Compiled by J. Delvigne*
*with contributions by P. Attema, W. de Neef, T.V. Sapelko, V. Stolba & D.A. Subetto*

## 3.1 INTRODUCTION | P. ATTEMA

In this chapter various aspects of the environment in the study area are discussed. In particular its geology, hydrology, soils and present land use are dealt with. A basic understanding of the environment and of (sub) recent patterns of land use aids in the explanation of the settlement patterns revealed by the survey and in hypothesizing on related forms of land use in the past (see Chapter 6).

Section 3.2 sets out with an introduction of earth science aspects pertaining to the limestone geology of the study area to which a blanket of dust (loess) was added during the last Ice Age. This is followed by a discussion of climate and natural vegetation. While the geology was mainly studied on the basis of literature and maps available to us, the discussion of climate and vegetation incorporates data from a pollen core taken in Lake Džarylgač in 2005 in a palaeo-environmental project directed by V. Stolba (reported on in Section 3.4). The chapter then moves on to a general discussion of the geomorphology, hydrology and soils in the study area.

In Section 3.3 the focus is on the survey transect. In this section a land type classification is presented compiled by J. Delvigne with W. de Neef on the basis of the former's fieldwork during the 2008 campaign. In the transect Delvigne discerns five land types: (a) the Plateaus/Uplands, (b) the peninsular Hillsides, (c) the Pediment, (d) the Coastal Lakes, and (e) the Lowland Ridge. These are systematically described and include observations in the field.

Part of the fieldwork for this classification consisted of hand augerings. These were carried out in transects in order to obtain soil profiles of the Pediment and Lowland Ridge land types. Also on selected archaeological sites hand augerings were carried out. These served mainly to establish the presence of possible subsurface archaeological deposits, but were also of help in the general reconstruction of soil cover in the transect. Soil augering data of the transects and their locations are presented in Appendices 3.1-4.

Finally, a brief description of (sub) recent land use is given providing insight in the environmental and socio-economic context in which the archaeological record nowadays is found (Section 3.4). Of known biases that affect the recovery of archaeological sites the contrasting surface visibility in the DSP landscape between the ploughed zone and the peninsular Hillsides is the most prominent factor (see Chapter 2.3.1). This is due to the on-going extensive ploughing of the Pediment and Lowland Ridge following the introduction of large scale farming during the Sovjet period. In the present survey this has led to a relatively high recovery rate of pottery scatters in this zone compared to the peninsular Hillsides and the Plateaus/Uplands.

The data that were used to write this chapter derive from the study of literature and maps as well as from field work. Since the available time was limited the picture we present here can only be preliminary. Nonetheless this chapter provides useful information for an evaluation of the settlement evidence found in the survey and its relation to the agricultural potential of the past landscape.

## 3.2 INTRODUCTION OF EARTH SCIENCE ASPECTS | J. DELVIGNE

### 3.2.1 Geology | J. Delvigne

In the research area, only limestone is found as bedrock (Fig. 3.1). The oldest limestone is Sarmatian (Middle and Upper Miocene) of age, followed by Meotian and Pontian limestone in the Upper Miocene. After uplift of the area, a planatian surface began to develop in the Early Pliocene, covered by a red weathering loam. Rapid uplift occurred after the deposition of Meotian and Pontian limestone. Following the minor folding that was present in the limestone, three low ridges were formed on the planation surface, running in a SW-NE direction. Differential tectonic movement at the end of the Pliocene and the beginning of the Pleistocene encouraged erosional processes, and during further uplift of the Plateau, gullies were cut, the weathering mantle was removed and, especially in the western parts, dissection took place (Embleton 1983, 387). The research area, situated in the N part of the Uplands, was only slightly dissected.

The Sarmatian limestone of the uplifted central Plateau area is almost horizontally bedded. Because slope angles are small, many dipslope situations, that is gentle slopes created by erosion of the tilted strata, occur.

Where deposits of sharp-edged fragments of limestone and shells are found along ancient coastlines of uplifted blocks, they have the characteristics of fan and delta deposits. These deposits mark locations where rivers flew out into the sea. These rivers cannot have been very long because hardly any rounding off of bedload material is found in these deposits. Because of the angular shape of their components, these deposits have a rather open structure, cemented together at their contact points.

The third geological unit in the research area is unconsolidated loess deposits (Eolian loam) dating from the last and possibly also from the second last Ice Age. This fine-textured material was blown here from more barren, cold desert surfaces that were present at that time in large parts of the continent to the N. This dust was then trapped in the more vegetated tundra fringe around the cold desert. Loess deposits have a characteristic yellow colour, show hardly any bedding and cover the landscape like a blanket. Its original thickness varied and depended to a large extent on the position of the receiving location in the wind field and the local small-scale relief. The leeward side of hill tops typically received a thicker blanket than the windward side. On the hill top itself less loess was deposited because of higher wind speeds there and a drier surface that is difficult for dust particles to adhere to.

Apart from the general loess 'blanket', a few more sizable bodies of loess were encountered that were used as quarries. Loess was probably used to even out farm floors and possibly in the more recent past to bake bricks. These bodies of loess can be over five m thick and for instance 50 m wide. They are interpreted as completely filled in parts of small valleys or of filled-in solution holes in the surface of the limestone. Both examples have been places of considerable local relief in a landscape with predominantly little relief.

The fourth geological unit comprises deposits associated with the sea in the Holocene period. They consist of lagoonal clays, and of the sands of spit barriers and dunes. In the research area they can be found in the Lake Panskoe – Lake Džarylgač region.

Also dating in the Holocene are small occurrences of river deposits in valley floors close to the sea. Some terraces may be distinguished in these deposits.

## 3.2.2 Geomorphology and hydrology | J. Delvigne

### 3.2.2.1 Geomorphology

The research area encompasses part of the central Uplands of the Tarchankut Peninsula and part of the coastal zone N of it, between Černomorskoe and Vodopojnoe.

The Uplands constitute an assemblage of well-rounded, not very high ridges, running in a WSW-ENE direction and surrounded by plateau-like surfaces. These surfaces probably represent a period of little or no upheaval in the general uplift of the area, causing the forming of a coastal terrace. The highest ridge in the area is the Tarchankut ridge SE of Černomorskoe with a crest at about 170 m altitude. The ridge that acts as the divide of the peninsula, E of Černomorskoe, has a crestline at about 130 m. Slopes in the Uplands are very gentle and reflect the slightly undulating bedding of the limestone, causing many dipslope-situations. On the limestone surfaces the initial stages of karst phenomena like sinkholes and small dolines can be found (Fig. 3.2).

The land type connecting the Uplands to the coastal zone is formed by the relative steep, sea-facing slopes that represent a period of rapid uplift in between the quiet period that caused the planation of the Plateau at about 130 m and a quiet period with little uplift of recent times. Some ravines are carved into the steepest parts of the slopes and disappear at the lower end of the slopes, where the Pediment begins.

The larger part of the coastal zone is occupied by a wide valley, running in an almost W-E direction and cut by the coastline at a small angle. Lakes and lagoons like Lake Panskoe, Jarylač Bay and Džarylgač Lake are found in this filled-in valley. Inland of the coastal zone, the wide valley narrows into a river valley with marked incised meanders.

NE of the village Mežvodnoe begins the low coastal ridge that separates the N side of the filled-in valley from the sea. Marine erosion accentuated the coastline here by cutting a cliff in the limestone rock.

3.2.2.2 Hydrology
Water in the steppe-climate of the region is supplied in the first place by rain fall with an average annual precipation between 350-650 mm. In spring, the soil underneath might still be frozen and impervious, giving rise to surface runoff. Some erosion by rivers as well as slope wash can therefore still take place in the present climate.

How much of the surplus of surface water enters the cracks in the limestone underneath will depend largely on the thickness of the loess cover. At a geological timescale, the limestone has not been above sea level for a very long time and will not have developed a large cave system underground. Nevertheless, the water in the soil may stay and flow there in two ways: in the open spaces in shellrock and in cracks in the rock, especially where the flow is confined by clay layers, as is the case in the lower parts of the Sarmatian limestone. There, lines of settlements sometimes mark the contact between clay and porous limestone where groundwater emerges. Because of its open structure, shell rock is a natural aquifer. The water carries enough oxygen to oxidize iron compounds, giving the shellrock its typical yellow colour. The relationship between rock type, rock structure and topography on the one hand and the location of springs and wells on the other, is rather complex and requires a separate study (Fig. 3.3). Most natural wet spots are stretches of valley floors where vertical erosion is blocked by a hard layer and small stones and fine earth are trapped. By digging out these stretches, the valley floor is brought even closer to the groundwater level and may become pools of open water, suitable to water cattle.

## 3.2.3 General aspects of soils | J. Delvigne & W. de Neef

Soils can be looked at as the types of loose material below the ground surface: the soil material, and as the typical vertical sequences of properties in this loose material: the soil profile.

3.2.3.1 Soil material
Where the land surface is made up of limestone, physical weathering will have attacked this rock and fragmented it into smaller pieces, acting to a large extent on natural bedding planes and folds. In the research area frost action will have been responsible for most of the fragmentation, especially before the deposition of loess in the last Ice Age. Further transport of these smaller pieces down the slope is made easier this way, just as subsequent solution of the rock fragments by the slightly acid rainwater and by organic acids in the soil. As far as solution is concerned, signs of this process are rather scarce, though some beginnings of karstic phenomena like sinkhole depressions were observed.

This dissolution of limestone may add small amounts of non-soluble impurities like clay and Fe- and Mn-oxides to the soil material. Another component of the soil material in the steppe-environment is organic matter that results from the biochemical break-down of dead organic (mainly plant) material. This component also plays an important role in the soil profile and will be discussed there. Finally, to the soil material on any rock type will be added small amounts of windblown dust (loess).

The Crimea lies near the southern fringe of the loess belt. Here, the original thickness of the loess blanket will have been less than two m (Velichko & Nechayev 1984). Loess has a loam texture. Mineralogically, the majority of the particles consists of quartz. Other components are clay, mica, Fe- and Mn-oxides. Fresh loess also contains carbonates and, due to the Fe-oxides, has a yellow-ochre colour (10YR 4/4-6/4). In the present climate, loess can only be weathered any further by solution of the carbonates and the vertical displacement of the Fe- and Mn-oxides. Loess is, however, susceptible to erosion by wind and, in a semi-arid climate especially, to slope wash by running water.

Weathered limestone and loess are the two basic types of autochthonous soil material on the Hillsides and Plateaus of the research area. However, in many places of this area a vertical sequence of soil materials was found that proved to be characteristic of certain parts of the area and may be called a third, partly autochthonous and complex, type of soil material. It can be described as a veneer of weathered limestone on top of a layer of no more than a few dm of loess that is resting on weathered limestone. This three-layered soil profile is sometimes called a triplex soil. The origin of this sequence must be that a thin layer of loess was, at the end of the last Ice Age, deposited on the existing, weathered Pediment. On the upper parts of the Pediment and the slopes above, hardly any loess was deposited. When slope processes restarted at the beginning of the Holocene, from these higher slopes weathered material of limestone was spread over the thin layer of loess on the lower slopes. Conclusions that can be drawn from this are:

(a) The lower boundary of the layer of loess represents the land surface that was present at the time that loess deposition began. Reddish colours in the soil material underneath the loess are signs of more intense soil formation, possibly dating back to a previous interglacial.

(b) The rather uniform, thin layer of loess on the Pediment indicates that the Pediment had, at the onset of loess deposition, virtually the same shape as the present Pediment. This also holds true for the end of the period of loess deposition. The shape of the transect over the Pediment apparently is not very sensitive to the process acting upon it.

(c) The slope process by which weathered limestone is deposited over a loess layer, forming a triplex soil profile, may be a process called solifluction. When snow melts in spring on slopes, where loess is not or no longer present, water may be prevented entering the subsoil because the subsoil is still frozen. The topsoil will become saturated and may slide downslope possibly covering a loess cover further down the slope. Soil sliding may resemble ice floes, when patches of top soil are held together by plant roots. Solifluction is common in a periglacial environment.

3.2.3.2 Soil profile

The zonal soil profile is the chernozem (USDA[30]: Chernozem, FAO[31]: Mollisol), an Ah-(B)-C profile with a thick (up to 1 m), dark and well-structured Ah horizon that typically has developed in lime-rich loamy and clayey soil materials. The high content of humic matter is caused by the dominance of macro-biological break-down of dead organic matter over microbiological destruction by bacteria because bacterial activity is hampered by very low temperatures in winter and dry conditions in summer. In vertical mixing of humic matter bioturbation plays an important role. In many profiles we may see the digging traces of hamster (in Russian *suslik*). On stable surfaces the topsoil may become leached of lime over a few centimeters up to a depth of a few decimeters. Lime may precipitate just below the leached horizon, creating a calcic horizon. The reprecipitated lime often looks like the mycelium of fungi or is concentrated in nodules. At the 'dry end' of the chernozem zone, the soils may contain some sodium on the exchange complex (solonetz) or may contain free salt (solonchak).

Where there is only a thin layer of dark soil material on top of light-coloured carbonate rock, the soil profile is called a rendzina (USDA: Inceptisol, FAO: Regosol).

Ploughing has two effects on the soil profile. It mixes the top 0.3 or 0.4 m of the soil profile. Ploughing also moves top soil material in the ploughing direction, especially where, due to small surface relief, rises are cut off and depressions filled.

### 3.2.4 The Soil Map of 1967 | J. Delvigne & W. de Neef

In 1967, a soil map of Northwestern Crimea was issued with emphasis on the type of ions on the exchange complex as well as on free Ca and Na ions (Fig. 3.1). Consequently, transitions exist between chernozems, solonetz and solonchak. Also, attention is paid to a varying humus content. Mainly soil associations are mapped, as the soil types distinguished vary according to their position in the landscape.

In an advanced stage of research it might be possible or useful to expand on the chemistry of the soil, for instance when quantitative estimates of agricultural yields are involved.

## 3.3 LAND TYPES AND SOIL OBSERVATIONS | J. DELVIGNE

### 3.3.1 Land type classification

One of the aims of DSP 2008 was to design a land classification that gives insight into the possible uses of the various parts of the landscape by (pre-)historic man. Because no detailed information on land properties was available to base such a map on, and time to collect that information in the field was limited, a different approach was applied.

In natural landscapes, the various earth science aspects of the landscape like rock type, landform and soil type are to a certain extent interrelated. One might make use of this fact by basing a land classification system mainly or exclusively on landforms, as these are easy to observe in the field and can quickly be mapped.

To choose such landforms or landform assemblages, a transect over the research area and surroundings was drawn on a topographic map, scale 1:45,000, with contour intervals of 4.34 m. Along this transect, five parts of the natural land-

30 United States Department of Agriculture.

31 Food and Agriculture Organization of the United Nations.

scape were distinguished and, after a field reconnaissance, were regarded representative of the natural variation in land types (Fig. 3.4). Starting at the highest point along the transect, in the middle of the Tarchankut Peninsula, and moving towards the coast to the N, all five land types are transversed:

- Plateaus/Uplands
- Peninsular Hillside
- Pediment
- Coastal Lake
- Lowland Ridge

The land types will be discussed below as to their general characteristics regarding lithology, landforms, position in the drainage system, and soil (material and profile). Also, some field observations and interpretations of soil augerings are presented. For each land type these will result in a list of properties that could have influenced the use of the land by (pre-)historic man in a positive or negative way, like soil thickness or groundwater depth. Depending on the time spent on this approach, the inventory will become more complete. In due course a land classification, based on a well-defined set of land properties, may be designed and used for mapping.

The inventory is based on the present situation. Some of the land properties that are taken into account, deal with probable differences in environment between prehistoric and present times: e.g. soil thickness.

## 3.3.2 General aspects of land type Plateaus/Uplands

At about 125 m altitude, there often is a plateau-like 'step' in the landscape on which a coherent pattern of shallow drainage channels is visible. Standing out in this plateau-landscape are small, SSW-NNE trending ridges, some with kurgans on their crests (Fig. 3.5).

The slopes are gentle. Nevertheless, denudation must have been active, because the soils are thin and the surface is smooth. The smooth surface is controlled by the top of a resistant limestone bed. Quite often the limestone is exposed, showing a dipslope situation and a soil cover generally not more than 0.3 m thick. It may well be that the bedrock is folded into a series of anticlines and synclines, as many hill ranges and valleys in the area have a SW-NE orientation. This orientation represents a major tectonic orientation in the Crimean Peninsula.

It is not known how thick a layer of loess the Plateau has received at the end of the last Ice Age (Valdai Ice Age). At the moment there usually is less than 0.3 m of loamy and stony soil material on the Plateau. We found one exception, above the head of a ravine on the Hillsides, where over two m of loess occurred (Section 3.3.2.1). At this place, there must have been a depression or cavity in the limestone surface. A possible overall loess cover of the Plateau may have been reduced to a few local occurrences. For the evaluation of the environment for prehistoric man, it is important to know whether or not a (large) loess cover was still present in prehistoric times. At the moment, the Plateau surface appears exhumed, without rills or gullies. The natural vegetation cover appears to be a good protection against rain drop and running water erosion, as the cover is composed of a mat of mosses, dead vegetation parts and living leaves that spread out over the soil surface. On the other hand, burrowing animals make heaps of loose soil material without such a cover.[32] Agricultural use, especially ploughing, even when it only means scratching the soil surface, will encourage soil erosion by wind and water.

In the western part of the Plateaus/Uplands, e.g. at Hill 12, limestone is found at the surface in the lower and steeper parts of the landscape and loess in the higher and flatter parts. Apparently, before loess was deposited, the limestone had a weathered, undulating surface showing some karstic phenomena. Clearly incised valleys were also present. At the end of the last Ice Age, a blanket of loess filled the V-shaped valley, creating gentle transitions into the Plateau landscape. Erosion after the Ice Age apparently laid bare parts of the incised valleys. This erosion must have been observed (and maybe caused) by prehistoric man. As a consequence of this erosion, slope angles in the upper, loess-covered parts of the landscape increased, causing some soil erosion by solifluction (Fig. 3.6). Although solifluction was observed by means of vegetated patches ('floes') of topsoil that are saturated with water and slide down the slope with bare patches in between, no rills or gullies were found as signs of superficial water erosion in the loess landscape, and deposition of washed-down

[32] We refer here to the rodent *Suslik*, which is any of the 13 species of Eurasian ground squirrels belonging to the genus *Spermophilus*.

loess material at the foot of slopes is also absent. At Hill 10, a pit was dug in loess where foundation stones of a Hellenistic house were found (**DSP07-H10-01** Feature 2; soil profile presented in Section 3.3.2.1). The depth of the stones did not indicate any soil erosion or soil accumulation, as the soil profile inside and outside the pit fits onto each other.

Is the reason for this difference with the more maritime climate zones of Western Europe the more limited role of slope wash in the research area, and that the little slope wash that does occur, is completely washed into the sea each time? Or is soil erosion here rather a matter of wind than of running water?

3.3.2.1 Hill 10

Hill 10, trial trench 10[33]

On the side of this trench the soil profile was described[34]:

| | |
|---|---|
| 0-0.28 | 10YR 3/2 (d), 10YR 2/2 (m) loam; 0-0.15 m decalcified |
| 0.28-0.96 | 10YR 4/3 (d), 10YR 3/3 (m) loam |
| 0.96 | Bottom of pit |

Hill 10-1, trial trench 1, pit 1

General depth of pit is 0.96 m. In the centre an augering was carried out of 0.85 m depth. In depth beneath the surrounding surface, the profile description is:

| | |
|---|---|
| 0-1.6 | 7.5YR 5/4 (m) loam |
| 1.6-1.8 | 10YR 5/4 (m) loam with grit |
| 1.8 | End, very hard |

3.3.2.2 Hill 12

On Hill 12, S of the site researched geophysically, two augerings were conducted at the location indicated by the survey team as a possible site. They had noticed that the soil in this area was darker, more "humic" than in the surrounding area, although this part of the slopes was unploughed. The vegetation in this landscape type consists of grass and herbs, no trees or shrubs grow here.

In the second layer in core 12, between 0.27 and 0.5 m below the surface small pottery fragments were detected. Being 0.23 m, this layer is not very thick, and seems intact. More augering is needed to find out what the extent of this layer is, and how it is related to surface finds.

3.3.2.3 Hill 13

This location is situated at the N end of a spur. On both the eastern and western side, the spur is bordered by ravines. On the spur are two small elevations of approximately 2.5 m, one W and one E. There is no agricultural activity on the slopes. Three augerings were carried out, one on each elevation and one in between. In the augering on the eastern elevation, the compact, heterogeneous layer below the topsoil is noteworthy. This may be a cultural layer as in it three pieces of bone were found, indicating that the hill became higher since its original formation. It is uncertain whether this was a natural or an artificial process. Further augerings are needed in order to determine the nature of layers 2 and 3.

The soil in the augering in the lower part between the two elevations was very hard and compact, making it impossible to continue augering beyond 1.4 m. In the third layer, intact plant remains were found.

3.3.2.4 Loess pocket on limestone plateau

The lithological control of the Plateau implies that the rim of the Plateau is sometimes visible as a small rock ledge. Where there is a ravine on the slope, above the rim there will be a saucer-shaped part of the Plateau where slope wash towards the ravine is stronger and the Plateau is slightly lower. In this limestone environment an augering (G1) was carried out to characterize the soil. A depth of 2 m was reached. The profile of this augering can be found in Appendix 3C. Because there are many small outcrops of limestone around, this must be a filled-in depression in the limestone landscape at the beginning of the loess-deposition; possibly an erosional or karstic land form.

33 J. Delvigne, P. Attema, T. Smekalova, A. Čudin, 12 May 2008.

34 Soil colours according to Munsell Soil Color charts, d, m soil in dry (d) or moist (m) state.

3.3.2.5 Quarry with fan-delta deposits (Fig. 3.4)
This quarry is located at the outer edge of a Plateau. The quarry is not very high (ca 6 m) and more than 100 m wide and long.

At various places in the wall, thin beds of rather coarse fragments of limestone were visible. Because of the sharp-edged, chaotic structure of the composing stones, there is not enough fine material to fill the open spaces between the stones and the total structure is rather open. The composing laminae dip at about 30-40 degrees; at their lower side, the laminae show a concave break of slope. The deposits are interpreted as delta deposits. In other places, the concave lower end of the laminae is missing. These deposits are interpreted as (fluvial) fan deposits. Both types of deposit indicate that once, when the sea level was higher or the land level lower, a river must have opened up into the sea at this point. This river cannot have been very long, as the limestone fragments that are transported to the coast by this river, are hardly rounded.

3.3.2.6 Stone circles on Plateau (Fig. 3.2)
On the flattish part of the Plateau, several stone circles had been recognized in 2007 and their origin attributed to human activity (possible Bronze Age tombs). The stone circles consist of a number of large, flat slabs of limestone, together roughly having a circular shape and a diameter of 4 to 10 m. The openings between adjacent slabs are only a few cm wide and have more the character of somewhat widened cracks than of independently bordering edges. This point of view even becomes more convincing where two thin slabs lie on top of each other. Openings between individual stones have the same shape in both layers, making it more likely that they are cracks that developed simultaneously in originally larger blocks. The circular appearance is not so much caused by a circular circumference but more by a roughly circular mass of dark fine earth, somewhere in the middle of the group of slabs that make up a 'stone circle'. It seems more likely that this heap of soil in the centre is pushed to the surface by digging hamster that are common animals here. Maybe this soil remains moist somewhat longer, causing a bit more dissolution. In the soil in the centre there is some enhanced dissolution, causing this soil to be somewhat 'cleaner' (free of limestone gravel).

On close inspection the origin of the stone circles as man-made features was thus refuted. They are now interpreted as a karst phenomenon (see also Chapter 4.4 for an excavated example of such a 'stone circle').

## 3.3.3 General aspects of land type Hillsides

The second land type is formed by the peninsular Hillsides between the Plateaus/Uplands on the inland side and the lower parts of the landscape on the coastal side, covering a vertical distance of some 70 m (Fig. 3.7). This vertical distance probably represents a period of relative strong uplift of the inland parts of the peninsula. Weathered limestone, moved down the slopes of the uplifted parts, must have washed away easily at the coastline, thus maintaining steeper slopes than inland. In fact, there even might have been cliffs off the coast, as is the case in present times. At the moment, slope angles generally vary around 10-12 % and the soils are thin. At irregular intervals, the Hillsides are dissected by ravines. Ravines are the places where springs may be found.

Erosional phenomena could well have developed on the Hillsides before the deposition of loess. Because of the significance of loess in land use, ravines should be checked for the occurrence of pockets of primary deposited loess and for loess that has been washed down from above. One such an occurrence was found, at quarry no. 1 (Section 3.3.4.3), where a layer of at least 4 m of loess was exposed. The absence of non-loess components and absence of stratification makes it likely to be a primary loess deposit. Possibly, it was deposited at the lee side of a ravine incision or coastal cliff, as only 100 m away, in quarry no. 2 (Section 3.3.4.3), limestone was found at the same level as the loess.

Ravines can also have been used to store drinking water for man and cattle. Both those that have been deepened as those that have only been dammed off, could have served such a purpose (Chapter 5.1.6). While driving around, both types were observed. From such observations it cannot be concluded, whether they are prehistoric or recent.

## 3.3.4 General aspects of land type Pediment

The Pediment is the land type that extends from the peninsular hillsides at a much smaller slope angle towards Lake Džarylgač (a typical slope angle being 5%) (Fig. 3.8). It is regarded as a 'wash'-surface, created both by erosional slope wash on weathered rock and depositional slope wash leaving behind washed-down material. Like on the Plateau, many places were observed where limestone rock constituted the surface of the Pediment or were only covered by a thin veneer of loose soil material. In this land type the ravines associated with the Hillsides above, usually become less incised and gradually disappear down the Pediment.

To gain more insight into the character of the Pediment, two locations were observed in more detail: the lower part of Hill 11-01 and survey field DSP08-F04. The former is situated on the higher part of the Pediment and is characterized by a ravine system cut into limestone rock. The bedding of the limestone is as usual almost horizontal but difficult to determine as the limestone is broken into irregular blocks and stones, rarely protruding much above the surface.

Where a ravine cuts through a more resistant limestone bed, the bed is exposed in a horseshoe shape, the opening pointing to the upper side of the ravine. When a slope surface is developed in limestone containing a more resistant bed, a more or less clear rim of limestone slabs is exposed. At a few places in the area in between rims, loess-like soil material was visible. In more stable positions, stonefree and over 0.15 m leached chernozem soils had developed.

Field DSP08-F04 is a large (2 km x 2 km) ploughed field at the lower side of the Pediment where 15 hand augerings were carried out along a transect from high to low to get an impression of the soil material and soil profile in a part of the landscape where one would expect accumulation of slope wash material. The description of the soil profiles along the transect is presented separately (Appendix 3A), the interpretation and results are presented here (Section 3.3.4.1).

#### 3.3.4.1 Interpretation of soil augerings along the transect across the Pediment

##### 3.3.4.1.1 Introduction

Along a transect across the major part of the Pediment on the S side of Lake Džarylgač, 15 hand augerings were carried out in a large surveyed field (DSP08-F04; Fig. 2.13). Already from a distance some whiter and some redder zones could be distinguished in this ploughed field. The transect and the augering points were chosen to include the centres of these zones. The Pediment surface along the transect is slightly concave and slopes down from 44 m altitude to sea level over a horizontal distance of about 1,300 m.

Aim of the augerings was to obtain information on the nature and variation of the soil material and soil profile and to gain understanding of their genesis and significance to agricultural use by prehistoric people.

##### 3.3.4.1.2 Soil material

In the more humid climate of Western Europe, a concave lower part of a slope is usually shaped by the downslope transport of soil material by the processes of slope wash and other types of mass movement. The thickness of this colluvium, with a maximum in the centre of the concave part, reflects that genesis of diffuse accumulation lower down the slope. As this colluvium often covers old soil profiles, it means that since the formation of the older soil profile, the supply of soil material from above has exceeded the downslope export of colluvium. This can often be attributed to human cultivation of the land.

Along the transect, the depth below the surface at which bedrock was reached, did not increase much down the slope, from 0.55 m on average over the seven upper augerings to 0.8 m over the seven lower augerings. This means that slope wash is not very strong. This is supported by the absence of rills. Not just the drier climate is responsible for less soil movement. The soil material here is lime-rich and contains quite a bit of organic material, creating a well-developed soil structure with good water-absorbing properties. Also, the human use of the land has not been as intense here as in Western Europe.

Bedrock underneath the soil material consists of limestone. From the soil profile descriptions it is learned that almost always there is in the profile a layer with a loam texture and a colour that has a strong resemblance to loess. It could well be a more or less *in situ* loess layer. An alternative explanation for the origin of this loam is as a leftover of weathered limestone. However, the weathering of limestone produces clay, not loam. The fact that, where the loam layer is thick, it has all the characteristics of loess (as in augering T10), makes it likely, that where the layer is thin, its origin is also loess with a possible admixture of weathered limestone. On this Pediment, being the upwind side of a hill, it is not surprising that the thickness of the loess layer is small. On the lee side, one would expect a thicker layer.

##### 3.3.4.1.3 Colour contrasts in ploughed fields

The colour contrasts between whiter and redder zones in this field that were observed from a distance, turned out to be zones with more limestone fragments at the surface (whiter zones) and zones where the loam in the topsoil is somewhat redder (7.5YR in stead of 10YR), respectively.

The whiter zones might have been slight rises in the weathered limestone bedrock. Ploughing of the field has flattened the surface, thereby removing the topsoil of rises in the surface and the weathered limestone. The redder zones probably are places where the loess is slightly deeper weathered.

3.3.4.1.4 Interpretation of soil augerings at two archaeological sites

Five augerings carried out in a row, were conducted across site DSP08-F04-07. This site is a ploughed scatter of Hellenistic pottery on the NW end of survey field DSP08-F04, W of an erosion ravine that runs towards Lake Džarylgač. The row of augerings runs N-S, at 25 m interval. The southernmost augering location lies between augering 14 and 15 in the long augering transect.

Noteworthy is, that in three augerings conducted in the core of the archaeological site, a very hard, compact layer is present above a softer layer of loamy clay. In augerings at either side of these, this layer was not detected. In the northernmost augering, the soft layer lies directly below the ploughed topsoil. In the southernmost augering, between augering 14 and 15, the hard layer is present above a loam layer which is neither plastic nor soft. The hard layer may owe its compactness to the precipitating of lime.

The limestone bedrock was reached at 1.1 and 1.83 m below the surface, respectively. The ploughed topsoil is 0.4 to 0.45 m thick. The intact layers between the topsoil and the bedrock are therefore 0.7 to 1.4 m thick. The hard, compact, yellowish-brown loam layer possibly is the finds layer of the archaeological site: in one case, archaeological remains were found in this soil layer. However, in none of the other augerings, archaeological remains were detected.

Two augerings were conducted at site DSP08-F04-17. This site was found in the same field as the previous site, but further to the S, and 300 m E of the augering transect. The site is also classified as a Hellenistic scatter. The soil at the find location is notably lighter in colour than the surrounding area. The site lies roughly in the middle of the field, W of an unploughed area. In the unploughed area no augerings were conducted, as the farmer was just starting to plough this part.

In one of the two augerings, the second, dark yellowish-brown layer may be interpreted as a finds or culture layer. Small pottery fragments were detected in this layer. More research through augering would be necessary to check, whether this layer is indeed the culture layer of the find location on the surface, and how far this layer extends.

The bedrock was reached at a deeper level than in the augerings at the N side of this field, at 1.48 and 2 m. The two augerings in the cross-area transect nearest by, 300 m to the W, had a soil thickness of 0.75 m (augerings T7 and T8).

3.3.4.2 Loess quarry next to limestone trench (Fig. 3.9)

On the way to the E Plateau, just after leaving the asphalt road, the main track is winding along the edge of a loess quarry with its base about 3 m below the track. 100 m to the S side of the track there is a trench exposing limestone beds up to a sharp, wavy boundary with loamy material on top. The quarry lies at the boundary of the land types Hillsides and Pediment.

In the walls of the quarry (quarry 1) the top two m are exposed:

| | |
|---|---|
| 0-20 | 10YR 4/3 (d) loam, well structured (loess) |
| 20-40 | 10YR 4/4 (d) loam, well structured |
| 40-60 | 10YR 4/4 (d) loam |
| 60-100 | 7.5-10YR 4/4 (d) loam |

Between 1 and 4 m below the surface, a network of round, tubular openings with a diameter of about 1 cm is found in the loam. On the inside, the tunnels are coated with grey-brown humic clay. Some smaller tunnels are completely filled with it. Lime migrates in the soil in this climate in a different way, as is shown by the occurrence of so-called lime-mycelium deeper than 0.5 m.

The slope above the quarry has an angle of 4%.

The trench 100 m away (quarry 2) is about 100 m long and about 5 m wide. In the middle it is about 3.5 m deep, sloping up to both ends. Probably it was dug to check the size of the loess body. In the seaside wall of the trench there is a sharp, wavy boundary between the loess-like deposits on top and the weathered limestone underneath. The loess-like layer might be a colluvial layer.

To learn about the landform that has been filled-in with loess and made the loess-quarries possible, one augering was carried out halfway between the track and the limestone trench (augering G2) and another on the other side of the track close to the track (augering G3). The descriptions of the profiles found in these augerings can be found in Appendix 3C. The soil profiles from these augerings unfortunately do not give enough information to decide on the landform present before the loess was deposited.

3.3.4.3 Red soils in ploughed field on the Pediment
Like the large field, where the transect runs through, there are other fields that show zones of redder colour, as well as darker zones. To check whether a redder soil is only a matter of redder loam, three augerings (G6, G7, and G8) were carried out in other 'red' zones. The soil profile descriptions of these augerings can be found in Appendix 3C.

### 3.3.5 General aspects of land type Coastal Lakes

Coastal Lakes are drowned parts of a river valley or former bays or lagoons (Fig. 3.10). Rivers, coastal currents and waves have delivered the sand that has helped building spits and bars that have closed them off and turned them into lakes. Depending on how and to what extent lakes are closed off, they can either be salty or contain fresh water. Usually, Coastal Lakes are not very deep, as quite a bit of the bed load of the river that opens into a lake, has been or is deposited there (Section 3.2.2). These sediments may be regarded as 'correlative sediments': periods of strong or little sedimentation in the low parts of the river system should correlate with periods of strong or little erosion in the higher parts of the catchment. This means that to a certain extent changes in environmental conditions can be derived from changes in the corresponding, correlated sediments. Periods of loess erosion (by surface wash) in the higher parts of the catchment may be recognized this way.

At least at the N side of the lake, wave action has formed a cliff in loess deposits that are over 5 m thick. Near the water line, the loess has been hardened by calcite crystals.

3.3.5.1 The Coastal Lake augerings (Fig. 2.13)
Augering T16 of the transect across the Lowland Ridge is similar to augering T15 on the other side of the Coastal Lake (transect across Pediment). The soil profile descriptions of both augerings can be found in Appendix 3A. As mentioned in Section 3.3.6.1, a thickness of 2 m of loess was observed here although the lakeside cliff nearby suggests that the sediment has a thickness of more than 3 m. This thick sediment in the lowest part of the research area may be the result of Eolian deposition of loam in a large valley, the present Lake Džarylgač. It is also possible that the loam was deposited at the bottom of the lake after being eroded off the slopes of the catchment during the Holocene. At the lakeside cliff, it could be observed that the lower layers of loam were reddish in colour. This colour may be the result of oxidation of iron compounds in the loam or in emerging groundwater.

3.3.5.2 The cliff on the northern lake shore (Fig. 3.10)
Over a distance of hundreds of meters along the N side of the lake, the cliff is about 5-6 m high. Middle part of it is well exposed:
– near the waterline there appears to be reddish stone. More probable it is loess loam cemented together
0.5-1 m above water level, 7.5YR 4/6 (m) clayey loam; many cementations by calcite-crystals
1-2.2 m above water level, ditto colour loam. Some pieces of limestone
1.5 m from this point in small pit, 10YR 4/3 (d) = 10YR 3/3 (m) loam

### 3.3.6 General aspects of land type Lowland Ridge/Coastal Cliff

The final land type in this classification consists of the Lowland Ridge (Fig. 3.11). It alternates with the Coastal Lakes and bays in building the coast, as the coast line cuts the parallel river valleys and Lowland Ridges at a small angle. The Lowland Ridges are neither high nor steep. At the coast marine erosion has cut a cliff, that can be more than 25 m tall and shows that the core of the ridge is composed of limestone, which at the cliff is covered by up to a few meters of loess. The Lowland Ridge can be regarded as developing autonomously.

To obtain an impression of the variations in soil materials and soil profiles across the ridge, a number of hand augerings was carried out along a transect. The results of these augerings are presented separately (Appendix 3B).

The ridge is slightly undulating. Because of ploughing, the undulations are visible by a regular alternation of dark and light zones, respectively showing the A and C horizon of the soil profile. Probably these horizons occur in weathered loess, in which case the undulations may be formed by wind. At the outskirt of the village Mežvodnoe a quarry (quarry no. 4) was found in at least 5 m of loess (Section 3.3.6.2). The loess may well be filling an erosional valley in the limestone ridge.

Has the presence of a loess pocket in any way influenced the choice of the settlement location? Like quarry no. 4, which is close to Mežvodnoe, quarry no. 1 is also close to a settlement (Section 3.3.4.3).

3.3.6.1 Interpretation of soil augerings along a transect across the Lowland Ridge (Fig. 2.13)

3.3.6.1.1 Introduction

A total of 14 augerings were conducted in a transect N of Lake Džarylgač, at 100 m intervals. Augering T16 was conducted at the lowest point, on top of the cliff at the lake (3 m above water level). Elevation difference with the highest augering location on top of the lowland hill, at augering T26, is 40 m. The northernmost augering location, near the coastal cliff on the Black Sea shore, was 30 m above sea level. The augerings are described in detail in Appendix 3B. Some general conclusions are presented here.

3.3.6.1.2 Soil material

The augering section shows a gradual decrease of soil depth between augerings 16 and 22. At augering T16, on top of the lake cliff, a depth of 2 m was reached without reaching underlying bedrock. Indeed, the cliff rose approximately 3 m above lake water level, and most likely continued below. Unfortunately, it was not possible to determine the depth of local bedrock on the lake shore. Contrastingly, bedrock was visible directly at the surface at augering location T22. The bedrock appears to be steeper at the S slope of the Lowland Ridge than at the N slope. The minimum depth of soil upslope implies that either less soil was deposited originally, or that more has eroded and moved downslope.

Augering T22 was conducted approximately 400 m S of the ridge top, in a fallow area partly overgrown with grass. In the section, this location appears to be a slight outcrop in the slope. With winds coming predominantly from the N, it is remarkable that a seemingly sheltered part of the ridge would be without Eolian loam sediment. In augering T26, a location on top of the coastal ridge and most exposed to northern winds, a 1.12 m thick loam sediment was observed. Therefore, it seems more likely that other erosional processes than Eolian movement of loam resulted in the uncovered bedrock.

South of sediment-less augering T22, the section shows an increase in sediment depth towards the lakeshore. Augering T18 is located slightly lower than the augerings to the N and S (19 and 17 respectively), in a depression of the slope. This may well be a pocket in the underlying limestone, however the loam fill has a similar depth as the soil on both sides.

The three augerings T23, T24 and T25, at 100 m from each other between the ridge top (T26) and the location where bedrock was observed at the surface (T22), yielded loam layers of 0.25 m or less. At augering T22, approximately 40% of the surface was covered with limestone fragments, and also augerings T24 and T25 were conducted at locations with many limestone parts on the surface. In augerings T23 and T24, the soil consisted of loam mixed with limestone, in augering T25, no limestone particles were observed. The many limestone parts on the surface at this part of the slope suggest that loess erosion has taken place, followed by limestone weathering (?). Limestone fragments must have moved over the loam from a point where weathered limestone was right at the soil surface.

At 40 m from the coastal cliff, the sediment was 0.7 m thick. Here the slope flattened out to an almost level surface. At the coastal cliff itself, a loam sediment several m thick was observed on top of the limestone rock.

3.3.6.1.3 Soil profile

On top of the coastal ridge, loam sediment with a thickness of 1.12 m was observed. Unfortunately, it is not known to what extent soils have been disturbed, when the asphalt road following the highest parts of the coastal ridge was built, yet augering 26 was conducted in an almond yard where no severe levelling seemed to have taken place. No stones were observed on the surface at this location, and a very dark top layer was present over lighter, undisturbed loam layers. The three augerings conducted on the N slope of the coastal ridge, from augering T26 down to the Coastal Cliff, all show comparable dark topsoil. The soils in these augerings are less thick than in augering T26. Augering T28 yielded a dark loam layer directly on top of the bedrock, in augering T27 a thin, lighter-coloured loam layer was present between the dark topsoil and bedrock. The very dark topsoil in augerings T26 through T29 suggests that circumstances must have been stable enough for soil formation. Indeed, at the almost level surface of augering T29, humic material formed a 0.37 m thick dark topsoil in a 0.7 m thick loam sediment. The colour of the topsoil in these four augerings is typical for chernozems.

None of the topsoils S of augering T22 are as dark as those observed on the N slope of the coastal ridge. In augering T16, the topsoil is brownish, as are those in augerings T18, T19, and T21. The underlying loam layers of augering T16, T17 and T18 have lighter colours typically seen in loess sediments. The soil samples seen in augerings T16 through T21 can be classified as underdeveloped chernozem soils, where humic material on the surface has not developed completely into a thick, dark top layer. Overall, it seems that darker topsoils occur in level parts of the research area.

3.3.6.1.4 Differing soil colour values

At augering location T19, a reddish loam layer was observed below a darker, brownish top layer. The soil sample was taken at a very gently sloping part of the coastal ridge. Together with loess quarry 4 (Section 3.3.6.2), this was the only location N of Lake Džarylgač where a 7.5YR Munsell value was measured. The reddish colour is either the result of post-depositional weathering of loess where it was in a stable-surface situation (where the profile development intensifies because there is no removal of material from the top) or it is a remnant of a pre-loess weathering of limestone, which has eroded away elsewhere. The augerings to the N and S on the same slope did not yield a similar red value.

3.3.6.2 Loess quarry 4 near Mežvodnoe (Fig. 3.12)

This quarry has a SW-NE orientation, the same as the coastline. The quarry bottom is a depression in the landscape, with a (dry) overflow towards the village.

The main wall (at a bend) of this quarry shows some 5 m of loess. The profile in estimated figures:

| | |
|---|---|
| 0-0.5 m | Greyish yellow, not structured |
| 0.5-2 m | 'Bleached' greyish yellow, moderately structured (some cracks) |
| 2-4 m | Most red colour (7.5YR 6/6), best structured |
| 4-4.5 m | Olive-green (10YR 6/6), massive |

Looking at the profile at the bend, the exposed part of the profile diminishes to the right. At the end is a sub-recent tile kiln.

3.3.6.3 Interpretation of augerings at a Bronze Age site on the Lowland Ridge

Two crossing rows of, altogether, 11 augerings were conducted at a Bronze Age site (DSP08-F06-01) found during the 2008 intensive survey of a ploughed field on the S coastal ridge slope. The main objective of these augerings was to establish how thick the sediment was at the site, and whether intact culture layers could be located. The soil profile descriptions can be found in the site entry in Appendix 1. The survey team observed that the surface at the site was more greyish than the surrounding ploughed field.

In nine of the 11 augerings at this site, bedrock was reached within 0.5 m below the surface. All of these augerings were conducted within the find concentration. The four augerings conducted outside the defined site borders all yielded a lighter brown, intact loam layer (10YR 5/4) underneath the humic topsoil. In augering S27, this layer was only 5 cm thick.

From these augerings, it can be concluded that the soil colour of the top soil is most likely influenced by human activity to the extent of the site boundary. Further research on the humic content and geochemical values of the top soil may add to our understanding of these activities, but this has not yet taken place. Unfortunately, this part of the coastal ridge was heavily ploughed and augerings at the site have not proven it likely for an intact finds layer to be present.

3.3.6.4 Soils of a Bronze Age site near Skalistoe (Coastal Cliff)

Four augerings were conducted at site DSP08-F03-02 (Skalistoe 2). The site is located in a shallow depression east of a ravine, which was heavily overgrown during the 2008 campaign. The field west of the ravine had not been mowed or ploughed in 2008 and was therefore not investigated by the survey team.

Two augerings were conducted in the grassy field west of the ravine, and two augerings within the site nucleus in the ploughed field. The objective of these augerings was to establish the thickness of the sediment and to see whether intact cultural layers containing archaeological indicators could be observed. The full augering descriptions can be found in the site description in Appendix 1.

The two augerings in the grassy field (10 and 60 m SW of the ravine) yielded very compact, hard loam soils containing limestone fragments. Depths of 30 and 70 cm respectively were reached, after which soil got too hard to penetrate, yet larger limestone parts suggest that bedrock was close.

Augering 3, within the site nucleus (10 m E of the ravine), yielded a similar profile with bedrock at 30 cm. Augering 4 (60 m E of the ravine), however, had a 116 cm thick soil profile. Here, the top 50 cm of loam showed no effervescences when tested, the underlying loess layer (50-116 cm) effervesced indicating a lime content. Bedrock was not reached due to the great compactness of the soil beyond 116 cm.

In contrast to site DSP08-F06-01, no variation in the colour of the top soil or humic content was observed to correspond with the pottery concentration. In none of the four augerings, archaeological indicators were recorded. Although augerings 1-3 showed shallow soils, the thickness of the sediment in augering 4 seems promising for possible intact layers in the eastern part of the site.

### 3.3.7 Triplex soils and landscape development

Loess is a wind deposit and is, contrary to water deposits, capable of covering a complete landscape. Towards the end of the last Ice Age, as the climate had become very dry, such a cover was probably formed over the limestone landscape of the Northwestern Crimean peninsula, thereby bringing slope processes to an end. At the beginning of the Holocene, the climate changed towards its present state, starting slope processes that are in accordance with the present climate. Although these slope processes have affected the post-loess landscape for about the same period of time (10,000 years), their effects on the landscape must have created typical local differences. From a theoretical point of view, the following effects on soils and landscape can be expected:

– Where a continuous layer of loess has been left behind, the top soil material remains loess, lying on top of the pre-loess limestone weathering profile. A landscape characterized by loess deposition will show rounded, gentle forms. Loess erosion is predominantly caused by slope wash and will result in rills and gullies. Some of the removed loess may have accumulated further down the slope, creating a layer of increased thickness and mixed composition.

– Where the loess has in patches or certain zones of hill slopes been removed, limestone weathering profiles should be found at the surface in the eroded patches. The limestone weathering products may spread out over the loess downslope of the patch, thereby creating a zone of mixed material or limestone weathering products on top of loess.

– Where the loess has completely been removed, the exposed land surface will be the pre-loess land surface. However, local landforms of strong relief like ravines or sinkholes, may still be completely filled by loess, creating a landscape in which the observer can be fooled by the landforms.

## 3.4 PALEO-LIMNOLOGICAL INVESTIGATIONS IN LAKE DŽARYLGAČ | V. STOLBA, D.A. SUBETTO & T.V. SAPELKO

The Džarylgač Lake, along with the adjacent lakes of Sasyk and Karlav as well as the Bakal' Lake to the NE and lakes Karadža and Kipčak to the SW, constitutes the Tarchankut group of the Crimean saline lakes.

A valuable summary of the former studies conducted at the lake in the 1920s by A.I. Dzens-Litovskij and others can be found in the volume *Soljanye ozera Kryma [Saline lakes of Crimea']* (Kurnakov et al. 1936), which still remains the basic reference work on the topic. In 2005, limnological investigations of Lake Džarylgač were carried out by an international team within the framework of a cross-disciplinary project 'Northern Black Sea in the 1st Millennium BC: Human history and climatic changes' (Stolba et al. 2006; Subetto, Sapelko & Stolba 2007; Subetto et al. 2007a; 2007b; 2008).

The lake with a total area of ca. 78 ha renders itself a submerged mouth of the Vodopojnoe Ravine, which, becoming gradually isolated from the sea by a sandbar, transformed into a lagoon and subsequently into the highly mineralized coastal lake. Its N and S shores are actively abrading. The over 600 m-wide sandbar, which separates the lake from the sea, consists of the oolitic limestone kernels of the Neogene period and shell debris. Along with the modern specimens the latter often includes the shells of extinct molluscs such as *Cardium tuberculatum*, *Tapes Colverti* etc. (Dzens-Litovskij 1933, 586; Longinov 1955, 158, 165; Zenkovič 1960, 144). The modern level of the lake is 0.4 m below that of the sea (- 0.7 m according to Longinov's measurements in 1945), the depth not exceeding 1.1 m.

The sampling of the one chosen site in the W part of the lake (489180 / 5047189, see Fig. 3.4) carried out in 2005 retrieved a 4.15 m-long sediment sequence which has been analysed for lithostratigraphy, pollen stratigraphy as well as diatoms and ostracods.

The bottom sediments are represented by grey, yellowish-grey and greenish-grey silts with a high content of mollusc shells. Unlike the sediments of Lake Saki studied by the same team in 2005, they do not feature a clearly expressed layered structure, a characteristic only observed in the upper 1.5 m section of the core. This made the correlation between the varve chronology and the absolute radiocarbon dating virtually impossible. The AMS C14 dates have been produced at the Radiocarbon Laboratory of the Institute of Physics and Astronomy, University of Aarhus, for various parts of the sequence. They place the beginning of marine sedimentation connected with the New Black Sea transgression, when the risen sea penetrated into the ravines and river valleys at around 7200-7050 cal BP, giving an average sedimentation rate of about 0.6 mm per year. Changes in the associations of ostracods observed along the retrieved sediment sequence and compared with the absolute radiocarbon dating allow the detailed reconstruction of the lake's history (Neustrueva et al. 2007; Subetto et al. 2007a). The transition from the open sea environment conditions to the lagoon is likely to have taken place around 5,590-5,350 BP, with the complete separation from the sea and the formation of saline lake to occur ca. 4,700 calendar yr BP.

The retrieved pollen data have been used to reconstruct the changes in vegetation and climate (Sapelko & Subetto 2007). Whereas the earlier phase of the Atlantic period was characterized by xerophytization of the steppes and lower share of tree species in conditions of warming climate, the later, more humid and cooler phase of this period was accompanied by expansion of forests, in which mixed oak stands (*Quercetum mixtum*) prevailed. The subsequent Subboreal is characterized by warmer and drier climate as well as by major appearances of crop cereals. The arboreal pollen spectra, in which *Pinus*, *Betula* and *Alnus* now preponderate, give way to *Chenopodiaceae* and *Artemisia* species, which mark the transition to arid and semi-arid steppe vegetation. Some increase in humidity is observable after 3,680±55 BP, as indicated by a rising proportion of tree pollen. The increase of tree pollen (20-40% of the assemblage), *Pinus*, *Picea* and *Abies* being predominant, also marks the transition to the Subatlantic period (2,575±41 BP), in which, as evidenced by pollen of crop cereals, the human factor plays an important role. The formation of modern vegetation takes place in the second half of this period.

## 3.5 (SUB) RECENT LAND USE | P. ATTEMA

The most prominent economic resource of the Crimean landscape is its potential for large scale agricultural production because of the fertility of its soils and the generally flat landscape. This potential was recognized by the Soviet planners, when Crimea was included in the Ukrainian Soviet Socialist Republic in 1954. Up to 1991, when the Autonomous Republic of the Crimea was formed, the main agricultural producers in Ukraine were collective and state farms (kolkhozes and sovchozes). Remains of abandoned Soviet farms, now often in ruinous state and used as quarries for building materials by local villagers, are ubiquitous in the study area as a testimony to the period (Fig. 3.13)

From our discussion of land types in the study area, it seems only logical that the Pediment and the Lowland Ridge were singled out as most suitable areas for large scale crop cultivation on tractor ploughed fields. This is because of their gentle relief, low stoniness, and fertile soils. In the current local agricultural economy following the land reforms of the 1990s, these areas are still under the plough in a rotation system (mainly cereals and sunflowers). For the survey this meant that as a rule no adjacent fields could be surveyed, adjacent fields being alternately ploughed.

However, also parts of the Hillsides and Plateaus/Uplands were subjected to large-scale farming as part of a more general trend in the former Soviet agricultural policy of pushing agriculture into zones less suitable for large-scale farming. Apparently this happened also in our study area. Air photos and geomagnetic maps clearly show traces of former land use, part of which can possibly be attributed to ultimately failed attempts at large scale farming of marginal parts of the Hillsides. Other traces visible on especially the geomagnetic maps may be remnants of smaller scale ploughing activity in recent times (Figs. 4.152-153, 4.235, 4.249).

Nowadays the Hillsides are for the greater part being used as pastureland, although on hills with sufficient soil cover crop cultivation continues. This is conform the current tendency in Ukrainian land use development, which shows a considerable decrease in arable land, followed by increases in pastureland and abandoned land (Kucher 2007, 6).

During the Soviet period, individual households played an insignificant role in the production of crops and were mainly operated by kolkhoz and sovchoz members on small land plots (Kucher 2007, 4). In the study area such small land plots are still exclusively found in the vicinity of the small villages that dot the landscape, while grazing takes place in the surrounding steppe. This is a settlement and land use pattern that continues that of the 19th century AD Tatar period and probably earlier. Where substantial cash crop cultivation takes place, this is nearly always on vast parcels of up to 2 km$^2$. From 1992 onwards, agricultural land in Ukraine increasingly went over to at first mostly collective and from 2000 on to private ownership (Kucher 2007, fig. 6).

A drawback of large-scale farming is soil erosion. According to Kucher (2007, 8) soil erosion is the most significant environmental issue in Ukrainian agriculture, while also a significant decrease in soil fertility has been noted. It is estimated that almost 40% of agricultural land is subject to water and wind erosion (see also Van Zon 2000). Indications of erosion were indeed observed both in the lower areas (Lowland Ridge and Pediment) and in the higher areas (Plateaus/Uplands, Hillsides).

From the perspective of site recovery we may conclude that post-war developments in land use have led to an increased visibility of the archaeological surface record in the Pediment and Lowland Ridge, while recent developments have reduced its visibility on the Hillsides and Plateaus/Uplands. This however only holds for settlement sites, not for the burial mounds (kurgans). In the former land types, settlements will typically surface as pottery scatters, in the latter land types, these can only be detected when there are stone structures present, although pottery scatters do occur sporadically. Because of their elevation, kurgans will have an equal chance to be detected in all land types. It follows that settlement sites, that in the archaeological record consist of pottery scatters only, are likely to be underrepresented on the Hillsides and Plateaus/Uplands, despite the adjusted methodology (see Chapter 2.3.1.2).

From the perspective of the evaluation of ancient land use the following aspects need to be considered:

– It is highly likely that the high estimations of soil erosion and substantial loss of soil fertility in Ukraine due to the negative impact of post-war large-scale farming practices also apply to our study area. This would mean that in the past, the soil cover on the whole may have been thicker and that soil fertility was higher. This may imply that the present-day contrast in the productive potential of crops between Lowland Ridge and Pediment on the one hand and Hillsides and Plateaus/Uplands on the other may have been less marked in the past, considering that today's requirements for mechanical ploughing to farm preferably only slightly undulating vast surfaces does not apply for past farming practices.

– Kucher (2007, 9) asserts that in the southern part of Ukraine, irrigation is essential for agriculture. As non-saline surface water is largely absent in the study area, the success of farming on a substantial scale in the past must have been heavily dependent on precipitation. With respect to the pre-war settlement pattern of 19th century AD Tatarian villages, it has been observed that their settlement locations coincide with those locations in the landscape where the groundwater level was within reach of man-made wells (Chapters 5.1.6). One of the central elements in these Tatar village communities is the communal well (Fig. 5.1). The hydrological situation indicates therefore that large-scale farming in the Crimea in the past, as supposed to have been undertaken during the Early Hellenistic period (see Chapter 6.3), may have been very vulnerable to periods of drought. Artificial and natural water pools in the ravines will have favoured extensive grazing on the Hillsides.

Salt needed for man and his animals could be extracted from the Coastal Lake, where salt works are recorded on the 1897 topographical map. The likelihood of salt extraction from these lakes in antiquity is discussed in Chapter 6.3.3. Detailed information on sub-recent topography and land use in the survey area is given in the catalogues of surveyed fields and hills (Chapter 4.4 and 4.5).

## CHAPTER 4

# The settlement data of the DSP landscape

*Compiled by K. Winther-Jacobsen*
*with contributions by P. Attema, P. Guldager Bilde, T. de Haas, S. Lancov, W. de Neef, T. Smekalova, V. Stolba & C. Williamson*

## 4.1 INTRODUCTION | P. GULDAGER BILDE

This chapter presents the settlement data collected in the course of the DSP fieldwork. It draws together various sources consisting of both published and yet unpublished work as well as DSP's own surveys. The chapter is ordered by landscape type. Section 4.2 focuses on the coastal zone between Černomorskoe to the W and Masliny to the E. It discusses settlement data from excavations and extensive surveys, in Russian referred to as *razvedki,* mainly carried out by Russian teams during the 20th century. Section 4.3 does the same for the major ravines dissecting the DSP study area. Apart from the major sites of the Černomorskoe Settlement and Panskoe I located at the outer ends of the ravines near the coast, the ravines have received little attention in past research. Unfortunately, investigations of this landscape feature by the DSP have necessarily remained limited due to the restricted time frame. They certainly merit further intensive investigations. Survey by the DSP teams was foremost focused on a transect cross-cutting the study area (Fig. 2.1). The settlement data resulting from the DSP survey are presented in Sections 4.4 (the Lowland Ridge and Pediment zone) and 4.5 (the Hillsides and Plateaus/Uplands). In combination, these data serve as the base material for our interpretations of the settlement and land use dynamics presented in Chapter 6.

## 4.2 THE COASTAL CLIFF/LOWLAND RIDGE: PAST AND CURRENT INVESTIGATIONS | P. GULDAGER BILDE WITH CONTRIBUTION BY T. SMEKALOVA & V. STOLBA

### 4.2.1 Introduction | P. Guldager Bilde

The coastal zone of the DSP survey area has been explored throughout most of the 20th century both by way of *razvedki*-type of surveys and by excavations (Fig. 1.1). Consequently, it was only investigated extensively in the limited period available by the DSP, foremost as a means to familiarize ourselves with the region. Due to the poor state of publication, the low accessibility of the few existing publications as well as the problem of language to a western audience, we have chosen to include a fairly detailed analysis of the known sites within the individual zones of the DSP survey area along with the new sites identified by the project (see Appendix 1).

Two major ancient settlements along the coast frame the survey area, the Černomorskoe Settlement (conventionally identified as Kalos Limen) to the W and Masliny to the E. In between these two sites is located the settlement of Panskoe I and along the shoreline are further situated a number of fortified and unfortified farmsteads as well as other types of sites.

The coastal zone was much more densely occupied in antiquity than is the case in later historical periods (Figs. 6.7-8, 6.10). In fact, just one of the post-antique sites is truly coastal (Meževodnoe) and the remaining localities are therefore discussed in the section on the coastal Lowland Ridge N of the Vodopojnoe Ravine/Džarylgač Lake.

The earliest site localized within this part of the DSP study zone is Jarylgač Severnoe, which dates to the Neolithic period (unpublished; mentioned in Blagovolin & Ščeglov 2002, 286, pl. 179.VI; Fig. 6.7). This is the only Neolithic site identified in the area, and therefore, we cannot evaluate, whether this location is typical for the period. However, it is noticeable that the site is in close vicinity of the Panskoe group of sites (ca. 500 m from Panskoe I).

Concerning the Bronze Age, the only coastal site in the area is the multi-period Baj Kijat 1, which was settled already from the Middle Bronze Age, and which was resettled even in the Hellenistic period (Fig. 4.1). It is located at a gentle slope facing the present coast sheltered by a small bay. 250 m further NW is located Baj Kijat 3, probably a Hellenistic farmhouse (Fig. 4.2). In general, Late Bronze Age sites are not found along the coast; instead, an inland location was preferred (see Chapter 6.2.1). The reason for the coastal location of Baj Kijat 1 may be a different site preference in the Middle Bronze Age. However, coastal erosion may also have had the effect that a number of sites have been irretrievably lost. This situation is for example suggested by the presence of three kurgan-lined roads aiming at two different bays immediately W of Bakal', which seems to suggest that the sites themselves are no longer preserved (See Chapter 6.6).

During the Late Classical-Early Hellenistic period, Greeks settled the coastline of the region. The local centre was the fortified Černomorskoe Settlement situated at the junction of the sea coast and the outflow of the Kelšeich Ravine (see below). Contemporary was the fort U7 of Panskoe I located on the coast of the Sasyk Ravine (see below). During the Hellenistic period, along the coastline proper were constructed a number of isolated fortified (Skalistoe 5, see below) and unfortified farmhouses (Mežvodnoe, Vetrenaja Bay). A number of localities found during the DSP extensive survey, where a small discrete scatter of finds were recorded, may have been farmhouses too (Baj Kijat 3, Burun Eli 5, Groty 2, Masliny 3, Skalistoe 6, 8, and 10; see Appendix 1). According to current knowledge based on published excavations, the only more complex settlement along the coast was Masliny 1-2, a group of fortified farmhouses (Figs. 4.3-5). However, tile was also collected at Groty 1, a possible farmhouse (Fig. 4.6). The sites Burun Eli 2 (Figs. 4.7-10), 3 and 4, Skalistoe 1-4 (Figs. 4.11-12), and perhaps 5 (see below) known prior to the DSP survey, have been interpreted as 'settlements' by their investigators. Revisited during the DSP extensive survey, they proved to be of small extension. In all likelihood, they must qualify as isolated farmhouses too. Found during the DSP extensive survey, Skalistoe 9 is a large scatter almost halfway between two ravines. It is possible that this site was more complex than just an isolated farmhouse. Of the mentioned sites, Groty 1, Masliny 1, Skalistoe 5 and 10 also had a Late Hellenistic component. In terms of finds and assemblage range, these sites (excluding Skalistoe 10) match Panskoe I-II.

### 4.2.2 Skalistoe 5 | T. Smekalova & V. Stolba

The site is situated on the coast, about 10 km NE of the present-day village of Mežvodnoe, on a high sea-cliff, on the right-hand slope of a wide ravine, close by a modern fishing station. Its N part is obliterated by the shore abrasion (Figs. 4.11-12). The features observable on the surface include two pronounced elevations in the E and W parts of the site, as well as traces of stone masonry exposed in the tracks of the modern dirt road running along the coast. Of the mentioned elevations, the former has a noticeable depression in the middle and is disturbed by four modern linear trenches joining at a right angle. The multiple finds of pottery, including roof-tiles, amphorai and plainware, recovered during the DSP extensive survey, come from the tracks of the road and from the cliff where the ancient material sticks out about 1 m below the modern top soil surface (GS 3027).

In 1973, trial excavations were conducted at the site by Z.Ch. Popandopulo as head of the detachment of the Kharkov University expedition. A trench measuring about 50 m$^2$ was tied to the edge of the cliff. The thickness of the cultural layer in it amounted to 0.40-0.50 m, with no architectural remains being recorded. The finds included numerous fragments of transport amphorai and handmade ware, as well as imported black-glaze pottery (fish-plates, bowls, kantharoi) and Chersonesean plainware (red-striped jugs, bowls and louteria). A preliminary date, late 4th through 3rd century BC, was suggested for the site on the basis of the amphora material (Latyševa 1974).

In May 2007, the site was re-visited by the DSP team during the extensive survey of the coastal zone. Except for a tiny corner at the very edge of the cliff, the remains of the 1973 trench are eroded away. This allows to roughly estimate the speed of the coast abrasion, which is significant. A closer examination of the E elevation, which was thought to be a tumulus disturbed by the Romanian military trenches from World War II, proved to be remains of a rectangular tower measuring ca. 9 x 9 m, the stone blocks of which have been removed for re-use elsewhere. The W elevation could be positively identified as an ash hill (see Chapter 5.1.5).

The archaeological sounding of the site was preceded by a detailed geomagnetic survey, which covered its whole area (Figs. 4.13-14). The geomagnetic map allows us to reconstruct the complete plan of this compound the corners of which are oriented to the main cardinal points. Its E corner is reinforced by a tower which from NE and SE is surrounded by a wide, ca. 50 m-long wall and a ditch. The latter occurs on the map as a clear positive anomaly which follows the trace of the wall on the outside. At the S corner of the farmhouse the traces of both the wall and the ditch disappear, possibly as an effect of destruction or rebuilding, to appear again SW of the ash hill as a ca. 15 m-long double (positive and negative) linear anomaly. The entire area embraced by the ditch amounts to ca. 1,550 m$^2$.

Inside the compound, about 4 m-wide rows of rooms built along the NW, SW and maybe SE sides of a courtyard can be distinguished. A small displacement in the orientation of these units and that of the tower as well as the fact that some of them are overlaid by an ash hill suggest more than one building period. A negative magnetic anomaly inside the courtyard can be an indication of the pavement or solid stone debris. The above-mentioned ash hill, the E periphery of which expands over the SW wall of the building, emerges on the magnetic map as an extensive positive anomaly (marked with green dashed line on Fig. 4.14). Its position SW of the building should possibly be explained by the prevailing wind directions.

In accordance with the geomagnetic map of the farmhouse, a 2 x 2 m trial trench was laid out in its western periphery where along with the architectural remains the geomagnetic survey indicated the presence of a large ash deposit (Figs. 4.15-20). The actual excavations and interpretation were made by S.B. Lancov and V.F. Stolba.

### 4.2.3 Conclusions on settlements on the Coastal Cliff/Lowland Ridge | P. Guldager Bilde

We may conclude that permanent settlement in the coastal zone, which started in the Neolithic period, intensified during the Middle and Late Bronze Ages. Whilst MBA sites may have preferred locations near the coast, as a rule, LBA sites are found inland off the coast. However, the low number of prehistoric sites as well as prograding coastal erosion is likely to leave us with a biased pattern.

So far, clear indications of Iron Age settlement lack for the coastal zone discussed here. The area seems only to have been resettled in the Late Classical and Early Hellenistic period, when a string of rural settlements appeared along the coast, some of which were fortified.

In Chapter 6 we will go into the socio-economic and historical significance of the observed settlement pattern.

## 4.3 THE RAVINES | P. GULDAGER BILDE WITH CONTRIBUTIONS BY P. ATTEMA, S. LANCOV, T. SMEKALOVA & V. STOLBA

A characteristic feature of the West Crimean landscape is the occurrence of long ravines, which, at their outflow, today have turned into salt- or brackish lakes (Fig. 2.3). The longest one is the Donuzlav Lake and Ravine, which stretches no less than 50 km inland and thus separates the Tarchankut Peninsula from southwestern Crimea. In antiquity, the sides of this ravine were densely settled (investigations of P.N. Schulz, cf. Lancov 2004, Fig. 2). We find a similar settlement pattern in the DSP study zone. Here we can observe two such ravines, the Kelšeich Ravine and the Džarylgač Lake/Vodopojnoe Ravine, stretching ca. 15 and ca. 25 km inland respectively, as well as the much smaller Sasyk Ravine extending only 10 km inland. The ravines provided the inhabitants with easy access to fresh water and fishing could have taken place in the lagoons, where also salt could have been extracted. In addition, the ravines gave the inhabitants shelter from the wind, a fact, which is appreciated even today (Fig. 4.21). The outer parts of the Kelšeich and Sasyk Ravines have been investigated by a number of Russian scholars and expeditions (Section 4.2 above), but the Džarylgač Lake/Vodopojnoe Ravine much less so. Due to the limited period available to the DSP fieldwork, the ravines have not been investigated by us. Here we can only make some general observations on their archaeology.

### 4.3.1 The Kelšeich Ravine | P. Guldager Bilde

Several river valleys have their mouth in the Uzkaja Bay (Fig. 4.22). Of these, the Kelšeich Ravine is the most prominent being deep, winding and much ramified. The topography of the Uzkaja Bay, as defined by the drowned mouth of several rivers, provided a sheltered location for the many sites, which dotted the coastline of the ancient as well as the present bay. The most important of these was the Černomorskoe Settlement, which was situated at the E side of the outer part of the ravine.

4.3.1.1 The Černomorskoe Settlement (Figs. 4.22-26)

Recent stratigraphical investigations have shown that the settlement was founded in the early 4th century BC. The excavators are of the opinion that the earliest settlers were Ionians (Užencev 2006, 21-23; Lantsov & Uzhentzev 2006, 701). Whether they came from Olbia, as suggested for nearby Panskoe I by A.N. Ščeglov, whether it was a joint Ionian settlement, or whether it was settled from originally Ionian Kerkinitis is a matter of dispute (Užencev 2006, 110-111;

Lantsov & Uzhentzev 2006, 701). Sparse building remains are reported from this phase (pits and mudbrick buildings on low stone socles). During the mid- or third quarter of the 4th century BC, Chersonesos expanded her territory along the coast at least up to Masliny (see Chapter 7.3). Thus, Chersoneseans took over and resettled the city. During the third quarter of the 4th century BC, the settlement obtained its basic layout (Užencev 2006, fig. 23). It was fortified with a well-built wall having several square towers constructed of mudbricks on an ashlar foundation. Inside the city were built rectangular blocks of houses (Lantsov & Uzhentzev 2006, fig. 11). In the first third of the 3rd century BC the city fortification was further strengthened with an additional tower and with so-called 'anti-ramming belts' around the towers (Fig. 4.25). Nevertheless, by the end of the first third of the 3rd century BC, part of the city's foundations seem to have been destroyed by fire. Apparently, some of the inhabitants abandoned the city and for a period, the occupation of the settlement was much diminished. During the last third of the century, the city was remodelled with a strong fortification called the 'citadel' located in the SW part of the city and taking up ca. 2,500 m$^2$ (Lantsov & Uzhentzev 2006, fig. 7). In the middle of the 2nd century BC, the Černomorskoe Settlement was destroyed in an enemy attack believed to be Late Scythians of Skiluros' Kingdom of Scythian Neapolis.

Life in the town continued for a substantial period. Nalivkina has noted layers of the 4th-2nd century BC as well as the 2nd-3rd and 8th-10th century AD (Nalivkina 1957, 271; Užencev 2006, 135) and the same was observed by Ščeglov (1967, 237; 1978, 20). The second half of the 2nd century BC-early 2nd century AD is attributed to the Late Scythian culture. A new type of irregular living houses is built inside the fortification of fieldstones (Užencev 2006, figs. 55-57) in addition to the numerous pit houses dug into the ground (Užencev 2006, figs. 51-52). It is the only locality within the DSP study zone, where evidence for occupation during the Roman period has been recorded.

To the E of the settlement is an immense ash hill (not investigated). The settlement was surrounded by at least three distinct groups of kurgans (Fig. 4.24): to the N and to the S and, with the largest group containing at least 34 kurgans, SE of the settlement (Kutajsov & Pridnev 2007). Of these, 15 kurgans have been excavated. The northern group may be connected with four farmhouses lying at the N outskirts of the fortified settlement (see below).

In antiquity, the settlement of Černomorskoe may have constituted the local centre of the Tarchankut district E of the Kelšeich Ravine with the settlement of Karadža as the local centre of the Tarchankut W of the same ravine (Fig. 1.1.15; see also Smekalova & Smekalov 2006, 238). From these two urban poles, local affairs could have been administered and at the same time they could have functioned as the main market place for exchanging local and foreign products.

The site is normally referred to as Kalos Limen ('the good harbour'). However, the proper ancient name of this locality is disputed (recent overview in Lantsov & Uzhentzev 2006, 698-699). Ancient sources inform us of two localities to be found N of Kerkinitis (present-day Evpatorija)[35]: Kalos Limen and Tamyrake. According to Fl. Arrianos "from Chersonesos to Kerkinitis is 600 stadia, and from Kerkinitis to Kalos Limen...another 700. From Kalos Limen to Tamyrake, 300" (*PPE* 19.5-20.1; Anon. PPE 83). Since there is ca. 75 km along the coast from Chersonesos to Kerkinitis, this measure must correspond to 600 stadia; accordingly, one stadion employed measures ca. 125 m. Thus, the distance from Kerkinitis to Kalos Limen should be 87.5 km. This corresponds precisely to the distance to Karadža, whereas the distance to Černomorskoe is ca. 115 km (920 stadia), thus not even corresponding to the mentioned distance to Tamyrake.[36]

Even though it is highly uncertain that the ancient fortified settlement of Černomorskoe can be identified as Kalos Limen, it is, as mentioned above, normally thus referred to in scientific and popular literature. This is probably due to the fact that until recently, Černomorskoe was the only one of the two candidates, the other one being Karadža,[37] that had been subjected to excavation. In the present publication we will restrain ourselves to use the modern name of the site.

#### 4.3.1.2 Farmhouses in the vicinity of the Černomorskoe Settlement

In the vicinity of the Černomorskoe Settlement, along the shore of the Uzkaja Bay are located at least nine ancient farmhouses (Figs. 4.23; 4.24). Most of them have never been subjected to archaeological investigation. Four are located on the northern outskirts of the fortified settlement (Užencev 2006, fig. 4, Ж 1-4). Presently, one is lying under water (Ж 1), and one can be seen on the beach (Ж 2; Figs. 4.23, 4.26). At the western promontory, at the entrance to the bay

[35] *IosPE* I$^2$ 352, 353, 401; Arrianos *PPE*, 19, 5; Anon. *PPE*, 57; Mela, *Chorogr.* 2.1.3; Strabon 7.4.2.

[36] Incidentally, Panskoe I is located 125 km (1,000 stadia!) from Evpatorija, and thus in fact a better candidate for Tamyrake. It would be strange, though, that this site should be mentioned, not the larger fortified town of the Černomorskoe Settlement. Thus, the proper identification of the sites must be regarded as problematical.

[37] Small scale excavations were conducted here in 2005-2006 under the direction of V.A. Kutajsov (unpublished).

have been localized two farm houses (Fig. 4.22 = Ščeglov 1967, fig. 2.10 and 11), and the site, where the Florya Pension in Ul. Revolucii is located must have been the place of yet another farm. At least, when excavating the plot in order to construct the house, a mouth of a well crafted in local lime stone was found (Fig. 4.27). The site has never been investigated archaeologically. Located on the opposite promontory, immediately SW of the settlement, another farmhouse has been referred to (Fig. 4.22 = Ščeglov 1967, 247, fig. 2.9; Hellenistic material is mentioned) and yet another one should be located NE of the settlement (Fig. 4.22 = Ščeglov 1967, fig. 2.8).

Some of the farmhouses around the Černomorskoe Settlement have been excavated, but very little information is at hand. One was excavated by Schulz in 1933 (Fig. 4.22 = Ščeglov 1967, 247, fig. 2.8), but the results were never published. Ščeglov mentions the presence of walls and pits. Of the two farms W of the settlement, stone walls are referred to as well as Hellenistic material such as Sinopean tiles, Chersonesean and Herakleian amphorai, and louteria of the 3rd-2nd century BC (Fig. 4.22 = Ščeglov 1967, 247, fig. 2.10-11). The farm shown in Ščeglov fig. 2.10 apparently continued in the Roman period, because Ščeglov also mentions finds of Red-glossed pottery. Also the Vetrenaja ("Windy") Bay farmhouse belongs to the immediate surroundings of the Černomorskoe Settlement (Figs. 4.28-4.30). This is the only farmhouse we have some information about (Ščeglov 1967, 249-256; fig. 2.5, 9; 1978, figs. 47-48; 2002a pl. 4.22; the structure can be found on the half-verst map of the 1890s too); see Appendix 1.

## 4.3.2 The Sasyk Ravine | P. Guldager Bilde with contribution by T. Smekalova, V. Stolba

The Sasyk Ravine (today Lake Panskoe) consists of the drowned parts of two seaside river valleys, which were penetrated by sea water during the glacio-eustatic New Black Sea transgression (Figs. 2.2, 2.5, 4.31). Geological and limnological descriptions of the lake and the Jarylgač Bay were carried out in the 1930-1940s by A.I. Dzens-Litovskij and V.V. Longinov (Dzens-Litovskij 1933, 585-595; Longinov 1955, 152-166). The sediments on the bottom of the lake were investigated by drilling in the 1970s within the framework of the Tarchankut expedition (Blagovolin & Ščeglov 2002, 285-302). The results of the mentioned geomorphological studies served as a basis for Ščeglov's reconstruction of the ancient landscape in the environs of the settlement of Panskoe I (Fig. 4.31). Until the 1980s, the lake was separated from the Jarylgač Bay by a massive sandbar, which was cut through, when the port of the "Černomorneftgaz" was constructed.

The ancient topography of the Sasyk Ravine, as reconstructed by A.N. Ščeglov (Fig. 4.31), mirrors at a much smaller scale that of the Kelšeich Ravine and its outflow in the Uzkaja Bay, and so does the settlement pattern (Fig. 4.22). Close to the Sasyk Ravine's outflow into the sea is located the major settlement of Panskoe I. Along the ravine's sides are at least three more ancient sites: Panskoe II, III (Fig. 4.32), and IV. Panskoe III and IV are farmhouses settled in the 4th century BC; Panskoe II was either a fortified farm or even a fort. It is located at a strategic point at the main thoroughfare between the Černomorskoe Settlement and the Vodopojnoe region, where a road branches off to Panskoe I (see Section 4.3.2.1 below and Chapter 6.6, crossing of Roads XII and XIV).

Undoubtedly, the Panskoe promontory between the lakes of Sasyk/Panskoe and Solenoe was an area densely settled even from an early time. On the tip of the promontory is located the Neolithic site Jarylgač Severnoe mentioned above (Section 4.2.1). Sparse Late Bronze Age elements came to light at the multi-period site of Jarylgač Vostočnoe located ca. 600 m E of Panskoe I, and a large kurgan of the same period is found 1.15 km S of the site, between Panskoe I and II. However, it is in the Late Classical-Early Hellenistic period that the area was most densely settled. Jarylgač Vostočnoe was thought by its investigators to have been a multiple-period settlement. However, in the Late Classical-Early Hellenistic period, it was probably an isolated farmhouse as Panskoe III and IV.

In the 18th century, a small village named Kolbaše was situated at the then inner end of the Sasyk Lake. It was marked as a ruin already on the map of 1817. In the 19th century salt was extracted from the lake.

### 4.3.2.1 Panskoe I (Figs. 4.33-37)

Panskoe I is one of the most significant and best-studied monuments of the Classical and Hellenistic period in Northwestern Crimea, and it served as the point of reference in the DSP survey area. The settlement, which extended over at least 4 ha, was located at a natural harbour at the mouth of the Sasyk Ravine. Today, the SW part of the settlement is submerged.

The foundation of the settlement took place in the late 5th century BC. The settlers came from an Ionian city, be it Olbia (Ščeglov 1987, 245-246; Chtcheglov 1992, 242-243) or Kerkinitis (Užencev 2006, 110-111; Lantsov & Uzhentzev 2006, 701). The material remains of this phase correspond to the lower stratigraphic horizon called Horizon B. Approximately from the middle of the 4th century BC, following a big fire, considerable changes in the material culture of the settlement, including a certain shift in the ceramic assemblage, in the composition of coin finds, changes in the language of local inscriptions, the appearance of a new building tradition, etc., can be observed (Ščeglov 1987, 246-247; Stolba 1991, 78-84; 2005a, 158-161). The nearest parallels to the new features come from Chersonesos and from other settlements in Chersonesos' rural territory. This phase in the settlement's existence corresponds to its upper stratigraphical horizon, called Horizon A. Most of the excavations at the site have been concerned with the upper stratigraphical horizon and accordingly were concerned with the site's Chersonesean phase. To this phase belong such completely investigated complexes as U6 (Fig. 4.34),[38] U13 and U14, as well as the archeologically unexplored or incompletely investigated remains of U1, U2, U3, U9, U10, and U12 (Figs. 4.35 and 4.36). The lower strata of Horizon B have only been recorded in the central part of the site (U7; Fig. 4.34)) and still remain insufficiently studied (Stolba 1991).

Around 270 BC, the life at the site ceased suddenly as an effect of an enemy attack, as testified by multiple finds of arrowheads and skeletons bearing signs of a violent death as well as by traces of a severe fire which destroyed the monumental building U6. A few fragments of mouldmade bowls recorded in U6 (Hannestad, Stolba & Hastrup 2002, 129-130, 142, no. B 144, pl. 73) and U2 (unpublished) as well as Chersonesean amphorai with stamps of the 2nd century BC (Ščeglov 1968, 213) seem to indicate a small-scale habitation during the second half of the 2nd century BC (cf. Hannestad 2005, 191), although the stratigraphic horizon of this period still has to be discovered.

In terms of urban layout, very little has been published. A first sketch plan was published by Wasowic (1972, fig. 12.I = Fig. 4.37). Originally, it was believed that the layout of the settlement was 'Hippodamic' (Wasowicz 1972, 215-216). From Wasowic's plan we are informed of the location of the buildings (U)1 to (U)11. A new sketch plan was published by Ščeglov in 1978 (1978, 47, fig. 17). This was made after U13 and U14 had been investigated, and after U6 had been brought to light, but before U7 was excavated. The position of what must originally have been U12 is indicated. This building was partly excavated in 1987, and erroneously published as U2 (see below). In 2002, the latest version of the sketch plan was included in the publication of U6 (Ščeglov 2002a, pl. 6 = Fig. 4.33). In this plan the overall lines of U7 are provided, and it is clearly to be seen that what now is called U7, originally was thought to be four individual houseblocks (U4, U5, U7, and U8).

In 2007 and 2008, the settlement of Panskoe I was surveyed geomagnetically by T.N. Smekalova assisted by A. Čudin (Figs. 4.33, 4.35 and 4.36). The purpose was to supplement the existing sketch plans in order to determine the extension and composition of the settlement's layout.

In 2007 were measured two large tracts S and W of U6 as well as a large tract E of U7 and a smaller one at the E outskirts of the settlement. In 2008 was measured a tract of land 35 m x 85 m E of U6. The results of these measurements allow us to complement the known picture of the settlement history of the site.

Unlike the majority of the other rural settlements of the region, which are represented by isolated fortified and unfortified farmhouses, the layout of Panskoe I is rather unique, as it represents an agglomeration of both well constructed, rectilinear farmhouses and other larger building structures.

Below follows a brief description of the individual houseblocks based on previous excavations supplemented with the results of the geomagnetic survey. The original numbering of the houseblocks has been retained.

U7 was the first structure to be built (Figs. 4.33-34). It was a fortified construction, square in plan (42 x 42 m) built on an ashlar foundation with walls of mudbricks with probably four round towers, three of which have been excavated (Ščeglov 1987, fig. 8). It had a single gate turned towards SE and the mainland. Its internal layout is almost completely unknown because it was overbuilt during the Chersonesean period. This first phase had two building horizons separated by a major fire (Ščeglov 1987, 245). During the Chersonesean period, the fortification was abandoned and inside and outside its walls 16-18 smaller farmhouses were constructed almost as a small village inside and immediately outside the former fortification (Ščeglov 1987, figs. 21 and 26; Stolba 1991).

Of the same orientation as U7 but built in the middle of the 4th century BC are U6, U9, and U10.

U6 This building is the only one fully published (Panskoye I 2002; Fig. 4.33). As U7, also this building is square in plan, but not as large (35 x 35 m). The building consists of an ashlar foundation with mudbrick walls; a number of tiles have been found. Its main feature is an immense courtyard surrounded by some 20-odd rooms. In a secondary phase one

[38] U=*usadba*, Russian for farm.

further row of rooms was added to the inner W and S side. Also this building had a single gate opening towards SW and thus, U6 flanked the road leading directly to/from U7. In U6's S corner (Room 12) was excavated a small house sanctuary devoted to the cult of Demeter and Dionysos-Sabazios (Panskoye I 2002, 47-50), and in the room added in front (Room 14) was an altar and a relief dedicated to Herakles (Panskoye I 2002, pl. 143.G1; see also Chapter 6.4). The geomagnetic measurements have not revealed further buildings SE of U6.

**U9+U10** This structure is located immediately S of U6 (Fig. 4.36). It was originally conceived of as two individual houses (U9 and U10), but since the S walls traced as part of U10 in Stolba, Hannestad & Ščeglov 1995, fig. 4.1 in fact belong to U11 (see below), U9 and U10 may have been part of the same building. Part of two rooms of U10 was excavated by a group of Danish archaeologists in 1987 (Stolba, Hannestad & Ščeglov 1995, fig. 4.2). The central part of the building is obscured by a neat round ash hill, which overlays the building. There may be a fairly large courtyard immediately W of the excavated rooms of U10. In a later phase a series of at least five deep rooms was added to the S part of the building.

A number of buildings follow a more western orientation. This applies to U2, U3, U8, and U13 as well as to the isolated house U7-15 (Ščeglov 1987, fig. 26) and a few rooms built over the S corner of U7.

**U2** U2 and U3 are located between two modern tracks (Fig. 4.35). U2 was fully exposed in the geomagnetic survey in 2007. The houseblock is ca. 30 m wide and 25 m long. It consists of probably two living units with two rows of rooms to the NW and one row of rooms to the S of a central courtyard.

**U3** The E side of U3 is exposed in the eastern track. This building block was fully revealed in the geomagnetic survey in 2007. It is a large square structure consisting of at least three houses with rows of rooms around smaller courtyards.

**U8** is lying under an immense ash hill and its features are therefore much obscured (Fig. 4.36). The geomagnetic data of 2007 revealed a series of square rooms to the N of a large courtyard. The orientation of this house is the same as the southernmost three rooms of a house built over U7. They may either belong to the same house or to a building of the same construction phase.

**U1** Part of this house was measured in the geomagnetic survey in 2008 (Fig. 4.33). It is situated NE of U2. It features one row of rooms to the N of a central courtyard and two rows of rooms to the W. It was in all likelihood an isolated house (c. 25 x 25 m). Its orientation differs from all other houses being oriented more to the W.

**U13** This house, which is partly submerged, is the northernmost of the houses of the settlement. It seems to be a free-standing house perhaps with a square plan with rooms surrounding a central courtyard. It was partly exposed in 1970 (Ščeglov, Saverkina & Glazunov 1971, 251) and in 1977 (Ščeglov et al. 1978; Sceglov 1987, fig. 26). According to the investigators the house covered an area of ca. 1,200 m². Preserved are three rooms and a round tower (?) and its date is late 4th-3rd century BC.

Two building blocks have yet another orientation, namely U11 and U14. If we consider the horizontal stratigraphy of the site, U11 and U 14 situated in the settlement's margins were probably the last ones to be built along with U1.

**U14** The building, which today is completely submerged, is square in plan, each side measuring 32-35 m. It was partly investigated in 1970 and 1971 (Ščeglov, Saverkina & Glazunov 1971, 251; Ščeglov et al. 1972, 342). It was dated to the late 4th-mid 3rd century BC by the investigators.

**U11** The complete plan of this building emerged as a result of the geomagnetic survey in 2007 (Fig. 4.36). The stone walls are very clearly visible as negative linear anomalies caused by stone walls on the magnetic map. The resulting structure is rectangular in plan ca. 15 x 20 m. All four sides of the building were occupied by rectangular rooms, ca. 5 x 5 m, around a rectangular courtyard. A weak positive anomaly off the centre can perhaps be interpreted as a well. In several rooms in the N and E side of the building strong positive anomalies testify to burning in the rooms. The building block was free-standing and separated from the building to the N by a passage of a few meters. In terms of plan and size, the building is very similar to U2-2 (Stolba, Hannestad & Ščeglov 1995, fig. 4.1). It should be noted that the walls traced as part of U10 in Stolba, Hannestad & Ščeglov 1995, fig. 4.1 in fact belong to U11, which also explains the two different orientations.

The geophysical data have shown that there are no further buildings to the S of U11. To the W and SW are two ash hills, which seem to respect the building's S corner. In general, a number of ash hills, which can partly be seen in the terrain and which are especially clear in the geomagnetic data dominate the site. Most of them are located W of the buildings and some overlay them. They attest to the troubled life in the settlement and to clean-up operations.

**U12** With the exclusion of the southernmost part, this building block was not covered in the geophysical survey, because heaps of soil left after excavation partly covered the site. It consists of at least four different houses built together. They were partly excavated in 1987 and erroneously published as U2.[39] One house is completely brought to light (published as U2-2) whereas of the house to the N, published as U2-3, is revealed only in its E part. Of the two houses to the E, published as U2-0 and 2-1, nothing is known. U2-2 and 2-3 both have two rows of rooms to the E, and at least U2-2 is in addition furnished with rooms to the N and S flanking a rectangular courtyard. The finds inside the building allow us to date it to the second half of the 4th-first third of the 3rd century BC (Stolba, Hannestad & Ščeglov 1995, 52).

Two strong positive magnetic anomalies to the E of U12 may have been caused either by furnaces or by pits filled with strongly magnetic material. Accordingly, there may have been some kind of production here in the outskirts of the settlement.

Due to the intensive interdisciplinary work carried out in Panskoe I, we are well informed about its mixed economy deriving from farming, animal husbandry, hunting and fishing (Januševič & Ščeglov 2002; Kasparov 2002).

300 m NW of the settlement itself is located an extensive tumular necropolis comprising over 60 burial mounds. Along with the burial mounds, the cemetery also contained a number of flat graves dug between the tumuli (Rogov 2011). The Tarchankut Expedition has investigated about one third of its extent (last year of excavation was in 1986). In the period from 1991 to 1993, the unexplored part of the necropolis was subject to intensive illicit digging.

Between the settlement and the necropolis runs a probably ancient road (Road XIV) leading towards the mainland. It follows the orientation of U7 (see also Chapter 6.6 below; Fig. 4.33).

The best parallel for this type of settlement is the one of Albeşti in the chora of Kallatis. Also here, several large rectangular farm complexes have been excavated, one of which was fortified (Rădulescu et al. 2000-2001; Buzoianu & Bărbulescu 2003; http://www.cimec.ro/scripts/arh/cronica/detaliu.asp?k=3058). The similarities may owe to the fact that Kallatis was a sister colony of Chersonesos.

### 4.3.3 The Vodopojnoe Ravine | P. Guldager Bilde

As mentioned above in Chapter 2.3.1.3, in 2007 as part of the brief visit to previously known sites in the DSP area the ravine E of Vodopojnoe was inspected by the DSP staff in the company of S.B. Lancov and S.D. Koltuchov. Investigations carried out by Koltuchov in the 1980s had located three sites in the northern branch of the ravine (Vodopojnoe 2-4) and one in the southern branch (Vodopojnoe 5). All four sites have been dated to the Late Bronze Age (unpublished; information S.G. Koltuchov; Fig. 6.7).

The Vodopojnoe Ravine splits off in two branches ca. 1 km E of Vodopojnoe (Fig. 2.2). Both branches stretch for more than 15 km further towards NE and SE. One further branch is located immediately S of Vodopojnoe at Novo Ul'janovka. This branch stretches for ca. 15 km towards the S while W of Vodopojnoe the ravine widens into the Džarylgač Lake. The Vodopojnoe Ravine was densely settled in Early Modern times. A number of foremost Tatarian villages are located within short distance from each other, primarily in the northern part of its three branches (see below, Chapter 6.2.3). Almost at the outflow to the sea at the N coast of the ravine is situated the modern town Mežvodnoe. Its present location is at the sea coast, but in antiquity, prior to the establishing of the sand bar stretching from the S coast of the ravine, the site was not located at the sea.

The present village of Vodopojnoe ("Watering place") is located at the major crossing over the ravine (Fig. 4.38). This

39 Stolba, Hannestad & Ščeglov 1995, 50-53; the position in the general plan in Ščeglov 2002a, pl. 6 is turned mirror-reversed and upside-down; correct position: see Stolba, Hannestad & Ščeglov 1995, fig. 4.1.

was probably also the case in antiquity, because several routes converge ca. 1 km E of the village (see below, Chapter 6.6, Roads I, II, III, IV, and X). Vodopojnoe seems to be an obvious place for an ancient settlement with plenty of water and a good location. A 'settlement' called Džarylgač is marked e.g. on a map of Ščeglov (2002, pl. 4.1.28 [wrongly given as Burun Eli] = Fig. 1.1.29), but this site was not verified in the DSP extensive survey in May 2007. An attempt to find the location of an isolated well, which is shown on the half-verst map SW of the village, also failed. The outskirts of the village are completely destroyed due to recent building activities. Moreover, numerous piles of modern soil and garbage are being pushed around in the same area. During our site visit, not a single ancient sherd was collected. The well was not located either, but a pump currently used for watering animals was found probably more or less where the well would have been. It is placed at the lowest point of a small ravine close to the lake. The water is slightly brackish, but still drinkable. At the S coast of the ravine, immediately S of Vodopojnoe is presently a modern 'suburb' to Vodopojnoe, Novo Ul'janovka.

### 4.3.4 Conclusions on the settlement in the ravines | P. Guldager Bilde

Above, we have discussed current evidence for settlement and land use in the ravines of the DSP study area on the basis of published and unpublished sources as well as on autopsy in the field during our field campaigns. In Western Crimea, the ravines in general appear to have been of major importance during all periods of human presence supplying water, shelter from the wind, and meadows for animals. In the study area we dealt with three major ravines, the Kelšeich Ravine, the Sasyk Ravine and the Vodopojnoe Ravine. A concise overview of past excavations at the sites of the Černomorskoe Settlement and Panskoe I respectively indicates the substantial scale these two settlements had. In Chapters 6 and 7 we will reflect upon the role of these settlements in the wider Crimean context.

Whilst all three ravines merit more intensive archaeological investigations in terms of their long-term histories of settlement and land use, our knowledge in the case of the Vodopojnoe Ravine is especially limited. Unfortunately, because of the fact that the DSP was shortened by a year, it was not possible for us to carry out systematic survey in this ravine. Our knowledge is therefore confined to the mentioned four Bronze Age settlements. The convergence of a number of nomadic routes in the ravine suggests that it was much frequented. It would therefore have been of considerable interest to learn more about this ravine's settlement pattern also in later periods.

## 4.4 THE LOWLAND RIDGE AND PEDIMENT ZONE | W. DE NEEF & C. WILLIAMSON WITH CONTRIBUTIONS BY P. ATTEMA, P. GULDAGER BILDE, T. DE HAAS & T. SMEKALOVA

This section presents the settlement data resulting from the pilot survey in 2006 and the intensive surveys carried out in the campaigns of 2007 and 2008 in ploughed fields located on the coast N of Lake Džarylgač in the land type defined in Chapter 3 as the Lowland Ridge and in ploughed fields to the S of Lake Džarylgač in the land type referred to as the Pediment. We will first present the summary data of the three campaigns separately (Sections 4.4.1 to 4.4.3) and then proceed with a discussion of known sites within the transect (Section 4.4.4) and new sites found in the DSP survey (Section 4.4.5). The latter sections take the form of a catalogue.

### 4.4.1 Summary data of the 2006 campaign

A pilot season was undertaken by P. Attema, P. Guldager Bilde and V. Stolba on the 25th to 27th of May 2006. One of the objectives was to make landscape observations and carry out test surveys in order to evaluate the potential of the Tarchankut landscape for an intensive survey project. The area surveyed in the campaign of 2006 consisted of four cultivated fields labelled DSP06 F1 to F4 (Fig. 4.39). The coordinates of the surveyed fields were recorded with a Garmin GPS and sketched on to a 1:25,000 map. Information concerning finds and observations were recorded in a note book. On the basis of the results of this pilot survey, in which evidence was found for small scatters of Hellenistic pottery in the ploughed fields S of Lake Džarylgač, the survey strategy was devised as employed in the surveys of 2007 and 2008 the results of which are discussed in the present publication. The mainly Hellenistic pottery collected during this survey was analysed in 2007 and incorporated in the database.

### 4.4.2 Summary data of the 2007 campaign

After one trial day, block survey started on the 7th of May and continued up to the 26th. During this period, team 1 spent 15.5 days block surveying, team 2 14.5 days. In total 1,939 blocks with a total surface of 579.2 ha (almost 5.8 $km^2$) were surveyed averaging 19.3 hectares per team per day.

Due to the varying number of surveyors, not all units were surveyed at the same coverage. While 91.3% of all units were surveyed at 20%, the other units were surveyed with a coverage varying between 14 and 28%.

The area surveyed in the campaign of 2007 consisted of eight cultivated fields labelled DSP07 F1 to F8 (Figs. 4.40-41).[40] Survey conditions in these fields were relatively uniform: part of Field 5 was ploughed in large peds, but all other fields had been recently ploughed and harrowed. In general, visibility was estimated to be relatively uniform as well: 13.1% of all units had excellent overall visibility, 60.3% good visibility and 21.7% average visibility. The distribution of finds (modern material excluded) over the survey area is displayed in Fig. 4.42. In this map, the densities have been corrected for coverage and unit size, but not for visibility.[41] This overview map, together with field observations and specific (combinations of) categories of finds was used as a basis to define sites (see Chapter 2.6). Only 5% lacked good visibility, while no areas with bad visibility were surveyed. The observed variation in visibility conditions is mainly due to the varying influence of weathering (mainly the amount of dust on the surface) and on the varying influence of shade (depending mainly on the weather and walking direction).

During the block survey circa 1,800 finds were collected. Pottery accounted for at least 73% of these, 6% were lithics and 21% were not determined. Within the ceramic finds, amphorai are the most common group, followed by handmade wares, tableware and utilities. Virtually all architectural finds were of recent date. These finds were supplemented by seven grab samples (16 finds) and 46 total samples (835 finds) of mixed composition. For a definition of the various sample types we refer to Chapter 2.3.1.1.1.

### 4.4.3 Summary data of the 2008 campaign

In 2008, a total of 1,872 blocks were surveyed with two teams. The total surface covered was 481.6 ha (4.8 $km^2$). This lower coverage of ploughed surface compared to the 2007 campaign was mainly due to a sudden case of campylobacter poisoning seriously crippling the teams. During the last three field days, the two teams working in the cultivated zone were reduced to one. Team 1 spent 14.5 days block surveying, team 2 spent 17.5 days covering on average 15 hectares a day.

In 2008, just as in 2007, not all units were surveyed at the same coverage, due to varying unit size or varying amount of field walkers. In 92% of the 1,872 units 20% of the surface was surveyed. The coverage in 6% of the units was 24%. Surface coverage in the remaining units varied between 17% and 28%.

The survey area of 4.8 $km^2$ consisted of six separate fields, numbered DSP08-F01 to F06 (Figs. 4.40-41). Field 4 was situated directly S of Lake Džarylgač, while the other five fields were all located on the coastal ridge N of the lake. The surveyed fields were all recently ploughed. Field 4 was partly ploughed into large peds, and part of Field 06 was harrowed during the survey. All other fields were harrowed before survey. Visibility conditions were relatively uniform, although weather conditions had some effect. The first week of the 2008 campaign was cloudy and rainy. Dust was washed away which affected the visibility of flints positively, yet at the same time the muddiness made walking and concentrating difficult. During the three subsequent weeks, the weather was generally sunny causing shade and dust. 11% of the surveyed units were described as having excellent overall visibility, 43% had good overall visibility, 34% had average visibility and 13% were below average (see Chapter 2.3.1.1.1). None of the surveyed units were classified as having a bad overall visibility.

During the intensive survey of the ploughed fields, 2,261 objects were collected. The distribution of finds (modern material excluded) over the survey area is displayed in Fig. 4.42. In this map, the densities have been corrected for coverage and unit size, but not for visibility. Most of these were pottery, 7% were lithics, including flint and stone tools like whetting stones and grinders, and 1% consisted of bone fragments. The finds classified as "indeterminable" (29% of the total amount) were all pottery of which the function could not be classified. These form the largest find group. Of the determinable finds, transport amphora fragments form the largest group (27%), followed by handmade pottery (25%).

40 Field 9 is not displayed in Fig. 4.1; it is located directly S of Field DSP07-5 and only two small blocks on the S side next to the lake have been surveyed (these areas are indicated on Fig. 4.1).

41 The low densities and the limited variation in visibility conditions make it unnecessary and perhaps even misleading to do so.

In total, seven iron objects were collected but after finds processing these were all discarded as (sub-) recent waste. In principle, modern ceramics were not collected.

### 4.4.4 Previously known sites in the Lowland Ridge and Pediment zone | P. Guldager Bilde

A number of sites have been found in the Lowland Ridge and Pediment zone within the surveyed area before the DSP (Fig. 4.43). The earliest sites are the quite extensive Late Bronze Age settlements of Burun Eli 1 and of Severnoe 1 and perhaps Vladimirovka 1. We are not aware of the precise location of the latter site. Burun Eli 1 and Severnoe 1 are both located on a low promontory at the confluence of two streams. They are connected with a 7 km long, straight kurgan-lined road (Road VI, see Chapter 6.6). Burun Eli 1 was an important multi-period site. In addition to its Late Bronze Age component, it had also a Hellenistic, a Late Hellenistic, a Medieval and an Early Modern component. Further inland on the N bank of the Džarylgač Lake is another Bronze Age site, Vodopojnoe 1 (see also DP07-F05), identified by Kolotuchin.

The same landscape zone was also the location of three Iron Age 'settlements' (Vladimirovka 2-4), however, again, we do not know, where precisely they are located, and they have never been published. Only Skalistoe 2 (see also DSP08-F03-02) and Severnoe 2 count as Hellenistic sites found prior to the DSP survey in this part of the landscape. In the DSP extensive survey, which took place primarily along the coast, inland sites were not at all investigated with the exception of near-coastal areas in the vicinity of already known sites. Thus was found Skalistoe 7, perhaps a Hellenistic farmhouse, in addition to the known sites.

Of the Tatarian villages located in this part of the landscape, two of them are near-coastal (Kula Šeich, Burun Eli) as is the outpost (*kordon*) N of Tubaka and the two farmsteads (*chutory*) around Kula Šeich and Skalistoe (e.g. DSP08-F03-01). Typically, they are drawn back from the coast being situated close to the bottom of a small ravine, where there is shelter from the wind and easy access to fresh water from a well.

### 4.4.5 Catalogue of surveyed fields

*DSP06-F01* (Fig. 4.39)
**Size** 1.3 km length.

**Investigations** Extensive survey. The field was walked in a rectangular strip with one pass back and forth during which pottery was collected. Coordinates were taken of all collected pottery.

**Landscape type** Plateaus/Uplands.

**Location** Field DSP06-F01, where the first test survey was carried out, is a rectangular strip in a much larger plot. It is located approximately 1 km W of the village Krasnosel'skoe, S of the asphalt road that runs to Olenevka (earlier named Karadža). The field is situated at the highest part of an inland ridge, a broad saddle, at about 120 m above sea level. The landscape slopes down gently to the W, in the direction of Olenevka, and to the N and S. From Krasnosel'skoe, the Kelšeich Ravine runs down N towards Černomorskoe.

**Cartography** On the half-verst map, the village of Krasnosel'skoe is called Kunan and it is inhabited by 84 Russians. Just 500 m SE of Kunan is a separate settlement, Čamak, home to 12 Tatarians. The buildings in the Russian village are larger and built along a central road, whereas the Tatarian settlement consists of smaller buildings with adjacent enclosures. About 2.5 km to the W, the ruins of a village called Džajlav are marked at the top of a ravine. On the half-verst map, several lines are depicted, parallel and perpendicular to this road. One of the parallel lines, a dashed stripe, runs through the survey field. The system of perpendicular and parallel lines may mark the remains of older land division lines in the landscape (Smekalova & Smekalov 2006, 235-240). From Kunan and Čamak, several tracks lead to the E, but not as systematically spaced as the ones in western direction. Approximately 300 m S of the present survey field, a mound on a hilltop is described with the name "K. [kurgan] Čolbat'oba", the height of this location is 64.1 sažens (132.68 m). To the NW of the field, another mound at 61.8 sažens (127.95 m) on a hilltop is described as "K. [kurgan] Uč'oba". These mounds mark both sides of the saddle.

On the mid-20th century map, the names of the kurgans are not mentioned anymore. The village of Kunan is now Krasnosel'skoe and has expanded to the S, the neighbouring Tatarian settlement of Čamak has turned to ruin. On the hilltop to the S of the survey field, two mounds are marked, whereas the hill to the NW has 10 mounds. The road to Olenevka and a parallel track to the S are the remains of the rectangular line system depicted on the half-verst map.

On the 1997 topographical map, two mounds are marked on the hilltop to the S with one star-shaped symbol. Nine of the 10 kurgans on the hill to the N are no longer mapped, only one large one is depicted at the elevation of 136.2

m. Interestingly, some kurgans further to the NW are still known by their specific names ("Kurgan Vysokaja Magoula", "Kurgan Čurokaja Magoula") known from the half-verst and mid-20th century maps.

**Survey Conditions** The rectangular strip had recently been ploughed in large peds. Visibility was even and good over the entire field. Harrowing and a little rain would have increased visibility to optimal survey conditions. Now dust and peds may be considered as slightly adverse.

**Sites** No sites were located in test Field DSP06-F01. In total, six pottery fragments were collected spread out over the field including an amphora body sherd, a jug handle and three fragments of orange-fired pottery. Also, the possible remains of land division walls were recorded at the southern end of the field (see Chapter 6.3.1).

**Relations to the surrounding area** South of the field, a kurgan cluster of at least nine kurgans were observed, one of which was larger than the others.

*DSP06-F02* (Fig. 4.39)

**Size** 700 m length.

**Investigations** Extensive survey. The field was walked in a rectangular strip with one pass back and forth during which pottery was collected. Coordinates were taken of all collected pottery.

**Landscape type** Pediment.

**Location** The second test survey field is a large field approximately 3 km E of Černomorskoe, on the SW side of the asphalt road leading along Lake Panskoe. The extensively surveyed area is the northernmost part of a field with an estimated length of 3 km N-S and a width of 1 km E-W The field was surveyed in one pass walking EW.

The field is situated on the eastern slope of a ridge extending from the Hillsides into the coastal ridge NE of Lake Panskoe. Just to the SE of the surveyed area, a shallow ravine leads from the Hillsides to the southwestern corner of the lake.

**Cartography** On the half-verst map, Field DSP06-F02 is located just S of a road. This is the same road that at present connects the town of Černomorskoe with Meževodnoe.

**Survey Conditions** The field was recently ploughed and harrowed, and therefore had optimal visibility.

**Sites** No sites were recorded in the surveyed part of the large ploughed field. Four orange-fired sherds were collected in the vicinity of a cut limestone block, but it is unclear what this represents. Spread out over the rest of the field, five more pottery sherds including one band handle were collected.

**Relations to the surrounding area** No relations to surrounding areas recorded.

*DSP06-F03* (Fig. 4.39)

**Size** Not recorded.

**Investigations** Extensive, non-systematic survey; pottery was observed but not taken.

**Landscape type** Lowland Ridge/Pediment.

**Location** Field DSP06-F03 is a field just S of the known Hellenistic settlement Panskoe II. It is located directly S of the asphalt road between Černomorskoe and Meževodnoe at the point where the asphalt road to Snežnoe runs off to the SE. This is the lowest part of the DSP research area, a few meters above sea level. Lake Panskoe is 50 m to the N.

**Cartography** The extensively surveyed part of Field DSP06-F03 is situated around a crossroads on the half-verst map. Here the road along Lake Panskoe crosses a track that connects the ruins of a village called Saja on the Hillside with the salt works on the eastern shore of Lake Panskoe. A small enclosure is depicted N of the field. On the slight ridges to the S and SW of the field, mounds are marked.

**Survey conditions** The NE corner of the field is not ploughed, and the terrain is bumpy.

**Sites** DSP06-F03-01 is close to Panskoe II. It is a scatter of ca. 40 m in diameter in a patch of greyish soil containing many stones and post-antique pottery. Finds related to Panskoe II continue for only 60 to 80 m and at places much less and there seems hardly any archaeological effect of this site on the surrounding terrain.

**Relations to the surrounding area** Panskoe I is located 1.6 km N-NW and there is a small group of kurgans just 450 m NE of Panskoe II.

*DSP06-F04* (Fig. 4.39)

**Size** Not recorded.

**Investigations** Extensive survey in two rectangular strips with two passes back and forth during which pottery was collected.

**Landscape type** Pediment.

**Location** The last test survey field of the 2006 campaign was located right on the southern shore of Lake Džarylgač, approximately 2.5 km N of the present-day hamlet of Zajcevo. Just S of the field, a dirt road leads from Zajcevo to Mežvodnoe crossing the sand bar that closes off Lake Džarylgač.

**Cartography** On the half-verst map, a track leads through the present survey field. It connects a mound W of the field right on the lakeshore with the village of Kipčak Učkuju 3 km to the SE. Another track leads off this track in a southern direction through the field. Directly S of the surveyed field, a mound is marked on a slight ridge. The ruins of a village called Džajlav are located approximately 1.1 km S of the field.

Neither the tracks nor the ruins are marked on the mid-20th century map, but the mound on the ridge S of the field is still there.

**Survey conditions** Field DSP06-F04 was recently ploughed and harrowed producing a good and even visibility. The field slopes down gently to the lakeshore. In the lower side of the field, nearest to the lake, soils were greyish in colour and contained much stone debris. The higher part of the field had more reddish soils.

**Sites** In the area with grey soils, a collection of amphora finds were made that in combination suggest the location of a Late Classical-Early Hellenistic site in this area (DSP06-F04-01).

**Relations to the surrounding area** DSP06-F04-01 is located 470 m due N of DSP07-F02-02, 870 m SE of DSP07-F02-01 and 1,140 m SE of DSP08-F04-03, which may all be contemporary.

*DSP07-F01* (Figs. 4.44-45)[42]

**Size** 38.4 ha

**Units** 1005-1083; 2001-2073.

**Investigations** Intensive survey in 50 x 50 m units except the corner ones. As it was expected to be dense with finds, this field was selected as the 'training grounds' for the 2007 field-walking teams. This was also the initial test site of the standard survey method using pre-mapped grids and GPS on handheld devices to establish unit boundaries (see Chapter 2.2); however, the northernmost units were set out with tape measure and measured in with Total Station.

**Landscape type** Lowland Ridge.

**Location** Field DSP07-F01 is located on a small peninsula between Lake Panskoe to the W and the Džarylgač Bay to the N and the E. The field is situated between the ancient sites of Panskoe I at its NW tip (Section 4.3.2.1), and Panskoe II to the SW.

The surveyed field is narrow at the northern end, where it skirts the lagoon and the site of Panskoe I, but widens towards the S. It slopes gently down towards Lake Panskoe in the W. The eastern field boundary follows the highest part of the strip between the lake and the bay. To the S, DSP07-01 borders on a ravine and the grassy strip that separates the field from the higher asphalt road connecting Černomorskoe and Mežvodnoe. W of the field is a grassy and marshy area close to the water separated by a dirt road running N-S from the main asphalt road.

**Cartography** On the half-verst map, there is no mention of ancient remains at the site Panskoe, but a small promontory is depicted. To the NE on a promontory along the coast are several five-point star symbols probably representing mounds in the area later known to be the necropolis of Panskoe I. A triangle marking the elevation of one of them is shown as 1.87 Ukrainian Height Units (one UHU is 2.07 m). Another mound is depicted directly SE of the field; this is the mound that is still visible today just outside Field 1.

A building is depicted near the eastern shore of Lake Panskoe (called Lake Sasyk on this map), at the border of the NW curve of the field boundary. Three roads lead to the building, one from the NW, another from the S, and the third from the E: these last two connect to the road further S, which coincides with the modern asphalt road between Černomorskoe and Mežvodnoe. A smaller track also leads to the structure from the N, following the curve along the lagoon and along the present survey field. Another track runs across the length of the promontory parallel to the lake and E of the field leading from the southern road to the northern coast, just W of the mounds; the upper eastern border of the field follows this track. Finally, a road follows the coastline joining the E-W road to the S. All of these roads and tracks have now vanished, except for the present asphalt road and the eastern shoreline road.

Lake Sasyk is a salt water lake (*solenoe*); three sides of its shores show salt works (*soljanoe promisl*). One of these is shown directly W of the surveyed field near the building. The eastern part of the lake is nowadays a salty marsh. N of the

[42] Fig. 4.40 works as a key to the period specific scales and legends of all the distribution maps.

lake on the sandbar that closes off the lake from the Black Sea, a small pier juts out into the sea (*prist.* for *pristan*, pier or wharf). This sandbar is now broken open giving Lake Panskoe and the harbour on the shore opposite of the ancient site, where the northern salt works once stood, access to the sea.

**Survey conditions** Field DSP07-01 had been recently ploughed, the plough lines ran E-W; this was also the walking direction of the intensive survey teams. The surface soil was brownish-red and included some limestone fragments with a few stonier areas spread out across the field, probably where the underlying limestone bedrock was turned up by the plough.

The northern part of the field contained some modern building debris including asbestos; the southern area also revealed construction material from the road. There was little vegetation, except in the southwestern part of the field. The days of the survey were sunny and windy. The surface visibility was good.

**Finds distribution** The highest densities of finds in Field DSP07-01 are located, not surprisingly, in the northern, narrow strip near Panskoe I, but also in the southern part of the field W of the large kurgan in the field to the E. Most of the finds collected are Late Classical-Early Hellenistic pottery fragments. In the northern section they form a 'halo' in conjunction with Panskoe I (see below, Relations with surrounding area). In the southern half of the field, the Late Classical-Early Hellenistic material is either connected to the kurgan in the barley field or to the destroyed kurgans (DSP07-F01-02 Features 1-3) discovered during the intensive survey or maybe both.

The survey produced no finds in this field from the Bronze Age or Byzantine periods. A single fragment was found in the middle of the field that could be dated to the Middle Ages, and two dispersed Early Modern sherds were found, over a kilometre apart. Some modern material was collected in two units in the southern area of the field.

**Sites** Two sites were identified in Field DSP07-F01. The northernmost site, DSP07-F01-01, is located in the narrow strip of the field directly E of Panskoe I near the highest part of the ridge between the lake and the bay. The finds in DSP07-F01-01 are concentrated around the middle section of the strip, across seven units; two units produced among the highest find densities recorded during the DSP survey seasons (Fig. 4.46). To the N and S of the site, find densities dropped notably. The distribution pattern is consistent with a site halo associated with Panskoe I. Finds consisted of Late Classical-Early Hellenistic pottery, mostly storage jars, but also tableware and handmade vessels and stone tools were found. One unit also yielded three pieces of worked flint of uncertain date. Within the concentration, a single green-glazed sherd dates to the Early Modern period.

Site DSP07-F01-02 is a kurgan group consisting of three separate features (kurgans). In the SE part of the field, a relatively high density of pottery fragments was found along the eastern field border (Fig. 4.47). Feature 1 was defined based on this and the proximity to the kurgan in the barley field just a few meters away (Fig. 4.48). The kurgan and this side of the field occupy the highest part of the ridge. The pottery concentration is most likely related to the kurgan, although a closer inspection of the mound itself yielded no finds. Most units along this field border yielded at least a few finds with one unit directly SE of the kurgan producing slightly more Late Classical-Early Hellenistic fragments as well as a single Early Modern sherd. Feature 2 is a ploughed scatter of stones and pottery most likely connected to a ploughed-out kurgan, which was excavated here by local archaeologists during the 1980s (pers. comm. V. Stolba). No noticeable elevation was detected in the field. The site is most likely also the same as the settlement Panskoe V identified by small-scale survey in 1993 (Randsborg 1994). The number of finds collected scarcely differs from the adjacent units: three sherds, two of which were Late Classical-Early Hellenistic, but the concentration and size of stones is consistent with a ploughed-out kurgan. Units N of the scatter yielded no finds; units to the E and S had similar densities. Located about 80 m to the S, Feature 3 is a similar scatter which probably also represents a destroyed kurgan. It is located at the border of four units, all of which yielded a few finds. Most of these were Late Classical-Early Hellenistic including handmade fragments. The scatter is located in the lower western part of the field. It is not known whether this kurgan was also excavated.

**Relations to the surrounding area** The large Late Classical-Hellenistic settlement Panskoe I is located 250 m NW of DSP07-F01-01 (see Chapters 1.4 and 4.2.1). SW of the field is the Hellenistic farmstead, Panskoe II (see DSP06-F03-01). In the field E of DSP07-01 is a large kurgan, just beyond the edge of the surveyed field.

A Neolithic site (Jarylgač Severnoe) was reported on the promontory NE of the field near the necropolis. This area was investigated in 2007 but was so overgrown that no systematic sampling could be undertaken.

*DSP07-F02* (Figs. 4.49-50)

**Size** 180.5 ha.

**Units** 1084-1351; 2074-2299.

**Investigations** Intensive survey in 60 x 60 m units except 60 x 100 m units along the eastern field border as well as geomagnetic survey.

**Landscape type** Pediment.

**Location** Field DSP07-F02 is located close to the southern shore of Lake Džarylgač in the lowest part of the Pediment zone. The field lies along the E side of a gravel road that crosses a causeway in the lake separating Lake Džarylgač from the smaller Lake Jarylgač SW and leads to Mežvodnoe further NW. The gravel road intersects with the asphalt road S of the field connecting Snežnoe and Novo Ul'janovka to the E; the hamlet of Zajcevo lies in the NW corner of this intersection.

Field DSP07-F02 is one of the largest fields surveyed during the project. It slopes down from the S and gently undulates towards the lake in the N some 30 m lower than the asphalt road. The northern corner of the field rises slightly taking the shape of an ear as it follows the curve from the lake to the inlet that separates it from Field DSP08-F04 to the E. The waterfront of this promontory is ringed with grass and shrubs. At a macro scale, this elevation is located at the northeastern end of a stretch of land cut by ravines to the SW and NE that empty into the lake. The northeastern ravine cuts across the field from the SW creating a low but increasingly discernible depression through the field as it deepens towards the inlet.

The northern part of DSP07-F02 is partly obscured from the lake and the southern shoreline road by the grassy strip of trees and shrubs. Towards the E is a lagoon or inlet of the lake crossed by a sandbar which continues the shoreline road further E and N of Field DSP08-F04. Field DSP08-F04 is immediately E of DSP07-02. A ravine defines the northwestern perimeter of the field emptying into a marsh close to the lake. Across this ravine and W of the marsh lies Field DSP06-F02. The western edge is adjacent to the gravel road to Mežvodnoe; except for Field DSP07-F04 this area was fallow. The southern end of DSP07-F02, some 350 m wide had not been tilled in 2007 and was thus excluded from survey.

**Cartography** Projected onto the half-verst map, DSP07-F02 is situated on the southern shore of Lake Džarylgač; this region may be marshier today than appears on the map. The two lagoons that frame the lake side of the field are shown as inlets cut off from Lake Džarylgač by sandbars. The elevation lines show the ravines that empty into both lagoons from the SW. The eastern lagoon receives run-off from two ravines, one cutting SW across Field DSP07-F02 and another running S-SE just within the boundary of DSP08-F04. The western lagoon receives runoff from a ravine that skirts the NW perimeter of DSP07-F02 and extends first S then SE, all the way from the Hillsides. The ravine originates around Hill 6 separating it from Hill 17. The modern shoreline track does not appear on the half-verst map. However, it does depict a road cutting across the northern corner of the field from the SE and continues across the sandbar of the western lagoon to the NW indicating firmer ground then than seems to be the case in this marshy area today. This road connects Džarylgač (modern Mežvodnoe) in the NW with the abandoned village of Kipčak Učkuju to the E.

The half-verst map shows another road that once crossed the middle of DSP07-F02 leading from Kipčak Učkuju to Ak Mečet (modern Černomorskoe) in the W. This middle road runs just N of an area marked on the map as the ruins of Džajlav. The ruins, which have now disappeared, are immediately W of the N-S ravine and E of a road parallel to this ravine continuing N to Džarylgač (modern Mežvodnoe). This middle road follows a straight line for at least 3 km through DSP07-F02 and to the W, up and down the ravines, running E-NE to W-SW. Exactly parallel to this middle road, and N of it by some 500 m is a dashed line on the half-verst map. The same mid-20th century map shows a second dotted row of trees perpendicular to the first; this one runs N-NW to S-SE and probably showing the boundaries of the present field. Though not discernible on the ground today, these lines as well as the old middle road have been interpreted as originally belonging to the field cadastre system set out by the Greek colonists (see Chapters 6.3.1 and 7.3.1).

S of DSP07-F02 is a road on the half-verst map that connects Kipčak Učkuju in the E with the abandoned village of Karlav in the W. This road crosses through an abandoned hamlet named Mussali SW of Field 2 immediately W of the modern hamlet of Zajcevo. Like Džajlav, Mussali is also located just W of the N-S ravine. The modern asphalt road lies further to the S.

The half-verst map also shows some stars within the boundaries marking mounds: one star is at the extreme eastern boundary measuring 21.6 m in elevation; two more stars appear roughly 800 m to the SW. The mid-20th century map only shows onc star with an elevation of 26 m. These mounds flank the road leading from Kipčak Učkuju to Mussali on the half-verst map; in fact the road seems oriented towards these mounds.

**Survey conditions** The surface of DSP07-F02 was reddish-brown with some very low vegetation, and on the whole it was moderately stony. Some very stony patches were observed in areas that did not seem related to surface finds. These occur most likely where the plough has cut into the bedrock bringing the chips to the surface. The field had been tilled E-W, and this was also the direction for field walking. The eastern 'ear' of the promontory in the N was initially left fallow, but it was surveyed later after having been ploughed. The days of survey of DSP07-F02 were sunny and clear. The ploughed surface was dry and very dusty. This had some negative effect on overall surface visibility, as had the limestone rocks. The low and sparse vegetation was hardly an obstruction except for longer cast shadows in the early morning light.

**Finds distribution** Finds are distributed evenly over the field in an extremely low density. Some adjacent units in the southern section yielded finds, whereas 'empty' units were mostly in the northern, middle and southwestern parts of the field.

By far the majority of the material collected dates to the Late Classical-Early Hellenistic period. These finds, including both handmade and wheelmade pottery are evenly spread across the field, with slightly higher densities in the southern part, and two concentrations in the northwestern and northeastern corners. No finds from the Bronze Age were collected in DSP07-F02 at all, nor were fragments dating from the Classical period. One possible Byzantine sherd was collected along the southern edge, and another possibly Medieval fragment was found more to the N. Early Modern finds were collected in four units in the southeastern quarter of the field. Flint was also regularly collected in this field. Low densities of modern fragments were found across the field, including some pottery, iron and asbestos. Some of this was collected, mostly in the middle and southern parts of the field. This may be due to field walkers still being unacquainted with the type of finds in the area, since this was only the second field surveyed.

**Sites** Although finds were scattered across the field, there were some noticeable concentrations at the northern and southern ends. Ultimately, five sites were identified.

At the upper NE section, DSP07-F02-01 was identified along the SE edge of the ear-shaped promontory (Figs. 4.51-52). The site consists of three discrete clusters of Late Classical-Early Hellenistic material (Features 1-3), including a fairly high amount of handmade pottery and some wheelmade sherds of both fine and coarsewares. Some flint was found further along the NE perimeter of the promontory. This area is very close to the overgrown part that leads towards the ravine and the lagoon separating this field from Field DSP08-F04 to the E. Much less material was found atop the elevation in the units to the NW. The highest densities were thus collected in two nonadjacent units in the SE corner with adjoining units containing no finds in the standard sample. This area affords good shelter from the northern wind coming from the lake; the site may extend further SE here, although a superficial examination of the grassy area beyond the field was inconclusive. Geomagnetic survey produced no structural anomalies except dipole ones suggestive of modern iron scraps.

In the opposite NW part of the field, five corner units showed relatively high densities of Late Classical-Early Hellenistic material from different functional groups including tablewares, amphorai and handmade material (DSP07-F02-02). Finds were dispersed along the corner units and the extent of the site was diffuse (Figs. 4.53-54). These finds may form the halo of a site outside the field boundary, but the thick vegetation precluded any further observations. The concentration was found in a similar setting as DSP07-F02-01 on an elevation near a ravine leading to the lake.

A small discrete concentration of Late Classical-Early Hellenistic handmade pottery, tableware, amphorae and stone tools was found upslope in the southern half of the field (DSP07-F02-03) (Fig. 4.55). It is located up on a slight slope yet is still somewhat sheltered and appears to be related to the ravine that empties into the eastern lagoon. The concentration is located on the border of two units. Adjacent units yielded only two Late Classical-Early Hellenistic fragments.

Further to the E were two small concentrations of Late Classical-Early Hellenistic tablewares, utility wares, amphorai and handmade pottery fragments located approximately 150 m apart (DSP07-F02-04) (Fig. 4.56). In between these, the units all yielded Late Classical-Early Hellenistic material suggesting that these finds may be the remains of a completely ploughed-out site. The two scatters (Features 1-2) and the adjacent units were defined as one site. A single Early Modern fragment was also collected in one of the units. Like DSP07-F02-04, this site is located on a slight slope but seemed less sheltered.

Nearby, a round, ploughed elevation was observed in the SE corner of the field (DSP07-F02-05) (Fig. 4.49-50). No finds were collected in the near vicinity, but a kurgan was depicted at exactly this location on the half-verst map.

No specific finds could unequivocally be associated with the cadastre system recorded across the middle of the field; material was found in units located directly on the divisions, yet this can easily be explained as part of the general pattern of dispersion.

**Relations to the surrounding area** DSP07-F02-01 and DSP08-F04-07 are located only 250 m apart on opposite sides of the ravine that separates their respective fields. Furthermore, these sites appear connected by the road shown on the half-verst map that stretched from Kipčak Učkuju across the ravine that separates the two sites and over the sandbar to the NW continuing on towards Džarylgač (modern Meževodnoe); perhaps this route was used in antiquity as well.

The proximity to open water will have been a location factor common to these sites. DSP07-F02-01 and DSP08-F04-07 are located near the ravine that empties into the lagoon between the fields. Roughly 1 km to the S, DSP07-F02-04 appears along the same ravine. Similarly, on the opposite side of the field is DSP07-F02-03 roughly 800 m W of DSP07-F02-01 quite near the ravine that empties into the lagoon NW of DSP07-F02. In the next field to the W, DSP07-F04, are two more sites related to the same ravine, DSP07-F04-01 and 02; although these are not contemporary with DSP07-F04-01, they do show the importance of water and/or shelter from the wind for settlement activity.

Site DSP07-F05 forms an exception to this pattern. It is located roughly 500 m E of DSP07-F02-01 and 400 m W of the kurgan (DSP07-F02-06). This kurgan is aligned with the long row of kurgans in Field DSP08-F04 further to the NE; these kurgans are visible on the horizon from DSP07-F02. It seems unrelated to the two isolated kurgans shown on the half-verst map (single kurgan on the mid-20th century map), which were not further investigated during the DSP.

DSP07-F02 is highly visible from the Hillsides to the S, e.g. from Hills 6 and 17, but also from across the lake as could be seen from the DSP08-F06.

*DSP07-F03* (Figs. 4.57-58)

**Size** 6.1 ha.

**Units** 1352-1364; 2300-2304.

**Investigations** Intensive survey in 60 x 60 m units except corner ones.

**Landscape type** Pediment.

**Location** This small triangular field lies some 3.5 km SE of Panskoe I and directly N of the modern village of Snežnoe across the asphalt road that leads to Vodopojnoe. It is located just S of Field DSP06-F01. Just beyond the NW edge of the field is a grassy kurgan mound, now reused as a cemetery and enclosed with a whitewashed stone wall (Fig. 4.59). The southwestern edge of the field borders on a dirt road. A row of trees lines the field to the E. Just S of Snežnoe and DSP07-F03, the Hillsides begin to rise more steeply towards the S. Immediately SW of the field is a ravine that separates Hills 19 and 24 further up and continues NW where it empties into the eastern edge of Lake Panskoe.

**Cartography** The village of Snežnoe is relatively new: no settlement is shown in this area on the half-verst map. On this map, DSP07-F03 is located N of a road at the foot of the Hillsides between the village of Karlav to the E and the ruins of a village called Saja to the W. Parallel tracks come down the hills towards the N-NW one of which runs through the field and passes SW of the cemetery kurgan. Both the tracks and the road from Karlav to Saja have been interpreted as part of the cadastre set out by the Greek colonists. It is unclear whether the parallel N-S ditches observable today may be related to this cadastre system or not. The kurgan reused as a cemetery by the inhabitants of Snežnoe is depicted on the map marked with an elevation of 11.4 sažens appearing on the mid-20th century map at 27.1 m.

Snežnoe appears on the mid-20th century map; the village of Karlav is renamed Čaïkino. A new cluster of buildings called Kostrovka was built on the location of previous Saja. The tracks coming down the hills have disappeared, and a new road runs from Čaïkino through Snežnoe towards the NW and Lake Panskoe. The cemetery-kurgan is still in use.

**Survey conditions** The field was ploughed recently and the surface soil was fine and somewhat powdery. The soil was reddish brown with some stony parts. DSP07-F03 was surveyed on a clear, sunny day. The overall visibility was good. Some parallel shallow ditches running N-S were observed in the central and eastern parts.

**Finds distribution** Of the 18 units in this field, eleven yielded archaeological finds. In all of these 11 units, Late Classical-Early Hellenistic pottery was collected. Highest densities were recorded in the unit bordering the cemetery-kurgan (Unit 2301) and in Unit 2304 (in the middle of the eastern field border). One unit in the southeastern corner yielded Early Modern material, and modern material was collected throughout the field, but particularly in the vicinity of the cemetery-kurgan.

**Sites** A group consisting of three ploughed-out kurgans observable as stony elevations were identified in the field (DSP07-F03-01) (Fig. 4.60). One of them, Feature 1, is located in the northwestern part of the field close to the cemetery kurgan; some amphora material could be associated with this. The other two, Features 2 and 3, lie parallel to the eastern field border. Between these two kurgans was a concentration of Late Classical-Early Hellenistic material, including handmade pottery and amphorai fragments, but also a black-gloss rim and an astragal.

**Relations to the surroundings area** None of the areas immediately surrounding DSP07-F03 were investigated, and no other kurgan groups are recorded in the direct vicinity. DSP07-H11-01 is located approximately 1 km to the SW and Panskoe II 2 km to the NW. No finds could be associated with the supposed cadastre line running through this area; in fact nearly all of the units it ran through were empty of finds.

*DSP07-F04* (Figs. 4.61-64)

**Size** 33.3 ha

**Units** 1365-1406; 2305-2364.

**Investigations** Intensive survey in 60 x 60 m units except corner ones.

**Landscape type** Pediment.

**Location** Field DSP07-F04 is located approximately 600 m S of Lake Džarylgač. The southern end of the field is rectangular and borders on a fallow field. The northern part extends to a sharp point in the NW where a gravel road separating this field from DSP07-F02 to the E and from DSP06-F02 to the N meets the tree-lined dirt road running N-S along the western side of the field. The gravel road is the main artery to Mežvodnoe across the lake.

**Cartography** On the half-verst map, the ridge in the western part of the field is clearly visible. At its northern end, a mound is marked with a round-star symbol. A track indicated by a dashed line leads N from this mound towards the lake. In the southern part of the field, another star symbol is shown on the eastern slope of the wide valley. It is possible that this marker is a mound related to the ruins of a village called Džajlav shown roughly 350 m S of the field. These ruins are just S of a road running from the southern shore of Lake Panskoe to the village Kipčak Učkuju. A little over 1 km SW of the northern mound are two more stars, one large and one smaller, indicating two mounds on a ridge W of the ruins of Džajlav; this ridge is the same one running through the western part of the field with the mound just beyond the northern border. Roughly 500 m to the NW of this is yet another mound next to a dashed line; this line is interpreted as stemming from the cadastre system set out by the ancient Greek colonists. The elevation lines reveal most of these mounds to be situated on prominent heights. The half-verst map shows a road W of DSP07-F04 running from the Hillsides in the S through the village of Mussali passing just W of the ruins of Džajlav and continuing N across the wide sandbar W of the lake to the village of Džarylgač (modern Mežvodnoe). Immediately to the E is another dashed line which ends at the boundaries of this field; this is also associated with the cadastre system (see above).

The mid-20th century map shows changes in the area. Mussali is now renamed Zajcevo. Kipčak Učkuju has been abandoned, and the road to Džajlav has disappeared along with Džajlav itself and the feature in the southern part of the field that was marked on the half-verst map with a star. The mound at the northern end of the ridge outside the northern field border is still present and marked at 10.5 m, but it has since disappeared. One of the two mounds further SW is shown at 22.2 m, and the mound to the W is shown at 15.4 m.

The wide road E of the field towards Mežvodnoe was already laid out in the half-verst map. The ancient cadastre system shown on the half-verst map has disappeared.

**Survey conditions** The field had recently been harrowed in small peds. The western section is distinctly higher than the eastern part of the field. In the mid-eastern area is a wide depression running roughly N-S through part of the ravine that empties into the lagoon at the lake's edge to the NE.

The soil colour changed from brown in the N to more reddish in the middle, to grey in the SE corner. Various stony patches occurred across the field, probably where underlying bedrock had been turned up by the plough. The stones in the northern part were especially distracting since from a distance they resembled sherds in colour and shape being reddish, small and rectangular. In the southern part of the field the quantities of straw, probably from fertilization, negatively affected surface visibility; also low vegetation regularly cropped up across the field. DSP07-F04 was surveyed on a sunny day and general overall visibility was moderate.

**Finds distribution** Most of the archaeological material has been found in the eastern, lower part of the field, whereas the higher western area was almost devoid of artefacts. Most of the finds were Late Classical-Early Hellenistic, but they occurred in low densities, and no distinct concentrations of Late Classical-Early Hellenistic material were identified. Bronze Age finds were collected in a small concentration within one unit near the northeastern corner of the field. No other finds from this period were collected in the rest of the field. Several Early Modern finds were collected in three units along the southern field border.

**Sites** Two sites were identified in Field DSP07-F04: the small concentration of Bronze Age material (DSP07-F04-01); and the Early Modern concentration along the southern field border (DSP07-F04-02).

Site DSP07-F04-01 is a small, diffuse scatter of handmade pottery fragments including part of a stone axe head (Fig. 4.65). This tool was found during standard intensive survey, while the pottery was collected when the area around the axe was sampled. Although the pottery fragments were all non-diagnostic, the axe head can be dated with certainty to the Bronze Age; the proximity of the pottery to the axe head suggests a contemporaneous date. The Bronze Age sites DSP08-F06-01 and DSP08-F03-02 across the lake both yielded similar broken stone axe heads. Some flint tools were collected 150 m S of DSP07-F04-01, and roughly 400 m to the N along the northern edge of Field DSP07-F02, but these could not be dated with any certainty. DSP07-F04-01 is located in a sheltered area in the ravine at the bottom of the slope.

DSP07-F04-02 is located along the southeastern border of the field also on the eastern side of the ravine (Figs. 4.66-67). It consists of a concentration of Early Modern utility and tablewares, associated with animal bones. A small elevation is visible, which coincides with the mound depicted on the half-verst map. While sampling the area, a total of

69 Early Modern fragments were collected and one Late Classical-Early Hellenistic amphora sherd. The adjacent units show similar low densities of Late Classical-Early Hellenistic finds. Two Early Modern fragments were collected in two nearby units. Although it may be possible that the Early Modern material postdates the mound on the same location, in which case the mound may be associated with the Late Classical-Early Hellenistic sherds, it is also possible that the mound is of Early Modern origin.

**Relations to the surrounding area** No other Bronze Age sites were identified on the Pediment S of the lake. DSP07-F04-02 is located roughly 400 m NE of the ruins of the village Džajlav.

*DSP07-F05* (Figs. 4.68-71)

**Size** 71.9 ha.

**Units** 1407-1526; 2365-2445.

**Investigations** Intensive survey in 60 x 60 m units except corner ones.

**Landscape type** Lowland Ridge.

**Location** This field is located N of Lake Džarylgač on the lower southern slope of the Lowland Ridge, roughly 500 m W of the present-day village of Vodopojnoe. DSP07-F05 spans a promontory that extends into the eastern branch of the lake some 600 m to the S; the field stretches from the eastern side of this promontory and continues W across the northern part of a large ravine that forms the main northern inlet in the wider part of the branch. The field is bordered to the N by a dirt road. Across this road is DSP07-F06. A minor dirt track runs along the southern border. Fallow land lies to the E, W, and S except for the southwestern corner which had just been deep-ploughed. DSP07-F05 is made up of three contiguous tilled areas, separated by narrow grassy strips. For the sake of reference, these were given one field number. The field slopes down to the S and SW with the southwestern corner being the lowest area of the field.

**Cartography** On the half-verst map present-day Vodopojnoe is called Kirleut and is located on the N shore of the eastern branch of the Vodopojnoe Ravine. This area must have been dry at times, as a track runs right across it. Kirleut is a crossroad with roads leading in all directions. The road to Džarylgač (present-day Mežvodnoe) still exists today and forms the northern field boundary of DSP07-F05. On the western edge of the village, a farm is depicted at the northern end of a ravine. Minor tracks lead from this building towards the W through DSP07-F05. A line of four mounds is marked on the ridge running down the slope to the lake; this area now coincides with the grassy strip separating the easternmost part of Field DSP07-F05. The southernmost of these mounds is marked at an elevation of 6.6 sažens (13.66 m). Another line of several mounds is depicted on a ridge across the ravine W of the field approximately 400 m from the field border; the second mound from the N of the lower group is marked at 5.9 sažens (12.21 m). The southern track extending from Kirleut to a lagoon further W runs directly along the two mounds marked with elevation levels on this map and they may have figured as points of orientation.

On the mid-20th century map, only one mound appears within the field boundary and perhaps two more just to the S marked at 15.5 m in elevation. The line of mounds W of the field now consists of three mounds on the lower ridge near the lake and a separate one on an elevation 130 m S of the dirt track to Mežvodnoe. This same map reveals that the village of Kirleut is now called Vodopojnoe. The farm on the western edge is marked as abandoned. A dirt track leads around it to the W and now forms the southern border of the field. Part of the field on the western side of the promontory is marked as a vineyard (with $-symbols). The western boundary of DSP07-F05 seems in reality closer to the ravine than would appear on this map, although a row of shrubs in the grassy area further W probably marks the western boundary of the field depicted on the mid-20th century map. The ravine is visible in the field today only as a very gentle depression.

**Survey conditions** The grassy strip that separates the two larger areas is located just W of the top of the promontory in the southwestern part of the field. The largest area is the rectangular field in the W. This field was ploughed and harrowed, and its surface visibility was fair except for some stony patches and shade. The soil is generally reddish-brown becoming a little lighter in colour towards the promontory. The two smaller eastern areas of this field had recently been deep-ploughed, but not harrowed. The surface consisted of very large peds, which made walking quite hazardous. Visibility here was below average, partly because of the shade created by the peds, partly because field walkers had to concentrate on staying upright. The days that DSP07-F05 was surveyed were sunny and hot.

**Finds distribution** Archaeological finds were collected throughout DSP07-F05. Most of these were Late Classical-Early Hellenistic. However, artefacts occurred only in low densities across the field, and widely dispersed. No clear concentrations from this period could be defined. The material consisted primarily of wheelmade pottery, including storage vessels and tablewares. Some handmade pottery was found as well as a few pieces of flint.

Pottery fragments dating to the Bronze Age period were collected in two units in the southwestern part of the field adjacent to DSP07-F05-02, and in one unit in the middle of the surveyed area.

Six units from different parts of the field yielded pottery, possibly from the Medieval period. Along the northern edge, in the E, moreover a high density of Early Modern finds was found. Other units in the eastern part also yielded Early Modern finds, but not adjacent to the concentration. Modern fragments were collected in the easternmost units, and in some dispersed units in the rest of the field. The modern material most likely comes from the nearby village of Vodopojnoe.

**Sites** Two sites were identified in DSP07-F05 neither of which coincides with the mounds shown on the historical maps. One grassy mound was recorded in the unploughed strip between the two easternmost parts of the field.

DSP07-F05-01 is located in the northeastern corner of the field (Figs. 4.72-73). It was found in the most deeply ploughed area where average visibility was poor; its full extent may therefore not have been recorded accurately. This site consists of a diffuse concentration of primarily Early Modern pottery fragments. Its location near the dirt road may be indicative of a roadside dump, especially since no ruins or buildings are recorded here on the historical map.

The identity of DSP07-F05-02 is based on a single find: a limestone stele with anthropomorphic form (Figs. 4.74-75). Similar limestone markers are known in the Black Sea area and are generally related to Pre- and Protohistoric burials. Although this particular stele was quite worn, its features were nonetheless recognizable. It was found in an elongated low rise spared by the plough on account of the stele. The stone was roughly 0.8-1 m across and about 0.2 m thick. A concentration of small stones was found around the stele, while neighbouring units had stones that were just a little smaller than this stele. Two of the Bronze Age sherds found in this field were located in units just to the SW of the anthropomorphic stele.

Soon after it was discovered and recorded, the stele was collected along with other large stones from this field by a villager with a tractor. Fortunately, most of the field had been surveyed already, and although other relatively large blocks were noticed, the anthropomorphic stone remained a singular find.

**Relations to the surrounding area** The Early Modern finds and the density of possibly Medieval finds both seem related to the dirt road at the northern edge of this field. Directly N of the dirt road in Field DSP07-F06, another site was identified which included mostly Late Classical-Early Hellenistic (and modern) material, but also some possibly Medieval finds (DSP07-F06-04). Although not contemporary, the proximity of these clusters and more diffused scatters indicates former activity in this area, and the road may have been ancient too.

Further N in Field DSP07-F06 was a pottery scatter including Late Classical-Early Hellenistic material, which may have come from a kurgan (DSP07-F06-02). This scatter is in line with the mounds shown to have existed on the half-verst map approximately 650 m N of the northernmost kurgan on the map.

The anthropomorphic stone, which likely belonged to a Bronze Age burial, appears to have been isolated. Besides, the two sherds found in the units to the SW, one other sherd was collected almost 500 m to the E. Also one Bronze Age sherd was collected near to the lake 700 m to the S at Vodopojnoe 1 (see Chapter 4.3.4), but no habitation sites could be defined.[43]

*DSP07-F06* (Figs. 4.76-77)

**Size** 45.3 ha.

**Units** 1527-1610; 2446-2534.

**Investigations** Intensive survey in 50 x 50 m units except corner ones.

**Landscape type** Lowland Ridge.

**Location** DSP07-F06 is located due N of DSP07-F05 and SE of DSP07-F07 on the southern slope of the Lowland Ridge. The village of Vodopojnoe lies about 500 m to the E, and Lake Džarylgač is 1.4 km to the S. The southern part of the field borders on the dirt road between Vodopojnoe and Mežvodnoe; remains of a small concrete building are located just within the southern border of the field. West of the field was a grassy strip with shrubs, separating this field from the recently ploughed field on the other side. The area N and E of the field lay fallow. Just beyond the NW corner are some skeletal concrete structures. A small track leads W to this structure from the northbound road from Vodopojnoe.

[43] These finds were collected at the southern edge of a large field that was initially targeted for survey as DSP07-F09. However this field had only been deep-ploughed in large peds. Since the experience in the eastern end of DSP07-F05 already proved such conditions to be too strenuous for effective surveying, the rest of this field was excluded from field-walking. The southern area close to the lake was only surveyed briefly after surveying DSP07-F06. DSP07-F09 is therefore not further discussed here.

The field gently slopes down to the S towards the lake. Some gentle undulations were observed; the dirt road along the southern border, for example, is not quite visible from the northern part of the field. The central area of the field is located atop the northern beginning of the promontory that continues S-SE into DSP07-F05, where the ridge functioned as a platform for the kurgans shown on the half-verst map.

**Cartography** The half-verst map shows the area of the field as being located just NW of the village of Kirleut (modern Vodopojnoe). It occupies the gentler northern part of the promontory that extends S down to the eastern branch of Lake Džarylgač, and is just N of the main road leading W from Kirleut to Džarylgač (modern Mežvodnoe).

A dashed line indicates a track that leads from this road to the NW cutting across the SW corner of the field. Some 300 m beyond the field, this track bends to the W parallel to the Kirleut-Džarylgač road running just below a group of kurgans near the top of the ridge (see also DSP07-F05); the track ends in DSP08-F05. This track is the only artificial feature shown within the boundaries of DSP07-F06. The modern field is located SW of a main road that leads from Kirleut to the NW. Just NE of the modern field, this road diverges into two sections leading to the two farms along the coast N of Fields DSP08-F02 and DSP08-F03.

The current northern border of the field is shown on the mid-20th century map as a row of trees lining a track running from the northbound Vodopojnoe-Skalistoe road in the E to the parallel N-S track that defines the western edge of DSP07-F07 and continues S (see DSP07-F05). The current eastern and western borders of the field are not depicted on this map although the western border does appear on the 1993 map, where it is adjacent to a vineyard in the W; a large orchard ran diagonally to the NE. The most recent map also shows a hydraulic system running through this field along the ridge continuing S through the eastern part of DSP07-F05, where it makes a turn towards Vodopojnoe; the concrete ruins NW of the field are the remains of a pumping station.

**Survey conditions** The soil colour is a fairly consistent reddish-brown with some stony patches. The field had been ploughed and harrowed. The days DSP07-F06 was surveyed were sunny and hot, and the surface was somewhat dry and dusty. Still, overall visibility was good.

**Finds distribution** Widely dispersed over the field archaeological material was collected but more in the southern half of the field than in the northern one. Most artefacts, however, were found in the middle area near the ridge, where some relatively high densities were recorded. Most of the finds date from the Late Classical-Early Hellenistic period. Low densities of possibly Medieval finds were recorded in six units, two in the northern part of the field, four in the southern. Relatively high densities of modern material were collected in units in the southern half of the field near the dirt road.

**Sites** Three sites were identified in DSP07-F06 based on find densities. All three were located in the mid-southern section of the field close to the ridge. DSP07-F06-01 is a ploughed-out kurgan and is characterized by a stony area with more vegetation than average on a slight elevation (Figs. 4.78-79). Some Late Classical-Early Hellenistic material was found nearby. Less than 100 m to the E is DSP07-F06-02 with a slightly higher density of Late Classical-Early Hellenistic material in one unit including both handmade and wheelmade vessels, as well as some flint (Figs. 4.78-79); both sites are in the middle of an area with a diffusion of Late Classical-Early Hellenistic finds. The higher concentrations of primarily Late Classical-Early Hellenistic material further S roughly 30 m from the dirt road were labelled site DSP07-F06-03 (Fig. 4.80). They consist of three main clusters of material found in two adjacent units consisting of some handmade but primarily wheelmade pottery, both storage vessels and tableware, and a stone tool. Also a single Early Modern fragment was collected here.

**Relations to the surrounding area** The location of the kurgan DSP07-F06-01 appears somewhat isolated, but it falls within the general alignment of the kurgans shown on the half-verst map on the ridge further down in DSP07-F05. The northernmost kurgan in this group is roughly 500 m S of DSP07-F06-01. In the opposite direction, roughly 2 km to the NW, is another group of kurgans on the coastal ridge at the location where this ridge begins.

DSP07-F06-02 is located near the kurgan. The finds date to the Late Classical-Early Hellenistic period, including a few handmade, tableware and utility ware sherds, imported amphorai from Chersonesos and Herakleia and a single flint.

DSP07-F06-03 is likely to be related to the widely dispersed scatter of Late Classical-Early Hellenistic material found across the field and in DSP07-F05 to the S. The range of finds is similar to DSP07-F06-02 although more pottery was collected and the variety of imports includes Chersonesos, Herakleia Sinope and the South Pontic area. It may also be associated with two dressed limestone blocks that were noted along the road nearby. During the survey of DSP07-F05, we witnessed the collection of these stones by locals.

*DSP07-F07* (Figs. 4.81-82)
**Size** 115.8 ha.

**Units** 1612-1911; 2536-2702.

**Investigations** Intensive survey in 50 x 50 m units except corner ones.

**Landscape type** Lowland Ridge.

**Location** DSP07-F07 is a large field on the coastal ridge sloping down from the highest part of the ridge towards the S. The northern end of the field borders on the main asphalt road from Meževodnoe to Razdolnoe. Immediately across the road is DSP08-F02, and to the SE is DSP07-F06. The field is bordered on all sides by fallow, grassy areas, except for one ploughed field to the W. The southern end of the field is a little over 2 km N of Lake Džarylgač. It is rectangular in shape except for the northeastern corner which follows the asphalt road towards the NE. The field spans the gentle ridge as it slopes down to the S through DSP07-F06 and –F05. The lower part of the field is quite flat.

**Cartography** On the half-verst map, a row of eight mounds is depicted with round stars in the northwestern corner of DSP07-F07; one is located just beyond the field border where the asphalt road now runs. One of the kurgans is marked with a triangle and an elevation of 24.54 sažens (50.79 m). This row of mounds is located at the apex of the coastal ridge aligned from N-NW to S-SE. The row is perpendicular to the ridge, which runs W-SW to E-NE. Interestingly, this direction corresponds to two more ridges that slope down from the coastal ridge to the N-NW and S-SE; these ridges are also dotted with kurgans all following the same alignment. Similar star-shaped symbols occur on the coastal ridge outside the field borders to the W and E; two more groups to the E show a similar alignment (the closest is the group in DSP08-F01), whereas the groups that follow the ridge down to the W-SW appear more clustered. Two more kurgans appear just beyond the northeastern border of DSP07-F07 on the coastal ridge.

Of the eight mounds shown on the half-verst map, three are still depicted on the mid-20th century map as are the two just outside the eastern border of the field. One within the field boundaries is marked as having an elevation of 53.9 m, the one to the E of 49.7 m. To the W, the closest kurgan is depicted one kilometre to the W on top of the ridge; one of the kurgans further E in DSP08-F01 is still shown. The field border is marked with rows of trees along the western, eastern and southern sides on the mid-20th century map.

The modern asphalt road from Meževodnoe that follows the highest part of the ridge does not yet appear on the half-verst map. A minor track marked with a dashed line runs parallel to the ridge but a bit further down the slope to the S crossing DSP07-F07 just S of the mounds. A main road that extends from the village of Kirleut (Vodopojnoe) to the SE crosses through the northeastern corner of this field and the eastern section of DSP08-F02 continuing towards a farm near the seacoast.

**Survey conditions** The soil was generally greyish-brown, which made the few stony patches in the field all the more pronounced. There was a large stony area in the middle of the field on the slope and another broad swath roughly 20 to 30 m wide, in the mid-northeastern part running diagonally from NW to SE i.e. the same direction as the ridge. These stony areas are very likely areas where the bedrock has been brought to the surface by the plough. Some shallow ditches were observed running N-S down the slope.

The field had been tilled and harrowed recently in small peds and had practically no vegetation. The days of the survey were hot, sunny days during which the soil was quite dry and dusty. This affected the otherwise optimal surface visibility.

**Finds distribution** The most noteworthy features in this field are the ploughed-out kurgans discovered in the northwestern units and one on the opposite side along the northeastern edge. These were observed as low elevations as well as in the form of clusters of small stones together with some pottery. For the rest, low densities of material appeared in units spread out over the field. The stony part in the middle of the field yielded however almost no finds at all.

Finds from the Late Classical-Early Hellenistic period, primarily wheelmade but some handmade in the northeastern section associated with the ploughed-out kurgan, were the most pervasive. Although there were no clear concentrations, material from this period was picked up in several consecutive units in the northwestern part and some in the SW. A broken stone axe head collected near one of the kurgans represents the only find from the Bronze Age.

Two units yielded finds possibly from the Byzantine period, one in the middle sloping part of the field and one in the flat southeastern corner of the field. Possibly Medieval sherds were collected in several dispersed units. Early Modern fragments were collected in four dispersed units in the northern part of the field, and three units in the southern end of the field. Relatively high densities of modern material were collected in all parts of the field.

**Sites** Two sites were defined in DSP07-F07. DSP07-F07-01 is the group of four, ploughed-out kurgans located in the northern part of the field towards the top of the coastal ridge (Fig. 4.83). The kurgans in this group are aligned in a N-S orientation. They are located in the northwestern corner of the field and coincide with the kurgans depicted on the half-verst map, a group that appears to have originally consisted of at least eight mounds.

DSP07-F07-02 is approximately 600 m E of DSP07-F07-01 right up against the northeastern edge of the field but probably extending beyond it a little (Fig. 4.84). All five kurgans have nearly vanished and were only discernable by stone scatters and slight elevations in the field that were at times only visible from a short distance.

Sherds were collected on and around the mounds but few were diagnostic. Late Classical-Early Hellenistic amphorai fragments appeared in the general area around DSP07-F07-01, Features 1, 3 and 4. Regular intensive survey around Feature 2 yielded no finds, but some amphora sherds were collected when the mound itself was total sampled. The broken stone axe was collected near Feature 3. Regular intensive survey of DSP07-F07-02 yielded some fragments of Late Classical-Early Hellenistic handmade and wheelmade pottery. This kurgan was not located on any map, but it is only some 150 m away from the two eastern kurgans shown on the half-verst map. These are probably represented by weak elevations observed in the grassy field to the E.

**Relations to the surrounding area** Another row of three kurgans was located in Field DSP08-F01 NE on the coastal ridge about 900 m E of DSP07-F07-01 sharing the same orientation. The group of kurgans to which DSP07-F07-01 belongs lies in the same general orientation as the kurgans that once dotted the ridge leading down S to Lake Džarylgač, two of which were identified in Fields DSP07-F06 and -05 respectively, as well as another group of five on a slope on the opposite side of the coastal ridge further down towards the sea just W of DSP08-F02.

The distance between the DSP07-F07-01 and the Late Classical-Early Hellenistic settlement DSP08-F02-01 is almost 2 km. This settlement may be related to the extensive scatter of mostly Late Classical-Early Hellenistic wheelmade fragments found throughout both fields. It is thus difficult to determine whether the material found on and around the kurgans actually belongs to them or is part of this wider scatter. The isolated Bronze Age axe head found near DSP07-F07-01 Feature 3 makes a dating in the Hellenistic period even less secure. Feature 3 is approximately 2.5 km NE of the large Bronze Age site DSP08-F06-01 and therefore too far away to connect the two.

Finally, some of the Early Modern and modern finds may be associated with the path that once cut diagonally across the northern section of the field just S of the kurgans and parallel to the ridge. No observations were made that could be related to the old Vodpojnoe-Skalistoe road that ran through the northeastern corner of this field. The wide stretches of stone were more to the W.

*DSP07-F08* (Figs. 4.85-86)

**Size** 86.7 ha.

**Units** 1912-1999; 2703-2802; 5000-5144.

**Investigations** Intensive survey in 50 x 50 m units except corner ones. Geomagnetic survey.

**Landscape type** Lowland Ridge.

**Location** DSP07-F08 is located on the southern slope of the Lowland Ridge. Its southern end is very close to the shore of Lake Džarylgač. To the N, the field borders on the dirt road lined with electricity masts between Mežvodnoe and Vodopojnoe. DSP08-F05 begins immediately across this road. E of this field and just N of the road across from DSP07-F08 is an abandoned concrete structure. Fields of barley lay E and W of DSP07-F08 partly separated from it by trees.

**Cartography** On the half-verst map the dirt road between the villages of Džarylgač (modern Mežvodnoe) and Kirleut (modern Vodopojnoe) is depicted. DSP07-F08 is S of this road on the western slope of the ridge that separates this area from DSP07-F05 further to the E. On top of this ridge is a row of seven mounds marked with round-star symbols. A separate mound is depicted further upslope.

A large lagoon cut off from Lake Džarylgač by a sandbar is located SW of DSP07-F08. A minor track runs from this lagoon to the village of Kirleut crossing the middle of the field continuing between the kurgans on the ridge and E through DSP07-F05. There is a clear E-W running depression across the middle of the field today with a stony patch at its western end which may correspond to this path. S of the field is a much smaller lagoon.

On the mid-20th century topographical map, the track has disappeared, but the small secondary lake is still there. The row of seven kurgans on the eastern ridge is reduced to three. Further upslope of the ridge just S of the dirt road, the separate kurgan is also depicted. The row of trees along the western field border is marked with small circles. In the SW corner just outside the field boundaries is a benchmark showing an elevation of 1.7 m.

The 1993 map shows a track following the shoreline to the W where it makes a sharp turn to the N through the field turning E after 300 m to continue around the vineyard on this side.

**Survey conditions** The field slopes gently down to the lake. Its eastern end follows the upper part of a ridge. The western end is lower, while the southwestern corner is the lowest part of the field. The surface in this area was quite stony. Smaller stony areas occur throughout the field. One of these in the southwestern part of the field had small pinkish stones which could be confused with pottery at first glance. Some slight ridges as well as low ditches were observed running N-S and

E-W. These were primarily in the northern part of the field. A few small concrete posts were found in the northwestern area as are typically associated with vineyards but it is unclear whether they originate from this field or from across the road.

The soil was generally reddish brown. The field had been ploughed and harrowed, and surface visibility was fine. The days during which the survey took place were sunny, hot and sometimes hazy, and the soil was a bit dusty.

**Finds distribution** Archaeological material was collected in units throughout the field, with the exception of the low southwestern part of the field, which was almost 'empty'. Most of the artefacts as well as the highest find densities were recorded in the northern highest part of the field. All of the units bordering the dirt road yielded finds.

The Late Classical-Early Hellenistic period was certainly predominant in the material record, as it was collected in units all across the field with higher densities in the northernmost part of the field.

Possibly Medieval pottery appeared in nine units in the southern half of the field with highest densities occurring on the eastern ridge. The origin of this material is unclear since the scatter was diffuse with no real concentrations. Two of the units were near the track on the half-verst map leading to the lagoon. Early Modern fragments were collected in three units in the northern part of the field and three units in the southeastern part. None of these were adjacent and no concentration can be observed.

Relatively high amounts of modern pottery were collected in the northern and southern part of the field often in several adjacent units. The occurrence in the northern part of the field of modern material is most likely associated with the concrete building on the opposite side of the dirt road. The remarkable occurrence of modern material in a N-S strip along the lagoon in the southwestern part suggests some recent activity near to the lagoon perhaps related to the shoreline track that crosses the field on the 1993 map.

**Sites** Two sites diverse in character were located at the northern end of the field. One consists of discrete concentrations of material that point to settlement activity (DSP07-F08-01), the other of some unusual elevations (DSP07-F08-02).

DSP07-F08-01 is a high density of material found in the northern units along the dirt road (Figs. 4.87-88). The soil here is greyer with occasional loose patches of stone. There appear to be two separate concentrations of archaeological material, one roughly 120 m from the northeastern corner and a second cluster about 180 m to the W. Both clusters are located along a low bank or 'ridge' (see below). This entire area contained a mixture of Late Classical–Early Hellenistic wheelmade pottery, handmade fragments, flint, and stone tools though there seemed to be a higher percentage of amphora fragments at the eastern end. Adjacent units to the S of both clusters form a halo with slightly lower find densities. Apart from pottery, stone tools, a whetting stone, part of a ground stone and flint also some Late Classical–Early Hellenistic tile fragments were found. A single sherd could be dated to the Bronze Age. DSP07-F08-01 is one of only two sites outside of Panskoe in the whole research area where tile was identified in the systematic survey. Because of its extent, DSP07-F08-01 was targeted for geomagnetic survey; however the amounts of concrete and wire and the proximity to the electric lines along the road prevented accurate readings.

DSP07-F08-02 consists of a system of low ridges in the northern part of the field of roughly E-W and N-S orientation (Fig. 4.89). Although the ridges were hardly visible in broad daylight, they were much clearer in the low light of early morning or late afternoon. Together these ridges form a system of somewhat askew blocks of circa 95 x 95 m and were observed for nearly a kilometre S of the dirt road, where they disappeared. The geomagnetic survey did not yield any results. The ridges were some 10 to 20 m in width and were occasionally cut through by shallow ditches roughly 1 m across. The ridges may be the remains of an ancient cadastral system, although this remains speculative; some pottery was found along these ridges, but the wide dispersion of material precludes any unequivocal contexts for these elevations.

The northernmost of the E-W ridges runs through the concentrations of DSP07-F08-01. The eastern cluster, in particular, is situated atop a particularly clear elevation which seemed more round in shape and was located in a patch of smaller stones that are often indicative of kurgans. However, this patch is less dense than normal, and the location of the elevation at the crossroads of the E-W and N-S ridge is peculiar.

**Relations to the surrounding area** A scatter of Late Classical-Early Hellenistic finds was recorded in DSP08-F05 just across the dirt road from DSP07-F08-01. Almost all of the finds in this concentration were Late Classical-Early Hellenistic except for one sherd which dated to the Bronze Age. Because the dirt road separates both concentrations, it is difficult to say if they in fact make up one larger concentration. Also the overgrown area around the concrete structure makes it impossible to determine whether any of the concentrations extend to the E. None of the fields N of DSP07-F08 revealed elevations similar to DSP07-F08-02.

The row of kurgans depicted on the topographical maps on the ridge to the E is located at about a kilometre SE of DSP07-F08-01. The separate northern kurgan lies about 500 m away to the E of the eastern part of the site. Two km

N of DSP07-F08-01, the half-verst map shows a loosely grouped cluster of kurgans at the apex of the ridge; these were probably visible from Field DSP07-F08 as are the trees lining the road along the coastal ridge today.

Directly across the lake is DSP08-F04 with several kurgans and settlements. Although not all of the lake is visible from everywhere in DSP07-F08 due to the angle of the slope, most of Field 4 is including the kurgans on it, and the slopes beyond.

*DSP08-F01* (Figs. 4.90-91)

**Size** 14 ha.

**Units** 5145-5166; 6000-6023; 8000-8015.

**Investigations** Intensive survey in 40 x 50 m units except corner ones. The field was used as a "training ground" to teach new field walkers the DSP methodology, and the inexperience of many field walkers at this point may have had a negative effect on sampling.

**Landscape type** Lowland Ridge.

**Location** DSP08-F01 is located N of Lake Džarylgač on the highest part of the Lowland Ridge between the lake and the coast. The field lies directly S of the asphalt road between Meževodnoe and Razdolnoe. To the E, the field borders on a dirt road that connects the asphalt road and the village of Vodopojnoe. To the S is a field of barley. The coast line is approximately 2.5 km to the N.

**Cartography** On the half-verst map, a track indicated by a dashed line lies S of the field. Another track leading S from Skalistoe farm ("xyT." is short for *chutor*) near the coast crosses the western part of the field. This track also runs through DSP08-F03. These tracks are no longer visible.

To the E of the field, a group of kurgans is depicted near a crossroads along the track S of DSP08-F01. In the field itself one mound is marked.

**Survey conditions** The field lies at the highest part of the coastal ridge and is almost flat. From this field there is a clear view towards Lake Džarylgač, the Pediment, Hillsides and Plateau in the S, as well as a view across the fields that slope down to the coast line in the N.

The soil had recently been ploughed in an E-W direction except for the field boundaries where the plough followed the field edge. The surface was somewhat dusty with scattered low vegetation. The western part of the field was stonier than the rest. A large grassy elevation spared by the plough was clearly visible and had already been identified on the map as a kurgan prior to the intensive survey.

DSP08-F01 was surveyed on a sunny but very windy day. Shadows on the ploughing peds, dust and stones affected the overall visibility.

**Find distribution** Most of the finds were recovered in the middle part of the field, and the northeastern part yielded no finds except in one unit. The same can be said of the western end of the field, which was largely empty except for one corner.

In one unit in the middle of the field a fragment dating from the Classical period was found. Late Classical-Early Hellenistic sherds were collected in units spread out over the field including several adjacent units in the central part of the field. Four possibly Medieval and Early Modern sherds were found SE of the kurgans. As there is no indication of Medieval or Early Modern activity nearby on either the historical maps or in the landscape, it is unclear where the finds from these periods came from.

**Sites** DSP08-F01-01 consists of a group of three kurgans aligned in a row running NW-SE (Figs. 4.92-93). Only the southernmost mound, Feature 3, is a grassy elevation. This is most probably also the mound marked by a star on the map of half-verst. One Late Classical-Early Hellenistic sherd was collected in the unit in which the grassy kurgan was located, as well as three possibly Medieval fragments. A closer inspection of the kurgan yielded no extra finds, but it did appear to have been partially dug out. Directly S of the grassy mound two slight elevations of lighter-coloured soil and stones were observed. These were interpreted as kurgans destroyed by agriculture (Features 1 and 2). There were very few finds around these kurgans. The unit directly to the E yielded eight fragments of Late Classical-Early Hellenistic pottery which are associated with the kurgans but dislocated by agricultural activities. The units to the N of the kurgans each yielded one Late Classical-Early Hellenistic sherd. As there are no finds from other periods, the three kurgans should most likely be dated to the Late Classical-Early Hellenistic period.

**Relations to the surrounding area** The closest archaeological sites are a single kurgan (DSP07-F07-02) approximately 900 m SW and a group of kurgans (DSP07-F07-01) another 600 m to the W. All the kurgans in DSP07-F07 and DSP08-F01 follow the same orientation, NW-SE. The kurgan group near the crossroads marked on the half-verst map lies at roughly the same distance in the opposite direction. Additional Late Classical-Early Hellenistic kurgans and/or settlements are located approximately 2.7 km to the S (DSP07-F06-01-3) and 1.8 km to the NW (DSP08-F02-01).

The nearest Early Modern settlements are the farms near Skalistoe approximately 2 km to the N such as DSP08-F03-01 and the village of Kirleut (modern Vodopojnoe) 2.9 km to the S.

*DSP08-F02* (Figs. 4.94-95)

**Size** 111.5 ha.

**Units** 6024-6225; 8016-8143; 8164-8209.

**Investigations** Intensive survey in 60 x 50 m units except for westernmost row of units (50 x 50 m) and corner ones.

**Landscape type** Lowland Ridge.

**Location** Field DSP08-F02 is located just N of DSP07-F07 in the coastal zone. To the S, the field is bordered by a ravine and a vegetation strip separating it from the asphalt road running from Meževodnoe to the E. To the W and E of the field, dirt roads extend from the asphalt road towards the coast. Just outside the northern edge of the field is a farm with sheds and a private graveyard.

**Cartography** On the half-verst map, a *chutor* (farm or small village) is located at approximately the same place as the present farm. Directly W of this farm are ruins named Kūla Šeich, perhaps the ruins of a tomb of a Turkish nobleman as suggested by the name. Another farm is located further to the NE. The farms are connected by two tracks one of which runs diagonally through DSP08-F02 to the SW. The other one runs S through DSP07-F07 down to a farm close to Kirleut (modern Vodopojnoe). Two tracks run parallel to the coastline, one directly N of the survey field and one diagonally through it. These features have disappeared. W of DSP08-F02, a row of four kurgans indicated by round-star symbols are shown on a ridge roughly 280 m away from the western edge of the field.

On the mid-20th century map, the farm N of the village is replaced by a large square complex under the Russian name Skalistoe. Several sites have been found in this area ranging in date from the Bronze Age to the Roman period (Section 4.2.3).

**Survey conditions** The asphalt road to the S of the field runs along the highest part of the Lowland Ridge. From here, the field slopes gently down towards the sea. In the northwestern part, wide ravines cut through the field diagonally from the SW to the NE, whereas in the northeastern part they run from the SE to the NW. The difference in elevation between the ravine bottoms and the surrounding field was up to 2 m. No augering was conducted in this field, but the ravines indicate that the sediment is at least 2 m thick here.

Ploughing and harrowing had taken place recently resulting in small peds. The ploughing direction was N-S as was the walking direction of the survey teams. Some stony patches were observed on the field surface. In the northern part of the field near the farm, straw and other vegetation occurred on the surface, most likely the result of field manuring.

The days during which the survey of DSP08-F02 took place were cloudy, windy and partly rainy. The surface visibility was good.

**Find distribution** The overall distribution map shows a wide dispersal of finds throughout the field with a concentration near the northern field boundary. Late Classical-Early Hellenistic material was collected in several units in all parts of the field. Although there was a slight concentration of handmade ware along the southern edge, higher densities were found in the N. Early Modern material was mostly found in the units near the northern concentration of finds, less in other parts of the field.

No finds from the Bronze Age or Classical period have been located in this field, and only two units at opposite ends of the field yielded finds which may date to the Byzantine period. However, there is a diffuse scatter of small numbers of possibly Medieval finds, including one unit with a somewhat higher density. Relatively high densities of modern fragments were recorded throughout the field.

**Sites** Only one site was identified. DSP08-F02-01 is located at the northern end of the field on a gentle NE slope (Figs. 4.96-97). Most of the material from this concentration dates from the Late Classical-Early Hellenistic period although some units also yielded Early Modern material. The core of the Late Classical-Early Hellenistic material was located in an area across three units, but there was also a significant "halo" of material in the adjacent units and a secondary concentration ca 100 m to the S. Straw and other refuse from the modern farm N of the field was spread out over the area suggesting that manuring or levelling had taken place. During this process, earlier remains may have been dislocated as well resulting in the "halo" around site DSP08-F02-01. The material includes transport- and utilitywares, parts of whetting stones and a stone sinker, probably used for weighing down a fishing net. The distance to the present coastal cliffs is approximately 700 m. The site may very well extend to the N into the grassy area around the modern farm. The Early Modern finds constitute most likely off-site material related to this farm and its predecessors. The local cemetery suggests that the location may have been in use for a longer period of time. A ravine runs diagonally NE-SW through DSP08-F02-01. It is possible that the ravine is a remnant of the track visible on the half-verst map; although its origin is unknown, it may have been used during the Medieval and Late Classical-Early Hellenistic periods as well.

**Relations to the surrounding area** Skalistoe 5 is the nearest Late Classical-Early Hellenistic settlement about 1.6 km to the NE. Several groups of kurgans are located on the coastal ridge approximately 2 km to the S (see DSP07-F07 and DSP08-F01) and at least some of these e.g. DSP08-F01-01 appear to be Late Classical-Early Hellenistic. The Early Modern farm drawn on the half-verst map is located approximately 1.3 km to the NE (DSP08-F03-01).

*DSP08-F03* (Figs. 4.98-101)

**Size** 43.4 ha.

**Units** 6234-6267; 8210-8311.

**Investigations** Intensive survey in 50 x 60 m units except corner ones. Geomagnetic survey.

**Landscape type** Lowland Ridge.

**Location** DSP08-F03 is the northernmost field surveyed. The field lies parallel to the coast, which is about 450 m to the N, and close to Skalistoe 5 (Section 4.2.3). The view towards Lake Džarylgač is blocked by the Lowland Ridge. The field is bordered on all sides by uncultivated land overgrown with grass. A track following the coastline runs N of the field. A fishing station is located towards the W at a low point along the cliff with a small jetty jutting out into the water. This is the modern location of Skalistoe, which is located next to the archaeological site Skalistoe 1.

**Cartography** The elevation model of the half-verst map clearly shows the ravine in the southwestern corner of the field. This ravine continues towards the coast where several tracks meet in the area of the modern fishing station. The NW corner of this field partly coincides with a farm named Skalistoe *chutor*, which is marked on the same map. From the farm four tracks lead in different directions. Two lead to a path running parallel to the coastal cliff. Another one leads to a farm located further SW (the farm N of DSP08-F02). The fourth track heads S to the village of Kirleut (modern Vodopojnoe) just W of DSP08-F02 continuing S towards DSP08-F01.

**Survey conditions** The field slopes down from the SE to the NW. About halfway is a ridge from where the field dips down more sharply to the W. The surface of this ridge is rich in stones probably ploughed up from the underlying bedrock. The lowest part of the field is in the southwestern corner. Here the soil was more greyish than in the rest of the field, where it was reddish-brown. Just outside the southwestern border of the field is an elevation overgrown with grass.

The field was ploughed and harrowed, and apart from some vegetation and stony patches the surface visibility was good. The days of survey were cold and stormy; extremely muddy soil and approaching whirlwinds were challenging conditions for fieldwork.

**Finds distribution** Most of the finds in Field DSP08-F03 were collected in the lower areas in the W. Although the higher eastern part also yielded material, find densities were much lower here.

Late Classical-Early Hellenistic pottery fragments were found all across the field, but a concentration is visible in the southwestern corner of the field. Additionally, all units in the westernmost row of units yielded Late Classical-Early Hellenistic fragments.

In the same southwestern corner was a marked concentration of Bronze Age material. Finds from this period are not dispersed over the field, but cover four adjacent units near the field boundary and one unit a bit upslope to the E. The highest density of Bronze Age material was found in one of the corner units at the northern edge of the grassy elevation.

No finds from the Classical period were collected in DSP08-F03. Two units in the western part of the field yielded material possibly from the Byzantine period. As no Byzantine finds were collected in the surrounding units, it is uncertain what these finds represents.

Finds possibly from the Medieval period were collected in nine units in the middle and western parts of the field. Early Modern fragments were also found in the middle and western parts, with one clear concentration in the northwestern corner of the field.

Relatively high densities of modern finds were recorded in the northwestern and southwestern parts of the field and in one unit upslope in the eastern part. These finds may be related to the nearby fishing station.

**Sites** Two sites were identified in Field DSP08-F03. DSP08-F03-01 is located in the northwestern part of the field (Figs. 4.102-104). Apart from a small Late Classical-Early Hellenistic component in the find assemblage, most of the collected material dates from the Early Modern period. The main concentration is located in the most northwestern unit and likely continues beyond the field boundaries where was (Skalistoe) *chutor* on the half-verst map. At present, there are no standing structures although the quantities of modern construction debris found here indicate more recent activity, although this predates the mid-20th century map. The assemblage consists of a high proportion of cookingwares, a category of finds that has not been identified at any other Early Modern site in the cultivated zone. This may indicate that the site was abandoned before metal pots became predominant.

DSP08-F03-02 is an artefact concentration at the southwestern corner of the field corresponding to the Bronze Age site Skalistoe 2, which was briefly investigated by by V.A. Kolotuchin in the 1970s (Figs. 4.105-107). A ravine marks the border of the ploughed field in this area, and the concentration was found along its edges. On the other side of this ravine was a grassy plateau which could not be surveyed systematically. It is likely that the site extends onto this Plateau, which seems to be an ideal settlement location; however, geomagnetic prospection was inconclusive. Six or seven weak positive anomalies were recorded that may be indicative of pit houses (Fig. 4.108), but further research is necessary. In the ploughed area around the ravine, the soil was much greyer and ashy than in the rest of the field and was quite stony as well. Augering showed the bedrock to be at a depth of about 0.6 m. DSP08-F03-02 has two components: a strong Bronze Age presence, which has the highest concentration in the unit immediately NE of the ravine, and a Late Classical-Early Hellenistic presence, which follows the course of the ravine.

**Relations to the surrounding area** So far, no other Bronze Age sites have been identified in the vicinity of Field DSP08-F03. A large Bronze Age site was identified about 5 km to the SW (DSP08-F06-01), and there may have been a Bronze Age kurgan roughly 5.5 km to the S (DSP07-F07-01 Feature 3). Neither of these sites would have been visible from DSP08-F03-02.

The Late Classical-Early Hellenistic component of this site is approximately 800 m to the S of the Late Classical-Early Hellenistic farmstead at Skalistoe 5, and roughly 1 km NE of the Late Classical-Early Hellenistic site DSP08-F02-01.

*DSP08-F04* (Figs. 4.109-112)

**Size** 224.2 ha.

**Units** 8312-8869; 6268-6626.

**Investigations** Intensive survey in 50 x 50 units except corner ones.

**Landscape type** Pediment.

**Location** This field is the largest of all the fields surveyed by the DSP. It is located on the southern shore of Lake Džarylgač. To the N, a strip of trees, a dirt track and a marshy area separate the field from the lake. The same dirt track continues W and E of the field along the strip of grass and trees bordering the field. The area E of the field beyond the track was overgrown with grass. The western edge of the field directly adjoins DSP07-F02.

The southern shoreline of the lake is characterized by inlets or lagoons. W of the field is a large lagoon and marsh separating it from DSP07-F02. Towards the E are two inlets created by ravines cutting through DSP08-F04. The large easternmost inlet is fed by the water of two ravines that from the S come down from Hills 16, 8, and 20; the western ravine begins along the W flank of Hill 16 and meanders down the middle of DSP08-F04 until it exits just S of the central eastern field boundary continuing E and emptying into the inlet. Wide ravines running N-S were also apparent in the northern lower part of the field probably also due to erosion. The southern end of both fields borders on the asphalt road between the villages of Snežnoe and Vodopojnoe.

**Cartography** Several features pertaining to this field are visible on the half-verst map. First of all, eight stars marking mounds are shown in a row from N to S roughly through the centre of the modern field. These mounds correlate with the mounds still visible today (see Sites). Directly E of the field is a village named Kipčak Učkuju. It encompasses several buildings, wells ("K." for *kalodeč*) and three windmills, two of which are located near the eastern edge of the field. The half-verst map also shows several roads passing through the village connecting it with Kurama Kostel' in the E and the villages of Mussali (modern Zajcevo) to the SW, Džarylgač (modern Mežvodnoe) to the W-NW, and Ak Mečet (modern Černomorskoe) farther W. These roads also pass through DSP08-F04, and some continue through DSP07-F02. Near the mounds, the road from Džarylgač (modern Mežvodnoe) splits into two routes heading E: one enters Kipčak Učkuju from the N and the other veers S and connects to Kipčak Učkuju from the W. Though no longer existing, some of these E-W roads have left their traces across the field as deepenings. Other roads just E of DSP08-F04 apparently follow the natural ravines, as they lead S from Kipčak Učkuju up the Hillsides towards the ridge. On the mid-20th century map, the village is depicted as a ruin by the Russian name of Majorovo.

**Survey conditions** The most remarkable aspect about Field DSP08-F04 is its sheer size: nearly 3 km in length and in some places 1 km across. It took two field teams a full week to survey it. The field is oblong, spreading out from S-SE to N-NW The surface is gently undulating and sloping down from the asphalt road towards the lake some 40 m lower.

Marked features are the mounds running N-S in a row from roughly the middle of the field towards its northeastern corner. On the largest mound (DSP08-F04-05 Feature 2) is a small beacon corresponding with the measuring point of 17.8 m on the mid-20th century map.

The soil of this vast field is quite varied. In the northern section, it was generally hard and brittle and had been ploughed in very large peds making walking difficult and visibility relatively low due to the strong contrasts in light and shade. The southeastern section had similar conditions though less extreme, while the southwestern part was generally much softer in texture and lighter in colour with much smaller peds. Stony sections appeared in various parts of the field most likely caused by ploughed up bedrock. There were also patches of grey in the otherwise brownish-red soil with the exception of an elevated strip in the northern half of the field, where it was distinctly red. A section of the field along the southeastern edge (1.8 km from the northeastern corner) had not been ploughed. The part of this area close to the road was planted with vines, and the part towards the centre of the field lay fallow, but had concentrations of larger stones probably collected from the field. The extreme southeastern corner was excluded from survey due to cultivation.

A row of augerings was taken along the western edge of the field as part of the larger cross-section through the research transect (Chapter 3.3) and more were carried out at two of the larger sites (DSP08-F04-03 and -07). Since these samples show that the loess sediment on top of limestone bedrock is generally about 1 m thick except for the area closest to the lake, some of the sites have intact stratigraphies.

**Finds distribution** Archaeological finds were collected throughout Field DSP08-F04 although they were more sparsely distributed across its northern part. Most of the material dates from the Late Classical-Early Hellenistic period. The distribution map shows a fairly even distribution of finds with a few concentrations.

The entire field yielded very little that could be dated with any certainty to the Bronze Age (DSP08-F04-04 Feature 3). Several pieces of flint were collected but they cannot be dated securely. A few finds from the Classical period were collected in this field at site DSP08-F04-03. Some possibly Byzantine and Medieval material was collected in units dispersed over the field. Finds possibly from the Medieval period were collected in three units near the lake, a unit near the kurgan row, and on the eastern field border near the location of the abandoned village of Kipčak Učkuju where also off-site scatters of Early Modern finds were collected. Relatively large amounts of modern material were collected across the entire field.

**Sites** A total of seven sites were defined in Field DSP08-F04. Five of these are interpreted as settlement locations, the other two as kurgan groups.

The three discrete features in DSP08-F04-01 have all been destroyed by agricultural activities. They were recognized in the field as slightly elevated stony scatters with lighter-coloured soil (Figs. 4.113-114). The soil of Feature 3 was very grey as opposed to the surrounding area, where the soil was distinctly reddish-brown. This feature may be interpreted as an ash heap. Ash heaps occur as refuse dumps near Tatarian households but are also well-known from the Late Classical-Early Hellenistic period (see discussion in Chapter 5.1.5). The diagnostic material found in the units around all three elevations is primarily Late Classical-Early Hellenistic although one louterion fragment of Sinopean provenance is dated to the Classical period. Apart from Sinope the handful of imported fragments originate in Chersonesos and Herakleia. A high density of mostly Late Classical-Early Hellenistic finds was located in the unit containing Features 2 and especially Feature 3. A piece of worked flint was collected near Feature 1, but it cannot be securely dated.

DSP08-F04-02 consists of two discrete clusters on slight elevations of lighter soil (Figs. 4.115-116). The clusters are located at the highest point of a ridge W of the row of kurgans. High densities of finds were collected here with concentrations of Late Classical-Early Hellenistic material. Around the site, find densities drop off sharply. Adjacent units yielded either singular Late Classical-Early Hellenistic finds or none at all. A large proportion of the pottery was handmade, but imported fragments of amphorai and tablewares originated in Chersonesos, Sinope, Herakleia, Thasos and possibly Mende as well as a possible louterion from Chersonesos.

A larger site, DSP08-F04-03, was identified in the northwestern corner of the field (Figs 4.117-118). A grab sample had been taken here in 2007, when the field was not ploughed. Immediately to the W is a small grove of trees in a grassy area; the site most likely extends in this direction. In the field itself a wide ravine probably caused by erosion borders the eastern edge of the site. The site is located at the northern end of a ridge just S of Lake Džarylgač. On the half-verst map, a water-carrying ravine runs W of this ridge. This ravine was, however, not well visible during the DSP survey seasons. The soil in the area of DSP08-F04-03 is dark greyish-brown, and less reddish than the rest of the field. The site has a distinct concentration of finds in one of the corner units, while densities drop off sharply in the units surrounding it forming a "halo" around the centre. All finds in and around the site are dated to the Late Classical-Early Hellenistic period. Almost no finds were collected E of the ravine. A large proportion of the pottery was handmade, but a fragment of a possible Attic black slip cup as well as imported fragments of amphorai and tablewares from Chersonesos, Sinope, Herakleia, Thasos and possibly Knidos were also collected. Five augerings were conducted across the units with the highest finds densities. The bedrock was reached within 1.10 and 1.85 m below the surface. No clear culture layer was discerned in any of the soil samples, yet a very hard, compact layer was present at 0.4 m below the topsoil in three augerings implying that at places archaeological features may be preserved.

A fourth settlement, DSP08-F04-06, was identified on account of high densities of finds in the units along the eastern edge near the abandoned village of Kipčak Učkuju (Figs. 4.119-121). Besides the Post-Medieval, Early Modern and modern fragments that one indeed would expect close to a near-recent village, also many Late Classical-Early Hellenistic finds were collected consisting mostly of wheelmade wares and some handmade. Most of the latter finds were recorded in a stony ravine running roughly NE-SW. The site is probably related to one of the roads close to Kipčak Učkuju following the ravine that leads further NE to an inlet of Lake Džarylgač. Being so close to the edge of the field, this concentration may well extend beyond the borders of the field to the E near or under the abandoned village. Some handmade pottery and a single fragment of a Herakleian import was recorded as well as some sherds of possible Medieval and of Early Modern date.

The fifth settlement, DSP08-F04-07, was identified by high densities of finds directly W of the unploughed area in the middle of the field (Figs. 4.122-123). A rather compact oblong elevation is visible in this S-central area extending from the E with its steepest sides in the S and N. The northern side of this 'ridge' borders on a depression due to a ravine leading NE over the field to the inlet of Lake Džarylgač. This depression is greener than the rest of the field, and the ploughed area of the site quickly turned green during survey due to water transport along this ravine. The elevation appears along the northern fork of the road leading SW from Kipčak Učkuju to Mussali on the half-verst map. Almost all the datable material collected here was Late Classical-Early Hellenistic consisting of both handmade and wheelmade pottery. Like DSP08-F04-03, this site also appears to have a "halo" of units with lower densities of Late Classical-Early Hellenistic material. A large proportion of the pottery collected was handmade as well as imported fragments of amphorai and tablewares from Herakleia, Sinope, Thasos and Chersonesos and fourteen pieces of flint. It is very likely that the site extends to the E in the unploughed area although a quick investigation of this area was inconclusive revealing only a few amphora fragments and clusters of stones, which could easily have been collected in the ploughed area and dumped there. Two augerings were conducted at this site. A distinct culture layer was found in one of these at a depth of 0.4-0.52 m below the surface containing small fragments of pottery. The limestone bedrock was reached at 2 m and at 1.48 m, respectively. Between the ploughed topsoil of 0.3 m and the underlying bedrock, the sediment was compact.

All kurgans in the field are aligned N-S and located on the highest ridge in the field from where the field clearly slopes down towards the W. They can be divided into two groups:

The mounds in the northernmost kurgan group, DSP08-F04-04, are visible in the field as grassy elevations or patches on the highest ridge of the field (Figs. 4.124-126). The soil around them was red, while the rest of the field was more brownish-red. This may have been an effect of limestone erosion. The five kurgans in this group coincide with the locations of the mounds depicted on the half-verst map. All five kurgans have been spared by the plough, although all are damaged. Kurgans, Features 3 and 4, had recently been robbed. In Kurgan 3, the robbers had left half a pot *in situ* in the western section of the burial chamber. This vessel could be dated to the Bronze Age, and the kurgan is the only certain feature from the Bronze Age in this field. No other finds from the Bronze Age were collected in the vicinity. In the unit to the NW, one Late Classical-Early Hellenistic sherd was collected. Five more Late Classical-Early Hellenistic sherds were collected about 100 m to the W. No other finds were collected in the vicinity of the five kurgans, with the exception of an Early Modern fragment near Kurgan 2.

The southern group of three mounds, DSP08-F04-05, is located on the southern end of the ridge (Figs. 4.127-128). The northernmost of these has been ploughed out in the past and is now left grassy (Feature 1). The southernmost one is likewise ploughed out although a small stony elevation is visible (Feature 3). The middle one is intact (Feature 2), and it is the highest in the field and therefore marked with a beacon. On the half-verst map, the road extending from Kipčak Učkuju to Ak Metčet (modern Černomorskoe) heads straight for this kurgan, skirting the southern flank to continue in a western direction; this is indicative of its continued function as an orientation marker. Several units in the vicinity of the three kurgans yielded finds, yet without clear concentrations even in the units which contained the mounds. Two units to the NW of Kurgan/Feature 1 each yielded one Early Modern sherd. Late Classical-Early Hellenistic fragments were collected in the three units around its southern perimeter. No finds were collected on or around the large Kurgan 2, but one Late Classical-Early Hellenistic sherd was picked up E of Kurgan 3.

**Relations to the surrounding area** In Field DSP07-F02, directly to the W of DSP08-F04, four settlement sites were identified. Near DSP08-F04-03 250 m across the marshy inlet of Lake Džarylgač is a small Late Classical-Early Hellenistic settlement (DSP07-F02-01). Both sites were connected by the ancient Road XII (Chapter 6.6).

The Late Classical-Early Hellenistic finds concentrations in the southern part of Field DSP07-F02 (-03 and -04) are located at a distance of 1.1 and 0.8 km to the SW of DSP8-F04-07. The single kurgan in DSP07-F02 (-05) lies higher up the ridge on which the row of kurgans in DSP08-F04 is located and is certainly a continuation of this row. The distance between DSP07-F02-05 and the southernmost mound in DSP08-F04 is 600 m.

To sum up on the location of sites in DSP08-F04 and DSP07-02, kurgans appear on the higher ridges, whereas possible habitation sites are generally located lower down closer to ravines or the lake (with DSP07-F02-04 as a possible exception). An excellent view is provided of this area from the Hillsides and vice versa.

*DSP08-F05* (Figs. 4.129-130)

**Size** 38.5 ha.

**Units** 8870-9034.

**Investigations** Intensive survey in 50 x 50 m units.

**Landscape type** Lowland Ridge.

**Location** This field is located roughly 1.5 km N of Lake Džarylgač on the southern slopes of the Lowland Ridge sloping down towards the S. South of the field is a dirt road connecting the villages of Vodopojnoe and Mežvodnoe. The same road separates DSP07-F05 from DSP07-F06 further E. On the N side of this road close to the southeastern corner of this field, is an abandoned concrete structure. DSP07-F08 lies across the road to the S. Two other dirt tracks border the field to the E and N separated from the ploughed area by a strip of grass. To the W is a vineyard partly overgrown in the southern part. To the E is a large ploughed field. Immediately to the N on the other side of the dirt track lies DSP08-F06.

**Cartography** On the half-verst map, a dashed line depicting a minor track starts in the northern section of DSP08-F05 and ends W of the village of Kirleut (modern Vodopojnoe). The (dirt) road between Kirleut and the village of Džarylgač (modern Mežvodnoe) is the same as the present dirt road just S of this field. E of the field a ridge can be seen extending S towards the lake. On this ridge, a row of mounds is depicted SE of the field. A separate mound is depicted further upslope to the N closer to the field. Directly E of the field, a cluster of four small stars marks the presence of low mounds. None of these features were observed during the DSP08 season.

The 1993 map indicates that this field was at one time part of the large vineyard that extends to the W.

**Survey conditions** DSP08-F05 had been ploughed and harrowed. The soil was reddish-brown. Two ravines probably caused by erosion ran parallel down the slope through the field. The days of survey were cloudy and partly rainy. Visibility was optimal.

**Find distribution** Of all the surveyed fields, this one yielded the lowest amount of archaeological material. The field was archaeologically speaking practically "empty," except for the units along the eastern and southeastern field boundary. A distinct concentration of Early Modern pottery fragments was encountered along the eastern field boundary. Lower densities of Late Classical-Early Hellenistic finds were spread out in the easternmost units and the southeastern field corner.

**Sites** DSP08-F05-01 consists of a low density concentration of Early Modern finds along the eastern field border, but too low to define the concentration as a site (Figs. 4.131-132). As it is located along on the outer edge of the field, the finds from DSP08-F05 may be the halo of a site located in the unploughed zone to the E. Some adjacent units yielded Early Modern material as well.

The Late Classical-Early Hellenistic finds collected in this part of the field likewise may constitute the halo of a site located outside the ploughed field (DSP07-F08-01 immediately to the S) (Figs. 4.131-132).

**Relations to the surrounding area** South of DSP08-F05 just across the dirt road is DSP07-F08-01, a settlement dated to the Late Classical-Early Hellenistic period and contemporary finds in the southeastern corner of DSP08-F05 most likely form part of the halo of this site. To the N, more Late Classical-Early Hellenistic sherds were collected in the northeastern section of DSP08-F06 (-03). The row of mounds depicted on the half-verst map is located approximately 350 m E of the northeastern edge of the field. There is also an isolated kurgan in the field to the N (DSP08-F06 (-02)).

*DSP08-F06* (Figs. 4.133-135)

**Size** 48 ha.

**Units** 9040-9240.

**Investigations** Intensive survey in 50 x 50 m units except corner ones.

**Landscape type** Lowland Ridge.

**Location** DSP08-F06 was the last ploughed field to be surveyed during the DSP seasons. It is a long narrow field located directly N and NW of DSP08-F05 located higher up the slope of the Lowland Ridge. It is the westernmost field surveyed N of Lake Džarylgač. In the N it is adjacent to a field that at the time of survey was overgrown with tall grass and wild flowers. Beyond this to the N is the main asphalt road running atop the coastal ridge from Meživodnoe, continuing on to Razdolnoe in the E. The field is bordered along the eastern side by the same small track that runs along DSP08-F05. At the western end of the field a large, partly ruined but still used agricultural complex overlooks the area.

**Cartography** The half-verst map shows no distinctive features within the field boundaries. There is no kurgan symbol at the location where now a grassy elevation with a central depression can be observed, most probably a looted kurgan, although there is a mound depicted on the modern 1993 map roughly 160 m NE of the field.

The modern dirt road between Vodopojnoe (Kirleut) and Meživodnoe (Džarylgač) is depicted S of the field as well as two minor tracks ending S of DSP08-F06 and heading towards the E and W. On the top of the coastal ridge N of the field, groups of mounds are marked on either side of the road.

**Survey conditions** The field had recently been ploughed and during the survey it was still being harrowed by two tractors in the northern part. Remains of concrete poles in the field indicate past vine growing. There still is a vineyard in the field to the SW and W of DSP08-F05. The surface of the field is gently undulating with the western part being slightly higher than the rest. Some areas in the middle section were quite stony. This is most likely due to the plough bringing the underlying limestone bedrock to the surface. These stony patches run N-S following the natural ridges; one is in the mid-western area and the other is concentrated more in the northern section. In the eastern part of the field were two wide ravines running N-S. These have probably formed as a result of downslope erosion from the coastal ridge. The soil was generally brownish in colour except for a grey area in the western part of the field. Here a large unploughed roughly circular area was observed just NW of the grey soil. The area is characterized by large stone blocks (>0.4 m) and a depression of about 1 m in the western end. Lack of time prevented us to survey the rest of the field W of this grassy area.

The largest part of the field was surveyed during overcast, windy days, when surface visibility was optimal. The westernmost units were surveyed during a sunny day, when shade affected visibility a bit.

**Finds distribution** Significant quantities of Bronze Age material were found in nine adjacent units in the central and western part of the field SE of the grassy area. Outside of this, very little Bronze Age material was found in the rest of the surveyed area.

A very diffuse distribution of Late Classical-Early Hellenistic material was found throughout the surveyed area of this field with a markedly higher concentration in one unit in the northeastern corner. A few Byzantine, Medieval, Early Modern, and Modern artefacts were collected in units dispersed throughout the field.

**Sites** Three sites were identified in Field DSP08-F06. The first, DSP08-F06-01, is the concentration of Bronze Age finds coinciding with the grey patch of soil (Figs. 4.136-138). As noted, the site is located in the higher western part of the field but not at its highest point. Instead, it is situated on the gentle slope going down towards the E in a somewhat sheltered position on the northern and western sides. A ravine caused by erosion leads down from the NW to the SE continuing just E of this site. The grey soil visible from afar was rather stony but also very soft and ashy with large quantities of bone. Finds here included high densities (in the hundreds) of handmade ware, but also some stone tools and large quantities of flint. One unit yielded 17 flint fragments in the standard survey alone. The Bronze Age finds were tightly related to the area of the grey soil, very few were collected outside this area. Although a few Late Classical-Early Hellenistic wheelmade fragments were found within the perimeter of the Bronze Age site, this most likely equals the off-site density that is typical for the rest of this field. 11 augerings were conducted at the site: seven in a N-S orientated row and four at right angles of this line. The soil on top of the limestone bedrock is very thin, between 0.3 and 0.5 m. This is about the depth of a tractor plough. This means that the original stratigraphy of the site will have been completely destroyed by ploughing. The augering conducted in the centre of the site yielded a pottery fragment and probably spots of ash at 0.35 m below the surface from a dark brown, damaged find layer. The bedrock was reached here at a depth of 0.45 m.

The second site, though not designated as such on the map, the grassy area to the NW (DSP08-F06-02) is tentatively interpreted as part of a kurgan due to its fairly prominent location, the elevation albeit slight (about 1 m), and the presence of large stones which would have been suitable for the construction of a burial chamber (Figs. 4.136-138). The pit, about 1 m deep, was located roughly 40 m W of the elevation, and is probably related to illegal excavation activities. Two Late Classical-Early Hellenistic sherds and two Early Modern sherds were found in the adjoining units, meaning that a dating for this kurgan remains insecure (visibility was very poor in the grass).

The third site, DSP08-F06-03, is the concentration of Late Classical-Early Hellenistic pottery consisting of eight amphora fragments, found in a single unit in the northeastern section of the field (Fig. 4.139). Since this unit is almost adjacent to the eastern edge of the field, it is possible that this concentration should be related to higher densities below the road or in the fallow field to the E.

**Relations to the surroundings** From DSP08-F06-01, one has a clear view down the slope to the E and towards Lake Džarylgač and beyond across DSP08-F04 and DSP07-F02 towards the Hillsides. DSP08-F06-01 is clearly visible from DSP07-F04-01 4.5 km to the SW. The Bronze Age site near the coast at Skalistoe, DSP08-F03-02 (Skalistoe 2), about 5 km to the NW is not visible since it is on the opposite side of the coastal ridge.

The Late Classical-Early Hellenistic concentration in the NE section as well as the dispersed off-site material may correspond to similar findings to the S along the eastern edge of DSP08-F05. These could be contemporary with the large Late Classical-Early Hellenistic site discovered in Field DSP07-F08, about 1 km to the S.

## 4.5 THE HILLSIDES AND THE PLATEAUS/UPLANDS | K. WINTHER-JACOBSEN WITH CONTRIBUTIONS BY P. ATTEMA, S. LANCOV, T. SMEKALOVA & V. STOLBA

### 4.5.1 Introduction

The Hillsides and the Plateaus/Uplands make up two separate land types according to the classification of Delvigne (see Chapter 3).[44] The Plateaus/Uplands zone is the relatively flat area above the Hillsides (Chapter 3.4.2). The Hillsides are the area between the Plateaus/Uplands and the Pediment, covering a vertical distance of some 70 m from approximately 125 m above sea level (Chapter 3.3.2) (Fig. 3.4). Generally, the slope angles vary around 10-12%. The slopes are dissected by ravines eroded by running water; however, according to Delvigne, the natural vegetation composed of a mat of mosses, dead vegetation parts and living leaves which spread out over the soil surface provides good protection against rain drop and running water erosion. Nevertheless, cultivation, especially ploughing, will have encouraged wind and water soil erosion.

Due to the vegetation, surface visibility is generally very low and only where burrowing animals, of which there is abundant evidence on the surface, make heaps of loose soil material, a glimpse of the subsoil is offered. The area is characterized by shallow soils mainly and today is predominantly used for grazing, but there are many pockets of deeper soils which have agricultural potential. The Plateaus/Uplands at the top of the ridge are currently cultivated, but changes in the vegetation and traces on the geomagnetic map clearly reveal that additional areas have been cultivated in the past, although this is difficult to date. Few of the roads appear on the old topographical maps from 1837/1865, as well as on the half-verst- and the mid-20th century maps. Generally, roads and dirt tracks are very "mobile" which may have to do with the location of fields. Cultivated areas on the Hillsides are indicated on the mid-20th century map, some of which are still exploited. The fields on the Hillsides tend to be laid out in long narrow strips following the contour of the slopes.

On the slopes, springs occur in the ravines, while the ravines themselves will have been used to store water by deepening or damming the reaches (Fig. 4.140). The water-bearing Sarmatian sediments which produce water of a very high quality are closest to the surface (less than 8 m below) at the bottom of the Hillsides and old maps reveal a series of Tatar villages located along this line. Today, there are only two settlements at the bottom of the Hillsides: Snežnoe and Zajcevo. At least one ancient settlement (DSP07-H11-01) was located at the bottom of the Hillsides.

Two post-depositional histories appear to dominate. Settlements tend to be located on deeper soils, but only in rare cases have the areas been cultivated pushing artefacts to the surface (e.g. DSP08-H11-01, DSP08-H12-01, DSP08-H21-01, Units 20041-20042). In most cases, there are few artefacts at the surface due to soil build-up and/or erosion.

In terms of pure surface observation, the slopes are characterised by three general types of alternating soils:

1. Large exposed patches of bedrock alternating with stone circles filled with soil often located along the edges of the promontories and on the floor of the ravines, but not exclusively so (Fig. 3.6);
2. Shallow soils with many tiny rock fragments suggesting the bedrock has been disturbed (Fig. 4.141);
3. "Deep" soils with no or little exposed bedrock, often, but not exclusively, located in shallow depressions (Fig. 4.142).

[44] When both zones are referred to the non-geomorphological term "the slopes" is applied.

As a rule, the actual depth of these soils may not exceed 0.2 m as indicated by the drilled cores observed in Units 20019-20020 (Fig. 4.143). Where the soil is thicker, the colours range from almost black to red. The deep pockets of red loess are most visible where they build up at the tip of promontories due to erosion (Fig. 4.144). Although settlements appear to be located on the thicker darker soils, the locations of kurgans are less clearly correlated to the actual surface. To this should be added a rich variety of flowers, mosses and grasses, which are also indicative of human activities. Thistles and low shrub almost always grow on settlements and in the kurgan chambers, but not exclusively so, while some flax and rape is usually also present on sites of more recent activities. No doubt a thorough botanical investigation would provide firm evidence for the correlation between certain types of vegetation and human activities.

Although clear settlement patterns emerged already in 2007, to test our own assumptions we persisted in a systematic investigation of the slopes in 2008. Consequently, areas surveyed in 2007 were recorded as hills whereas predetermined survey units were laid out in 2008. Survey on the slopes in 2007 involved systematic intensive, extensive and *razvedki*-style survey, and recording of the results was inconsistent. As explained in the chapter on methodology (see Chapter 2.3.1.2.1), in 2008 certain restraints were placed on our research strategy by the local authorities forcing us to focus on a prioritized list of areas; however, as we were also instructed to survey larger coherent areas, we chose to survey long lines running roughly N-S from the bottom of the Hillsides to the Plateaus/Uplands and a few lines running E-W using the windmill foundations as focal points. Because of this, the lines cut across geomorphologically diverse units such as promontories and ravines. Since the weather changed several times while surveying due to the size of the units, specific weather conditions were not recorded. In 2008, we surveyed a total of 16 lines in 32 units including 37 windmill foundations and 24 km of projected cable- and road lines. Furthermore, we did a complete survey of seven sites, two of which were identified during the 2007 campaign.

## 4.5.2 Previously known sites in the area

The half-verst and the mid-20th century maps reveal only tracks and dirt roads and kurgans on the slopes (Figs. 2.4 and 2.5). In fact, prior to the 2007 campaign of the DSP, no ancient settlements had ever been identified on the slopes, only kurgans. We have verbal information from S. Koltuchov on controlled excavations of a number of kurgans on the slope carried out during the Northwestern Crimea Expedition (e.g. DSP07-H19-01 kurgans 1-3) (Chapter 5.3).

On the Pediment, along the bottom of the Hillsides, is a series of current villages (Snežnoe and Zajcevo) and abandoned ones (Aldermen, Aldermenskaja Skala, Saja/Kostovka, Karlav/Čaikino and Mussali, Kipčak-Učkuju and Kurama Kostel') as well as properties (*nostrovka*) marked on the maps (Chapter 4.3.4). The location of these villages is guided by the fresh water accessibility (Chapter 3.2.3.2). On the maps, the villages are connected with a road in the Uplands running E-W by dirt roads and tracks cutting across the Hillsides. These roads and tracks tend not to follow the ridges of the individual promontories, but rather shallow depressions. Some of the tracks are still visible on the slopes where some have multiplied into broad lanes of tracks laying side by side.

## 4.5.3 Catalogue of surveyed hills and units (Figs. 4.145-146)[45]

*Hill 1* (Figs. 4.147-148)[46]

**Investigations** Partial intensive survey, geomagnetic survey, two trial trenches.

**Landscape type** Hillsides.

**Location** Short wide promontory in the upper part of the Hillsides.

**Cartography** On the mid-20th century map the southwestern part of Hill 1 is indicated as cultivated.

**Finds distribution** All finds were associated with the site.

**Sites** Concentration of pottery (GS3043, GS3045-3046 and GS3049-3060) associated with rectangular stone structure recorded in the northern end of the slope on the western side (DSP07-H01-01) (Figs. 4.149-150). The pottery consists of mostly indefinable wheelmade sherds, but there are also two fragments of Sinopean transport amphorai and a handmade sherd of Late Classical-Early Hellenistic date. The geomagnetic survey confirmed the presence of a settlement consisting

[45] Please note that survey conditions are not noted for the slopes, because they were often surveyed over several days.

[46] Fig. 4.40 works as a key to the legends of all the distribution maps.

of several buildings, household pits and a double pen (Figs. 4.151-152). From Google Earth can clearly be distinguished a significant ash concentration at the promontory's northern end (Fig. 4.148). An aerial photo of the hill revealed that the long linear negative and positive anomalies which appear to cross under the site are consistent with a recent agricultural episode (Fig. 4.153). The same phenomenon was observed on the geomagnetic maps of Hills 6, 8 and 18 (Figs. 4.201, 4.235; 4.251). Material from a circular household pit (Feature 18) of dark ashy phosphorous soil with small stones including fragments of Chersonesean, Thasian and Sinopean amphorai, multiple handmade sherds and two jug fragments confirmed the Late Classical-Early Hellenistic date (Figs. 4.154-155). The topsoil also included a Herakleian amphora fragment apart from sherds of similar type to what was recovered from the pit. Another trial trench in Feature 5 revealed a layer of stones on an ashy bed, a possible floor. The finds consisted of mostly handmade Late Classical-Early Hellenistic sherds but also three Bronze Age fragments (Figs. 4.156-158). Unfortunately, fragments of the same vessels were found in different strata, and it was impossible to establish a relative chronology. The number of structures and the double pen suggests a fair-sized settlement associated with livestock rearing although not all the structures may be contemporary. The artefacts suggest a date in the Late Classical-Early Hellenistic period, possibly preceded by a Bronze Age settlement at the same site. The elements of the site appear to be very similar to DSP08-H08-01 (Fig. 4.251) dated to the Bronze Age and DSP07-H02-01 (Fig. 4.163). Further S, three kurgans and a linear elevation that seemed to be connected to at least one kurgan were observed, but not recorded.

**Relations to the surrounding area** DSP07-H01-01 is located 1 km due N of the settlement on Hill 12 (DSP08-H12-01) (Fig. 4.296). There is a group of kurgans located only 300 m W of the settlement across the ravine.

*Hill 2* (Figs. 4.159-160)

**Investigations** *Razvedki*, partial extensive and geomagnetic survey.

**Landscape type** Hillsides, Plateaus/Uplands.

**Location** Short wide promontory in the upper part of the Hillsides.

**Cartography** On the mid-20th century map there are three strips of cultivated land indicated on Hill 2.

**Finds distribution** Seven grab samples (GS3077-3082, 3096) were collected on Hill 2. All were assumed to be related to the site, having been collected in relation to structures visible on the surface, but no coordinates were recorded; however, due to the flint tool mentioned in the notes, GS3081 can be placed on the western tip of the promontory. The total grab sample consists of 111 Late Classical-Early Hellenistic wheelmade sherds including Herakleian, Thasian and Chersonesean transport amphora fragments, one Late Classical-Early Hellenistic handmade sherd, 14 Bronze Age handmade sherds, two flint tools and a piece of bone.

**Sites** Various structures were observed on the lower northern part of the slope during the surveys (Figs. 4.161-162). DSP07-H02-01 consists of an area of very ash-rich soil located on the higher western tip of the promontory. A flint tool was collected here as well as nine handmade sherds of the Bronze Age and a piece of bone (GS3081). Geomagnetic survey revealed a double pen (Fig. 4.163). DSP07-H02-02 located 145 m to the NE consisted of three stone structures with double walls of standing slabs on the lower tip of the promontory (Fig. 4.164). Geomagnetic survey revealed a group of four angular structures with one to two rooms as well as one with a possible courtyard. DSP08-H02-03 is a double faced wall of large stones damming the eastern ravine (Fig. 4.140; 4.159; 4.161). The number of structures and the double pen suggests a fair-sized settlement associated with livestock rearing although not all the structures may be contemporary. The artefacts suggest a date in the Late Classical-Early Hellenistic period, preceded by a Bronze Age settlement at the same site. The elements of the site appear to be very similar to DSP08-H08-01 (Fig. 4.251) and DSP07-H01-01 (Fig. 4.149).

**Relations to the surrounding area** There are several kurgans located on the ridge of Hill 17 (see Units 20003, 20005 and 20006) approximately 500 m NE. Approximately 650 m due N is another settlement (see Hill 7).

*Hill 3* (Fig. 4.165)

**Investigations** *Razvedki* and partial extensive survey.

**Landscape type** Hillsides.

**Location** Wide promontory in the lower part of the Hillsides.

**Cartography** A symbol indicating a pit is marked on both the half-verst and mid-20th century maps.

**Finds distribution** Some pottery (GS3041-3042, 3047) and features were mapped, especially in the lower northwestern part. Most of these features belong to Early Modern/Modern, perhaps Tatarian structures as suggested by two tile fragments.

**Sites** No sites recorded (see also Units 20033-34).

**Relations to the surrounding area** The area with the possible Tatarian structures is located due S of the village Karlav on the half-verst map. According to the mid-20th century map, this village had been abandoned.

*Hill 4* (Figs. 4.166-167)
**Investigations** Partial intensive survey.
**Landscape type** Hillsides.
**Location** Small promontory in the upper part of the Hillsides. Dominated by exposed bedrock and stone circles.
**Cartography** No features indicated.
**Finds distribution** All finds associated with sites.
**Sites** In the northern lower part, several rectangular structures were mapped. DSP07-H04-01 consists of four rectangular stone structures roughly associated with a small cluster of five indefinable wheelmade sherds tentatively dated to the Late Classical-Early Hellenistic period (GS3068-3072) (Figs. 4.168-169). DSP07-H04-02 is a small settlement located in a shallow ravine on the northern edge of Hill 4 (Figs. 4.170-171). A single sherd (GS3075) of possibly Medieval date was associated with rubble and several pits were indicated by the vegetation. DSP07-H04-03 consists of a patch of intense vegetation suggesting a subsurface structure associated with a single indefinable wheelmade sherd (GS3074) (Fig. 4.172).
**Relations to the surrounding area** No specific relations observed.

*Hill 5* (Figs. 4.173, 4.167)
**Investigations** Partial intensive survey.
**Landscape type** Hillsides, Plateaus/Uplands.
**Location** Small promontory in the upper part of the Hillsides. Dominated by exposed bedrock and stone circles.
**Cartography** No features indicated.
**Finds distribution** No finds recorded.
**Sites** No sites identified.
**Relations to the surrounding area** No specific relations observed.

*Hill 7* (Fig. 4.174)
**Investigations** *Razvedki* and geomagnetic survey.
**Landscape type** Hillsides, Plateaus/Uplands.
**Location** Hill 7 is a long promontory running from the Uplands to the Pediment. See also Units 20064-20065.
**Cartography** On the half-verst map there is a track running S from Karlav splitting into two. On the mid-20th century map is a strip of cultivated land in the shallow area between Hills 25 and 7 and a track from Karlav/Čaïkino leads to it.
**Finds distribution** Two grab samples were recorded (GS3062-3063) both associated with the sites. More finds were collected in June 2007 reported as Finds Lists 9 and 10 in the original report but never integrated into the official finds recording system.
**Sites** DSP07-H07-01 consists of two rectangular stone structures associated with four jug fragments of indeterminable date (2007 FL09) as well as five Late Classical-Early Hellenistic amphora and handmade sherds and pieces of bone (Fig. 4.175).

Geomagnetic survey revealed a rectangular structure with two rooms with a possible pit or oven inside as well as several household pits and a possible track leading to the entrance of the house (Fig. 4.176). DSP07-H07-02 consisted of a possible linear feature associated with seven amphora fragments and a thin-walled jug fragment of indeterminable date (2007 FL10). Geomagnetic survey revealed another rectangular structure of two rooms, one L-shaped as well as a round structure possibly bordered by posts (Fig. 4.177). For DSP08-H07-03-04, see Units 20064-20065.
**Relations to the surrounding area** Approximately 650 m due S is the settlement on Hill 2 (see Hill 2).

*Hill 9* (Fig. 4.178)
**Investigations** *Razvedki* and geomagnetic survey.
**Landscape type** Hillsides.
**Location** Hill 9 is a long promontory running from the Uplands to the Pediment.
**Cartography** On the half-verst map there is a property marked on the bottom tip of Hill 9 and it was crossed by a track from Saja/Kostovka running SW but ending at an unmarked location on the Hillsides. On the mid-20th century map there is a track running on the western slope of Hill 9 to a strip of cultivated land.

**Finds distribution** Three grab samples GS3088-90 were recorded in this area, but unfortunately without coordinates. They consist of a mixed batch including two possibly Late Classical-Early Hellenistic handmade sherds, nine fragments of a possibly Medieval vessel, one Early Modern fragment and five indefinable wheelmade sherds. More finds were collected on Hill 9 in June 2007 reported as Finds List 14 in the original report but never integrated into the official finds recording system. All finds are assumed to be associated with the sites.

**Sites** On the eastern and western edge of the promontory towards the lower end linear stone walls of standing blocks were recorded (Fig. 4.179). On the eastern edge geomagnetic survey revealed a rectangular structure with a possible courtyard (DSP08-H09-01) (Fig. 4.180). On the western edge geomagnetic survey revealed another stone structure (DSP08-H09-02) (Fig. 4.181). According to Smekalova, the sites were Tatarian, which cannot be confirmed by the pottery collected.

**Relations to the surrounding area** No specific relations observed.

*Hill 10* (Fig. 4.182)

**Investigations** *Razvedki* and geomagnetic survey.

**Landscape type** Hillsides.

**Location** Hill 10 is a small, roughly N-S oriented promontory.

**Cartography** No features indicated.

**Finds distribution** Three grab samples (GS3084, GS3086-3087) were collected on Hill 10 and 18, but the coordinates were not recorded. These consist of seven Late Classical-Early Hellenistic sherds including three transport amphorai, one possibly from Herakleia, a tableware fragment and three handmade sherds one of which may be Bronze Age and a piece of flint. More finds were collected on Hill 10 in June 2007 reported as Finds List 13 in the original report but never integrated into the official finds recording system. All finds are associated with the sites.

**Sites** DSP07-H10-01 is located near the tip of the promontory (Fig. 4.183). Four well-preserved, rectangular stone structures built with double walls of standing blocks were recorded on the surface, and also noted on satellite photos (Figs. 4.184-185). Geomagnetic survey revealed four rectangular stone structures and several household pits (Fig. 4.186). Feature 2 consisted of at least six rooms around a courtyard. Trial excavation on the NE perimeter of Feature 2 revealed a double wall constructed of field stones and roughly cut boulders packed with rubble in-between (Figs. 4.187-191). Sitting directly on the soil, the wall consisted of at least two courses; the upper one had collapsed creating the debris inside the structure. The original thickness of the wall must have been ca. 0.8-0.82 m, the currently preserved height is 0.32 m, and it can be clearly traced on the surface beyond the excavated area. No traces of mud brick were found. Below the stone debris were found several large fragments of Chersonesean transport amphorai. No other pottery was associated with the wall. Trial excavation at Feature 5 included two circular household pits (Figs. 4.192-194). The northern pit (Pit 1) was roughly piriform with concave bottom, its filling was homogeneous. A large flag stone was found at the bottom. Pit 1 was associated with fragments of Chersonesean and Sinopean transport amphorae as well as handmade pottery, animal bones, cereals and burnt mud brick. The eastern pit (Pit 2) is of the same shape as Pit 1 but much deeper (Fig. 4.194). To a depth of 0.8 m, the upper fill was dark brown soil while the lower fill was densely packed with stone boulders. These two layers were associated with a few fragments of Chersonesean, Sinopean and possibly Herakleian amphorai, handmade pottery, joining fragments of black-glossed tableware, a fragmentary whetstone, animal bones, Helix shells and cereals. The bottom of the pit consisted of an ashy layer with slightly more finds of Chersonesean and Herakleian amphorai fragments, handmade pottery including a restorable handmade pot, animal bones, Helix shells and cereals. Similar types of pottery as well as several jug fragments were found in the fill above the pits. The pottery dated both the features to the late 4th-early 3rd century BC. A single wheelmade sherd did not come from an amphora. A large stone was found blocking a burrow of some animal at the bottom of Pit 2 and this may have been the reason for abandoning the pit and filling it with stones. Sherds of the same vessels were found over 0.2-1.3 m suggesting that the pit was filled in one episode. The date of the bowl (FL 9 no. 5) suggests that the pit was given up in the early 3rd century BC which is consistent with the Chersonesean amphora rim (FL 5 no. 1). Including the structures on Hill 18 (DSP07-H18-01), the number of structures and the range of finds suggest a sizeable settlement. The building technique suggests this settlement is closely related to DSP08-H12-01 (Figs. 4.299-300).

**Relations to the surrounding area** DSP07-H10-01 is located across a shallow ravine only 135 m E of DSP07-H18-01; the sites are likely to be associated (Fig. 4.200). Two small groups of kurgans are located 600 m to the NW and 300 m to the NE.

*Hill 11* (Fig. 4.195)

**Investigations** *Razvedki* and geomagnetic survey.

**Landscape type** Hillsides.

**Location** Long promontory running from the Uplands to the Pediment.

**Cartography** On the mid-20th century map there is a track running on the western slope of Hill 11 to a strip of cultivated land, and two kurgans are marked some distance apart on the upper part of the hill.

**Finds distribution** All finds are associated with the site.

**Sites** At DSP07-H11-01 located at the bottom of Hill 11 (the Pediment), large grab samples were collected after ploughing (GS3083 and GS3093) (Figs. 4.196-197). The range of finds is equivalent only to Panskoe I including Late Classical-Early Hellenistic Herakleian, Chersonesean, Sinopean, Thasian and possibly Chian transport amphorai, fragments of tile, table- and utility ware from Sinope, grey tableware, handmade pottery of the Late Classical-Early Hellenistic period and (possibly Middle and) Late Bronze Age both as well as four lithics. The stones on the surface seem natural with the possible exception of a line of stones on the E bank of the shallow ravine which may be a check dam (Fig. 4.198). Geomagnetic survey revealed no preserved architectural remains (Fig. 4.199).

**Relations to the surrounding area** DSP08-H11-02 is located approximately 1 km due N at the bottom of the slope (see Units 20050-20055) (Fig. 4.277; 4.289).

*Hill 18* (Fig. 4.200)

**Investigations** *Razvedki* and geomagnetic survey.

**Landscape type** Hillsides.

**Location** The eastern spur of Hill 18 a long wide promontory.

**Cartography** No features indicated.

**Finds distribution** Three grab samples (GS3084, GS3086-3087) were collected on Hills 10 and 18, but the coordinates were not recorded. They consisted of seven Late Classical-Early Hellenistic sherds including three transport amphorai, one possibly from Herakleia, a tableware fragment and three handmade sherds as well as a piece of flint. All finds are associated with the sites.

**Sites** Four possible stone structures were visible on the surface, one at least partly built with standing blocks (Feature 3) (Fig. 4.201). Geomagnetic survey revealed at least four structures and household pits. For interpretation, see Hill 10.

**Relations to the surrounding area** DSP07-H18-01 is located across a shallow ravine only 135 m W of DSP07-H10-01 (Fig. 4.200), and the sites are likely to be associated. Two small groups of kurgans are located 600 m to the NW and 300 m to the NE.

*Hill 19* (Figs. 4.202, 4.148)

**Investigations** *Razvedki*, partial intensive, topographical and geomagnetic survey and trial excavation.

**Landscape type** Hillsides.

**Location** Long promontory running from the Pediment to the Uplands.

**Cartography** Eight kurgans are marked on the half-verst map (fig. 4.203); seven in a row near the Uplands and a solitary one on a large plateau further N. On the mid-20th century map the solitary kurgan, five of the six kurgans and an additional two on another Plateau even further N are marked. On the map produced by S. Koltuchov 18 kurgans are marked mainly clustered around the already known groups.

**Finds distribution** GS3033-3039, GS3099-3100. Except for GS3036, the grab samples are all associated with the sites. The total number is 10 sherds, mostly transport amphorai whose type could however not be identified. Nine of these are tentatively dated to the Late Classical-Early Hellenistic period. See also Units 20036-20037.

**Sites** Along the ridge of Hill 19, 31 kurgans were identified during the topographical survey (for the individual kurgans, see Appendix 1). They appear in eight distinct groups based on intermediary distance (Figs. 4.204-227). Within each group, diameters and heights of kurgans may vary. Their present form, however, has in all cases been altered by excavation, looting and/or erosion. It is thought that all kurgans would have had a so-called krepis wall that would have held the soil of the mound in place. If true, this would mean that all kurgans mapped by us, originally were smaller in diameter, as we must assume that soil from the mound has washed over the krepis wall to form the sloping mounds that we see today. Indeed, in a number of kurgans wall remains were noted at the surface forming a circle halfway through the present mounds (most significantly Kurgan 10). Rather than as carefully constructed retaining walls, these appear however in most cases (possibly with the exception of Kurgan 10) as loosely placed stone circles. Other kurgans show no sign of a krepis wall or stone circle whatsoever at the surface. As far as the grave architecture is concerned, it was possible to reconstruct the orientation on the basis of still protruding head stones (e.g. Kurgan 8 and 10) based on the evidence from the excavated kurgans (see Section 4.4.1 and Kurgan 1, 2 and 4). Their orientation corresponds more or less with Kurgans 1 and 2.

Allowing for considerable changes over time to have happened to the original morphology of the kurgans on this slope, we nonetheless may note differences within clusters that are unlikely to have been caused by changes that have occurred after erection of the mound. The cluster comprising Kurgans 1-4, for instance, has three larger and/or higher (1-3) kurgans and one smaller and lower specimen (4). This height/diameter differentiation applies to all other groups except for the cluster made up of Kurgans 16 and 17 both of which show approximately the same diameters and height in their present form.

On basis of the excavations of the Northwestern Crimea Expedition, it is assumed that although not many datable artefacts had been found in the chamber graves, the graves date between the late 5th and late 4th century BC. Such a chronology seems now confirmed by ceramic finds from a trial trench dug during the 2007 campaign approximately 10 m SE of Kurgan 3 at a spot where a clear anomaly had turned up in the geomagnetic map (Fig. 4.204). As far as the dating of this feature is concerned, we should, however, realize that no stratigraphic relationship was established between the trial trench and the kurgan. The sherds found in the trial trench SW of Kurgan 3 may tentatively be connected to ritual practice: it is possible that feasting occurred around the burial mounds, and the amphora sherds in the trial trench could testify to this (pers. comm. T. Smekalova). It is reported that artefacts collected from the tombs in the mounds in the 1970s excavations consisted of iron knives and arrow heads. It is possible that the excavated mounds had been robbed of other materials prior to the excavation (pers. comm. S. Koltuchov).

**Relations to the surrounding area** The kurgans on the southern part of the ridge of Hill 19 (Groups 1 and 2) are located less than 200 m W of DSP08-H12-01, although they are not at present visible in the viewscape of the site. Group 3 is located 350 m due N of DSP08-H12-01. Group 4 is located across the ravine 300 m W of DSP07-H01-01. Groups 6-8 are located in the Uplands 1.3 km S of DSP08-H12-01 and 1 km SE of DSP08-H13-01. See also Units 20035-20036 (Fig. 4.270).

*Hill 27* (Fig. 4.228)
**Investigations** Extensive survey.

**Landscape type** Hillsides.

**Location** Low promontory on the eastern edge of the survey area.

**Cartography** The map of Koltuchov indicates a settlement in this area.

**Finds distribution** No finds.

**Sites** Site on the NW edge of Hill 27 indicated by large roughly square patch of intense vegetation with cut-away corners towards the N (DSP08-H27-01).

**Relations to the surrounding area** It seems probable that this site was associated with the exploitation of the cultivated area to the N at the lower end of the Hillsides. A very similar site, DSP08-H15-01, is located 400 m to the E (Fig. 4.264)

*Hill 28* (Fig. 4.229)
**Investigations** Extensive survey.

**Landscape type** Hillsides.

**Location** Wide promontory in the lower part of the Hillsides.

**Cartography** The map of Koltuchov indicates a settlement in this area.

**Finds distribution** No finds.

**Sites** Site by the edge of the steep NE ravine consisting of one roughly square structure and two possible pits suggested by vegetation (DSP08-H28-01).

**Relations to the surrounding area** No specific relations observed.

*Units 20000-20001 (Hill 26)* (Fig. 4.230)
**Size** 2.7 ha.

**Investigations** Intensive survey.

**Landscape type** Hillsides, Plateaus/Uplands.

**Location** The line runs 2 km roughly N-S from the road between Zajcevo and Novaja Ul'janovka at the bottom of the Hillsides to the Uplands running parallel with the road to Krasnaja Poljana on either side. It passes Foundation 49 on the western side of the road where it is crossed by Units 20029-30. The area is characterised by earth works and garbage dumping related to the road.

**Cartography** The road was constructed after 1955. On the half-verst map, four kurgans are marked while the map of Koltuchov shows two kurgans on the western side of the road and six on the eastern side of the road.

**Finds distribution** All finds are associated with sites.

**Sites** The road cuts a group of kurgans, three on the eastern side (DSP08-H26-02) and one possibly two on the western side (DSP08-H26-01) (Figs. 4.231-232). The easternmost one on the western side and the three smaller ones on the eastern side are either very much destroyed or possibly spoil heaps. A total of seven sherds were recorded in relation with the kurgans on both sides of the road. Among these are two Herakleian amphora fragments dated to the Late Classical-Early Hellenistic period (see also Units 20029-30 and GS3200).

Approximately 300 m S of the kurgans on the western side of the road, Unit 20000 cuts across a small promontory onto a larger one, Hill 6. Here and immediately to the E, four handmade sherds were recorded (GS3201-3) associated with DSP07-H06-01 (see Units 20002 and 20007).

**Relations to the surrounding area** The group of kurgans is located in an area where the slope becomes steeper to the N while it flattens out to the S. This is only 450 m due W of the settlement on Hill 6 (see Units 20002 and 20007).

*Units 20002 and 20007 (Hills 6 and 17)* (Figs. 4.233-234)

**Size** 8.4 ha.

**Investigations** Intensive and geomagnetic survey.

**Landscape type** Hillsides.

**Location** The line runs just under 1 km roughly E-W from the Krasnaja Poljana road to Foundation 54. The line cuts across Hill 6, a medium-sized promontory oriented broadly E-W, onto Hill 17 which is a large promontory running from the Uplands to the bottom of the Hillsides. The northeastern side of Hill 6 was characterised by patches of exposed bedrock and stone circles.

**Cartography** On the half-verst and mid-20th century maps, dirt roads from Karlav and Mussali meet on Hill 17 and continue together SE. The road from Mussali existed already on the 1865 map. Another dirt road from Mussali also on the 1837/1865 and half-verst maps runs SW, whereas a road running between the southeastern and southwestern road no longer existed on the half-verst map. On the mid-20th century map is a long strip of cultivated land to the S. At the time of survey, the cultivated area extended all the way to the Zajcevo-Krasnaja Poljana road.

**Finds distribution** Apart from a single sherd (20007/1) recorded on Hill 17, all finds were associated with sites. More finds were collected on Hill 6 in June 2007 reported as Finds List 8 in the original report but never integrated into the official finds recording system.

**Sites** In June 2007 geomagnetic survey revealed a double pen and a possible small stone structure close by in the central part of Hill 6 (DSP07-H06-01) (Fig. 4.235). The satellite image of Google Earth clearly shows a significant ash accumulation over the pen (Fig. 4.234). The site is crossed by long linear negative and positive anomalies at regular intervals. The same phenomenon was observed on Hill 1 where aerial photography has revealed the lines to be consistent with a recent agricultural episode (Fig. 4.153). A grab sample collected at the start of the season consisted of one handmade sherd (GS3085) and furthermore, a stone with a circular depression was recorded (Fig. 4.236). A total of nine sherds and a piece of flint were recorded on Hill 6 and immediately to the E in 2007 and 2008. Grab samples from 2008 consisted of four handmade sherds (GS3201-3203). The piece of flint and five sherds were recorded in Units 20002 and 20007. One of the sherds was a Late Classical-Early Hellenistic handmade one, and for two of the sherds a Medieval date was tentatively proposed. Based on the double pen, the site was interpreted as a settlement tentatively dated to the Late Classical-Early Hellenistic period that was involved in livestock rearing.

On the eastern edge of Hill 17 on a wide spur, an ancient settlement was identified by S. Lancov & S. Koltuchov in 2007, but no surface finds or features were recorded by the survey in 2008 apart from a single possibly Early Modern sherd (20007/1). Geomagnetic survey revealed a cluster of possible archaeological anomalies in the central part of Hill 17 (DSP08-H17-01), and a single indefinable body sherd (GS3204) may be associated with this site (Figs. 4.237-238).

**Relations to the surrounding area** There is a group of kurgans (DSP08-H26-01-02) 450 m due E of the settlement on Hill 6 (see Units 20001).

*Units 20003 and 20006 (Hill 17)* (Figs. 4.239, 4.234)

**Size** 5.1 ha.

**Investigations** Intensive survey.

**Landscape type** Hillsides.

**Location** The line runs approximately 600 m roughly N-S from Foundation 54 on either side of and parallel to a dirt track on the ridge of Hill 17.

**Cartography** Five kurgans are marked on the 1837 map on the eastern side of a roughly N-S running dirt road from Mussali. The half-verst map however shows only four kurgans along the same road. On the mid-20th century map a track that appears to be the same track runs roughly N-S on the western side of three kurgans. Six kurgans are marked on the map by Koltuchov with the track running through it. Also on the mid-20th century map is depicted a long strip of cultivated land to the SE. Nowadays, the cultivated area has extended all the way to the Zajcevo-Krasnaja Poljana road.

**Finds distribution** All finds are associated with sites.

**Sites** A piece of flint and four sherds including a Thasian amphora handle (also GS3205-6) were recorded in relation to a group of five kurgans located on the western side of the track (DSP08-H17-02) (Fig. 4.240).

**Relations to the surrounding area** The group of kurgans is located 500 m W of DSP07-H06-01 (Fig. 4.233) and approximately 800 m NE of DSP07-H01-01 (Fig. 4.147), but none of these sites are located within the viewscape of the kurgans.

*Units 20004-20005 (Hill 17)* (Figs. 4.241, 4.234)
**Size** 2.7 ha.

**Investigations** Intensive and geomagnetic survey.

**Landscape type** Hillsides.

**Location** The line runs approximately 300 m roughly E-W from the dirt track (see Units 20003 and 20006) to Foundation 59 on a short wide spur on the western edge of Hill 17.

**Cartography** On the 1837/1865 and the half-verst maps, a dirt road from Mussali runs SW across Hill 17.

**Finds distribution** No finds recorded.

**Sites** Unit 20005 bypassed a single kurgan (DSP08-H17-03) on the S side close to the dirt track 375 m S of the southernmost kurgan of DSP08-H17-02.

Towards the western tip of the spur, which is dominated by patches of exposed bedrock, a stone circle was excavated by the project in 2007 (Fig. 4.242). The context was empty of finds, and indeed the geological observations done in 2008 (see Chapter 3.3.2.6) determined that these very common features are natural phenomena caused by limestone erosion (karst). Of course, this does not rule out cultural uses, but so far only one tiny sherd has been recorded in areas of stone circles (see Unit 20010).

**Relations to the surrounding area** The spur lies adjacent to Hill 2 with settlements DSP07-H02-01-03 (Fig. 4.159).

*Units 20008-20009 (Hills 16, 8 and 20)* (Fig. 4.243)
**Size** 18.6 ha.

**Investigations** Intensive and geomagnetic survey.

**Landscape type** Hillsides, Plateaus/Uplands.

**Location** The line runs approximately 1.5 km in a more or less N-S direction from Foundation 42 to Foundation 44 in the Uplands. The line starts on Hill 19 at Foundation 42 towards the S and crosses over to Hill 8 following the eastern edge of this past Foundation 43 until it crosses over to Hill 20 with Foundation 44 in the Uplands. South of Foundation 44 the Uplands are cultivated. Vegetation and evidence of field clearance indicates that the cultivated zone started approximately 300 m further N at some time prior to the present, which is consistent with observations made in Units 20010-20012 (Fig. 4.244).

**Cartography** On the mid-20th century map, the Uplands were cultivated as is currently the case. The map of Koltuchov marks a settlement on the lower part of Hill 8 which was identified by the project in June 2007 (DSP07-H08-01) and another site on the eastern edge of Hill 8 further to the S in an area of stone circles.

**Finds distribution** Only Early Modern sherds were recorded on the Hillsides: 16 fragments of a glazed jar were found unrelated to other observations (**Cat. 822**), and a single sherd was noted approximately 70 m S of the stone circle site identified by Koltuchov. All other finds were associated with sites.

**Sites** On the S slope of Hill 16 N of Hill 8 several shallow banks were observed (DSP08-H16-01). In the same area geomagnetic survey revealed a rectangular bank open towards the W (Fig. 4.245). Inside is a square structure associated with a circular structure and another small square structure. No surface finds were recorded on this slope, but this may be due to erosion on the slope. The architecture suggests a settlement associated with the exploitation of the landscape possibly of more recent date.

At the bottom of Hill 16 where the line crosses over to the western edge of Hill 8, two possible stone foundations were recorded in a shallow area with dark soil and high grasses which belonged to the southern part of DSP07-H08-01 (see Units 20013-15) (see also Fig. 4.249).

On the eastern edge of Hill 8, the line crossed a shallow area where the map by Koltuchov identified a site consisting of stone circles, a heap of small stones and a small mound of soil. In 2008 we recorded an L-shaped elevation associated with a possible spoil heap. These earth works are likely to be related to a possible pipeline or cable visible as a shallow elevation crossing the area. The results of the geomagnetic survey support the predominance of modern activities in this area.

In the Uplands of Hill 20, the small cluster of finds was interpreted as evidence for the location of a Late Classical-Early Hellenistic site located in the cultivated field to the S (DSP08-H20-01). This included two Late Classical-Early Hellenistic handmade fragments, a possibly Late Classical-Early Hellenistic transport amphora fragment and two pieces of flint.

**Relations to the surrounding area** The settlement on Hill 16 would be within the viewscape of the settlements on the lower part of Hill 8 (DSP07-H08-01 and DSP08-H08-02) (Fig. 4.248). DSP08-H20-01 is associated with finds in Units 20011 (Fig. 4.247).

*Units 20010 and 20012 (Hill 20)* (Fig. 4.246)
**Size** 15.9 ha.

**Investigations** Intensive and geomagnetic survey.

**Landscape type** Hillsides, Plateaus/Uplands.

**Location** The line runs approximately 1.4 km in a more or less N-S direction from Foundation 41 to Foundation 39 in the Uplands. South of Foundation 39, the Uplands are cultivated. Vegetation and evidence of field clearance indicates that the cultivated zone started approximately 300 m further N at some time prior to the present, which is consistent with observations made in Units 20008-9 and 20011 (Fig. 4.244).

**Cartography** On the mid-20th century map the Uplands were cultivated as is currently the case. The area believed to have been previously cultivated is not indicated as such on the mid-20th century map suggesting a partly abandoned extension of the cultivated area since 1955. The map of Koltuchov indicates two sites associated with stone circles.

**Finds distribution** One tiny possibly red-slipped sherd was recorded at one of the stone circle sites identified by Koltuchov near Foundation 40. The condition of the sherd and the character of the area suggest erosion as a likely cause for its location in this particular area. It was the only sherd found in relation to stone circles, and although cultural use of these natural features cannot be completely rejected, there is no positive evidence to suggest it.

A possible Late Classical-Early Hellenistic amphora body fragment was recorded in the previously cultivated area in the Uplands.

**Sites** No sites identified.

**Relations to the surrounding area** No specific relations observed.

*Unit 20011 (Hill 20)* (Fig. 4.247)
**Size** 3.7 ha.

**Investigations** Intensive and geomagnetic survey.

**Landscape type** Uplands.

**Location** The line runs approximately 400 m roughly E-W between Foundations 39 and 44 parallel to a large cultivated field in the Uplands. Deep soils dominate, and the vegetation as well as evidence for field clearance suggests that this area was cultivated at some time in the past.

**Cartography** The mid-20th century map indicates that the western half of the surveyed Uplands were cultivated. The area believed to have been previously cultivated is not indicated as such on the mid-20th century map suggesting a partly abandoned extension of the cultivated area since 1955.

**Finds distribution** A piece of flint and two sherds, one Early Modern, the other recent, were recorded on the edge of the cultivated field.

**Sites** No sites identified.

**Relations to the surrounding area** The flint is probably associated with the finds in the southern end of Unit 20009 (DSP08-H20-01) (Fig. 4.243).

*Units 20013-20018 (Hill 8)* (Fig. 4.248)
**Size** 11.5 ha.

**Investigations** Intensive and geomagnetic survey, trial excavation.

**Landscape type** Hillsides.

**Location** The line runs approximately N-S, the units form a block in the lower northern part of Hill 8 although they stop some 270 m short of the tip of the promontory. The western edge of the lower part of Hill 8 is grassy and has deep soils while the central and eastern side is dominated by patches of exposed bedrock. The vegetation suggests that the tip of the promontory may have been cultivated in the recent past.

**Cartography** The map of Koltuchov indicates a settlement in this area which was identified by the project in June 2007 (DSP07-H08-01).

**Finds distribution** Four clusters of finds were recorded all associated with architecture. Three of the clusters were located along the western edge of Hill 8 and the last one on the eastern edge. More finds were collected on Hill 8 in June 2007 reported as Finds Lists 11-12 in the original report but never integrated into the official finds recording system. Two Early Modern tableware sherds were recorded in the southern half of Unit 20017 (no. 1). These were however not associated with any archaeological structural remains.

**Sites** The distribution of pottery suggests at least two distinct scatters. The three clusters of finds on the western edge are associated with four stone foundations and several pits that can be recognized by intense vegetation on top of them (DSP07-H08-01) (Figs. 4.249-250). The entire area covers 2 ha. The northernmost cluster, a discrete scatter of eight sherds included three handmade Bronze Age sherds as well as Late Classical-Early Hellenistic ones (20013-1-2, 20014-1-2 and 20015-1-3). A grab sample (2007-FL11) from the broader area from 2007 included 20 handmade sherds dated to the Bronze Age. Geomagnetic survey revealed a double pen (Feature 2), and finds from the trial excavation, handmade pottery and numerous animal bones, suggested a Bronze Age date for this structure. The wall of the double pen was constructed as a double wall of large blocks filled-in with rubble (Figs. 4.251-254). The width could not be reconstructed. In the western corner of the trench, in stratum 4 was a small concentration of small stones associated with a concentration of animal bones. The middle, less discrete cluster of pottery consisted of six sherds most of them handmade Late Classical-Early Hellenistic sherds and a piece of flint located approximately 80 m to the SW (20015-4-10 and 20016-4). A grab sample (2007-FL12) from the broader area surveyed in 2007 included 13 handmade sherds dated to the Bronze Age. Geomagnetic survey revealed a complex of rectangular stone structures with multiple rooms as well as circular structures. A trial trench at Feature 7 revealed a collapsed stone wall in a layer of ashy soil which also contained the majority of the artefacts excavated, handmade pottery and lithics dated to the Bronze Age as well as animal bones (Figs. 4.255-256). The orientation of the wall is N-S, and it is 0.8-1 m wide. The wall is sitting on an ashy floor where a flint knife was found, several fragments of handmade pottery and animal bones (FL bag 31). The southernmost small cluster of three sherds, one Early Modern and two handmade Late Classical-Early Hellenistic ones, was located 55 m further SW (20008-9, 20016-1-3). Geomagnetic survey revealed three rectangular and more irregular stone structures with multiple rooms, pits and possible enclosure walls. The number of structures and the double pen suggest a settlement of substantial size associated with livestock rearing although not all the structures may be contemporary. The artefacts suggest a date in the Bronze Age. The range of vessel types is impossible to estimate based on the small handmade fragments. The elements of the site appear to be very similar to DSP07-H01-01 and DSP07-H02-01 which both were occupied during the Late Classical-Early Hellenistic period as well as sometime during the Bronze Age. The Late Classical-Early Hellenistic element at Hill 8 is however much more spread out over the hill.

Two Early Modern sherds found on the eastern edge are associated with a square stone foundation (DSP08-H08-02) (Figs. 4.257-258).

**Relations to the surrounding area** The settlements on the lower Hill 8 would be within the viewscape of the settlement on Hill 16 (DSP08-H16-01) (Fig. 4.243).

*Units 20019-20020 (Hill 16)* (Fig. 4.259)
**Size** 8 ha.

**Investigations** Intensive and geomagnetic survey.

**Landscape type** Hillsides.

**Location** The line runs approximately 800 m roughly N-S from Foundation 46 to 47 on the western part of Hill 16. A large cultivated field runs parallel to the upper half of the line on the western side. Just SW of Foundation 46, there was an approximately 12 m long shallow elevation oriented N-S, but no finds were recorded. At both windmill foundations, three cores were drilled the day before our survey and five of them were left on the surface (e.g. Fig. 4.260). According to these, the soil is no more than 0.2 m deep.

**Cartography** A strip of cultivated land parallel to the upper half of the line on the western side is marked on the mid-20th century map. The map of Koltuchov marks a group of five kurgans some distance W of the line, but only one was observed by the survey in 2008.

**Finds distribution** A single piece of flint was recorded just N of Foundation 46.

**Sites** No sites identified.

**Relations to the surrounding area** From the ridge of this hill is a fine view of DSP07-F02 and DSP08-F04.

*Units 20021 and 20024 (Hill 14)* (Fig. 4.261)
**Size** 5.6 ha.

**Investigations** Intensive and geomagnetic survey.

**Landscape type** Hillsides, Plateaus/Uplands.

**Location** The line runs approximately 800 m roughly N-S from Foundation 31 to 32. The line follows a shallow ravine with deep soils into the Uplands.

**Cartography** The map of Koltuchov indicates four kurgans at some distance W of the line.

**Finds distribution** A single indefinable sherd was noted.

**Sites** The line passed the western edge of a Bronze Age site (DSP08-H14-01), but no finds were recorded (see Units 20043-20048).

**Relations to the surrounding area** See Sites.

*Units 20022-20023 (Hills 14 and 21)* (Fig. 4.262)
**Size** 4.6 ha.

**Investigations** Intensive and geomagnetic survey.

**Landscape type** Hillsides, Plateaus/Uplands.

**Location** The line runs approximately 500 m roughly E-W from Foundation 32 to 34 crossing a N-S running dirt track to run parallel to another dirt track on the southern side. The area is dominated by deep soils, and the vegetation and evidence for field clearance suggests that the area was cultivated at some point in the past. Some modern material is found along both tracks.

**Cartography** The half-verst- and mid-20th century maps reveal both a N-S running dirt road from Kurama Kostel' in the S while also an E-W running track is marked on the mid-20th century map that appears to wind between strips of cultivated land in the Uplands.

**Finds distribution** No finds recorded.

**Sites** Just W of Foundation 34, a large angular patch of high vegetation and thistles indicate the presence of a structure probably of more recent date although no structures are visible on the half-verst- or mid-20th century maps (DSP08-H21-01) (Fig. 4.263). A piece of iron was embedded in the soil which may have been a plough fragment caught on a stone although no large stones were visible. The site must have been occupied between 1897 and 1955 or more likely after 1955.

**Relations to the surrounding area** It seems probably that this site was associated with the exploitation of the cultivated area to the S in the Uplands.

*Units 20025-20028 (Hill 15)* (Fig. 4.264)
**Size** 3.6 ha.

**Investigations** Intensive survey.

**Landscape type** On the transition between Hillsides and Pediment.

**Location** Running E-W, the units form a block on the shallow northern part of Hill 15 near the bottom of the Hillsides. Deep soils dominate. There is a dirt track running E-W just to the N.

**Cartography** On the 1955 map, the area to the S is cultivated. The map by Koltuchov indicates a settlement.

**Finds distribution** No finds recorded.

**Sites** The large angular patch of intense vegetation on the western slope of Hill 15 suggests the presence of a structure confirmed by near contemporary building debris (DSP08-H15-01) (Fig. 4.265). The site must have been occupied between 1897 and 1955 or more likely after 1955 as it is not indicated on either of these maps.

**Relations to the surrounding area** It seems probable that this site was associated with the exploitation of the cultivated area to the N at the lower end of the Hillsides. A very smiliar site, DSP08-H27-01 is located 400 m to the W (Fig. 4.228).

*Units 20029-20030 (Hill 26)* (Fig. 4.266)
**Size** 4.1 ha.

**Investigations** Intensive and geomagnetic survey.

**Landscape type** Hillsides.

**Location** The line runs approximately 575 m roughly N-S between Foundations 48 and 49 on Hill 26 cutting across the road between Zajcevo and Krasnaja Poljana and Units 20000-1. The area near the road is characterised by earth works and garbage dumping related to the road. Generally, this side of the hill is dominated by deeper soils.

**Cartography** The road was constructed after 1955. On the half-verst map four kurgans are marked while the map of Koltuchov shows six on the eastern side of the road and an undated structure 160 m northeast of Foundation 48.

**Finds distribution** All finds were associated with sites.

**Sites** For DSP08-H26-02, see Unit 20001 (Fig. 4.230; 4.232). On the site marked on the map of Koltuchov NE of Foundation 48 on the northeastern edge of the promontory, two patches of low bushes revealed the outline of two structures, the larger one measuring 120 m$^2$. No finds were recorded. The vegetation was similar to the kind recorded on structures at DSP08-H12-01.

**Relations to the surrounding area** See Unit 20001 (Fig. 4.230).

*Units 20031-20032 (Hill 25)* (Fig. 4.267)
**Size** 11.7 ha.

**Investigations** Intensive and geomagnetic survey.

**Landscape type** Hillsides.

**Location** The line runs approximately 1.3 km roughly N-S between Foundations 60 and 63 on Hill 25. Deeper soils dominate the northern part of the line while thinner soils and exposed patches of bedrock characterise the southern part.

**Cartography** On the half-verst map runs a dirt road from a farm between Karlav and Mussali SW on the western slope of Hill 25 where it crosses another dirt road from Karlav coming over Hill 3. The road from Karlav (here named Čaikino) is marked in the 1955 map running in the shallow ravine between Hills 3 and 25, as well as a strip of cultivated land in the shallow ravine between Hills 25 and 17. The map by Koltuchov indicates two kurgans between Foundations 60 and 61.

**Finds distribution** All finds were associated with sites.

**Sites** Midway between Foundations 60 and 62, the line crossed three kurgans. One indefinable body sherd was found just S of the kurgans, but no datable finds were associated. The southernmost kurgan was confirmed by geomagnetic survey (DSP08-H25-01). Approximately halfway between Foundations 62 and 63 a small mound with some very large stones in it was identified. An indefinable wheelmade sherd was recorded 8 m NE of the kurgan, but no datable finds were associated. Geomagnetic survey confirmed this is a kurgan as well as identified another one to the NW (DSP08-H25-02) (Fig. 4.268).

**Relations to the surrounding area** Just S of Foundation 63, a settlement was identified in Units 20064-20065 (DSP08-H07-03-04) (Fig. 4.304). East of the settlement, three kurgans were observed.

*Units 20033-20034 (Hill 3)* (Fig. 4.269)
**Size** 0.6 ha.

**Investigations** Intensive survey.

**Landscape type** Hillsides.

**Location** Running N-S, the units form a block basically covering Foundation 61. Deep soils dominate.

**Cartography** To the SW there is a symbol indicating a pit on both the half-verst- and the mid-20th century maps.

**Finds distribution** A fragment of modern tile was grab-sampled just W of the line.

**Sites** No sites were identified (see also Hill 3).

**Relations to the surrounding area** No specific relations observed.

*Units 20035-20036 (Hills 19 and 12)* (Fig. 4.270)
**Size** 2.3 ha.

**Investigations** Intensive and geomagnetic survey, augering.

**Landscape type** Hillsides, Plateaus/Uplands.

**Location** The line zigzags 2.6 km roughly N-S between Foundations 68 and 71 on Hill 19 and 12. In the area of Foundation 70, the line crossed a series of earth works associated with an abandoned well. There is plentiful modern garbage in the wider area as well as earth works associated with pipelines. The well post-dates the 1955 map. Still operating, there is another one like it close to the Zajcevo-Krasnaja Poljana road.

**Cartography** The half-verst map marked a line of seven kurgans near the Uplands and a single one on a long plateau approximately 1 km N of this. These are also marked on the 1955 map except that the line of kurgans is reduced to six. This map also marks two kurgans towards the bottom of the hill. Also on the 1955 map are tracks running from Čaikino and Snežnoe to a strip of cultivated land in the shallow area between Hills 24 and 19. The Koltuchov map marks a quarry just N of Foundation 69 and 11 kurgans between Foundations 70 and 71 as well four kurgans on the long plateau to the N. There are also three kurgans N of Foundation 68.

**Finds distribution** 22 sherds of an Early Modern pot (**Cats. 568-569**) were recorded midway between a group of kurgans (DSP08-H19-02) and a settlement in the Uplands (DSP08-H12-01). A handmade sherd was found just N of Foundation 71 (20035/10). All other finds were associated with sites.

**Sites** In the area of Foundation 69 and S towards Foundation 70, the line crossed an area of exposed bedrock of which several were quarried (DSP08-H19-09) (Figs. 4.271-272). An Early Modern tableware rim, a piece of flint and a ground stone were recorded in this area (**Cat. 570** and 20035/2-3). Another Early Modern tableware sherd found 100 m to the S may be associated with other quarries in the wider area.

Just N of Foundation 70, the line bypassed a group of four kurgans to the NE. This is close to an abandoned well, where two sherds were recorded (20036/19-20) although they may be part of the modern rubbish associated with this area.

11 ancient sherds were recorded around a group of four kurgans S of Foundation 70 (DSP07-H19-03) (Fig. 4.207). They consisted of Late Classical-Early Hellenistic amphora fragments with among them a Herakleian fragment (20035/5-6 and 20036/14-18).

In the Uplands, the line by-passed seven kurgans positioned roughly in line (DSP07-H19-01-02) (Figs. 4.205-206). Due E of these, a patch of exposed bedrock with evidence for quarrying was recorded as well as low shrubs that suggested the location of four structures. The latter were associated with four sherds and a piece of flint (**Cat. 537**, 20035/7-8, 20036/11 and GS3208). Also associated is a cluster of 11 sherds and a piece of flint (20035/9 and 20036/1-10) recorded in the upland zone in a shallow depression with dark loose soils and high grassy vegetation associated with DSP08-H12-01 (see Units 20061-20063). Most of these sherds are of amphorai including a Sinopean, a Herakleian and a possible Thasian fragment. Geomagnetic survey clearly indicates that this area was cultivated at some point in the nearer past, but only the remains of one structure were found preserved below the surface (Figs. 4.302-303). In a core drilled in this area, tiny pottery fragments were detected in a cultural layer 0.23 m deep (see DSP08-H12-01 Auger S14-S15).

**Relations to the surrounding area** The quarrying in the southern part of Hill 19 may be associated with Snežnoe. The structures and pottery on the upper eastern slope of Hill 19 form the western edge of DSP08-H12-01, a large settlement spreading across two smaller promontories to the E (see Units 20061-20063) (Fig. 4.296). It is tempting to associate the kurgans on the upper part of Hill 19 with either of the two large settlements to the E (Figs. 4.148 and 4.296). The kurgans on the southern part of the ridge of Hill 19 are located less than 200 m W of DSP08-H12-01, although they are not at present visible in the viewscape of the site. The kurgans N of Foundation 70 are located only 300 m W of the settlement across the ravine on Hill 1.

*Units 20037-20038 (Hill 11)* (Fig. 4.273)

**Size** 5.8 ha.

**Investigations** Intensive and geomagnetic survey.

**Landscape type** Hillsides.

**Location** The line runs 725 m roughly N-S between Foundations 86 and 87. The line follows a shallow ravine crossing an area of stone circles.

**Cartography** On the 1955 map there is a strip of cultivated land to the N and W.

**Finds distribution** No finds recorded.

**Sites** No sites identified.

**Relations to the surrounding area** No specific relations observed.

*Units 20039-20040 (Hill 24)* (Fig. 4.274)
**Size** 2.1 ha.

**Investigations** Intensive and geomagnetic survey.

**Landscape type** Hillsides.

**Location** The line runs 2 km in an approximately N-S direction between Foundations 75 and 78. The line follows the western slope of Hill 24, a long promontory characterised by larger areas of exposed bedrock and stone circles.

**Cartography** A single kurgan is marked on the half-verst map in the lower part of the hill, while the 1955 map shows two kurgans at this location. The map of Koltuchov records eight kurgans in the same location and an area of stone circles between Foundations 75 and 76. Both the half-verst- and mid-20th century map show a track running roughly N-S on the western slope of Hill 24 probably linking several strips of cultivated fields marked in the 1955 map directly W of the track.

**Finds distribution** Apart from one sherd recorded (20040/5) SE of Foundation 78, there is a scatter of three sherds and two pieces of flint (**Cat. 574**, 20039/1-2 and 20040/2,4) recorded between the two kurgan groups crossed by this line. In terms of post-depositional processes, these fragments are most likely to have come from the southern group (DSP08-H24-02).

**Sites** Two groups of kurgans and a structure associated with a possible quarry were recorded. The northern group of eight kurgans is located NE of Foundation 77 (DSP08-H24-01) (Fig. 4.274). In this area there is also evidence for quarrying, which is the only instance of a close geographical correlation between kurgans and a quarry. A single amphora sherd was recorded in one of the kurgans (GS3217).

The southern group of five kurgans is located NE of Foundation 76 (DSP08-H24-02). Four of these kurgans were confirmed by geomagnetic survey (Fig. 4.274). In this general area, 12 sherds were recorded, apart from a possible Late Classical-Early Hellenistic amphora fragment all of unknown date (20039/3-9 and 20040/1). Nine of these sherds were collected in an area characterised by loose dark soil W of the kurgans.

Past Foundation 76, the line crossed an area of exposed bedrock and stone circles. Here vegetation suggested the location of a structure and a possible Late Classical-Early Hellenistic transport amphora rim (**Cat. 575**) was found just S of it (DSP08-H24-03). A possible quarry was located 85 m E of the structure (Figs. 4.275-276).

**Relations to the surrounding area** Apart from Hill 19, Hill 24 has the highest number of kurgans recorded.

*Units 20041-20042 (Hill 23)* (Fig. 4.277)
**Size** 2.4 ha.

**Investigations** Intensive and geomagnetic survey.

**Landscape type** Hillsides.

**Location** The line zigzags 2.1 km roughly N-S between Foundations 79 and 82. The line follows the western side of Hill 23, a long promontory characterised by large areas of exposed bedrock and stone circles.

**Cartography** Both the half-verst- and mid-20th century map show a track running roughly N-S on the western slope of Hill 23, probably linking several strips of cultivated fields marked in the 1955 map directly W of the track. The Koltuchov map marked an area of stone circles between Foundations 80 and 81. SE of this are three kurgans.

**Finds distribution** A small diffused cluster consisting of four possibly Late Classical-Early Hellenistic sherds, one of which is a Sinopean amphora fragment, three indefinable and one Early Modern sherd (20041/1-5 and 20042/3-4) were recorded on the northern side of Foundation 80. 20041/1-5 came from an area of dark loose soil and high grasses similar to conditions in the southern part of DSP08-H12-01; however, no structures were recorded. This type of soil is usually associated with cultivation on the slopes. Post-depositional processes may have produced the cluster by moving artefacts to the surface. A visit to a small promontory adjacent to this area did not produce any additional finds. An Early Modern sherd was recorded just N of Foundation 82 (20041/7).

**Sites** Eight kurgans were recorded between Foundations 80 and 81 in an area of exposed bedrock and stone circles (DSP08-H23-01) (Fig. 4.277). Only a single sherd of undeterminable date was recorded (**Cat. 573**).

Two sherds, of which one was possibly Late Classical-Early Hellenistic (20041/6 and 40042/1), were recorded between Foundations 81 and 82 (DSP08-H23-02). Here the geomagnetic survey revealed the remains of an angular structure probably associated with a curved feature approximately 70 m to the NW (Fig. 4.277).

**Relations to the surrounding area** The cluster of pottery on the lower northern part of the hill may be associated with the large settlement located 260 m to the W (DSP08-H11-02).

*Units 20043-20048 (Hill 14)* (Fig. 4.278)
**Size** 9 ha.

**Investigations** Intensive and geomagnetic survey, trial excavation.

**Landscape type** Transition between Hillsides and Pediment.

**Location** The block of units runs N-S on the lower northern part of Hill 14 near the bottom of the Hillsides. Deep dark soil dominated by high dense grasses affecting surface visibility.

**Cartography** The site was identified by the project in June 2007 and marked on the map of Koltuchov.

**Finds distribution** A discrete concentration of pottery was recorded on the border of Units 20047 and 20048 within a more diffuse pottery scatter spreading from 20045 to 20048. Additionally, a Late Classical-Early Hellenistic amphora handle and an Early Modern sherd (**Cat. 539** and 20046/1) were recorded 140 m to the S, also a piece of flint (20048/4) was recorded 140 m to the N.

**Sites** A site was supposed in June 2007 when a grab sample of 34 handmade sherds was collected (DSP08-H14-01). Although rocks were visible on the surface, no clear structures could be made out (Fig. 4.279). The core of this site is represented by 16 sherds of which 15 are handmade (20047/2-3 and 20048/2-3); however, due to the grab sampling the spatial distribution is not reliable. Three of the handmade sherds were dated to the Bronze Age, but the date of the remainder may be Late Classical-Early Hellenistic. Geomagnetic survey revealed six possible structures (Figs. 4.280-281). Trial excavation at Feature 2 confirmed the presence of a structure with a small square stone cist dated to the Bronze Age (Figs. 4.282-284). Below the topsoil was a sterile erosion layer which had helped to protect the archaeological remains. The archaeological remains sit in a grey ashy layer associated with handmade pottery and animal bones. In the northern half of the trench was a layer of field stone debris. In the southern half was a rectangular stone cist (1.2 x 0.9 m) constructed of small slabs: 3-5 cm thick and 5-10 cm high, sitting directly on the bedrock. In the middle of the cist is a natural depression filled with dense loam and rubble, but empty of artefacts. The more diffused pottery scatter consisted of two Late Classical-Early Hellenistic transport amphora fragments, one Sinopean, a possibly Byzantine transport amphora fragment and a Late Classical-Early Hellenistic handmade fragment (20045/1-2, 20048/1 and 3).

**Relations to the surrounding area** It is possible that the site was associated with the exploitation of the cultivated area to the N at the bottom of the Hillsides.

*Units 20049 and 200058-60 (Hill 13)* (Fig. 4.285)
**Size** 4.8 ha.

**Investigations** Intensive and geomagnetic survey, augerings.

**Landscape type** Plateaus/Uplands.

**Location** Running N-S, the units form a block on Hill 13, a small promontory. There is evidence for agriculture on the southern part of the promontory, and although the soil appears deep there are also areas of exposed bedrock. The tip of the promontory is flanked by two small peaks with stone structures on top of them.

**Cartography** The site marked on the map of Koltuchov was identified by S. Lancov and S. Koltuchov in the autumn of 2007. According to the map, it is a settlement of the Bronze to Early Iron Age.

**Finds distribution** The pottery is found on the southern part of the promontory, 100 m S of the flanking peaks. It consisted of a possible Herakleian tile fragment, two Late Classical-Early Hellenistic transport amphora fragments, one Thasian, a Late Classical-Early Hellenistic handmade sherd, an Early Modern sherd, a modern sherd and five indefinable sherds (**Cat. 538**, 20049/1-3, 20059/1-2 and 20060/1-4).

**Sites** Remains of six possible structures were visible on the surface of the peaks and 100 m S of the tip on the central part of the promontory, but no pottery was recorded in the vicinity (DSP08-H13-01) (Figs. 4.286-287). Geomagnetic survey recorded the remains of only three rectangular structures (Features 2, 3 and 6), one of which is on the eastern peak (Feature 2) (Fig. 4.288) Augerings on the peaks revealed no secure cultural layers. Based on the rectangular structures the site was interpreted as a possible settlement.

**Relations to the surrounding area** It is possible that the site was associated with the exploitation of the cultivated area to the S in the Uplands. The ravines cutting Hill 13 from the surrounding landscape continue N to settlement DSP08-H11-02 (see Units 20050-20055) (Fig. 4.277). Kurgan Group 7 is located in the Uplands 1 km SE of DSP08-H13-01.

*Units 20050-20055 (Hill 11)* (Fig. 4.289)
**Size** 11.6 ha.

**Investigations** Intensive and geomagnetic survey, trial excavation.

**Landscape type** Hillsides.

**Location** Running N-S, the units form a block on the eastern slope of Hill 11, a long promontory. The eastern slope of Hill 11 has a long narrow spur. Towards the northern tip, the soil is deeper but towards the S, the soil becomes gradually thinner.

**Cartography** No features indicated.

**Finds distribution** All finds are associated with the site spread over an area of 7 ha. The finds of the intensive survey tend to cluster in the northwestern part of the promontory spreading onto the eastern slope of the main body of Hill 11 over an area of approximately 3 ha. Unfortunately, two large grab samples (GS3094 and GS3218) were taken prior to the systematic survey, which has corrupted the spatial distribution of the clustering.

**Sites** The discrete cluster of artefacts associated with architectural remains was identified as a site spreading over 2.6 ha (DSP08-H11-02) (Fig. 4.290). Remains of ten possible stone structures are visible on the surface. The architecture seems concentrated at some distance S of the tip as was the case with the settlement on Hill 8 (DSP07-H08-01) (Fig. 4.248). Of course this may have been caused by erosion and soil build-up at the tip, which appears to have been the case at DSP08-H12-01 (see Units 20060-20063) (Fig. 4.296). Three additional structures are located 85 m to the N on the E slope on the main body of Hill 11. A total of 104 sherds and three lithics were collected from DSP08-H11-02. Apart from a single possible Bronze Age sherd, the pottery consists of Late Classical-Early Hellenistic transport amphorai from Herakleia, Chersonesos, Sinope, Thasos and possibly Chios, a Sinopean louterion and handmade sherds as well as many unidentified wheelmade body fragments of red clay some of which are Early Modern. The most closely dated artefact is a Sinopean amphora fragment dated ca 350-250 BC (**Cat. 494**). One amphora fragment shows traces of repair (**Cat. 497**). Geomagnetic survey revealed eleven poorly defined anomalies indicating possible structures up to 30 to 70 m apart and household pits (Fig. 4.291). Trial excavation at Feature 9 revealed an irregularly shaped pit house with a S entrance dated to the Late Classical-Early Hellenistic period (Figs. 4.292-294). The diameter of the main pit is 2.8 m. Above the pits was an ashy layer associated with fragments of Chersonesean and Sinopean amphorai, handmade pottery, tableware as well as a jug of possibly Samian origin and animal bones. The loamy sand-filling of the pits was associated with Chersonesean, Sinopean, Thasian and Herakleian amphora fragments, handmade pottery, tableware and animal bones as well as a fragmented stone slab of possible southeastern import. The number of structures and the range of finds suggest a settlement of substantial size dated to the Late Classical-Early Hellenistic period. The architecture makes this settlement unique among the sites identified so far.

**Relations to the surrounding area** DSP08-H11-01 is located approximately 1 km due S at the bottom of the slope (Fig. 4.195). The ravine E of the site continues S to settlement DSP08-H13-01 (see Units 20049, 20058-20060) (Fig. 4.195). Across the ravine approximately 260 m to the E was recorded a small cluster of ceramics (see Units 20041-20042) (Fig. 4.277).

*Units 20056-20057 (Hill 22)* (Fig. 4.295)

**Size** 15.4 ha.

**Investigations** Intensive and geomagnetic survey.

**Landscape type** Hillsides, Plateaus/Uplands.

**Location** The line zigzags 1.6 km roughly N-S between Foundations 92 and 94. The line roughly follows the ridge of Hill 22, a long promontory. Patches of exposed bedrock and stone circles are mainly observed in the southern part between Foundations 92 and 93. Towards the Uplands, the soil becomes visibly darker.

**Cartography** A single kurgan is marked on the half-verst map in the Uplands while on the map of Koltuchov six kurgans appear. This map also indicates three kurgans NE of Foundation 93 and another one to the SE.

**Finds distribution** The only pottery recorded in these units is six fragments of an Early Modern cooking pot (**Cat. 572**) collected S of Foundation 93 NW of a lone kurgan.

**Sites** Northeast of Foundation 93, a group of three kurgans was recorded (DSP08-H22-01) and to the SE a single kurgan (DSP08-H22-02) (Fig. 4.295). In the Uplands were a quarry as well as seven kurgans and remains of more recent structures revealed as patches of intense vegetation and stone foundations (DSP08-H22-03). This is the only recorded instance of a near recent settlement being located adjacent to ancient kurgans on the slopes. No finds were collected in this area, and the geomagnetic survey revealed a number of irregular ditches of different sizes (Fig. 4.295).

**Relations to the surrounding area** No specific relations observed.

*Units 20061-20063 (Hill 12)* (Figs. 4.296, 4.148)

**Size** 6.7 ha.

**Investigations** Intensive and geomagnetic survey, augering.

**Landscape type** Plateaus/Uplands.

**Location** Running N-S, the units form a block on the small promontory of Hill 12 spreading E onto another small promontory (see also Units 20035-36). Deep soils dominate the area especially to the S where the soil is dark and loose and dominated by high grasses. The site cuts across two shallow ravines with relatively steep banks.

**Cartography** No features are indicated.

**Finds distribution** All finds are associated with the site, which spreads out over a total area of 6 ha (see also 20035-20036). Finds tend to cluster in the S (20035-20036) on the tip of Hill 12 (GS3209-3216) and on the western slope of the small promontory adjacent to Hill 12 (20063). The cluster on the tip is probably the result of post-depositional processes of soil accumulation and subsequent erosion. The cluster in the S may likewise be the result of post-depositional processes. Because of cultivation of this area in the recent past, artefacts have been ploughed out into the surface.

**Site** DSP08-12-01 is located near the tip of Hill 12, but the architecture spreads onto the eastern slope of Hill 19 and the western slope of a small promontory E of Hill 12 (Figs. 4.297-298). At least 14 stone structures are visible on the surface, some covered with shrub (Figs. 4.299-301). At least one of these was constructed with double walls built with standing blocks. Including the artefacts from 20035-36, a total of 53 sherds (20035/7-10, 20036/1-9, 20036/11-12, 20061/1-4, 20062/1-7, 20063/3-9, 20063/11, GS 3209-3216) and six lithics (20036/10, 20063/2, 20063/10 and GS3208) were collected from DSP08-H12-01. The pottery consists of Late Classical-Early Hellenistic handmade pottery and transport amphorai from Herakleia, Sinope, and possibly Thasos and Chios as well as some unidentified wheelmade fragments of red clay. One handmade sherd may be dated to the Bronze Age and for two amphora sherds, a Byzantine and Medieval date has been tentatively suggested. Geomagnetic survey suggests the presence of 11 architectural complexes some consisting of one to three rooms, but at least four complex multi-room structures, and multiple household pits (Figs. 4.302-303). Several attempts at augering inside the structures were unsuccessful due to stones. The number of stone structures of relative complexity and the range of finds suggest a settlement of a substantial size of the Late Classical-Early Hellenistic period, although not all of the structures may be contemporary. If the curved feature in the northeastern part of the settlement is a double pen, the architectural elements appear to be consistent with DSP07-H01-01, DSP07-H02-01 and DSP07-H08-01; however, only one Bronze Age sherd was identified and the technique of constructing the foundation of standing blocks has not been observed at the Bronze Age sites.

**Relations to the surrounding area** DSP08-H12-01 is located 1 km due S of the settlement on Hill 1 (DSP07-H01-01) (Fig. 4.270). The kurgans on the southern part of the ridge of Hill 19 (Groups 1-2) are located less than 200 m W of DSP08-H12-01, although they are not at present visible in the viewscape of the site (Fig. 4.270). Kurgan Group 4 is located 350 m due N, and Groups 6-8 are located in the Uplands 1.3 km to the S.

*Units 20064-20065 (Hill 7)* (Fig. 4.304)
**Size** 1.4 ha

**Investigations** *Razvedki*, intensive and geomagnetic survey.

**Landscape type** Transition between Hillsides and Plateaus/Uplands.

**Location** The line runs approximately 150 m N-S in a lower area between Hills 25 and 7 near the Uplands just S of Foundation 63. The soil is deep and vegetation dominated by high thick grasses accounting for a very poor visibility. Only 100 m to the S the cultivated zone in the Uplands begins.

**Cartography** On the half-verst map, the area is crossed by two dirt roads running SW from Karlav. On the mid-20th century map, the area is indicated as cultivated. The Koltuchov map indicates three kurgans to the NE.

**Finds distribution** The original survey in June 2008 produced only five tiny sherds (20064/1 and 20065/1-4). The low number was probably due to poor visibility. A later visit when the vegetation was dry produced a grab sample of six sherds around the largest ash hill (GS3221), four of which were Early Modern, but also one handmade sherd. All finds are associated with the site.

**Sites** At least three rectangular stone structures and three possible ash hills are visible at DSP08-H07-03 (Fig. 4.305). Additionally, geomagnetic survey revealed a number of household pits and a cluster of three fireplaces (Figs. 4.306-307). The architecture and finds suggests an Early Modern settlement.

**Relations to the surrounding area** There is a dam (DSP08-H07-04) consisting of two roughly L-shaped earth banks in the shallow ravine 110 m SW of DSP08-H07-03 and probably associated with the site (Fig. 4.308); however, just 130 m to the W is a small scatter of sherds associated with the settlement on Hill 1 (DSP07-H01-01) (Fig. 4.147). It seems probable that DSP08-H07-03 was associated with the exploitation of the cultivated area to the S in the Uplands.

CHAPTER 5

# The material culture of the DSP landscape

*Compiled by P. Guldager Bilde*
*with contributions by P. Attema, S. Handberg, M.E. van Kruining,*
*M.J.L.Th. Niekus & K. Winther-Jacobsen*

In this chapter we shall present the material culture of the DSP landscape in terms of settlement and funerary architecture, pottery, and lithics, and though not material culture proper, the chapter ends with a brief discussion of the osseous remains found in three trial trenches. A more detailed discussion of the burial architecture, foremost the kurgans, is included in Chapter 6.5.

## 5.1 SETTLEMENT ARCHITECTURE | P. GULDAGER BILDE WITH CONTRIBUTION BY K. WINTHER-JACOBSEN

As part of the DSP, various settlement types featuring stone walls made of local limestone blocks were inventoried (see Appendix 1). Structures of cut and sometimes dressed blocks are, however, only found at the larger sites, such as the Černomorskoe Settlement, farmhouses around Černomorskoe including the Vetrenaja Bay farm, Panskoe I-IV (see Chapter 4.3.2), Skalistoe 5 (see Chapter 4.2.2), and Masliny 1. On the Hillsides, in many cases only stone slabs are visible on the surface revealing the presence of quite another type of architecture. Geomagnetic surveys conducted at a number of sites with such slabs have provided a reliable picture of both their architecture and layout. Below, the following features are discussed; pit houses, houses(?) with stone foundations, household pits, double pens and ash hills. Of the pit houses, only a single specimen was excavated, but the presence of further pit houses is suggested from geomagnetic survey, and they are known from excavations in other places. T. Smekalova has suggested that a circular anomaly (DSP07-H07-02 Feature 2) with an approximate diameter of 18 m is the imprint of a nomadic tent, a *jurt*. However, without excavation this cannot be verified. The presence of household pits and (double) pens for rearing animals are vivid testimonies to mixed farming at individual Hillside settlements. A final category of architecture to be discussed is burial structures. They were constructed from stone, mudbrick, and turf.

### 5.1.1 Pit houses | P. Guldager Bilde

Pit houses or dugouts, in Russian *zemljanki*, are huts partially dug into the ground. It is a widespread type of architecture especially in the northern hemisphere above 32° northern latitude. In the 20th century, they were common especially in South Russia between the rivers Dniepr and Bug (e.g. Feist 1913, 125), and many are the examples that exiles to the *gulag* lived in such dugout pit houses (e.g. Appelbaum 2003, 190). In antiquity, particularly in the Northern Pontic region, they were commonly employed by indigenous tribes as well as by Greek settlers. Because they are dug into the ground, temperature and humidity remain the same throughout the year, and in addition, they are well sheltered from the wind. Thus, less energy is required to maintain stable temperatures inside, which must have been of importance in windy Crimea.

On Hill 11 an assemblage of pit houses was located. One of the houses was fully excavated (DSP07-H11-02 Feature 9). It measures 5.7 x 4.1 m with a maximum depth of 1.27 m and with the entrance towards S. The house had several successive floor levels. Finds from the excavation make it clear that the house dates to the Late Classical-Early Hellenistic period (see Appendix 1). At Skalistoe 2, six or seven weak positive anomalies are possibly the traces of pit houses. Previously, this site was excavated by V.A. Kolotuchin, on which occasion house foundations of field stones were brought to light (Kolotuchin 1973; 1987). The settlement phases are dated to the Late Bronze Age as well as to the Hellenistic period, and we have no means of ascertaining to which period the pit houses belong.

At our current level of knowledge, they have been sparingly identified in the Tarchankut. Pit houses were excavated in the Late Bronze Age settlement of Baj Kijat 1 (Kolotuchin 2003, fig. 48) and Hellenistic pit houses have been excavated in Masliny, where they were connected with a number of household pits and an ash hill outside (to the E of) the Masliny (1) Settlement (Ščeglov 1978, 82-83). Similar structures have been found at Tarpanči (Ščeglov 1978, 82-83) and in the Černomorskoe Settlement. In the latter, they were connected with Late Scythian culture of the Late Hellenistic and Roman period (Užencev 2006, figs. 51-52).

## 5.1.2 Houses (?) with stone foundations | P. Guldager Bilde

In Hillside settlements, stone slabs were frequently encountered on the surface, which allude to the presence of stone architecture. Two types of walls have been observed: single- and double-faced built walls of local slabs and boulders and filled in with small-sized rubble. The stone is local limestone, mostly unworked or occasionally slightly chipped. Many of the structures appear to have had one or two rooms.

| Rectangular structures | | |
|---|---|---|
| Size category | $m^2$ | Number of structures |
| 1 | up to 49 | 28 |
| 2 | 50-99 | 18 |
| 3 | 100-149 | 9 |
| 4 | 150-199 | 10 |
| 5 | 200-249 | 5 |
| 6 | 250-299 | 4 |
| 7 | 300 and more (up to 1,320) | 8 |

Table 5.1. Number of structures according to size category.

| **Rectangular structures** | | | | |
|---|---|---|---|---|
| **Site** | **Size category** | **Feature no.** | **m2** | **Date range** |
| DSP07-H01-01 | 1 | 4 | 30 | BA, LC-EH |
| | | 17 | 30.4 | BA, LC-EH |
| | | 15 | 36 | BA, LC-EH |
| | | 12 | 44.8 | BA, LC-EH |
| | 2 | 3 | 51 | BA, LC-EH |
| | | 9 | 70 | BA, LC-EH |
| | | 13 | 88 | BA, LC-EH |
| | 3 | 14 | 103.5 | BA, LC-EH |
| | 4 | 10 | 152.5 | BA, LC-EH |
| | | 6 | 153 | BA, LC-EH |
| | | 16 | 157.5 | BA, LC-EH |
| | | 2 | 176 | BA, LC-EH |
| | 5 | 11 | 200 | BA, LC-EH |
| | 6 | 5 | 270 | BA, LC-EH |
| | 7 | 1 | 371.3 | BA, LC-EH |
| | | 7 | 486.4 | BA, LC-EH |

| Rectangular structures | | | | |
|---|---|---|---|---|
| **Site** | **Size category** | **Feature no.** | **m2** | **Date range** |
| DSP07-H02-01 | 1 | 2 | 28 | BA |
| DSP07-H02-02 | 1 | 2 | 49 | BA, LC-EH |
| | 2 | 1 | 99 | BA, LC-EH |
| | 3 | 3 | 100 | BA, LC-EH |
| | 4 | 5 | 160 | BA, LC-EH |
| | | 4 | 196 | BA, LC-EH |
| DSP07-H04-01 | 1 | 2 | 36 | LC-EH |
| | | 1 | 48 | LC-EH |
| | 3 | 4 | 109.25 | LC-EH |
| | 5 | 3 | 221 | LC-EH |
| DSP07-H04-02 | 6 | 2 | 280.5 | EM? |
| DSP07-H06-01 | 2 | 2 | 54 | ? |
| DSP07-H07-01 | 6 | 2 | 260 | BA?/LC-EH |
| DSP07-H07-02 | 4 | 1 | 150 | BA?/LC-EH |
| DSP07-H08-01 | 1 | 12 | 22.5 | BA, LC-EH? |
| | | 4 | 42 | BA, LC-EH? |
| | | 10 | 49 | BA, LC-EH? |
| | 2 | 3 | 60 | BA |
| | | 6 | 87.5 | BA, LC-EH? |
| | 3 | 7 | 115 | BA, LC-EH? |
| | | 1 | 136.5 | BA |
| | | 11a | 144 | BA, LC-EH? |
| | 4 | 9 | 165 | BA, LC-EH? |
| | 7 | 11b | 696 | BA, LC-EH? |
| | | 8 | 1320 | BA, LC-EH? |
| DSP07-H09-01 | 2 | 1 | 75 | EM |
| DSP07-H09-02 | 3 | 1 | 105 | EM? |
| DSP07-H10-01 | 1 | 6 | 35 | LC-EH |
| | | 4 | 36 | LC-EH |
| | 2 | 3 | 70 | LC-EH |
| | 7 | 2 | 340 | LC-EH |
| DSP07-H11-02 | 1 | 18 | 30 | LC-EH |
| | | 11 | 35.75 | LC-EH |
| | | 13 | 36 | LC-EH |
| | | 8 | 45 | LC-EH |
| | | 5 | 48 | LC-EH |
| | 2 | 12 | 63 | LC-EH |
| | | 9 | 70 | LC-EH |
| | 4 | 16 | 180 | LC-EH |
| | 5 | 14 | 225 | LC-EH |

| Rectangular structures | | | | |
|---|---|---|---|---|
| Site | Size category | Feature no. | m2 | Date range |
| DSP07-H18-01 | 2 | 3 | 50 | LC-EH |
| | 3 | 5 | 117 | LC-EH |
| | 5 | 2 | 216 | LC-EH |
| | 6 | 4 | 255 | LC-EH |
| DSP07-H22-03 | 2 | 8 | 68.25 | Modern |
| DSP08-H07-03 | 1 | 5 | 16.5 | EM? |
| | | 4 | 20 | EM? |
| | | 6 | 40 | EM? |
| | 2 | 3 | 63 | EM? |
| DSP08-H08-02 | 1 | 1 | 16 | EM |
| DSP08-H12-01 | 1 | 11 | 49 | LC-EH? |
| | 2 | 6 | 64 | LC-EH? |
| | | 3 | 88 | LC-EH? |
| | | 7 | 90 | LC-EH? |
| | 3 | 8 | 126 | LC-EH? |
| | 4 | 13 | 162 | LC-EH? |
| | 5 | 4 | 202.5 | LC-EH? |
| | 7 | 5 | 836 | LC-EH? |
| | | 2 | 912 | LC-EH? |
| DSP08-H13-01 | 1 | 1 | 6 | BA?/ LC-EH |
| | | 3 | 16 | BA?/ LC-EH |
| | | 4 | 16 | BA?/ LC-EH |
| | | 6 | 30 | BA?/ LC-EH |
| | | 5 | 36 | BA?/ LC-EH |
| DSP08-H16-01 | 2 | 3 | 52.5 | ? |
| | 7 | 2 | 720 | ? |

Table 5.2. Distribution of rectangular structures at DSP sites according to size.

Based on the measurements of the individual features in the site catalogue (Appendix 1), it has been possible to obtain an overall idea of the size variation of the structures.[47] Ca. 82 structures have been measured. We may conclude that over half of the buildings are less than 100 m$^2$ (Table 5.1). Considering the fact that they are structures in a rural setting, this is hardly surprising. However, there is a fair number of larger buildings with an area between 100 and 300 m$^2$, 28 in total as well as up to eight large to very large, complex buildings. It is difficult to form an opinion about how regular or irregular the individual structures are, because their current state of preservation make them appear more irregular than they probably are. In all likelihood, they are all living units. It is interesting to observe that at three sites, which we may consider complex, all size categories were encountered (DSP07-H01-01, DSP07-H08-01, DSP08-H12-01). The former is a site with Bronze Age as well as Late Classical-Early Hellenistic components, whereas the last two are exclusively dated to the Late Classical-

[47] Please note that the measures are obtained from the geomagnetic maps. It is likely that the area of the individual structures is a bit too large, because the outside of the walls are measured, and this outside border may be the representation of stones, which have fallen away from the house. Nevertheless, since the same method is used concerning all of the houses, the measurements can, at any rate, be compared.

Early Hellenistic period. It should also be noted that at site DSP07-H10-01, where a category 7 structure was noted, roof tiles were found as the only Hillside site (Appendix 2A-B). The latter structure was measured geomagnetically and with the Total Station. In all likelihood, this is a farmhouse with at least four rooms around two sides of a courtyard.

In one of the excavated stone-built structures, a hearth was discovered (DSP08-H14-01 Feature 2). It is constructed of stone slabs 3-5 cm thick, 5-10 cm high measuring 1.20 x 0.9 m. It is placed directly on the bedrock and thus pertains to the earliest phase of the structure. The finds made inside the hearth date to the Late Bronze Age (**Cat. 555-564**). Similar hearths were found in other Late Bronze Age settlements, such as Baj Kijat 1 (Kolotuchin 2003, figs. 41-46, 47.2).

## 5.1.3 Household pits | P. Guldager Bilde

A number of settlements featured household pits (Table 5.3). They could be seen as patches in the terrain of more intensive vegetation and they were verified in the geomagnetic survey. Household pits occur in clusters (50-200 m$^2$) near buildings and one or a few isolated pits were identified inside buildings; they also occur in larger clusters at some distance away from any architecture (200-500 m). Three pits were excavated (DSP07-H01-01 Feature 18, DSP07-H10-01 Feature 5). The mentioned pits are roughly circular in plan and piriform in section with a concave bottom. One is only 0.47 m deep with a diameter of the mouth of 0.86 m and a maximum diameter of 0.96 m. The second one has the same shape; it is 0.76 m deep with a 0.9 m diameter of the mouth and a maximum diameter of 1.47 m. In the bottom level were found grains of cereals.

| Site | Feature | | Approx. area m$^2$ | Date | Excavation |
|---|---|---|---|---|---|
| DSP07-H01-01 | 18 | Large cluster of ca. 32 pits | 202 | LC-EH? | Yes |
| DSP07-H01-01 | 20 | Large cluster of pits | 268 | LC-EH? | |
| DSP07-H04-02 | 1 | Cluster of pits | ? | Early Modern? | |
| DSP07-H07-01 | 4 | Cluster of ca. 12 pits | 84 | BA?/LC-EH | |
| DSP07-H08-01 | 5 | Cluster of pits? | 560 | BA, LC-EH? | |
| DSP07-H10-01 | 5 | Cluster of pits | 55 | LC-EH | Yes |
| DSP07-H10-01 | 7 | Cluster of pits | 47 | LC-EH | |
| DSP07-H11-02 | 10 | Cluster of pits | 197 | LC-EH | |
| DSP07-H11-02 | 15 | Cluster of pits? | 168 | LC-EH | |
| DSP07-H11-02 | 17 | Cluster of pits? | 76 | LC-EH | |
| DSP07-H11-02 | 6 | Diffused cluster of pits | 200 | LC-EH | |
| DSP07-H18-01 | 1 | Cluster of pits | 80 | LC-EH | |
| DSP07-H18-01 | 6 | Cluster of pits | 95 | LC-EH | |
| DSP08-H12-01 | 12 | Large cluster of pits | 300 | LC-EH/? | |
| DSP08-H12-01 | 14 | Large cluster of pits | 500 | LC-EH/? | |

Table 5.3. Distribution of pits at DSP sites in terms of number, size, and chronology.

All three excavated pits yielded finds of the Hellenistic period. This is in concordance with the general picture that pits on all locations are foremost associated with Hellenistic surface finds. The presence of storing facilities shows that the settlements in all likelihood were permanent or semi-permanent.

## 5.1.4 Double pens | P. Guldager Bilde

In four individual localities double pens were recorded during the geomagnetic surveys (DSP07-H01-01 Feature 8, DSP07-H02-01 Feature 1, DSP07-H06-01 Feature 1, DSP07-H08-01 Feature 2; Table 5.4). The last-mentioned was further investigated by means of a 2 x 2 m trial trench. This provided an idea about the construction of these pens. It had a broad double stone wall built of large blocks, which was filled-in with rubble.

The double pen has been and still is in use among herdsmen in regions of Central, Southern, and Eastern Europe such as Ukraine, Romania, Bulgaria, Greece, Hungary, Slovakia, and Serbia (Vuia 1964, 60; Podolak 1987; Chang 2000; Chang & Tourtellotte 1993, 256, fig. 6). In Slavonic language, the small pen is called *strunga* and its function is to gather the sheep prior to milking, and when they are milked they are let into the larger enclosure connected with the *strunga*, where they also sleep. Frequently, the pens are made of perishable materials, such as brushes or sticks, leaving no traces for survey archaeologists to discover. However, they could also be constructed from field stones as is the case with the ones found in the DSP study zone. Their significant size and quite monumental construction technique makes it likely that they were intended for more or less permanent, long-term use.

| | Location of *strunga* | Approximate Ø | |
|---|---|---|---|
| DSP07-H01-01 Feature 8 | W, upslope | 21.5 m | 38 m |
| DSP07-H02-01 Feature 1 | S, upslope | 28 m | 48 m |
| DSP07-H06-01 Feature 1 | SW, upslope | 28 m | 45 m |
| DSP07-H08-01 Feature 2 | SW, upslope | 22 m | 30 m |

Table 5.4. Location of *strunga* at DSP sites.

According to Podolák, when pens are placed on a slope, the *strunga* is always uphill in order to avoid accumulations of dung and urine in the milking area thus keeping the milking area clean. This is also the case with the four double pens found in the DSP survey.

In connection with other settlements on the slopes were localized further curvilinear structures, which may also be pens (DSP08-H12-01 Feature 9 Ø ca. 20 m; DSP08-H16-01 Feature 4, last-mentioned concentric and thus perhaps a double pen).

The excavation of DSP07-H08-01 Feature 2 provided a date in the Late Bronze Age for this particular pen. Late Bronze Age sherds were also collected in the vicinity of DSP07-H01-01 Feature 8 and DSP07-H02-01 Feature 1, but in connection with DSP07-H01-01 Feature 8 also Hellenistic pottery was found. Taken into account that this type of pen is known throughout time and space in animal-rearing societies it is problematic to couple them exclusively to the Bronze Age period, even though this is the date suggested by the majority of our find assemblages.

## 5.1.5 Ash hills | P. Guldager Bilde

Even though ash hills (in Russian *zol'niki*) are not an architectural element proper, they are, nevertheless, a significant feature of the settled DSP landscape and therefore they merit a brief discussion in this section. In general, ash hills are characteristic of Northern Black Sea sites. In the DSP study zone we find them connected with all periods of activity. Ash hills are common in settlements of the Late Bronze Age Sabatinovka culture (Gerškovič 1999, 43-45; Chişinău 2005). A large ash hill was found at Baj Kijat 1 and a similar ash hill was also observed at one of the two major Bronze Age settlements identified during the survey (DSP08-F06-01). In addition, several of the Bronze Age sites on the slopes were associated with ashy soil (DSP07-H01-01, DSP07-H02-01 and DSP07-H08-01). Ash hills were also connected with sites of the Hellenistic period such as the Černomorskoe Settlement, Burun Eli 4, Masliny 2, Panskoe I, Skalistoe 5, and DSP08-F04-01 Feature 3, and they are especially characteristic of the Early Modern, Tatarian sites, e.g. Kurama Kostel', Panskoe II, and DSP08-H07-03 Feature 1 and 2. The ashy soil of the ancient and Early Modern settlements can be clearly distinguished on Google Earth (e.g. Figs. 4.148, 4.167, 6.11).

In scholarly literature, their function has been debated. Currently, many researchers are of the opinion that the Greco-Roman *zol'niki* are to a large extent the remains of cultic activity (e.g. Rusanova 1998; Vinogradov, Butjagin & Vachtina 2003, 817). This may be the case in some instances, an example being, e.g., the large Cape Takil' ash hill in eastern Crimea, which is located far from any settlement (Vinogradov, Butjagin & Vachtina 2003). However, in the DSP study zone, no ash hill can readily be connected with cultic activity. On the contrary, in most cases they are most probably to be understood as dumps of garbage and other refuse including ash from the hearths. Perhaps brushes and shrubs were used as fuel, when houses were heated and food was prepared. These types of fuel produce much more ash, because when compared with proper wood the calorific value is low wherefore large volumes are needed in order to obtain the required heat. In addition, ash hills may represent clean up of a larger area after a fire, which is probably the case with some of the ash hills in Panskoe I and with the ash hill over part of the Skalistoe 5 farmhouse.

### 5.1.6 Hydrological installations (wells and dams) | P. Guldager Bilde & K. Winther-Jacobsen

Three types of hydrological installations have been recorded: wells, dams and possible check dams, but no actual cisterns were identified. The most prominent well is the Early Modern stone well in the village of Sneznoe, a village occupied after 1897 (Fig. 5.1). The Early Modern villages such as Snežnoe tend to be located along the geographical line where the water-bearing Sarmatian limestone bedrock is closest to the surface and consequently where access to the highest quality water was most accessible (Smekalova 2007, 83-85). Due to the short duration of the project and their main locations in the occupied villages, the wells were not systematically investigated. The 1890s map indicates the location and number of wells in each of these villages with a K (that is, *kolodec*, a drawing well) or with a dotted circle. There may have been difference between the two, since at some sites both occur (Table 5.5).

| | Dotted circle | K (*kolodec*) | Dotted circle with K |
|---|---|---|---|
| Coastal zone/Lowland Ridge (from W to E) | | | |
| Ak Mečet | | 2 | |
| Šeichlar | | 2 | |
| Burun Eli | 7 | | |
| Tubaka | 3 | | |
| Baj Kijat | 3 | 1 | |
| Časilmaj/Čigiltaj | 3 | | |
| Kirgiz Kazak | | 1 | |
| N branch of the Vodopojnoe Ravine (from W to E) | | | |
| Kirleut | 11 | 1 | |
| Kurman Adži | 7 | | |
| Tok Džol' | 1 | | |
| Otus | 4 | | |
| Dauldžar' | 2 | 1 | |
| Kul' Džakin | 1 | | |
| Baim | 1 | 1 | |
| Tatarian Čongurči | 1 | | |
| Foot of Hillsides (transition between Pediment and Hillsides, from W to E) | | | |
| Aldermenskaja Skala | | | 1 |
| Karlav | | | 3 |
| Musali | | | 1 |
| Kipčak Učkuju | | | 1 |
| Kurama Kostel' | | | 2 |

Table 5.5. List of wells occurring in the Tatar villages according to the map of the 1890s.

The toponym of Kipčak Učkuju means the Kipčak village of three wells (Jankowski 2006, 1125-1126). The village is known at least since the time of Mukhin's map of 1817; on the 1890s map, only one well is noted.

Two types of dams were recorded: one consisting of stone walls constructed across a small ravine with relatively steep sides (DSP08-H02-03). The walls are double faced and constructed of standing slabs. The bedrock behind the dam was flat, which is the natural structure of this local type of limestone. The other type of dam consists of two L-shapes earth banks constructed across a shallow ravine on the top of the ridge (DSP08-H07-04). No finds were directly associated with either structure, but the latter is spatially associated with DSP08-H07-03, a small Early Modern settlement, and the former is spatially associated with a Bronze and Late Classical-Early Hellenistic settlement (DSP07-H02-01-02).

A course of stone on the shallow bank of a ravine at the bottom of the Hillsides associated with DSP07-H11-01 may be the remains of a check dam (Fig. 4.198). However, it was not associated with finds and may also be associated with Early Modern and Modern activities at the same location.

## 5.2 FUNERARY ARCHITECTURE | P. ATTEMA

In the campaigns of 2007 and 2008, one of the teams mapped and documented kurgans in the survey area on Hill 19. Here we discuss their morphology and architecture. On basis of this investigation, a catalogue was compiled containing standard descriptions, photographs, a schematic plan and two cross-sections perpendicular to each other. At kurgans where geomagnetic surveys were carried out, maps and interpretation of the results were added in the catalogue entries. In all, 31 kurgans were recorded with Total Station in order to obtain digital terrain models (DTM's) of each individual kurgan and in relation to each cluster. Geomagnetic surveys were carried out at nine of these (Fig. 4.204-227).

The kurgans on the surveyed slope appear in eight distinct groups based on intermediary distance (DSP07-H19-01-06, DSP08-H19-07-08). Within each group, diameters and height of kurgans varied. Their present form, however, will in all cases have been altered by excavation, looting and/or erosion. It is thought that all kurgans would have had a so-called krepis wall that would have held the soil of the mound in place. If true, this would mean that all kurgans mapped by us, originally were smaller in diameter, as we must assume that soil from the mound has washed over the krepis wall to form the sloping mounds that we see today. As can be noted in the descriptions in the catalogue, at a number of kurgans, indeed, remains of a possible krepis wall were noted at the surface (most significantly DSP07-H19-03 Kurgan 10), although they nowadays rather appear as stone circles and not as carefully constructed retaining walls. But even allowing for considerable changes over time to have happened to the original morphology of the kurgans on this slope, we may note differences within clusters. The cluster comprising DSP07-H19-01 Kurgans 1-4, for instance, has three larger and higher (1-3) and one clearly smaller and lower specimen (4). This height/diameter differentiation applies to all other groups except for the cluster made up of DSP07-H19-05 Kurgans 16 and 17 both of which show approximately the same diameters and height in their present form. As far as the grave architecture is concerned, we must rely on the information furnished in excavations carried out during the North-West Crimean expedition (pers. comm. S. Koltuchov). In the campaigns between 1970 and 1973, three tombs were excavated (DSP07-H19-01 Kurgans 1-3). Two of these (1 and 2) showed the mounds to have an approximately E-W orientated stone chamber that could be entered by means of a N-S orientated dromos. Chamber and dromos are constructed of standing orthostats as well as flat-lying stones. From the same orientation that was adopted by the mound builders for the chambers in Kurgans 1 and 2 we may deduce that the kurgans were either constructed at the same time, or that chamber and/or dromos remained visible so that the next kurgan could be placed in line with an existing one and could be reopened for additional burials. For some kurgans it was possible to reconstruct the orientation on basis of still protruding headstones (DSP07-H19-03 Kurgans 8 and 10). These correspond with Kurgans 1 and 2. The described type and its orientation is said to be typical for Northwestern Crimea and is interpreted as a family tomb where multiple burials took place.

On basis of the said excavations, it is assumed that, although not many datable artefacts had been found in the chamber graves, these graves date between the late 5th and late 4th century BC. Such a dating seems now confirmed by ceramic finds from a trial trench dug during this campaign SW of DSP07-H19-01 Kurgan 3 at a spot where a clear anomaly had turned up in the geomagnetic survey. As far as the dating is concerned, we should realize that no stratigraphical relationship was established between test pit and kurgan. Reportedly, the artefacts collected from the mounds in the excavations during the 1970s consisted of iron knives and arrow heads. Apparently, the excavated mounds had already been robbed of other materials. The sherds found in the test pit may tentatively be connected to ritual practice: it is possible that feasting occurred around the burial mounds, and the amphora sherds in the test pit could testify to this (pers. comm. T. Smekalova).

# 5.3 POTTERY | S. HANDBERG & K. WINTHER-JACOBSEN

## 5.3.1 Introduction | S. Handberg

The total number of pottery fragments pertaining to all the chronological periods represented in the DSP ceramic material, excluding modern finds, amounts to 9,725 fragments. In addition, the 10 excavations undertaken on the Hillsides produced 1,198 fragments (Table 5.6). All rims, bases and handles as well as fragments with characteristic decoration were recorded as diagnostic. The total share of diagnostic pieces among the survey finds makes up 820 fragments, i.e. 8.4%, the corresponding number of diagnostic pieces from the excavations constitutes 158 fragments, i.e. 13.2%.

| Excavation | Number of frs. |
|---|---|
| Hill 01-1 trench 1 | 124 + 2 oven frs. |
| Hill 01-1 trench 2 | 19 |
| Hill 19 kurgan 3 | 9 |
| Skalistoe 5 | 540 + 1 oven fr. |
| Hill 10-1 trench 1 | 108 + 1 oven fr. |
| Hill 10-2 trench 1 | 28 |
| Hill 8 trench 3 | 20 |
| Hill 8 trench 4 | 101 |
| Hill 14-1 trench 5 | 98 |
| Hill 11-2 trench 6 | 151 + 1 oven fr. |

Table 5.6. Distribution of pottery fragments from the DSP excavation areas.

The chronological distribution of the pottery makes clear that the majority of the finds fall within the Late Classical and Early Hellenistic periods (Table 5.7). Hellenistic finds datable to after the 3rd century BC are very few and although some fragments have been tentatively dated to the Byzantine and Medieval periods, there is a general lack of pottery in the survey area until the Early Modern period, i.e. the 18th-19th century AD. The second largest concentration belongs to the prehistoric period. A fairly large number of Middle and Late Bronze Age fragments were recognized, although a Bronze Age date for 746 of these is not entirely secure (see below, Section 5.3.2.1).

| **Period** | **Approx. Dates** | **% of pottery** |
|---|---|---|
| MBA-LBA | 18th-12th century BC | 19.9 |
| LC-EH | 4th-3rd century BC | 65.5 |
| Byzantine/Medieval | 6th-13th century AD | 0.3/1.3 |
| Early Modern | 18th-19th century BC | 5.1 |
| Unknown date | ? | 9.6 |

Table 5.7. Chronological phases with distribution of the DSP pottery.

As described in Chapter 2.4, all finds were recorded according to their primary function and divided into 'function groups': '*architecture*', '*transport amphorai*, '*tableware*', '*handmade pottery*', '*utility ware*', and '*lithics*'. The three groups '*tableware*', '*utility ware*' and '*transport amphorai*' correspond roughly to the tableware, processing and storage functional groups employed on some Mediterranean surveys.[48] These groups were supplemented by an additional category termed '*indeterminable*' that contains fragments, which could not be securely placed in any of the functional groups. Because no other architectural remains were found during the survey, '*architecture*' consists exclusively of fragments of roof tiles. Due to the difficulties in determining the specific functions of the fragmented handmade pottery, it was decided to catalogue these simply as '*handmade pottery*'. The '*utility ware*' group includes mostly pottery used in households, such as *louteria*, basins and cookingware, but storage vessels like *pithoi* were also included in this group, however, the numbers are very small. The distinction between transport amphorai and storage vessels is deceptive, as amphorai in the Northern Black Sea area are often reused for the storing of for instance grain as was the case in Panskoe I U6.[49]

The overall distribution of the survey pottery according to functional groups (Table 5.8) shows a remarkable preponderance of handmade pottery with 4,140 recorded fragments. The second largest group is the transport amphorai with 1,870 fragments. The tableware amounts to 697 fragments, 252 of which belong to the Early Modern period, whereas only a very limited number of utility ware, mostly of Early Modern date, was identified.

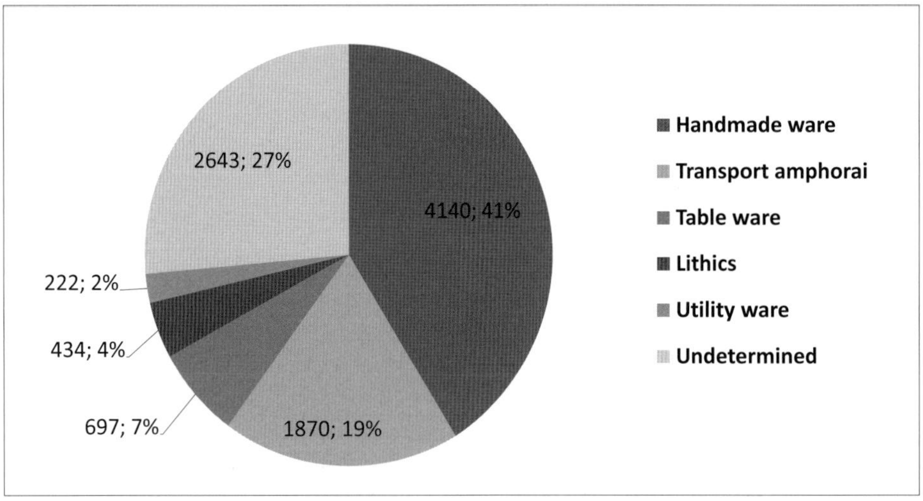


Table 5.8. Overall distribution of the survey pottery according to functional groups, all periods.

Considering the absolute sizes of the different groups it is important to keep in mind the large chronological span of the finds, which covers the period from the Middle Bronze Age ca. 1800 BC to the Early Modern pottery of the 18th-19th centuries AD. The predominance of the handmade pottery is to a large extent caused by two factors; first of all, the prehistoric pottery exclusively consists of handmade pottery and the two major Bronze Age sites DSP08-F03-02 (Skalistoe 2) and DSP08-F06-01 contributed with a total of 1,423 fragments. Secondly, the general low firing temperature and hence friable composition of the handmade fabrics increases the breakage rate of these vessels.[50]

The ascription of fragments to the different function groups was in most cases rather straight forward, although the number of fragments in the *indeterminable* group is substantial with 27%. The handmade pottery was easily identifiable among the wheelturned pottery, whereas the dating of this group is more problematic (see more below, Section 5.3.3.5.1). A characteristic feature of the composition of the DSP pottery assemblage is the lack of coarseware fabrics, which are found in abundance on surveys in the Mediterranean. Taken together, the handmade pottery and transport amphorai constitute 62% of the collected Late Classical/Early Hellenistic material. The large group of undetermined fragments only includes wheelmade pottery and consists almost exclusively of fragments of levigated red or reddish brown clay sometimes with mica or pyroxene inclusions.

48 For the development of the typology based on the Northern Keos Survey data, see Whitelaw 1998; for the Laconia Survey, see Shipley 2002; for the Boeotia Survey, see Bintliff, Howard & Snodgrass 2007, 39-42.

49 Cf. Januševič & Ščeglov 2002, the Chersonesan amphora Cat. Ad 10.

50 Firing temperatures above 800° Celsius have, however, been recorded for some of the handmade pottery from Panskoe I U/6, see Kovnurko 2002, 320-321 (group 2).

## 5.3.2 The Middle and Late Bronze Age pottery | S. Handberg

The Late Bronze Age period of the Crimea is characterised by several 'cultures', the most prominent of which are the later period of the Srubnaja 'culture', the Sabatinovka and the Belozerka 'cultures'. It was previously thought that the Sabatinovka 'culture' of the lower Dnieper area and the Crimea developed out of the Srubnaja 'culture'. Recently, this assumption has been generally abandoned, and the Sabatinovka 'culture' is now seen as an autonomous group (Gerškovič 1999, esp. 75-77). It is roughly divided into three periods, an early period that overlaps with the later period of the Srubnaja 'culture', a middle 'independent' period and a late phase, which overlaps with the Late Bronze Age/Early Iron Age Belozerka 'culture' (Toščev 2007, 194-195).

Thus, the Srubnaja and the Sabatinovaka 'cultures' are now viewed as more or less synchronous groups occupying somewhat different geographical areas. In addition to these, a few groups are defined in terms of pottery assemblages like the so-called 'merhwulstkeramik' (*Mnogovalikovoj keramiki*), and the Kamenka-Livencovka group from eastern Crimea. The differentiation of the material assemblages, especially the ceramic material, has proven particularly difficult and ascriptions are still contested (Toščev 2007, 173 with further references).

Most of the pottery recovered during the survey is related to either the Sabatinovka or the Kamenka-Livencovka groups. The material cultural expression of the Sabatinovka group, especially the pottery, shares features with the Noua 'culture' of the western Black Sea area and 'cultures' of western Siberia.[51] However, recent opinions regard it as having developed out of the 'Mehrwulstkeramik' tradition. The Kamenka-Livencovka group in Crimea is believed to have played a major role in the formation of the Sabatinovka pottery, and it has even been proposed that the group first originated here.[52] The recognition that there was a close link between the 'Mehrwulstkeramik' and the Sabatinovka pottery departs from earlier opinions that the two groups were unrelated.[53]

The scholarly literature on the Bronze Age of the Northern Black Sea area and the Eurasian steppe zone in general is extensive, even so, no single coherent classification of Late Bronze Age ceramics exists. Ceramic classifications are still largely based on eponymous sites, like the Sabatinovka site in the Lower Dnieper, the Kamenka-Livencovka group on the Kerč Peninsula and the Early Iron Age Kizil Koba Cave in the Chatyr-Dag Mountain in the Crimean Foothills. Most problematic is the lack of a clear typological differentiation between the different ceramic assemblages and the inconsistencies in the use of terminology and the taxonomy for the different types of ceramics. The general low level of rigorous taxonomic uniformity in descriptions of especially pottery, but also burial types has blurred the borders of the specific expressions of the material culture associated with the different 'cultural' groups.[54] The characteristic features of the Sabatinovka pottery group is not very clear from the existing literature as few synthetic works exist. A recent synthetic account and typology has been proposed by J.P. Gerškovič, which has been followed in the DSP classification to the extent it has been possible to ascribe fragments to specific types (Gerškovič 1999). The lack of common identifying traits for the Sabatinovka pottery is partly caused by the lack of closed deposits and stratigraphy. Also, with the exception of Kirovo, none of the relevant sites for the later period have been extensively excavated (Toščev 2007, 197). Similarities among the groups are profound and V.A. Kolotuchin has noted the difficulties in differentiating between the different 'cultural' groups of Bronze Age Crimea on account of their material culture (Kolotuchin 2003, 10). The Late Bronze Age period in the Crimea was most likely permeated by a 'cultural' osmosis. Thus, the development can only be traced in broad terms.

In the Late Bronze Age period the Northwestern Crimea and the Tarchankut Peninsula in particular, is characterised by settlements of the Sabatinovka and the Belozerka 'cultures' best represented by the two larger settlement of Baj Kijat and Burun Eli not far to the NE from the survey area (Kolotuchin 2003). In fact, almost all settlements of the Late Bronze Age in the Crimea are now ascribed to either the Sabatinovka or Belozerka 'cultures' (Toščev 2007, 175 with further references). However, the Bronze Age pottery from the DSP has close connections to the eastern Crimea with several parallels to the material from Kirovo on the Kerč Peninsula as well. A connection between the western and eastern Crimea is furthermore supported by the unique finds of moulds for a sickle of similar type found at both Burun Eli and Kirovo (Toščev 2007, 191). The above-ground, stone constructions for houses as well as the ash hills are typical of settlements of

---

[51] For a good short overview with references to the relevant literature, see Parzinger 2006, 536-539.

[52] See, e.g., Toščev 2007, 173 with further references, but see also Gerškovič 1999, 79.

[53] Leskov & Merpert 1967, 177-180. For recent discussions of the origin of the Srubnaja cultural group and its relationship to the Sabatinovka group, see Kolotuchin 2003, 6-10; Gerškovič 1999, esp. 75-77.

[54] See e.g. Kolotuchin 2003, 41-42 and Gerškovic 1999, 47-50 for this problem.

the Sabatinovka 'culture' and both features were identified at Bronze Age sites in the DSP survey area (see Sections 5.1.2 and 5.1.5 above).[55]

5.3.2.1 The DSP Bronze Age pottery

A total of 1,701 fragments of Bronze Age pottery were found during the survey. The majority of the prehistoric pottery belongs to the Late Bronze Age period, i.e. ca. 16th-12th century BC, and only 98 fragments could be dated to the Middle Bronze Age exclusively. However, the distinction between Middle and Late Bronze Age pottery is not always straightforward and 63 fragments have been dated broadly to the Middle or Late Bronze Age.

The very fragmentary state of the material from the survey poses a considerable problem in comparing individual pieces to better preserved, published examples.[56] Some general observations on the Bronze Age pottery can be made. The clay of the Middle and the Late Bronze Age is often described as having a black core caused by lack of oxidization and a low firing temperature. Sand and lime particles as well as organic material are typically added or naturally occurring in the clay.[57] The bodies of vessels of the Middle and Late Bronze Age are sometimes covered with horizontal striations created by tools used to dress the surfaces.[58] Smoothed or polished surfaces can be found during the Middle and Late Bronze Age, but are perhaps more frequent in the Middle Bronze Age. During the initial recording of the survey pottery, we encountered the problem of securely identifying Bronze Age fragments and especially body fragments. This problem arose because Late Classical and Early Hellenistic handmade and wheelturned pottery often appeared on the same sites as diagnostic Bronze Age material.[59] However, as more diagnostic Bronze Age fragments emerged, it became increasingly easy to distinguish between the Late Classical/Early Hellenistic handmade pottery and the Bronze Age pottery. Some features, such as the striations seen on **Cats. 326** and **454** and the smoothed or polished surfaces on **Cats. 268**, **668** and **674** became more and more prominent and were never observed on the diagnostic fragments belonging to the later periods. It was also observed that the wall thickness of the handmade fragments generally tends to be larger in the Bronze Age pottery. The most characteristic general feature of the Bronze Age pottery is, however, the fabric. The core of the Bronze Age fragments is almost always dark grey or black. The texture often appears somewhat granular, and there is usually a relatively high inclusion of sand. The colour of the surface of the pottery provides another indication of a Bronze Age date. Whereas the Late Classical and Early Hellenistic handmade pottery often has red or reddish brown surfaces (see Section 5.3.3.5 below), the Bronze Age pottery predominantly has a greyish or sometimes black surface. In some cases the surface is light, generally ranging from pink (5YR 8/4) to a pale brown (10YR 7/4), but the colour of the surface can vary from light to grey on the same fragment as seen for instance on **Cat. 780**. Even with these criteria, 746 fragments have only tentatively been allocated a Bronze Age date (marked as *Bronze Age?* in the finds lists).

Some characteristic Bronze Age fragments among the DSP finds can be singled out. A few fragments (**Cats. 295**, **298** and **451**) have a characteristic yellow surface, which is smoothly polished. This group finds clear parallels in the late Middle Bronze Age pottery from Planerskoe I, where they are often decorated with impressed cord decoration (Toščev 2007, 110). A similar decoration is seen on the DSP fragment **Cat. 298**). The use of impressed cord decoration was especially favoured in the Kamenka-Livencovka group and is often found on fragments at Kirovo in layers contemporary to the Planerskoe fragments (Gerškovič 1999, 78; Leskov 1970, fig. 3). The lower layers from Planerskoe I and Kirovo as well as the settlement of Kamenka, and hence the Kamenka-Livencovka group, are ascribed to the Catacomb 'culture', i.e. the Middle Bronze Age (Toščev 2007, 143, 160). However, the practice of using impressed cord decoration continued into the Late Bronze Age period (Gerškovič 1999, 79).

Fragments with deeply incised furrows like **Cats. 672** and **779** are similar to finds from the upper layer at Planerskoe I and typical of the post-Catacomb period, i.e. the transition from Middle to Late Bronze Age. Similar pieces are also found at Kirovo and Kamenka in the eastern Crimea (Leskov 1970, fig. 3.5-6, 12-14; Rybalova 1974, fig. 14.1,2). A few

[55] Gerškovič 1999, 43-45. For more generally on the Bronze Age ash hills, see Chişinău 2005.

[56] See Kolotuchin 2003, 41 for the problem of identifying fragments from settlements.

[57] Cf. e.g. Toščev 2007, 100 and 153 for a description of the clay fabric from Planerskoe I and Kimmerik.

[58] Rice 2005, 136-141 for the technique. See, e.g., Gerškovič 1999, 8 and 48 and Toščev 2007, 129, 190 for examples of this practice.

[59] A similar problem of dating the handmade pottery was encountered at the site of Bezymyannaya near Chersonesos, where the handmade pottery has variously been suggested to date from the Hellenistic period until the eighth or ninth century AD, see Arthur 2000, 46-47.

fragments with heavily combed surfaces (e.g. **Cats. 287** and **665**) have been found during the survey, and a cup with similar decoration comes from the robbed kurgan 3 in field DSP08-F04-04 (**Cat. 250**). Clear parallels for this type of decoration are found at Planerskoe I, and it is most typical of the Middle Bronze Age.

The largest group of the diagnostic Late Bronze Age fragments found during the survey belongs to J.P. Gerškovič's 'Type T' (*Töpfe*) or "Type VT" (*Vorratsbehälter*) of the Sabatinovka group (Gerškovič 1999, 50-59). Both of these types are characterized by tall pots with out-turned rim and a rather broad neck. There is often a horizontal relief moulding on the exterior below the rim, which is either left plain or furnished with impressed or incised decoration. Since the diameter of the rim is the decisive factor for discriminating between the two types, ascriptions are difficult when dealing with fragments. Although with minor variations in the shape of the rim, the same types are also common at the two larger nearby Late Bronze Age sites of Baj Kijat and Burun Eli (See e.g. Kolotuchin 2003, fig. 49-52).

5.3.2.2 The shift to Belozerka pottery

The transition from the Late Bronze Age period to the Iron Age is connected with the emergence of the Belozerka 'culture'. Although distinguishing between burial mounds of the Sabatinovka and the Belozerka 'cultures' is problematic (Toščev 2007, 191), the Belozerka pottery shares some distinctive features like the tall cylindrical necks of many of the larger vessels and the rounded or slightly biconical lower parts (Parzinger 2006, 538). The Belozerka pottery is present on the Tarchankut Peninsula at Baj Kijat (Kolotuchin 2003, 42-44, 120-121 fig. 53-54). Chronologically, the beginning of the Belozerka 'culture' is traditionally dated to the 12th century BC continuing until the 10th or 9th century BC. No fragments among the DSP pottery can positively be associated with the Belozerka pottery.

## 5.3.3 Pottery of historical antiquity | S. Handberg

Almost all the finds datable to the period of Greek presence in the survey area can be dated to the Late Classical and Early Hellenistic period (see above, Section 5.3.1). In total, 6,322 fragments can be dated to this period. Thus, the main period of activity in the historical period in the survey area was the 4th century BC. This chronologically rather narrowly defined period is confirmed by the finds from the excavations (Appendix 1). The functional distribution is presented in Table 5.9.

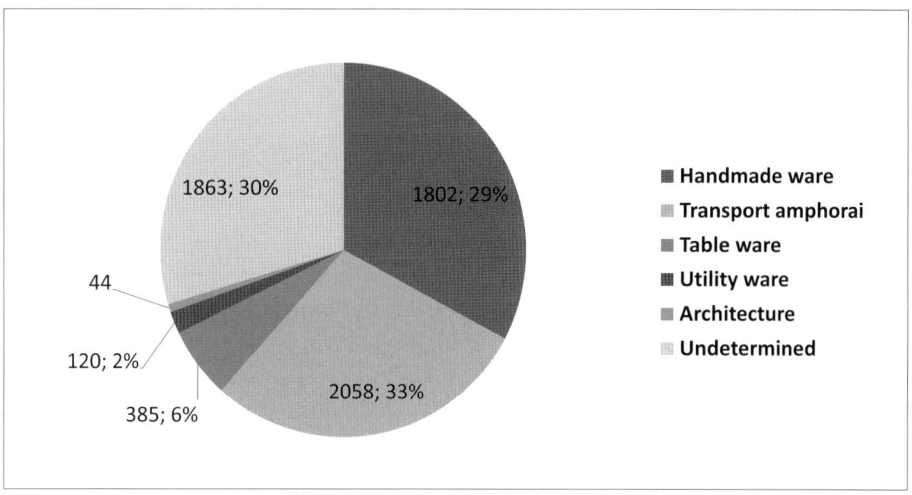


Table 5.9. Distribution of the survey pottery according to functional groups, historical periods.

Unfortunately, comparable information on the distribution of pottery in the rural landscape of the entire western Crimea is scarce.[60] Although 5th century BC finds have been recorded in the DSP area, the number is negligible compared to the evidence from the 4th century BC (see Section 5.3.3.6). On the Herakleian Peninsula in the home chora of Chersonesos,

[60] Kutajsov 2006 has a discussion of the chora of Kerkinitis, but without any mention of the distribution of pottery. The surveys in the Chersonesan home *chora* are generally unpublished, see Zubar' & Kravčenko 2003.

the majority of the pottery dates after the middle of the 4th century BC and continues into the 3rd century BC, but in some cases, as in the building on land plot 5, the assemblage is confined to the 4th century BC.[61] However, a substantial occupation from the late 5th and first half of the 4th century BC is recognized at Panskoe I (Chapter 4.3.2.1). Apart from the different functional groups presented below, mention should also be made of a terracotta egg (**Cat. 778**) found in 2008 in Field 2 and apparently not associated with a site. A similar terracotta egg was found in room 13 in building complex U6 at Panskoe I and the excavators suggested that it could be associated with the worship of Asklepios or Hygieia but the egg was also used in Dionysian rituals.[62]

5.3.3.1 Transport amphorai

The most reliable type of pottery in respect to dating is the transport amphorai. Out of the 1,802 identified amphora fragments datable to the Late Archaic through Hellenistic periods, four major provenances can be recognized: Herakleia Pontike, Chersonesos, Sinope and Thasos (Table 5.10). Apart from these, a few other less prominent groups are present too; those are Chian, Mendean, and Knidian productions. However, a substantial number of fragments could not be attributed to known production centres (Table 5.11). In total, 21 different amphora fabrics were recorded although most groups only included a small number of fragments. The largest group of amphora fragments with an unknown provenance consists of fragments of red fabric. 322 fragments were ascribed to this group although the group is not strictly speaking homogenous and includes fragments both with and without quartz and feldspar (but no inclusions of mica or pyroxene).

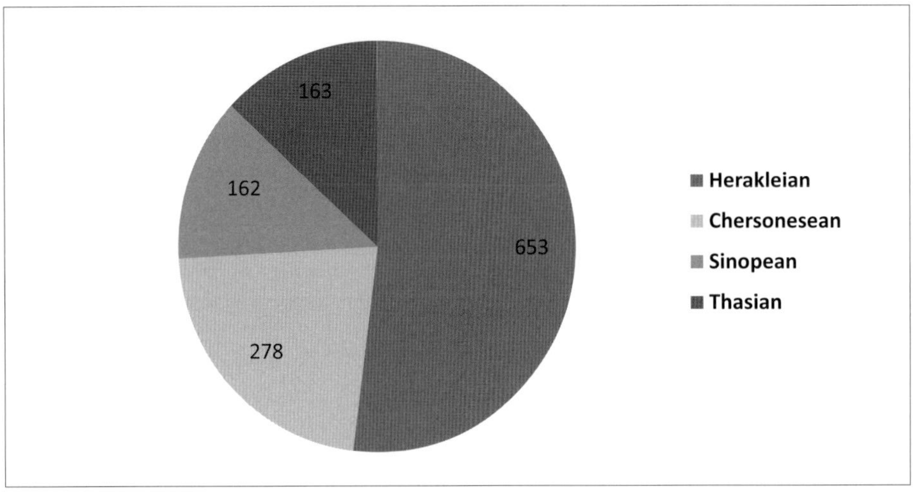


Table 5.10. Distribution of major amphora types of known production place.

[61] For land plot 5, see Nikolaenko 2003, 136; see also more generally Carter et al. 2000, 714.

[62] See Ščeglov 2002b, 221 Cat. G20 and for the same terracotta in the same publication Rogov 2002, 274, Cat. M22. For dedications of eggs to Dionysos see Plut. Quaest. Symp. II.3.1 (discussed in Harrison 1992, 627-628).

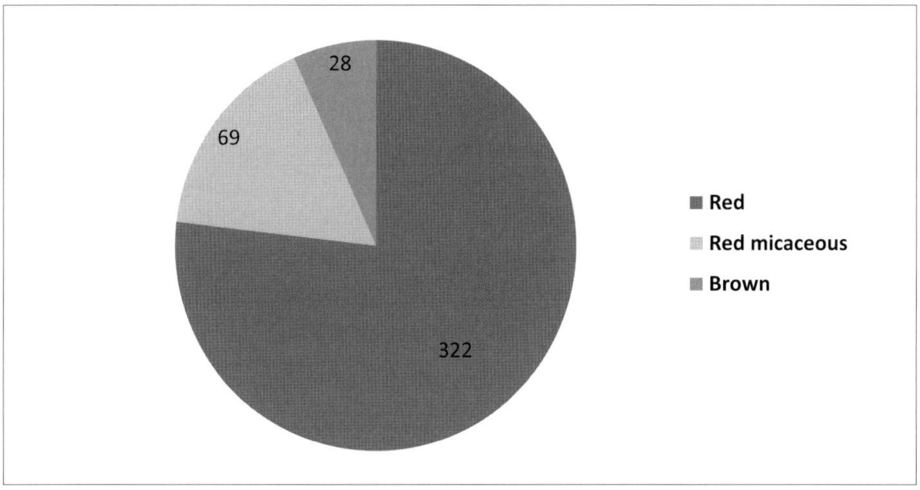


Table 5.11. Distribution of amphorai of unknown production place according to the largest fabric groups.

No formal criteria for the selection of amphora wall fragments were established. Amphorai from Panskoe I stored in the Museum in Černomorskoe and amphorai in the collection of the Museum in Evpatorija were studied prior to the survey and served as reference material. In addition, morphologically distinctive pieces observed among the surface material from several sites on the Tarchankut Peninsula, which were visited during the survey campaigns, complemented the museum collections.

5.3.3.1.1 Amphora fabrics

Several of the amphora fabrics are distinctive compared to other classes of pottery. The Sinopean amphorai are recognizable by the presence of abundant black volcanic pyroxene,[63] the Thasian amphorai are characterized by a reddish fabric and a high number of mica particles (Whitbread 1995, 165-197). Although the colour range of the Herakleian amphorai varies from brown and light-brown to a light-red, most fragments are identifiable on account of the almost ubiquitous gritty surfaces and lesser number or absence of black pyroxene inclusions.[64] Wall fragments of Chersonesean amphorai are more difficult to positively identify, as there seems to be substantial variability in the fabric, but a reddish colour and a higher amount of lime inclusions are characteristic of the majority of the diagnostic pieces (see Whitbread 1995, 238-242). Chian amphorai are not very common in the DSP survey area, but a number of fragments have been ascribed to this production. Wall fragments of Chian amphorai were particularly difficult to identify due to the fact that the fabric is rather 'anonymous' without characteristic inclusions. Furthermore, the fabric possibly changed over time (Whitbread 1995, 135-153; Lawall 1998, 78 with further references).

When compared to the amphora assemblage in Panskoe I U6, the large number of Herakleian amphora fragments found on the survey is surprising. At U6 Chersonesean amphorai are the prevalent type with the maximum number of examples amounting to 168 compared to a maximum of eight examples of imported Herakleian amphorai (Kac et al. 2002, 104-107, table 1). The Herakleian amphora fragments found during the survey have all been dated to the 4th century BC since the production peaked during this century, but also because none of the Herakleian stamps found at Panskoe I U6 can be dated to the 3rd century BC (Kac et al. 2002, 105). The higher number of Herakleian amphorai may suggest a more intense activity in the survey area before the last quarter of the 4th century BC when Chersonesean amphora became the predominant amphora type. A similar pattern with Herakleian amphorai appearing earlier than the Chersonesean amphorai has been observed in Olbia (Lawall 2010, 378, Period 3, summary).

332 diagnostic amphora fragments were collected during the survey. If we compare the relative numbers of wall fragments of the four larger provenances with the relative number of diagnostic fragments (Table 5.12), we see that wall frag-

[63] For analysis of the Sinopean amphora fabric, see Whitbread 1995, 234-244.

[64] M. Lawall et al. present detailed descriptions of the fabrics of stamped Herakleian amphorai from the lower city (Sector NGS) in Olbia Sector NGS, see Lawall et al. 2010, Cat. L55-, L57-L58, L111-L112. See also the characterisation of Herakleian amphorai from Panskoe I U/6 (Ščeglov & Selivanova 2002).

ments in the total collection of amphorae appear to be overrepresented in relation to the diagnostic fragments. Especially the number of Herakleian wall fragments seems to be overrepresented, which suggests that some of the wall fragments recorded as Herakleian may, in fact, not be Herakleian at all.

| Amphora type | Diagnostic frs. | Undiagnostic frs. | Ratio of diag/undiag frs. |
|---|---|---|---|
| Herakleian | 20.48% | 35.7% | 1:1.74 |
| Chersonesean | 10.04% | 14.98% | 1:1.49 |
| Sinopean | 6.32% | 8.73% | 1:1.38 |
| Thasian | 6.02% | 8.70% | 1:1.44 |

Table 5.12. Percentages and ratios of diagnostic and undiagnostic fragments of selected transport amphora types.

5.3.3.2 Architectural ceramics

The almost complete absence of roof tiles in the survey area is noteworthy, since this is contrary to the evidence from sites in the Mediterranean and many of the farmhouses in the Chersonesean home *chora* (e.g. Bintliff, Howard & Snodgrass 2007, 39; Nikolaenko 2003; Saprykin 2003). At Farmhouse 151 on the Herakleian Peninsula, the fragments of roof tiles amounted to 5% of the total assemblage but roof tiles seem generally to have been removed from sites when they went out of use (Carter et al. 2000, 725 see also Pettegrew 2001). The lack of roof tiles might also be explained by the use of thatched roofs for dugout structures, which were especially common in the northwestern Black Sea area in the Archaic and Classical periods and are known from Chersonesos, Kerkinitis and in the DSP study zone (Section 5.1.1).[65] Certainly, the near absence of roof tiles is in clear contrast to the farmhouse at Vetrenaja Bay, where they were abundant (Ščeglov 1967, 253 fig. 14). Even though a few roof tiles were found in one room of the building complex U6 at Panskoe I, buildings at that site though apparently quite monumental in layout were seemingly not tiled and the tiles found were even re-used from another building suggesting that roof tiles were a scarce commodity in this area in the Early Hellenistic period (Kac et al. 2002, 101). Only three fragments of Corinthian type pan tiles were found during the test excavation at Skalistoe 5 (**Cats. 733-734**). Two were of Chersonesean and one of Sinopean productions suggesting a use of mixed tiles as at Panskoe I U6.

5.3.3.3 Tablewares

385 fragments of tableware have been dated to the Early Hellenistic period. The identification of different shapes was a continuous problem in the recording of the tableware as well as their decorations. When decorated fragments were found, only very little of the slip, gloss or decoration had normally been preserved. The use of gloss or slip to discriminate between open and closed vessels could therefore in most cases not be strictly followed. Ascriptions were based mainly on the estimated size of the vessels as well as on the shape and curvature of the fragments. Sometimes shoulder fragments of jugs or the lower parts of cups could be identified with a higher degree of certainty. Table 5.13 shows the proportional distribution of fragments according to major shape groups. Jugs constitute the majority of the identifiable shapes, which is also the case in the assemblage of local/regional tableware from Panskoe I U6. The production of the characteristic jugs with banded decoration at Chersonesos is well known since a 3rd century BC workshop producing them has been excavated (Borisova 1958). The assemblage from building complex U6 at Panskoe I and a cistern in Chersonesos provide the best datable contexts for the Chersonesean jugs, which give us a date in the late 4th and first three quarters of the 3rd centuries BC (Zolotarev 2005). Fragments belonging to the group of closed vessels were identified by the presence of clearly visible interior throwing marks and in a few cases the preservation of an exterior slip. Many fragments in the group of closed vessels are likely to belong to jugs as well.

65 For Chersonesos, see Zolotarev 1998 and more generally Tsetskhladze 2004.

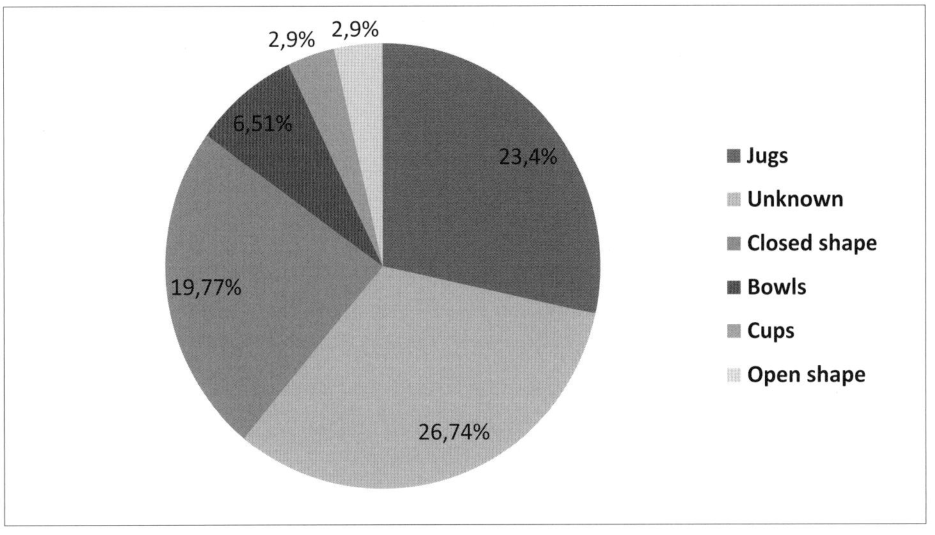


Table 5.13. Distribution of tablewares according to shape groups.

The relatively low number of tableware (6%) is perhaps surprising since tableware is generally frequent on Hellenistic sites in Greece sometimes amounting to half of all the finds (Bintliff, Howard & Snodgrass 2007, 40-41).

However, the percentage of tableware from DSP is only slightly lower than the percentage of fineware from a rural settlement identified in the Methana survey where the percentage of fineware amounts to less than 10% (Foxhall 2004, 261-262, fig. 12). A generally low number of tableware has also been noted for farmhouses in the Chersonesean home *chora*, although the percentage at Farmhouse 151 amounts to 26% of the total ceramic finds, an amount comparable to the tablewares from Panskoe I and Skalistoe 5 sites in the DSP survey area (see Section 5.3.4.2) (Carter et al. 2000, 714, 725).

#### 5.3.3.4 Utility wares

Only 120 fragments of utility ware of this period were identified. A combination of the thickness, inclination and estimated diameter of the vessel became a guiding rule in the ascription of body fragments to the '*utility ware*' group. If fragments of this combination did not seem likely to belong to an amphora, they were catalogued as utility ware. Special attention was paid to fragments with clear soot deposits or burn marks as well as those with a ledge for a lid (i.e. cookingware). However, only very few examples (nine) could be identified. Most of these come from field DSP07-01 close to the settlement of Panskoe I and probably ploughed out of that site. From the diagnostic fragments, it is clear that the utility wares closely resemble the common types from Panskoe I U6. The only recognisable types among the diagnostic fragments are the *louterion*/mortar, the *pithos* and the cooking pot. The very few identified fragments of Sinopean *pithoi* are remarkable because this was the preferred storage vessel in the home *chora* of Chersonesos, where among other places, it has been identified at the farmhouses on land plots 6, 9, and 132 (Nikolaenko 2003; Nikolaenko 2001, 57, fig. 2.58). Closer to the survey area, they are present in large quantities at the farmhouse at Vetrenaja Bay as well as in lesser numbers at Panskoe I U6, while an unspecified number was recorded at Masliny (Ščeglov 1967; Kac et al. 2002, 103-104; Latyševa 1978, 60).

#### 5.3.3.5 Handmade pottery

Like the utility ware, the handmade pottery found during the survey greatly resembles the handmade pottery from Panskoe I U6, but with the important exception of one type, possibly a bowl. 2,058 fragments of handmade pottery can be dated to the Late Classical and Early Hellenistic periods, but only 100 (4.9%) diagnostic fragments have been recorded. This high ratio of undiagnostic fragments is probably due to the high breakage rate of the handmade pottery as well as a lack of handles on handmade pottery. Out of the 100 diagnostic fragments datable to the Late Classical and Early Hellenistic periods, 69 fragments belong to a type of pot with bulbous body, narrow neck and out-turned rim defined as 'Type 1' by V.F. Stolba for the handmade pottery at Panskoe I (Stolba 2002, 183-184). Among the handmade pottery from Panskoe I U6, this type accounts for 67% of all the handmade pottery. Not much of the handmade pottery from

Chersonesos has been published,[66] but two recent studies give some idea as to the repertoire of the handmade pottery there. L.A. Kavalevskaja and I. Ju. Suchanova have published the handmade pottery from Farmhouse 338 in the south-eastern part of the Herakleian Peninsula, and S.N. Senatorov published some 6th-4th century BC assemblages from the city itself (Kovalevskaja-Suchanova 2005; Senatorov 2003). At Farmhouse 338, handmade pottery constitutes one of the largest ceramic groups, and fragments of approximately 70 vessels, most likely datable to the Hellenistic period, have been identified. There the pot with bulbous body, narrow neck and out-turned rim is also the most common type and can be compared to the DSP material. A similar type is also common at Chersonesos, where assemblages datable to both the Late Archaic period and the 4th century BC contained fragments of this characteristic shape. Handmade tableware shapes such as cups and bowls are difficult to recognize when dealing with fragments only, especially because their surfaces were often not treated in any way, i.e. polished or burnished, but resemble the surfaces of the larger pots. This phenomenon has been observed at both Panskoe I and Chersonesos (Stolba 2002, 186; Senatorov 2003, 19). However, it was possible to identify 12 fragments of bowls among the diagnostic material. A few fragments (**Cats. 198-199** and **215**) belong to the most common type of bowl in Panskoe I U6 with in-turned rim (Stolba 2002, 186 'Type 9'). However, rim fragments of another type with a projecting lip, which find no parallels at U6 or Chersonesos, have been identified (e.g. **Cats. 374**, **688**, **696** and **792**). They might belong to a type of bowl known from eastern Crimea during the 3rd-2nd century BC (Maslenikov 1995, fig. 20.2). Only three pieces have been identified as belonging to cups and one larger fragment comes from a decorated jug of the so-called Kizil Koba pottery (**Cat. 9**). The best example of a handmade cup with part of the handle preserved (**Cat. 643**) was found close to Panskoe I and resembles a cup from Chersonesos found in a context datable to the second half of the 5th – first half of the 4th century BC (Senatorov 2003, fig. 4.18). The Kizil Koba jug fragment was also found in the 'halo' of Panskoe I. It is decorated with incised and impressed geometric patterns similar to a published example from U6 (Stolba 2002, Cat. D104).

#### 5.3.3.5.1 Dating the handmade pottery

One of the most controversial issues in dealing with the handmade pottery is the question, whether a more precise chronology is possible. Comparisons with Late Archaic and Classical handmade pottery from Chersonesos have already made clear the difficulties involved in attempts at more precise dates. The general conformity in the typology of the handmade pottery of this period is corroborated by evidence from the recent excavations near General'skoe West on the shore of the Asov Sea in eastern Crimea (Abramov & Maslenikov 2005). Here a sealed layer containing some amphora and handmade fragments was found. The amphorae were predominantly imported from Chios, possibly from Thasos and Mende and could all be dated within the 5th century BC. As pointed out by the excavators, some of the handmade pottery found in the layer is typologically similar to 4th and 3rd century BC handmade pottery from the steppe region (Abramov & Maslenikov 2005, 15). In fact, it is also similar to 'Type 1' from Panskoe I U6.[67]

Indigenous sites have been identified on the Herakleian Peninsula dating to the 6th and 5th century BC, i.e. contemporary with the early "Ionian" settlement there.[68] An attempt to date the handmade pottery more exactly is therefore crucial to the question, whether an indigenous population was present on the Tarchankut Peninsula before the arrival of the Greek settlers. As much of the handmade pottery is very similar and undergoes few typological changes at least from the Archaic through the Early Hellenistic periods, it is difficult to obtain a narrow date for the handmade pottery of the DSP survey. However, almost all the DSP handmade pottery has been dated to the Early Hellenistic period due to its clear resemblance to the handmade pottery from Panskoe I U6. The dating of undiagnostic wall fragments is even more difficult. From the beginning of the project, it was unclear how to differentiate between wall fragments of the Bronze Age and fragments of the Early Hellenistic period. The main problem was caused by the fact that Hellenistic material was often found on sites that contained Bronze Age pottery (Section 5.3.2.1).

---

66 See the comments by Kavalevskaja-Suchanova 2005, 219.

67 Stolba 2002, 183-184. Similar handmade pottery is also known from a 5th century BC pit near Patraeus, see Lomtadze 1994, fig.3.2-4.

68 See Stolba 2008, 140 for a brief overview of the indigenous presence at Chersonesos. For a catalogue of indigenous sites on the Herakleian peninsula, see Kravčenko 2005.

5.3.3.5.2 Handmade fabrics

In view of the problems concerning the date of the handmade pottery, it would obviously be worthwhile to use different fabrics as dating criteria, as has already been advocated for and employed on other survey projects (e.g. MacDonald 1995; Attema & van Oortmerssen 1997-1998.). This task is severely hampered by a lack of existing detailed publications on handmade fabrics and the scarcity of closely datable contexts in the region. Apart from a detailed petrological thin-section analysis of the handmade pottery from Panskoe I U6, we only have cursory descriptions at our disposal. The analysis of the U6 assemblage demonstrated at least six different fabric compositions (Stolba 2002, 180-183 and the appendix by G. M.Kovnurko (2002)). The main tempers consist of different kinds of sands, quartz and several different magmatic minerals, small inclusions of crushed pottery, limestone and occasionally mica. Sand is one of the most commonly added or naturally occurring inclusions of which four different compositions could be recognised. Only one of the compositions (sand composition 3) appears to be associated with the local shell-sand deposits containing small amounts of quartz, feldspars and limestone. The remaining three types of sands containing magmatic and volcanic minerals appear foreign to the region.

The characterisation of the DSP handmade pottery relied exclusively on a visual examination, which excluded the possibility of differentiating between the different sand compositions. However, the following inclusions were identified: limestone, quartz, feldspar, shells, mica, smaller grit as well as occasional inclusions of grog. The fragments of handmade pottery were sorted into three main fabric groups; a sandwich fabric, a red fabric and a coarse fabric, which may correspond to different firing techniques (Table 5.14). All these groups are also found at Panskoe I U6, but as is the case with the Panskoe pottery, a high degree of heterogeneity is characteristic of the DSP handmade pottery and overlaps do occur. No discernable correlations between fabric and shapes are recognizable.

| Fabric | Description |
|---|---|
| Sandwich fabric | *This fabric is characterised by having a light-red (10R7/8), red (5YR 5/6), reddish-brown (5YR 6/3) or pale-brown (2.5Y 6/2) surface and a grey or black core caused by a lack of oxidisation during firing. In a few cases the surfaces may be greyish or black. The surfaces of the fragments are often powdery to the feel. The fabric is remarkably pure and rather dense. A small amount of quartz is found in approximately one third of the fragments within this group. Mica is rare* |
| Red fabric | *The colour range within this group is large ranging from red (2.5YR 5/8) over yellow (10YR 7/6) to pale-brown (10YR 8/4). Common to the group is the generally well oxidized firing. Mica occurs slightly more often in this group* |
| Coarse fabric | *This is numerically the smallest group. Almost all fragments within it have a greyish core. The colour of the surfaces is similar to fragments of the sandwich and red fabrics. The fabric is more porous than the previous groups and generally contains more limestone particles and more importantly many small crushed shells. Quartz minerals occur in roughly half of the fragments* |

Table 5.14. Handmade fabrics.

The rather inconclusive situation of the handmade fabrics from the DSP does not provide much scope for a narrower dating, which is hardly surprising for surface finds. However, two observations regarding the DSP handmade pottery's relation to the previously mentioned handmade pottery from Chersonesos and eastern Crimea should be mentioned. With the exception of a few fragments, all the handmade pottery from Farmhouse 338 dated to the Hellenistic period has a reddish or brownish surface. Contrary to this, the late 6th-5th century BC assemblages of handmade pottery from Chersonesos and General'skoe West are described as having black or greyish surfaces. This might indicate a general change in firing techniques sometime in the early 4th century BC, however, this observation must remain hypothetical.

5.3.3.6 The question of the pre-Chersonesean phase

The almost complete absence of 5th century BC material has already been noted (Section 5.3.3). Since the area is believed to have been under the control of Olbia from the late 5th to around the middle of the 4th century BC (Chapters 4.2.1, 4.3.2.1, and 7.3), it is worthwhile to elaborate on the sporadic occurrence of early pottery in the DSP area. Greek pottery from Western Crimea is generally scarce before the beginning of the 5th century BC, and especially the northwestern part seems completely devoid of Greek pottery before the 4th century BC (e.g. Vnukov 2001, 152-154). The nearby Greek Černomorskoe Settlement was not founded earlier than the early 4th century BC (Chapter 4.3.1.1; Ščeglov 1978, 20), whereas there is some pottery belonging to the late 5th and first half of the 4th century BC at both the necropolis and building complex U7 at Panskoe I.[69] Banded "Ionian" pottery of the 6th and especially the 5th century BC is fairly well represented in Chersonesos and Kerkinitis, but no comparable examples of this early banded ware was collected in the survey (Zolotarev 1993; Kutajsov 2004, 79 fig. 59-60). Only one fragment (**Cat. 789**) appears similar, but this piece comes from the Early Hellenistic assemblage at Skalistoe 5 (Chapter 4.2.2). Since figure-decorated fineware is completely absent and glazed pottery uncommon among the collected pottery, it is difficult to compare the occurrence of these two groups in the 6th and 5th centuries BC at Chersonesos and Kerkinitis to the finds from the rural hinterland of Panskoe. In fact, figure-decorated fineware pottery is rarely found in rural contexts (Foxhall 2004).

Two rim fragments of transport amphorai (**Cats. 96** and **704**) belong to the Late Archaic period. The profile of the rims is a variant of Zeest's 'Samian' type, which is reported to amount to as much as 30% of the amphora finds from the chora of Olbia (Cook & Dupont 1998, 178-182, fig. 23.12b). A handle fragment (**Cat. 117**) could conceivably belong to an Archaic Samian amphora, a type also known at Kerkinitis (Kutajsov 2004, 70). Among the diagnostic fragments are a few amphorai which can probably be dated in the first half of the 4th century BC. Five belong to Herakleian amphorai (**Cats. 140**, **192** and **782-784**) and four are of Thasian fabric (**Cats. 23-24** and **785-786**).

Two amphora fragments can be dated to the 5th century BC (**Cats. 223** and **787**). One fragment of a Chian amphora with bulging neck (**Cat. 787**) from field DSP08-04 can be securely placed in the second quarter of the 5th century BC. Chian amphorai with bulging necks, similar to the DSP fragment are not uncommon in the northwestern Black Sea area and are among other places found at Chersonesos, Kerkinitis, and Olbia (Kerkinitis: Kutajsov 2004 69-70, fig. 37; Chersonesos: Zolotarev 1993, 14-15, pl. XV, nos. 1-7; Olbia: Lejpunskaja 1981, 1986, 463-465; Lawall et al. 2010, 364-365). Another rim fragment (**Cat. 223**) resembles the early 5th century BC rims on Thasian amphorai, but is clearly not of Thasian fabric. A wall fragment with a thin horizontal band (**Cat. 703**) might also belong to a 5th century BC amphora. A single fragment of the lower part of a black-glazed cup, presumably a skyphos, with concentric circles on the underside possibly of Attic manufacture (**Cat. 247**) could also belong in the 5th century BC. However, considering the poor quality of the drawing of the concentric circles and the somewhat matt appearance of the glaze it seems more likely that the fragment should be dated in the 4th century BC. The Chian amphora and the black-glazed cup fragments were found less than 100 m apart at site DSP08-F04-03. Considering that 5th century BC pottery is found on the site, a neck fragment of a jug with grooved neck (**Cat. 249**) is interesting because of the possibility of a 5th century BC date for this fragment as well. This type of jug was popular in the Greyware production of the northwestern Black Sea area in the late 5th century BC. However, **Cat. 249** is of a red fabric, and the type of grooves bear more resemblance to a Greyware jug from Belozerskoe dated to the late 4th century BC (Bylkova 2005, fig. 6.3). It is clear that these 5th century BC finds constitute a very small group compared to the mass of material of the Late Classical and Early Hellenistic period. We may also note a conspicuous absence of certain ceramic types typical of the 5th century BC. Thus, apart from the few examples already cited, there is a pronounced lack of the most common 5th century BC types known from the surrounding area. The Chian bulging neck amphorai are very common in Kerkinitis, Chersonesos and Olbia (Kerkinitis: Kutajsov 2004 69-70, fig. 37; Chersonesos: Zolotarev 1993, 14-15, pl. XV, nos. 1-7; Olbia: Lejpunskaja 1981, 1986, 463-465; Lawall et al. 2010, 364-365). So too are Samian amphorai, red and grey Lesbian amphorai and the late 5th century BC type of Mendean amphora (Kutajsov 2004, 70; Lawal et al. 2010). Also noteworthy is the absence of Greyware pottery, which is abundantly present in the period from the late 5th and first half of the 4th century BC at building complex U7 at Panskoe I. Greyware is ubiquitous at the settlements in the Olbian chora of this period, and it is generally believed that the Greyware found at Panskoe I was imported from Olbia.[70] The lack of Greyware in the rural hinterland therefore also appears significant. In addition, we may note that no pottery of the 5th century BC or earlier Taurian Kizil Koba pottery was found. This further strengthens the supposition that the DSP survey ceramic assemblage within the Greek period is basically confined to the Late Classical and Early Hellenistic periods.

---

[69] For finds from the necropolis see Rogov 1989; Monachov & Rogov 1990a, 1990b; Rogov & Tunkina 1998; Rogov 2011. Building complex U7 still remains largely unpublished, but see Stolba 1991.

[70] Cf. E.g. Hannestad, Stolba & Blinkenberg 2002, 131. See, however, the reservations put forward by Bujskich concerning the Olbian origin of the Greyware (2006, 46).

5.3.3.7 Reuse of Late Classical-Hellenistic pottery
A small group of pottery fragments (16 in all) found during the survey had already in Antiquity been re-used for various purposes. A limited number of fragments, three in total, were used as weights, presumably loom- or net-weights (Table 5.15); one larger neck fragment of an amphora (**Cat. 226**) shows several cuts on the exterior surface and possibly served as a cutting board;[71] finally a larger group of fragments of transport amphorai, mostly consisting of handles and a few toes, were employed as a kind of polishing tool. Concerning the latter, one or both of the ends of the amphora handles and the undersides of the toes have a faceted and smoothed area with multidirectional striations, which shows that these pieces must have been re-used for an extended period of time. Similar pottery fragments with polished surfaces in addition to polishing tools of bone and stone have been found at the site of Zavetnoe 5 near Akra in eastern Crimea while they are also reported from the chora of Theodosia.[72] It is difficult to associate these tools with any specific task. They cannot readily be connected to agricultural activities, and most of the tools were found in connection with identified sites both in the Pediment and on the Hillsides (Table 5.16) and might therefore be associated with household activities where they could have been used as pestles. Alternatively they could have been used as polishing tools in pottery manufacture.[73]

| Weights | Associated site |
|---|---|
| **Cat. 788** | DSP07-F01-02 |
| **Cat. 158** | DSP08-F02-01 |
| **Cat. 497** | DSP08-H11-02 |
| **Polishing tools** | Associated site |
| **Cat. 644**, **647** and **652** | Panskoe II |
| **Cat. 464** and **466** | DSP07-H11-01 |
| **Cat. 141** | DSP08-F01-01 |
| **Cat. 143**, **149** and **158-159** | DSP08-F02-01 |
| **Cat. 781** | near DSP08-F02-01 |
| 6226/01/1 | Skalistoe 2 |
| **Cat. 226** (cutting board), **231** | DSP08-F04-03 |
| **Cat. 527** | DSP08-H11-06 |
| 6111/01/1 | Off-site |

Table 5.15. Reuse of pottery, Late Classical and Hellenistic periods.

## 5.3.4 A note on site assemblages | S. Handberg & K. Winther-Jacobsen

5.3.4.1 The Bronze Age site assemblages
All the Bronze Age pottery is handmade, and the extremely low degree of specialisation of fabric and shape makes it difficult to comment on the composition of the site assemblages (Table 5.15). At Skalistoe 2, four sherds out of a total of 191 sherds came from closed vessels and one sherd was identified as a cooking vessel. The rest was identified as coming from pots. 37 lithics, mostly chipped stones, were collected at the site, but flint was common at later sites too and the Bronze Age site overlaps a Late Classical-Early Hellenistic site. At DSP08-F06-01, 11 sherds came from bowls, and one each from a cooking vessel, a frying pan and an open shape. The remaining 1,057 sherds were identified as coming from pots. 108 lithics, mostly chipped stones, were collected at the site. Although only part of Skalistoe 2 could be surveyed,

[71] The cuts could have been caused by ploughing, but the fact that the cuts are placed closely together and that similar cuts have not been observed on other fragments from the same place, makes this seem unlikely.
[72] I am grateful to Sergej Solovyov for sharing information with me about the finds from Zavetnoe 5, see Solov'ev & Šepko 2006, 58 fig. 184. For the finds from the chora of Theodosia, see Gavrilov 2006, 257-258.
[73] See Handberg 2011, for more on the argument that these tools could have been used in pottery manufacturing.

the estimated sizes of the scatters of the two sites were not dissimilar, but the densities of sherds were. The sherds were not weighed so it is not possible to determine the possible significance of post-depositional processes. The different densities may also reflect length of occupation, but both sites were active during the Middle and the Late Bronze Age. Another process affecting the size of the samples is the actual collection. In the case of Skalistoe 2, the team leader realized that the team seemed to be collecting the standard sample (ST) less meticulously, because they knew they were doing a total sample (TS) afterwards. The TS to ST pottery sample ratios suggest that this was indeed the case: At Skalistoe 2 the TS to ST ratio was 1:2.4 at DSP08-F06-01 it was 1:4.3. At Skalistoe 2, 27 out of 37 lithics were collected in the ST sample, whereas at DSP08-F06-01 only 16 out of 108 lithics came from the ST sample. Many tiny chips of flint were collected at DSP08-F06-01, which suffered from a low visibility and required the more intense sampling method. At Skalistoe 2, 45% of the site area was total sampled, at Skalistoe 2 it was 36%, which is not dissimilar enough to result in such a large difference in the lithics ratios. It is possible that the tiny flints reflect a site activity difference between the two sites, perhaps a small workshop, even if no nodes were recorded.

| Sites | Survey pottery | | | Excavation pottery | Total number sherds | ST to TS pottery ratio | ST to TS lithics ratio | Site size |
|---|---|---|---|---|---|---|---|---|
| | ST | TS | GS | | | | | |
| DSP08-F04-03 | 526 | 179 | 0 | 0 | 705 | 1:0.3 | 1:1 | 4,810m$^2$ |
| DSP08-F04-07 | 207 | 320 | 0 | 0 | 527 | 1:1.5 | 1:3.7 | 1 ha |
| DSP07-F01-01 | 215 | 255 | 0 | 0 | 470 | 1:1.2 | - | |
| DSP07-F08-01 | 121 | 220 | 1 | 0 | 341 | 1:1.8 | 1:0.9 | 2.25 ha |
| DSP08-F04-02 | 109 | 207 | 0 | 0 | 316 | 1:1.9 | 1:1 | 2,460 m$^2$ |
| DSP08-F02-01 | 63 | 72 | 0 | 0 | 135 | 1:1.14 | 1:1 | 2.9 ha |
| DSP07-F06-all sites | 59 | 61 | 7 | 0 | 120 | 1:1.03 | 2:0 | - |
| Skalistoe 5 | 0 | 0 | 136 | 540 | 676 | - | - | |
| DSP08-H11-02 | 86 | 0 | 32 | 138 | 256 | - | - | 2.6 ha |
| DSP07-H01-01 | 0 | 0 | 15 | 141 | 156 | - | - | |
| DSP07-H10-01 | 0 | 0 | 27* | 106 | 133* | - | - | |

Table 5.16. Quantities of sherds collected at selected sites by different collection strategies, and site sizes. * including finds from site DSP07-H18-01.

5.3.4.2 The Late Classical-Early Hellenistic site assemblages

10 of the sites identified by the DSP have produced relatively substantial quantities of Late Classical and Early Hellenistic pottery (Table 5.16). However, different collection strategies were applied including standard, total and grab samples (GS) as well as excavation resulting in quantities not immediately comparable. In order to include as many sites as possible working with standard samples would be preferable; however, the standard samples collected on the slopes are too small to be statistically significant. On the Pediment, total samples are generally larger than standard samples and for comparison of surface finds, the total samples should also be the most reliable (Table 5.17).

| | Type of sample | TR | HW | TW | IN | UT | LI |
|---|---|---|---|---|---|---|---|
| DSP08-F04-07 | TS | 35.2 | 34.5 | 0.7 | 27.2 | 0.0 | 2.4 |
| DSP07-F01-01 | TS | 22.4 | 24.4 | 28.3 | 19.3 | 4.7 | 0.8 |
| DSP07-F08-01 | TS | 6.0 | 77.0 | 12.0 | 1.0 | 0.0 | 3.5 |
| DSP08-F04-02 | TS | 14.6 | 70.4 | 0.5 | 13.6 | 0.5 | 0.5 |
| DSP08-F04-03 | TS | 28.2 | 28.8 | 0.5 | 38.0 | 0.0 | 0.0 |
| Skalistoe 5 | Excavation | 26.0 | 3.0 | 44.0 | 25.0 | 0.0 | 2.0 |
| DSP08-H11-02 | Excavation | 47.0 | 43.0 | 10.0 | 0.0 | 0.0 | 0.0 |
| DSP07-H01-01 | Excavation | 18.0 | 69.0 | 10.0 | 3.0 | 0.0 | 0.0 |
| DSP07-H10-01 | Excavation | 35.0 | 58.0 | 5.0 | 0.0 | 0.0 | 2.0 |

Table 5.17. TS and excavated samples proportions in percent of find classes at selected sites.

Transport amphorai, handmade pottery, tablewares, and undetermined wares are represented at all the selected sites, whereas utility wares are absent from three out of four sites on the Pediment. The most differentiated distribution within a single class of finds is tablewares ranging from 0.5% to 28.3% in the surface assemblages on the Pediment and Lowland Ridge (ratio 1:56.6), although distribution within the undetermined wheelmade class is also rather extreme (ratio 1:38). Among the selected sites, lithics are only absent from DSP08-F04-03, DSP07-H01-01, DSP07-H10-01 and DSP08-H11-02, but common at a probably contemporary site located only 200 m away (DSP07-F02-01). Most of the lithics collected in the DSP area are imported flints either in the form of fragmented tools or chips, but 40 ground stone tools have been identified all except for two found on the Pediment and Lowland Ridge. The find spots of 30 of the ground stones are clearly associated with sites. Although the majority of lithics cannot be dated stylistically (Section 5.4), the fairly consistent appearance of flint at Late Classical-Early Hellenistic sites in the DSP area is evidence to the continued use of chipped stone tools in the historical period. Also at the contemporary sites Kavkaz and Energik in southwestern Crimea, flints were encountered (Zubar' & Kravčenko 2003, 190). These sites are indigenous sites belonging to the Kizil Koba culture, and topographically, they are located at the perimeter of the Chersonesean home chora.

On the Pediment, only the excavated sites provide architectural evidence: Panskoe I and Skalistoe 5 both represent a well-known type of fortified settlements. Of course the post-depositional processes and collection strategies applied to the investigation of the sites are not similar. The excavated finds from the trial trench at Skalistoe 5 represent a depositionally restricted assemblage as well as post depositional processes and it is mainly suitable for comparison with other excavated assemblages. The surface assemblage at DSP07-F01-01 is part of the 'halo' of Panskoe I representing secondary rubbish disposal and/or possibly manuring on the in-field and post-depositional processes (Bintliff, Howard & Snodgrass 2007, 23-26) and as such, Panskoe I provides a point of reference for the surface assemblages. The remaining surface site assemblages from the Pediment were collected on the actual sites representing the full range of depositional and post-depositional processes.

Although there is some resemblance between the compositions of the two assemblages from known fortified settlements, the dissimilar collection strategies suggest that this is likely to be a random pattern. However, the two assemblages include very high proportions of tablewares, 28.3% and 44% respectively. This is in clear contrast to the remaining selected surface sites, where tablewares range between 0.5% and 12%. In terms of range, the tablewares are less surprising: mostly Chersonesean jugs predominate, but cups, including kantharoi, bowls and fish-plates are also present at both sites, and at DSP07-F01-01 the base of a probable lekythos was found. The low amount of tablewares on the minor sites in the Pediment zone might suggest that the sites represent smaller agricultural establishments, possibly with smaller sheds or shelters.[74] This predominance of tablewares at Skalistoe is not a phenomenon entirely explained by collection strategies. The three Late Classical-Early Hellenistic excavated sites on the slopes have produced very different assemblages with DSP07-H01-01 and DSP07-H10-01 being dominated by handmade wares and DSP08-H11-02 producing 43% handmade wares

[74] For the possibility that the *klision* known from ancient texts might refer to such smaller rural establishments see Osborne 1985, 21.

and 47% transport amphorai. The assemblages from DSP07-H01-01 and DSP07-H10-01 consist of an amalgamation of the finds from two different trenches each including household pits and stone structures, whereas DSP08-H11-02 is a pit house. Consequently, DSP07-H01-01 and DSP07-H10-01 consist of depositional and post-depositional finds, while the DSP08-H11-02 assemblage is mainly post-depositional.

Two of the selected sites on the Pediment produced predominantly handmade wares: DSP07-F08-01 and DSP08-F04-02 ranging between 70.4% and 82.2%. Another group defined by the 25-35% proportions of handmade wares consists of DSP07-F01-01, DSP08-F04-03 and DSP08-F04-07 including also DSP08-F02-01. The fact that three of the sites appearing in different groups are located in DSP08-F04 and consequently share their post-depositional history supports the behavioural meaningfulness of the pattern. Unfortunately, there are not other similarities to confirm the pattern.

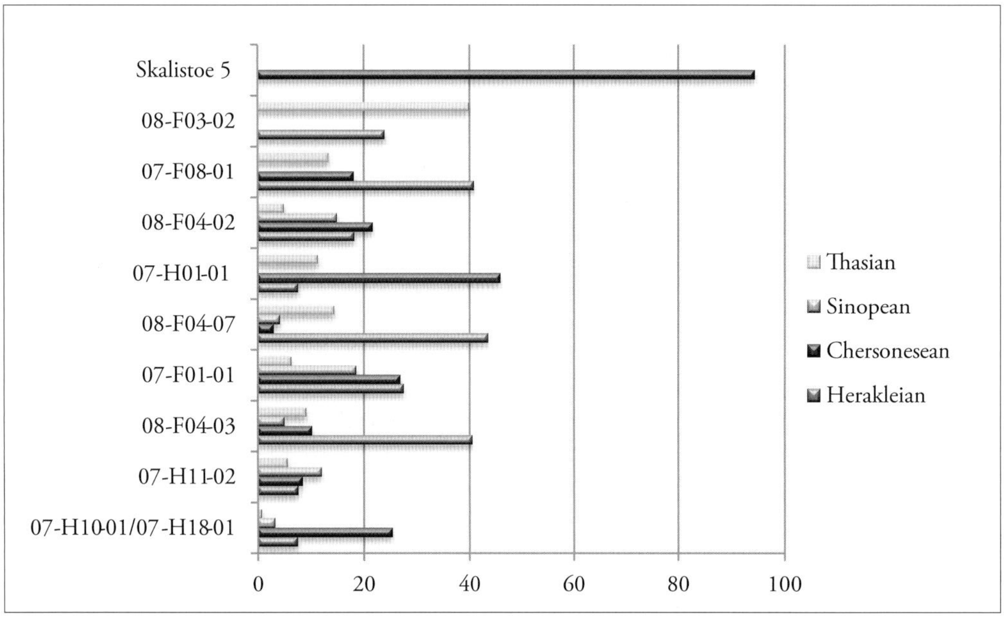


Table 5.18. Distribution of the four dominant transport amphora types at selected sites.

The preponderance of Herakleian, Chersonesean, Sinopean and Thasian productions within the transport amphorai has already been mentioned (Table 5.10). Comparing the distribution of the four amphora types at selected sites, it is clear that Herakleian amphorai are dominant on three of the sites (DSP08-F04-03, DSP08-F04-07, DSP07-F08-01). We may also note that Chersonesean amphorai are not nearly as predominant, as might have been expected, since there are roughly twenty times as many Chersonesean as Herakleian amphorai at Panskoe I U6. Surprisingly, the share of Chersonesean and Herakleian amphora in the assemblage from the 'halo' of Panskoe I U6 DSP07-F01-01 is almost the same. Another interesting picture emerges, if we compare the representation of Chersonesean amphora on the hill sites DSP07-H10-01/H18-01, DSP07-H01-01 and the fortified settlement at Skalistoe 5 to the overall distribution of amphora. The Chersonesean amphorai are much better represented on the hill sites and at Skalistoe 5 than on other sites in the Pediment, which might indicate a chronological difference between the hill sites and Skalistoe 5 and other sites in the Pediment, which is not otherwise traceable (see also the discussion in Section 5.3.3.1.1*)*. Finally, the share of Sinopean and Thasian amphorai is relatively constant on all the sites except for Skalistoe 2, which is mostly likely a result of the very low number of amphorai (0.43%) from this site.

The above-mentioned, close proximity between DSP07-F02-01 and DSP08-F04-03, only 200 m currently across a lagoon, and the highly specialised distribution of pottery at DSP07-H02-01 with 55% transport amphorai suggest that these two sites may belong to the same settlement. Although the short chronological span does suggest contemporaneity, the chronology is too coarse to be certain. In terms of range of finds, only 12 Chian amphora fragments were collected in the entire area, two of these at DSP07-F02-01 and another one on the margin of DSP08-F04-03; however, the latter fragment is part of the evidence for a possible pre-Chersonesean phase at DSP08-F04-03 (Section 5.3.3.6). All the Chian amphora fragments were collected on the S side of the lake, five out of 12 from DSP08-F02.

The TS of DSP07-F02-01 and DSP08-F04-03 both have high percentages of undetermined wheelmade wares, 23% and 38% respectively, and it seems prudent to speculate a little on the identity of this class. Undetermined wheelmade wares consist of wheelmade body sherds of unrecognisable shape, unknown fabrics, and consequently, uncertain date. It seems most likely that many of these sherds came from utility types of the historical period of which no diagnostic fragments were found. Given the overall proportions of the periods in the DSP area, many of these sherds should be Late Classical-Early Hellenistic, and another large group probably dates to the Early Modern period. The percentage of undetermined wheelmade wares at the selected surface sites ranges from 1% at DSP07-F08-01 to 42% at DSP07-F06-01 to 03. DSP07-F06-01 to 03 are all located near the old road from Vodopojnoe and Meževodnoe. A cluster of Early Modern finds on the opposite side of the road (DSP07-F05-01) is suggested to be the result of rubbish dumped along the road. It appears likely that some of the very high proportions of undetermined wheelmade wares collected at DSP07-F06-01-03 reflect the same discard behaviour. The two Late Classical-Early Hellenistic sites DSP08-F04-06-07 both have high percentages of undetermined wheelmade wares, and these two sites are located near the abandoned village of Kipčak Učkuju, a similar case can be made for DSP08-F02-01 located next to a farm built sometime between 1899 and 1955. At DSP08-F04-03, a few finds have been dated to the Medieval to Modern period and it seems likely that the high proportion of undetermined wheelmade wares includes sherds of this period.

## 5.3.5 Medieval to Modern pottery | K. Winther-Jacobsen

### 5.3.5.1 Introduction

Publications of Medieval to Modern pottery are still uncommon, and when they appear, they tend to deal only with glazed wares (e.g. Zographou-Korre 1995; 2000). Hayes' (1992) publication of the pottery from Saraçhane in Istanbul is a rare exception, but although this has provided useful parallels, there is an obvious difference between the pottery found in the capital city of an empire as opposed to tiny backwater settlements. Additionally, Vroom (2003) has published two studies of Medieval to Modern pottery from the Aegean. At 'Kalos Limen', 8th-10th centuries and later Medieval phases have been reported, but very few finds have been published (Nalivkina 1957; Ščeglov 1978, 36-37; Guldager Bilde et al. 2008, 134). Krasil'nikov (1976; 1979) has published articles on pottery of the Saltovo-Majak culture of the 8th-10th centuries, and Jakobson (1979) has published a book on Medieval pottery from Southern Crimea. Also in Southern Crimea, the University of Texas, Austin project at Chersonesos has excavated several Medieval contexts of the Early (6th-7th centuries) and Late (11th-13th centuries) Byzantine periods. Preliminary publications of some of the pottery are posted on the internet (Arthur 2000; Andreeva 2003; Rabinowitz 2005). Generally, the state of publication has not changed since Arthur (2000, 48) commented on the quality of illustrations and descriptions of pottery and the availability of the literature. *Historical and Archaeological Investigations in Azov and the Lower Don in 2005* (in Russian) provides much needed new references for pottery of the Medieval period (e.g. Dmimrienko & Maslovski 2006).

A total of 160 sherds are tentatively dated to the Byzantine and/or Medieval period compared to 502 sherds of the Early Modern period. 35 of the Byzantine and/or Medieval finds were diagnostic compared to 98 diagnostic Early Modern sherds resulting in a similar ratio of diagnostic to undiagnostic around 1:4. Generally, the sherds dated to the Byzantine and/or Medieval period were poorly preserved, and no close parallels have been found for any of them. The date is based on the fabrics (different from the fabrics known from the other periods e.g. **Cat. 809**) and the appearance of high handles on amphorai (e.g. **Cat. 821**) and deep mechanical corrugation (e.g. **Cat. 823**). No tablewares have been recorded in the DSP area to support the existence of a Byzantine or Medieval phase in the hinterland of Panskoe.

Generally, the glazed wares from the DSP area suggest that the majority of the Medieval to Modern pottery should be dated to the 18th-20th century. The style of pottery is consistent with what is termed Late Turkish in the Mediterranean area and bears witness to the close relations to the ceramic traditions of the Ottoman Empire. Well-known monochrome or bichrome green (both pale and dark) and yellow glazes dominate, as well as brown glazes. Only two fragments of polychrome painted glazed wares were collected. The most conspicuous artefact of this period is a mid-19th century stemmed, green-glazed candlestick found off-site in DSP08-F05 in an area with only near-site level densities. As is the case in the Aegean, pottery from this late period has not been treated stratigraphically in Crimea (Vroom 2003, 73). There is rich ethnographic material from the 18th century at the Khans palace at Bakčisaraj, but the exhibition includes little utility ware as is the case with the local and regional museums of the Tarchankut at Černomorskoe and Evpatorija.

The old topographical maps of the area probably provide us with the most precise chronological indicator for finds of such recent date. DSP08-F03-01 is marked as a hamlet on the 1897 map, but no longer existed on the 1955 map providing us with a 1955 *terminus ante quem*. This assemblage consists of a high proportion of cookingwares, a category which has not been identified at any other Early Modern sites in the cultivated zone. One of the two polychrome fragments (**Cat. 188**) and the single fragment with circular combing (**Cat. 171**) also came from this site. DSP07-F04-02 is a very interesting cluster because of the absence of glazed fragments and high number of animal bones. The site is located in a slightly elevated area on the southern edge of the field. The elevation may indicate the presence of a buried structure, and the 1897 map shows a kurgan in this area. No other elevations were observed in the area. The finds may be associated with Džajlav located approximately 400 m to the S on the 1817 map and already a ruin by 1837. Despite the absence of more easily datable sherds, the style and fabrics of pottery are consistent with the pottery collected at Kurama Kostel'. A few Hellenistic amphora body sherds may have come from the kurgan. None of the remaining ceramic clusters in the cultivated zone (DSP07-F05-01, DSP08-F05-01) are indicated on any of the maps. Consequently, if they are indeed sites, they were either abandoned before 1897 or too short-lived to be indicated on the maps. Kurama Kostel', which was only grab sampled, is marked on both the 1897 and the 1955 map, but has since been abandoned and demolished. The absence of clay cooking pots at Kurama may be explained by the transfer to metal cooking pots, while the site was occupied.

Finds were only collected at two Early Modern sites on the slopes: two sherds were collected around a possible stone foundation on the eastern edge of Hill 8 (DSP08-H08-02 and 11 sherds at a small settlement on the Plateau (DSP08-H07-03). Due to the few finds, the settlements cannot be securely dated, but all the diagnostic Medieval to Modern sherds collected on the slopes appear to belong in the 18th-20th century. Of course, the small clusters in the cultivated zone may represent similar settlements; however, they may also reflect different behaviours such as garbage dumping. DSP07-F05-01 is located S of the road to Meževodnoe approximately 600 m from the western edge of the village of Vodopojnoe, which existed already in 1817. Among the artefacts collected were several fragments of glass, faience and porcelain, a metal spoon and an iron knife blade indicating a fairly recent date. None of the maps indicate any structures in the area, and the abraded state of the finds suggests a dump site. The cluster is located perpendicular to the road, and it is tempting to interpret it as the ploughed-in remains of refuse discarded by the roadside. The Early Modern element at DSP08-F05-01 appears to be too insignificant to suggest the existence of a site. It may belong to the halo of a site in the adjacent field, but no such site is marked on any of the maps. Another possible Early Modern halo or off-site scatter was recorded at the Hellenistic site DSP08-F04-06. This site coincides with the eastern edge of the village Kipčak Učkuju, which is first indicated on the 1817 map and marked as a ruin on the 1967 map.

### 5.3.5.2 Utility wares, tablewares and transport amphorai

The Early Modern utility and tablewares form a very homogeneous group. The majority is made of a fine sandy fabric with some tiny to small lime inclusions and possibly tiny feldspar and quartz, although sometimes slightly more and slightly larger particles occur as well as longish voids. Occasionally large red particles, possibly grog, are present. The wares are very hard fired and of high quality, fired to a light or yellowish red. The study of this group revealed two characteristics. The surface of the unglazed vessels is often of a slightly different colour than the core. This is probably the result of working intensively with a very wet surface which brings out the fine clay silicate. Sometimes the interior surface feels almost soapy (e.g. **Cat. 637**). Under oxidized firing conditions, this produces a slightly different colour on the surface, a self-slip. Some sherds have an exterior matt, dark-red to brown to slightly purplish "self-slip", which tends to come off in stripes, creating a pattern of striation.

A few sherds have been identified as transport amphorai of the Byzantine to Medieval period based on the quality, morphology and decoration, although no specific parallels have been found due to the poor state of preservation. Three body sherds of large well-made vessels have corrugated or ribbed bodies (**Cats. 810, 812**, and **823**, see Section 5.3.5.2.2). Additionally, some handle fragments appear to come from transport amphorai of fabric types not belonging to the Late Classical-Hellenistic period (e.g. **Cats. 284, 809, 819**, and **821**). It is not the individual types of inclusions, which set these fragments apart from the remaining material, but the fabric mixture and manufacture.

#### 5.3.5.2.1 Morphology

Out of 19 base fragments, only three are ring-bases the remaining 16 are flat. All the ring-bases are glazed on the inside (**Cats. 619** and **630**), but only one on the outside as well (**Cat. 286**). Glaze is only preserved on three of the flat base fragments (**Cats. 113** and **185**) and only one on the inside as well (**Cat. 114**); however on one of the fragments with exterior glaze, it stopped well above the base (**Cat. 113**). Since most fragments are small, such a style of glazing would not be observable on the preserved sherds very often. The flat bases often flare a little towards the bottom (e.g. **Cats. 104, 113-114, 403, 569** and **794**), which is typical of early 19th century pottery e.g. from Didymoteichon (Bakirtzis 1980, pl. IV). Diameters of the flat bases appear to fall into two groups: 7 cm and 10-13 cm, although many of the fragments were too small for the diameter to be estimated reliably. The flat bases come from vessels with almost cylindrical or slightly out-tapering lower bodies, the smaller ones probably from jugs and the larger ones probably from jars. Zographou-Korre (2000, 280 no. A138) dated glazed jars of this type to the 19th century. The jugs (**Cats. 106**, **109**, **618**, **626-628**) probably belong to the spouted type so common among Ottoman pottery both glazed and unglazed (Bakirtzis 1980, fig. 10 right; Hayes 1992, 342-342, pl. 46.f-j, fig. 106; Zographou-Korre 2000, 348 no. B94a). At Saraçhane the type dates back to the mid 16th to mid 17th century; however, there was very little Late Turkish material at this site (Hayes 1992, 334-339) and Zographou-Korre (2000, 348) dated the glazed type to the 19th century AD. At Saraçhane the spouted jugs appear in contexts with wide mouthed basins (Hayes 1992, 342-343, pl. 46.a-d) which have also been collected in the DSP area (**Cats. 105**, **108** and **631**). Interior wheel marks are usually prominent on the body fragments.

#### 5.3.5.2.2 Decoration

As mentioned above, monochrome or bichrome green and yellow glazes dominate, but there are also some brown-glazed pieces. The green and brown glazes come in several shades from pale (e.g. **Cats. 175** and **820**) to bright (e.g. **Cats. 171** and **179**), but the yellow glaze is very homogeneous (e.g. **Cat. 114**). Only two fragments of polychrome painted glazed wares were collected (**Cats. 188** and **811**), both from open shapes probably dishes only glazed on the inside. **Cat. 188** from DSP08-F03-01 is decorated with concentric dark-orange lines on a cream-coloured background. **Cat. 811** is decorated with blue petals framed with a cream-coloured line on a rose-coloured background. It was collected off-site in DSP08-F04 SW of Kipčak, a Tatarian village abandoned before 1955 except for a single house. One interior and exterior green glazed body sherd is decorated with a moulding with impressed strokes (**Cat. 813**).

Additionally, a few sherds with mechanical deep combing (**Cats. 812** and **823**), crisp horizontal combing with 2-5 grooves per set (e.g. **Cats. 402**, **638-639**), a few grooves (e.g. **Cat. 621**) and rare soft ribbing (e.g. **Cats. 637** and **810**) mainly on the shoulder or neck are seen. Combing and corrugation are two common styles of decoration in the Mediterranean both known particularly from storage vessels and transport amphorai of the Late Roman and Early Byzantine periods (Robinson 1959, 6; see Chapter 5.3.3.2). Many of these common types of amphorai also occur in Crimea (Jakobson 1979, figs. 1-3, 12-13), and the two decorative styles continued during the Medieval period (Jakobson 1979, figs. 33-34, 43, 68-69). Jars and jugs with corrugated or combed shoulder and/or neck are associated with the Saltovo-Majak culture dated to the 8th to 10th century (Krasil'nikov 1976, fig. 3; Ščeglov 1970, fig. 2; Andreeva 2003, fig. 1A). Similar decoration with narrow bands of crisp combing is common on unglazed jugs and amphorai from Turkish layers at Thebes, Didymoteicho and Saraçhane (Hayes 1992, fig. 106; Zographou-Korre 2000, 348 no. B94a; Vroom 2003, 179-180, fig. 6.46 no. W 38.Ex2). Only two sherds are decorated with combing, which does not follow the shape of the vessel: **Cat. 171** comes from a bichrome green- and yellow-glazed vessel decorated with horizontal combing just under the rim and circular patterns below. One shoulder fragment of a utility vessel collected off-site (**Cat. 818**) is decorated with what appears to be a horizontal set of combing and a wavy set below.

#### 5.3.5.2.3 Cookingwares

Two types of cooking vessels have been collected: a pot with a high collar rim (**Cats. 167-169**, **175**, **178**, **180-181**, **184**, and **572**) and a possible casserole or pot with everted rim with a ledge for a lid (**Cats. 170**, **174**, **176**, **182** and **186**). The cookingwares are made of very pale brown clay (10YR 8/3) covered with a matt dark-red to black slip (5YR 6/4). The sandy fabric is coarse and gritty compared to that of the utility wares. It is tempered with sparse, rounded tiny quartz and what appears to be small clay pellets. The majority of the diagnostic vessels were collected at DSP08-F03-01, which suggests 1955 as a *terminus ante quem*. The collar-rimmed type of cooking pot has a long history in the Mediterranean. It survived during the Medieval period (e.g. Vroom 2005, 104-105, fig. 18.4) and its common appearance in a glazed variant at DSP08-F03-01 gives further evidence to the longevity of this shape.

### 5.3.5.3 Other

Two diagnostic fragments of pipes for smoking (**Cats. 634**, **799**) and three body sherds (**Cat. 162**, 1581/01/01 and 8019/01/1) were collected, one associated with Kurama Kostel'. **Cat. 799** belongs to the Eastern Mediterranean *chibouk* type used with a long tube and a mouth piece (Hayes 1992, 391-395). This type of pipe was made at many production centres from the 17th into the 20th century. The fabric and shape of **Cat. 799** is reminiscent of a pipe photographed by Vroom (2005, 172-175, fig. 15.1) and another one from Saraçhane of the late 17th century (Hayes 1992, 343, 393 type III, pl. 50.e top). The fragmented pipe (**Cat. 634**) from Kurama Kostel' is morphologically very plain and undecorated.

A completely flat sherd with a carefully- and prefiring-made hole at the centre (**Cat. 115**) may be a lid. The near contemporary date of the remaining sherds in this homogeneous assemblage (DSP07-F05-01) indicates a much later date than suggested by the parallels (see Section 5.3.5.1). A much larger lid of a somewhat similar type has come from a 14th century context at Azak/Azov (Dmimrienko & Maslovski 2006, fig. 10.8). However, the sherd may also have come from a vessel similar to a Byzantine casserole with central hole in the bottom from Saraçhane (Hayes 1992, pl. 12 no. E).

The mid-19th century candlestick similar to one found at Saraçhane (Hayes 1992, 337, fig. 143, no. 6) came from DSP08-F05-01, a possible Hellenistic site with a minor Early Modern element.

## 5.4 LITHICS | M.E. VAN KRUINING & M.J.L.TH. NIEKUS

During the Džarylgač Survey Project nearly 100 pieces of flint, mostly artifacts, and approximately 70 worked or modified stones were collected. This amounts to ca. 1.5% of the total number of finds. Nearly all flint artifacts and stones were found during the surveys, either as single finds or as small scatters. Only a few lithic artifacts were found during the excavation of one of the trial trenches at DSP07-H08-01, a Bronze Age site. In this section several aspects of the finds will be discussed including the characteristics of the raw material and the age of the collected artifacts. It was decided to refrain from an in-depth discussion of the finds for a number of reasons. First of all, the basic technological and typological descriptions that were compiled during the survey are sketchy, incomplete and at times inconsistent. Since it was not possible to take the finds abroad for more detailed study we have not been able to check the classifications that were provided obviously introducing a source of error. Second, expertise on Stone Age artifacts from the study area is limited and publications in English are scarce, hampering an interpretation of the finds. Third, nearly all artifacts are surface finds and their interpretative potential is limited.[75] In the following paragraphs we will focus mainly on the flint artifacts.

### 5.4.1 Raw material

Most of the flint is brown, greyish brown to (dark) grey in colour, sometimes with grey specs or small fossils (Bryozoa?). The texture is fine to medium grained but transparent flint with a glassy texture is also present. The majority of artifacts appear to be quite fresh but several pieces display white patina, possibly the result of contact with plant roots (Stapert 1976) or soil formation processes. Concretions of lime and iron stains, the latter probably due to ploughing, have also been recorded. Unfortunately, little information is available on primary sources of flint and other lithic raw materials in Crimea except for the south and southwest, regions that are well known for their Palaeolithic sites. The bedrock in the research area consists of limestone (see Chapter 3) while more inland siliceous rocks are present. The cortical surfaces that are present on quite a few artifacts are weathered and abraded which suggests that the nodules were collected from secondary raw material sources such as gravel-bearing fluviatile deposits or glacial deposits.

### 5.4.2 Flint artifacts

For 63 flint artifacts a more detailed classification is presented in Table 5.19. By comparing the database with photographs that are available for a number of finds it became clear that quite a few classifications are erroneous. For example, several artifacts that are classified as flakes are in fact blades, one of the scrapers is a surface retouched arrowhead, a retouched blade is a backed point and so on. Part of these problems could be resolved by combining categories such as flakes and blades, by reclassifying artifacts and by introducing "new" types, for example "bifacial tools" and "gun-flints". Nevertheless, some inconsistencies remain because photographs have not been taken of all finds or they do not provide enough detail for a more accurate classification. The relatively high number of retouched tools (N=41 or 65%) is most likely due to these classification problems in combination with collector bias, i.e. retouched tools and larger artifacts such as cores are more easily recognised and picked up in the field than small chips, simple flakes and other waste products of flint working.

[75] See for discussion Lewarch & O'Brien 1981 and Smit 2010.

| | H08-01 | F03-01 | F03-02 | F04-02 | F04-03 | F04-04 | F04-07 | F06-01 | Slope | Single find | **Total** |
|---|---|---|---|---|---|---|---|---|---|---|---|
| **Debitage** | | | | | | | | | | | |
| Flakes/blades | | 1 | 1 | | | | 1 | 1 | 1 | 6 | **11** |
| Cores | | | | | | | 1 | 1 | | 9 | **11** |
| | | | | | | | | | | | |
| **Retouched tools** | | | | | | | | | | | |
| Scraper | | | 1 | | | | | 4 | 1 | 5 | **11** |
| Knife | 1 | | | | | | | 3 | 1 | | **5** |
| Bifacial tool | | | | | | | | 2 | | | **2** |
| Retouched blade | | | | 1 | | | | | | | **1** |
| Borer/reamer | | | | | | | | 4 | | | **4** |
| Arrowhead | | | | | | 1 | | 2 | 1 | | **4** |
| Microlith/point | | | | | 1 | | | 2 | | | **3** |
| Flaked tool | | | | | | | | | | 1 | **1** |
| Tool fragment, indet | | | | | | | | 4 | 1 | 3 | **8** |
| Gun-flints | | | | | | | | | | 2 | **2** |
| | | | | | | | | | | | |
| **Total** | 1 | 1 | 2 | 1 | 1 | 1 | 2 | 23 | 5 | 26 | 63 |
| | | | | | | | | | | | |
| **Dating of site** | BA/LC-EH | LC/EH | BA/LC-EH | LC/EH | LC/EH | Unknown | LC/EH | BA/LC-EH? | | | |

Table 5.19. Classification of 63 flint artifacts from the Džarylgač Survey Project, arranged by surface site (indicated by "F"), the trial trench (site H08-01), slope and single finds. The age of the site, based on characteristics of the pottery, is indicated as well.

### 5.4.3 Dating

Without thorough knowledge of the prehistory of the Crimea, it is nearly impossible to date the finds, especially when dealing with flakes and other undiagnostic debris such as at sites DSP08-F03-02/Skalistoe 2 (but see the paragraph on the stone finds below) and DSP08-F04-03. These sites cannot be dated more precisely than "Stone Age" or "Bronze Age". For most of the single finds, it is equally impossible to date more precisely.

The presence of (possibly retouched) regular flint blades at sites DSP08-F03-01 and DSP08-F04-02 may indicate a Late Upper Palaeolithic or Mesolithic age. The same is true for a possible backed point of Gravette-type (length ca. 2.5 cm) at site DSP08-F04-03. Among the single finds there are also blades (6388/01, 8275/01 & 8703/01) and a single platform core (Fig. 5.2) with a length of ca. 7 cm for the production of blades (1579/02). At least one of the single finds (9050/01) is not prehistoric but far more recent. This trapezoid flint object, measuring ca. 2.7 x 2.6 cm, is a so-called gun-flint (type C) which were used in flintlock pistols or muskets from the 16th or early 17th century onwards (Fig. 5.3). The DSP-find is made from a dark grey to black flint and originates from Brandon in the UK, one of the largest production centres for gun-flints (Clarke 1935). The gun-flint must be dated to the second half of the 19th century, more specifically the period of the Crimean-war (1853-1856). During this war Brandon supplied millions of gun-flints to the Turkish Army every year. The heavily damaged edges of the flint testify to it actually being used. A flint artefact with a similar shape and damaged edges but somewhat smaller (8759/01) might also be a gun-flint. It is not made from Brandon flint but most probably from locally available raw material. Whether it is a gun-flint or a prehistoric scraper – these types resemble each other – is unclear.

At site DSP08-F04-04, a fragment of a bifacially worked flint arrowhead was found. The fragment – a large part of the tip is missing – has a maximum length and width of 2.4 and 2.3 cm respectively and is 0.4 cm thick. The arrowhead should be dated in the Neolithic or Bronze Age. A similar type of artifact (9202/10) was also found at site DSP08-F06-01. The maximum length, width and thickness of this broken artifact are 2.2, 1.9 and 0.5 cm. At this site, dated in the

Bronze Age, a few other bifacially worked tools (9208/1 and -/2) were found as well, but whether these are fragments of arrowheads or daggers is not clear. Both fragments are relatively thick (0.7 and 0.8 cm) compared to the arrowheads. The other artifacts mainly consist of simple flakes, cores, angular debris and several retouched pieces/scrapers. The absence of a blade technology and the *ad hoc* character of the assemblage seems to fit well with a dating in the Bronze Age.

In the trial trench at feature 7 at the Bronze Age site DSP07-H08-01 an asymmetrically shaped artifact made from brown-greyish flint was found with the following measurements: 5.0 cm (length), 2.3 cm (width) and 0.7 cm (thickness). The artefact is bifacially worked and can be described as a "knife" (Fig. 5.4).

### 5.4.4 Stone tools

According to the database, the following types of stone tools were recovered: 32 whetstones, five possible whetstones, nine grinding stones, six possible grinding stones, six axes, one possible axe and one stone tool without further specification. Fragments of perforated battle-axes have been found at some of the sites listed in Table 5.19, including DSP08-F06-01 and DSP07-H08-01 (Fig. 5.5). These battle-axes with cylindrical perforations most likely date to the Bronze Age (or perhaps a little bit earlier), which is in agreement with the classification of the pottery. The few undiagnostic flint artifacts from site DSP08-F03-02/Skalistoe 2 may date to the Bronze Age as well in view of a battle-axe fragment found at the same location. The whetstones and grinding stones are likely to be later prehistoric (Neolithic and/or Bronze Age) as well.

### 5.4.5 Conclusion

During the Džarylač Survey Project a modest amount of flint artifacts and stone tools was found. Artifacts dating to the Middle Palaeolithic, which are known from other parts of the Crimea, have not been found. The oldest artifacts (a core, a point and a number of blades) date to the Late Upper Palaeolithic and/or the Mesolithic but whether these single finds are part of hunter-gatherer campsites could not be established. More than one-third of the flint artifacts can be dated to the Bronze Age in view of their association with pottery and the presence of for example arrowheads and other bifacially worked tools, knives and perforated battle-axes. The *ad-hoc* character of the flakes and cores and the grinding stones also fit well in a Bronze Age settlement context. A Brandon gun-flint, a relic of the Crimean-war, is the most recent artefact.

## 5.5 OSTEOLOGICAL DATA | M.E. VAN KRUINING

This section considers the osteological finds from three trial trenches, DSP07-H01-01, DSP07-H10-01, and Skalistoe 5 (see Appendix 1). Bones found during the field surveys are not discussed here, since not all were collected. The preservation is generally poor, and it is difficult, if not impossible, to determine whether they are part of an archaeological context, or were left behind in more recent times.

From the three trenches among other finds, animal remains have been recovered that were preserved well enough to be studied. The animal remains have been found at different stratigraphic levels in the trenches, but these have not been connected with distinct periods and therefore, the bones from each site were treated as one sample. The archaeozoological remains can provide information on the diet of the inhabitants of the area from the Bronze Age until the Late Classical-Early Hellenistic period.

Only a small part of the excavated bones was preserved well enough to be identified to the skeletal element and the species or group of species. Other aspects were in some cases visible, such as marks on the bones caused by gnawing, weathering, preparation of food or disease. First the results for the individual sites shall be presented, then an overall picture will be drawn.

### 5.5.1 DSP07-H01-01

On site H01-01, two trenches were excavated and labelled Feature 5 and Feature 18. Both features are dated to the Late-Classical and Early Hellenistic Periods. Some of the finds from Feature 5 can also be dated to the Bronze Age. From both trenches, in total 87 fragments of bone were excavated. Of these only 23 fragments or 26%, could be classified. The majority of the bones were too fragmented and/or too worn to warrant identification. Among the indentified bones, three species could be recognized: cattle, sheep/goat and horse. The difference between the bones of sheep and goat is small and in this case not obvious (Schmid 1972). They are therefore grouped in one combined category. Remains of sheep/goat are the most numerous, but there is not a big difference in number compared to cattle and horse (Table 5.20). Most of the bones (eight; 35%) originate from the lower part of the leg, such as metacarpi/metatarsi and phalanges. The horse bones present were only those from the lower part of the leg. Another category with 21.7% (five in number) are teeth.

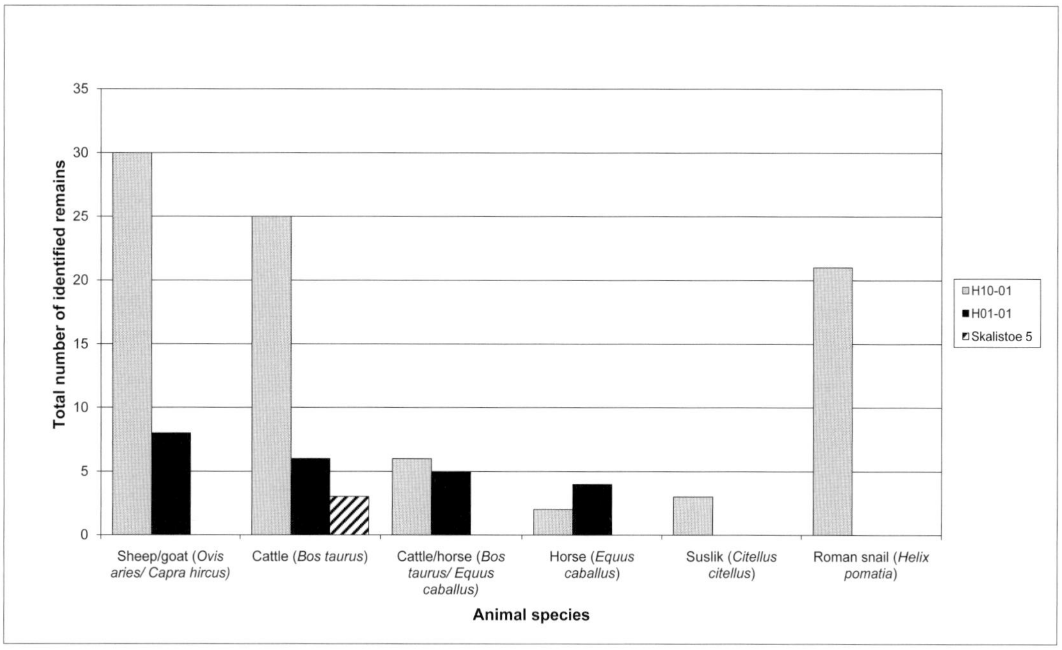


Table 5.20. Distribution of identified animal remains.

A lot of bones show marks of weathering or gnawing. Therefore, it is likely that the bones laid on the surface for quite some time, instead of being buried immediately. The gnaw marks are an indirect indication of the presence of dogs (or other canines) on the site. On two bones cut marks are present. One of them is a phalanx I (foot bone) from a horse. This might be caused by removal of the skin from the bone. Although horses are normally not kept for their meat, this one probably might (at least partly) have been eaten.

## 5.5.2 Skalistoe 5

Only three identifiable bone fragments were excavated on the Late Classical-Early Hellenistic site of Skalistoe 5. All three are fragmented and heavily worn cattle bones.

## 5.5.3 DSP07-H10-01

The preservation at the Late Classical-Early Hellenistic site H10-01 was better than at the H01-01 site and therefore 68% of the animal remains could be identified (88 of in total 129 bones) to species and skeletal element. The other bones were too fragmented and/or too weathered to be identified.

Among the 88 identified animal remains, five animal species are present: cattle, sheep/goat, horse, suslik (Eurasian ground squirrel, *Citellus citellus*) and Roman snail (*Helix pomatia*). In some cases it was not possible to classify a bone as either horse or cattle. These bones are therefore taken as one category. Like the H01-01 site, the sheep/goat remains are the most numerous together with those of cattle (Table 5.21).

Horses, however, are poorly represented (4.5%). Among the mammal remains, teeth again form a large category, namely 33.8% (22 teeth) of the identified remains. These are also the only remains found from horse. The other skeletal elements are evenly distributed among both sheep/goat and cattle. These include 26 fragments (46.4%) from the vertebral column and the chest (vertebrae and costae). Other skeletal elements present are mainly those from the lower part of the leg (10.8%).

Almost all bones are worn to very worn. Other phenomena such as cut marks and burning were observed on a few bones and teeth. Cut marks are present on four sheep/goat bones (costae, astragalus and humerus), two on bones of cattle (costae), one on a bone which is either cattle or horse (vertebra) and five on bones that are unidentifiable to a certain animal species. These marks are indications of butchering. Next to that, three burned teeth of sheep/goat are present.

Apart from remains of cattle, sheep/goat and horse, Roman snails and bones from suslik were found. The snails are edible and are a common find in archaeological contexts in for instance Hellenistic southern Greece (Prummel 1996; Reinders & Prummel 2003). This is not the case with suslik. The suslik is a burrowing rodent which is commonly found in the area. The suslik bones found during the excavation are hardly worn compared to the other bone material. They do not exhibit any cut mark or traces of burning. Although susliks are edible, it looks like all bones originate from one animal that died in recent times and does not belong to the archaeological record. In the excavated feature also a foxhole was found. This even stresses the probable recent origin of the suslik bones.

## 5.5.4 Discussion and conclusion

The animal remains from the three excavated sites are badly preserved and therefore difficult to identify. At the Skalistoe 5 site only three cattle bones were found. This collection is considered to be too small to be compared to the other two sites. From the bones that could be analysed, only a narrow animal spectrum came to light. Apart from suslik that was probably not eaten and Roman snails that were not systematically collected during the excavations, the remains of three species were found. Sheep/goat and cattle are the most numerous and horse is only present with a few remains. This is most clear on the H10-01 site. On this site horse bones are far less numerous than those of sheep/goat and cattle. The function of horse might have been different on this site or in this period. Gnaw marks on bones are an indirect indication of a fourth mammal present on the sites: dog. All the bones were excavated on sites associated with animal husbandry. Pigs probably were not a part of the food economy. This is also pointed out by Marčenko & Vinogradov (1989) who state that the major part of animals in Greek colonies is acquired from local tribes. Therefore, in both Greek and Scythian archaeological contexts, predominance is visible of sheep/goat and pigs are completely absent.

On a small number of bones cut marks were found, including one on a horse bone, which may be a sign of butchering. Other bones are (partly) burned, which may be an indication of food preparation. The mammal remains found mainly consist of teeth, bones from the lower part of the leg and parts of the torso. The bones that were preserved well enough to be analysed do, however, more or less exclude fragments of the long bones like the upper part of legs. Those are, especially when fragmented, difficult to determine and were therefore left out of the analysis. The skeletal distribution as stated above is for that reason probably not entirely representative. Despite the small amount of identified animal remains, they do give a glimpse of the food economy people depended on in the Bronze Age and Late Classical-Early Hellenistic period.

| | Sheep/goat (*Ovis aries/ Capra hircus*) | Cattle (*Bos taurus*) | Cattle/horse (*Bos taurus/ Equus caballus*) | Horse (*Equus caballus*) | Suslik (*Citellus citellus*) | Roman snail (*Helix pomatia*) |
|---|---|---|---|---|---|---|
| H10-01 | 30 | 25 | 6 | 2 | 3 | 21 |
| H01-01 | 8 | 6 | 5 | 4 | | |
| Skalistoe 5 | | 3 | | | | |

Table 5.21. Overview over identifiable faunal remains.

The findings of the DSP are in strong contrast to those from Panskoe I (Kasparov 2002), where horse constituted the vast majority of the bones (58.3%). At this site, horse meat certainly was eaten (Kasparov 2002, 332). However, pig was not attested either, whereas the earth-hare (jerboa, *Allactaga jaculus* Pall.) constituted a fair part of the bones, and was probably hunted for meat (13.6%).

CHAPTER 6

# Prehistory and history of the DSP landscape

*Compiled by P. Guldager Bilde*
*with contributions by P. Attema, T. de Haas, W. de Neef,*
*T. Smekalova, C. Williamson & K. Winther-Jacobsen*

When initiating the project, its delimitation was determined as the area roughly between Černomorskoe to the W and Masliny to the E. The southern border of the study area was the edge of the Plateau, which constitutes the spine of the peninsula. Thus, the overall study zone is roughly a trapeze measuring 5 (NS to the W) x 25 (EW to the S) x 15 (NS to the E) x 30 km (along the coast). Because the project was cancelled prior to its completion, some parts of this zone were not investigated at all. This holds true for the area between Černomorskoe and Panskoe and for the Plateau itself. In addition, in 2008 due to the wish of our Ukrainian partners, the study zone was further restricted to a 12 km wide N-S transect located roughly between Mežvodnoe and Vodopojnoe, but covering all environmental zones of the region.

## 6.1 FROM SITE CLASSIFICATION TO SITE INTERPRETATION | P. ATTEMA, T. DE HAAS, W. DE NEEF & C. WILLIAMSON

### 6.1.1 Introduction

In order to discuss the long term settlement organization in the DSP landscape, we present in this paragraph a classification of the sites mapped in the transect, based on their descriptions in the catalogue. This site classification will then be the basis of the period-by-period maps that, in combination with the landscape classification, will serve a functional interpretation of settlement organization in the wider area of study. Archaeological sites are defined in the DSP as *all artefact scatters and structural features that are indicative of past human activity at circumscribed locations in the landscape, provided these remains are detected 'in situ'* (Chapter 2.5). Fig. 6.1 shows that archaeological sites have been attested in all land types. In the following we will try to establish whether there are any correlations between site classes (and related human activities in the landscape) and land types as discussed in Chapter 3. To this end we will first present the site classification to subsequently show in the period-by-period maps how site types may correspond with land types. The resulting interpretation of settlement organization in the DSP survey area will then be evaluated in light of the regional settlement pattern in Sections 6.2.1 and 6.2.3.

The *first* step in devising the classification presented here entailed the characterization of the archaeological data and their attribution into defined classes. The substantial environmental differences and related post-depositional histories between the higher and lower areas in the landscape are determining factors in the appearance of archaeological sites and the ways in which they may be investigated. Because of this we decided to distinguish two classifications: one for the cultivated area featuring the landscape types Lowland Ridge/Coastal Cliff, Coastal Lakes, Pediment and one for the higher parts in the landscape featuring the landscape types Hillsides and Plateaus/Uplands.

### 6.1.2 The cultivated area

Most sites in the cultivated area (Lowland Ridge/Coastal Cliff, Coastal Lakes, Pediment) appeared as (multiple) surface scatters of artefacts in the ploughed terrain, although isolated artefacts did occur. A minority of sites was (also) visible as an elevated feature in the landscape (for instance kurgan, cadastre or ash heap). Although magnetometer surveys at Panskoe I gave evidence for new structures, we have no information as of yet on possible subsurface archaeology revealing plans of (stone) structures at other sites in the cultivated area. In the first column in the first table of Appendix 1, all sites found are listed with a short description based on the catalogue entries in the second column. In the third column the attribution to one of the following classes is given:

- Part of a large complex settlement (scatters related to Panskoe I)
- Large settlement (> 1 ha, multiple scatter)
- Small settlement (ca. 0.5 ha, single or multiple scatter)
- Non-settlement (< 0.1 ha, single scatter)
- Funerary area (ploughed out and/or intact kurgans)
- Land divison
- Isolated find

The map (Fig. 6.2) shows the distribution of the various classes over the landscape.

## 6.1.3 The slopes

Most settlement sites in the higher parts (Hillsides and Plateaus/Uplands) are noticed by their stone features or by marked differences in the vegetation, while kurgans appear as clear elevations. Artefact scatters are rare and generally only appeared in the occasional ploughed areas. Many sites were subjected to magnetometer surveys yielding images of the plans of the stone structures and, when wider geomagnetic prospection was conducted, the extent of the settlements to which they belong. Therefore more settlement classes could be discerned compared to the cultivated area. In the first column of the first table of Appendix 1 all sites are listed with in the second column a short description based on the catalogue entries. In the third column the attribution to one of the following classes is given:

- Large complex settlement (multiple stone structures, pits, pens, > 2 ha)
- Small complex settlement (multiple stone structures, pits, pens, < 1 ha)
- Large simple settlement (multiple simple stone structures > 2 ha)
- Small simple settlement (multiple simple stone structures < 1 ha)
- Isolated double pen
- Isolated stone structure
- Funerary area (groups of kurgans)
- Quarry
- Dam
- Crop marks
- Natural phenomenon

The map (Fig. 6.3) shows the distribution of the various classes over the landscape.

## 6.1.4 General observations

From the list of site classes compiled for the cultivated areas and the higher parts of the landscape we observe the following:

- Animal pens have only been discovered on the Hillsides, both isolated and in combination with settlements. In the cultivated area no pastoral structures were identified.
- Settlements on the Hillsides all feature stone foundations, but indications for the use of stone as a building material were very rare in the cultivated area. Sites here consist of pottery scatters only (excepting the Hellenistic coastal farmhouses). Although this might be due to the lack of geomagnetic observations in the ploughed area, the significant absence of stone construction materials in the pottery scatters would indicate that less durable materials were used here.
- Settlements on the Hillsides are often associated with so-called household pits that were presumably used for storing foodstuffs. No such pits have been identified in the cultivated area. This is likely due to the lack of geomagnetic observations in the ploughed area. Against this argument would plead the general absence of surface discolorations associated with the small pottery scatters and the fact that in the rare instances that geomagnetics were carried out on pottery scatters, no subsurface archeology was found, and finally that augerings with the exception of two sites did

not reveal the presence of clear cultural layers. This would indicate that the cultural layers associated with the pottery scatters have been homogenized by ploughing.

- Possible land divisions were only observed in the cultivated area (cadastre) and were visible as elevations in the landscape (earth banks). Although on the Hillsides several 'lines' in the landscape were observed on the geomagnetic maps (Fig. 4.152-153, 4.201, 4.235, 4.249), these cross-cut older features and most probably are of recent date.
- Kurgans appear both in the high- and cultivated parts of the landscape.
- Field conditions allowed to record off-site pottery only for ploughed fields. In the cultivated areas, where survey was carried out exclusively in ploughed fields, off-site pottery was consistently present in the fields and as such can be used as an indicator of past land use (manuring, discard). In the higher parts the recording of single sherds or very low densities of sherds are not necessarily indicative of off-site, but did in some cases signify sites hidden below the vegetation revealed by geomagnetic survey.

## 6.1.5 Period-by-period maps

The *second* step consisted of the compilation of period-by-period maps indicating the distribution of site classes over the landscape types. Because the sites that appeared during survey were predominantly Bronze Age, Late Classical-Early Hellenistic and some Early Modern, the discussion here is focused on these periods; no sites from the Medieval period were identified, although sherds from this period were collected in off-site contexts. We will start out with describing the map depicting the Bronze Age sites.

6.1.5.1 Bronze Age sites in the DSP transect (Fig. 6.4)
The Lowland Ridge has:

- Large settlements. They were classified as 'simple' because we do not know anything about their internal structure due to the absence of good geomagnetic data. The sites measure 1.2 (DSP08-F03-02 (Skalistoe 2)) and 1.9 ha (DSP08-F06-01). These sites are associated with variations in elevation and soil composition.
- Funerary areas (kurgans). These however do not show a clear spatial relationship with the settlements.

The Hillsides have:

- Two large complex settlements. They were classified as 'complex' because they consist of various types of stone structures, at times associated with pits and pens. We have to be cautious however as the dating of the agglomerations is based on a few observations only. Not all structures within an agglomeration need therefore be of Bronze Age date, although the consistency of the structures makes this likely.
- Small complex settlements. Three clusters on one hill, the others are 'isolated'. Such small settlements were not identified on the Lowland Ridge.
- Isolated double pens, probably corresponding to settlements not yet found in the survey.
- No clear funerary areas; the kurgans so far are thought to be all Late Hellenistic-Early Classical and the stone circles appeared to be natural phenomena, not megalithic tombs.

6.1.5.1.1 Functional interpretation
While the larger sites on the Lowland Ridge are, due to the availability of arable soils and proximity of the Lake Džarylgač, most plausibly interpreted as the permanent habitations of communities involved in mixed farming, hunting and fishing, the settlement remains on the Hillsides rather point to pastoral activities. Whether the settlements and animal pens on the Hillsides were used on a semi-permanent basis by seasonally returning groups of pastoralists *or* were settled all-year-round by pastoralists/farmers *or* were the facilities of people living on the coastal ridge, we are unable to say. The many pits at some of the sites – if we accept their interpretation as storage pits – may point to permanent settlement or at least to prolonged stays. However, ethnographic parallels should instruct us as to whether pastoralists seasonally returning to the same spot would have built such extensive durable stone structures and facilities as were mapped in the geophysical survey (see Chapter 5.1.4). Unlike the off-site material from the Late Classical-Early Hellenistic period, the off-site material from the Bronze Age could not be related to sites; therefore this should be interpreted as belonging to sites that have either not been located or have already disappeared altogether.

6.1.5.1.2 Chronological interpretation
Permanent settlement on the Lowland Ridge started in the Middle Bronze Age and continued into the Late Bronz Age. It is unclear when in the Bronze Age the construction of the pastoralist settlements began.

6.1.5.2 Late Classical/Early Hellenistic sites in the DSP transect (Fig. 6.5)
The Lowland Ridge/Pediment has:

- Three large settlements (possibly hamlets or farms as they consist of more than one pottery scatter), one small settlement and two non-settlements (possibly storage facilities?).
- A large complex settlement (Panskoe I).
- Possibly one more large settlement.
- A number of small settlements.
- Funerary areas.

The Hillsides have:

- Large complex settlements (pens, numerous stone house structures, household pits).
- Large simple settlements (e.g. agglomerate of >10 simple structures).
- Small complex settlements (e,g, three structures and pits) and small simple settlements (isolated farmhouses).
- Funerary areas.

Plateaus/Uplands[76] have:

- Funerary areas.

6.1.5.2.1 Functional interpretation
The largest pottery scatters are found on the Lowland Ridge. They are best interpreted as hamlets consisting of a number of rural dwellings made of non-durable materials (wooden construction, mudbrick, wattle and daub, straw). They lack any signs of stone architecture. The Pediment is characterized by small pottery scatters, again without any sign of durable building materials. These small scatters may represent single huts or sheds. Sites near the Coastal Lakes may have been fishing stations or involved in salt extraction, sites inland may be associated with arable farming. The consistent pattern of off-site materials around sites would be expected if we accept the manuring hypothesis, which could correspond with the presence of animal pens or simply household waste (e.g. Wilkinson 1982; 1989; Bintliff and Snodgrass 1988; Bintliff 1992). Larger single scatters on the Pediment may just as well represent small farmsteads. Settlements may hypothetically be related to groups of kurgans (Fig. 4.270). Most sites on the survey on the lowland however seem to represent rather poor dwellings.

The Hillsides are characterized above all by small villages and hamlets, although isolated farms also occur. Here almost all the evidence is based on stone structures and not on sherd scatters. As on the Lowland Ridge and Pediment, also on the Hillsides most sites seem to represent rather basic dwellings. There is however a limited number of more complex house plans with an internal articulation. These sites may be comparable to the non-fortified farmsteads along the coast that are known from the topographic survey (see Section 6.2.2) Kurgans occur on separate hills and are clearly separated from the rural settlements and also here it is possible to model which funerary areas would have belonged to which settlements. On account of the presence of animal pens in the larger simple settlements, it is likely that they were primarily inhabited by pastoralists. On the other hand, farming likely took place as well, as is indicated by the complex architectural structure on DSP07-H18-01.

6.1.5.2.2 Chronological interpretation
The majority of the sites date within the relatively short time span between mid-4th century BC and ca. 270 BC.

[76] However, survey has been very limited here.

6.1.5.2.3 LC/EH in sum

Taken together the site data for the Late Classical-Early Hellenistic period point to a complex rural site typology with, speaking in hierarchical terms, Panskoe I as a complex fortified settlement at the top. Other rural site types are fortified and non-fortified farmsteads along the coast (as recorded in the topographic survey), hamlets inland on the Lowland Ridge, isolated small farmsteads and sheds on the Pediment, villages, hamlets and farmsteads on the Hillsides. The overall economy in the cultivated areas is likely to have been geared towards agricultural production and fishing/salt extraction, while on the Hillsides pastoralist activities will have been a more prominent feature in the economy. This settlement organization developed quickly, making a marked impact on the landscape, but was also quickly abandoned.

6.1.5.3 Early Modern sites in the DSP transect (Fig. 6.6)

The Lowland Ridge/Pediment has:

- Three small settlements.
- One non-settlement (dump?).

The Hillsides have:

- Three small simple settlements.
- Two isolated stone structures.
- One dam.

6.1.5.3.1 Functional interpretation

The sites in the cultivated areas are all characterized by small scatters with few indications of stone architecture. It is difficult to tell whether these scatters belonged to villages or isolated structures. Some may correspond to larger nearby settlements known to have been in the area based on the cartography.

Although these sites may be small settlements, they could also be temporary or seasonal structures; the single non-settlement site is located close to a road and may have been a waste dump.

On the Hillsides, the isolated stone structures may have served as facilities related to the villages in the vicinity. The simple settlements may also be related to these villages. One site may have been a settlement of several small houses (DSP07-H04-02) but this is difficult to interpret due to the heavy vegetation. Dams in the ravines are relatively common, and the one found during the survey may have been related to the small settlement close by.

Off-site material from the Early Modern period was found in nearly every field surveyed. In field DSP08-F04 this may be related to the Early Modern settlement at Kipčak, The heavier concentration near DSP08-F05-01 indicates Early Modern activity near this Late Classical-Early Hellenistic site.

6.1.5.3.2 Chronological interpretation

The material from this period is relatively poorly known, relying primarily on the presence of glazed pottery, and is thus difficult to date with any precision. Therefore, it is not possible yet to interpret possible chronological relationships or developments in the Early Modern period for the area studied in the DSP.

## 6.2 The settled landscape | P. Guldager Bilde & K. Winther-Jacobsen

In this section we will place the observations on the settled landscape within the transect as found in the systematic survey (Section 6.1) in the wider perspective of the DSP study area.

In the cultivated zone, prehistoric sites are associated with relatively high density discrete clusters of artefacts sometimes associated with soil changes and elevations. In the past, some of these coastal sites have been excavated. On the slopes, the sites are associated with relatively low density discrete and diffused clusters of artefacts associated with soil changes. Structural remains associated with a few artefacts and distinctive vegetation indicate spatially well-defined soil changes.

### 6.2.1 Prehistoric period (Fig. 6.7)

Based on the methodology and the post-depositional history of the individual sites, six different categories of sites can be distinguished: 1) Structural remains of settlements visible on the surface associated with a few artefacts and distinctive vegetation to indicate spatially well-defined soil changes and relatively low density discrete and diffused clusters of artefacts such as DSP07-H06-01, DSP07-H08-01, DSP07-H14-01, possibly DSP07-H01-01 and DSP07-H02-01, whereas it is difficult to form an opinion about DSP07-H07-1-2, DSP07-H09-01 and DSP08-H13-01 due to sampling methods. 2) High density discrete clusters of artefacts sometimes associated with soil changes and elevations interpreted as settlements such as DSP08-F03-02 and DSP08-F06-01. 3) Structural remains of settlements associated with a Bronze Age element but dominated by finds from later periods such as DSP08-H12-01 and DSP07-H18-01. 4) Kurgans associated with Bronze Age finds such as DSP08-F04-4 (Feature 3). 5) Isolated artefacts possibly associated with burials such as DSP07-F04-01 (axe) and DSP07-F05-02 (stele). 6) Previously identified Prehistoric sites within the survey area such as Baj Kijat 1-2, Burun Eli 1, Jarylgač Severnoe, and Skalistoe 2, where excavations have been carried out. 7) Sites, where Prehistoric material has previously been recorded by means of *razvedki*-type of survey, e.g. the Vodopojnoe 1-5 sites. The overall topographical distribution suggests that settlements are located in all the ecological zones of the DPS landscape even if the coastal zone is only represented by one settlement, namely Baj Kijat 1. The settlements in the cultivated zone on the N side of the Džarylgač Lake such as DSP08-F06-01, Skalistoe 2, Burun Eli 1 and Severnoe 1 are all substantial sites. Assemblage studies of the DSP sites suggests that DSP08-F06-01 may have been associated with lithics production – at least when compared with Skalistoe 2. The chronology is very coarse. Both DSP sites have Middle and Late Bronze Age elements, and all the sites may have been occupied intermittently; nevertheless, the spatial distribution is approximately equidistant. The area on the N side of the Džarylgač Lake is broken by a roughly E-W running ridge surface if generally very smooth and there is no current conclusive evidence to suggest that the localities were sheltered from the dominating winds, except Skalistoe 2, which is located in a shallow depression.

Notably, we have not been able to locate any Bronze Age settlements on the Pediment of the S side of the Džarylgač Lake. One secure Bronze Age kurgan (DSP08-F04-4 Feature 3) and one possible burial of the same general period (DSP07-F04-01) have been identified in the same zone. Moreover, the two possible Bronze Age burials identified on the N side of the lake (DSP07-F07-01 (Feature 3) and DSP07-F05-02) are also not located anywhere near contemporary settlements, and the location of kurgans during this period may be determined by other criteria.

A number of settlements have been identified on the slopes all belonging to the first category of settlements mentioned above. Even if structural remains have been identified at all the sites, the Bronze Age component in the form of ceramics is not always firm. All the sites were reoccupied during the Late Classical-Early Hellenistic period, and the precise association between the architecture and the Bronze Age ceramics is tentative. The location of the settlements on the slopes appears to be guided by dominating winds as the majority is sheltered by higher hills on the NW side. Furthermore, generally the settlements are all located towards the tip of a promontory near a ravine which provided the easiest access to fresh water in the form of wells as well as dams. Of course, the ravines also provided shelter for the animals.

The same topographical features are found in the Vodopojnoe Ravine where at least four Bronze Age settlements have been identified (Vodopojnoe 2-5). Apart from these topographical amenities, Vodopojnoe 2-3 are also located at geostrategic points namely at the major crossings of the ravine tying these sites into the system of long distance routes (see Chapter 6.6).

In order to compare approximate population density, we will assume that the settlements were contemporary and continuously occupied. Even compared to the excavated sites on the slopes, the finds densities N of the lake are disproportionally higher. Also the settlements on the N side of the Džarylgač Lake are located at ca. 4-5 km apart; whereas on the slopes and around the Vodopojnoe Ravine, the settlements are located much closer between 0.7 and 2 km apart. The topographical variety of the DSP area provides the basis for a mixed agropastoral economy. The absence of settlements on the Pediment S of the lake suggests that this zone was reserved for a different purpose. Held up against the relatively closer settlement pattern on the slopes, it is possible that the people settled there cultivated the Pediment S of the lake while grazing their flocks on the slopes. Of course, if the settlements were occupied intermittently, there is a whole range of possibilities.

## 6.2.2 Greco-Roman period (Fig. 6.8)

Based on the methodology and the post-depositional history of the individual sites, the following different categories of sites can be distinguished: 1) Structural remains of settlements visible on the surface associated with a few artefacts and distinctive vegetation to indicate spatially well-defined soil changes and relatively low density, discrete and diffused clusters of artefacts such as DSP07-H01-01, DSP07-H06-01, DSP07-H07-01, DSP07-H10-01/H18-01, DSP08-H11-02, DSP08-H12-01, and possibly DSP07-H02-01, DSP07-H08-01, and DSP07-H14-01, whereas it is difficult to form an opinion about DSP08-H13-01 due to sampling methods. 2) High density discrete clusters of artefacts sometimes associated with soil changes and elevations interpreted as settlements such as DSP07-F08-01, Skalistoe 2, DSP08-F04-03, and DSP08-F04-07. 3) Relatively low density, discrete clusters of artefacts sometimes associated with soil changes and elevations interpreted as settlements such as DSP07-F02-01, DSP07-F02-02, DSP07-F02-03, DSP07-F02-04, DSP07-F06-02-03, DSP08-F02-01, DSP08-F04-01-02, DSP08-F04-06, DSP08-F05-01, DSP08-F06-03 and DSP08-H20-01. 4) Individual sherds associated with subsurface structures such as DSP08-H17-01 and DSP08-H23-01-02. 5) Kurgans associated with Greco-Roman finds such as DSP07-F01-02, DSP07-F03-01, DSP07-F07-01, and probably DSP08-F01-01, DSP08-H17-02, DSP07-H19-03-04, DSP08-H24-01, DSP08-H26-01-02. 6) Previously identified Greco-Roman site within the survey area such as Černomorskoe and associated farms, Panskoe I-V, Groty 1, Skalistoe 1, 3-4, Baj Kiat 1 and 3, Burun Eli 1-5, Jarylgač, Jarylgač Vostočnoe, Masliny 1-2, Meževodnoe, Severnoe 2 and possibly Vodopojnoe 2.

The overall topographical distribution suggests that settlements are located in all the ecological zones of the DPS landscape. However, unlike the Bronze Age, a large number of the previously known sites are located on the coast. Furthermore, a large number of previously known sites are located around the Kelšeich and Sasyk Ravines, whereas the Bronze Age sites were located on the Vodopojnoe Ravine. At least one of the sites in the Vodopojnoe Ravine has a Hellenistic element, but the area was not thoroughly investigated, so it is difficult to form an opinion as to the settlement pattern in this ravine after the prehistoric periods and before the Early Modern times.

Only kurgans are attested on the Coastal Ridge N of the Džarylgač Lake, but there appear to be two groups of settlements in this zone: two sites of the third category (DSP08-F02-01, Skalistoe 2) located 600-700 m from the present coast and close to several coastal sites (Skalistoe 1, 5 and 8). Skalistoe 2 is located in the same area as the Bronze Age site Skalistoe 2. A small cluster of sites appear to be located along the old road between the modern settlements at Meževodnoe and Vodopojnoe (DSP07-F08-01, DSP08-F05-01, DSP08-F06-03, DSP07-F06-01-03). Unlike the Bronze Age, the cultivated zone on the S side of the Džarylgač Lake is spatially, densely occupied (DSP06-F04-01, DSP07-F02-01-04, DSP08-F04-01-03 and DSP08-F04-06-07). A long chain of kurgans of unknown date range is located in the same area, at least one of which dates back to the Bronze Age. A Late Archaic amphora rim was collected near the southern-most of these (DSP07-F02-05). DSP07-H11-01 is located at the foot of the Hillsides where there is a string of 19th century villages. 850 m to the NE is a small chain of contemporary kurgans (DSP07-F03-01). Immediately NW of this group is a large kurgan reused as a cemetery by the nearby village of Snežnoe (Fig. 6.9). According to the traditional typology a kurgan of this size would be dated to the Bronze Age.

Several of the sites on the slopes are new (DSP07-H04-01, DSP07-H10-01, DSP08-H11-02, DSP08-H20-01, DSP08-H23-02, DSP08-H24-03 and probably DSP08-H12-01). However, the location of a small majority of Hellenistic settlements sites on the slopes coincides with Bronze Age settlements (DSP07-H01-01, DSP07-H02-02, DSP07-H06-01, DSP07-H07-01, DSP07-H08-01, DSP08-H13-01, DSP08-H14-01). The slopes are characterised by the numerous kurgan chains running roughly N-S, the most impressive of which is the chain on Hill 19 consisting of at least 31 kurgans (see Appendix 1). The kurgans have been looted one time or other. In some cases, a few Hellenistic sherds and flint tools have been collected around the kurgans probably left by looters. Apart from the officially excavated ones (DSP07-H19-01 Kurgans 1-3 and 10), the kurgans are almost impossible to date. The excavation of a strong anomaly near Kurgan 3 produced an amphora toe of the mid-4th century BC. Generally, the kurgans and settlements are located on different hills, and there appears to be no direct spatial relationship between the two.

In order to compare approximate population density, we will assume that the settlements were contemporary and continuously occupied. The combination of densities, assemblage composition and geographical size implies the existence of a settlement hierarchy. One group is defined by the presence of tile and tablewares combined with high density level: DSP07-F08-01, DSP08-F04-03 and DSP08-H11-01. To this group should be added the Černomorskoe settlement, Panskoe I and II, and possibly Groty 1 and Masliny 1. Tiles are in general rare on the Tarkhankut Peninsula even at Panskoe I and significantly, single tile fragments have been collected from all these sites. Apart from Panskoe I (and II), tablewares are rare at the sites identified by survey, but tablewares also always occur at the sites where tiles have been collected. Otherwise assemblage compositions distributions are diverse, possibly due to the relatively small size of the surface assemblages. The spatial distribution of these settlements is dispersed at a distance no closer than 3 km. The combined elements characterising this group of sites suggests that they are multiple settlements as has been proven the case by excavation at Černomorskoe, Panskoe I and Masliny 1.

Another group is defined by the absence of tile, almost absence of tablewares and relatively lower overall densities (DSP08-F02-01, Skalistoe 2, DSP08-F06-03, DSP08-F05-01, DSP07-F06-01-03, DSP07-F02-01-04, DSP08-F04-01-02, and DSP08-F04-06-07) and possibly Groty 2. In the group probably belong slope sites DSP07-H04-01, DSP08-23-02, DSP08-H24-03, DSP07-H01-01, DSP07-H02-02, DSP07-H07-01, DSP07-H06-01, DSP07-H08-01, DSP08-H20-01, DSP08-H14-01. Due to different post-depositional processes surface densities are very low on the slopes making direct comparative analysis with sites in the cultivated zone impossible. On the other hand, the number of structures recorded at some of these sites is substantial, but since they occupy sites also inhabited during the Bronze Age it is impossible to date the individual structures for certain based on surface observations alone. The proportion of handmade wares and flint tools appears not to be correlated to the hierarchical structure.

Structural elements define two additional groups: DSP08-H11-02 is a unique site consisting of pit houses, and although stone structures appear to be visible on the surface these could not be definitely identified on the geomagnetic map. DSP07-H10-01/18-01 and DSP08-H12-01 both have multiple stone structures visible on the surface and subsurface, but tiles have not been collected distinguishing these sites from the group mentioned above. Of course, finds densities are very low since the sites have not been ploughed.

Another consistent pattern which appears to divide the settlements into two groups is the occurrence of large clusters of pits located at sometimes a short distance away from the structures (DSP07-H01-01, DSP07-H07-01, DSP07-H10-01/H18-01, DSP08-H11-02, DSP08-H12-01). Of course these features have only been identified at sites where geomagnetic survey were carried out, and they appear to be absent at a number of hills where extensive areas have been surveyed (DSP07-H02, DSP07-H08, DSP08-H13-H14). The area investigated on Hill 6 is too small to postulate absence. On DSP07-H01-01 the cluster of pits was located in the southern part and the excavated pit revealed only Hellenistic material. The remaining sites where clusters of pits have been recorded are all predominantly Hellenistic. The fact that clusters of pits were not recorded on Hill 8 and 14 – the major Bronze Age settlements on the slopes – would suggest that this is not a phenomenon associated with this period.

The chronology is only as precise as the locally available ceramic typologies. Consequently, the period between 400 and 270 BC is well attested due to the publication of Panskoe I. A few finds appear to precede this period (Chapter 5.3.3.6). The earliest finds are two Late Archaic amphora rim fragments collected near DSP07-F02-06 and at Skalistoe 2. At the latter site a 5th century BC amphora sherd was also collected. Additionally, a small cluster of 5th and early 4th century pottery was collected along the present S bank of Lake Džarylgač at and around DSP07-F04, DSP07-F02-01-02, and DSP08-F04-01-03. These finds suggest that the sites mentioned are earlier or at the latest contemporary with the foundation of Panskoe I. The bulk of Greco-Roman material collected dates to the mid/late 4th to early 3rd century. No more than 12 Greco-Roman finds post-date 270 BC. Eight of these sherds were collected at the larger settlements where also tiles were collected as well as higher proportions of tablewares (Panskoe I-II and DSP07-F01-02, DSP07-F08-01, DSP08-F04-03 and Groty 1). This Late Hellenistic phase is documented at Černomorskoe and Masliny 1 and of course at Panskoe I. Two Late Hellenistic sherds were collected at the multi-period (Bronze Age and Late Classical – Early Hellenistic) settlement DSP08-F03-02, where a Late Archaic sherd was also found. The last two Late Hellenistic sherds were collected at Skalistoe 10 and in DSP08-F06.

### 6.2.3 Medieval and Early Modern period (Fig. 6.10)

The Early Modern settlement history can be studied from surface finds, existing and abandoned sites as well as a series of historical maps. There are only two towns of some size, both located close to but sheltered from the sea: Černomorskoe and Mežvodnoe. Apart from two/three coastal/near coastal farms in that area and presumably replaced by Skalistoe by 1955 (e.g. DSP08-F03-02), there are no Medieval to Early Modern sites between the major settlements at Mežvodnoe and Vodopojnoe on the N side of the Džarylgač Lake. The fishing station currently located next to the ancient site Skalistoe 5 is an even more recent addition to the landscape.

Until 1944, the DSP study zone was relatively densely settled with foremost Tatar villages. These are located, where there is the easiest access to fresh water: at the lower slopes in the ravines and at the foot of the slopes of the ridge of the Džangul' Uval, where the water-bearing horizon of the Sarmatian clays is closest to the surface (Smekalova 2007, 83-85).

Thus they are situated as three strings of pearls running roughly E-W. The northernmost consists of (from W to E): Burun Eli, Tubaka, Baj Kijat (present village Vladimirovka), Sakaja, Časilmaj/Čigiltaj (present village Dalekoe), Tanabaj, and Kirgiz Kazak. The middle one is to be found in the N branch of the Vodopojnoe Ravine E of Kirleut/Vodopojnoe (again from W to E): Kubaran, Kurman Adži, Tok Džol', Otus (present village Zorjanoe), Dauldžar' (present village Groznoe), Kul' Džakin, Baim, Tatarian Čongurči and Russian Čongurči (present village Zadornoe). In the middle branch of the ravine are located the villages: Buzak and Džamal' (present village Kirovskoe) with Gadži Gazan in a southern side branch. In the inner part of the S branch of the ravine we find Ak-Baš. The southern string of settlements consists of: (from W to E) Evan(?),[77] Aldermen, Aldermenskaja Skala, unnamed village ruin, Saja, (Koš) Ateš (Kostrovka), Snežnoe, Karlav (Čaikino), Musali (Zajcevo), Džajlav, Kipčak Učkuju/Majorovo, Kurama Kostel' (Malyševka), Kostel', Kurama, Kabla Kurama, and Syrt Kurama. None of these sites were surveyed systematically and intensively. Apart from the eastern part of Kurama Kostel', which was grab sampled, the remaining Early Modern sites are either currently occupied or left as grassy ruins with very low visibility (Fig. 6.11). A discrete cluster of finds (DSP07-F04-02) must probably be associated with Džajlav. This assemblage is particularly interesting because it is the only one of the Early Modern sites where no glazed wares were collected, and according to the maps this site was a ruin already in 1837.

Not all settlements were occupied at the same time (Table 6.1). In addition, today there are by far fewer settlements than was the case in the 19th and even in the first third of the 20th century. We can follow the villages' demographic development through the maps of the 19th century (Chapter 6.2.3), and from a head count in 1806. From the half-verst map of the 1890s we are also informed of the ethnic composition of the individual towns and villages. The earliest more detailed and reliable map is Major General Semjon Mukhin's *Vojennaja topografičeskaja karta poluostrova Kryma* in 10 sheets. It is a topographical map made for military purposes in the scale of 1:168 000. The survey was done in 1816 and the map was printed the following year (published in English, French and Italian editions; see also Jankowski 2006, 11-12). The second map is by Betev made in 1837 (see Chapter 2.2.1). Very valuable are the handwritten emendations made to the map from a second survey made in 1865. Demidoff's map compiled in 1837 (published in 1853) is included as a supplement. Very important is also the half-verst map of the late 1890s. Last to be mentioned is the pre-WWII map of 1931 revised in 1941, which presents the situation at the eve of the Tatar deportation.

According to the half-verst map, in the late 19th century the two larger towns were inhabited by Russians alone. Of the first string, Tatars were the only inhabitants with the exception of Baj Kijat and Kirgiz Kazak, which had a mixed settlement of Tatars and Russians; in the 1931 map Tanabaj is mentioned as *nemeckij* (German). Of the villages in the Vodopojnoe Ravine, only in the two settlements at either end of the N branch lived Russians. In Kirleut/Vodopojnoe the Russians lived side by side with the Tatars, whereas at the end of the ravine a small Russian village was built with the same name as the nearby Tatar village Čongurči; Russians were also the exclusive inhabitants of the village of Gadži Gazan in an isolated side branch off the middle branch of the ravine. Finally, in the southern string the village of Kostel', which was in ruin already in 1837, was Tatarian and the very small settlement of Musali was Russian. The other villages we have information about consisted of a mix of Russian and Tatar inhabitants.

[77] Only mentioned in the map of Demidoff 1853, but an area of ca. 100 x 100 m with a different colour can be seen on the map of Google Earth, which may mark this location.

| Name in 19th century | Recent name | 1790 | 1817 Mukhin | 1837a Demidoff | 1837b Betev | 1865 Betev rev. | 1890s | 1931 | 1941 | 1955 1:25,000 | 1:100,000 | Number of households | | | | | Ethnicity of inhabitants[1] |
|---|---|---|---|---|---|---|---|---|---|---|---|---|---|---|---|---|---|
| | | | | | | | | | | | | 1806[2] | 1817 | 1837 | 1865 | 1890s | |
| *Coastal zone* | | | | | | | | | | | | | | | | | |
| Šeichlar[3] | Černomorskoe | | | | X | X | X | X | X | | | | | 26 | 82 | 104 | Russian |
| [N] Šeichlar | Černomorskoe | | | | X | cem. | | | | | | | | 14 | | | |
| Ak Mečet[4] | Černomorskoe | X | X | X | X | X | X | X | | X | X | | | | | | |
| Takil'[5] | Černomorskoe | | | | X | | | | | | | | | 6 | | | |
| *Cultivated zone N of Džarylgač Lake/Vodopojnoe Ravine* | | | | | | | | | | | | | | | | | |
| Kula Šeich | | | X | X | X | X | ruin | | | X | OTF | | | 10 | 1 | | |
| Kula Šeich *chutor* | | | | | | | X | | | | | | | | | | |
| [Skalistoe] *chutor* | Skalistoe | | | | | | X | X | X | | | | | | | | |
| Atajčik[6] | | | X | X | | | ruin | | | | | | | | | | |
| Bornel/Burun Eli[7] | | | X | X | X | X | X | | | FS | FS | | 8 | 4 | 12 | 20 | Tatarian |
| Tuaka/Tubaka | Rovnoe | | X | X | X | X | X | | | X | Ur. | | | 11 | 6 | 23 | Tatarian |
| Kutaj | | | X | | | | | | | | | | | | | | |
| (Köp or Baj) Kijat[8] | Vladimirovka | | X | X | X | X | X | | | X | X | | 30 | 24 | 11 | 46 | Tatarian and Russian |
| Baj Kijat *chutor* | | | | | | | X | | | | | | | | | | |
| Sakab/Sakaja | Morozova | | X | ruin | ruin | ruin | | X | X | | | | | | | | |
| Čagaltaj/ Čagiltaj/ Čigiltaj/ Čegoltaj[9] | Dalekoe | | X | X | X | X | X | | | X | X | | | 7 | 11 | 30 | Tatarian |
| Tananbaj[10] | Najdenovo | | | X | X | X | X | | | | Ur. | | | 25 | 17 | 44 | Tatarian[11] |
| Kirgiz (Kazak) | Kazaube | | | X | | | X | | | | | | | | | 34 | Tatarian and Russian |
| *Džarylgač Lake/Vodopojnoe Ravine* | | | | | | | | | | | | | | | | | |
| Erilgač/Jarylgač /Džarylgač[12] | Mežvodnoe | | X | X | X | X | X | | | X | X | | | 8 | 50 | 48 | Russian |
| Kirleut[13] | Vodopojnoe | | X | X | X | X | X | | | X | X | | 45 | 43 | 33 | 83 | Tatarian and Russian |
| Kubaran | | | X | | ruin | | | | | ruin | | | | | | | |
| Kurman Adži/ Romašino | Uročišče Romašino | | X | X | X | X | X | X | X | X | Ur. | 8[14] | 11 | 10 | 30 | 50 | Tatarian |
| Tok Džol'[15] | Kolodežnoe | | X | X | X | X | X | X | X | ruin | OTF | | | 14 | | 23 | Tatarian |
| Otuz[16] | Starosel'e / Zorjanoe | | X | X | X | X | X | X | X | X | X | | 34 | 16 | 10 | 32 | Tatarian |
| Dauldžar[17] | Groznoe | | | | X | X | X | X | X | X | OTF | | | 26 | 27 | 36 | Tatarian |
| Kul' Džakin' | Ochotniki | | | | | | X | X | X | | | | | | | 32 | Tatarian |
| Baim[18] | Tichaja | | | | | | X | X | X | | | | | | | 31 | Tatarian |
| Čongurči[19]Tat. | Jaroslavka | | | | | | X | X | X | | | | | | | 34 | Tatarian |
| Čongurči Russ. | Zadornoe | | | | | | X | X | X | | | | | | | 31 | Russian |

| Name in 19th century | Recent name | 1790 | 1817 Mukhin | 1837a Demidoff | 1837b Betev | 1865 Betev rev. | 1890s | 1931 | 1941 | 1955 1:25,000 | 1:100,000 | Number of households | | | | | Ethnicity of inhabitants[1] |
|---|---|---|---|---|---|---|---|---|---|---|---|---|---|---|---|---|---|
| | | | | | | | | | | | | 1806[2] | 1817 | 1837 | 1865 | 1890s | |
| Assan/(G)adži Gazan | | | X | X | ruin | ruin | | X | X | | | 2[20] | 2 | | | 11 | Tatarian (1806); Russian (1890s) |
| (Kastel) Bozak/ Buzak/Buzav[21] | Dedovo | | X | X | ruin | ruin | | X | X | | | | | | | | |
| Džamal' (Adži)[22] | Nizovka/ Kirovskoe | | X | X | X | X | X | X | X | X | X | 37[23] | 44 | 42 | 25 | 40 | Tatarian |
| | Novo Uljanovka | | | | | | | X | X | X | X | | | | | | |
| Ak Baš[24] | Vjačeslavka | | X | | X | ruin | X | X | X | X | Ur. | 3[25] | 3 | 1 | 1 | 19 | Tatarian |
| *Sasyk Lake/Ravine* | | | | | | | | | | | | | | | | | |
| Kolbaše | | | | | ruin | ruin | | | | OTF | | | | | | | |
| *Cultivated zone S of Džarylgač Lake/Vodopojnoe Ravine* | | | | | | | | | | | | | | | | | |
| Evan | | | | X | | | | | | | | | | | | | |
| Aldermen | | | X | X | X | X | ruin | | | | | | | 8 | | | |
| Aldermenskaja Skala | Uročišče Aldermenskaja Skala | | | | X | X | X | X | X | Ur. | Ur. | | | 8 | | | |
| Unnamed village ruin B (1890s) | | | | | | | ruin | | | | | | | | | | |
| Saja[26] | | | X | X | ruin | ruin | ruin | ruin | ruin | X | | | | | | | |
| (Koš) Ateš | Kostrovka | | X | X | ruin | ruin | ruin | X | X | X | | | | | | | |
| | Snežnoe | | | | | | | X | X | X | X | | | | | | |
| Karlav | Čaikino | | X | X | X | X | X | X | X | X | | | | 19 | 10 | 22 | Tatarian and Russian |
| Musala/Musali | Zajcevo | | X | X | X | X | X | X | X | X | X | | | 4 | | 4 | Russian |
| Džajlav[27] | | | X | | ruin | ruin | ruin | | | | | | | | | | |
| Uč Kuju Kipčak /Kipčak Učkuju[28] | Majorovo/ Uročišče Majorov | | X | X | X | X | X | X | X | ruin | Ur. | 15[29] | | 22 | 10 | 40 | Tatarian and Russian |
| Karama/ Kurama Kostel' (Kastel')[30] | Malyševka | | X | | X | X | X | | | X | OTF | | | 26 | 15 | 40 | Tatarian and Russian |
| Kostel' (Kastel') | | | | | ruin | ruin | ruin | | | | | 10[31] | | | | | Tatarian |
| Kurama | | | | | | | ruin | ruin | ruin | | | | | | | | |
| Kabla Kurama | | | | | ruin | ruin | ruin | ruin | ruin | | | | | | | | |
| Syrt Kurama[32] | | | | | ruin | ruin | ruin | | | | MTF | | | | | | |

Table 6.1
Overview over Early Modern and modern settlement history. Ur. = *Uročišče* (small, wild wood); OTF (cattle farm), PTF (chicken farm); FS = fishing station

1 Korosteleva & Obcinnikova 2002, appendix 4 and half-verst map of 1890s.
2 Korosteleva & Obcinnikova 2002, appendix 4.
3 Meaning 'of the sheiks'.
4 Meaning 'white mosque' (Jankowski 2006, 183-184).

5 Meaning 'deserted steppe'.
6 Ata = father; land owned or inhabited by this person (Jankowski 2006, 226-227).
7 Meaning 'Cape of El(i)', the name of a people (Jankowski 2006, 347).
8 Baj means a wealthy man; Kijat is a Mongolian and Turkish ethnonym (Jankowski 2006, 367).
9 Čigil is the ethnic name of nomadic tribe (Jankowski 2006, 404-405).
10 Tana + baj meaning rich owner of a cow (Jankowski 2006, 1025).
11 In 1931 map mentioned as nemeckij (German).
12 Jaryl means split or broken in two; gač is suffix: a lake split in two (Jankowski 2006, 514).
13 Kirleut is the name of Mongol tribe (Jankowski 2006, 604-606).
14 26 males, 19 females.
15 Tok = filled; džol' = road, so filled road (Jankowski 2006, 1077-1078).
16 Male name, anthroponym (Jankowski 2006, 768-769).
17 Davul = storm or hurricane; džar = ravine (Jankowski 2006, 436-437). The Russian name is a direct translation.
18 Baim = Baj, a rich or wealthy man (Jankowski 2006, 258).
19 Čongurči = Čongurov, family name (Jankowski 2006, 258).
20 4 males, 5 females.
21 Ethnonym, boz = calf (Jankowski 2006, 333-334).
22 Džamal' = beauty, Adži = Hadji, person who has been on pilgrimage to Mecca (Jankowski 2006, 496-497).
23 150 males, 145 females.
24 Ak Baš = white head (Jankowski 2006, 173-174).
25 16 males, 9 females
26 Pen for sheep or cattle (translation: Jankowski 2006, 949).
27 Summer residence or summer pasture (translation: Jankowski 2006, 495).
28 Uč = three, kuju = wells (Jankowski 2006, 1125-1126.
29 72 males, 62 females.
30 Kastel'/Kostel' = castle, stronghold (relations to kurgans: Jankowski 2006, 606).
31 40 males, 28 females.
32 Syrt = northeastern (Jankowski 2006, 53).

In almost all cases can we observe an increased number of households from the first half of the 19th century to the end of the same century. The head count of 1806 (Korosteleva & Obcinnikova 2002, appendix 4) gives not only the number of farms, but also the number of individuals and their gender. In this count, 50 farms are numbered as well as the 308 males and 268 females living in them, in total 576 persons (with an average number of 11.5 persons per farm).

The best preserved Early Modern site on the slopes is the settlement on the ridge top (DSP08-H07-03). It consists of at least three rectangular stone structures surrounded by small ash hills, household pits, a cluster of fire places as well as a dam. The settlement was probably associated with the exploitation of the cultivated area on the top of the ridge. Due to the few finds, the settlement cannot be securely dated, but the diagnostic Medieval to Modern sherds collected on the slopes appear to belong in the 18th-20th century. Several "Modern" single settlements were also identified on the slopes which were not indicated on any of the maps. The identity and date was suggested by vegetation and architectural remains, e.g. DSP08-H15-01, DSP08-H22-03, DSP08-H21-01 and DSP08-H27-01, but due to very poor visibility no artefacts were collected. All of these sites are located on the edge of the Plateau/Uplands or at the bottom of the Hillsides suggesting they were associated with the cultivation of the land in these areas as well as the resources of the Hillsides.

## 6.3 The productive landscape | P. Guldager Bilde, P. Attema & K. Winther-Jacobsen

The DSP study zone is an area with many different ecological niches, which allowed the region's inhabitants to have a mixed, sustainable economy. However, the geology and hydrology makes it best suited for pastoralism and mixed agro-pastoralism. Periods of intensive cultivation are mainly the result of external forces occupying the region, be that the Greeks of the Late Classical and Hellenistic periods, the Russian Empire and later the Soviet Union. Characteristic of these are that decisions on cultivation are made according to standards developed for different ecological regions. Because the Tarchankut is an area of marginal lands intensive cultivation was bound to fail. In the Hellenistic period after the initial success in a period of climatic optimum, a severe crisis pushed the area off the edge wherefore after only half a century of intensive cultivation this was completely abandoned. The same was the case after the collapse of the Soviet Union in 1991. Last-mentioned can easily be read in the landscape in the form of ruined collective farms, which today serve as mining ground for building materials. In the beginning of the new millennium economy in the region is again mainly agro-pastoral supplemented with some fishing in addition to the touristic facilities of the coastal towns, foremost Černomorskoe.

### 6.3.1 Agriculture

As discussed in Chapter 7.3.1 below, large parts of the Chersonesean chora was divided into regular plots (Fig. 6.12). This was also the case with the Tarchankut Peninsula. In this section we shall briefly discuss the possible cadastral system of the region NE of the Kelšeich Ravine, which corresponds to the wider study area of the DSP.

During the preliminary survey in May 2006, several plot division walls preserved as straight earthen banks were measured in the western part of the Tarchankut Peninsula (Figs. 6.13-14). The location of the walls prove beyond doubt that the orthogonal system of tracks, unrelated to the Early Modern settlement structure, which can be seen on the half-verst map of the 19th century on the western tip of the Tarchankut Peninsula, follows the ancient cadastre as also suspected by Ščeglov, Smekalova and Smekalov. Closer to the Karadža Settlement, the division wall taking its departure near the settlement itself can be followed as a dry stone wall over a considerable stretch (Fig. 6.15).

The presence of a cadastral system around the Černomorskoe Settlement has long been assumed. In his 1967 article on the rural surroundings of this settlement, Ščeglov proposed the identification of a land division system in the immediate surroundings of the town with Vetrenaja Bay being furthest away (Ščeglov 1978, 87 fig. 43; Chtcheglov 1992, 118-120; 2002, pl. 4.2; Fig. 4.22). According to Ščeglov, the divided land was located within a ca. 2 km radius from the town, and he was of the opinion that to the NE, the divided land did not continue on the E side of the small ravine having its mouth where the Vetrenaja Bay farm is located (Fig. 4.22). On the contrary, he identified a low earthen rampart, which he believed protected the farmland. Thus, he recognized four contiguous blocks of land stretching from Vetrenaja Bay towards SW as well as small part of a block close to the site of the supposed Florya Pension farm house (a corner of the division wall was found under the market square (Ščeglov 1967, fig. 5, ca. 250 m from the pension) and some minor patches NE of the settlement (Ščeglov 1967, fig. 2.4-7 and 3 respectively). Today, the Florya site is completely built over and plot divisions can no longer be verified. However, the space images available from Google Earth makes it possible to identify blocks of land division on the entire promontory W of Ščeglov's blocks 6 and 7 (Fig. 6.16). He identified ca. 60 plots each measuring 210 x 420 m (Chtcheglov 1992, 118) or 362 x 200 m (Ščeglov 1967, 243). The plots were marked by a low wall of field stones. In some cases, we can even study the further sub-division of the plots into strips of 10, 20 or 30 m width. There are several kurgans close to the Vetrenaja Bay farm. These can be earlier than the Hellenistic period; however, we cannot exclude that they belonged to the landed aristocracy.

The orthogonal system of tracks, again unrelated to the contemporary settlement structure, which we find NE of the Kelšeich Ravine on the half-verst map of the late 19th century, can be similarly assumed to mirror the ancient cadastral system, as we saw concerning the western part of the Tarchankut Peninsula. The tracks indicate that the today small promontory stretching W of the Sasyk Ravine was also part of the divided land (Fig. 6.17). It should be noted that the three parallel roads visible in the map of Google Earth (Road XIIIa-c; see below Section 6.6) are of the 20th century. Though only the southeasternmost one of them (XIIIc) is retained today, the fields still follow their orientation. 20th century agriculture on the promontory most likely has obscured all remains of the ancient cadastral system. Panskoe III (and IV?) is a farm located in the divided land similar to the farm of Vetrenaja Bay. Also on this promontory a fair number of kurgans is located.

It is likely that the land S of the Sasyk and Vodopojnoe Ravines and N of the last-mentioned was also divided into regular plots. Ščeglov hints at the existence of cadaster S of Panskoe and on the Meživodnoe Peninsula (Chtcheglov 1992, fig. p. 251; 2002, pl. 4.3.A), but this is not otherwise described. Faint lines can be observed on the Meživodnoe Peninsula in the air photos of 1978, but nothing shows up on Google Earth. The lines S of Panskoe mentioned by Ščeglov are even more difficult to identify, and they cannot be distinguished from Google Earth either. Yet again, if we trace the cadastral system in the tracks of the half-verst map, we find that it followed the orientation of the coastline of the Sasyk Ravine and hence of the E-W Road XII, which Ščeglov considered as ancient (1987, fig. 2.I; see below Section 6.6). Thus, the reconstructed cadastral system NE of the Kelšeich Ravine follows one and the same orientation. However, since this is also the orientation of the modern fields, modern landuse may obscure traces of the ancient one.

In Field DSP07-08 were found remains of what may perhaps be interpreted as internal plot divisions. Over an area of ca. 47 ha, a grid of soil banks was visible on the surface (Fig. 6.18). However, the overall orientation of this grid does not coincide with the one of the supposed overall system of the region mentioned above. It is therefore likely that we deal with traces of more recent agriculture.

It should be noted that the orientation of the plot demarcation grid of the agricultural territory W and NE of the Kelšeich Ravine respectively followed two slightly different orientations (Fig. 6.12). Surely, this is due to the fact that the gridding respected the major landscape features of the two different regions, but it may also support the theory that there seemingly were two central places in the Tarchankut Peninsula: the Karadža and the Černomorskoe Settlements respectively.

Smekalova & Smekalov mention that on aerial photos of 1977 traces of rectangular plots can be seen near Masliny (Smekalova & Smekalov 2006, 239). However, the only kind of 'parcelling' we can see in this area, are the irregular 'parcels' devoid of burials, which were created by the nomadic migratory routes (Sections 6.5 and 6.6). In all likelihood, they may have served for grazing animals rather than for farming. The tracks of the half-verst map does moreover not give reason to assume the presence of cadastral systems this far north.

### 6.3.2 Fish

A large number of sites were located by the shore of lakes and the sea, and it is therefore likely that fish constituted a significant part of the ancient inhabitants' daily diet. However, no fish bones are recorded from sites in the area, so this must remain a hypothesis.

### 6.3.3 Salt

Salt was one of the major raw materials of the Crimea and tens of thousands of tons were produced annually in the 19th century (Holderness 1923, 89; Lyall 1825, 230). In the DSP area salt was harvested from salt vats in the shallow Sasyk and Džarylgač Lakes, as can be gleaned from the half-verst map of the 1890s (Fig. 6.17). Whether salt was also collected here in antiquity is a matter of speculation, but it is likely.

### 6.3.4 Quarries

Exposed bedrock with preserved cut marks is evidence for limestone quarrying on Hills 19, 22 and 24. Extensive quarrying in the S part of Hill 19 (DSP08-H19-01) may be associated with the village of Snežnoe 1.5 km to the SW. Although specific relations are impossible to demonstrate, the remaining quarries appear to be associated with sites: approximately midway between a line of kurgans (DSP08-H19-03) and DSP08-H12-01, 78 m NW of DSP08-H22-03 and 85 m E of DSP08-H24-03. 30 m NW of DSP08-H11-02 Feature 1 is a patch of exposed bedrock (Feature 15) which may be a quarry, but there were no obvious tool marks. Evidence for recent lime stone quarrying were observed in several locations as well as evidence for loess quarrying (see Chapter 3.3).

## 6.4 THE SACRAL LANDSCAPE | P. GULDAGER BILDE

During the DSP survey, the only find that can be related to religious activities of the inhabitants of the region was a terracotta egg (**Cat. 778**) found in 2008 in Field 2 and apparently not associated with a site. It may have come from a ploughed-out grave, which was not otherwise observed. A similar terracotta egg was found in Panskoe I U6 room 13 (Rogov 2002, 274 M 22, pls. 149, 175). However, many local limestone reliefs found in the Chersonesean chora with depiction of Herakles bear witness to his status as the mythological Doric founder of the colony.[78] In addition can be mentioned a club found in the Vetrenaja Bay farmhouse (Ščeglov 1967, fig. 16). Similar finds have come to light closer to Chersonesos: two Herakles' clubs in terracotta were unearthed in an outside niche in the central tower of farmhouse

[78] *Mežvodnoe* farmhouse: Vysotskaja 1968 [reclining]; *Panskoe I* U6, room 14: Panskoye I 2002, pl. 143.G1 [standing with club]; *Kulčuk*: found in 2008 [reclining], S.B. Lancov, personal communication; *Kerkinitis*: Kutajsov 2004, 90, fig. 88.1 [reclining]; *Mojnak Lake* farmhouse: Kutajsov 2004, 90, fig. 88.2 [reclining]; *Lake Saki* (farmhouse?): Kutajsov 2004, fig. 90.1 [seated]; *Čaika*: flanking tower room R: Kutajsov 2004, fig. 90.4 [standing with club], farmhouse 3 room 10: Popova & Kovalenko 1996, fig. 3-4 [standing with club], *Čaika*, farmhouse: Karasev 1966 [reclining].

151 in the Herakleian Peninsula (Carter et al. 2000, 727) and in a farmhouse at the Evpatorija lighthouse too (Kolesnikov 1984, 81 [not seen]).

On the Tarchankut Peninsula only one sanctuary has ever been found, which attests to public worship. This is located at the cliffs of Džangul' NE of the Karadža Settlement (unpublished; Ščeglov 2002a, pl. 4.1.17; Fig. 1.1.17). A small group of finds, which were located *ex situ*, were excavated by the Tarchankut Expedition in 1983 with a few sherds picked up in 1984. The finds are presently for the larger part kept in the storerooms of the Institute of the History of Material Culture (Russian Academy of Sciences), St. Petersburg. They consist of handmade and wheelmade pottery (amphorai account for almost 50%), glass, and sea shells. The finds have a ca. 500 years dating horizon from the 4th century BC to at least the 1st century AD. Of great importance are the numerous graffiti, which show that the sanctuary was devoted to a goddess by the name of Targa, and one is a joint dedication to Targa and Achilleus, the main god of Roman Olbia. Though the graffiti are incised in Greek language, the goddess is not of the Greek pantheon. In all likelihood she is a goddess of the Hellenized 'barbarian' population of the region. However, the sanctuary must also have attracted the Greek settlers, so it is likely that it was a common arena for the local inhabitants notwithstanding their ethnic background.

## 6.5 THE FUNERARY LANDSCAPE | P. ATTEMA & T. SMEKALOVA

### 6.5.1 Kurgans in Western Crimea: an introduction | T. Smekalova

Thousands of kurgans are scattered throughout the steppe of the Northern Black Sea littoral. They break the visual monotony of the steppe and provide a sense of monumentality to the landscape. Even though they belong to different periods and vary in dimension, relative position, and the number of burials inside, there are certain regularities in their arrangement within the landscape possibly determined by various historical conditions.

The word *kurgan* is of Turkish origin. Originally kurgan meant a hill, burial hill, fortress, settlement or town. Owing to Russian scholarship it has been established in scientific literature as a technical term for a barrow mound. From Russian it was introduced into other Western languages, e.g. German.

The earliest kurgans occurred in the Northern Black Sea area during the Middle and Late Chalcolithic period, when societal change from hunting and gathering to animal husbandry and agriculture took place. During this period the earliest social and economical differentiation became visible. In all likelihood, it was pastoralism that gave the impulse to the culture of constructing barrows. The only way to make the ancestral graves noticeable in the steppe landscape was to heap a mound above the burial. Later kurgans probably imitated those of the past. Given that the migrations of the population took place over vast territories, kurgans, which originally were exclusively burial monuments, immediately became noticeable signs in a space otherwise devoid of any reference points.

Kurgans were built mostly of turf blocks. To construct the largest Scythian kurgans, it was necessary to remove the soil layer from a fairly large area. Thus, the famous Čertomlyk kurgan was erected from turf blocks measuring 15 × 25 × 20 cm; according to various calculations, the area of turf removed for its construction was 35 to 75 ha (Murzin & Rolle 1989, 93; Alekseev, Murzin & Rolle 1991, 171; Klamm, Viebrok & Meyer 1991, 306). Kurgans of the Chalcolithic and Bronze Age periods, as well as some Scythian barrows, e.g. the Melitopol' Kurgan, were occasionally built of excavated soil, clay rolls or primitive mud bricks (Terenožkin 1955, 30; Mozolevskij & Polin 2005, 249).

Constructing a kurgan demanded considerable collective efforts. Therefore the emergence of the earliest kurgans must have corresponded to a stage in the development of society, when fairly large groups of kinsmen had been established in order to enable the organization of such works. The construction of kurgans may have constituted a bonding link, which united the entire group. These collectives appeared due to the necessity of labour division after the transition from food gathering to food production. During the same period, surplus production first emerged, which enabled the collective to spend its efforts on non-productive goals. Using the primitive digging tools of the Mesolithic and Chalcolithic epoch, a collective body of 20-25 members was in 1-2 days able to build a fairly small barrow of soil 0.5-1 m high, 8-20 m in diameter, with a volume of 45-50 m$^3$ (Belov & Ljasko 1991, 29; Semenov & Korobkova 1983, 84-88). A minor Scythian barrow, 1.5-2 m high, 18 m in diameter and with a volume of the mound between 60 and 115 m$^3$ could have been constructed in 1-1.5 weeks time by a fairly small team also (Mozolevskij & Polin 2005, 253). When kurgans up to 3-4 m high with stone or wooden structures were built, undoubtedly much greater labour investment had to be spent and correspondingly larger working teams had to be attracted. But even this scale of work cannot be compared with that of the construction of the huge Scythian kurgans of the 4th century BC for which, according to some calculations, 1,000-1,500 persons were working simultaneously for a week or two (Mozolevskij & Polin 2005, 252-253).

The surface of a newly built kurgan "revived" immediately after its completion recovering from the wounds and scars afflicted by the cutting of the turf blocks. Soon it would become part of the general soil development processes. This fact made the kurgans inseparably linked with the natural landscape providing simultaneously their long term preservation and stability. Only man himself was able to disturb a kurgan otherwise proof against any natural effects such as earthquakes, showers, animal or plant activities.

Kurgans are mostly situated so as to be best viewed from all (or at least more) sides, i.e. at margins of watersheds, on dominating hills or in other prominent places. This makes them into landscape markers that function as reference points organizing the wide space of the steppe. The resulting patterns of reference points may have served various purposes.

It is normally assumed that kurgans arranged as long chains in the steppe marked ancient roads or traditional routes of seasonal migrations of nomadic tribes (e.g. Boltrik & Fialko 1987; Boltrik 1990). Inter-visibility must have been consciously aimed at so that from any kurgan the next one could be viewed. Under steppe conditions lacking any other reference points, the kurgans were a kind of "road mark" ensuring movement during migration. During long-distance movement of nomads with their herds it was of vital importance to follow certain routes with localities of good grazing and watering. The chains of kurgans must have marked those routes, the deviation from which could have been dangerous for an entire tribe. One of the longest and straightest kurgan chains can be traced from the sea coast of the Džarylgač Bay towards the Dnieper at Kakhovka. The appearance of that route hardly seems to have been accounted for due to the peculiarities of the relief, because the route in places passes absolutely flat terrain devoid of any elevation. According to V.S. Ol'khovskij the kurgans, which compose this "highway", belonged to the Jagorlyk-Kalančak tribe (to the S) and the Podneprovskaja tribe (to the N), two groups kindred to each other (Ol'chovskij 1991, 135). Presumably, this kurgan chain indicates the seasonal N-S migrations of the Scythians from the Dnieper region. The route is exceptionally convenient because it crosses no water or other obstacles on its way and passes through rich grazing and watering territories. It is not without reason that in a map of the late 19th century numerous stockyards and wells are marked there. We cannot rule out that one of the salt routes used both in antiquity and in Polovtsy and Kiev Russian times went along this route (Boltrik 1990, 39). Chains and clusters of kurgans may also have marked particular territories and their boundaries.

Despite the fact that for two centuries kurgans have invariably attracted the interests of scholars, no overall work concerning their spatial and chronological distribution has ever been made. The accelerated destruction due to agricultural activities and looting undermines the scientific base represented by these archaeological monuments – the commonest in the region under consideration. Therefore, scholars lacking this scientific base miss the possibility of arriving at the correct historical conclusions concerning the life of the people who once were living here. Identification of kurgans using detailed historical maps, investigation by remote sensing methods in addition to field survey are able to provide us with a real picture of the present-day state of these archaeological monuments. In addition, a comparison of topographic maps and the archive evidence on the kurgans will allow us to consider their preservation for at least the last 120 years and to assess the anthropogenic factor influencing their state and number.

The present general introduction is part of an overall study by T. Smekalova of territorial and chronological patterns in the arrangement of kurgans within the landscape of the Northern Black Sea littoral entitled 'Communicating power, a study of kurgan burials'.[79] This study is based on the study of cartographic evidence, geophysical methods and archaeological survey. Today, primarily results of remote earth sounding are widely used for investigating large areas: space photographs of high resolution and maps of scanning the earth's surface. In addition, historical aerial photos preserved in the aerophotoarchives of the Institute of Archaeology in Moscow as well as at other scientific institutions and production companies are now easily available. In addition, it is now possible to study detailed topographic maps, both old and new ones of high resolution, preserved in archives because now most of these materials have been declassified. The importance of examination of large-scale maps was already understood by B.N. Mozolevskij, who fulfilled the meticulous work of identifying areas of distribution of large kurgans in maps. He has revealed barrows from 8 to 15 m high in the area from the Moločnaja River to the E up to the Southern Bug River to the W and based on these studies has developed a concept of the ethnical geography and political structure of Scythia in the 4th century BC (Mozolevskij & Polin 2007, 46; Terenožkin & Mozolevskij 1988, 179-222).

---

79 http:www.pontos.dk/research/ra_6/research/ra_6/ra-6b-communicating-power-a-study-of-kurgan-burials-1

In this study, a number of detailed historical maps are employed. In 1836-1838, the Corps of Military Topographers under the direction of F.P. Schubert and D.D. Oberg triangulated the Crimea. Their work resulted in a map based on the topographic survey of Colonel Betev (Fig. 2.2). His map of 1837 clearly shows the relief, the main and minor roads and the numerous contemporary villages. Furthermore, it also showed water sources and wells as well as a fairly large number of kurgans. The proportions of the peninsula practically coincide with those of the modern maps. In 1865 a reconnaissance was conducted on the basis of the maps of 1837 within the frame of large-scale works of 1845-1870, when entire Poland and 22 provinces of central and southern Russia, including the province of Taurica, were surveyed in the scale of one verst (1 verst corresponds to 1.067 km). The map of 1865 shows a greater number of roads, settlements, kurgans and wells. The basis of the three-verst map issued in 1865 was also based on Betev's survey (Fig. 2.2).

Since the early 1870s, surveys almost completely ceased in the internal provinces of Russia and were continued only in the border regions, including the Crimea, but now conducted in half-verst scale (1:21,000; Fig. 2.3). In the half-verst military topographic map of the Crimea issued in the late 19th-early 20th century, an enormous amount of kurgans is marked with height specifications for the largest mounds. Such serious attention was paid to kurgans because they were viewed as strategic military points. In addition, water sources and wells were drawn in detail and roads and paths, including the abandoned ones, are shown. The relief of the landscape was rendered with contour lines at an interval of 2 sažens (c. 4.3 m). Crimea's proportions correspond exactly to the present ones, so objects identified in this map may easily be transposed into new maps or compared with air and space photos.

The volume of information represented in the half-verst map is comparable with the data of another detailed map, namely that in the scale of 1:25,000, which was reconnoitred in the 1950s and revised in the 1980s on the basis of the results of air-photography (Fig. 2.5). The contour lines in that map are drawn at an interval of 5 m. In that map, kurgans are also shown but their number is considerably less than in the half-verst map. The heights of the largest kurgans are specified. More kurgans with names are depicted in the western part of the Tarkankhut Peninsula on the half-verst map, but E of Černomorskoe, no kurgans are mapped with names.

The half-verst map of the Crimea of the late 19th century has been chosen as the main source for composing a general map with distribution of kurgans. Although some mounds, particularly the smaller ones, are not shown in that map, nevertheless, the percentage of such omissions is relatively low in comparison with the number of kurgans that are presented. Moreover, it can be hoped that this percentage is uniform throughout the whole Crimea, thus allowing for a comparative analysis of the distribution of kurgans of the entire Crimea.

The scanned sheets of the half-verst map were assembled by means of MapInfo software. Kurgans were marked in a special layer with circles of different sizes depending on the dimensions of the kurgans. In the half-verst map, near mounds over 1 sažen high (1 sažen = 2.1336 m) their approximate height (in sažens) is noted. These larger barrows are marked with larger circles, while smaller kurgans are signed with small circles. Not surprisingly, the latter kurgans are much more numerous than the larger ones. It has proved possible to identify a total of 15,500 kurgans in the half-verst map of the Crimea.

The analysis based on the half-verst map defined three areas with particularly high number of kurgans. These are: the E part of the Kerch Peninsula, the Tarkhankut Peninsula and northern foothills of the Crimean Mountains. The number of kurgans in the areas exceeds many times that in the remaining central part of the Crimea.

In the Tarkhankut Peninsula alone the total number of kurgans as evidenced by cartography is at least 5,000. However, it seems as if the region of Lake Džarylgač in a certain sense is peripheral to the Tarkhankut Peninsula, because W of the Kelšeich Ravine a significantly larger concentration of kurgans has been recorded. In the Tarchankut the largest kurgans are situated mostly on the crests of watersheds and they are spaced at certain intervals equal to the limits of vision. Kurgan chains are well discernible along the crest of the Džangul' Uval rising up to 137 m S of Lake Džarylgač. They are also found on the N bank of the lake along the top of the Plateau rising 20 m above sea level. Numerous smaller kurgans form compact clusters, which are arranged on the slopes created between the small ravines running from the main ridges to the lake and to the sea. The small kurgans are especially abundant on the S shore of the Tarchankut Peninsula, on the slopes coming down towards the coastal valley, as well as in the area of the Džangul' Uval, where it slopes down towards Lake Džarylgač.

## 6.5.2 Kurgans in the DSP area | P. Attema

A specific aim in the DSP survey programme was to analyze the spatial relationship between burial mounds and contemporary settlement organization in the transect area. Both in the 2007 and 2008 campaign work focused on one particular slope of the Hillsides featuring several distinct groups of kurgans, a small number of which had been previously excavated (Hill 19). Although all the kurgans on this particular slope had already been generically recorded on the half-verst map, they had not been surveyed in detail nor was their exact geographical location known. Moreover, it appeared that several kurgans still needed to be mapped (Chapter 5.2).

Although several anomalies were recorded, only one case merited a trial trench excavation. The trench did not, however, yield convincing evidence for cultic activity nor were any fossa graves detected. The geomagnetic survey generally covered the same areas as the digital elevation models, allowing us to superimpose the layers in GIS for further analysis.

## 6.6 THE LANDSCAPE OF PATHS | P. GULDAGER BILDE

Because the Western Crimea has been and still is a marginal area for agriculture, in many patches, the nomadic landscape can be recognized even today. The most readily visible remains are the numerous kurgans, which were dealt with in detail in Section 6.5 above. Moreover, in many places the actual paths of nomadic migration can be distinguished in the maps of Google Earth. Since these paths are frequently lined with kurgans and/or cross hubs with clusters of kurgans, and since they rarely coincide neither with modern routes, nor with routes documented on historical maps, their antiquity can hardly be disputed (Fig. 6.19).

The Tarchankut Peninsula is one of the Crimean regions most densely occupied by kurgans (c. 5,000 according to the investigations of T.N. Smekalova).[80] Alone in the area between Bakal', Černomorskoe and the end of the Donuzlav Lake, thus the region immediately surrounding (and including) the DSP study zone, roughly 1,500 kurgans can be distinguished with the help of historical maps and Google Earth. These kurgans are the physical remains of the nomadic, the semi-nomadic and settled population of the region from at least the Bronze Age until the present day.[81]

As already discussed by Smekalova (Section 6.5.1), kurgans are not dispersed randomly in the landscape, on the contrary. Moreover, kurgans and roads are intricately linked; it is therefore necessary briefly to sketch the types of spatial arrangement of kurgans in the landscape in order to obtain an idea of the landscape of paths.

Kurgans are normally found in one of the following four types of spatial configuration:

(a) In chains on watersheds (e.g. Fig. 4.148)
(b) In clusters around settlements (e.g. Fig. 4.24)
(c) In clusters at hubs, where several routes converge (e.g. Figs. 6.23 and 6.30)
(d) In straight lines along roads

Ad (a) Kurgans are located in chains following larger or smaller ridges and crests in the landscape. In the DSP area there is a major E-W chain on the ridge N of the Džarylgač Lake as well as along the N edge of the Plateau S of the lake. In addition, there are smaller chains running downhill N-S on the ridges separating the small ravines N and S of the lake. A good example is the kurgan chain described in Section 6.5.2. Even in the ploughed zone, these kurgan chains can be seen on barely perceptible watersheds. Because the kurgan chains follow the watersheds they are rarely placed in straight lines.

The position on a ridge made the kurgans highly visible from more sides. There may have been paths along the watersheds, but these would mainly have served local movement. It is more likely that the kurgans on watersheds marked particular territories. A common feature of traditional peoples worldwide is how they conceive of nature in terms of a geographical boundary, in particular a watershed. This is described by Berkes et al (1998) based on a large ethnographic material. According to them, the watershed unit forms the basis for a people's or a group's self-understanding ('we are the people of NN River') and the watershed thus forms the easily recognizable boundary of a territory, be that of hunting, grazing, living etc. As an example, they mention how NW Pacific chiefs' territory run from "mountain top to mountain-top" (Berkes et al. 1998, 410). It is conceivable that something similar is the case in the DSP study zone.

Ad (b) It is a characteristic element of cities and settlements of the Northern Black Sea region that they featured sig-

[80] It should be noted that this considerable sum is derived from digitizing the Russian half-verst map of the 1890s; however, when compared with the number of kurgans readily visible from Google Earth, the number of kurgans is yet significantly higher.

[81] It is quite a common phenomenon that kurgans are used for contemporary cemeteries, as we can gather e.g. from a kurgan near Snežnoe (Figs. 4.59; 6.9) and one at Hub II.5.

nificant kurgan necropoleis immediately outside the inhabited area (Chapter 4.3.1.1 on the Černomorskoe Settlement, Chapter 4.3.2.1 on Panskoe I). It has habitually been taken for granted that they are evidence of barbarians living in the settlements. We will not enter this discussion here. In the half-verst map of the 1890s are marked as kurgans a number of clusters around the deserted Tatar villages of Baj Kijat, Tanabaj, Kirgiz Kazak, Dauldžar, Kul Džakin and Čongurči. They are not included in the present section, because in many cases, they are more likely sub-recent ash hills rather than kurgans. Especially in the maps of Google Earth, it is easy to observe the difference in how they appear when compared with kurgans proper: their margins are blurred and the colour deviates from the one of the kurgans.

Ad (c) Kurgan clusters, that are not located in the vicinity of settlements, normally mark hubs where several routes meet. A good example is the Superhub V.1/VI.3 (Fig. 6.30). Hubs marked by kurgans may have constituted sites of temporary camps in use during long-distance movement.

Ad (d) In many places the actual paths lined with kurgans can be distinguished on Google Earth. Visually two different types can be noted:

Type A: the path is preserved as a dark line (black, brown or deep green) surrounded by an ample very light margin at each side, which differs from the general colour of the terrain. This type of path is normally rectilinear and it runs over considerable distances. Surely, a path, which left such an imprint, and which has withstood later times' agriculture, must have been well trodden and probably used over considerable time. It could be interesting to verify such a path through excavation.

Type B: the path is preserved as a black line, more diffused in the margins and without the bordering light margins of Type A. This type of path normally runs over shorter distances and it tends to be more irregular in its orientation. Probably this type of path was of more local use with less heavy 'traffic' and perhaps also more short-lived. Good examples can be found around the immense Bakal' Settlement.

How visible these paths are on the individual sheet of the Google Earth map depends on the time of the year, the satellite picture was taken, as well as of the crops on the ground. It is necessary to employ the feature "Historical pictures" in order to catch as many features as possible.

An attentive study of the maps available on Google Earth compared with data concerning location of kurgans and roads in the Russian half-verst map of the 1890s allow us to distinguish the following overall pattern (Fig. 6.19): First of all, there is a significant difference between the areas W and E of Vodopojnoe in terms of number and orientation of roads. The area E of Vodopojnoe features a large number of roads, whereas to the W features few roads. The shift corresponds nicely to the pattern, we saw concerning the presence of farmland laid out in regular plots (Fig. 6.12). Let us first discuss the area E of Vodopojnoe, where no cadastre was detected. In this part of the Tarchankut, most of the roads run NE-SW following the overall contour of the peninsula. Two roads are long-distance roads. Road I brings the Tarchankut Peninsula in connection with the Perekop region; Road II connects the crossings of the Vodopojnoe Ravine and of the Donuzlav Lake and Ravine respectively. Roads III to VI represents medium-range traffic between the area of Bakal' and Vodopojnoe, whereas Roads VII to IX are major connecting roads running NW-SE. Road X connects Vodopojnoe with the intensely settled area between Medvedevo and Ozerovka at the coast of the Donuzlav Lake.[82] Road XI runs at the N edge of the Plateau bordering the Hillside settlements further N.

W of Vodopojnoe is the farmland, which was divided into regular plots. The main artery is Road XII running almost E-W, which most likely links the area between the Černomorskoe Settlement and the zone S of Vodopojnoe. It is likely that in antiquity, this was also a main thoroughfare, and it probably constituted the longest of the roads in the cadastre system. Road XIIIa-c are three recent parallel roads running parallel throughout the length of the small peninsula N and W of the Sasyk Ravine. Road XIV is a track leading to the settlement of Panskoe I. Road XVa-b are two short, parallel roads. Road XVI is the main route of the 19th century. It connects the area W and E of Vodopojnoe.

Between hubs, the road runs in a straight line, whereas it may slightly change orientation between hubs. This shows beyond doubt that the kurgans were used as markers of orientation in the steppe. Because so few kurgans have been excavated, we have little means to date them and accordingly to date the roads. It is therefore uncertain whether all of the mentioned routes existed at the same time. In some cases it can be seen how the road was shifted some dozen meters. Such roads were hardly contemporary. But the overall pattern is that the location of roads was rather stable. Beneath is a brief characterization of the 16 most important roads. With the exception of Roads XIII and XVI, which belong to the

[82] For the settlement in the mentioned area, see Lantsov & Uzhentzev 2007, fig. 2.

20th and 19th century respectively, it is likely that all the remainder are ancient. Nevertheless, some of them had a long life; thus, Road XI also occurs on the half-verst map of the late 19th century. The same is the case with Road XII, which even today in its westernmost part is covered by the modern asphalt road, T0107.

*Road I* (Figs. 6.20-21)
Straight SW-NE road between Hub I.1/III.1 and the hub leading to Bakal' (Hub I.5). The distance between the two hubs is ca. 25 km. The road continues beyond Bakal' leading towards Perekop, where it can be distinguished as a line of kurgans, the path itself not being visible. Hub I.1/III.1 is a superhub, where several major roads converge at one of the Vodopojnoe crossings (Fig. 6.21). Three further hubs are located with regular distances at the road: HI.1 to H1.2: 4.5 km; HI.2 to HI.3: 4.5 km; HI.3 to HI.4: 7.5 km; HI.4 to HI.5: 7.5 km. Between Hubs I.1 and I.5 the path itself is clearly visible over long tracts of land. Kurgans are mostly located as clusters around hubs. The multi-period settlement of Vodopojnoe 2 (Late BA; LC-EH?; Early Modern) is located near Hub I.1/III.1. Road IX crosses Road I in Hub I.2/IX.3 (Fig. 6.22). At the same hub the road has two further forks towards SW and SE. At Hub I.3, which is located at a cluster of kurgans centred on one of the very large kurgans, the road branches towards N and S. From Hub I.4 (Fig. 6.23), which is a superhub, a number of roads converge and a straight road aims at Hub III.3/VIII.3. Also this hub is centred at a kurgan cluster There are several NW-SE and SW-NE roads in the immediate vicinity.

*Road II* (Figs. 6.24-25)
Partly straight W-E road between Hubs II.1/X.1 and II.5 with a distance of ca. 17.5 km between the hubs. The road is located on the plateau between the two main branches of the Vodopojnoe Ravine. The distance between hubs varies. From HII.1/X.1 to HII.2: 2.7 km; HII.2 to HII.3: 1 km; HII.3 to HII.4: 3.3 km; HII.4 to HII.5: 9.6 km. Hub II.2 is marked by a kurgan cluster with at least 10 kurgans. It is crossed by a road leading to the Bronze Age settlement of Vodopojnoe 4 1.5 km to the NE and to the crossing opposite another Bronze Age settlement of Vodopojnoe 5 1.5 km to the SW. At Hub II.3 located at one of the very large kurgans, the road forks off towards NE. Hub II.4 is a superhub, where several roads converge (Fig. 6.25); most notably the Road II is here crossed by a straight road between the Tatar villages of Buzak and Dauldžar, but since it does not occur on any of the historical maps, this road must be ancient. Also this hub is marked by one of the very large kurgans. Between Hubs II.1/X.1 and II.4 the actual path can be followed over long tracts of land. Between Hubs II.4 and II.5 the route can only be traced by means of a few kurgans. Hub II.5 is a superhub located 1 km S of (Russian) Čongurči and thus in the very end of the Vodopojnoe Ravine. From this hub, which is marked by a kurgan cluster including a large kurgan used as a modern cemetery, the road branches off in different directions aiming for the main crossing of the Donuzlav Ravine as well as for the end of the same ravine. The road also branches off in a more northern direction heading for the Perekop area.

*Road III* (Figs. 6.26-27)
A significant route stretches for 25 km between the Vodopojnoe crossing and the mouth of the Bakal' Ravine. Either end constituted a major pole in the landscape. Road III started/ended at the same Vodopojnoe crossing as Road I (HI.1/III.1; Fig. 6.21). The other end of the road is immediately opposite the much extended multi-period settlement of Bakal' discovered in the *razvedki* by P.N. Schulz in 1948 (unpublished; Lancov 2004, fig. 1.82), which was probably one of the most important settlements in the entire region. The road crosses a number of hubs. At the first hub (HIII.2) the road is crossed by a short path linking Road I and III by the multi-period settlement of Vodopojnoe 3 (Late BA; LC-EH; Early Modern). It is also connected with a straight road ending at the E outskirts of present day Vodopojnoe. 1.3 km further NE the road is crossed by Road IX. The cross is not marked in particular, and it is therefore possible that the two roads were not contemporary. At Hub III.3 a road branches off towards SW and several cross roads cut Road III in the vicinity. At the location of the hub a well is marked on the half-verst map. In the late 19th century a sheepfold was stationed there; in all likelihood the well has greater antiquity. Hub III.4 located between Čigiltaš/Dalekoe and Tanabaj is a superhub, where a number of roads cross (Fig. 6.27). From here a straight road crosses Road I. The hub is marked by a small kurgan cluster. Hub III.5 is a cross road marked again by a small kurgan cluster. A straight road leading from Hub I.4 ends in this hub, provided the road continues the straight line. Finally at Hub III.6 a large number of roads converge. This is an area of intense traffic. The distances between the hubs vary and probably not all hubs have been recognized. From HI.1/III.1 to HIII.2: 2.3 km; HIII.2 to HIII.3: 7.5 km; HIII.3 to HIII.4: 2.5 km; HIII.4 to HIII.5: 8.1 km; HIII.5 to HIII.6: 2.6 km. The actual path can be followed over long stretches of land. In places it is lined with kurgans.

*Road IV* (Fig. 6.28)
This road runs N of and parallel to Road III at a distance of up to 1.2 km. It is quite irregular in its layout. It starts/ends immediately above Vodopojnoe 2, mentioned several times above. It possibly ends in Hub III.5. It is crossed by a number of minor roads leading to the sea. In Hub IV.2 the road is crossed by Road IX. Its maximum length is 21.3 km. The actual path can be followed over some stretches; it is partially lined by kurgans.

*Road V* (Figs. 6.29-30)
Straight road connecting Hubs V.1/VI.3 and III.6/V.2. In the central part of this road, the actual path can be identified. The road is lined by a few kurgans. The total length of this road is 15.3 km.

*Road VI* (Figs. 6.30-31)
Straight road between the Bronze Age settlements of Burun Eli 1 and Severnoe 1. The road continues beyond both sites. Beyond Burun Eli 1 it may have joined the line of kurgans on the watershed of the ridge N of Lake Džarylgač; beyond Severnoe 1 it continued to the coast of the sea. Its length is at least 14 km. The E end of the road may mark a (Bronze Age?) locality, which has been eroded away by the sea. Traces of the road itself are preserved. The road is lined by a discrete number of kurgans. Hubs are located with short distances. From HVI.1/IX.1 to HVI.2: 3 km; HVI.2 to HV.1/VI.3: 1.3 km; HV.1/VI.3 to the sea: 6.3 km. At Hub V.1/VI.3 is a kurgan cluster with at least 10 kurgans (Fig. 6.30). From this hub several cross roads branches out. From Hub VI.2 a road connects Road VI and III crossing Road IV. In all likelihood, Road VI can be dated to the Late Bronze Age, the date of the two settlements it connects – if not as early as the Middel Bronze Age, because a very large kurgan in the kurgan cluster of Hub V.1/VI.3 seems to overlay the road line as well as of yet a road branching out in a more southern direction.

*Road VII* (Fig. 6.19)
Short transversal road between Hub VII.1 and I.5/VII.2. The road probably continues across the Bakal' Ravine ending at the sea. Its total distance is 10.2 km. Hub VII.1 is a superhub, because here converge at least eight roads. Its centre is a very large kurgan and around it is a kurgan cluster of at least six more kurgans. Hub I.5/VII.2 is also a superhub where several roads converge. Also this is marked with a discrete kurgan cluster. In addition, midway between the two hubs is one further kurgan cluster.

*Road VIII* (Fig. 6.32)
Short, straight transversal road linking Hub III.4/VIII.1 with Hub VIII.2. The distance between the hubs is 6.8 km. At the latter the road forks and continues further 1.9 km SW towards Kul Džakin and 2.8 km towards SE. Both roads end with a kurgan. The actual path can be followed over a considerable distance.

*Road IX* (Fig. 6.33)
Long-distance road running in a straight line from Burun Eli 2 to at least Hub II.4/IX.4 but probably on to the Donuzlav Ravine, where it seemingly used the same crossing as is used toady. The distance is at least 13 km, but more likely ca. 30 km. The road may have been early, because at some point it was superposed by another route running E-W crossing it and seemingly obliterating the former passage 2 km E of Vodopojnoe 3. The actual path can be followed in some areas. Hub VI.1/IX.1 is marked by a couple of kurgans. In Hub IV.2/IX.2 it crosses Road IV, and in Hub I.2/IX.3 it crosses Road I, whereas it joins or crosses Road II in Hub II.4/IX.4.

*Road X* (Figs. 6.34-35)
Road leading from one of the Vodopojnoe crossings at Hub II.1/X.1 towards SE probably continuing in a straight line as far as the area between Medvedevo and Ozerovka at the coast of the Donuzlav Lake. At Hub X.2 the road forks in a more northern direction, which may ultimately have ended in Hub XI.1. The actual paths are visible and a discrete number of kurgans line the roads.

*Road XI* (Fig. 6.36)
This road is a curving road aiming at the N edge of the Plateau. It is very weak in the satelite images og Google Earth. There is one hub (Hub XI.1), which is a superhub, because here a number of roads converge. At this hub is a kurgan cluster with at least one large kurgan. Road XI defines the N edge of the Plateau. It must have been an ancient road,

because it is lined with kurgans. However, it also occurs on the late 18th century map. The Plateau is practically devoid of kurgans and further traces of roads.

*Road XII* (Fig. 6.37)
Straight E-W road connecting the area immediately S of the Černomorskoe Settlement with the area S of Vodopojnoe. The road is easily distinguishable in the zone S and SE of Panskoe, and it passes immediately S of the fortified site of Panskoe II, which may have controlled the road. Further W it is either underlying the modern asphalt road, T0107, or is shifted ca. 100 m to the further N, because a track close to the S shore of the Sasyk Lake is of the same orientation. The preserved stretch is at least 4 km long. It crosses a number of kurgan chains, but it does not appear to be kurgan-lined itself. This road was noted by A.N. Ščeglov, and he considered it ancient (1987, fig. 2.I). Several stretches of it are also indicated on the 1890s map (the stretch S of Lake Sasyk; stretch N of Karlav and Kipčak Učkuju). It may mark a major artery in the divided farmland of the Černomorskoe Settlement.

*Road XIIIa-c* (Fig. 6.17)
Three parallel, straight roads located on the small peninsula constituting the W side of the Sasyk Ravine. They are without doubt of the first half of the 20th century, because they do not occur on the half-verst map of the late 19th century but instead on the mid-20th century map in scale 1:25,000 (Fig. 2.5). Only the southernmost of the three still exists.

*Road XIV* (Figs. 6.38-39)
This road was noted already by the Tarchankut Expedition (Ščeglov 1978, fig. 17; Ščeglov 1987, fig. 2.I [erroneously given as a straight road]; Ščeglov 2002a, pl. 6). It is a small, local track branching off from Road XVI immediately E of the fortified site of Panskoe II. The first km runs NNE; after this point, it turns sharply to a NNW direction leading to the sea shore between the settlement to the W and its necropolis to the E. The length of this part of the road is 1.1 km, and the overall length accordingly 2.1 km. On its way, close to Panskoe V it passes immediately W of a kurgan cluster formed around a large Bronze Age kurgan as well as closer to Panskoe I it passes two isolated kurgans. If we accept Ščeglov's reconstruction of the ancient topography, we may have an explanation for the sharp bend of the road, namely that it followed the side of the ancient ravine (Ščeglov 1987, fig. 2.I).

*Road XVa and b* (Fig. 6.40)
Two parallel tracks running from the Hillsides NNE to the crossing of the Sasyk Ravine. The length is ca. 2.3 km. The eastern one is lined by several kurgans. Their date is difficult to ascertain, but since they do not occur on any historical or modern maps they are probably ancient.

*Road XVI* (Fig. 6.41)
In the 19th century, the main road of the DSP area was an inland route leading from Ak Metčet via Bakal' to the mainland. In Mukhin's map of 1817 the major villages are marked as Ak Metčet → Aldermen → Karlav → Učkuju Kipčak → Kubaran → (Baj) Kijat → Sakab → Čagaltaj → Tanabaj. The same route is given in A. de Demidoff's map from 1837 and in S.A. Mitchell's map from 1859,[83] whereas the indication of roads and tracks in Betev's map of 1837 with revision in 1835 is considerably more detailed showing also the internal connection between the villages. And they also make it clear that tracks were highly mobile, not necessarily remaining in the same place but changing with changes in settlement pattern. The modern main road (T0107) follow the line of the 19th century main road from slightly W of Vladimirovka (Baj Kijat) to Bakal'. However, this is not the case between Černomorskoe and Vladimirovka, where today the road is led via Mežvodnoe. The actual road can be faintly distinguished between Evan and Aldermen (Fig. 6.41) and between Aldermenskaja Skala and Saja/Kostrovka. It is noteworthy that the main road of the 19th century does *not* coincide with the nomadic routes and it is equally noteworthy that this road does only sparingly show up in the maps of Google Earth indicating that traffic in antiquity must have been much more 'heavy' than in the 19th century.

---

83 *A Complete Map of the Crimea Showing the Military and Carriage Roads with Distances from Various Points of the Western Coast From English & Russian Surveys.* Philadelphia. Available at http://www.davidrumsey.com/luna/servlet/detail/RUMSEY~8~1~36464~1200831:A-Complete-Map-of-the-Crimea-Showin.

Summing up

The routes mentioned above are tangible testimonies of mobility and man's interaction with the landscape, and the preserved and reconstructed paths reveal an intricate web of movement. In all likelihood, the watershed units as defined by the kurgan chains may characterize zones of micro-mobility of a sedentary or semi-sedentary population, for example the zone in which a flock of animals could be moved during a day. The rectilinear roads on the other hand bear witness to movement over longer distances. It is normally assumed that these are nomads' routes of seasonal migration thus of macro-mobility. Their location in the landscape is very rational being foremost concerned with finding the easiest crossing at the beginning of the ravines. Incidentally, it appears that the survey area is located at the junction, where several types of mobility patterns can be observed. As we have already seen, there seems to be a relation between the area, where the cadastre can be reconstructed and the number of paths. Moreover, there is a similar change in the way the kurgans were spatially arranged, because kurgan chains, which are so common in our area and further W in the Tarchankut Peninsula, are rarely found to the E and NE of the Džarylgač Lake.

It is likely that the settlements found on the slopes are more or less permanent villages constituting the winter quarters sheltered from the winds. They may have been inhabited throughout the year by inhabitants practicing agro-pastoralism. In that case, some members of the family probably went to summer pastures perhaps in the foothills or the *jajlas* of the Crimean Mountains ca. 100 km and more to the SW. In the 19th century, they attracted shepherds with their flocks from as far away as Romania (Cordova, Rybak & Lehman 2001, 372). Thus, in a comparative perspective, the distance from the DSP study area to the Crimean Mountains is slight when discussing transhumance, because we know from ethnographical studies how flocks can be moved 500 km or more.[84]

[84] Eurasian steppe in Kazakhstan: Chang 2006, 197; Altai: Suttie & Reynolds (eds.) 2003.

CHAPTER 7

# The broader Tarchankut and Western Crimean context

*Compiled by P. Guldager Bilde*
*with contribution by S. Handberg & K. Winther-Jacobsen*

The DSP study zone is an artificial entity. Therefore, in order to understand the historical development in our area we need to contextualize our study zone within a wider geographical and historical context. The DSP area is located on the N coast of the Tarchankut Peninsula and it makes up a fraction only of the peninsula. Through the centre of the Tarchankut Peninsula runs a low ridge (height up to 178 m a.s.l.), which stretches E-W from Karadža almost to the end of the Donuzlav Balka. The peninsula is divided in two by the deep and much ramified Kelšeich Ravine running NS from Černomorskoe to the E-W ridge. Though not insurpassable, this ravine, nevertheless, seems to have bisected the peninsula in two distinct micro regions. In historic times, these two micro regions may have been oriented towards each a larger fortified settlement: the settlements of Karadža and Černomorskoe respectively. In turn, in the Late Classical, Hellenistic and perhaps the Roman period the entire region was part of Chersonesos' distant chora, whereas in the Medieval period until the early 18th century the Tarchankut was ruled by the Crimean Khanate. In the Early Modern period the region was foremost settled by the Nogaj Tatars. After Catharine the Great's annexation of Crimea from the Ottoman Empire in 1783, Russians and other Europeans were encouraged to settle in the region. Nevertheless, as late as 1944, when the Tatars were collectively exiled, the Tarchankut was dominated by Tatar villages. Today, the area is thinly settled. The local district town is Čornomorske/Černomorskoe (2001, population 11,709).

## 7.1 PREHISTORY | S. HANDBERG

There is very little evidence for occupation on the Tarchankut Peninsula and the western Crimea in the Middle Bronze Age period and with the exception of Baj Kijat and possibly Skalistoe 2, settlements of this period are lacking altogether on the Tarchankut. A few burials from Dalekoe and Mežvodnoe have been associated with the 'Mehrwulstkeramik' 'cultural' group of the post-Catacomb period, i.e. the Late Middle Bronze Age period, but corresponding settlements have not been located. In view of the scarcity of permanent settlements on the peninsula prior to the Late Bronze Age period, the sporadic finds of Middle Bronze Age pottery from Skalistoe 2 and DSP08-F06-01, which can be interpreted as settlement sites, appear to be a significant new contribution to the historical reconstruction of the settlement pattern in the area.

Although 10 Late Bronze Age settlement sites had previously been identified on the Tarchankut Peninsula, excavations have only been undertaken at the three settlements of Burun Eli, Baj Kijat and Skalistoe 2. These previously recognised Late Bronze Age sites on the Tarchankut are clustered around the Džarylgač and Jarylgač lakes in the immediate vicinity of the DSP survey area and the geographical location on hills near ravines and watersheds of the Late Bronze Age sites identified during the survey are similar to the position of the known settlements in the area as well as the Late Bronze Age settlements known from the foothills of the Crimean mountains to the S.[85]

The pre-existing observations concerning the pattern of Bronze Age kurgans, which are isolated from the settlements, was confirmed by the DSP survey, where no kurgans were found associated with the major Bronze Age settlements such as Skalistoe 2 and DSP08-F06-01 with the possible exception of the kurgan group with a robbed Late Bronze Age kurgan DSP08-F04-04 (Feature 3). This kurgan belongs to the secondmost common type of Late Bronze Age burial type with a stone lined cist grave.[86]

[85] Kolotuchin 1996, 24.
[86] Tošhev 2007, 180.

For the largest settlements, Baj Kijat and Burun Eli, V.A. Kolotuchin has proposed several short occupational phases (four in the case of Baj Kijat and several at Burun Eli).[87] Given the nature of survey material, short term occupational differences cannot be recognised. Even so, the monumental double pen on DSP07-H08 combined with Middle and Late Bronze Age pottery suggest a continued occupation. The find of a fragment of a stone axe adds additional importance to the settlement on Hill DSP07-H08. Such stone axes are more often found in burials as for instance in Kurgan no. 2/1982 at Krulovka in the Pervomajev region 70-80 km to the E of the DSP survey area.[88] This kurgan has a circular stone foundation over 30 m in diameter and we may consider the possibility that the inner circle of the double pen was originally a kurgan which was later re-used as part of a milking pen. Out of approximately one thousand analysed bones from the two years of excavations at Baj Kijat 52% of the osteological material could be identified as belonging to cattle.[89] The four double pens identified on the hills during the DSP survey provide new additional evidence for the rearing of dairy cattle. Here it might also be mentioned that remains of pigs appear to be absent in the Late Bronze Age in and around the DSP survey area, whereas they are represented on other Late Bronze Age settlements in Crimea and the Kherson region.[90]

The emergence of the last phase of the Bronze Age, the so-called Belozerka 'culture', has been recognised at Baj Kijat 1 and possibly at Skalistoe 2, but following the Sabatinovka 'culture' there is a remarkable decline in the settlements in the northern part of Crimea. Already at this time, in the last centuries of the second millennium BC, most settlements of the Belozerka 'culture' are found in the mountainous region in southern Crimea between Sevastopol and Belogorsk.[91] The migration has traditionally been connected with climatic changes when the whole Eurasian steppe became progressively more arid.[92] This migration is also reflected in the material culture retrieved during the DSP survey since no material could be ascribed to the Belozerka 'culture'.

## 7.2 IRON AGE | S. HANDBERG

The Early Iron Age in Crimea is associated with the transition from the Belozerka 'culture' to the Kimmerian, Taurian and Kizil-koba 'cultural' groups. The distinction in the material culture of the various groups, or tribes, has been widely discussed but clear differences in the ceramic repertoire in the transitional period are lacking.[93] Following the Belozerka period, the Tarchankut Penisula became completely abandoned and the Early Iron Age settlements were moved to the mountainous southern area. There the continuity from the Late Bronze Age to the Early Iron Age has been observed at several settlement sites.[94] The most characteristic pottery of the Early Iron Age, the so-called Kizil-Koba pottery with incised combed decoration, does not appear to pre-date the 8th century BC.[95] Kizil-Koba pottery of the 6th century BC is known from both Chersonesos and its home chora, but similar material is not known from the Tarchankut.[96]

In past research, it was believed that before the settling of the Greeks, the region was not at all settled (e.g. Ščeglov 2002a, 15), but alone frequented by an occasional nomad, as witnessed by a few kurgans dated to the 5th century BC left behind in the landscape (e.g. Ščeglov 1978, fig. 8.I). In the 1970s and 1990s the North-Crimean Expedition excavated a number of kurgans in the DSP study zone: five kurgans near Vodopojnoe; four kurgans over Snežnoe, four kurgans near Meživodnoe, and two kurgans near Dalekoe. Unfortunately they have never been published. According to the opinion of S.G. Koltuchov, who participated in the excavation, they all date to the second half of the 5th-beginning of the 4th century BC.[97] If this date can be upheld all of them may predate the Greek settling of the region. The few finds of Greek

87 Kolotuchin 2003, 28-29
88 Kolotuchin 2003, 69 fig. 2.4.
89 Kolotuchin 2003, 41.
90 Osteological remains of pigs are known from e.g. Družnoe 2 (18.5%) and Dremailovka (6.5%), see Kolotuchin 1996, 28 and Sekerskja 2000.
91 Kolotuchin 1996, 68, 87.
92 For short overviews, see Bokovenko 2004 and Makhortykh 2004.
93 See Chrapunov 1995; For the ceramics also e.g. Kolotuchin 1996, 69.
94 See Kolotuchin 1996, 19-24.
95 Vlasov 1997.
96 For Chersonesos, see Senatorov 2003. For Kizil-koba occupation on the Herakleian Peninsula see Kravčenko 2005.
97 See also Kolotuchin 1996, 65-66 for a short discussion of the Snežnoe as well as another kurgan group at Donuslav.

Late Archaic and Classical transport amphorai found in the DSP survey area might be related to the presence of a limited indigenous population in the area before the arrival of the Greeks (see Chapter 5.3.3.6).

## 7.3 GREEK AND ROMAN ANTIQUITY | P. GULDAGER BILDE

Whereas the landscape of the nomads and the agro-pastoralists predominantly was one of 'path' and accordingly of 'space' (Chapter 6.6), the landscape of the Greeks was one of 'place', in which the settlement or the individual farmhouse marked a zone of permanence and stability. In the Late Classical and Hellenistic period, the DSP study zone became a sub-region of the Chersonesean chora. The period of the maximum expansion of the Chersonesean state was the brief interval from the mid- to second half of the 4th century BC to ca. 270 BC, thus not much more than half a century. During this period, a considerable number of fortified and unfortified settlements and farms were constructed between Chersonesos and at least as far as Masliny, ca. 220 km by the sea following the coast.

Tradition has it that Chersonesos was founded by Dorian Greeks from the S Pontic city Herakleia Pontike in 422/1 BC, but archaeological finds prove Greek presence around 100 years earlier, and a batch of *ostraka* from the late 6th and early 5th century BC points to a joint colonization by Pontic settlers from Ionian Sinope and Doric Herakleia (Zolotarev 1996). One further polis, 'Old Chersonesos ... razed to the ground' was mentioned by Strabon (7.4.2). This city is by most scholars placed on the fortified Majačnyj (Lighthouse) Peninsula at the Kazač'ja Bay. On this site, the earliest Greek material is contemporary with that found in the city at Karantinnaja (Quarantine) Bay (Ščeglov 1997). However, Strabon's words about the location of this city between Cape Parthenion (Cape Fiolent) and Symbolon Limen (Balaklava) cannot be mistaken, because Cape Fiolent is exactly 12.5 km S of the city at Karantinnaja Bay corresponding to the 100 stadia mentioned by Strabon. Accordingly, there were at least two *poleis* as well as the settlement of the Majačnyj Peninsula in the area. Old Chersonesos has not been convincingly located and it may have been eroded away completely. Therefore, we also ignore the date of its foundation. The settlement of the Majačnyj Peninsula is at our present knowledge contemporary with that of Chersonesos. However, from a topographical point of view, its location in the landscape corresponds better to the one preferred by the earliest Greek settlers, namely on a narrow peninsula, which was fortified across its narrowest point as for example in Orgame, the Sinoe-"Insula Lupilor" Settlement, and the Berezan' Settlement, so by analogy, this may have constituted the earliest of the Chersonesean settlements having the same relation to Chersonesos as did the Berezan' Settlement to Olbia.

Old and New Chersonesos were not the only *poleis* of Western Crimea: also Karkinitis/Kerkinitis was mentioned as a polis (FGrHist 1 fr. 184 = Steph. Byz. 360.1). As pointed out above (Chapter 4.3.1.1), it was located 600 stadia from Chersonesos, wherefore it is commonly identified as the settlement W of modern Evpatorija. Here dugouts and pottery of late 6th and early 5th century BC have been found as well as a fortification wall of the 5th century BC (Vnukov 2001, 152-153). This city minted its own coins. The earliest ones were cast coins in the shape of arrow heads and fish/dolphins of the 5th century BC some with the city name KA inscribed (Kutajsov 1996). This type of monetary practice finds its closest parallels in contemporary Olbia. During the second half of the 5th century BC, also more ordinary cast coins were issued, these too with city name K or KA (Kutajsov 1996, fig. 6). For a short period in the mid-4th century BC Kerkinitis minted further coins inscribed KARK/KARKI (Stolba 1996). It is likely that Kerkinitis was an independent Ionian polis until this period. Thereafter, it was seemingly incorporated into the territory of the Chersonesean state, as we can gather from the Chersonesean civic oath of ca. 300 BC (*IosPE* I$^2$ 401) probably as a dependent polis (Avram, Hind & Tsetskhladze 2004, 946).

The influence of Olbia in Western Crimea has been much debated. Several scholars are of the opinion that the coast of northwestern Crimea was outright part of Olbian territory in the 5th until mid-4th century BC (Rusjaeva 1986, 57 note 145; Ščeglov 1986, 166 [not seen]; Zolotarev 1986; Chtcheglov 1992, 244-245; Vinogradov 1989; Ščeglov 2002a, 16-17). However, this historical reconstruction has not been unanimously accepted. On the contrary, some scholars suggest that Kerkinitis before being subjected to Chersonesos may have settled the Tarchankut (e.g. Lantsov & Uzhentzev 2006, 701).

### 7.3.1 The Chersonesean chora in the Late Classical and Early Hellenistic period

If the character of the settling of the region in the Archaic and Classical periods is up to debate, there is unanimous agreement that during the brief interval between ca. 350 and 325 BC the entire western and northwestern Crimea was incorporated into the Chersonesean state as its distant chora. Whether this was the result of a regular war between the two Greek poleis or whether it was a peaceful takeover is a matter of scholarly dispute (overview in Lantsov & Uzhentzev 2006, 701). The material remains witnessing this process attest to the spread of a unitary culture in terms of building style and design as well as a common spiritual culture centred on the Dorian founder hero, Herakles (Ščeglov 1978, 125; see also Chapter 6.4).

It is debated, how far N the Greek-controlled territory extended. Common agreement has it that it extended no further than Masliny (Lancov 2004, 59 with references). However, it is well known that three or four (Scythian?) settlements are found further 40 km N of Bakal' (Portovoe: *razvedki* of A.A. Ščepinskij, cf. Ščepinskij & Čerepanova 1974, 62-63; Borisovka and Kumovka: *razvedki* of M.K. Zinov'ev, cf. Zinov'ev 1957, 325; Lancov 2004, fig. 2.83-86). Since Greek as well as Scythian material is found on the mentioned sites, Lancov is of the opinion that the Chersonesean distant chora continued thus far (Lancov 2004, 62; Lantsov & Uzhentzev 2007, 681-682). The largest site in the region by far seems to be the one at Bakal', which is located ca. 15 km NE of Masliny. This site extends over an area of more than 1,000 x 500 m. Certainly, much more investigation is needed in order to obtain a firm understanding of the settlement structure E of Masliny. In any case, the DSP survey area was part of the distant chora of the administrative polis, be that Olbia or Chersonesos.

We are well informed of the settlement hierarchy not just from archaeological fieldwork but also from ancient literary and epigraphic sources. In this period, Chersonesos was the polis, which controlled the dependent polis of Kerkinitis and the town of Kalos Limen. The presence of forts (τείχη) is mentioned in three Chersonesean inscriptions (*IosPE* I² 352, 353, 401). The land is called the chora (χώρας) (*IosPE* I² 401) as well as the productive plain (πεδίον) (*IosPE* I² 401, 418).

The settling of the chora had two main effects: the introduction of a number of new species such as grapevine (*vitis*), almond (*amygdalus*), and perhaps walnut (*juglans*), and a variety of fruit trees. But at the same time, a considerable deforestation of the region took place. Both can be read in the pollen diagrams, which have been produced by Cordova and Lehman (2003; 2006).

It is well known that the Chersonesean territory was divided into regular plots in several different phases. We have already discussed the evidence for cadasters on the Tarchankut Peninsula in Chapter 6.3.1. Until recently, the Chersonesean cadaster system was well preserved, especially on the Herakleian Peninsula in the near or home chora of Chersonesos (Ščeglov 1980; Carter 2006; Nikolaenko 2006; Smekalova & Smekalov 2006). Clear traces of property divisions over large tracts of land in the Tarchankut Peninsula testify to the substantial scale of cultivation of the steppe also in Chersonesos' distant chora. This was first discovered by A.N. Ščeglov through the study of aerial photographs in the 1970s (Ščeglov 1980). Recently T.N. Smekalova has returned to the issue (Smekalova & Smekalov 2006, 235-240). The mentioned studies have made it clear that significant contiguous stretches of land foremost on the flat inland Plateau W of the Kelšeich Ravine and also small patches of land on the S coast around Ojrat and Džan Baba were thus divided (Ščeglov 2002a, pl. 4; Smekalova & Smekalov 2006, fig. 14).

### 7.3.2 The crisis of the 270s BC

The first third of the 3rd century BC was a watershed in the life of the entire northern Black Sea region. A series of interlinked crises resulted in a near collapse of the local cultures:

a) A climate crisis (Alexeev 2003, 416; Stolba 2005b and c; Užencev 2006, 115-116)
b) Abandonment of the chorai in the entire northern Black Sea area; cessation of Chersonesean wine production (or at least of stamping the amphorai)
c) An economic crisis resulting in the change of weight standards, adding of base metals to alloys, overstriking and countermarking, and cessation of striking of gold (Stolba 2005b and c). The economic crisis may be reflected in the stop of import of Attic Black Gloss pottery and the presence of many repairs on pottery

Because the North Pontic landscape is a marginal zone for Greeks as well as for the nomadic tribes, production was hit hard by changing climate. Thus grain became a scarce commodity, as we may gather from the Chersonesean civic oath from ca. 300 BC (*IosPE* I² 401). In the oath, the citizens are obliged to swear that "neither shall I sell grain suitable for exportation which comes from the plain, nor export grain from the plain to another place, except to Chersonesos". The pressure on the resources also led to intensified diplomatic contacts between the Bosporan Kingdom and Egypt during the first half of the 3rd century BC perhaps in order to secure import of grain from Egypt.

Ecological crises are well-known push factors for migration and in all likelihood, the 3rd century crisis led to movement of nomadic tribes. Thus, already around 300 BC, the Classical culture of Scythia Magna collapsed (Alexeev 2003, 416) probably pushed by the arrival of a new group of nomads, the Sarmatians in the area. It is believed that many Greek sites were destroyed by nomadic attacks during this period, e.g. Panskoe I, as witnessed by burnt layers in most of the Greek sites.

Even though the worsened climatic conditions may have prevailed for more than a century, the chorai of many Greek cities, especially of the Bosporan Kingdom were resettled already around 230 BC, many with fortifications that illustrate that the threat from outsiders was real.

### 7.3.3 Late Hellenistic period

In the DSP study zone, after the collapse of the agricultural system in the 270s BC, it is only in the Late Hellenistic period the region was populated again. However, settling was significantly limited and the evidence exclusively comes from coastal sites (Černomorskoe Settlement, Masliny; a few pieces of Mouldmade bowls from Panskoe I also attest to some activity in the second half of the 2nd century BC; possibly the Skalistoe sites 5, 8 and 10; possibly Baj Kijat 2). It is generally believed that Scythians took over the Greek settlement sites in the Late Hellenistic period turning them into to coastal strongholds. This goes for Masliny and the Černomorskoe Settlement as well as Kulčuk, Tarpanči, Ojrat, and Beljaus, and even Greek farms, such as Bol'šoj Kastel' are assumed to have been taken over by the Scythians.

### 7.3.4 Roman period

The Roman period is scarcely known in the Tarchankut Peninsula. In the DSP study zone, it is only in Černomorskoe that activity in the Roman period is recorded (see above, Chapter 4.3.1.1).

## 7.4 MEDIEVAL PERIOD | P. GULDAGER BILDE & K. WINTHER-JACOBSEN

At Černomorskoe, 8th-10th centuries phases associated with the Saltovo-Majak culture linked to the Khazars as well as later Medieval phases have been reported, but scarce evidence has been published (Nalivkina 1957; Ščeglov 1978, 36-37; Guldager Bilde et al. 2008, 134). None of the pottery collected by the DSP can be securely dated to this period (see Chapter 5.3.5.1). Kurgan burials persist, even though the presence of moundless catacomb graves indicates that the burial rites have not been restricted to this practice only. One such burial has been unearthed around 1990 at the settlement of Panskoe I (unpublished).

The ancestors of (most of) the Crimean Tatars were the Medieval tribes of the Kuman/Kipčak (in Rus annals called Polovetses). These were nomadic and semi-nomadic tribes living in the 10th to the 13th century occupying a significant territory in the Eurasian steppe, stretching from N of the Aral Sea westward to the region N of the Black Sea as far as Central Europe. According to H. Jankowski, the bulk of Crimean toponyms and oikonyms are Kipčak (2006, 31). They are the witnesses of sedentarisation of the area in the Medieval period. In the DSP study zone, the presence of the Kipčak is alluded to in the toponym Kipčak-Učkuju, which is the name of one of the deserted Tatar villages S of the Džarylgač Lake.[98] Currently, the finds from Černomorskoe and the surviving toponyms are the only evidence for the Medieval occupation of the DSP study zone. As was the case with the material of the 8th-10th centuries, none of the pottery collected by the DSP can be dated securely to the 11th-13th centuries (see Chapter 5.3.5.1). However, none of the Tatarian villages identified on the maps such as Kipčak-Učkuju were the object of intensive studies mainly due to current occupation and the short duration of the project. The tendency to occupation continuity witnessed on the slopes suggests that the Medieval villages may be located on the sites of the Early Modern Tatarian ones. These tend to be located like a string of pearls along the geographical line where access to the highest quality water was the easiest, and high quality water for watering the animals would have been a vital resource to the nomadic and semi-nomadic peoples of the Medieval period.

[98] There is one further settlement at the N coast of the Tarchankut Peninsula with the name of Kipčak (also Hellenistic settlement at this site: Lancov 2004, fig. 1.64). The highest hill N of Krasnosel'skoe is equally called Kipčak Hill.

The Kipčak buried their dead under kurgans. At Dalekoe a Medieval kurgan has been excavated by S.G. Koltuchov and others (Koltuchov et al. 1994, 114-115 [not seen]). Stone statues (*baba*'s) were used to mark graves. In Černomorskoe Museum two fragmentary *baba*'s from the region are exhibited (Kutajsov 2005). According to V.I. Pavlenkov fragments of two similar statues were recorded from Mežvodnoe (Pavlenkov 1986, 294 [not seen]).

## 7.5 EARLY MODERN PERIOD | P. GULDAGER BILDE

In 1783, Catharine the Great annexed the Crimea from the Ottoman Empire, whereby Russia gained access to the Black Sea. At that time, the region was becoming gradually depopulated (Lyall 1825, 340). Catherine therefore consolidated her new acquisition with the settling of Russian, Ukrainian, German, Bulgarian and other colonists. Already by 1863, the immigrants outnumbered the Tatar population.

The annexation of Crimea by the Russian Empire signified a fundamental change of the Crimean landscape: new cities and towns were constructed, new roads were made and a shift to more intensive horticulture with focus on vine and tobacco was carried out (Williams 2001, 64). P.S. Pallas claimed that extensive deforestation took place, which led to an exposure to cold winds from the N (Pallas 1801, 382). Some deforestation also took place between 1812 and 1913, when the area of forests was reduced from 361,000 ha to 335,000 ha (Cordova, Rybak & Lehman 2001, 372). The Tatars were by the government incited to cultivate the land, but they are "of all the colonists, far the worst cultivators; and are still much addicted to grazing large flocks and herds and numerous studs of mares. Under the encouragement of government, agriculture has, nevertheless, much increased amongst them; arising in a great degree from the high price of wheat..." (Holderness 1823, 141). In the DSP area, a colony was planted W of the mouth of the Kelšeich Ravine with regular house plan (see Betev's map of 1837). The preceding but still existing Tatar villages Ak Metčet (the 'white mosque') and Šeichlar extended along the E side of the ravine from the small village of Takil' to immediately S of the ancient settlement. By the end of the century it was completely taken over by Russians.

After the deportation of the Tatars in 1944, the Turkish names of their villages were supplanted by Russian names (Jankowski 2006, 1). The Crimean Tatars were officially rehabilitated in 1967, and from 1988, they began returning to Crimea. In the Ukrainian census of 2001, 12.7% of the population in the Tarchankut Peninsula is Tatar.[99] Today, most of the Tatar villages in the DSP area remain deserted. However, especially in Vodopojnoe and Mežvodnoe a significant number of Tatars are living today (Fig. 4.38).

[99] http://en.wikipedia.org/wiki/File:Etnics.svg accessed 9 May 2009.

# Appendix 1
# Site catalogue

*P. Attema, P. Guldager Bilde, W. de Neef & K. Winther-Jacobsen*
*with contributions by S. Lancov, D. Pilz, T. Smekalova, V. Stolba & C. Williamson*

The sites discovered within the DSP area of study are listed here, together with those previously known. Site names beginning with 'DSP' were generally in areas subject to intensive and systematic survey techniques unless otherwise stated. 'F' indicates ploughed fields, 'H' indicates hillside. Throughout, the following abbreviations for chronological periods are used: BA = Bronze Age; MBA = Middle Bronze Age; LBA = Late Bronze Age; LC = Late Classical; EH = Early Hellenistic; LH = Late Hellenistic; EM = Early Modern (18th-19th century). 'FL' are the official finds lists in which finds from trial trenches were recorded on the request of our Ukrainian partners. Now, the lists prevail in the archives of the institutes of archaeology of the Ukrainian Academy of Sciences in Kiev and Simferopol' respectively.

## Overview of sites

This table presents an overview of the sites, with period and general type, correlated with the landscape zone.

| Site name | Period | Type | Landscape zone |
|---|---|---|---|
| (DSP06-F03-01- see Panskoe II) | | | |
| DSP06-F04-01 | LC-EH | Settlement? | Pediment |
| DSP07-F01-01 | LC-EH | Settlement | Lowland ridge/Coastal cliff |
| (DSP07-F01-02 – see Panskoe V) | | | |
| DSP07-F02-01 | LC-EH | Settlement | Pediment |
| DSP07-F02-03 | LC-EH | Settlement | Pediment |
| DSP07-F02-04 | LC-EH | Settlement | Pediment |
| SP07-F02-05 | Unknown | Kurgan | Pediment |
| DSP07-F03-01 | LC-EH? | Kurgan group | Pediment |
| DSP07-F04-01 | BA | Settlement | Pediment |
| DSP07-F04-02 | LC-EH? Early Modern | Kurgan? and Settlement | Pediment |
| DSP07-F05-01 | Early Modern | Settlement | Lowland ridge/Coastal cliff |
| DSP07-F05-02 | BA? | Kurgan / ritual | Lowland ridge/Coastal cliff |
| DSP07-F06-01 | LC-EH | Kurgan | Lowland ridge/Coastal cliff |
| DSP07-F06-02 | LC-EH | Settlement? | Lowland ridge/Coastal cliff |
| DSP07-F06-03 | LC-EH | Settlement | Lowland ridge/Coastal cliff |
| DSP07-F07-01 | BA, LC-EH | Kurgan group | Lowland ridge/Coastal cliff |
| DSP07-F07-02 | Unknown | Kurgan | Lowland ridge/Coastal cliff |
| DSP07-F08-01 | LC-EH | Settlement | Lowland ridge/Coastal cliff |
| DSP07-F08-02 | Unknown | Cadastral system | Lowland ridge/Coastal cliff |
| DSP08-F01-01 | Unknown | Kurgan group | Lowland ridge/Coastal cliff |
| DSP08-F02-01 | LC-EH | Settlement | Lowland ridge/Coastal cliff |
| DSP08-F03-01 | Early Modern | Settlement | Lowland ridge/Coastal cliff |

| Site name | Period | Type | Landscape zone |
|---|---|---|---|
| (DSP08-F03-02 see Skalistoe 2) | | | |
| DSP08-F04-01 | LC-EH | Settlement | Pediment |
| DSP08-F04-02 | LC-EH | Settlement | Pediment |
| DSP08-F04-03 | LC-EH | Settlement | Pediment |
| DSP08-F04-04 | Unknown | Kurgan group | Pediment |
| DSP08-F04-05 | Unknown | Kurgan group | Pediment |
| DSP08-F04-06 | Modern | Settlement | Pediment |
| DSP08-F04-07 | LC-EH | Settlement | Pediment |
| DSP08-F05-01 | LC-EH | Settlement? | Lowland ridge/Coastal cliff |
| DSP08-F06-01 | BA | Settlement | Lowland ridge/Coastal cliff |
| DSP08-F06-02 | Unknown | Kurgan | Lowland ridge/Coastal cliff |
| DSP08-F06-03 | LC-EH | Settlement | Lowland ridge/Coastal cliff |
| DSP07-H01-01 | BA, LC-EH | Settlement | Hillsides |
| DSP07-H02-01 | BA, LC-EH? | Double-pen | Hillsides |
| DSP07-H02-02 | BA, LC-EH | Settlement? | Hillsides |
| DSP08-H02-03 | Unknown | Dam | Hillsides |
| DSP07-H04-01 | LC-EH | Settlement | Hillsides |
| DSP07-H04-02 | Early Modern? | Settlement | Hillsides |
| DSP07-H04-03 | Unknown | Structure? | Hillsides |
| DSP07-H06-01 | BA/LC-EH? | Settlement | Hillsides |
| DSP07-H07-01 | BA?/LC-EH | Settlement | Hillsides |
| DSP07-H07-02 | BA?/LC-EH | Settlement | Hillsides |
| DSP07-H07-03 | Early Modern? | Settlement | Hillsides |
| DSP07-H07-04 | Early Modern? | dam | Hillsides |
| DSP07-H08-01 | BA/LC-EH? | Settlement | Hillsides |
| DSP07-H08-02 | Early Modern | Rectangular stone structure | Hillsides |
| DSP07-H09-01 | Early Modern | Stone structure | Hillsides |
| DSP07-H09-02 | Early Modern? | Stone structure | Hillsides |
| DSP07-H10-01 | LC-EH | Settlement | Hillsides |
| DSP07-H11-01 | LC-EH | Settlement | Hillsides |
| DSP08-H11-02 | LC-EH | Settlement | Hillsides |
| DSP08-H12-01 | LC-EH? | Settlement | Hillsides |
| DSP08-H13-01 | BA?/LC-EH | Settlement? | Hillsides |
| DSP08-H14-01 | BA/LC-EH | Settlement | Hillsides |
| DSP08-H15-01 | Modern | Farmstead | Hillsides |
| DSP08-H16-01 | Unknown | Settlement | Hillsides |
| DSP08-H17-01 | Unknown | Storage site? | Hillsides |
| DSP08-H17-02 | LC-EH? | Kurgan group | Hillsides |
| DSP08-H17-03 | Unknown | Kurgan | Hillsides |
| DSP08-H18-01 | LC-EH | Settlement | Hillsides |
| DSP08-H19-01 | Unknown | Kurgan group | Hillsides |
| DSP08-H19-02 | Unknown | Kurgan group | Hillsides |
| DSP08-H19-03 | Unknown | Kurgan group | Hillsides |

| Site name | Period | Type | Landscape zone |
|---|---|---|---|
| DSP08-H19-04 | LC-EH? | Kurgan group | Hillsides |
| DSP08-H19-05 | Unknown | Kurgan group | Hillsides |
| DSP08-H19-06 | Unknown | Kurgan group | Hillsides |
| DSP08-H19-07 | Unknown | Kurgan group | Hillsides |
| DSP08-H19-08 | Unknown | Kurgan group | Hillsides |
| DSP08-H19-09 | Modern? | Area of quarries | Hillsides |
| DSP08-H20-01 | LC-EH | Settlement? | Hillsides |
| DSP08-H21-01 | Modern | Settlement | Hillsides |
| DSP08-H22-01 | Unknown | Kurgan group | Hillsides |
| DSP08-H22-02 | Unknown | Kurgan | Hillsides |
| DSP08-H22-03 | Modern/? | Settlement/Kurgan group | Hillsides |
| DSP08-H23-01 | Unknown | Kurgan group | Hillsides |
| DSP08-H23-02 | LC-EH? | Settlement? | Hillsides |
| DSP08-H24-01 | LC-EH? | Kurgan group | Hillsides |
| DSP08-H24-02 | Unknown | Kurgan group | Hillsides |
| DSP08-H24-03 | LC-EH? | Settlement? | Hillsides |
| DSP08-H25-01 | Unknown | Kurgan group | Hillsides |
| DSP08-H25-02 | ? | Kurgan group? | Hillsides |
| DSP08-H26-01 | LC-EH? | Kurgan group? | Hillsides |
| DSP08-H26-02 | LC-EH? | Kurgan group? | Hillsides |
| DSP08-H27-01 | Modern? | Farmstead | Hillsides |
| DSP08-H28-01 | Modern? | Storage site? | Hillsides |
| Baj Kijat 1-2 | MBA-LBA/LC-EH | Settlement/ash hill | ? |
| Baj Kijat 3 | LC-EH/Early Modern | Farmhouse?/settlement? | ? |
| Burun Eli 1 | LBA/Late Scythian/Med | Settlement | Lowland ridge/Coastal cliff |
| Burun Eli 2 | LC-EH | Settlement or Farmhouse? | Lowland ridge/Coastal cliff |
| Burun Eli 3 | LC-EH | Settlement or Farmhouse? | Lowland ridge/Coastal cliff |
| Burun Eli 4 | LC-EH/Early Modern | Settlement/ash hill | Lowland ridge/Coastal cliff |
| Burun Eli 5 | LC-EH/Early Modern | Settlement or Farmhouse? | Lowland ridge/Coastal cliff |
| Černomorskoe settlement | Cl-Roman/LR/Medieval | Fortified city | Lowland ridge/Coastal cliff |
| Černomorskoe farm 1 | LC-EH | Farmhouse | Lowland ridge/Coastal cliff |
| Černomorskoe farm 2 | LC-EH | Farmhouse | Lowland ridge/Coastal cliff |
| Černomorskoe farm 3 | LC-EH/Roman | Farmhouse | Lowland ridge/Coastal cliff |
| Černomorskoe farm 4 | LC-Hellenistic | Farmhouse | Lowland ridge/Coastal cliff |
| Černomorskoe farm 5 | LC-EH? | Farmhouse | Lowland ridge/Coastal cliff |
| Černomorskoe farm 6 | Unknown | Farmhouse | Lowland ridge/Coastal cliff |
| Černomorskoe farm 7 | Unknown | Farmhouse | Lowland ridge/Coastal cliff |
| Černomorskoe farm 8 | Unknown | Farmhouse | Lowland ridge/Coastal cliff |
| Černomorskoe farm 9 | Unknown | Farmhouse | Lowland ridge/Coastal cliff |
| Džarylgač | LC-EH | Settlement? | Lowland ridge/Coastal cliff |
| Groty 1 | LC-EH/LH | Unfortified settlement | ? |
| Groty 2 | LC-EH | Farmhouse | ? |
| Jarylgač | LC-EH | Settlement? | Lowland ridge/Coastal cliff |

| Site name | Period | Type | Landscape zone |
|---|---|---|---|
| **Jarylgač Severnoe** | Neolithic | Settlement | Lowland ridge/Coastal cliff |
| **Jarylgač Vostočnoe** | LBA/LC-EH/Medieval | Settlement? | Lowland ridge/Coastal cliff |
| **Kula Seich** | Early Modern? | Unknown | Lowland ridge/Coastal cliff |
| **Kurama Kostel** | LC-EH/Early Modern | Settlement? Kurgan? | Pediment |
| **Masliny 1 (Vladimirovka)** | LC-EH/LH | Fortified settlement | Lowland ridge/Coastal cliff |
| **Masliny 2** | LC-EH | Settlement | Lowland ridge/Coastal cliff |
| **Masliny 3** | LC-EH | Farmhouse? | Lowland ridge/Coastal cliff |
| **Meživodnoe** | LC-EH | Unfortified? farmhouse | Lowland ridge/Coastal cliff |
| **Panskoe I** | LC-EH/LH | Settlement | Lowland ridge/Coastal cliff |
| **Panskoe II / DSP06-F03-01** | LC-EH/LH | Settlement | Lowland ridge/Coastal cliff |
| **Panskoe III** | LC-EH | Farmhouse | Lowland ridge/Coastal cliff |
| **Panskoe IV** | LC-EH | Farmhouse | Lowland ridge/Coastal cliff |
| **Panskoe V / DSP07-F01-02** | LC-EH/LH | Settlement? Kurgans? | Lowland ridge/Coastal cliff |
| **Severnoe 1** | LBA/Early Modern | Settlement | Lowland ridge/Coastal cliff |
| **Severnoe 2** | LC-EH | Settlement? | Lowland ridge/Coastal cliff |
| **Skalistoe 1** | LC-EH | Settlement? | Lowland ridge/Coastal cliff |
| **Skalistoe 2 / DSP08-F03-02** | BA/LC-EH | Settlement | Lowland ridge/Coastal cliff |
| **Skalistoe 3-4** | LC-EH | Settlement? | Lowland ridge/Coastal cliff |
| **Skalistoe 5** | LC-EH/LH | Fortified farm/ash hill | Lowland ridge/Coastal cliff |
| **Skalistoe 6** | LC-EH | Farmhouse?/settlement? | Lowland ridge/Coastal cliff |
| **Skalistoe 7** | LC-EH | Farmhouse? | Lowland ridge/Coastal cliff |
| **Skalistoe 8** | LC-EH | Farmhouse? | Lowland ridge/Coastal cliff |
| **Skalistoe 9** | LC-EH | Settlement?/Farmhouse? | Lowland ridge/Coastal cliff |
| **Skalistoe 10** | LC-EH/LH | Farmhouse? | Lowland ridge/Coastal cliff |
| **Veterenaja Bay** | LC-EH | Farmhouse | Lowland ridge/Coastal cliff |
| **Vladimirovka** | LBA | Settlement? | Lowland ridge/Coastal cliff |
| **Vodopojnoe 1 / DSP07-F09-01** | LBA | Settlement | Lakeshore |
| **Vodopojnoe 2** | LBA/LC-EH?/ Early Modern | Settlement/kurgan/farmhouse? | Lowland ridge/Coastal cliff |
| **Vodopojnoe 3A-B** | LBA/LC-EH/ Early Modern | Settlement | Lowland ridge/Coastal cliff |
| **Vodopojnoe 4** | LBA | Settlement? | Lowland ridge/Coastal cliff |
| **Vodopojnoe 5** | LBA | Settlement? | Lowland ridge/Coastal cliff |

## Documentation

The data on each site are presented below including basic topographical description and the method of investigation (see Chapter 2). Sites DSP08-H19-01-03 and 05-08 consists of kurgans which were primarily measured for 3D rendering using the Total Station. The finds from the sites are discussed and analyzed in Chapter 5; the pottery from each site is tabulated in Appendix 2A and selected individual sherds are catalogued in Appendix 2B. Coordinates are given in UTM.

## DSP06-F03-01 see Panskoe II

| **DSP06-F04-01** | | *Coordinates (centre)*<br>490107/5045265 | | *Fig.*<br>4.39 |
|---|---|---|---|---|
| | *Interpretation*<br>Settlement? | *Estimated size*<br>Unknown | *Date range*<br>LC-EH | |

*Topography*
DSP06-F04-01 is located in the SE corner of Field 4 about 500 m S of Lake Džarylgač

*Investigations*
**Extensive survey** The area with grey soils associated with amphora sherds suggests the presence of a Hellenistic site

*Surface finds*
1 Herakleian amphora sherd, 1 Mediterranean amphora sherd (Thasos, Mende or Aegean?), 1 Mendean amphora toe, 1 amphora toe (origin?), 1 body sherd of Herakleian fabric, 1 Hellenistic sherd, 1 flat base glazed on the interior, 1 possibly "Medieval" body sherd (i.e. Post-antique), 1 S coast amphora sherd (Sinope?), 1 unidentified sherd (not retained and therefore not included in the catalogue)

*Site interpretation*
Periphery of Hellenistic settlement?

| **DSP07-F01-01**<br>**See also Panskoe I** | *Core units*<br>1005-14, 2001-09 | *Sample density*<br>20% | *Coordinates (centre)*<br>485486/5044536 | *Figs.*<br>4.44-46, 4.33-37 |
|---|---|---|---|---|
| | *Adjacent units*<br>– | *Sample density* | | |
| | *Interpretation*<br>Settlement | *Estimated size*<br>250 x 250 m | *Date range*<br>LC-EH | |

*Topography*
**DSP07-F01-01** is located in a narrow ploughed strip in the N part of the field, directly W of the excavation location of Panskoe I. This is the highest part of the ridge between Lake Panskoe and the Bay of Džarylgač

*Investigations*
**Extensive survey** Grab sample 1004
**Intensive survey** 50 x 50 m. High-density scatter of a wide range of pottery
(Geomagnetic survey was carried out at Panskoe I, directly W of the survey site, but not in the surveyed field)

*Site interpretation*
The E halo of the previously known Hellenistic settlement Panskoe I (Chapter 4.3.2.1)

## DSP07-F01-02 see Panskoe V

| **DSP07-F02-01** | *Core units*<br>2164-65, 2205-06 | *Sample density*<br>20% | *Coordinates (centre)*<br>491044/5045225 | *Figs.*<br>4.49-52 |
|---|---|---|---|---|
| | *Adjacent units*<br>2166, 2194-95 | *Sample density*<br>20% | | |
| | *Interpretation*<br>Settlement | *Estimated size*<br>1 ha | *Date range*<br>LC-EH | |
| **Feature 1** | *Core units*<br>2205 | *Sample density*<br>20% | *Coordinates (centre)*<br>491073/5045272 | |
| | *Interpretation*<br>Structure? | *Estimated size*<br>250 m$^2$ | *Date range*<br>LC-EH | |
| **Feature 2** | *Core units*<br>2206, 2165 | *Sample density*<br>20% | *Coordinates (centre)*<br>491074/5045232 | |
| | *Interpretation*<br>Structure? | *Estimated size*<br>590 m$^2$ | *Date range*<br>LC-EH | |
| **Feature 3** | *Core units*<br>2164 | *Sample density*<br>20% | *Coordinates (centre)*<br>491003/5045213 | |
| | *Interpretation*<br>Structure? | *Estimated size*<br>360 m$^2$ | *Date range*<br>LC-EH | |

*Topography*
**DSP07-F02-01** is located on a local peak in the NE ear-shaped promontory of the surveyed field

*Investigations*
**Intensive survey** 50 x 50 m units; total samples 2164/02, 2205/02, 2206/02. A ploughed scatter with three discrete clusters

| **Geomagnetic survey**<br>T. Smekalova<br>*Dimensions*<br>Non-systematic search | *Description of measuring conditions*<br>Many dipole anomalies |
|---|---|

*Interpretation of features* The data revealed no structures

*Site interpretation*
The combination of three discrete pottery concentrations with relatively high densities of transport, utility, table and handmade wares, suggests that this location was in use as a settlement. The three separate clusters may be the remains of separate buildings or households

<table>
<tr><td rowspan="3">DSP07-F02-02</td><td>Core units<br>2100-02, 2117-19, 2146-47</td><td>Sample density<br>20%</td><td>Coordinates (centre)<br>490199/5044946</td><td>Figs.<br>4.49-50, 4.53-54</td></tr>
<tr><td>Adjacent units<br>2104</td><td>Sample density<br>20%</td><td></td><td></td></tr>
<tr><td>Interpretation<br>Settlement</td><td>Estimated size<br>Unknown</td><td>Date range<br>LC-EH</td><td></td></tr>
<tr><td colspan="5">Topography<br>DSP07-F02-02 is located in the NW part of the field on a slight elevation overlooking a ravine, which runs towards Lake Džarylgač</td></tr>
<tr><td colspan="5">Investigations<br>Intensive survey 50 x 50 m units. TS 2101/02 (13 x 13 m in W part of Unit 2102). A discrete, relatively high-dentity scatter of pottery in five corner units yielding pottery from different functional groups including tablewares, amphorai and handmade material. The pottery frs. may be part of the halo or perimeter of a larger site located outside the field border. Therefore, the extent of the site remains unknown</td></tr>
<tr><td colspan="5">Site interpretation<br>The combination of several functional pottery groups in this scatter suggests the presence of a settlement</td></tr>
</table>

<table>
<tr><td rowspan="3">DSP07-F02-03</td><td>Core units<br>1233-34</td><td>Sample density<br>20%</td><td>Coordinates (centre)<br>490839/5044023</td><td>Figs.<br>4.49-50, 4.55</td></tr>
<tr><td>Adjacent units<br>1228, 1230, 1259, 1261</td><td>Sample density<br>20%</td><td></td><td></td></tr>
<tr><td>Interpretation<br>Settlement</td><td>Estimated size<br>330 m$^2$</td><td>Date range<br>LC-EH</td><td></td></tr>
<tr><td colspan="5">Topography<br>DSP07-F02-03 is located approximately 1.4 km S of Lake Džarylgač. The site is located uphill, but seems a bit sheltered between undulating parts of the field</td></tr>
<tr><td colspan="5">Investigations<br>Intensive survey 50 x 50 m units; TS 1233/02 and 1234/02. A small, discrete scatter of handmade pottery, amphorai, tableware and stone tools (whetting stones, grinding stones) was found</td></tr>
<tr><td colspan="5">Site interpretation<br>The presence of several types of pottery in combination with stone tools suggests that this is a small settlement site</td></tr>
</table>

<table>
<tr><td rowspan="3">DSP07-F02-04</td><td>Core units<br>1134, 1181</td><td>Sample density<br>20%</td><td>Coordinates (centre)<br>491358/5043919</td><td>Figs.<br>4.49-50, 4.56</td></tr>
<tr><td>Adjacent units<br>1089, 1103-04, 1120, 1122, 1132-33, 1150-51, 1162, 1180, 1192</td><td>Sample density<br>20%</td><td></td><td></td></tr>
<tr><td>Interpretation<br>Settlement</td><td>Estimated size<br>Unknown</td><td>Date range<br>LC-EH</td><td></td></tr>
<tr><td rowspan="2">Feature 1</td><td>Core units<br>1181</td><td>Sample density<br>20%</td><td>Coordinates (centre)<br>491303/5043818</td><td></td></tr>
<tr><td>Interpretation<br>Structure?</td><td>Estimated size<br>?</td><td>Date range<br>LC-EH</td><td></td></tr>
<tr><td rowspan="2">Feature 2</td><td>Core units<br>1134</td><td>Sample density<br>20%</td><td>Coordinates (centre)<br>491331/5044034</td><td></td></tr>
<tr><td>Interpretation<br>Structure?</td><td>Estimated size<br>?</td><td>Date range<br>LC-EH</td><td></td></tr>
<tr><td colspan="5">Topography<br>DSP07-F02-04 is located uphill, S of Lake Džarylgač, in the pediment zone</td></tr>
<tr><td colspan="5">Investigations<br>Intensive survey 50 x 50 m units; TS 1134/02-03. Two small discrete scatters of Hellenistic tablewares, utility wares, amphorai and handmade pottery frs. 150 m apart. Relatively high densities of archaeological material recorded in the adjacent units</td></tr>
<tr><td colspan="5">Site interpretation<br>The combination of a high-density area surrounded by a halo of lower densities suggests that this is the location of a completely ploughed-out site. This site may be interpreted as a settlement based on the combination of different functional pottery groups</td></tr>
</table>

<table>
<tr><td rowspan="3">DSP07-F02-05</td><td>Core units<br>1091, 1100</td><td>Sample density<br>20%</td><td>Coordinates (centre)<br>491719/5043831</td><td>Figs.<br>4.49-50</td></tr>
<tr><td>Adjacent units<br>1090, 1125</td><td>Sample density<br>20%</td><td></td><td></td></tr>
<tr><td>Interpretation<br>Kurgan</td><td>Estimated size<br>630 m$^2$</td><td>Date range<br>Unknown</td><td></td></tr>
<tr><td colspan="5">Topography<br>DSP07-F02-05 is located in the pediment zone, S of Džarylgač Lake.</td></tr>
<tr><td colspan="5">Investigations<br>Intensive survey 50 x 50 m units. A round, ploughed-out elevation in the SE corner of the surveyed field with a diameter of about 25 m. No finds associated with the feature</td></tr>
<tr><td colspan="5">Site interpretation<br>On the half-verst map, a kurgan is marked on the exact location of the ploughed-out elevation</td></tr>
</table>

| **DSP07-F03-01** | *Core units* 1361-62, 2301, 2304 | *Sample density* 20% | *Coordinates (centre)* 487536/5041802 | *Figs.* 4.57-60 |
|---|---|---|---|---|
| | *Adjacent units* 1353, 2300, 2302 | *Sample density* 20% | | |
| | *Interpretation* Group of kurgans | *Estimated size* 27 x 75 m | *Date range* LC-EH? | |
| **Feature 1** | *Core units* 2301 | *Sample density* 20% | *Coordinates (centre)* 487497/5041882 | |
| | *Interpretation* Kurgan | *Estimated size* 160 m² | *Date range* LC-EH? | |
| **Feature 2** | *Core units* 2304 | *Sample density* 20% | *Coordinates (centre)* 487576/5041798 | |
| | *Interpretation* Kurgan | *Estimated size* 360 m² | *Date range* LC-EH? | |
| **Feature 3** | *Core units* 1361-1362 | *Sample density* 20% | *Coordinates (centre)* 487569/5041737 | |
| | *Interpretation* Kurgan | *Estimated size* 270 m² | *Date range* LC-EH | |

*Topography*
**DSP07-F03-01** is located on the lower extension of a hillside

*Investigations*
**Intensive survey** 50 x 50 m units. TS 2304/02 (on Feature 2). Three ploughed-out elevations associated with a scatter of stones

*Site interpretation*
Group of kurgans based on the shape and size of the elevations

| **DSP07-F04-01** | *Core units* 1400 | *Sample density* 20% | *Coordinates (centre)* 490131/5044597 | *Figs.* 4.61-65 |
|---|---|---|---|---|
| | *Adjacent units* 1394, 1399, 1401, 2361-63 | *Sample density* 20% | | |
| | *Interpretation* Settlement | *Estimated size* 26 m² | *Date range* BA | |

*Topography*
**DSP07-F04-01** is located at the bottom of the hillside, in a wide ravine running towards Lake Džarylgač. The lake is approximately 900 m to the N

*Investigations*
**Intensive survey** 50 x 50 m units; TS 1400/03 and 1400/04 (each of 5 x 20 m around the find spot of the partial stone axe). A small diffuse scatter of handmade pottery frs. and part of a stone axe

*Site interpretation*
Small Bronze Age settlement?

| **DSP07-F04-02** | *Core units*<br>1388 | *Sample density*<br>20% | *Coordinates (centre)*<br>490117/5044233 | *Figs.*<br>4.61-64, 4.66-67 |
|---|---|---|---|---|
| | *Adjacent units*<br>1386-87, 1389-90, 1404-06, 2356 | *Sample density*<br>20% | | |
| | *Interpretation*<br>Kurgan? and Settlement | *Estimated size*<br>780 m² | *Date range*<br>LC-EH? EM | |

*Topography*
**DSP07-F04-02** is located in the pediment zone, on the E side of a wide ravine running towards Lake Džarylgač. Lake Džarylgač is approximately 1250 m to the N

*Investigations*
**Intensive survey** 50 x 50 m units; TS 1388/02-07 (each 7 x 50 m N-S). Scatter of pottery and animal bones associated with small ploughed-out elevation.

*Site interpretation*
Early Modern settlement, possibly re-use of a kurgan. On the half-verst map, a kurgan is marked on this exact location, and the ruins of a village called "Džajlav" are depicted 400 m to the SW

| **DSP07-F05-01** | *Core units*<br>1502-05 | *Sample density*<br>20% | *Coordinates (centre)*<br>494117/5048736 | *Figs.*<br>4.68-73 |
|---|---|---|---|---|
| | *Adjacent units*<br>1480, 1482-84, 1499-1501, 1506, 1519, 1522 | *Sample density*<br>20% | | |
| | *Interpretation*<br>Settlement | *Estimated size*<br>? | *Date range*<br>EM | |

*Topography*
**DSP07-F05-01** is located on the S slope of the coastal ridge, approximately 500 m W of the present-day village of Vodopojnoe, and 1 km NW of Lake Džarylgač. It lies on the E part of a slight ridge extending towards the lake

*Investigations*
**Intensive survey** 60 x 60 m units; TS 1502/02 (6 x 50 m N-S in the E part of Unit 1502). A diffuse scatter of pottery was recorded in a deep-ploughed field S of the dirt road between Vodopojnoe and Mežvodnoe

*Site interpretation*
Small Early Modern settlement or road dump. On the half-verst map, no ruins or buildings are documented at this location

| **DSP07-F05-02** | *Core units*<br>2404 | *Sample density*<br>20% | *Coordinates (centre)*<br>493428/5048271 | *Figs.*<br>4.68-71, 4.74-75 |
|---|---|---|---|---|
| | *Adjacent units*<br>2403, 2422, 2427 | *Sample density*<br>20% | | |
| | *Interpretation*<br>Kurgan / ritual | *Estimated size*<br>720 m$^2$ | *Date range*<br>BA? | |

*Topography*
**DSP07-F05-02** is located on a small elongated elevation on the W slope of a ridge extending towards Lake Džarylgač

*Investigations*
**Intensive survey** 60 x 60 m units. A diffused area with many limestone frs. One large, damaged anthropomorphic stone was observed, later to be removed by a farmer clearing his field of stones. Although no pottery was collected within the stony patch around the stele, handmade pottery dated to the Bronze Age were collected in two units SW of the stony area

*Site interpretation*
Antropomorphic stelai are known as grave markers in the Bronze Age around the Black Sea. The patch represents the remains of a ploughed-over Bronze Age kurgan

| **DSP07-F06-01** | *Core units*<br>1607 | *Sample density*<br>20% | *Coordinates (centre)*<br>493970/5049021 | *Figs.*<br>4.76-79 |
|---|---|---|---|---|
| | *Adjacent units*<br>1587, 1608-09, 2525 | *Sample density*<br>20% | | |
| | *Interpretation*<br>Kurgan | *Estimated size*<br>1,210 m$^2$ | *Date range*<br>LC-EH | |

*Topography*
**DSP07-F06-01** is located on the S slope of the coastal ridge, on an elevation extending towards Lake Džarylgač

*Investigations*
**Extensive survey** Grab sample 1607/02 in the stony area
**Intensive survey** 50 x 50 m units. A relatively large elevated area associated with a scatter of pottery and stones

*Site interpretation*
Kurgan?

| **DSP07-F06-02** | *Core units*<br>1579 | *Sample density*<br>20% | *Coordinates (centre)*<br>494058/5049035 | *Figs.*<br>4.76-78 |
|---|---|---|---|---|
| | *Adjacent units*<br>1558, 1578, 1580 | *Sample density*<br>20% | | |
| | *Interpretation*<br>Settlement? | *Estimated size*<br>470 m$^2$ | *Date range*<br>LC-EH | |

*Topography*
**DSP07-F06-02** is located on the S slope of the coastal ridge, on the W part of a ridge extending towards Lake Džarylgač

*Investigations*
**Extensive survey** Grab sample 1579/02 within the stony elevated area
**Intensive survey** 50 x 50 m units. A ploughed scatter on a small relatively stony elevation

*Site interpretation*
The presence of amphora frs. in combination with handmade pottery suggests that this may have been a settlement

| **DSP07-F06-03** | *Core units*<br>2476-77, 2519-20 | *Sample density*<br>20% | *Coordinates (centre)*<br>493857/5048856 | *Figs.*<br>4.76-77, 4.80 |
|---|---|---|---|---|
| | *Adjacent units*<br>– | *Sample density*<br>20% | | |
| | *Interpretation*<br>Settlement | *Estimated size*<br>690 m² | *Date range*<br>LC-EH | |
| **Feature 1** | Core unit<br>2477 | *Sample density*<br>20% | *Coordinates (centre)*<br>493841/5048852 | |
| | *Interpretation*<br>Settlement | *Estimated size*<br>280 m² | *Date range*<br>LC-EH | |
| **Feature 2** | Core unit<br>2519 | *Sample density*<br>20% | *Coordinates (centre)*<br>493879/5048858 | |
| | *Interpretation*<br>Settlement | *Estimated size*<br>280 m² | *Date range*<br>LC-EH | |
| **Feature 3** | Core unit<br>2519 | *Sample density*<br>20% | *Coordinates (centre)*<br>493891/5048847 | |
| | *Interpretation*<br>Settlement | *Estimated size*<br>130 m² | *Date range*<br>LC-EH | |

*Topography*

**DSP07-F06-03** is located in the S part of the surveyed field, just N of the dirt road between Vodopojnoe and Mežvodnoe. The site is located on the top part of an elevation sloping down from the coastal ridge towards Lake Džarylgač

*Investigations*

**Intensive survey** 50 x 50 m units; TS 2477/02 (25 x 25 m in the SE corner of Unit 2477); 2519/02 (25 x 50 m in the S half of Unit 2519); 2520/02 (6 x 50 m in the N half of Unit 2520). A large, ploughed-out, stony, slightly elevated area. Within the stony area, three discrete find clusters consisted of amphora and handmade pottery frs. and a fragmented ground stone tool. One pottery fr. may date to the Medieval period. *Adjacent units* yielded no finds

*Site interpretation*

The combination of utility and transport wares and a tool suggest that the three clusters may have formed a settlement

| **DSP07-F07-01** | *Core units*<br>2586, 2610-13, 2636-39 | *Sample density*<br>20% | *Coordinates (centre)*<br>492978/5051041 | *Figs.*<br>4.81-83 |
|---|---|---|---|---|
| | *Adjacent units*<br>2587, 2640, 2661-62 | *Sample density*<br>20% | | |
| | *Interpretation*<br>Group of kurgans | *Estimated size*<br>– | *Date range*<br>BA, LC-EH | |
| **Feature 1** | Core unit<br>2586 | *Sample density*<br>20% | *Coordinates (centre)*<br>492960/5051144 | |
| | *Interpretation*<br>Kurgan | *Estimated size*<br>440 m$^2$ | *Date range*<br>LC-EH? | |
| **Feature 2** | *Core units*<br>2610-11 | *Sample density*<br>20% | *Coordinates (centre)*<br>492978/5051096 | |
| | *Interpretation*<br>Kurgan | *Estimated size*<br>520 m$^2$ | *Date range*<br>LC-EH? | |
| **Feature 3** | *Core units*<br>2612-13, 2636-37 | *Sample density*<br>20% | *Coordinates (centre)*<br>492975/5051026 | |
| | *Interpretation*<br>Kurgan | *Estimated size*<br>1,140 m$^2$ | *Date range*<br>BA, LC-EH | |
| **Feature 4** | *Core units*<br>2638-39 | *Sample density*<br>20% | *Coordinates (centre)*<br>493001/5050941 | |
| | *Interpretation*<br>Kurgan | *Estimated size*<br>1,620 m$^2$ | *Date range*<br>LC-EH? | |

*Topography*

**DSP07-F07-01** is located in the NW part of the surveyed field on the highest part of the coastal ridge. On the half-verst map, a row of eight mounds is depicted at this location. On the 1955 map, only three of these are depicted

*Investigations*

**Intensive survey** 50 x 50 m units; TS 2586/02 (23 x 38 m over Feature 1); 2610/02 (20 x 25 m in the E half of Unit 2610); 2611/02 (10 x 25 m in the W half of Unit 2611); 2612/02 (23 x 36 m in the S half of Unit 2612) and 2613/02 (13 x 20 m in the SW corner of Unit 2613); 2639/02 (covering Feature 4). Four ploughed-out elevations associated with relatively many stones. Hellenistic amphora frs. were collected on Feature 3, as was part of stone axe. This tool can be dated to the Bronze Age, yet no contemporary pottery was collected. More Hellenistic amphora frs. were collected in the general area

*Site interpretation*

A group of kurgans based on the size and shape of the elevations and the presence of mounds at this location on the half-verst map

| **DSP07-F07-02** | *Core units*<br>2572-73 | *Sample density*<br>20% | *Coordinates (centre)*<br>493550/5051217 | *Figs.*<br>4.81-82, 4.84 |
|---|---|---|---|---|
| | *Adjacent units*<br>2552 | *Sample density*<br>20% | | |
| | *Interpretation*<br>Kurgan | *Estimated size*<br>470 m² | *Date range*<br>Unknown | |

*Topography*
**DSP07-F07-02** is located on the highest part of the coastal ridge

*Investigations*
**Intensive survey** 50 x 50 m units; TS 2572/02 and 2573/02 (together a block of 22 x 35 m over the elevation). A stony, ploughed-out elevation. Handmade pottery frs. were collected in the general area of the elevation. These cannot be dated with certainty

*Site interpretation*
Kurgan based on the size and shape of the elevation

| **DSP07-F08-01** | *Core units*<br>2705-10, 2715-20, 8991-92, 9019 | *Sample density*<br>20% | *Coordinates (centre)*<br>492102/5048591 | *Figs.*<br>4.85-88 |
|---|---|---|---|---|
| | *Adjacent units*<br>– | *Sample density*<br>20% | | |
| | *Interpretation*<br>Settlement | *Estimated size*<br>ca.250 x 90 m | *Date range*<br>LC-EH | |

*Topography*
**DSP07-F08-01** is located S of the dirt road between Vodopojnoe and Mežvodnoe, on the S slope of the coastal ridge. To the E, a slight ridge extends S towards Lake Džarylgač. SW of the site, a lagoon is cut off from the larger salt lake by a low barrier. The distance to this lagoon is 880 m

*Investigations*
**Intensive survey** 50 x 50 m units; TS 2706/02 (6 x 35 m in the E end of Unit 2706); 2707/02 (25 x 45 m in the W half of Unit 2707); 2709/02 (10 x 50 m in the E half of Unit 2709); 2710/02 (7 x 30 m in the E half of Unit 2710); 2719/02 (50 x 10 m in the N end of Unit 2719). This large concentration of tile, handmade pottery, table- and utility wares, amphora frs. and stone tools (whetting stone, ground stone, flint), but no clear features; the elevated strips that form site DSP07-F08-02 continue in the core of the present site

*Site interpretation*
A settlement dating to the Hellenistic period, based on the combination of pottery and stone tools. This is the only site outside Panskoe I where a part of a Hellenistic tile was found suggesting that the site may have been occupied permanently over a longer period of time

<table>
<tr><td rowspan="3">DSP07-F08-02</td><td>Core units<br>N part of field<br>DSP07-F08</td><td>Sample density<br>20%</td><td>Coordinates (centre)<br>492003/5048362</td><td>Fig.<br>4.89</td></tr>
<tr><td>Adjacent units<br>–</td><td>Sample density<br>20%</td><td></td><td></td></tr>
<tr><td>Interpretation<br>Cadastral system</td><td>Estimated size<br>unknown</td><td>Date range<br>Unknown</td><td></td></tr>
<tr><td colspan="5">Topography<br>DSP07-F08-02 is located in the N half of ploughed field DSP07-F08 running almost N-S and E-W</td></tr>
<tr><td colspan="5">Investigations<br>Intensive survey 50 x 50 m units. A system of low ridges in a ploughed field delimiting lots of approximately 80 x 90 m. The ridges occur about halfway to the S slope of the coastal ridge. They were almost completely destroyed, but are visible at the end of the day, when low sun created long shadows. Finds were generally dispersed throughout Field DSP07-F08; see DSP07-F08-01 and Chapter 4.4</td></tr>
<tr><td colspan="2">Geomagnetic survey<br>T. Smekalova<br>Dimensions 50 x 90 m</td><td colspan="3">Description of measuring conditions<br>Many dipole anomalies</td></tr>
<tr><td colspan="5">Interpretation of features The data revealed no structures</td></tr>
<tr><td colspan="5">Site interpretation<br>Possible remains of a cadastral system</td></tr>
</table>

| **DSP08-F01-01** | *Core units*<br>5153-54, 6009-10, 6016 | *Sample density*<br>20% | *Coordinates (centre)*<br>4943328/5051717 | *Figs.*<br>4.90-93 |
|---|---|---|---|---|
| | *Adjacent units*<br>– | *Sample density*<br>20% | | |
| | *Interpretation*<br>Group of kurgans | *Estimated size*<br>– | *Date range*<br>Unknown | |
| **Feature 1** | *Core units*<br>5153-54, 6009-10 | *Sample density*<br>20% | *Coordinates (centre)*<br>494315/5051746 | |
| | *Interpretation*<br>Kurgan | *Estimated size*<br>750 m$^2$ | *Date range*<br>Unknown | |
| **Feature 2** | *Core units*<br>6009-10, 6016 | *Sample density*<br>20% | *Coordinates (centre)*<br>494327/5051717 | |
| | *Interpretation*<br>Kurgan | *Estimated size*<br>1,090 m$^2$ | *Date range*<br>Unknown | |
| **Feature 3** | *Core units*<br>6016 | *Sample density*<br>20% | *Coordinates (centre)*<br>494353/5051671 | |
| | *Interpretation*<br>Kurgan | *Estimated size*<br>510 m$^2$ | *Date range*<br>Unknown | |

*Topography*
The elevation is located on the highest part of the coastal ridge and is the northernmost in a row of three mounds. On the half-verst map, one mound is marked at this point, most probably the highest, southernmost mound

*Investigations*
**Intensive survey** in 40 x 50 m and 70 x 50 units. Two ploughed-out, low elevations of lighter-coloured soil than the rest of the ploughed field and a high grassy mound. The two units in which Features 1 and 2 are located yielded no finds, but in the unit directly to the E of Feature 1, eight Hellenistic sherds were collected. It is uncertain whether these are associated with the features. One Hellenistic fr. and three Medieval sherds were collected around Feature 3, but it is uncertain whether these are associated with the mound

*Site interpretation*
Group of kurgans based on the size and shape of the elevations, and the presence of a mound at this location on the half-verst map

| **DSP08-F02-01** | *Core units*<br>8177-78, 8181 | *Sample density*<br>20% | *Coordinates (centre)*<br>493131/5053076 | *Figs.*<br>4.94-97 |
|---|---|---|---|---|
| | *Adjacent units*<br>8175-76, 8179, 8182-84 | *Sample density*<br>20% | | |
| | *Interpretation*<br>Settlement | *Estimated size*<br>ca. 2.9 ha | *Date range*<br>LC-EH | |

*Topography*
**DSP08-F02-01** is located at the N end of the surveyed field, near a farmstead. This place has been in use over a longer period of time: on the half-verst map, both a farm and ruins of a village are depicted nearby. The site is located near the Black Sea coast, at the flat end of the N slope of the coastal ridge. A broad ravine was observed E of the site, which may be the remains of an old road or track. The site may extend to the N, in the uncultivated area around the farm

*Investigations*
**Intensive survey** 50 x 50 m units; TS 8177/02 (5 x 50 m in the N end of Unit 8177) and 8178/02 (5 x 50 m in the S end of adjacent Unit 8178). A large discrete scatter of transport, table- and utility wares, handmade pottery and stone tools, as well as a stone sinker

*Site interpretation*
The combination of pottery and stone tools indicates that this site was a settlement. The stone sinker and the location close to the sea suggest that part of the site's subsistence consisted of fishing

| **DSP08-F03-01/ Skalistoe *chutor*** | *Core units*<br>6234 | *Sample density*<br>20% | *Coordinates (centre)*<br>493929/5054031 | *Figs.*<br>4.98-104 |
|---|---|---|---|---|
| | *Adjacent units*<br>6230-33, 6235-40, 6252-56 | *Sample density*<br>20% | | |
| | *Interpretation*<br>Settlement | *Estimated size*<br>? | *Date range*<br>EM | |

*Topography*
**DSP08-F03-01** is located in the NW corner of the surveyed field at the E side of a horseshoe-shaped depression near the coastal cliff at the end of a ravine. On the 1897 map, a farm (Skalistoe *chutor*) is depicted just outside the field border, and two wells are marked a bit further to the NW, near the coastline

*Investigations*
**Intensive survey** 50 x 60 m units; TS 6234/02. Discrete scatter of pottery

*Site interpretation*
The combination of pottery correlating with the location of a farm on the map suggests that the site was a settlement. For the Hellenistic element, see Skalistoe 2

## DSP08-F03-02, see Skalistoe 2

<table>
<tr><td rowspan="3">DSP08-F04-01</td><td>Core units<br>8468-71, 8510-11</td><td>Sample density<br>20%</td><td>Coordinates (centre)<br>492102/5045538</td><td>Figs.<br>4.109-114</td></tr>
<tr><td>Adjacent units<br>8470, 8509, 8516</td><td>Sample density<br>20%</td><td></td><td></td></tr>
<tr><td>Interpretation<br>Settlement</td><td>Estimated size<br>2,380 m²</td><td>Date range<br>LC-EH?</td><td></td></tr>
<tr><td rowspan="2">Feature 1</td><td>Core units<br>8468-8471</td><td>Sample density<br>20%</td><td>Coordinates (centre)<br>492132/5045581</td><td></td></tr>
<tr><td>Interpretation<br>Settlement</td><td>Estimated size<br>1,370 m²</td><td>Date range<br>LC-EH?</td><td></td></tr>
<tr><td rowspan="2">Feature 2</td><td>Core units<br>8510</td><td>Sample density<br>20%</td><td>Coordinates (centre)<br>492092/5045511</td><td></td></tr>
<tr><td>Interpretation<br>Settlement</td><td>Estimated size<br>610 m²</td><td>Date range<br>LC-EH?</td><td></td></tr>
<tr><td>Feature 3</td><td>Core units<br>8510-8511</td><td>Sample density<br>20%</td><td>Coordinates (centre)<br>492089/5045480</td><td></td></tr>
<tr><td></td><td>Interpretation<br>Ash heap</td><td>Estimated size<br>400 m²</td><td>Date range<br>LC-EH?</td><td></td></tr>
<tr><td colspan="5">Topography<br>DSP08-F04-01 is located at the N end of a slight ridge, the extension of one of the hillsides. It is located in the pediment area 750 m S of Lake Džarylgač</td></tr>
<tr><td colspan="5">Investigations<br>Intensive survey 50 x 50 m units. TS 8510/02-05. A scatter of pottery on a slight elevation in grey soil. Two additional slight elevations of light-coloured soil, but no finds could be directly associated. A relatively high density of Hellenistic frs. in the general area may be associated with the site</td></tr>
<tr><td colspan="5">Site interpretation<br>The combination of the ash heap, pottery and two additional discrete clusters of finds suggest that this was a small settlement</td></tr>
</table>

| **DSP08-F04-02** | *Core units*<br>8499-8500, 8520-21 | *Sample density*<br>20% | *Coordinates (centre)*<br>491823/5045461 | *Figs.*<br>4.109-112, 4.115-116 |
|---|---|---|---|---|
| | *Adjacent units*<br>8480, 8498, 8501, 8519 | *Sample density*<br>20% | | |
| | *Interpretation*<br>Settlement | *Estimated size*<br>2,460 m² | *Date range*<br>LC-EH | |
| **Feature 1** | *Core units*<br>8499-8500 | *Sample density*<br>20% | *Coordinates (centre)*<br>491823/5045461 | |
| | *Interpretation*<br>Settlement | *Estimated size*<br>1,270 m² | *Date range*<br>LC-EH | |
| **Feature 2** | *Core units*<br>8520-21 | *Sample density*<br>20% | *Coordinates (centre)*<br>491811/5045423 | |
| | *Interpretation*<br>Settlement | *Estimated size*<br>1,190 m² | *Date range*<br>LC-EH | |

*Topography*
**DSP08-F04-02** is located on a local ridge top. Lake Džarylgač is approximately 580 m to the NW

*Investigations*
**Intensive survey** 50 x 50 m units; TS 8500/02-06 and 8521/01-02. A ploughed scatter around two discrete, relatively high-density clusters of transport, utility and handmade wares and flint on slight elevations in lighter-coloured soil

*Site interpretation*
The combination of two discrete pottery concentrations with relatively high densities of pottery and flint suggests that this location was a settlement. The two separate clusters may be the remains of two buildings or households, although more research is needed to assess this assumption

| **DSP08-F04-03** | *Core units*<br>8531, 8533-34,<br>8568-70, 8591-94 | *Sample density*<br>20% | *Coordinates (centre)*<br>4913123/5045157 | *Figs.*<br>4.109-112, 4.117-118 |
|---|---|---|---|---|
| | *Adjacent units*<br>8530, 8567, 8595,<br>8637-38 | *Sample density*<br>20% | | |
| | *Interpretation*<br>Settlement | *Estimated size*<br>4,810 m² | *Date range*<br>LC-EH | |
| **Feature 1** | *Core units*<br>8533-34, 8568-69 | *Sample density*<br>20% | *Coordinates (centre)*<br>491349/5045180 | |
| | *Interpretation*<br>Structure | *Estimated size*<br>1,350 m² | *Date range*<br>LC-EH | |
| **Feature 2** | *Core units*<br>8332-33, 8569-70 | *Sample density*<br>20% | *Coordinates (centre)*<br>491287/5045136 | |
| | *Interpretation*<br>Structure | *Estimated size*<br>3,460 m² | *Date range*<br>LC-EH | |

*Topography*
**DSP08-F04-03** is located on either side of a shallow ravine which runs towards the lake. At the present, this is a shallow, marshy secondary lake, separated from the proper lake by a low natural sand barrier. The distance to this marsh is approximately 200 m. The W scatter is located on a low promontory just S of Lake Džarylgač

*Investigations*
**Extensive survey** GS 3005
**Intensive survey** 50 x 50 m units; TS 8532/02, 8534/02-03, 8568/02, 8570/02-05. Two ploughed scatters of transport, utility and table wares, one fr. of black-gloss, handmade pottery, reused sherds and flint

**Augering** Five augerings in two rows across the site. The hard loam layer in Augers 2, 3 and 4 is probably a culture layer. Finds: One pottery fr. found in the topsoil of Auger 4

| Augering S1: 491278/5045157 | |
|---|---|
| 0-0.42 m | Topsoil, loam; 10YR 3/2. Plant remains |
| 0.42-0.72 m | Loam, plastic, intact layer; 10YR 5/4 |
| 0.72-1.10 m | Loam, plastic; 10YR 5/4. Limestone inclusions |
| Augering S2: 491278/5045179 | |
| 0-0.42 m | Topsoil, loam; 10YR 3/2 |
| 0.42-0.75 m | Loam, very hard, limestone inclusions; 10YR 5/4 |
| 0.75-1 m | Loam, very hard, compact. No inclusions; 10YR 5/4 |
| 1-1.70 m | Loamy clay, soft, plastic, no inclusions; 10YR 5/4 |
| 1.70-1.83 m | Loamy clay, soft, plastic, limestone inclusions; 10YR 5/4 |
| 1.83 m- | End of auger. Bedrock |
| Augering S3: 491280/5045132 | |
| 0-0.45 m | Topsoil, loam,10YR 3/2. Humus |
| 0.45-1.20 m | Loam, very hard, compact, intact layer. Few limestone inclusions; 10YR 5/4 |
| 1.20-1.40 m | Loamy clay, soft; 10YR 5/4. Small stone inclusions |
| Augering S4: 491282/5045107 | |

| 0-0.41 m | Topsoil, loam, humus; 10YR 3/2. One pottery fr. in core |
|---|---|
| 0.41-0.80 m | Loam, compact, hard, intact; 10YR 5/4 |
| 0.80-1.05 m | Loam, compact, hard, intact. Limestone shells, 10YR 5/4 |
| 1.05-1.30 m | Loamy clay, soft, plastic, small lime inclusions; 10YR 5/4 |
| 1.30 m- | End of auger. Bedrock |
| Augering S5: 491285/5045058 | |
| 0-0.40 m | Topsoil, loam,10YR 3/2 |
| 0.40-0.90 m | Loam, compact, hard, intact layer; 10YR 5/4 |
| 0.90-1.15 m | Loam; 10YR 5/4 |
| 1.15 m- | End of auger. Bedrock |
| *Site interpretation*<br>The combination of finds, density and size of the two scatters suggests that this is a settlement | |

| **DSP08-F04-04** | *Core units*<br>8707, 8737-38, 8745, 8771, 8784, 8859 | *Sample density*<br>20% | *Coordinates (centre)*<br>492100/5044905 | *Figs.*<br>4.109-112, 4.124-126 |
|---|---|---|---|---|
| | *Adjacent units*<br>– | *Sample density*<br>20% | | |
| | *Interpretation*<br>Group of kurgans | *Estimated size*<br>– | *Date range*<br>Unknown | |
| **Feature 1** | *Core units*<br>8707 | *Sample density*<br>20% | *Coordinates (centre)*<br>492135/5045064 | |
| | *Interpretation*<br>Kurgan | *Estimated size*<br>440 m² | *Date range*<br>Unknown | |
| **Feature 2** | *Core units*<br>8737-38 | *Sample density*<br>20% | *Coordinates (centre)*<br>492137/5045007 | |
| | *Interpretation*<br>Kurgan | *Estimated size*<br>350 m² | *Date range*<br>Unknown | |
| **Feature 3** | *Core units*<br>8745, 8771 | *Sample density*<br>20% | *Coordinates (centre)*<br>492099/5044905 | |
| | *Interpretation*<br>Kurgan | *Estimated size*<br>230 m² | *Date range*<br>BA | |
| **Feature 4** | *Core units*<br>8784 | *Sample density*<br>20% | *Coordinates (centre)*<br>492079/5044831 | |
| | *Interpretation*<br>Kurgan | *Estimated size*<br>250 m² | *Date range*<br>Unknown | |
| **Feature 5** | *Core units*<br>8859 | *Sample density*<br>20% | *Coordinates (centre)*<br>492090/5044783 | |
| | *Interpretation*<br>Kurgan | *Estimated size*<br>160 m² | *Date range*<br>Unknown | |

*Topography*
**DSP08-F04-04** is located on top of a ridge extending towards Lake Džarylgač. The soil on top of the ridge is distinctly red, whereas the lower parts of the surveyed field are brownish

*Investigations*
**Intensive survey** 50 x 50 m units. Five grassy elevations surrounded by red soil. Feature 3 had been robbed recently. Part of a pot was found *in situ* in the W section of the robbers' trench. This feature is the fourth from the N in a cluster. Early Modern pottery frs. were found in the general area

*Site interpretation*
Group of kurgans based on the size and shape of the elevations, all marked on the half-verst map. Probably associated with DSP08-F04-13

| **DSP08-F04-05** | *Core units*<br>6581, 6602, 6617, 8805-06, | *Sample density*<br>20% | *Coordinates (centre)*<br>491960/5044468 | *Figs.*<br>4.109-112, 4.127-128 |
|---|---|---|---|---|
| | *Adjacent units*<br>6582, 6616, 6618, 8804 | *Sample density*<br>20% | | |
| | *Interpretation*<br>Group of kurgans | *Estimated size*<br>– | *Date range*<br>Unknown | |
| **Feature 1** | *Core units*<br>8805-06, 6617 | *Sample density*<br>20% | *Coordinates (centre)*<br>491987/5044547 | |
| | *Interpretation*<br>Kurgan | *Estimated size*<br>3,010 m$^2$ | *Date range*<br>Unknown | |
| **Feature 2** | *Core units*<br>6602 | *Sample density*<br>20% | *Coordinates (centre)*<br>491960/5044468 | |
| | *Interpretation*<br>Kurgan | *Estimated size*<br>490 m$^2$ | *Date range*<br>Unknown | |
| **Feature 3** | *Core units*<br>6581 | *Sample density*<br>20% | *Coordinates (centre)*<br>491925/5044409 | |
| | *Interpretation*<br>Kurgan | *Estimated size*<br>280 m$^2$ | *Date range*<br>Unknown | |

*Topography*

**DSP08-F04-05** is located on top of a ridge extending towards Lake Džarylgač. The soil on top of the ridge is distinctly red, whereas the lower parts of the surveyed field are brownish

*Investigations*

**Intensive survey** 50 x 50 m units. A large round grassy patch (Feature 1), a grassy elevation (Feature 2) and a ploughed-out slight elevation (Feature 3) all surrounded by red soil. On top of Feature 2, an iron beacon was built

*Site interpretation*

Group of kurgans based on the size and shape of the grassy patches, all marked on the half-verst map. Probably associated with DSP08-F04-08

| **DSP08-F04-06** | *Core units*<br>6521-22, 6525-26 | *Sample density*<br>20% | *Coordinates (centre)*<br>492454/5044421 | *Figs.*<br>4.109-112, 4.119-121 |
|---|---|---|---|---|
| | *Adjacent units*<br>6523, 6527 | *Sample density*<br>20% | | |
| | *Interpretation*<br>Settlement | *Estimated size*<br>minimum 9,000 m$^2$ | *Date range*<br>LC-EH and EM | |

*Topography*

**DSP08-F04-06** is located on a gentle slope at the E edge of the surveyed field, near the abandoned village of Kipčak Učkuju. A ravine may be the remains of one of the roads leading to the village as depicted on the half-verst map

*Investigations*

**Intensive survey** 50 x 50 m units; TS 6526/02-06. A ploughed scatter of table, utility and transport wares. Most finds were collected in a stony ravine running roughly NE-SW. The concentration may extend to the E

*Site interpretation*

The combination of pottery suggests that this is a settlement with two components of the Hellenistic and Early Modern period

| **DSP08-F04-07** | *Core units*<br>6492-94 | *Sample density*<br>20% | *Coordinates (centre)*<br>492104/5044144 | *Figs.*<br>4.109-112, 4.122-123 |
|---|---|---|---|---|
| | *Adjacent units*<br>6475-80, 6490-91, 6495, 6506-07 | *Sample density*<br>20% | | |
| | *Interpretation*<br>Settlement | *Estimated size*<br>minimum 1 ha in a 2.5 ha halo to the N, W, and S | *Date range*<br>LC-EH | |

*Topography*
**DSP08-F05-01** is located at the E edge of the surveyed field

*Investigations*
**Intensive survey** 50 x 50 m units; TS 6492/02-06, TS 6493/02-03. A high-density scatter of utility, table and transport wares, handmade pottery, stone tools and worked flint on a slight elevation of lighter-coloured soil. The site may well extend to the E outside the survey area

**Augering** 2 augerings. Layer 2 in Auger 2, containing pottery frs., is interpreted as a culture layer. Finds: Small indeterminate frs. of pottery in Layer 2 of Augering 2

| Augering S6: 492094/5044232 | |
|---|---|
| 0-0.30 m | Topsoil, loam,10YR 3/2 |
| 0.30-1.20 m | Loam, compact, hard. Very few limestone inclusions; 10YR 5/4 |
| 1.20-2 m | Loam, plastic,10YR 5/4. Small brown clayey inclusions |
| 2 m- | End of auger. Bedrock |
| Augering S7: 492111/5044145 | |
| 0-0.40 m | Topsoil, loam; 10YR 3/2. Mottled |
| 0.40-0.52 m | Loam, not hard; 10YR 4/3. Several small pieces of pottery |
| 0.52-1.48 m | Loam, plastic. Limestone and brown stone inclusions; 10YR 4/4 |
| 1.48 m- | End of auger. Bedrock |

*Site interpretation*
The combination of pottery, stone tools and worked flint suggests that is a settlement, possibly related to site DSP07-F08-01

| **DSP08-F05-01** | *Core units*<br>9024-25 | *Sample density*<br>20% | *Coordinates (centre)*<br>492119/5048936 | *Figs.*<br>4.129-131 |
|---|---|---|---|---|
| | *Adjacent units*<br>9013, 9026, 9028 | *Sample density*<br>20% | | |
| | *Interpretation*<br>Settlement | *Estimated size*<br>Unknown | *Date range* LC-EH and EM | |

*Topography*
**DSP08-F05-01** is located at the E edge of the surveyed field

*Investigations*
**Intensive survey** 50 x 50 m units. A ploughed scatter of table and utility wares

*Site interpretation*
The presence of table- and utility wares suggests a settlement likely extending into the unploughed field to the E

| **DSP08-F06-01** | *Core units*<br>9200-03, 9207-09, 9218-19 | *Sample density*<br>20% | *Coordinates (centre)*<br>491200/50493218 | *Figs.*<br>4.133-138 |
|---|---|---|---|---|
| | *Adjacent units*<br>9210, 9217 | *Sample density*<br>20% | | |
| | *Interpretation*<br>Settlement | *Estimated size*<br>1.9 ha | *Date range*<br>BA | |
| **Feature 1** | *Core units*<br>9207-9209, 9219 | *Sample density*<br>20% | *Coordinates (centre)*<br>491195/5049294 | |
| | *Interpretation*<br>Settlement | *Estimated size*<br>2,860 m² | *Date range*<br>BA | |
| **Feature 2** | *Core units*<br>9201-02 | *Sample density*<br>20% | *Coordinates (centre)*<br>491242/5049301 | |
| | *Interpretation*<br>Settlement | *Estimated size*<br>1,100 m | *Date range*<br>BA | |

*Topography*

**DSP08-F06-01** is located on a hillside that gently slopes down to the E. The site is not located at the highest part of the ridge and seems therefore sheltered on the N and W sides. An erosion ravine leads down from the NW to the SE, just E of the site.

*Investigations*

**Intensive survey** 50 x 50 m units; TS 9200/02-07; 9201/02-11; 9202/02-09; 9203/02; 9207/02-03; 9209/02; 9219/02. Two grey patches of rather stony, but also soft and ashy soil associated with a high-density, discrete scatter of handmade pottery (including decorated frs.) stone tools and large quantities of worked flint and bone material. Find densities dropped significantly outside the grey patch

**Augering** 11 augerings in two crossing rows across the grey patch

| Augering S17: 491126/5049287 | |
|---|---|
| 0-0.50 m | Topsoil, loam, humic, appears ashy; 10YR 4/2 |
| 0.5 m- | End of auger. Bedrock |
| Augering S18: 491151/5049292 | |
| 0-0.45 m | Topsoil, loam, humic, appears ashy; 10YR 4/2 |
| 0.45 m- | End of auger. Bedrock |
| Augering S19: 491177/5049296 | |
| 0-0.35 m | Topsoil, loam, humic. Lime stone inclusions; 10YR 2/2 |
| 0.35 m- | End of auger. Bedrock |
| Augering S20: 491188/5049302 | |
| 0-0.15 m | Topsoil, loam, dry; 10YR 4/2 |
| 0.15-0.45 m | Loam; 10YR 4/2. Ash inclusions, one piece of handmade pottery at 0.35 m |
| 0.45 m- | End of auger. Bedrock |
| Augering S21: 491209/5049307 | |
| 0-0.30 m | Topsoil, loam; 10YR 3/2. Small pottery frs., limestone inclusions |
| 0.30 m- | End of auger. Bedrock |
| Augering S22: 491243/5049317 | |

| | |
|---|---|
| 0-0.30 m | Topsoil, loam; 10YR 3/2. Small pottery frs., limestone inclusions |
| 0.30 m- | End of auger. Bedrock |
| Augering S23: 491277/5049320 | |
| 0-0.42 m | Topsoil, loam, humic; 10YR 2/2. Limestone inclusions |
| 0.42-0.60 m | Loam, plastic, limestone inclusions; 10YR 5/4 |
| 0.60 m- | End of auger. Bedrock |
| Augering S24: 491186/5049319 | |
| 0-0.30 m | Topsoil, loam, many limestone inclusions; 10YR 4/2 |
| 0.30 m- | End of auger. Bedrock |
| Augering S25: 491182/5049354 | |
| 0-0.57 m | Topsoil, loam, humic; 10YR 2/2 |
| 0.57-0.72 m | Loam, plastic, many limestone inclusions; 10YR 5/4 |
| 0.72 m- | End of auger. Bedrock |
| Augering S26: 491190/5049285 | |
| 0-0.48 m | Topsoil, loam, limestone inclusions; 10YR 3/2 |
| 0.48 m- | End of auger. Bedrock |
| Augering S27: 491193/5049256 | |
| 0-0.30 m | Topsoil, loam, humic. Some limestone inclusions; 10YR 4/2 |
| 0.30-0.35 m | Loam, slightly lighter colour, limestone inclusions |
| 0.35 m- | End of auger. Bedrock |
| *Site interpretation*<br>The combination of high pottery densities, stone tools, large quantities of worked flint and bone material suggests that this was a settlement | |

| **DSP08-F06-02** | *Core units*<br>9226-27 | *Sample density*<br>20% | *Coordinates (centre)*<br>491078/5049324 | *Figs.*<br>4.133-135 |
|---|---|---|---|---|
| | *Adjacent units* | *Sample density*<br>20% | | |
| | *Interpretation*<br>Kurgan | *Estimated size*<br>2,500 m² | *Date range*<br>Unknown | |
| *Topography*<br>**DSP08-F06-02** is located in the E part of a larger unploughed, grassy patch, on top of a slight ridge | | | | |
| *Investigations*<br>**Intensive survey** 50 x 50 m units. A grassy elevation associated with large stones. No finds were collected on the actual elevation, but two possibly LC-EH sherds were collected in the area around it | | | | |
| *Site interpretation*<br>Kurgan based on the size and shape of the elevation | | | | |

| **DSP08-F06-03** | *Core units*<br>9045 | *Sample density*<br>20% | *Coordinates (centre)*<br>492082/50496233 | *Figs.*<br>4.133-135, 4.139 |
|---|---|---|---|---|
| | *Adjacent units*<br>9044, 9049-50 | *Sample density*<br>20% | | |
| | *Interpretation*<br>Settlement | *Estimated size*<br>Unknown | *Date range*<br>LC-EH | |

*Topography*
**DSP08-F06-03** is located at the NE edge of the field

*Investigations*
**Intensive survey** 50 x 50 m units. A small scatter of amphora frs. which probably extends into the unsurveyed field to the E

*Site interpretation*
The presence of amphora frs. suggests a settlement location, although this may be the perimeter of a site in the unploughed field to the E

| **DSP07-H01-01** | *Core units*<br>H01 | *Coordinates*<br>4895420/5040220 | | *Figs.*<br>4.147-158 |
|---|---|---|---|---|
| | *Adjacent units*<br>H03, H12, H19 and H25 | | | |
| | *Interpretation*<br>Settlement | *Estimated size*<br>minimum 300 x 110 m | *Date range*<br>BA, LC-EH | |
| **Feature 1** | *Core units*<br>H01 | *Coordinates*<br>489542/5040379 | | |
| | *Interpretation*<br>Stone structure | *Estimated size*<br>35.7 x 10.4 m | *Date range*<br>BA, LC-EH | |
| **Feature 2** | *Core units*<br>H01 | *Coordinates*<br>489546/5040357 | | |
| | *Interpretation*<br>Stone structure | *Estimated size*<br>16 x 11 m | *Date range*<br>BA, LC-EH | |
| **Feature 3** | *Core units*<br>H01 | *Coordinates*<br>489539/5040325 | | |
| | *Interpretation*<br>Stone structure | *Estimated size*<br>8.5 x 6 m | *Date range* | |
| **Feature 4** | *Core units*<br>H01 | *Coordinates*<br>489555/5040319 | | |
| | *Interpretation*<br>Stone structure | *Estimated size*<br>5 x 6 m | *Date range*<br>BA, LC-EH | |
| **Feature 5** | *Core units*<br>H01 | *Coordinates*<br>489478/5040306 | | |
| | *Interpretation*<br>Stone structure | *Estimated size*<br>18 x 15 m | *Date range*<br>BA, LC-EH | |
| **Feature 6** | *Core units*<br>H01 | *Coordinates*<br>489470/5040277 | | |
| | *Interpretation*<br>Stone structure | *Estimated size*<br>16 x 9.5 m | *Date range*<br>BA, LC-EH | |
| **Feature 7** | *Core units*<br>H01 | *Coordinates*<br>489498/5040293 | | |
| | *Interpretation*<br>Stone? structure | *Estimated size*<br>25.6 x 19 m | *Date range*<br>BA, LC-EH | |
| **Feature 8** | *Core units*<br>H01 | *Coordinates*<br>489530/5040291 | | |
| | *Interpretation*<br>Double pen | *Estimated size*<br>Diameter ca. 21.5/38 m | *Date range*<br>BA, LC-EH? | |

| | | | | |
|---|---|---|---|---|
| **Feature 9** | *Core units*<br>H01 | *Coordinates*<br>489530/5040275 | | |
| | *Interpretation*<br>Stone structure | *Estimated size*<br>10 x 7 m | *Date range*<br>BA, LC-EH | |
| **Feature 10** | *Core units*<br>H01 | *Coordinates*<br>489550/5040283 | | |
| | *Interpretation*<br>Stone structures? | *Estimated size*<br>30.5 x 5 m | *Date range*<br>BA, LC-EH | |
| **Feature 11** | *Core units*<br>H01 | *Coordinates*<br>489561/5040268 | | |
| | *Interpretation*<br>Stone structure | *Estimated size*<br>20 x 10 m | *Date range*<br>BA, LC-EH | |
| **Feature 12** | *Core units*<br>H01 | *Coordinates*<br>489543/5040260 | | |
| | *Interpretation*<br>Stone structure | *Estimated size*<br>8 x 5.6 m | *Date range*<br>BA, LC-EH | |
| **Feature 13** | *Core units*<br>H01 | *Coordinates*<br>489545/5040237 | | |
| | *Interpretation*<br>Stone structure | *Estimated size*<br>Ca. 8 x 11 m | *Date range*<br>BA, LC-EH | |
| **Feature 14** | *Core units*<br>H01 | *Coordinates*<br>489489/5040193 | | |
| | *Interpretation*<br>Stone structure | *Estimated size*<br>11.5 x >9 m | *Date range*<br>BA, LC-EH | |
| **Feature 15** | *Core units*<br>H01 | *Coordinates*<br>489544/5040172 | | |
| | *Interpretation*<br>Stone structure | *Estimated size*<br>7 x 6 m | *Date range*<br>BA, LC-EH | |
| **Feature 16** | *Core units*<br>H01 | *Coordinates*<br>489545/5040139 | | |
| | *Interpretation*<br>Stone structure | *Estimated size*<br>21 x 7.5 m | *Date range*<br>BA, LC-EH | |
| **Feature 17** | *Core units*<br>H01 | *Coordinates*<br>489555/5040118 | | |
| | *Interpretation*<br>Stone structure | *Estimated size*<br>7.6 x 4 m | *Date range*<br>BA, LC-EH | |
| **Feature 18** | *Core units*<br>H01 | *Coordinates*<br>489532/5040128 | | |
| | *Interpretation*<br>Large cluster of pits | *Estimated size*<br>17 x 22 m | *Date range*<br>LC-EH? | |
| **Feature 19** | *Core units*<br>H01 | *Coordinates*<br>489509/5040261 | | |
| | *Interpretation*<br>Structure? | *Estimated size*<br>26.5 x 28 m | *Date range*<br>BA, LC-EH | |

| | | | | |
|---|---|---|---|---|
| **Feature 20** | *Core units*<br>H01 | *Coordinates*<br>489445/5040331 | | |
| | *Interpretation*<br>Large cluster of pits | *Estimated size*<br>30 x 15 m<br>*Date range*<br>LC-EH? | | |

*Topography*
**DSP07-H01-01** is located on the NW side of Hill 1, a short wide promontory located in the upper part of the hillsides

*Investigations*
**Intensive survey** Cluster of pottery (GS 3043, 3045-46, 3049-60) associated with rectangular stone structure with double walls of standing slabs in an area with ashy soil (Feature 17). Most of the surface pottery was collected N, S and W of Feature 18 and most likely came from the household pits.

| | |
|---|---|
| **Geomagnetic survey**<br>T. Smekalova<br>*Dimensions* 90 x 270 + 100 x 20 m | *Description of measuring conditions*<br>Good |

*Interpretation of features* Multiple linear and curve linear structures and features are visible in the data situated 25-40 m apart. In the S part are five single or double-room stone structures with possible hearths in three of them (Features 15-17). Feature 16 consists of three such, possibly overlapping structures of different orientations. Directly W of these is a cluster of 33 household pits (Feature 18). 67 m to the N is part of a possibly rectangular, stone structure (Feature 4). The N end of the surveyed area appears to consist of a large number of regular and irregular stone structures located around a double pen. The double pen consists of a circular enclosure inside a larger one (Feature 8). Some structures overlap the pen (Features 9-10). Features 3-4, 9-13 consists of individual or groups of single or double-room structures. Features 2 and 5-6 are multiple structures of irregular shape. Feature 1 is probably also a multiple-room structure, but the shape appears to be rectangular and three large pits are located inside the building. Feature 17 is a roughly horseshoe-shaped structure open towards the SW with an indication of possible post holes or bases for posts. Feature 19 is a large poorly preserved feature. NW of Feature 5 is another large cluster of household pits (Feature 20). The long linear anomalies which appear to cut under the site (equipment lifted when crossing over the large stones) are remains of recent agricultural processes

**Excavation** P. Guldager Bilde, S. Lancov, E.Weistra, interpreted by S. Lancov

| | | | |
|---|---|---|---|
| **Feature 5** | *Coordinates*<br>489472/5040297 | *Trench dimensions*<br>2 x 2 m | *Features*<br>Field stones on floor/collapsed wall? |

Description of stratigraphy by layer: 0-1,1 m
1) Ashy topsoil: 0-0.30.5 m (FL 6:1-4)
2) Ashy yellow loam with field stones and rock frs. on a ashy bed: 0.3-0.85 m (FL 6:5-10)
3) Loose yellow loam: 0.45-0.85 m (FL 6:11-15)
4) Hard yellow loam: 0.85-1.1 m (FL 6:16-18)

| | | |
|---|---|---|
| *Finds*<br>Pottery, bone and shells | *Date range*<br>BA and LC-EH | *Interpretation*<br>Stone structure |

| | | | |
|---|---|---|---|
| **Feature 18** | *Coordinates*<br>489556/5040112 | *Trench dimensions*<br>3.2 x 2.9 m | *Features*<br>Circular pit |

Description of stratigraphy by layer: 0-1.2 m
1) Top soil: 0-0.2 m (FL 4)
2) Looser yellow loam with rock frs.: 0.15-0.25 m (FL 2)
3) Sterile yellow loam: top level 0.25 m
Circular pit of dark looser soil, 0.9 m deep. Top D: 0.95 m. Bottom D: 1.15-1.2 m
1) Looser yellow loam with rock frs.: 0.25-0.4 m (FL 3)
2) Dark ashy phosphorous soil with small stones. Top level 0.4-0.8 m (FL 1 and 5)
3) Sterile yellow loam. Top level 0.55 m

| | | |
|---|---|---|
| *Finds*<br>Pottery, animal bones | *Date range*<br>LC-EH | *Interpretation*<br>Household pit |

*Site interpretation*
A complex settlement including at least 16 roughly rectangular, single and multiple-room, stone structures and a double pen associated with the keeping of live stock. The settlement includes at least two phases: a Bronze Age phase and a Hellenistic phase. The structures overlapping the double pen suggest that this was not in use during both phases

| **DSP07-H02-01** | *Core units*<br>H02 | *Coordinates*<br>490946/5040687 | | *Figs.*<br>4.159-163 |
|---|---|---|---|---|
| | *Adjacent units*<br>H07, H17 and H25 | | | |
| | *Interpretation*<br>Double pen | *Estimated size*<br>Diameter 28/48 m | *Date range*<br>BA, LC-EH? | |
| **Feature 1** | *Core units*<br>H02 | *Coordinates*<br>490908/5040623 | | |
| | *Interpretation*<br>Double pen | *Estimated size*<br>Diameter 28/48 m | *Date range*<br>BA | |
| **Feature 2** | *Core units*<br>H02 | *Coordinates*<br>490950/5040717 | | |
| | *Interpretation*<br>Stone structure? | *Estimated size*<br>Ca. 7 x 4 m | *Date range*<br>BA | |

*Topography*
**DSP07-H02-01** is located on a small promontory on the NE edge of Hill 2, a short wide promontory located in the upper part of the hillsides

*Investigations*
**Extensive survey** Area with ash-rich soil. In this general area, GS 3077-82 were collected, but their precise location was not recorded. According to the original report, handmade Bronze Age sherds and a flint tool were collected in this area, which suggests that 3081 was sampled at DSP07-H02-01. From Google Earth can clearly be distinguished a significant ash concentration between two small streams at the promontory's northern end (Fig. 4.159)

| **Geomagnetic survey**<br>T. Smekalova<br>*Dimensions* 65 x 700 + 30 x 30 m | *Description of measuring conditions*<br>Good |
|---|---|

*Interpretation of features* The data revealed a double pen consisting of a circular enclosure inside a larger one. There appears to be a small stone structure ca. 3 m N of the double pen. The long linear anomalies are remains of recent agricultural processes

*Site interpretation*
A double pen associated with the keeping of live stock during the Bronze Age possibly also associated with the Bronze Age?/ Hellenistic settlement DSP07-H02-02

<table>
<tr><td rowspan="3">DSP07-H02-02</td><td>Core units<br>H02</td><td colspan="2">Coordinates<br>490782/5040925</td><td>Figs.<br>4.159-162, 4.165</td></tr>
<tr><td>Adjacent units<br>H07, H17 and H25</td><td></td><td></td><td></td></tr>
<tr><td>Interpretation<br>Settlement?</td><td>Estimated size<br>100 x 70</td><td>Date range<br>BA and LC-EH</td><td></td></tr>
<tr><td rowspan="2">Feature 1</td><td>Core units<br>H02</td><td>Coordinates<br>490760/5040975</td><td></td><td></td></tr>
<tr><td>Interpretation<br>Stone structure</td><td>Estimated size<br>11 x 9 m</td><td>Date range<br>BA and LC-EH</td><td></td></tr>
<tr><td rowspan="2">Feature 2</td><td>Core units<br>H02</td><td>Coordinates<br>490814/5040936</td><td></td><td></td></tr>
<tr><td>Interpretation<br>Stone structure</td><td>Estimated size<br>minimum 7 x 7 m</td><td>Date range<br>BA and LC-EH</td><td></td></tr>
<tr><td rowspan="2">Feature 3</td><td>Core units<br>H02</td><td>Coordinates<br>490799/5040930</td><td></td><td></td></tr>
<tr><td>Interpretation<br>Stone structure</td><td>Estimated size<br>10 x 10 m</td><td>Date range<br>BA and LC-EH</td><td></td></tr>
<tr><td rowspan="2">Feature 4</td><td>Core units<br>H02</td><td>Coordinates<br>490766/5040917</td><td></td><td></td></tr>
<tr><td>Interpretation<br>Stone structure</td><td>Estimated size<br>14 x 14 m</td><td>Date range<br>BA and LC-EH</td><td></td></tr>
<tr><td rowspan="2">Feature 5</td><td>Core units<br>H02</td><td>Coordinates<br>490809/5040890</td><td></td><td></td></tr>
<tr><td>Interpretation<br>Stone structure</td><td>Estimated size<br>16 x 10 m</td><td>Date range<br>BA and LC-EH</td><td></td></tr>
<tr><td colspan="5">Topography<br>DSP07-H02-02 is located on the N tip of Hill 2, a short wide promontory located in the upper part of the hillsides</td></tr>
<tr><td colspan="5">Investigations<br>Extensive survey Three stone structures with double walls of standing slabs were observed, but not recorded. In the same general area GS 3077-82 were collected, but their precise location was not recorded. According to the original report, Hellenistic handmade and wheelmade sherds were collected in this area</td></tr>
<tr><td colspan="2">Geomagnetic survey<br>T. Smekalova<br>Dimensions ca. 130 x 60 m</td><td colspan="3">Description of measuring conditions<br>The bedrock is close to the surface</td></tr>
<tr><td colspan="5">Interpretation of features Five, roughly rectangular, stone structures visible in the data situated 25-40 m apart. Two single-room ones (Features 2-3), a double-room one (Feature 5) and a multiple-room one with four pits inside (Feature 1). Feature 4 appears somewhat irregular in shape reconstructed as an ante-building open towards the N with a possible line of post holes or bases for posts</td></tr>
<tr><td colspan="5">Site interpretation<br>Small Bronze Age?/Hellenistic settlement probably associated with the keeping of live stock as indicated by DSP07-H02-02 which appears to have been originally constructed in the Bronze Age</td></tr>
</table>

| **DSP08-H02-03** | *Core units*<br>H02 | *Coordinates*<br>491007/5040710 | | *Figs.*<br>4.159-163 |
|---|---|---|---|---|
| | *Adjacent units*<br>H07, H17 and H25 | | | |
| | *Interpretation*<br>Dam | *Estimated size*<br>Length 10-15 m | *Date range*<br>Unknown | |
| *Topography*<br>In the E ravine of Hill 2, a short wide promontory located in the upper part of the hillsides | | | | |
| *Investigations*<br>**Extensive survey** Remains of double wall of large stone blocks blocking the ravine | | | | |
| *Site interpretation*<br>Dam possibly related to the settlement on Hill 2 (see DSP07-H02-01-02) | | | | |

| **DSP07-H04-01** | *Core units*<br>H04 | *Coordinates*<br>486244/5038519 | | *Figs.*<br>4.166-169 |
|---|---|---|---|---|
| | *Adjacent units*<br>H05, H09 and H22 | | | |
| | *Interpretation*<br>Settlement | *Estimated size*<br>minimum 6,900 m$^2$ | *Date range*<br>LC-EH | |
| **Feature 1** | *Core units*<br>H04 | *Coordinates*<br>486269/5038555 | | |
| | *Interpretation*<br>Rectangular stone structure | *Estimated size*<br>8 x 6 m | *Date range*<br>LC-EH | |
| **Feature 2** | *Core units*<br>H04 | *Coordinates*<br>486289/5038549 | | |
| | *Interpretation*<br>Rectangular stone structure | *Estimated size*<br>6 x 6 m | *Date range*<br>LC-EH | |
| **Feature 3** | *Core units*<br>H04 | *Coordinates*<br>486194/5038518 | | |
| | *Interpretation*<br>Rectangular stone structure | *Estimated size*<br>17 x 13 m | *Date range*<br>LC-EH | |
| **Feature 4** | *Core units*<br>H04 | *Coordinates*<br>486177/5038478 | | |
| | *Interpretation*<br>Rectangular stone structure | *Estimated size*<br>11.5 x 9.5 m | *Date range*<br>LC-EH | |
| *Topography*<br>**DSP07-H04-01** is located on the tip of Hill 4, a small promontory in the upper part of the hillsides. The SW edge of the promontory was currently cultivated | | | | |
| *Investigations*<br>**Intensive survey** Small cluster of pottery (GS 3068-3072) roughly associated with four rectangular stone structures | | | | |
| *Site interpretation*<br>Small settlement? | | | | |

<table>
<tr><td rowspan="3">DSP07-H04-02</td><td>Core units<br>H04</td><td colspan="2">Coordinates<br>486536/5038488</td><td>Figs.<br>4.166-167, 4.170-171</td></tr>
<tr><td colspan="3">Adjacent units<br>H05, H09 and H22</td><td></td></tr>
<tr><td>Interpretation<br>Settlement</td><td>Estimated size<br>Unknown</td><td>Date range<br>EM?</td><td></td></tr>
<tr><td rowspan="2">Feature 1</td><td>Core units<br>H04</td><td>Coordinates<br>486545/5038538</td><td></td><td></td></tr>
<tr><td>Interpretation<br>Cluster of pits</td><td>Estimated size<br>?</td><td>Date range<br>EM?</td><td></td></tr>
<tr><td rowspan="2">Feature 2</td><td>Core units<br>H04</td><td>Coordinates<br>486531/5038431</td><td></td><td></td></tr>
<tr><td>Interpretation<br>Stone structure</td><td>Estimated size<br>16.5 x 17 m</td><td>Date range<br>EM?</td><td></td></tr>
<tr><td colspan="5">Topography<br>DSP07-H04-02 is located on the NE edge of Hill 4 spreading into the shallow ravine on the NE side of the hill, which is a small promontory in the upper part of the hillsides</td></tr>
<tr><td colspan="5">Investigations<br>Extensive survey Several pits (Feature 1) indicated by vegetation and rubble associated with a single sherd (GS 3075)<br>Intensive survey Stone foundation (Feature 2), but no finds associated</td></tr>
<tr><td colspan="5">Site interpretation<br>Small Early Modern settlement</td></tr>
</table>

<table>
<tr><td rowspan="3">DSP07-H04-03</td><td>Core units<br>H04</td><td colspan="2">Coordinates<br>486899/5037905</td><td>Figs.<br>4.166-167, 4.172</td></tr>
<tr><td colspan="3">Adjacent units<br>H05, H09 and H22</td><td></td></tr>
<tr><td>Interpretation<br>Structure?</td><td>Estimated size<br>0.04 ha</td><td>Date range<br>Unknown</td><td></td></tr>
<tr><td colspan="5">Topography<br>DSP07-H04-03 is located at the S end of Hill 4, a small promontory in the upper part of the hillsides. The SW edge of the promontory was cultivated</td></tr>
<tr><td colspan="5">Investigations<br>Intensive survey Patch of vegetation indicating subsurface structures associated with one sherd (GS 3074)</td></tr>
<tr><td colspan="5">Site interpretation<br>Storage site?</td></tr>
</table>

<table>
<tr><td rowspan="3">DSP07-H05-01</td><td>Core units<br>H05</td><td colspan="2">Coordinates<br>486580/5037209</td><td>Fig.<br>4.173</td></tr>
<tr><td>Adjacent units<br>H04, H05, H22</td><td colspan="2"></td><td></td></tr>
<tr><td>Interpretation<br>Structure? Settlement?</td><td>Estimated size<br>7 x 4 m</td><td>Date range<br>Unknown</td><td></td></tr>
<tr><td colspan="5">Topography<br>DSP07-H05-01 is located at the S end of Hill 5, a small promontory in the upper part of the hillsides</td></tr>
<tr><td colspan="5">Investigations<br>Intensive survey Possible structure made of large stones. No finds</td></tr>
<tr><td colspan="5">Site interpretation<br>Settlement site? An ash accumulation at the N tip of the promontory can clearly be distinguished on Google Earth (Fig. 4.167)</td></tr>
</table>

<table>
<tr><td rowspan="3">DSP07-H06-01</td><td>Core units<br>H06, 20002 and 20007</td><td colspan="2">Coordinates<br>491658/5041721</td><td>Figs.<br>4.233-236</td></tr>
<tr><td>Adjacent units<br>H17 and H26</td><td colspan="2"></td><td></td></tr>
<tr><td>Interpretation<br>Settlement</td><td>Estimated size<br>minimum 70 x 90</td><td>Date range<br>BA/LC-EH?</td><td></td></tr>
<tr><td rowspan="2">Feature 1</td><td>Core units<br>H06</td><td>Coordinates<br>491674/5041721</td><td></td><td></td></tr>
<tr><td>Interpretation<br>Double pen</td><td>Estimated size<br>Diameter 28 and 45 m</td><td>Date range</td><td></td></tr>
<tr><td rowspan="2">Feature 2</td><td>Core units<br>H06</td><td>Coordinates<br>491666/5041755</td><td></td><td></td></tr>
<tr><td>Interpretation<br>Stone structure</td><td>Estimated size<br>9 x 6 m</td><td>Date range</td><td></td></tr>
<tr><td colspan="5">Topography<br>DSP07-H06-01 is located on the ridge of Hill 6, a small promontory in the upper part of the hillsides</td></tr>
<tr><td colspan="5">Investigations<br>Extensive survey Pottery (GS 3085, 2007 FL 8) associated with a stone structure. A stone block with a small circular depression possibly from a post was recorded<br>Intensive survey A rather diffused cluster of pottery and flint (20002/01-02, 20007/02-03, 3201-03)</td></tr>
<tr><td colspan="2">Geomagnetic survey<br>T. Smekalova<br>Dimensions 100 x 100 m</td><td colspan="3">Description of measuring conditions<br>Good</td></tr>
<tr><td colspan="5">Interpretation of features The data revealed a double pen consisting of a circular enclosure inside a larger one (Feature 1) associated with a small irregular, stone double-room structure to the NE (Feature 2). The long linear anomalies which cut across the site are remains of recent agricultural processes</td></tr>
<tr><td colspan="5">Site interpretation<br>Small settlement associated with the keeping of livestock probably occupied during the Bronze Age</td></tr>
</table>

| **DSP07-H07-01** | *Core units*<br>H07 | *Coordinates*<br>490573/5041333 | | *Figs.*<br>4.174-176 |
|---|---|---|---|---|
| | *Adjacent units*<br>H02, H17 and H25 | | | |
| | *Interpretation*<br>Settlement | *Estimated size*<br>minimum 44 x 24 m | *Date range*<br>BA?/LC-EH | |
| **Feature 1** | *Core units*<br>H07 | *Coordinates*<br>? | | |
| | *Interpretation*<br>Stone structure | *Estimated size*<br>? | *Date range*<br>BA?/LC-EH | |
| **Feature 2** | *Core units*<br>H07 | *Coordinates*<br>490567/5041340 | | |
| | *Interpretation*<br>Stone structure | *Estimated size*<br>20 x 13 m | *Date range*<br>BA?/LC-EH | |
| **Feature 3** | *Core units*<br>H07 | *Coordinates*<br>490569/5041320 | | |
| | *Interpretation*<br>Road/path | *Estimated size*<br>minimum 37 m | *Date range*<br>BA?/LC-EH | |
| **Feature 4** | *Core units*<br>H07 | *Coordinates*<br>490582/5041324 | | |
| | *Interpretation*<br>Cluster of pits | *Estimated size*<br>84 m$^2$ | *Date range*<br>BA?/LC-EH | |

*Topography*
**DSP07-H07-01** is located on the E edge of the lower part of Hill 7, a long slope running from the top ridge to the cultivated zone

*Investigations*
**Extensive survey** Two rectangular stone structures with double walls built with standing slabs associated with nine sherds, five of Hellenistic date (GS 3062-63 and 2007 FL 9)

| **Geomagnetic survey**<br>T. Smekalova<br>*Dimensions* 50 x 65 m | *Description of measuring conditions*<br>Good |
|---|---|

*Interpretation of features* Rectangular, double-room stone structure with one interior pit and four exterior ones (Feature 2). A road leads up to the building on the W side (Feature 3). Due S is a cluster of 12 pits (Feature 4). The long linear anomalies which pass on the SE side of the site are remains of recent agricultural processes

*Site interpretation*
Small settlement possibly associated with DSP07-H07-02

<table>
<tr><td rowspan="3">DSP07-H07-02</td><td>Core units<br>H07</td><td colspan="2">Coordinates<br>490168/5041704</td><td>Figs.<br>4.174, 4.177</td></tr>
<tr><td>Adjacent units<br>H02, H17 and H25</td><td colspan="2"></td><td></td></tr>
<tr><td>Interpretation<br>Settlement</td><td>Estimated size<br>minimum 140 x 30 m</td><td>Date range<br>BA?/LC-EH</td><td></td></tr>
<tr><td rowspan="2">Feature 1</td><td>Core units<br>H07</td><td>Coordinates<br>490108/5041684</td><td></td><td></td></tr>
<tr><td>Interpretation<br>Stone structure</td><td>Estimated size<br>10 x 15 m</td><td>Date range<br>BA?/LC-EH</td><td></td></tr>
<tr><td rowspan="2">Feature 2</td><td>Core units<br>H07</td><td>Coordinates<br>490215/5041750</td><td></td><td></td></tr>
<tr><td>Interpretation<br>Jurt?</td><td>Estimated size<br>Diameter 18 m</td><td>Date range<br>BA?/LC-EH</td><td></td></tr>
<tr><td rowspan="2">Feature 3</td><td>Core units<br>H07</td><td>Coordinates<br>490217/5041714</td><td></td><td></td></tr>
<tr><td>Interpretation<br>Fence</td><td>Estimated size<br>minimum 35 m</td><td>Date range<br>BA?/LC-EH</td><td></td></tr>
<tr><td colspan="5">Topography<br>DSP07-H07-02 is located N of DSP07-H07-01 on the E edge of the lower part of Hill 7, a long slope running from the top ridge to the cultivated zone</td></tr>
<tr><td colspan="5">Investigations<br>Extensive survey Intense vegetation and multiple-room, stone structure built with standing slabs associated with eight sherds of amphorai and jugs (2007 FL 10)</td></tr>
<tr><td colspan="2">Geomagnetic survey<br>T. Smekalova<br>Dimensions 30 x 60 and 50 x 50 m</td><td colspan="3">Description of measuring conditions<br>Some dipole anomalies</td></tr>
<tr><td colspan="5">Interpretation of features Rectangular, multiple-room, stone structure associated with metal debris (Feature 1). 50 m NE is a round structure made up of 17 to 20 post-holes or bases for posts with four pits inside, possibly a jurt (Feature 2). To the S is a long linear anomaly possibly a fence (Feature 3)</td></tr>
<tr><td colspan="5">Site interpretation<br>Small settlement possibly associated with DSP07-H07-01</td></tr>
</table>

| DSP08-H07-03 | *Core units*<br>20064-65 | *Coordinates*<br>490257/5040112 | | *Figs.*<br>4.304-307 |
|---|---|---|---|---|
| | *Adjacent units*<br>H02, H17 and H25 | *Date range*<br>EM? | | |
| | *Interpretation*<br>Settlement | *Estimated size*<br>5,500 m² | *Date range*<br>EM? | |
| Feature 1 | *Core units*<br>20064 | *Coordinates*<br>490260/5040151 | | |
| | *Interpretation*<br>Ash hill | *Estimated size*<br>Diameter ca. 23 m | *Date range*<br>EM? | |
| Feature 2 | *Core units*<br>20065 | *Coordinates*<br>490208/5040141 | | |
| | *Interpretation*<br>Ash hill | *Estimated size*<br>Diameter 17 m | *Date range*<br>EM | |
| Feature 3 | *Core units*<br>20064 | *Coordinates*<br>490277/5040139 | | |
| | *Interpretation*<br>Rectangular stone structure | *Estimated size*<br>9 x 7 m | *Date range*<br>EM? | |
| Feature 4 | *Core units*<br>20065 | *Coordinates*<br>490226/5040128 | | |
| | *Interpretation*<br>Rectangular stone structure | *Estimated size*<br>8 x 2.5 m | *Date range*<br>EM? | |
| Feature 5 | *Core units*<br>20064 | *Coordinates*<br>490254/5040134 | | |
| | *Interpretation*<br>Rectangular stone structure | *Estimated size*<br>5.5 x =/>3 m | *Date range*<br>EM? | |
| Feature 6 | *Core units*<br>20064 | *Coordinates*<br>490249/5040127 | | |
| | *Interpretation*<br>Rectangular stone structure | *Estimated size*<br>8 x 5 m | *Date range*<br>EM? | |
| Feature 7 | *Core units*<br>20064 | *Coordinates*<br>490241/5040117 | | |
| | *Interpretation*<br>Group of hearths | *Estimated size*<br>38 m² | *Date range*<br>EM? | |
| Feature 8 | *Core units*<br>20064 | *Coordinates*<br>490276/5040120 | | |
| | *Interpretation*<br>Rectangular? stone structure | *Estimated size*<br>Length 17 m | *Date range*<br>EM? | |

<table>
<tr><td rowspan="2">Feature 9</td><td>Core units<br>20064-20065</td><td>Coordinates<br>490263/5040098</td><td></td><td></td></tr>
<tr><td>Interpretation<br>Rectangular earth bank</td><td>Estimated size<br>Length >105 m</td><td>Date range<br>EM?</td><td></td></tr>
<tr><td colspan="5">Topography<br>DSP08-H07-03 is located in the central part of Hill 7 on the top of the ridge</td></tr>
<tr><td colspan="5">Investigations<br>Extensive survey Ash hill associated with a single tiny sherd and several stone structures. A later visit yielded a small cluster of pottery on and around the ash hill (GS 3221)<br>Intensive survey Two ash hills (Features 1-2) and three possible rectangular stone structures of double walls (Features 3-4 and 8) were recorded associated with a few tiny sherds (20064-65)</td></tr>
<tr><td colspan="2">Geomagnetic survey<br>Eastern Atlas<br>Dimensions 100 x 78 m</td><td>Description of measuring conditions<br>Some dipole anomalies</td><td></td><td></td></tr>
<tr><td colspan="5">Interpretation of features Archaeological features are noticeable in the NW corner, SW of the ash hill. Three rectangular buildings with one or two rooms show up as weak, white (negative) anomalies (Features 4-6). Three strong possible anomalies with a white "halo" are most likely former fireplaces (Feature 7). Several pits most likely of archaeological origin are located in and around the structures, but also random pattern of small pits. To the S, the data are less disturbed, probably due to a thicker soil horizon. The whole site is pervaded by linear features that either belong to roads/paths or agricultural patterns (Feature 9). A circular feature in the centre is of modern origin</td></tr>
<tr><td colspan="5">Site interpretation<br>Early Modern settlement probably associated with DSP08-H07-04 located 110 m to the SW. The geomagnetic survey did not cover the entire site which consists of a total of five possible stone structure and two ash hills</td></tr>
</table>

<table>
<tr><td rowspan="3">DSP07-H07-04</td><td>Core units<br>H07</td><td colspan="2">Coordinates<br>490105/5040058</td><td>Figs.<br>4.304, 4.308</td></tr>
<tr><td colspan="3">Adjacent units<br>H02, H17 and H25</td><td></td></tr>
<tr><td>Interpretation<br>Dam</td><td>Estimated size<br>48 x 30.5 m</td><td>Date range<br>EM?</td><td></td></tr>
<tr><td colspan="5">Topography<br>DSP07-H07-04 is located in a shallow ravine at the W side of Hill 7, on the top of the ridge, 110 m SW of DSP07-H07-03</td></tr>
<tr><td colspan="5">Investigations<br>Extensive survey Two interlocking L-shaped earth banks, but no artefacts associated</td></tr>
<tr><td colspan="5">Site interpretation<br>A dam which is probably associated with DSP08-H07-03</td></tr>
</table>

| **DSP07-H08-01** | *Core units*<br>20008-9, 20013-16 | *Coordinates*<br>493206/5042276 | | *Figs.*<br>4.248-256 |
|---|---|---|---|---|
| | *Adjacent units*<br>H16 and H20 | | | |
| | *Interpretation*<br>Settlement | *Estimated size*<br>2.8 ha | *Date range*<br>BA, LC-EH? | |
| **Feature 1** | *Core units*<br>20013 | *Coordinates*<br>493290/ 5042399 | | |
| | *Interpretation*<br>Stone structure | *Estimated size*<br>10.5 x =/>13 m | *Date range*<br>BA | |
| **Feature 2** | *Core units*<br>20013-14 | *Coordinates*<br>493252/5042355 | | |
| | *Interpretation*<br>Double pen | *Estimated size*<br>Diameter 22 and 30 m | *Date range*<br>BA | |
| **Feature 3** | *Core units*<br>20013 | *Coordinates*<br>493289/5042342 | | |
| | *Interpretation*<br>Stone structure | *Estimated size*<br>6 x 10 m | *Date range*<br>BA | |
| **Feature 4** | *Core units*<br>20014 | *Coordinates*<br>493231/5042368 | | |
| | *Interpretation*<br>Stone structure | *Estimated size*<br>10.5 x 4 m | *Date range*<br>BA, LC-EH? | |
| **Feature 5** | *Core units*<br>20014-15 | *Coordinates*<br>493251/5042375 | | |
| | *Interpretation*<br>Cluster of pits? | *Estimated size*<br>560 $m^2$ | *Date range*<br>BA, LC-EH? | |
| **Feature 6** | *Core units*<br>20015 | *Coordinates*<br>493176/5042225 | | |
| | *Interpretation*<br>Stone structure | *Estimated size*<br>12.5 x =/> 7 m | *Date range*<br>BA, LC-EH? | |
| **Feature 7** | *Core units*<br>20016 | *Coordinates*<br>493132/5042237 | | |
| | *Interpretation*<br>Stone structure | *Estimated size*<br>11.5 x 10 m | *Date range*<br>BA, LC-EH? | |
| **Feature 8** | *Core units*<br>20015-16 | *Coordinates*<br>493151/5042205 | | |
| | *Interpretation*<br>Complex stone structure/structures | *Estimated size*<br>60 x 22 m | *Date range*<br>BA, LC-EH? | |
| **Feature 9** | *Core units*<br>20016 | *Coordinates*<br>493126/5042211 | | |
| | *Interpretation*<br>Stone structure | *Estimated size*<br>11 x 15 m | *Date range*<br>BA, LC-EH? | |

| | | | | |
|---|---|---|---|---|
| **Feature 10** | *Core units*<br>20016 | *Coordinates*<br>493114/5042179 | | |
| | *Interpretation*<br>Stone structure | *Estimated size*<br>7 x 7 m | *Date range*<br>BA, LC-EH? | |
| **Feature 11** | *Core units*<br>20015-16 | *Coordinates*<br>493101/5042128 | | |
| | *Interpretation*<br>Complex stone structure | *Estimated size*<br>16 x 9 and 22 x 29 m | *Date range*<br>BA, LC-EH? | |
| **Feature 12** | *Core units*<br>20015 | *Coordinates*<br>493128/5042104 | | |
| | *Interpretation*<br>Stone structure | *Estimated size*<br>5 x 4.5 m | *Date range*<br>BA, LC-EH? | |

*Topography*
**DSP07-H08-01** is located along the W edge of Hill 8, a short shallow promontory between two high promontories on the lower part of the hillsides

*Investigations*
**Extensive survey** A cluster of Bronze Age pottery (GS 3095, 2007 FL 11-12) associated with stone structures in an area of ashy soil with intense vegetation
**Intensive survey** One discrete cluster of pottery (20013/01-02, 20014/01-02 and 20015/01-03) associated with two stone structures (Features 2 and 4) and six possible pits indicated by intense vegetation (Feature 5) all within a larger area of diffused intense vegetation of 0.87 ha. Another less discrete cluster of pottery and flint is located approximately 100 m to the SW (20015/04-10 and 20016/04). Further 100 m to the SW are the remains a stone structure (Feature 11) associated with a small cluster of finds (20016/01-03) in a larger area of greener, grassier vegetation

| | |
|---|---|
| **Geomagnetic survey**<br>T. Smekalova<br>*Dimensions* 1.25 ha | *Description of measuring conditions*<br>Some dipole anomalies |

*Interpretation of features* The data revealed a complex site with several types of stone structures. In the N end is a large double pen consisting of a circular enclosure inside a larger one (Feature 2) associated with two irregular structures (Features 1 and 3). Feature 4 to the E of the double pen is obscured by dipole anomalies. 120 m to the S is a complex cluster of rectangular (Features 6-7 and 9) and irregular structures mostly of different orientation, although one long wall links a cluster of walls (Feature 8). The long linear anomalies which cross the site are remains of recent agricultural processes

| | |
|---|---|
| **Geomagnetic survey**<br>Eastern Atlas<br>*Dimensions* 210 x 48 m | *Description of measuring conditions*<br>Rare dipole anomalies |

*Interpretation of features* Rectangular structure with pits inside and around it 80 m SW of Feature 8 (Feature 10). Small cluster of small pits 15 m S of this. Another complex further S consisting of a rectangular multi-room structure associated with curved walls to the SW and NW (Feature 11). Large irregular pits inside and around the northernmost room and the small curved structure NE of this. A small square structure 9 m to the SW (Feature 12)

**Excavation** V. Stolba, S. Lancov and P. Gecko.

| | | | |
|---|---|---|---|
| **Feature 2** | *Coordinates*<br>493248/5042340 | *Trench dimensions*<br>2 x 3 m | *Features*<br>Broad/collapsed stone wall |

Description of stratigraphy by layer: 0-0.75 m
1) Topsoil: 0-0.2 m (FL bag 18)
2) Loose dark soil: 0.2-0.3 m (FL bags 19-20)
3) Layer of ashy soil with a broad, double stone wall constructed from large blocks and filled with rubble: top level 0.1 m. Bottom level: 0.4 m. Thickness: 1.6 m. Preserved H: 0.35 m
4) Yellowish loam with concentration of small stones in the W corner: 0.3-0.65 m. Concentration of animal bones (FL bags 21-26 and 32)
5) Sterile loam: top level 0.45-0.65 m

| *Finds* | *Date range* | *Interpretation* |
|---|---|---|
| Handmade pottery, animal bones | BA | Double pen |

| **Feature 7** | *Coordinates* | *Trench dimensions* | *Features* |
|---|---|---|---|
| | 493134/5042240 | 2 x 2 m | Collapsed stone wall |

Description of stratigraphy by layer: 0-0.3 m
1) Topsoil: 0-0.2/0.25 m (FL bag 27)
2) Layer of ashy soil with collapsed stone wall: top level: 0.1 m. Bottom level: 0.3 m. Numerous animal bones, handmade pottery frs. and a flint blade (FL bags 28-31)
3) Sterile loam, top level: 0.3 m below surface

| *Finds* | *Date range* | *Interpretation* |
|---|---|---|
| Handmade pottery, micro-lithic, stone axe, animal bones | BA | Settlement |

*Site interpretation*
Complex Bronze Age settlement associated with the keeping of livestock although the structures may not all be contemporary

| **DSP08-H08-02** | *Core units* | *Coordinates* | | *Figs.* |
|---|---|---|---|---|
| | 20018 | 493372/5042379 | | 4.248, 4.257-258 |
| | *Adjacent units* 20013-17, H16 and H20 | | | |
| | *Interpretation* | *Estimated size* | *Date range* | |
| | Rectangular stone structure | 4 x 4 m | EM | |

*Topography*
**DSP07-H08-02** is located along the E edge of Hill 8, a short shallow promontory between two high promontories on the lower part of the hillsides

*Investigations*
**Intensive survey** Possible stone structure associated with two Early Modern sherds (20018/1-2)

*Site interpretation*
Early Modern storage site?

<table>
<tr><td rowspan="3">DSP07-H09-01</td><td>Core units<br>H09</td><td colspan="2">Coordinates<br>485116/50401354</td><td>Figs.<br>4.178-180</td></tr>
<tr><td>Adjacent units<br>H04-05 and H22</td><td colspan="2"></td><td></td></tr>
<tr><td>Interpretation<br>Stone structure</td><td>Estimated size<br>15 x 5 m</td><td>Date range<br>EM</td><td></td></tr>
<tr><td colspan="5">Topography<br>DSP07-H09-01 is located on the E edge on the lower part of Hill 9, a long slope running from the ridge top to the cultivated zone</td></tr>
<tr><td colspan="5">Investigations<br>Extensive survey Stone structure built with standing slabs associated with intense vegetation and pottery (2007 FL 14). The W wall is at least 15 m long, the N one at least 5 m long. The finds recorded as GS 3088-3090 and 2007 FL 14 possibly belongs either to this site or to DSP07-H09-02.</td></tr>
<tr><td colspan="2">Geomagnetic survey<br>T. Smekalova<br>Dimensions 1 ha</td><td colspan="3">Description of measuring conditions<br>Some dipole anomalies</td></tr>
<tr><td colspan="5">Interpretation of features The data revealed no clear features</td></tr>
<tr><td colspan="5">Site interpretation<br>Early Modern settlement?</td></tr>
</table>

<table>
<tr><td rowspan="3">DSP07-H09-02</td><td>Core units<br>H09</td><td colspan="2">Coordinates<br>484702/5040385</td><td>Figs.<br>4.178, 4.181</td></tr>
<tr><td>Adjacent units<br>H04-05 and H22</td><td colspan="2"></td><td></td></tr>
<tr><td>Interpretation<br>Stone structure</td><td>Estimated size<br>14 x 7.5 m</td><td>Date range<br>EM?</td><td></td></tr>
<tr><td colspan="5">Topography<br>DSP07-H09-02 is located on the W edge on the lower part of Hill 9, a long slope running from the ridge top to the cultivated zone</td></tr>
<tr><td colspan="5">Investigations<br>Extensive survey Stone structure associated with cluster of pottery. The finds recorded as GS 3088-3090 and 2007 FL 14 possibly belong either to this site or to DSP07-H09-01 (see DSP07-H09-01)</td></tr>
<tr><td colspan="2">Geomagnetic survey<br>T. Smekalova<br>Dimensions 31 x 55 m</td><td colspan="3">Description of measuring conditions<br>Some dipole anomalies</td></tr>
<tr><td colspan="5">Interpretation of features The data revealed a rectangular stone structure. No interpreted map</td></tr>
<tr><td colspan="5">Site interpretation<br>Early Modern settlement?</td></tr>
</table>

| **DSP07-H10-01** | *Core units* H10 | *Coordinates* 482497/5037941 | | *Figs.* 4.182-194 |
|---|---|---|---|---|
| | *Adjacent units* H18 | *Date range* LC-EH | | |
| | *Interpretation* Settlement | *Estimated size* minimum 120 x 115 m | *Date range* LC-EH | |
| **Feature 1** | *Core units* H10 | *Coordinates* 482446/5037977 | | |
| | *Interpretation* Cluster of pits | *Estimated size* 48 m² | *Date range* LC-EH | |
| **Feature 2** | *Core units* H10 | *Coordinates* 482514/5037943 | | |
| | *Interpretation* Stone structure | *Estimated size* 20 x 17 m | *Date range* LC-EH | |
| **Feature 3** | *Core units* H10 | *Coordinates* 482477/5037944 | | |
| | *Interpretation* Stone structure | *Estimated size* 10 x 7 m | *Date range* LC-EH | |
| **Feature 4** | *Core units* H10 | *Coordinates* 482507/5037921 | | |
| | *Interpretation* Stone structure | *Estimated size* 7 x 6 m | *Date range* LC-EH | |
| **Feature 5** | *Core units* H10 | *Coordinates* 482513/5037925 | | |
| | *Interpretation* Cluster of pits | *Estimated size* 55 m² | *Date range* LC-EH | |
| **Feature 6** | *Core units* H10 | *Coordinates* 482447/5037924 | | |
| | *Interpretation* Stone structure | *Estimated size* minimum 7 x 5 m | *Date range* LC-EH | |
| **Feature 7** | *Core units* H10 | *Coordinates* 482435/5037915 | | |
| | *Interpretation* Cluster of pits | *Estimated size* 47 m² | *Date range* LC-EH | |

*Topography*
**DSP07-H10-01** is located on the tip of Hill 10, a small promontory towards the lower part of the hillsides

*Investigations*
**Extensive survey** Four rectangular stone structures with double walls occasionally employing standing blocks (Features 2-4 and 6), associated with a cluster of pottery (2007 FL 13, GS 3084, 3086-87), although these grab samples may also be associated with DSP07-H18-01 approximately 50 m to the SW. Feature 2 appears to be a house with multiple rooms and a possible court yard

| **Geomagnetic survey** T. Smekalova *Dimensions* 0.75 ha | *Description of measuring conditions* Some dipole anomalies in the SW |
|---|---|

<table>
<tr><td colspan="5">Interpretation of features The data revealed four rectangular stone structures (Features 2-4 and 6), of which Feature 2 has multiple rooms with three pits and a possible courtyard. There are also three clusters of household pits (1, 5 and 7). Feature 5 is associated with Features 4 and 6 possibly with 7. A single pit is associated with Feature 3. The long linear anomalies are remains of recent agricultural processes</td></tr>
<tr><td colspan="5">Augering One core was drilled from the bottom of Feature 5, Pit 1</td></tr>
<tr><td>0.96 m</td><td colspan="4">Top of auger. Bottom of the pit</td></tr>
<tr><td>0.96-1.81 m</td><td colspan="4">Loam with a thin layer of limestone crumble above the rock surface</td></tr>
<tr><td>1.81 m</td><td colspan="4">End of auger. Limestone bedrock</td></tr>
<tr><td colspan="5">Excavation V. Stolba and P. Gecko</td></tr>
<tr><td>Feature 2</td><td>Coordinates<br>482518/5037946</td><td>Trench dimensions<br>2 x 2 m</td><td colspan="2">Features<br>Double wall constructed of roughly chipped stones with packing of small stones in between</td></tr>
<tr><td colspan="5">Description of stratigraphy by layer: 0-0.4 m<br>1) Topsoil: 0-0.2 m. Double stone wall clearly discernable on the surface. 0.8-0.82 m thick. Preserved H: 0.32 m. Bottom level: 0.28-0.3 m<br>2) Layer of stone debris W of the wall: 0.2-0.4 m (FL 12)<br>3) Dark brown fill SW of the wall: 0.2-0.4 m<br>4) Sterile loam: top level 0.4 m</td></tr>
<tr><td colspan="3">Finds<br>Amphora frs. and helix shell</td><td>Date range<br>LC-EH</td><td>Interpretation<br>Settlement</td></tr>
<tr><td>Feature 5</td><td>Coordinates<br>482511/5037924</td><td>Trench dimensions<br>2 x 2 m<br>extended 0.6 x 1 m E and 0.7 x 1.1 m N</td><td colspan="2">Features<br>Two circular pits</td></tr>
<tr><td colspan="5">Description of stratigraphy by layer: 0-1.81 m<br>1) Topsoil: 0-0.28 m (FL 1, 3 and 8)<br>2) Upper fill of yellowish loam: top level 0.28 m (FL 2 and 4)<br>3) Control pit on the S edge of the trench: depth 0.60-0.70 m (FL 7)<br>Circular roughly piriform pit with concave bottom (1): 0.96 m deep (measured from topsoil surface). Max D: 0.96 m<br>1) Darker soil with admixture of rubble: 0.50/0.60-0.80 m (FL 5-6)<br>2) Large flag stone (0.50 x 0.40 m) on the floor of the pit: top level 0.80 m. Bottom level 0.96 m<br>Circular piriform pit with concave bottom (2): 1.36 m deep (measured from topsoil surface). Max D: 1.47 m<br>1) Upper fill of dark brown soil: 0.20-60 m (FL 9)<br>2) Fill of densely packed rubble: 0.60-1.15 m (FL 10)<br>3) Fill below stone packing: 1.15-1.30 m. An almost complete vessel was found in this layer, but frs. came from depth 0.40-1.30 m (FL 11)<br>4) Same as above but with ashy inclusions: 1.30-1.36 m. A fox hole blocked with stones towards the bottom on the S side. Mudbrick found on the bottom of the pit<br>5) Sterile loam 1.36-1.81 m<br>6) Bedrock: top level 1.81 m</td></tr>
<tr><td colspan="5">Finds<br>Amphora, handmade, plainware and black-glazed frs., stone tools, animal bones, helix shells and cereal</td></tr>
<tr><td colspan="5">Site interpretation<br>Small settlement including one large structure with multiple rooms and a courtyard, two smaller structures and several household pits. Probably associated with DSP07-H18-01, located just across a shallow ravine 140 m to the W</td></tr>
</table>

<table>
<tr><td rowspan="3">DSP08-H11-01</td><td>Core units<br>H11</td><td colspan="2">Coordinates<br>486890/5041136</td><td>Figs.<br>4.195-199</td></tr>
<tr><td>Adjacent units<br>H22, H23</td><td colspan="2">Date range<br>LC-EH</td><td></td></tr>
<tr><td>Interpretation<br>Settlement</td><td>Estimated size<br>minimum 100 x 80 m</td><td>Date range<br>LC-EH</td><td></td></tr>
<tr><td rowspan="2">Feature 1</td><td>Core units<br>H11</td><td>Coordinates<br>486915/5041146</td><td></td><td></td></tr>
<tr><td>Interpretation<br>Check dam?</td><td>Estimated size<br>3 m</td><td>Date range<br>LC-EH</td><td></td></tr>
<tr><td colspan="5">Topography<br>DSP07-H11-01 is located on thin soils on the E side at the bottom of Hill 11, a long slope running from the top of the ridge to the cultivated zone</td></tr>
<tr><td colspan="5">Investigations<br>Extensive survey A large cluster of pottery and ground stone tools (GS 3083 and 3093) associated with possible remains of check-dams in the shallow ravine on the E side of Hill 11</td></tr>
<tr><td colspan="2">Geomagnetic survey<br>T. Smekalova<br>Dimensions 50 x 80 and 70 m</td><td colspan="3">Description of measuring conditions<br>Poor. Many dipole anomalies</td></tr>
<tr><td colspan="5">Interpretation of features No identifiable structures or features preserved</td></tr>
<tr><td colspan="5">Site interpretation<br>Apart from the possible check-dam, any remains of architecture appear to have been removed. The pottery suggests the presence of a relatively large Hellenistic settlement</td></tr>
</table>

| **DSP08-H11-02** | *Core units*<br>20050-20055 | *Coordinates*<br>487265/5040066 | | *Figs.*<br>4.289-294 |
|---|---|---|---|---|
| | *Adjacent units*<br>H13 and H22-23 | *Date range*<br>LC-EH | | |
| | *Interpretation*<br>Settlement | *Estimated size*<br>2.6 ha | *Date range*<br>LC-EH | |
| **Feature 1** | | *Coordinates*<br>487254/5040179 | | |
| | *Interpretation*<br>Quarry? | *Estimated size*<br>Length 2.5 m. | *Date range*<br>LC-EH | |
| **Feature 2** | | *Coordinates*<br>487320/5040167 | | |
| | *Interpretation*<br>Structure | *Estimated size*<br>? | *Date range*<br>LC-EH | |
| **Feature 3** | | *Coordinates*<br>487251/5040150 | | |
| | *Interpretation*<br>Structure | *Estimated size*<br>? | *Date range*<br>LC-EH | |
| **Feature 4** | | *Coordinates*<br>487285/5040145 | | |
| | *Interpretation*<br>Stone structure | *Estimated size*<br>Length 17 m | *Date range*<br>LC-EH | |
| **Feature 5** | | *Coordinates*<br>487356/5040133 | | |
| | *Interpretation*<br>Stone structure | *Estimated size*<br>4 x 12 m | *Date range*<br>LC-EH | |
| **Feature 6** | | *Coordinates*<br>487297/5040114 | | |
| | *Interpretation*<br>Diffused cluster of pits | *Estimated size*<br>200 m$^2$ | *Date range*<br>LC-EH | |
| **Feature 7** | | *Coordinates*<br>487313/5040088 | | |
| | *Interpretation*<br>Structure | *Estimated size*<br>? | *Date range*<br>LC-EH | |
| **Feature 8** | | *Coordinates*<br>487224/5040063 | | |
| | *Interpretation*<br>Stone structure | *Estimated size*<br>5 x 9 m | *Date range*<br>LC-EH | |
| **Feature 9** | | *Coordinates*<br>487266/5040063 | | |
| | *Interpretation*<br>Stone structure and pit house | *Estimated size*<br>10 x 7 and 8 x 5 m | *Date range*<br>LC-EH | |

| | | | | |
|---|---|---|---|---|
| **Feature 10** | | *Coordinates* 487300/5040059 | | |
| | *Interpretation* Cluster of pits | *Estimated size* 197 m² | *Date range* LC-EH | |
| **Feature 11** | | *Coordinates* 487314/5040060 | | |
| | *Interpretation* Stone structure | *Estimated size* 5.5 x 6.5 m | *Date range* LC-EH | |
| **Feature 12** | | *Coordinates* 487396/5040041 | | |
| | *Interpretation* Stone structure | *Estimated size* 7 x 9 m | *Date range* LC-EH | |
| **Feature 13** | | *Coordinates* 487172/5040033 | | |
| | *Interpretation* Stone structure | *Estimated size* 4.5 x 8 m | *Date range* LC-EH | |
| **Feature 14** | | *Coordinates* 487224/5040030 | | |
| | *Interpretation* Stone structure | *Estimated size* 15 x 15 m | *Date range* LC-EH | |
| **Feature 15** | | *Coordinates* 487260/5040022 | | |
| | *Interpretation* Cluster of pits? | *Estimated size* 168 m² | *Date range* LC-EH | |
| **Feature 16** | | *Coordinates* 487301/5040018 | | |
| | *Interpretation* Stone structure and two pits? | *Estimated size* 10 x 18 m | *Date range* LC-EH | |
| **Feature 17** | | *Coordinates* 487232/5040009 | | |
| | *Interpretation* Cluster of pits? | *Estimated size* 76 m² | *Date range* LC-EH | |
| **Feature 18** | | *Coordinates* 487387/5039980 | | |
| | *Interpretation* Stone structure | *Estimated size* 5 x 6 m | *Date range* LC-EH | |

*Topography*
**DSP08-H11-02** is located on the E side of Hill 11, a long slope running from the top of the ridge to the cultivated zone. The site spreads over an area subdivided by three very shallow ravines

*Investigations*
**Extensive survey** A scatter of pottery (GS 3094 and 3218) associated with area of concentrated intense vegetation (see below)
**Intensive survey** A diffused scatter of pottery (20050-55) associated with 10 stone structures in an area of diffused intense vegetation. There is an area of more concentrated intense vegetation to the N on the tip of the promontory (Feature 2). A possible quarry is located to the NW (Feature 15). Traces of three possible stone structures were found on another promontory approximately 80 m to the NW along the E edge of the ravine

| **Geomagnetic survey**<br>T. Smekalova<br>*Dimensions* 3.5 ha | *Description of measuring conditions*<br>Few dipole anomalies |
|---|---|

*Interpretation of features* The data revealed 11 poorly defined anomalies on either side of a shallow ravine. A strong anomaly was chosen for excavation. The long linear anomalies which cut cross the site are remains of recent agricultural processes

**Excavation** S. Lancov and P. Gecko

| **Feature 9** | *Coordinates*<br>487276/5040055 | *Dimensions*<br>5.7 x 4.1 m | *Features*<br>Concentric pits |
|---|---|---|---|

Description of stratigraphy by layer: 0-1.47 m
1) Topsoil: 0-0.2 m
2) Grey ashy layer with many frs. of bones and pottery: 0.2-0.45 m
3) Irregular pit of loamy sand with several floors. Top level: 0.4 m. 1st floor: top level 0.73-0.78 m. 2nd floor top level 0.98-1.17 m. Bottom level: 1.4-1.47 m
4) Sterile loam: top level 0.45-1.47 m

| *Finds*<br>Pottery, animal bone | *Date range*<br>LC-EH | *Interpretation*<br>Pit house |
|---|---|---|

*Site interpretation*
The investigations suggest that the settlement consisted of at least 14 structures dispersed around two shallow ravines. Some of the structures are visible as possible stone structures on the surface (Features 4-5, 8-14, 16 and 18), but at least Feature 9 is associated with a pit house and this may be the case at several features e.g. Features 5, 11 and 16. There appear to be four, possible, large clusters of pits (Features 6, 10, 15 and 17), two of which are relatively closely associated with individual structures (Features 10-11 and Features 14 and 17). The stone structures are poorly preserved, but appear to be less complex than the contemporary stone structures from DSP08-H12-01 and DSP07-H10-01/18-01

| DSP08-H12-01 | *Core units* 20035-36, 20061-63 | *Coordinates* 489515/5039038 | | *Figs.* 4.296-303 |
|---|---|---|---|---|
| | *Adjacent units* H1 and H19 | | | |
| | *Interpretation* Settlement | *Estimated size* 3.6 ha | *Date range* LC-EH/? | |
| **Feature 1** | *Core units* 20036, 20061 | *Coordinates* 489424/5039054 | | |
| | *Interpretation* Wall | *Estimated size* Length 8.5 m | *Date range* LC-EH/? | |
| **Feature 2** | *Core units* 20036 | *Coordinates* 489439/5039053 | | |
| | *Interpretation* Stone structure | *Estimated size* 38 x 24 m | *Date range* LC-EH/? | |
| **Feature 3** | *Core units* 20036 | *Coordinates* 489429/5039003 | | |
| | *Interpretation* Stone structure | *Estimated size* 8 x 11 m | *Date range* LC-EH/? | |
| **Feature 4** | *Core units* 20061-62 | *Coordinates* 489510/5039063 | | |
| | *Interpretation* Stone structure | *Estimated size* 13.5 x 15 m | *Date range* LC-EH/? | |
| **Feature 5** | *Core units* 20061-62 | *Coordinates* 489498/5039017 | | |
| | *Interpretation* Complex stone structure | *Estimated size* 38 x 22 m | *Date range* LC-EH/? | |
| **Feature 6** | *Core units* 20063 | *Coordinates* 489576/5039075 | | |
| | *Interpretation* Stone structure | *Estimated size* 8 x 8 m | *Date range* LC-EH/? | |
| **Feature 7** | *Core units* 20063 | *Coordinates* 489592/5039080 | | |
| | *Interpretation* Stone structure | *Estimated size* 10 x 9 m | *Date range* LC-EH/? | |
| **Feature 8** | *Core units* 20063 | *Coordinates* 489601/5039063 | | |
| | *Interpretation* Stone structure | *Estimated size* 9 x 14 m | *Date range* LC-EH/? | |
| **Feature 9** | *Core units* 20063 | *Coordinates* 489594/5039031 | | |
| | *Interpretation* Curved wall | *Estimated size* Diameter ca. 20 m | *Date range* LC-EH/? | |

| | | | | |
|---|---|---|---|---|
| **Feature 10** | *Core units*<br>20062-63 | *Coordinates*<br>489571/5039012 | | |
| | *Interpretation*<br>Stone structure | *Estimated size*<br>Length 9 m | *Date range*<br>LC-EH/? | |
| **Feature 11** | *Core units*<br>20062-63 | *Coordinates*<br>489560/5038998 | | |
| | *Interpretation*<br>Stone structure | *Estimated size*<br>7 x 7 m | *Date range*<br>LC-EH/? | |
| **Feature 12** | *Core units*<br>20063 | *Coordinates*<br>489584/5038992 | | |
| | *Interpretation*<br>Large cluster of pits | *Estimated size*<br>0.03 ha | *Date range*<br>LC-EH/? | |
| **Feature 13** | *Core units*<br>20063 | *Coordinates*<br>489572/5038966 | | |
| | *Interpretation*<br>Stone structure | *Estimated size*<br>18 x 9 m | *Date range*<br>LC-EH/? | |
| **Feature 14** | *Core units*<br>20062 | *Coordinates*<br>489530/5038962 | | |
| | *Interpretation*<br>Large cluster of pits | *Estimated size*<br>0.05 ha | *Date range*<br>LC-EH/? | |
| **Feature 15** | *Core units*<br>20036, 20061 | *Coordinates*<br>489447/5038978 | | |
| | *Interpretation*<br>Long wall? | *Estimated size*<br>Length 111 m | *Date range*<br>LC-EH/? | |
| **Feature 16** | *Core units*<br>20036, 20061 | *Coordinates*<br>489451/5038888 | | |
| | *Interpretation*<br>Stone structure | *Estimated size*<br>16 x? m | *Date range*<br>LC-EH/? | |
| **Feature 17** | *Core units*<br>20035 | *Coordinates*<br>489387/5039098 | | |
| | *Interpretation*<br>Quarry? | *Estimated size*<br>Length 7 m | *Date range*<br>LC-EH/? | |
| *Topography*<br>**DSP08-H12-01** is located on Hill 12, a small low promontory just N of the ridge top. The site spreads across Hill 12, the W edge of which is cut by a ravine, onto the W slope of Hill 19 | | | | |

*Investigations*

**Intensive survey** 13 structures indicated by slight elevations covered by intense vegetation and linear stone features within a large area of diffused scatter of pottery of mostly probably Late Classical-Early Hellenistic date and flint (20035/08, 20036/10-12, 20061-63 and GS 3208). Generally, the vegetation on the structures in the W part of the site is low shrub (Features 1-5 and 15) whereas the vegetation on the structures in the E part is grassy (6-8, 10-11 and 13). Visible in Feature 5 is a line of stones and a corner in the central part of the elevation, 9 m long. Feature 11 is a square stone structure with double walls built with standing slabs. Feature 1 is associated with an amphora handle possibly of Medieval date and a piece of flint (20036/12 and GS 3208). Feature 4 is associated with an amphora and a handmade sherd both of Hellenistic date (20062/02). Feature 6 is associated with two sherds, one a possible Late Bronze Age handmade one (20063/07-08). The general distribution of finds includes two discrete scatters: a ceramic scatter on the tip of the middle promontory (GS 3209-16) and another one including flint approximately 50 m to the SW in the area of Feature 16 (20035/09-10 and 20036/01-09). The soil in this area is very dark and the grasses more intensive suggesting recent cultivation. There is a possible quarry on the W slope of Hill 19 (Feature 17). Long earth banks in the S and SE part of the site probably associated with modern earth works

| **Geomagnetic survey**<br>Eastern Atlas<br>*Dimensions* 405 x 250 m | *Description of measuring conditions*<br>Rare dipole anomalies. Plough marks dominate the SW corner |
|---|---|

*Interpretation of features* 13 architectural complexes as well as numerous pits are visible in the data. Some structures such as Features 1, 4, 6-8, 10-11 and 16 may consist of single- and double-room structures, although some of them may belong to the same building e.g. Features 6-8. Features 2, 5 and 13 are more complex, multi-room structures. Feature 9 may be a circular pen although there appears to be a corner of a building inside it. Large irregular pits are mostly associated with the architecture, but small regular pits appear in clusters some distance away from any architecture especially on the E side of the settlement (Features 12 and 14). Possible remains of a long wall oriented roughly E-W S of Features 2 and 4 (Feature 15). Linear features derive from topography, dams and/or agriculture. The dominating orientation is NNW-SSE. Terrain steps up to 0.7 m show in the data as negative anomalies, they either result from ravines or recently dug pits. A large field with NE-SW-orientated plough marks is noticeable in the SW part where Feature 16 is located, and there is a road along its NE edge. Few pieces of metal debris (dipole anomalies) are spread over the site

**Augering**

In the S part of DSP08-H12-01, two augerings were conducted in the area of Feature 16. The second layer in Auger 11, between 0.27 and 0.50 m below the surface, was interpreted as a culture layer due to the detection of small pottery frs. This layer was not very thick, 0.23 m in this sample, but it seemed intact. At the centre of DSP08-H12-01, six attempts at augering were conducted with structures visible on the surface, but at all locations, stone was met within 0.25 m below the surface (Auger 13). Finally two augerings were conducted at the centre of two magnetometric anomalies on the lower E slope of Hill 19. In both augers, the second layer was very dark and organic and contained small ceramic inclusions, while there were no signs of burnt soil or charcoal. Both layers can be interpreted as pit fills. Auger 16 was conducted on Hill19, where Eastern Atlas had detected another magnetic anomaly. At this location, no clear pit fill could be detected. The second layer was heterogeneous but did not contain archaeological indicators

| Auger S11: 489425/5038854 | |
|---|---|
| 0-0.50 m | Topsoil, loam, loose, organic material; 10YR 2/2 |
| 0.50-0.85 m | Loam, compact, some stone inclusions, plastic, organic material; 10YR 4/4 |
| 0.85-1.22 m | Loam, compact, few stone inclusions, plant remains at 0.90 m, homogeneous; 10YR 4/3 |
| 1.22-1.55 m | Clayey loam, brown, plastic, limestone inclusions; 10YR 4/3 |
| 1.55-1.80 m | Loam, some lime shells, loose; 10YR 6/4 |
| 1.80 m - | End of auger. Limestone bedrock |
| Auger S12: 489409/5038812 | |
| 0-0.27 m | Topsoil, chernozem; 10YR 2/2, loam, organic material, loose |
| 0.27-0.50 m | Loam, compact, heterogeneous, organic material, a few small pottery frs. included; 10YR 4/3 |
| 0.50-1.05 m | Loam, compact, plastic, homogeneous, some lime shells, small stone inclusions; 10YR 5/4 |

| | |
|---|---|
| 1.05 m - | End of auger. Limestone bedrock |
| Auger S13: 489562/5038993 | |
| 0-0.25 m | Topsoil |
| 0.25 m- | End of auger |
| Auger S14: 489430/5039036 | |
| 0-0.32 m | Topsoil, loam, organic material; 10YR 2/2 |
| 0.32-0.65 m | Loam, heterogeneous, stone inclusions, small pottery frs.; 10YR 3/2 |
| 0.65-0.85 m | Loam, compact, not plastic; 10YR 5/4 |
| 0.85 m- | End of auger |
| Auger S15: 489467/5038989 | |
| 0-0.30 m | Topsoil, loam, loose, organic material; 10YR 2/2 |
| 0.30-0.60 m | Loam, very organic, very hard, heterogeneous, stone inclusions, small shell frs., small pottery frs.; 10YR 2/1 |
| 0.60-0.80 m | Loam, plastic but hard, some limestone inclusions; 10YR 4/4 |
| 0.80 m- | End of auger |
| Auger S16: 489524/5039040 | |
| 0-0.20 m | Topsoil, loam, organic material, loose; 10YR 3/2 |
| 0.20-0.55 m | Loam, with brown spots, compact, some lime inclusions; 10YR 4/3 |
| 0.55-0.80 m | Loam, plastic, stone and limestone inclusions; 10YR 5/4 |
| *Site interpretation*<br>Hellenistic settlement consisting of 11 buildings and numerous household pits. Simple structures with clusters of single-rooms dominate but some clusters may represent structures with multiple rooms. On the NE side of the settlement is a possible circular pen of a diameter of about 25 m, even if there appears to be a corner of a building inside it. Household pits cluster near the buildings but also some distance away from any architecture, especially on the E side of the settlement. Possible remains of a long wall oriented roughly E-W just S of the main settlement area may indicate the edge of the settlement on this side. In this case, the structure in the ploughed field to the S situated almost 100 m away may not belong with the settlement. The date of the pottery collected in this area was inconclusive and it may be of a later date. However, the E-W-running wall is located on the edge of the recently ploughed area and post-depositional processes would have destroyed remains of architecture in the surface. There are no traces of the S building on the surface. Possibly, the settlement is associated with the group of kurgans on the ridge of Hill 19 immediately to the W although these are not visible from the site | |

| **DSP08-H13-01** | *Core units*<br>20049, 20058-60 | *Coordinates*<br>487837/5037883 | | *Figs.*<br>4.285-288 |
|---|---|---|---|---|
| | *Adjacent units*<br>H11, H22-23 | | | |
| | *Interpretation*<br>Settlement? | *Estimated size*<br>2 ha | *Date range*<br>BA?/ LC-EH | |
| **Feature 1** | *Core units*<br>20058 | *Coordinates*<br>487841/5037972 | | |
| | *Interpretation*<br>Possible stone structure | *Estimated size*<br>2 x 3 m | *Date range*<br>BA?/ LC-EH | |
| **Feature 2** | *Core units*<br>20049 | *Coordinates*<br>487876/5037974 | | |
| | *Interpretation*<br>Possible stone structure | *Estimated size*<br>Length 5 m | *Date range*<br>BA?/ LC-EH | |
| **Feature 3** | *Core units*<br>20049 | *Coordinates*<br>487858/5037891 | | |
| | *Interpretation*<br>Possible stone structure | *Estimated size*<br>4 x 4 m | *Date range*<br>BA?/ LC-EH | |
| **Feature 4** | *Core units*<br>20058 | *Coordinates*<br>487836/5037871 | | |
| | *Interpretation*<br>Possible stone structure | *Estimated size*<br>4 x 4 m | *Date range*<br>BA?/ LC-EH | |
| **Feature 5** | *Core units*<br>20058 | *Coordinates*<br>487836/5037861 | | |
| | *Interpretation*<br>Possible stone structure | *Estimated size*<br>6 x 7 m | *Date range*<br>BA?/ LC-EH | |
| **Feature 6** | *Core units*<br>20058 | *Coordinates*<br>487834/5037850 | | |
| | *Interpretation*<br>Possible stone structure | *Estimated size*<br>6 x 5 m | *Date range*<br>BA?/ LC-EH | |

*Topography*
**DSP08-H13-01** is located on Hill 13, a small shallow promontory on the ridge top. Towards the N, the tip of the promontory is flanked by two small peaks

*Investigations*
**Extensive survey** Koltuchov site 15
**Intensive survey** Linear stone features suggesting possible stone structures on either peaks (Features 1-2) and four possible stone structures approximately 75 m S from the tip of the promontory (Features 3-6). The 10 sherds collected (20049, 20059-60) were found mainly S of the possible structures, slightly uphill

| **Geomagnetic survey**<br>Eastern Atlas<br>*Dimensions* 366 x 95 m | *Description of measuring conditions*<br>Prominent plough marks |
|---|---|

*Interpretation of features* The site is dominated by geological and agricultural patterns. Geological patterns show up through a great variation in magnetic data, agricultural patterns through a lot of parallel anomalies. Built structures as well as pits were detected in the N part. There is a roughly rectangular structure with several large irregular pits to the N and S on the E peak (Feature 2). 70 m due S are two rectangular structures with irregular pits inside (Features 3 and 6)

| **Augering** Three augurings were undertaken, two on each peak and one in between. The difference in elevation between the hilltops (Auger 8 and 10) and the lower part in between was approximately 2.50 m. Noteworthy in Auger 8 is the compact heterogeneous layer below the topsoil. No archaeological indicators were detected in this layer, but it may have been a cultural layer. In the layer below, three pieces of bone were found, indicating that the hill has become higher since the deposition. It is uncertain whether this was a natural or an artificial process. The soil in Auger 9 was very hard and compact, making it impossible to continue after 1.40 m. In the third layer, intact plant remains were found. | |
|---|---|
| Auger S8: 487873/5037985 (top of E peak) | |
| 0-0.60 m | Topsoil, loam, loose, not ploughed. Humus; 10YR 5/3 |
| 0.60-1.25 m | Loam, compact, dark stone inclusions, not homogeneous; 10YR 4/4 |
| 1.25-2.10 m | Loam, many limestone shells; 10YR 5/4. Three frs. of bone at 1.50 m below surface |
| 2.10 m- | End of auger |
| Auger S9: 487858/5037980 (next to track, between two hills) | |
| 0-0.50 m | Loam, dark brown, loose, humus, not ploughed; 10YR 3/2 |
| 0.50-0.90 m | Loam, very compact, plastic; 10YR 3/2 |
| 0.90-1.10 m | Loam, very compact, inclusions, limestone, small plant remains; 10YR 4/3 |
| 1.10-1.40 m | Loam, loose, light brown, limestone shells and inclusions; 10YR 4/3 |
| 1.40 m- | End of auger. Soil too hard to continue |
| Auger S10: 487842/5037972 (top of W peak) | |
| 0-0.07 m | Topsoil, loam, loose, not ploughed; 10YR 3/2 |
| 0.07-0.50 m | Loam, loose, no inclusions; 10YR 4/4 |
| 0.50-0.90 m | Loam, compact, limestone inclusions; 10YR 4/4 |
| 0.90-1.05 m | Loam, less compact, limestone shells, small pieces of granite? 10YR 5/4 |
| *Site interpretation*<br>Settlement? | |

| | | | | |
|---|---|---|---|---|
| **DSP08-H14-01** | *Core units*<br>20045-48 | *Coordinates*<br>495341/5042577 | | *Figs.*<br>4.278-284 |
| | *Adjacent units*<br>20021-22, 20043-44, H15 | | | |
| | *Interpretation*<br>Settlement | *Estimated size* | *Date range*<br>BA/ LC-EH | |
| **Individual phases**<br>**BA** | *Core units*<br>20047-48 | *Coordinates*<br>495341/5042577 | | |
| | *Adjacent units*<br>20021-22, 20043-46, H15 | | | |
| | *Interpretation*<br>Settlement | *Estimated size* | *Date range*<br>BA | |
| **Individual phases**<br>**LC-EH** | *Core units*<br>20045-48 | | *Coordinates*<br>495341/5042577 | |
| | *Adjacent units*<br>20021-22, 20043-44, H15 | | | |
| | *Interpretation*<br>? | *Estimated size*<br>? | *Date range*<br>LC-EH | |
| Feature 1 | | *Coordinates*<br>495320/5042622 | | |
| | *Interpretation*<br>Stone structure? | *Estimated size*<br>? | *Date range*<br>BA | |
| Feature 2 | | *Coordinates*<br>495310/5042595 | | |
| | *Interpretation*<br>Stone structure with stone cist | *Estimated size*<br>? | *Date range*<br>BA | |
| Feature 3 | | *Coordinates*<br>495326/5042548 | | |
| | *Interpretation*<br>Stone structure? | *Estimated size*<br>? | *Date range*<br>BA | |
| Feature 4 | | *Coordinates*<br>495335/5042522 | | |
| | *Interpretation*<br>Stone structure? | *Estimated size*<br>? | *Date range*<br>BA | |
| Feature 5 | | *Coordinates*<br>495300/5042507 | | |
| | *Interpretation*<br>Stone structure? | *Estimated size*<br>? | *Date range*<br>BA | |
| Feature 6 | | *Coordinates*<br>495316/5042484 | | |
| | *Interpretation*<br>Stone structure? | *Estimated size*<br>? | *Date range*<br>BA | |

*Topography*
**DSP08-H14-01** is located on a wide ravine between two promontories at the bottom of the hillsides. The area was probably ploughed in the recent past as indicated by the loose, dark soils

*Investigations*
**Extensive survey** S. Lancov, S. Koltuchov possibly in autumn 2007
**Intensive survey** A discrete scatter of handmade sherds was concentrated in an area approximately 50 m wide, between lines 20047 and 20048 in an area of high grasses. However, only three of these could be dated with certainty to the Bronze Age (20047-02); there is a more diffused scatter of Hellenistic material and possibly later e.g. a reused amphora handle (20043, 20045, 20047-02, 20048)

| **Geomagnetic survey**<br>T. Smekalova<br>*Dimensions* 2.5 ha | *Description of measuring conditions*<br>Some dipole anomalies |
|---|---|

*Interpretation of features* Weak anomalies of six possible structures on either side of a shallow ravine. The long linear anomalies which cut cross the site are remains of recent agricultural processes

**Excavation** S. Lancov and P. Gecko

| **Feature 2**<br>495304/5042593 | *Coordinates* | *Trench dimensions*<br>2 x 2 m | *Features*<br>Rectangular stone cist of stone slabs and possible collapsed wall |
|---|---|---|---|

Description of stratigraphy by layer: 0-0.85 m
1) Topsoil empty of archaeological artefacts: 0-0.20 m
2) Brown loam with tiny rock frs. empty of archaeological artefacts: 0.20-0.50/60 m
3) Stone debris: top level 0.34 m. Bottom level: 0.79 m (FL 3-6)
4) Ashy layer: top level 0.50-0.60 m. Stone cist: 1.2 x 0.9 m built of stone slabs 3-5 cm thick, 5-10 cm high directly on the bedrock. Natural shallow depression in the centre filled with loam and rubble (FL 1-2)
5) Bedrock: top level 0.79-0.85 m

| *Finds*<br>Handmade pottery, animal bones | *Date range*<br>BA | *Interpretation*<br>Stone structure with stone cist |
|---|---|---|

*Site interpretation*
BA settlement possibly consisting of six stone structures; however, the area appears to have been resettled during the LC-EH period, and some of the structures may date to this period

| **DSP08-H15-01** | *Core units*<br>20026 | *Coordinates*<br>495464/5042459 | | *Figs.*<br>4.264-265 |
|---|---|---|---|---|
| | *Adjacent units*<br>20025, 20027-28. H14 | | | |
| | *Interpretation*<br>Farmstead | *Estimated size*<br>54 x 41 m | *Date range*<br>Modern | |

*Topography*
**DSP08-H15-01** is located on the W slope of Hill 15, a wide low promontory at the bottom of the hillsides

*Investigations*
**Extensive survey** Koltuchov site 36
**Intensive survey** Large angular patch of intense vegetation. No pottery directly associated, but modern building debris

*Site interpretation*
Near contemporary settlement

<table>
<tr><td rowspan="3">DSP08-H16-01</td><td>Core units<br>20008-9</td><td colspan="2">Coordinates<br>493049/5042275</td><td>Figs.<br>4.243, 4.245</td></tr>
<tr><td colspan="3">Adjacent units<br>H08</td><td></td></tr>
<tr><td>Interpretation<br>Settlement</td><td>Estimated size<br>0.4 ha</td><td>Date range<br>Unknown</td><td></td></tr>
<tr><td rowspan="2">Feature 1</td><td>Core units<br>20009</td><td>Coordinates<br>493069/5042257</td><td></td><td></td></tr>
<tr><td>Interpretation<br>Ditches?</td><td>Estimated size<br>Diameter ca. 11 m</td><td>Date range<br>Unknown</td><td></td></tr>
<tr><td rowspan="2">Feature 2</td><td>Core units<br>20008</td><td>Coordinates<br>493037/5042247</td><td></td><td></td></tr>
<tr><td>Interpretation<br>Stone structure</td><td>Estimated size<br>24 x 30 m</td><td>Date range<br>Unknown</td><td></td></tr>
<tr><td rowspan="2">Feature 3</td><td>Core units<br>20008</td><td>Coordinates<br>493021/5042232</td><td></td><td></td></tr>
<tr><td>Interpretation<br>Stone structure</td><td>Estimated size<br>7.5 x 7 m</td><td>Date range<br>Unknown</td><td></td></tr>
<tr><td rowspan="2">Feature 4</td><td>Core units<br>20008-9</td><td>Coordinates<br>493053/5042247</td><td></td><td></td></tr>
<tr><td>Interpretation<br>Enclosure</td><td>Estimated size<br>100 x =/>100 m</td><td>Date range<br>Unknown</td><td></td></tr>
</table>

*Topography*
**DSP08-H16-01** is located on the SE slope of Hill 16, a high short promontory near the bottom of the hillsides

*Investigations*
**Intensive survey** Three long shallow elevations running E-W at irregular intervals. No finds

**Geomagnetic survey**
Eastern Atlas
*Dimensions* 630 x 53 m

*Description of measuring conditions*
Metal debris is randomly spread over the site

*Interpretation of features* Square structure with multiple internal walls surrounded on three sides by large irregular pits (Feature 2). SW of this is a rectangular structure with small pits inside and large irregular pit outside the S wall (Feature 3). NE of Feature 1 are two concentric U-shaped ditches. This entire complex is surrounded by shallow banks on three sides. Open towards the W (Feature 4). Additional E-W running bank on the E side. Linear anomalies close to the archaeological findings could be due to farming

*Site interpretation*
Agricultural settlement? Date unknown

| DSP08-H17-01 | *Core units*<br>20002 | *Coordinates*<br>491253/5041648 | | *Fig.*<br>4.233, 4.237-238 |
|---|---|---|---|---|
| | *Adjacent units*<br>20007, H06-07 | | | |
| | *Interpretation*<br>Storage site? | *Estimated size*<br>0.1 ha | *Date range*<br>Unknown | |
| Feature 1 | *Core units*<br>20002 | *Coordinates*<br>491259/5041642 | | |
| | *Interpretation*<br>Circular anomaly | *Estimated size*<br>Diameter 16.5 m | *Date range*<br>Unknown | |
| Feature 2 | *Core units*<br>20002 | *Coordinates*<br>491207/5041645 | | |
| | *Interpretation*<br>Circular anomaly | *Estimated size*<br>Diameter 6.5 m | *Date range*<br>Unknown | |

*Topography*
**DSP08-H17-01** is located on the central part of Hill 17, a long wide promontory running from the top of the ridge to the cultivated zone

*Investigations*
**Intensive survey** No anomalies were observed on the surface
A grab sample (GS 3204) was collected 60 m N of the site

| **Geomagnetic survey**<br>Eastern Atlas<br>*Dimensions* 205 x 200 m | *Description of measuring conditions*<br>Metal debris is spread over the area |
|---|---|

*Interpretation of features* The area is pervaded by road anomalies crossing NW-SE. The highest density is noticeable along the linear anomalies in the E. Some weak circular patterns (Features 1-2) are visible in the NE corner as well as some possible anomalies displaying cavities – these features could be of archaeological origin (e.g. kurgans, pits)

*Site interpretation*
Possible archaeological features, e.g. kurgans or pits

| **DSP08-H17-02** | *Core units*<br>20003, 20006, H17 | *Coordinates*<br>491131/5041648 | | *Figs.*<br>4.239-240 |
|---|---|---|---|---|
| | *Adjacent units*<br>H06-07 | | | |
| | *Interpretation*<br>Group of kurgans | *Estimated size*<br>– | *Date range*<br>LC-EH? | |
| **Feature 1** | *Core units*<br>H17 | *Coordinates*<br>491034/5041477 | | |
| | *Interpretation*<br>Kurgan | *Estimated size*<br>Diameter 21 m | *Date range*<br>LC-EH? | |
| **Feature 2** | *Core units*<br>H17 | *Coordinates*<br>491082/5041433 | | |
| | *Interpretation*<br>Kurgan | *Estimated size*<br>Diameter 17 m | *Date range*<br>LC-EH? | |
| **Feature 3** | *Core units*<br>20003 | *Coordinates*<br>491132/5041398 | | |
| | *Interpretation*<br>Kurgan | *Estimated size*<br>Diameter 34 m | *Date range*<br>LC-EH? | |
| **Feature 4** | *Core units*<br>20003 | *Coordinates*<br>491152/5041372 | | |
| | *Interpretation*<br>Kurgan | *Estimated size*<br>Diameter 19 m | *Date range*<br>LC-EH? | |
| **Feature 5** | *Core units*<br>20003, 20006 | *Coordinates*<br>491216/5041318 | | |
| | *Interpretation*<br>Kurgan | *Estimated size*<br>Diameter 15 m | *Date range*<br>LC-EH? | |

*Topography*
**DSP08-H17-02** is located on the central part of Hill 17, a long wide promontory running from the top of the ridge to the cultivated zone

*Investigations*
**Intensive survey** Five grassy elevations associated with five sherds of which one is a Thasian amphora and one is probably Early Modern and a piece of flint (20003/01-04 and GS 3205-6)

*Site interpretation*
Group of kurgans

<table>
<tr><td rowspan="3">DSP08-H17-03</td><td>Core units<br>H17</td><td colspan="2">Coordinates<br>491235/5040946</td><td>Fig.<br>4.241</td></tr>
<tr><td>Adjacent units<br>20004-5, H06-07</td><td colspan="2"></td><td></td></tr>
<tr><td>Interpretation<br>Kurgan</td><td>Estimated size<br>Diameter 10 m</td><td>Date range<br>Unknown</td><td></td></tr>
<tr><td colspan="5">Topography<br>DSP08-H17-03 is located on the central part of Hill 17 300 m S of DSP08-H17-02</td></tr>
<tr><td colspan="5">Investigations<br>Extensive survey Grassy elevation located S of the intensively surveyed area</td></tr>
<tr><td colspan="2">Geomagnetic survey<br>Eastern Atlas<br>Dimensions 180 x 100 and 80 x 50 m</td><td colspan="3">Description of measuring conditions</td></tr>
<tr><td colspan="5">Interpretation of features The kurgan shows up in the magnetic data as a circular and very strong positive anomaly. Linear anomalies deriving from tracks</td></tr>
<tr><td colspan="5">Site interpretation<br>Kurgan</td></tr>
</table>

| **DSP07-H18-01** | *Core units*<br>H18 | *Coordinates*<br>482307/5038002 | | *Figs.*<br>4.200-201 |
|---|---|---|---|---|
| | *Adjacent units*<br>H10 | | | |
| | *Interpretation*<br>Settlement | *Estimated size*<br>=/>100 x 50 m | *Date range*<br>LC-EH | |
| **Feature 1** | *Core units*<br>H18 | *Coordinates*<br>482311/5038049 | | |
| | *Interpretation*<br>Cluster of household pits | *Estimated size*<br>80 m² | *Date range*<br>LC-EH | |
| **Feature 2** | *Core units*<br>H18 | *Coordinates*<br>482280/5038041 | | |
| | *Interpretation*<br>Stone structure | *Estimated size*<br>=/>18 x 12 m | *Date range*<br>LC-EH | |
| **Feature 3** | *Core units*<br>H18 | *Coordinates*<br>482301/5038034 | | |
| | *Interpretation*<br>Stone structure | *Estimated size*<br>5 x 10 m | *Date range*<br>LC-EH | |
| **Feature 4** | *Core units*<br>H18 | *Coordinates*<br>482310/5037989 | | |
| | *Interpretation*<br>Complex stone structure | *Estimated size*<br>15 x 17 m | *Date range*<br>LC-EH | |
| **Feature 5** | *Core units*<br>H18 | *Coordinates*<br>482342/5037963 | | |
| | *Interpretation*<br>Stone structure | *Estimated size*<br>=/>9 x 13 m | *Date range*<br>LC-EH | |
| **Feature 6** | *Core units*<br>H18 | *Coordinates*<br>482328/5037957 | | |
| | *Interpretation*<br>Cluster of household pits | *Estimated size*<br>95 m² | *Date range*<br>LC-EH | |

*Topography*
**DSP07-H18-01** is located on the small E promontory of the broad Hill 18 in the lower part of the hillsides

*Investigations*
**Extensive survey** Pottery (see DSP07-H10-01) associated with four possible stone structures (Features 2-5). Feature 4 is at least partly built with standing slabs

| **Geomagnetic survey**<br>T. Smekalova<br>*Dimensions* 50 x 110 and 20 x 25 m | *Description of measuring conditions*<br>Good |
|---|---|

*Interpretation of features* Four rectangular stone structures: Features 2 and 4-5 have multiple rooms, Feature 4 with three pits inside while Feature 3 appears to be a single-room structure. Features 3 and 5 and Features 2 and 4 appear to have similar orientation. Two clusters of household pits are located to the N (Feature 1) and E of Feature 5 (Feature 2). There are also a few diffused pits. The long linear anomalies are remains of recent agricultural processes

*Site interpretation*
Small settlement including three small structures and several household pits. Probably associated with DSP07-H10-01, located just across a shallow ravine 140 m to the E

| **DSP07-H19-01** | | *Coordinates*<br>489257/5038946 | | *Figs.*<br>4.204-205,<br>4.213-214 |
|---|---|---|---|---|
| | *Interpretation*<br>Kurgan group | *Estimated size*<br>– | *Date range*<br>Unknown | |

*Description*
This group consists of four kurgans, three of which were excavated in 1971-1972. In all cases the earthen embankments of the kurgans are reflected in the magnetic field as positive anomalies, because the material of the soil is slightly magnetic

*Investigations*
**Geomagnetic survey**
T. Smekalova
There are three local positive magnetic anomalies SE and NW kurgan 3, which are caused, possibly, by the magnetic soil filling of the natural (?) depressions in the non-magnetic limestone bedrock. There are no other anomalies, which could be created by archaeological features in the space between the kurgans there

| **Kurgan 1** | *Coordinates (centre)*<br>489259/5038884 | *Diameter*<br>20 m | *Status*<br>Excavated (and restored to original height?) |
|---|---|---|---|
| | *Architecture*<br>Chamber tomb in situ | *Surface finds*<br>None | *Geomagnetic survey*<br>T. Smekalova |

*Description*
Kurgan 1 is a high mound that is gradually sloping on its E and W side and more steeply on its N and S side. It has been excavated and plans have, according to Koltuchov (pers. comm.), been published. The architecture consists of a chamber tomb situated in the central part of the mound that is reasonably well preserved. It has a narrow entrance on the S side lined by huge flat-lying limestone slabs that give out on a substantial chamber lined by three huge, still-standing orthostats. Presumably, the head of the deceased was located at the W end of the trench where the largest orthostat was placed. The chamber itself measures 2.80 in length. Length from entrance to the outer lateral back wall of the chamber is 3.10 m. The mound itself has numerous small stones at its surface. Halfway between top and foot of the mound larger blocks are visible that may be the remains of a stone circle. If so, this circle would be situated approximately 4 m inward from the lower outline of the mound. The mound itself measures approximately 20 m in diameter. Recently, a number of stones have been removed from the mound. No artefacts were observed at the surface. The orientation of the chamber itself is 90°, i.e. E-W

*Interpretation of geomagnetic features*
The round earthen embankment of the kurgan is visible as a positive anomaly.

| **Kurgan 2** | *Coordinates (centre)*<br>489261/5038924 | *Diameter*<br>Ca. 18 m | *Status*<br>Excavated (and restored to original height?) |
|---|---|---|---|
| | *Architecture*<br>Chamber tomb in situ | *Surface finds*<br>None | *Geomagnetic survey*<br>T. Smekalova |

*Description*
Kurgan 2 is a high mound that is gradually sloping on its S side and more steeply on its N, W and E side. The mound has been excavated and plans have, according to Koltuchov (pers. comm.), been published. Large stones from the grave chamber and/or krepis wall or stone circle have been deposited NW of the mound after excavation. In the central part of the mound is the grave chamber. Like Kurgan 1, it has a narrow entrance lined with flat-lying stones, while the chamber itself is lined with large orthostats. The entrance is orientated S-N which is almost the same orientation as Kurgan 1. All over the surface of the mound, many small limestones are visible and halfway the slope is quite a number of large blocks that seem in situ (remains of a stone circle?). The grave architecture of the chamber repeats that of Kurgan 1 as to its construction. The length of chamber of Kurgan 2 is ca. 2.80 and its width is between 1.40 and 1.50 m. The E flat-lying stone of the chamber of Kurgan 2 has been displaced about 0.4 m. The width of the entrance is 0.55-0.6 m. These are the same dimensions as Kurgan 1. The publications will have to show how the chambers were closed. Both Kurgan 1 and 2 have a large slab lying loose at the entrance (apparently replaced by the excavators). Apparently, this is a closing or covering slab. Its orientation is 100° i.e. slightly S of E-W

<table>
<tr><td colspan="4">Interpretation of geomagnetic features<br>The earthen embankment of the kurgan is clearly visible as a positive anomaly its soil being slightly magnetic. Its form, however, shows up as triangular. The contour of the embankment is visible best in the grey-scale map.</td></tr>
<tr><td rowspan="2">Kurgan 3</td><td>Coordinates (centre)<br>489257/5038965</td><td>Diameter<br>22-24 m</td><td>Status<br>Excavated</td></tr>
<tr><td>Architecture<br>Remains of chamber tomb not in situ, stone circle?</td><td>Surface finds<br>Unnumbered grab sample collected by I. Florja incl. glazed cup</td><td>Geomagnetic survey<br>T. Smekalova</td></tr>
<tr><td colspan="4">Description<br>The mound is comparable in size and height with Kurgans 1 and 2. It has been excavated but no grave architecture has been preserved. On top of the mound, limestone slabs lie about indicating that this mound had a grave chamber like Kurgans 1 and 2. Halfway up the mound some large blocks are visible in situ that may be the remains of a stone circle. Unlike the other two kurgans, this mound has not been reconstructed to its original height.</td></tr>
<tr><td colspan="4">Interpretation of geomagnetic features<br>The earthen embankment of the kurgan is clearly visible as a positive anomaly. The contour of the embankment and grave architecture is visible best in the grey-scale map. Anomalies to the NE, NW and SW are interpreted as pits or depressions filled with magnetic soil</td></tr>
<tr><td colspan="4">Excavation S. Lancov, S. Koltuchov</td></tr>
<tr><td>Geomagnetic anomaly</td><td>Coordinates<br>489270/5038955</td><td>Trench dimensions<br>2 x 2 m</td><td>Features<br>Two large stones but no clear features</td></tr>
<tr><td colspan="4">Description of stratigraphy by layer, reconstruction based on FL 7:<br>1) Top soil: 0-0.55 m (FL 7:1-3)<br>2) The second stratum starts at 0.55 m, and two large stones as well as an amphora fr. was found on top of it (FL 7:4)<br>3) Bedrock. Top level: 0.85 m (FL 7:5)</td></tr>
<tr><td>Finds<br>Pottery</td><td>Date range<br>LC-EH</td><td colspan="2">Interpretation<br>Remains of ritual activity?</td></tr>
<tr><td rowspan="2">Kurgan 4</td><td>Coordinates (centre)<br>489270/5039016</td><td>Diameter<br>18-19 m</td><td>Status<br>Excavated or looted</td></tr>
<tr><td>Architecture<br>No visible remains</td><td>Surface finds<br>None</td><td>Geomagnetic survey<br>T. Smekalova</td></tr>
<tr><td colspan="4">Description<br>This lowest mound among the line-up of four mounds has been excavated in the centre. Here some large limestone blocks lie about. All over the surface of the mound there are many small limestones visible. This mound has no grave architecture left. Only one stone maybe still in situ (part of wall of a cist or a chamber?)</td></tr>
<tr><td colspan="4">Interpretation of geomagnetic features<br>The earthen embankment of the kurgan is just visible as a positive anomaly. Its precise form is, however, hard to make out</td></tr>
<tr><td colspan="4">Site interpretation<br>Group of kurgans</td></tr>
</table>

<table>
<tr><td rowspan="2">DSP07-H19-02</td><td>Core units</td><td>Coordinates<br>489280/5039151</td><td></td><td>Figs.<br>4.206, 4.215-216</td></tr>
<tr><td>Interpretation<br>Kurgan group</td><td>Estimated size<br>–</td><td>Date range<br>Unknown</td><td></td></tr>
<tr><td colspan="5">Description<br>Three kurgans belong to the middle group. Here again as on the geomagnetic maps of the previous group, there are no ground burials in the space between the kurgans. The stone circles, surrounded the earthen embankments are reflected in the magnetic field as weak negative anomalies. In the mappings several strong signals were recorded caused by iron objects. There are some local positive anomalies to the N of the kurgan 7, which one could connect with natural depressions in the bedrock, filled with soil. Linear anomalies from the modern earthen road are visible at the NE corner of the plot rendered in yellow</td></tr>
<tr><td rowspan="2">Kurgan 5</td><td>Coordinates (centre)<br>(489290/5039117)</td><td>Diameter<br>11-13 m</td><td colspan="2">Status<br>Excavated/looted</td></tr>
<tr><td>Architecture<br>Present, but not in situ; grave chamber? Stone circle?</td><td>Surface finds<br>none</td><td colspan="2">Geomagnetic survey<br>T. Smekalova</td></tr>
<tr><td colspan="5">Description<br>Low mound with on its E side some blocks preserved that may indicate the presence of a stone circle. In the centre lie a number of large blocks about, one of which is of the orthostat slab type. This would indicate the presence of a grave chamber with cist and/or dromos. The mound has been excavated or looted. The slopes of the mound are covered with typical limestone rubble</td></tr>
<tr><td colspan="5">Interpretation of geomagnetic features<br>The round earthen embankment of the kurgan is visible as a positive anomaly</td></tr>
<tr><td rowspan="2">Kurgan 6</td><td>Coordinates (centre)<br>489272/5039150</td><td>Diameter<br>16 m</td><td colspan="2">Status<br>Excavated/looted</td></tr>
<tr><td>Architecture<br>Remains of a grave chamber, stone circle</td><td>Surface finds<br>None</td><td colspan="2">Geomagnetic survey<br>T. Smekalova</td></tr>
<tr><td colspan="5">Description<br>Kurgan 6 is located ca. 20 m N of Kurgan 5. There is evidence for a krepis wall in the form of large blocks that can be made out at the surface at ca. 3 m inwards from where the slope of the kurgan starts. Part of the grave chamber is preserved. It has an E-W orientation with one large orthostat at its W end and two orthostats perpendicular to it. There is no evidence for a dromos, although this may well have existed. Where the entrance can be postulated, there is now rubble and soil. There is a possibility that, if the tomb was looted (and not excavated), the robbers concentrated on the grave chamber itself. The chamber has a width between 1.08 and 1.9 m. Judging from the position of a stone in situ, which may have been part of the back wall of the chamber, the length of the chamber may have been 3.40 m at maximum. The orientation of the grave chamber is 50°, i.e. N of E-W</td></tr>
<tr><td colspan="5">Interpretation of geomagnetic features<br>The round earthen embankment of the kurgan can be made out. Within the interpreted perimeter appears a strong signal probably caused by an iron object</td></tr>
<tr><td rowspan="2">Kurgan 7</td><td>Coordinates (centre)<br>489300/5039194</td><td>Diameter<br>15-16 m</td><td colspan="2">Status<br>Excavated/looted?</td></tr>
<tr><td>Architecture<br>Stone circle?</td><td>Surface finds<br>None</td><td colspan="2">Geomagnetic survey<br>T. Smekalova</td></tr>
<tr><td colspan="5">Description<br>Kurgan 7 is the northernmost in this row. It is a large low mound with in its centre a slight depression (attempt at looting?). Outside the perimeter of the mound lie a number of large blocks about, apparently from the mound. Some blocks at ca. 2-2.5 m inwards from the outer perimeter may be evidence for the existence of a stone circle/krepis wall. The mound has a fairly regular form and does not seem to have been extensively damaged or looted</td></tr>
</table>

*Interpretation of geomagnetic features*
The round earthen embankment of the kurgan can be made out. To the W, N and NE, three anomalies occur interpreted as pits or depressions filled with magnetic soil

*Site interpretation*
Group of kurgans

| **DSP07-H19-03** | *Core units*<br>20035 | *Coordinates*<br>489267/5039496 | | *Figs.*<br>4.207, 4.217-218 |
|---|---|---|---|---|
| | *Adjacent units*<br>20036, H01, H03, H12 and H24 | | | |
| | *Interpretation*<br>Group of kurgans | *Estimated size* | *Date range*<br>LC-EH? | |

*Topography*
**DSP07-H19-03** is located on Hill 19, a long slope running from the top of the ridge to the cultivated zone

*Investigation*
**Intensive survey** Line of four grassy elevations just N of Hill 12 (for detailed descriptions, see kurgan catalogue). 12 sherds collected in the area including three Hellenistic transport amphora frs. (20035/05-06 and 20036/14-18)
**Topographical survey** The N group of barrows consists of three kurgans. The northernmost kurgan has been excavated, which is reflected in the magnetic field as a circular positive anomaly. The two remaining kurgans give us only weak circular negative anomalies, most probably, from the stone walls surrounding the embankments. Again, there are no anomalies, which could be caused by ground burials in the space between the barrows. There are linear anomalies from the modern earth road at the NW part of the plot

| **Kurgan 8** | *Coordinates (centre)*<br>489294/5039404 | *Diameter*<br>18 m | *Status*<br>Looted |
|---|---|---|---|
| | *Architecture*<br>Grave chamber | *Surface finds*<br>None | *Geomagnetic survey*<br>T. Smekalova |

*Description*
Large mound with on its slopes the characteristic limestone rubble exposed in a very dense packing. Its top is flat and it appears that the grave chamber has been plundered and damaged. A probable head stone, however, still protrudes from the mound and seems to be in situ. The orthostat slabs lining the chamber have, however, been removed and lie tilted and out of place. On the N slope, a few exposed blocks lie in situ and may indicate the position of a stone circle. These blocks are situated ca. 3.50 m inwards from the outer boundary of the mound. For this mound it is not possible to give dimensions of the grave chamber and there is no sign of an entrance of the dromos-type. Where it can be postulated, no digging has been taken place. This grave may have been of the grave chamber and lateral dromos type. The orientation of the grave chamber is 70°, i.e. slightly N of E-W

*Interpretation of geomagnetic features*
Of this kurgan only a very weak circular anomaly could be made out. To the S, various strong signals indicate the presence of (recent) iron objects

| **Kurgan 9** | *Coordinates (centre)*<br>489284/5039477 | *Diameter*<br>16 m | *Status*<br>Excavated/looted |
|---|---|---|---|
| | *Architecture*<br>Possibly a grave chamber? | *Surface finds*<br>None | *Geomagnetic survey*<br>T. Smekalova |

<table>
<tr><td colspan="4">Description<br>Kurgan 9 is a rather flat low mound, apparently looted and/or excavated. On the surface there lie many limestone blocks about that are not in situ. The lower slopes are characterized by a thin cover of limestone rubble. From the central part of the mound protrudes one orthostat that may constitute the N wall of the grave chamber. Based on the orientation of this orthostat (if), the orientation of this grave chamber would be 60°. The position of the orthostat has been measured in. Although some stones lie about, this mound has no clear indications for the existence of a stone circle/krepis wall</td></tr>
<tr><td colspan="4">Interpretation of geomagnetic features<br>Of this kurgan only a weak circular anomaly could be made out. Within it a strong signal indicates the presence of a (recent?) iron object</td></tr>
<tr><td rowspan="2">Kurgan 10</td><td>Coordinates (centre)<br>489257/5039511</td><td>Diameter<br>16-20 m</td><td>Status<br>Excavated (excavators unknown)</td></tr>
<tr><td>Architecture<br>Remains of a grave chamber, possibly a dromos type entrance, stone circle / krepis</td><td>Surface finds<br>None</td><td>Geomagnetic survey<br>No</td></tr>
<tr><td colspan="4">Description<br>Kurgan 10 was excavated by archaeologists: it is, however, not known when or by whom. The outer perimeter of the mound seems untouched and still has the characteristic surface cover of limestone rubble. This mound has the best evidence for the existence of a krepis. A regular circular pattern of stones is exposed at the surface and was measured in stone by stone. There also seems to be another stone circle, but this is not certain as many blocks lie loose at the surface and they may have been removed from the grave chamber itself. The architecture of the grave chamber itself consists of four orthostats that, save for one in the N wall and the possible "head" stone itself, are tilted, either in- or outwards. Two stones probably indicate an entrance of dromos type, which, as is usual, would have been located in a N-S orientation. The orientation of the grave chamber is 70°</td></tr>
<tr><td colspan="4">Interpretation of geomagnetic features<br>(Figs. 45-47) The earthen embankment of this excavated kurgan appears clearly on the geomagnetic map. To the NE, a pit or depression and to the W linear anomalies of the modern track. To the W, just outside the embankment, appears a strong signal of a (probably recent) iron object</td></tr>
<tr><td rowspan="2">Kurgan 11</td><td>Coordinates (centre)<br>489260/5039558</td><td>Diameter<br>13 m</td><td>Status</td></tr>
<tr><td>Architecture<br>None visible</td><td>Surface finds<br>None</td><td>Geomagnetic survey<br>No</td></tr>
<tr><td colspan="4">Description<br>Kurgan 11 is a low, fairly small mound. It is covered at the surface with a dense packing of limestone rubble. Some large blocks lie loose about and two are protruding from the mound. These do, however, not seem to indicate the grave chamber. The mound looks as if it is untouched having no depression at its centre</td></tr>
<tr><td colspan="4">Site interpretation<br>Group of kurgans</td></tr>
</table>

| **DSP07-H19-04** | *Core units* | *Coordinates*<br>489110/5040063 | | *Figs.*<br>4.208, 4.219-220 |
|---|---|---|---|---|
| | *Interpretation*<br>Group of kurgans | *Estimated size* | *Date range*<br>LC-EH? | |

*Description*
Group IV consists of two pairs of two kurgans. The group is located at a distance of ca. 400 m downslope from Group III. No geomagnetic maps were made of this group, and consequently we have no information on any associated buried features

| **Kurgan 12** | *Coordinates (centre)*<br>489085/50400231 | *Diameter*<br>13 m | *Status* |
|---|---|---|---|
| | *Architecture*<br>Grave chamber? | *Surface finds*<br>None | *Geomagnetic survey*<br>No |

*Description*
Kurgan 12 is a low mound of ca. 13 m in diameter. It has a moderately dense packing of limestone rubble at the surface. In the centre, the mound has a depressed area, indicating that it may have been topped off by looters. If so, this has not been done recently, however. There are three stones at the surface that may indicate the existence of a grave chamber. One may be an orthostat and possible "head' stone, a second one could constitute one of the orthostats lining the N back wall of the chamber. The 'head' stone does not seem to lie in situ, the one that possibly lines the N wall may be in situ and as such might furnish the orientation

| **Kurgan 13** | *Coordinates (centre)*<br>489138/5040027 | *Diameter*<br>11-13 m | *Status* |
|---|---|---|---|
| | *Architecture*<br>Stone circle? | *Surface finds*<br>None | *Geomagnetic survey*<br>No |

*Description*
Kurgan 13 is a low mound with the typical limestone rubble at the surface in a quite dense packing. In the centre there lies a heap of limestone blocks. Some large blocks may indicate the existence of a stone circle/krepis wall wall. This mound has no indicators for a grave chamber

| **Kurgan 14** | *Coordinates (centre)*<br>489098/5040097 | *Diameter*<br>10 m | *Status* |
|---|---|---|---|
| | *Architecture*<br>None | *Surface finds*<br>None | *Geomagnetic survey*<br>No |

*Description* Kurgan 14 is a small low mound, covered by a dense limestone rubble packing. No signs of orthostats protruding from the mound or blocks indicating the existence of an inner stone circle

| **Kurgan 15** | *Coordinates (centre)*<br>489131/5040114 | *Diameter*<br>21 m | *Status*<br>Possibly intact |
|---|---|---|---|
| | *Architecture*<br>Grave chamber? Stone circle? | *Surface finds*<br>None | *Geomagnetic survey*<br>No |

*Description*
Kurgan 15 is a large, rather low mound, covered with the typical limestone rubble in a quite dense packing. In the centre there is a depression (looted?). Part of a large stone is protruding from the W part of the central area of the mound, probably an orthostat. It is tilted, but seems a clear indicator of a chamber or cist. There are some flat blocks surfacing ca. 4 m inward from the outer perimeter that may indicate a stone circle/krepis wall (or rather inner stone circle)

*Site interpretation*
Group of kurgans

| **DSP07-H19-05** | *Core units* | *Coordinates*<br>488620/5040904 | | *Figs.*<br>4.209, 4.221 |
|---|---|---|---|---|
| | *Interpretation*<br>Kurgan group | *Estimated size*<br>– | *Date range*<br>Unknown | |
| *Description* This is a group of two kurgans on the lower end of the hill situated close together. | | | | |
| **Kurgan 16** | *Coordinates (centre)*<br>488597/5040918 | *Diameter*<br>16 m | *Status*<br>Excavated/looted? | |
| | *Architecture*<br>Grave chamber? Stone circle? | *Surface finds*<br>None | *Geomagnetic survey*<br>No | |
| *Description*<br>Kurgan 16 is a rather low mound covered with a dense packing of small limestone rubble. The kurgan has a depression in the middle. At ca. 3.50 m inwards from the perimeter there are stones visible that may indicate the existence of an inner stone circle. If so, then we would in this particular case deal with orthostats as part of the stone circle. Near the centre is a large orthostat that may indicate the existence of a grave chamber or cist | | | | |
| **Kurgan 17** | *Coordinates (centre)*<br>488643/5040889 | *Diameter*<br>18 m | *Status*<br>Possibly intact | |
| | *Architecture*<br>Stone circle? | *Surface finds*<br>None | *Geomagnetic survey*<br>No | |
| *Description*<br>Kurgan 17 is a large low mound covered with rather sparse small limestones. The centre of the mound is slightly depressed, but there is not much debris at the surface. The mound may, therefore, be intact. Inwards of the mound here and there limestone blocks are visible that may indicate the existence of a stone circle. There are no stones visible in the centre indicative of a grave chamber. In fact the mound may be untouched | | | | |
| *Site interpretation*<br>Group of kurgans | | | | |

| **DSP07-H19-06** | *Core units* | *Coordinates* 489220/5037665 | | *Figs.* 4.210, 4.216, 4.222-223 |
|---|---|---|---|---|
| | *Interpretation* Kurgan group | *Estimated size* – | *Date range* Unknown | |

*Description* This is a group of 5 kurgans on the edge of the plateau of which 18 to 21 are more or less aligned. A larger kurgan, no. 22 lies excentric to these. Amidst the remains of a (sub)-recent structure were mapped.

| **Kurgan 18** | *Coordinates (centre)* 489200/5037726 | *Diameter* 13 m | *Status* |
|---|---|---|---|
| | *Architecture* Stone circle? | *Surface finds* None | *Geomagnetic survey* No |

*Description*
Kurgan 18 has abundant vegetation on top. It has a sparse packing of limestone rubble at the surface, although this may be partially hidden by the grass. There are no signs of orthostats protruding from the central part of the mound. Some large blocks at ca. 3 m inwards from the mound's perimeter may indicate a stone circle, but this is far from clear and thus rather hypothetical. The S slope seems damaged

| **Kurgan 19** | *Coordinates (centre)* 489208/5037689 | *Diameter* 15 m | *Status* |
|---|---|---|---|
| | *Architecture* Grave chamber? Stone circle? | *Surface finds* None | *Geomagnetic survey* No |

*Description*
Kurgan 19 is a fairly high, medium large mound with in the centre a protruding stone, maybe a head stone or a large stone lining one of the grave chamber/cist walls. On the S side large stones may indicate the existence of a stone circle

| **Kurgan 20** | *Coordinates (centre)* 489240/5037659 | *Diameter* 13 m | *Status* |
|---|---|---|---|
| | *Architecture* No visible remains | *Surface finds* None | *Geomagnetic survey* No |

*Description*
Kurgan 20 is a medium large, low mound covered with a sparse limestone packing. No sign of stones indicating the existence of grave architecture. A large stone on the E side may indicate the existence of an inner stone circle, but this is very hypothetical

| **Kurgan 21** | *Coordinates (centre)* 489313/5037641 | *Diameter* 11-16 m | *Status* Intact |
|---|---|---|---|
| | *Architecture* Stone circle | *Surface finds* None | *Geomagnetic survey* No |

*Description*
Kurgan 21 is a medium large, low mound, covered with a sparse lime stone rubble packing. The mound has a very regular form and there are no signs of a grave chamber or cist architecture. The mound seems untouched. At ca. 2.5 m. from the perimeter inwards there are large blocks that quite convincingly form a circle

| **Kurgan 22** | *Coordinates (centre)* 489177/5037615 | *Diameter* 20-22 m | *Status* Looted/excavated? |
|---|---|---|---|
| | *Architecture* None visible | *Surface finds* None | *Geomagnetic survey* No |

| *Description*<br>Kurgan 22 is a large, high mound, with a depression in the central part (excavated? looted? or just an attempt at?). No clear remains of grave architecture are visible at the surface, although there are some large blocks lying about. There is quite some vegetation on the mound through which one can discern the limestone rubble covering the surface of the mound. No signs at the surface of the existence of an inner stone circle |
| --- |
| *Site interpretation*<br>Group of kurgans |

| **DSP08-H19-07** | *Core units* | *Coordinates*<br>489046/5037633 | | *Figs.*<br>4.211, 4.224-225 |
|---|---|---|---|---|
| | *Interpretation*<br>Kurgan group | *Estimated size*<br>– | *Date range*<br>Unknown | |

*Description* This is a group of 3 kurgans on the edge of the plateau, two of which (23 and 24) are located relatively close to each other, 25 is located at some distance of the latter two. Amidst of them were found wall remains of (sub)recent date.

**Wall remains near DSP08-H19-07**
Also was mapped what at first seemed to be a looted kurgan but which appeared to be the remains of a rectangular structure. Several parts of wall are visible at the surface at a substantial elevation above ground level. This may indicate that there are several layers of stones on top of each other. The structure is orientated E-W over its length axis. The blocks join without cement. Wall blocks have flat sides when facing towards the exterior or interior. A number of blocks (all of local hard limestone) have been cut. These measure 50 x 20 cm, others are irregular in form. The approximate dimensions of the structure are 6 x 4 m. Associated with the walls were two terracotta bricks that seem fairly recent. This is most likely a sub-recent structure. The brick remains were collected as grab sample 3092 "brick found in stone structure near kurgan 25"

| **Kurgan 23** | *Coordinates (centre)*<br>489075/5037659 | *Diameter*<br>14 (NS) – 15 (EW) m | *Status*<br>Intact |
|---|---|---|---|
| | *Architecture*<br>Loose boulders | *Surface finds*<br>None | *Geomagnetic survey*<br>No |

*Description*
This is a medium low mound that seems still intact. In its central part, some large boulders lie at the surface (40 x 30 cm max.). On the sides of the kurgan, smaller stones surface, but there is no dense packing visible. The soil at the surface is blackish and quite loose. It is covered by regular steppe vegetation. The mound lies S of the track on which it borders

| **Kurgan 24** | *Coordinates (centre)*<br>489034/5037610 | *Diameter*<br>14 m | *Status*<br>Disturbed |
|---|---|---|---|
| | *Architecture*<br>No remains visible | *Surface finds*<br>None | *Geomagnetic survey*<br>No |

*Description*
Low mound 'smeared out' towards the NW side where the slope is very gentle. Towards the SE there is an unnatural outflow of soil which renders the normally circular form of a kurgan irregular. It seems therefore that part of the top of the mound has been moved aside. There is, however, no robber's pit visible in the middle. On the sides of the mound, medium large (40 x 30 cm) limestones surface (as with Kurgan 23), but no dense packing can be seen. This mound has a brownish soil covered by regular steppe vegetation. No signs of recent disturbance

| **Kurgan 25** | *Coordinates (centre)*<br>488911/5037657 | *Diameter*<br>21 m | *Status*<br>Disturbed |
|---|---|---|---|
| | *Architecture*<br>No remains visible | *Surface finds*<br>None | *Geomagnetic survey*<br>No |

*Description*
This is the highest and by far the largest mound of the group. It slopes evenly on all sides, but in the middle it is concave with a convex centre. Apparently, it has been topped off and a circular trench was dug into the mound. Occasionally fairly large stones surface on the sides of the mound (ca. 50 x 40 cm). The N slope shows a reasonably dense packing of smaller stones. In the trench dug in the top there are a number of recent wall blocks with cement attached lying about. These are not of local limestone. Possibly the mound has been reused for some purpose unknown to us (war trench?). The larger part of the sides of the mound are covered by normal steppe vegetation and do not seem to be disturbed

*Site interpretation*
Group of kurgans

| **DSP08-H19-08** | *Core units* | *Coordinates* 489984/5037665 | | *Figs.* 4.212, 4.225-227 |
|---|---|---|---|---|
| | *Interpretation* Kurgan group | *Estimated size* – | *Date range* Unknown | |

*Description* This is a group of five kurgans on the edge of the plateau, four of which were placed close to each other, one lies excentric to the group

| **Kurgan 26** | *Coordinates (centre)* 490047/5037676 | *Diameter* 13 m EW x 14 m NS | *Status* Looted |
|---|---|---|---|
| | *Architecture* No remains visible | *Surface finds* None | *Geomagnetic survey* No |

*Description*
This is a low mound of 13 m E-W and 14 m N-S. Its sides are covered with a rather dense packing of small limestones (20 x 20 cm), some loose on the surface. The kurgan has been looted: the inner part is completely removed to a depth of ca. 60 cm. The inner side gives a good view on the packing of stones with brownish loose soil in between. The slopes of the mound have normal steppe vegetation and seem not to have been disturbed. Most stone debris has been thrown out onto the E slope of the mound. In the vicinity of the mound more digging has taken place. Especially towards the N, various pits are visible

| **Kurgan 27** | *Coordinates (centre)* 489979/5037618 | *Diameter* 17 m | *Status* Looted |
|---|---|---|---|
| | *Architecture* No remains visible | *Surface finds* None | *Geomagnetic survey* No |

*Description*
This is a large low mound with very gentle slopes. There are a few stones at the surface. A clandestine pit of ca. 1 x 1 m has been dug into the top of the mound on the W side. This is clear from the different vegetation. A smaller pit may have been dug in the E part of the mound, but this is not clear

| **Kurgan 28** | *Coordinates (centre)* 489960/5037576 | *Diameter* 23 m | *Status* Disturbed |
|---|---|---|---|
| | *Architecture* Possible traces of a grave chamber | *Surface finds* None | *Geomagnetic survey* No |

*Description*
This is a large and once high mound. A trench had been dug from the S slope into the mound removing the entire centre (war trench?). In the middle are large stones lying about, most out of situ. These may indicate that the mound originally had a grave chamber, but there is no clear evidence. The trench dug into the mound continues towards the S where it seems to have destroyed another mound (which was measured in with the Total Station as Kurgan 31)

| **Kurgan 29** | *Coordinates (centre)* 489946/5037548 | *Diameter* Between 16 and 17 m | *Status* Limited disturbance |
|---|---|---|---|
| | *Architecture* Possible remains | *Surface finds* None | *Geomagnetic survey* No |

*Description*
This is a medium high mound. It is somewhat concave in the middle, indicating only limited digging. In the centre large stones with flat sides are visible, possibly an indication of grave architecture. The sides of this mound have a dense packing of limestones of ca. 20 x 20 cm, though some stones are larger

| **Kurgan 30** | *Coordinates (centre)* 489904/5037556 | *Diameter* Ca. 13 m | *Status* Disturbed |
|---|---|---|---|
| | *Architecture* No remains visible | *Surface finds* None | *Geomagnetic survey* No |

<table>
<tr><td colspan="4">Description<br>This is a low mound of which the W half has been removed. It has a quite dense packing of stones preserved on its sides. Some large stones surface at the top. There is, however, no sign of grave architecture</td></tr>
<tr><td rowspan="2">Kurgan 31</td><td>Location<br>Directly S of Kurgan 28</td><td>Diameter<br>Unknown</td><td>Status<br>Destroyed</td></tr>
<tr><td>Architecture<br>None visible</td><td>Surface finds<br>None</td><td>Geomagnetic survey<br>No</td></tr>
<tr><td colspan="4">Description<br>This kurgan has been excavated for the larger part; it could only be recognized through the presence of low 'left-overs' of its sides that show the dense stone packing typical for kurgans. Some large stones lie about. This kurgan may have been destroyed during WW II (pers. comm. A. Kajarka)</td></tr>
<tr><td colspan="4">Site interpretation<br>Group of kurgans</td></tr>
</table>

| **DSP08-H19-09** | *Core units*<br>20035-36, H19 | *Coordinates*<br>488583/5040350 | | *Figs.*<br>4.271-272 |
|---|---|---|---|---|
| | *Adjacent units*<br>H01, H03, H12 and H24 | | | |
| | *Interpretation*<br>Area of quarries | *Estimated size*<br>– | *Date range*<br>Modern? | |
| **Feature 1** | *Core units*<br>H19 | *Coordinates*<br>488541/5040406 | | |
| | *Interpretation*<br>Quarry | *Estimated size*<br>Length 66 m | *Date range*<br>Modern? | |
| **Feature 2** | *Core units*<br>20035 | *Coordinates*<br>488588/5040340 | | |
| | *Interpretation*<br>Quarry | *Estimated size*<br>Length 23 m | *Date range*<br>Modern? | |
| **Feature 3** | *Core units*<br>20035-36 | *Coordinates*<br>488670/5040279 | | |
| | *Interpretation*<br>Quarry | *Estimated size*<br>Length 83 m | *Date range*<br>Modern? | |
| **Feature 4** | *Core units*<br>H19 | *Coordinates*<br>488771/5040277 | | |
| | *Interpretation*<br>Quarry | *Estimated size*<br>Length 28 m | *Date range*<br>Modern? | |
| **Feature 5** | *Core units*<br>20036 | *Coordinates*<br>4887670/5040199 | | |
| | *Interpretation*<br>Quarry | *Estimated size*<br>Length 31 m | *Date range*<br>Modern? | |
| **Feature 6** | *Core units*<br>20036 | *Coordinates*<br>488929/5040093 | | |
| | *Interpretation*<br>Quarry | *Estimated size*<br>Length 27 m | *Date range*<br>Modern? | |
| *Topography*<br>**DSP08-H19-09** is located on Hill 19, a long slope running from the top of the ridge to the cultivated zone | | | | |
| *Investigations*<br>**Extensive survey** Koltuchov site 49b<br>**Intensive survey** Area of quarrying spread over 500 m. A single Early Modern sherd was collected in the area (20035/01) | | | | |
| *Site interpretation*<br>Probably recent quarrying | | | | |

<table>
<tr><td rowspan="3">DSP08-H20-01</td><td>Core units<br>20009</td><td colspan="2">Coordinates<br>493719/5041094</td><td>Fig.<br>4.247</td></tr>
<tr><td colspan="3">Adjacent units<br>H8 and H21</td><td></td></tr>
<tr><td>Interpretation<br>Settlement?</td><td>Estimated size<br>–</td><td>Date range<br>LC-EH</td><td></td></tr>
<tr><td colspan="5">Topography<br>DSP08-H20-01 is located on Hill 20 on the top of the ridge just on the edge of a large ploughed field in an area which was cultivated in the recent past based on the vegetation</td></tr>
<tr><td colspan="5">Investigations<br>Intensive survey Small scatter of pottery and flint (20009/01-04 and 20011/03)</td></tr>
<tr><td colspan="5">Site interpretation<br>Possibly the edge of a settlement located in the cultivated field to the S</td></tr>
</table>

<table>
<tr><td rowspan="3">DSP08-H21-01</td><td>Core units<br>H21</td><td colspan="2">Coordinates<br>494502/5041672</td><td>Figs.<br>4.262-263</td></tr>
<tr><td colspan="3">Adjacent units<br>20022-23</td><td></td></tr>
<tr><td>Interpretation<br>Settlement</td><td>Estimated size<br>37 x 40 m</td><td>Date range<br>Modern</td><td></td></tr>
<tr><td colspan="5">Topography<br>DSP08-H21-01 is located on Hill 21 on the top of the ridge just N of an area which was cultivated in recent times based on the vegetation</td></tr>
<tr><td colspan="5">Investigations<br>Intensive survey Square patch of high, intense vegetation. No finds, but a piece of iron embedded in the soil may be a plough fr. caught on a stone although no large stones are visible</td></tr>
<tr><td colspan="5">Site interpretation<br>Near contemporary settlement</td></tr>
</table>

| **DSP08-H22-01** | *Core units*<br>20056-57, H22 | *Coordinates*<br>487072/5038333 | | *Fig.*<br>4.295 |
|---|---|---|---|---|
| | *Adjacent units*<br>H04-5, H09, H11 and H13 | | | |
| | *Interpretation*<br>Group of kurgans | *Estimated size* | *Date range*<br>Unknown | |
| Feature 1 | *Core units*<br>H22 | *Coordinates*<br>487008/5038394 | | |
| | *Interpretation*<br>Kurgan | *Estimated size*<br>Diameter 6.5 m | *Date range*<br>Unknown | |
| Feature 2 | *Core units*<br>20056 | *Coordinates*<br>487023/5038376 | | |
| | *Interpretation*<br>Kurgan | *Estimated size*<br>Diameter 10 m | *Date range*<br>Unknown | |
| Feature 3 | *Core units*<br>20056 | *Coordinates*<br>487079/5038336 | | |
| | *Interpretation*<br>Kurgan | *Estimated size*<br>Diameter 9 m | *Date range*<br>Unknown | |
| Feature 4 | *Core units*<br>20056-57 | *Coordinates*<br>487097/5038319 | | |
| | *Interpretation*<br>Kurgan | *Estimated size*<br>Diameter 16.5 m | *Date range*<br>Unknown | |
| Feature 5 | *Core units*<br>20056 | *Coordinates*<br>487092/5038286 | | |
| | *Interpretation*<br>Kurgan | *Estimated size*<br>Diameter 17 m | *Date range*<br>Unknown | |

*Topography*
**DSP08-H22-01** is located on the upper part of Hill 22, a long promontory running from the top of the ridge to the cultivated zone

*Investigations*
**Intensive survey** Rough line of five grassy elevations. A single inconclusive body fr. (20057/02)

*Site interpretation*
Group of kurgans

<table>
<tr><td rowspan="3">DSP08-H22-02</td><td>Core units<br>20057</td><td colspan="2">Coordinates<br>487118/5037938</td><td>Fig.<br>4.295</td></tr>
<tr><td colspan="3">Adjacent units<br>20056, H04-5, H09, H11 and H13</td><td></td></tr>
<tr><td>Interpretation<br>Kurgan</td><td>Estimated size<br>0.02 ha</td><td>Date range<br>Unknown</td><td></td></tr>
<tr><td colspan="5">Topography<br>DSP08-H22-02 is located on Hill 22, 350 m S of the southernmost kurgan of DSP08-H22-01</td></tr>
<tr><td colspan="5">Investigations<br>Intensive survey Single grassy elevation. No finds immediately associated, but six frs. of Early Modern cooking pot (20057/01) were collected 65 m to the NE</td></tr>
<tr><td colspan="5">Site interpretation<br>Kurgan</td></tr>
</table>

<table>
<tr><td rowspan="3">DSP07-H22-03</td><td>Core units<br>20056-57</td><td colspan="2">Coordinates<br>487205/5037507</td><td>Fig.<br>4.295</td></tr>
<tr><td colspan="3">Adjacent units<br>H04-5, H09, H11 and H13</td><td></td></tr>
<tr><td>Interpretation<br>Settlement/group of kurgans</td><td>Estimated size</td><td>Date range<br>Modern/?</td><td></td></tr>
<tr><td rowspan="2">Feature 1</td><td>Core units<br>20057</td><td>Coordinates<br>487090/5037590</td><td></td><td></td></tr>
<tr><td>Interpretation<br>Quarry</td><td>Estimated size<br>Length 10.5 m</td><td>Date range<br>Unknown</td><td></td></tr>
<tr><td rowspan="2">Feature 2</td><td>Core units<br>20056</td><td>Coordinates<br>487186/5037610</td><td></td><td></td></tr>
<tr><td>Interpretation<br>Kurgan?</td><td>Estimated size<br>Diameter 7 m</td><td>Date range<br>Unknown</td><td></td></tr>
<tr><td rowspan="2">Feature 3</td><td>Core units<br>20057</td><td>Coordinates<br>487205/5037564</td><td></td><td></td></tr>
<tr><td>Interpretation<br>Kurgan?</td><td>Estimated size<br>Diameter 10.5 m</td><td>Date range<br>Unknown</td><td></td></tr>
<tr><td rowspan="2">Feature 4</td><td>Core units<br>20057</td><td>Coordinates<br>487194/5037534</td><td></td><td></td></tr>
<tr><td>Interpretation<br>Kurgan?</td><td>Estimated size<br>Diameter 15.5 m</td><td>Date range<br>Unknown</td><td></td></tr>
<tr><td rowspan="2">Feature 5</td><td>Core units<br>20057</td><td>Coordinates<br>487159/50375217</td><td></td><td></td></tr>
<tr><td>Interpretation<br>Structure?</td><td>Estimated size<br>44 x 47 m</td><td>Date range<br>Modern</td><td></td></tr>
</table>

| | | | | |
|---|---|---|---|---|
| **Feature 6** | *Core units*<br>20056-57 | *Coordinates*<br>487165/5037490 | | |
| | *Interpretation*<br>Kurgan? | *Estimated size*<br>Diameter 7 m | *Date range*<br>Unknown | |
| **Feature 7** | *Core units*<br>20056 | *Coordinates*<br>487222/5037529 | | |
| | *Interpretation*<br>Kurgan? | *Estimated size*<br>Diameter 26.5 m | *Date range*<br>Unknown | |
| **Feature 8** | *Core units*<br>20056-57 | *Coordinates*<br>4872120/5037494 | | |
| | *Interpretation*<br>Structure | *Estimated size*<br>10.5 x 6.5 m | *Date range*<br>Modern | |
| **Feature 9** | *Core units*<br>20057 | *Coordinates*<br>487198/5037477 | | |
| | *Interpretation*<br>Kurgan? | *Estimated size*<br>Diameter 19 m | *Date range*<br>Unknown | |
| **Feature 10** | *Core units*<br>20057 | *Coordinates*<br>487213/5037447 | | |
| | *Interpretation*<br>Structure? | *Estimated size*<br>Diameter 7.5 m | *Date range*<br>Modern | |
| **Feature 11** | *Core units*<br>20056-57 | *Coordinates*<br>487219/5037424 | | |
| | *Interpretation*<br>Structure? | *Estimated size*<br>29 x 24.5 m | *Date range*<br>Modern | |
| **Feature 12** | *Core units*<br>20057 | *Coordinates*<br>487232/5037407 | | |
| | *Interpretation*<br>Kurgan? | *Estimated size*<br>Diameter 14 m | *Date range*<br>Unknown | |

*Topography*
**DSP07-H22-03** is located on the top of the ridge of Hill 22, a long promontory running from the top of the ridge to the cultivated zone

*Investigations*
**Intensive survey** Three square patches of high, intense vegetation (Feature 5, 10-11) and a stone foundation (Feature 8) among six grassy elevations. No finds associated.

| | |
|---|---|
| **Geomagnetic survey**<br>Eastern Atlas<br>*Dimensions* 53 x 740 + 101 x 100 m | *Description of measuring conditions*<br>Many dipole anomalies |

*Interpretation of features* The surveyed area crosses only part of the site and none of the features are clearly visible due to the many dipole anomalies caused by metal or other debris. At the southern edge of the site several pits were recorded as well as possible ditches

*Site interpretation*
Near contemporary settlement associated with a group of kurgans

| **DSP08-H23-01** | *Core units*<br>20041-42, H23 | *Coordinates*<br>487796/5039387 | | *Fig.*<br>4.277 |
|---|---|---|---|---|
| | *Adjacent units*<br>H11, H13 and H24 | | | |
| | *Interpretation*<br>Group of kurgans | *Estimated size* | *Date range*<br>Unknown | |
| **Feature 1** | *Core units*<br>20041-42 | *Coordinates*<br>487801/5039653 | | |
| | *Interpretation*<br>Kurgan | *Estimated size*<br>Diameter 10.5 m | *Date range*<br>Unknown | |
| **Feature 2** | *Core units*<br>20042 | *Coordinates*<br>487731/5039586 | | |
| | *Interpretation*<br>Kurgan | *Estimated size*<br>Diameter 11 m | *Date range*<br>Unknown | |
| **Feature 3** | *Core units*<br>20042 | *Coordinates*<br>487784/5039505 | | |
| | *Interpretation*<br>Kurgan | *Estimated size*<br>Diameter 10.5 m | *Date range*<br>Unknown | |
| **Feature 4** | *Core units*<br>H23 | *Coordinates*<br>487787/5039461 | | |
| | *Interpretation*<br>Kurgan | *Estimated size*<br>Diameter 9.5 m | *Date range*<br>Unknown | |
| **Feature 5** | *Core units*<br>20042 | *Coordinates*<br>487759/5039441 | | |
| | *Interpretation*<br>Kurgan | *Estimated size*<br>Diameter 10.5 m | *Date range*<br>Unknown | |
| **Feature 6** | *Core units*<br>H23 | *Coordinates*<br>487884/5039335 | | |
| | *Interpretation*<br>Kurgan | *Estimated size*<br>Diameter 13 m | *Date range*<br>Unknown | |
| **Feature 7** | *Core units*<br>H23 | *Coordinates*<br>487843/5039307 | | |
| | *Interpretation*<br>Kurgan | *Estimated size*<br>Diameter 14 m | *Date range*<br>Unknown | |
| **Feature 8** | *Core units*<br>H23 | *Coordinates*<br>487887/5039277 | | |
| | *Interpretation*<br>Kurgan | *Estimated size*<br>Diameter 10.5 m | *Date range*<br>Unknown | |
| **Feature 9** | *Core units*<br>20042 | *Coordinates*<br>487837/5039229 | | |
| | *Interpretation*<br>Kurgan | *Estimated size*<br>Diameter 14 m | *Date range*<br>Unknown | |

| *Topography* **DSP08-H23-01** is located on the W edge of Hill 23, a long promontory running from the top of the ridge to the cultivated zone | |
|---|---|
| *Investigations* **Intensive survey** Nine grassy elevations recorded in an area of thin soils. A single sherd which could not be dated was collected (20042/02) | |
| **Geomagnetic survey** Eastern Atlas *Dimensions* 700 x 53 m | *Description of measuring conditions* Some dipole anomalies in the S |
| *Interpretation of features* The southernmost kurgan of the group (Feature 9) was detected at the E margin, close to the centre of the surveyed area. | |
| *Site interpretation* Group of kurgans | |

| **DSP08-H23-02** | *Core units* 20041-42 | *Coordinates* 4879211/ 5038947 | | *Fig.* 4.277 |
|---|---|---|---|---|
| | *Adjacent units* H11, H13 and H24 | | | |
| | *Interpretation* Settlement? | *Estimated size* 2,000 m² | *Date range* LC-EH? | |
| **Feature 1** | *Core units* 20042 | *Coordinates* 487889/5039003 | | |
| | *Interpretation* Curved bank/wall? | *Estimated size* Length 57 m | *Date range* LC-EH? | |
| **Feature 2** | *Core units* 20041-42 | *Coordinates* 487946/5038909 | | |
| | *Interpretation* Rectangular structure? | *Estimated size* 40 x =/>13 m | *Date range* LC-EH? | |

| *Topography* **DSP08-H23-02** is located along the W edge of Hill 23, a long promontory running from the top of the ridge to the cultivated zone | |
|---|---|
| *Investigations* **Intensive survey** No structures were observed on the surface only two sherds one of which is Hellenistic (20041/06 and 20042/01) | |
| **Geomagnetic survey** Eastern Atlas *Dimensions* 700 x 53 m | *Description of measuring conditions* Some dipole anomalies in the S |
| *Interpretation of features* Approximately in the middle of the surveyed area some several meters long linear anomalies were detected (Feature 2). These patterns seem to be structures with irregular pits inside. A slightly curved black and white linear anomaly is noticeable (Feature 1). It is of unknown origin and could be archaeological | |
| *Site interpretation* Settlement? | |

| **DSP08-H24-01** | *Core units*<br>20040, H24 | *Coordinates*<br>488137/5040149 | | *Fig.*<br>4.274 |
|---|---|---|---|---|
| | *Adjacent units*<br>20039, H19 and H23 | | | |
| | *Interpretation*<br>Group of kurgans | *Estimated size* | *Date range*<br>LC-EH? | |
| Feature 1 | *Core units*<br>20040 | *Coordinates*<br>488073/5040208 | | |
| | *Interpretation*<br>Kurgan | *Estimated size*<br>Diameter 14.5 m | *Date range*<br>LC-EH? | |
| Feature 2 | *Core units*<br>20040 | *Coordinates*<br>488085/5040171 | | |
| | *Interpretation*<br>Kurgan | *Estimated size*<br>Diameter 15 m | *Date range*<br>LC-EH? | |
| Feature 3 | *Core units*<br>H24 | *Coordinates*<br>488153/5040160 | | |
| | *Interpretation*<br>Kurgan | *Estimated size*<br>Diameter 15.5 m | *Date range*<br>LC-EH? | |
| Feature 4 | *Core units*<br>H24 | *Coordinates*<br>488215/5040151 | | |
| | *Interpretation*<br>Kurgan | *Estimated size*<br>Diameter 15.5 m | *Date range*<br>LC-EH? | |
| Feature 5 | *Core units*<br>H24 | *Coordinates*<br>488212/5040106 | | |
| | *Interpretation*<br>Kurgan | *Estimated size*<br>Diameter 13 m | *Date range*<br>LC-EH? | |
| Feature 6 | *Core units*<br>20040, H24 | *Coordinates*<br>488176/5040100 | | |
| | *Interpretation*<br>Kurgan | *Estimated size*<br>Diameter 11 m | *Date range*<br>LC-EH? | |
| Feature 7 | *Core units*<br>H24 | *Coordinates*<br>488153/5040100 | | |
| | *Interpretation*<br>Kurgan | *Estimated size*<br>Diameter 13 m | *Date range*<br>LC-EH? | |
| Feature 8 | *Core units*<br>H24 | *Coordinates*<br>488189/5040088 | | |
| | *Interpretation*<br>Kurgan | *Estimated size*<br>Diameter 12 m | *Date range*<br>LC-EH? | |

*Topography*
**DSP08-H24-01** is located in the lower part on the W edge of Hill 24, a long narrow promontory running from the top of the ridge to the cultivated zone

*Investigations*
**Intensive survey** Eight grassy elevations associated with a Hellenistic amphora sherd (GS 3217) collected from Feature 3. A Hellenistic amphora fr. and a piece of flint (20039/01, 20040/03-04) were collected approximately 100 m uphill

<table>
<tr><td rowspan="3">DSP08-H25-01</td><td>Core units<br>20031, H25</td><td colspan="2">Coordinates<br>489691/5041081</td><td>Fig.<br>4.267</td></tr>
<tr><td colspan="3">Adjacent units<br>20032, H01-03 and H07</td><td></td></tr>
<tr><td>Interpretation<br>Group of kurgans</td><td>Estimated size</td><td>Date range<br>Unknown</td><td></td></tr>
<tr><td rowspan="2">Feature 1</td><td>Core units<br>H25</td><td>Coordinates<br>4896709/5041101</td><td></td><td></td></tr>
<tr><td>Interpretation<br>kurgan</td><td>Estimated size<br>Diameter 20.5 m</td><td>Date range<br>Unknown</td><td></td></tr>
<tr><td rowspan="2">Feature 2</td><td>Core units<br>20031, H25</td><td>Coordinates<br>489694/5041069</td><td></td><td></td></tr>
<tr><td>Interpretation<br>kurgan</td><td>Estimated size<br>Diameter 10 m</td><td>Date range<br>Unknown</td><td></td></tr>
<tr><td rowspan="2">Feature 3</td><td>Core units<br>20031</td><td>Coordinates<br>489716/5041048</td><td></td><td></td></tr>
<tr><td>Interpretation<br>kurgan</td><td>Estimated size<br>Diameter 17.5 m</td><td>Date range<br>Unknown</td><td></td></tr>
<tr><td colspan="5">Topography<br>DSP08-H25-01 is located on Hill 25, a narrow promontory in the lower part of the hillsides</td></tr>
<tr><td colspan="5">Investigations<br>Intensive survey Three grassy elevations associated with a single sherd of unknown date (20031/01)</td></tr>
<tr><td colspan="2">Geomagnetic survey<br>Eastern Atlas<br>Dimensions 550 x 75 m</td><td colspan="3">Description of measuring conditions<br>Dipole anomalies and large irregular, natural anomalies</td></tr>
<tr><td colspan="5">Interpretation of features The northernmost kurgan (Feature 1) is detectable on the SW edge of the upper third. Only central anomalies/the chambers? are detected in Features 2-3</td></tr>
<tr><td colspan="5">Site interpretation<br>Group of kurgans</td></tr>
</table>

| **DSP08-H25-02** | *Core units*<br>20031 | *Coordinates*<br>489850/5040589 | | *Figs.*<br>4.267-268 |
|---|---|---|---|---|
| | *Adjacent units*<br>H01-03 and H07 | | | |
| | *Interpretation*<br>Group? of Kurgans | *Estimated size*<br>– | *Date range*<br>Unknown | |
| **Feature 1** | *Core units*<br>20031 | *Coordinates*<br>489836/5040591 | | |
| | *Interpretation*<br>Kurgan | *Estimated size*<br>Diameter 7 m | *Date range*<br>Unknown | |
| **Feature 2** | *Core units*<br>20031 | *Coordinates*<br>489864/5040586 | | |
| | *Interpretation*<br>Kurgan | *Estimated size*<br>Diameter 6.5 m | *Date range*<br>Unknown | |
| *Topography*<br>**DSP08-H25-02** is located on Hill 25, a narrow promontory in the lower part of the hillsides | | | | |
| *Investigations*<br>**Intensive survey** Small grassy elevation, but no finds associated | | | | |
| **Geomagnetic survey**<br>Eastern Atlas<br>*Dimensions* 75 x 550 m | | *Description of measuring conditions*<br>Dipole anomalies and large irregular, natural anomalies | | |
| *Interpretation of features* Kurgan (Feature 1) clearly detectable on the SW edge. Strong positive anomalies S of it | | | | |
| *Site interpretation*<br>Kurgans | | | | |

<table>
<tr><td rowspan="3">DSP08-H26-01</td><td>Core units<br>20000, H26</td><td colspan="2">Coordinates<br>491715/5042001</td><td>Figs.<br>4.230-231</td></tr>
<tr><td>Adjacent units<br>H06 and H16</td><td colspan="2"></td><td></td></tr>
<tr><td>Interpretation<br>Group of kurgans?</td><td>Estimated size<br>–</td><td>Date range<br>LC-EH?</td><td></td></tr>
<tr><td rowspan="2">Feature 1</td><td>Core units<br>H26</td><td>Coordinates<br>491652/5042007</td><td></td><td></td></tr>
<tr><td>Interpretation<br>Kurgan</td><td>Estimated size<br>Diameter 9 m</td><td>Date range<br>LC-EH?</td><td></td></tr>
<tr><td rowspan="2">Feature 2</td><td>Core units<br>20000</td><td>Coordinates<br>491794/5042004</td><td></td><td></td></tr>
<tr><td>Interpretation<br>Kurgan?</td><td>Estimated size<br>Diameter 13.5 m</td><td>Date range<br>LC-EH?</td><td></td></tr>
<tr><td colspan="5">Topography<br>DSP08-H26-01 is located on the lower part of Hill 26, a wide promontory in the lower part of the hillsides</td></tr>
<tr><td colspan="5">Investigations<br>Intensive survey Two grassy elevations associated with a possible Herakleian amphora fr. (GS 3200) collected by Feature 1</td></tr>
<tr><td colspan="5">Site interpretation<br>Group (?) of kurgans</td></tr>
</table>

| **DSP08-H26-02** | *Core units*<br>20001, 20030, H26 | *Coordinates*<br>491930/5041989 | | *Figs.*<br>4.230, 4.232 |
|---|---|---|---|---|
| | *Adjacent units*<br>20000, 200029, H06 and H16 | | | |
| | *Interpretation*<br>Group of kurgans | *Estimated size* | *Date range*<br>LC-EH? | |
| **Feature 1** | *Core units*<br>20001 | *Coordinates*<br>491901/5042028 | | |
| | *Interpretation*<br>Kurgan | *Estimated size*<br>Diameter 12.5 m | *Date range*<br>LC-EH? | |
| **Feature 2** | *Core units*<br>H26 | *Coordinates*<br>491929/5042027 | | |
| | *Interpretation*<br>Spoil heap? | *Estimated size*<br>Diameter 8.5 m | *Date range*<br>LC-EH? | |
| **Feature 3** | *Core units*<br>20001 | *Coordinates*<br>491924/5042006 | | |
| | *Interpretation*<br>Spoil heap? | *Estimated size*<br>Diameter 8.5 m | *Date range*<br>LC-EH? | |
| **Feature 4** | *Core units*<br>20001 | *Coordinates*<br>491919/5041993 | | |
| | *Interpretation*<br>Spoil heap? | *Estimated size*<br>Diameter 10.5 m | *Date range*<br>LC-EH? | |
| **Feature 5** | *Core units*<br>20001 | *Coordinates*<br>491929/5041974 | | |
| | *Interpretation*<br>Kurgan | *Estimated size*<br>Diameter 15.5 m | *Date range*<br>LC-EH? | |
| **Feature 6** | *Core units*<br>20001, 20030 | *Coordinates*<br>491970/50419769 | | |
| | *Interpretation*<br>Kurgan | *Estimated size*<br>Diameter 20 m | *Date range*<br>LC-EH? | |

*Topography*
**DSP08-H26-02** is located on the lower part of Hill 26, a wide promontory in the lower part of the hillsides

*Investigations*
**Intensive survey** Three grassy elevations associated with heaps of rock frs. Five Hellenistic sherds associated with the area (20001/01, 20029/01 and 20030/01), three of which came from just N of Feature 6

*Site interpretation*
Group of kurgans

| **DSP08-H27-01** | *Core units*<br>H27 | *Coordinates*<br>495868/5042411 | | *Fig.*<br>4.228 |
|---|---|---|---|---|
| | *Adjacent units*<br>H15 | | | |
| | *Interpretation*<br>Farmstead | *Estimated size*<br>1,000 m² | *Date range*<br>Modern? | |

*Topography*
**DSP08-H27-01** is located on the NW edge of Hill 27

*Investigations*
**Extensive survey** Koltuchov site 37. Large roughly square patch of intense vegetation with cut-away corners towards the N

*Site interpretation*
Near contemporary farmstead

| **DSP08-H28-01** | *Core units*<br>H28 | *Coordinates*<br>492099/5042234 | | *Fig.*<br>4.229 |
|---|---|---|---|---|
| | *Adjacent units*<br>H26 and H16 | | | |
| | *Interpretation*<br>Storage site? | *Estimated size*<br>170 m² | *Date range*<br>Modern? | |

*Topography*
DSP08-H28-01 is located on the steep NE edge of Hill 28

*Investigations*
**Extensive survey** Koltuchov site 52. Roughly square elevation covered with intense shrub and two small patches of shrub. Same type of shrub as covers the W structures of DSP08-H12-01

*Site interpretation*
Storage site? Date unknown

| **Baj Kijat 1 (and 2)**[98] | | *Coordinates*<br>500187/5058205 | | *Figs.*<br>4.1-2 |
|---|---|---|---|---|
| | *Interpretation*<br>Settlement; large ash hill | *Estimated size*<br>300 x 700 m | *Date range*<br>Middle BA; Late BA; LC-EH | |
| *Topography*<br>On a gentle slope facing the sea | | | | |
| *Investigations*<br>Excavations in 1973-1974 (A.A. Šepinskij) and 1991-1995 (V.A. Kolotuchin). The excavation alone concerned the Bronze Age settlement. Visited during the DSP **Extensive survey** 17.05.07; GS 3009<br>*Additional information* The excavations have brought to light a number of rectangular one-room houses built in field stones. The houses follow the same orientation. Inside the houses were hearths. The area of the houses is 30-56 $m^2$ | | | | |
| *Site interpretation*<br>Multi-period settlement | | | | |
| *Bibliography*<br>Kolotuchin 1997; Kolotuchin 2003, 31-39, 68 fig. 1.27; Kolotuchin 2007 | | | | |

| **Baj Kijat 3** | | *Coordinates*<br>499979/5058379 | | *Fig.*<br>4.2 |
|---|---|---|---|---|
| | *Interpretation*<br>Farmhouse? settlement? | *Estimated size*<br>ca. 150 x 250 m | *Date range*<br>LC-EH; EM | |
| *Topography*<br>Located on flat terrace facing the sea; to the E a small ravine | | | | |
| *Investigations*<br>Found during the DSP **Extensive survey** 17.05.07; GS 3010 | | | | |
| *Site interpretation*<br>Farmhouse or settlement? | | | | |
| *Bibliography*<br>– | | | | |

100 When extensively surveyed in 2007 by the DSP team, the site was divided as Baj Kijat 1 and 2, but since the excavator includes the entire area as one site, the distinction has been given up. Accordingly there is no Baj Kijat 2.

| **Burun Eli 1** | | *Coordinates*<br>496838/5053761 | | |
|---|---|---|---|---|
| | *Interpretation*<br>Settlement | *Estimated size*<br>3 ha | *Date range*<br>Late BA, 'Ancient';<br>Late Scythian;<br>Medieval[99] | |
| *Topography*<br>Situated on tongue at confluence of two streams | | | | |
| *Investigations*<br>Excavated by V.A. Kolotuchin (1988-1990). The excavation alone concerned the Bronze Age settlement.<br>*Additional information* Found are a number of rectangular one-room houses built in field stones as well as at least one semi-dugout. The houses by and large follow the same orientation. Inside the houses were hearths. The largest house (sector I house I, Kolotuchin 2003, fig. 36) has an area of 34-36 m². The excavator distinguished four different building periods | | | | |
| *Site interpretation*<br>Multi-period settlement | | | | |
| *Bibliography*<br>Kolotuchin 2003, 28-31, 68 fig. 1.26; Lancov 2004, fig. 1.77-78 | | | | |

| **Burun Eli 2** | | *Coordinates*<br>496857/5056469 | | *Figs.*<br>4.7-10 |
|---|---|---|---|---|
| | *Interpretation*<br>Settlement or farmhouse? | *Estimated size*<br>ca. 100 x 100 m | *Date range*<br>LC-EH | |
| *Topography*<br>Located on terrace immediately facing the sea; partly eroded away | | | | |
| *Investigations*<br>Found by the North-Crimean Expedition in 1987; unexcavated. Visited during the DSP **Extensive survey** 17.05.07; GS 3011<br>*Additional information* In the soil profile eroded by the sea a pavement consisting of large, flat pebbles can be seen (Fig. D). A road leading from the site can clearly be seen as a dip in the terrain. It is directed towards a kurgan (Fig. E) | | | | |
| *Site interpretation*<br>Settlement or farmhouse? | | | | |
| *Bibliography*<br>Lancov 2004, 62; Lantsov & Uzhentzev 2007, 681 | | | | |

101 The find of multi-period potsherds is mentioned by Kolotuchin (2003, 28).

<table>
<tr><td rowspan="2">Burun Eli 3</td><td></td><td colspan="2">Coordinates<br>497953/5057070</td><td></td></tr>
<tr><td>Interpretation<br>Settlement or farmhouse?</td><td>Estimated size<br>ca. 100 x 30 m</td><td>Date range<br>LC-EH</td><td></td></tr>
<tr><td colspan="5">Topography<br>Situated on terrace immediately facing the sea</td></tr>
<tr><td colspan="5">Investigations<br>Found by the North-Crimean Expedition in 1987; unexcavated</td></tr>
<tr><td colspan="5">Site interpretation<br>Settlement or farmhouse?</td></tr>
<tr><td colspan="5">Bibliography<br>Lancov 2004, 62; Lantsov & Uzhentzev 2007, 681</td></tr>
</table>

<table>
<tr><td rowspan="2">Burun Eli 4</td><td></td><td colspan="2">Coordinates<br>495871/5055721</td><td></td></tr>
<tr><td>Interpretation<br>Settlement; ash hill</td><td>Estimated size<br>50 x 50 m</td><td>Date range<br>LC-EH; EM</td><td></td></tr>
<tr><td colspan="5">Topography<br>Located at cliff close to the sea</td></tr>
<tr><td colspan="5">Investigations<br>Found in the 1970s by the North-Crimean Expedition. Visited during the DSP Extensive survey 17.05.07, GS 3021 (derives from the ash hill)</td></tr>
<tr><td colspan="5">Site interpretation<br>Unfortified settlement? with ash hill</td></tr>
<tr><td colspan="5">Bibliography<br>Ščeglov 1978, fig. 8.44</td></tr>
</table>

<table>
<tr><td rowspan="2">Burun Eli 5</td><td></td><td colspan="2">Coordinates<br>495360/5055308</td><td></td></tr>
<tr><td>Interpretation<br>Farmhouse?; settlement?</td><td>Estimated size<br>ca. 100 x 50 m</td><td>Date range<br>LC-EH; EM</td><td></td></tr>
<tr><td colspan="5">Topography<br>Situated at cliff facing the sea</td></tr>
<tr><td colspan="5">Investigations<br>Found during the DSP Extensive survey 17.05.07; GS 3022</td></tr>
<tr><td colspan="5">Site interpretation<br>Farmhouse? or settlement?</td></tr>
<tr><td colspan="5">Bibliography<br>–</td></tr>
</table>

| **Černomorskoe Settlement** (traditionally identified as Kalos Limen) | | *Coordinates* 477655/5040468 | | *Figs.* 4.23-27 |
|---|---|---|---|---|
| | *Interpretation* Fortified city | *Estimated size* 200 x 140 m | *Date range* Early 4th c. BC-early 2nd c. AD; late 3rd-early 4th c. AD; 8th-10th c. AD | |

*Topography*
At the NE outskirts of the modern district town of Černomorskoe by the Uzkaja ('Narrow') Bay is located a small fortified harbour city. Its location, which is drawn back and thus sheltered from the sea, is at the E side of the outer part of the Kel'šejch Ravine. The topography of the site is somewhat altered since antiquity. In early antiquity the city was probably surrounded by water to the W (as today) as well as to the S. At some point, the stream of the Kel'šejch Ravine has been cut off from the sea by a narrow sand barrier thus creating a small brackish lagoon. Today, part of the ancient town is under water

*Investigations*
Scientific excavation has taken place in the ancient settlement since 1929 (L.A. Moiseev; for a summary of the excavation history with references, see Užencev 2006, 10-15; Lantsov & Uzhentzev 2006, 699-705). P.N. Schulz worked here in 1933 and in 1948 as did K.E. Grinevič in 1946. The Tauro-Scythian Expedition excavated at the site in 1948 and 1950 (M.A. Nalivkina, G.D. Smirnov) and in 1959 a joint Soviet-Polish expedition worked here (V.F. Gajdukevič, K. Michalowski; M.L. Bernhard). In 1962-1963 the Tarchankut Expedition excavated in the site (A.N. Ščeglov). After a hiatus in the investigations, in 1988 the site became the study object of the West-Crimean Expedition under the direction of V.A. Kutajsov. Recently, the first monograph of the site was published (Užencev 2006)

*Site interpretation*
Multi-period fortified city

*Bibliography*
Ščeglov 1978, fig. 8.35; Ajbabin 2003, fig. 4.105; Lancov 2004, fig. 1.66; Užencev 2004; Lantsov & Uzhentzev 2006, 699-705; Užencev 2006; Kutajsov & Pridnev 2007

| **Černomorskoe farm 1** | | *Coordinates (approx.)* 477816/5040932 | | |
|---|---|---|---|---|
| | *Interpretation* Farmhouse | *Estimated size* ? | *Date range* LC-EH | |

*Topography*
By Uzkaja Bay, NE of the Černomorskoe Settlement

*Investigations*
Excavated by P.N. Schulz in 1933
*Additional information* Ščeglov mentions the presence of walls and pits

*Site interpretation*
Farmhouse?

*Bibliography*
Ščeglov 1967, fig. 2.8

<table>
<tr><td rowspan="2">Černomorskoe farm 2</td><td></td><td>Coordinates (approx.)<br>476624/5040055</td><td></td><td></td></tr>
<tr><td>Interpretation<br>Farmhouse</td><td>Estimated size<br>?</td><td>Date range<br>LC-EH</td><td></td></tr>
<tr><td colspan="5">Topography<br>Located by Uzkaja Bay, on the tip of the central promontory between two branches of ravine SW of the Černomorskoe Settlement</td></tr>
<tr><td colspan="5">Investigations<br>A.N. Ščeglov, year unknown</td></tr>
<tr><td colspan="5">Site interpretation<br>Farmhouse?</td></tr>
<tr><td colspan="5">Bibliography<br>Ščeglov 1967, 247, fig. 2.9</td></tr>
</table>

<table>
<tr><td rowspan="2">Černomorskoe farm 3</td><td></td><td>Coordinates (approx.)<br>475805/5039923.</td><td></td><td></td></tr>
<tr><td>Interpretation<br>Farmhouse</td><td>Estimated size<br>?</td><td>Date range<br>LC-EH, Roman</td><td></td></tr>
<tr><td colspan="5">Topography<br>Located at S side of promontory W of Uzkaja Bay</td></tr>
<tr><td colspan="5">Investigations<br>A.N. Ščeglov, year unknown<br>Additional information Stone walls are referred to by Ščeglov as well as Hellenistic material such as Sinopean tiles, Chersonesean and Herakleian amphorai, and louteria of the 3rd-2nd century BC</td></tr>
<tr><td colspan="5">Site interpretation<br>Multi-period farmhouse?</td></tr>
<tr><td colspan="5">Bibliography<br>Ščeglov 1967, fig. 2.10</td></tr>
</table>

<table>
<tr><td rowspan="2">Černomorskoe farm 4</td><td></td><td>Coordinates (approx.)<br>475940/5040170</td><td></td><td></td></tr>
<tr><td>Interpretation<br>Farmhouse</td><td>Estimated size<br>?</td><td>Date range<br>LC-EH to 2nd c. BC</td><td></td></tr>
<tr><td colspan="5">Topography<br>Located at S side of promontory W of Uzkaja Bay</td></tr>
<tr><td colspan="5">Investigations<br>A.N. Ščeglov, year unknown<br>Additional information Stone walls are referred to by Ščeglov as well as Hellenistic material such as Sinopean tiles, Chersonesean and Herakleian amphorai, and louteria of the 3rd-2nd century BC</td></tr>
<tr><td colspan="5">Site interpretation<br>Farmhouse?</td></tr>
<tr><td colspan="5">Bibliography<br>Ščeglov 1967, fig. 2.11</td></tr>
</table>

| **Černomorskoe farm 5** | | *Coordinates*<br>476373/5039509 | | |
|---|---|---|---|---|
| | *Interpretation*<br>Farmhouse | *Estimated size*<br>? | *Date range*<br>LC-EH? | |
| *Topography*<br>Located by Uzkaja Bay, on the central promontory between two branches of ravine SW of the Černomorskoe Settlement (Ul. Revoljucii, Florya Pension) | | | | |
| *Investigations*<br>The site has never been investigated archaeologically<br>*Additional information* The mouth of a well crafted in local limestone was found during construction work of the pension (Fig. 4.27). The pension is located just 250 m S of the area, where a corner of a cadastre division wall was found under the market square (Ščeglov 1967, fig. 5) | | | | |
| *Site interpretation*<br>Farmhouse | | | | |
| *Bibliography*<br>– | | | | |

| **Černomorskoe farm 6** | | *Coordinates (approx.)*<br>477584/5040659 | | |
|---|---|---|---|---|
| | *Interpretation*<br>Farmhouse | *Estimated size*<br>? | *Date range*<br>? | |
| *Topography*<br>By the Uzkaja Bay ca. 200 m N of the settlement; currently located under water | | | | |
| *Investigations*<br>No investigations are known to us | | | | |
| *Site interpretation*<br>Farmhouse? | | | | |
| *Bibliography*<br>Užencev 2006, fig. 4, Ж 1 | | | | |

| **Černomorskoe farm 7** | | *Coordinates (approx.)*<br>477644/5040708 | | |
|---|---|---|---|---|
| | *Interpretation*<br>Farmhouse | *Estimated size*<br>? | *Date range*<br>? | |
| *Topography*<br>By the Uzkaja Bay ca. 250 m N of the settlement; the ruin is visible on the beach | | | | |
| *Investigations*<br>No investigations are known to us | | | | |
| *Site interpretation*<br>Farmhouse? | | | | |
| *Bibliography*<br>Užencev 2006, fig. 4, Ж 2 | | | | |

| **Černomorskoe farm 8** | | *Coordinates (approx.)*<br>477443/5040832 | | |
|---|---|---|---|---|
| | *Interpretation*<br>Farmhouse | *Estimated size*<br>? | *Date range*<br>? | |
| *Topography*<br>Located at the Uzkaja Bay ca. 400 m NW of the settlement | | | | |
| *Investigations*<br>No investigations are known to us | | | | |
| *Site interpretation*<br>Farmhouse? | | | | |
| *Bibliography*<br>Užencev 2006, fig. 4, Ж 3 | | | | |

| **Černomorskoe farm 9** | | *Coordinates (approx.)*<br>477497/5040934 | | |
|---|---|---|---|---|
| | *Interpretation*<br>Farmhouse | *Estimated size*<br>? | *Date range*<br>? | |
| *Topography*<br>Located at the NE promontory of Uzkaja Bay ca. 500 m NW of the settlement | | | | |
| *Investigations*<br>No investigations are known to us | | | | |
| *Site interpretation*<br>Farmhouse? | | | | |
| *Bibliography*<br>Užencev 2006, fig. 4, Ж 4 | | | | |

| **Džarylgač** | | *Coordinates*<br>? | | |
|---|---|---|---|---|
| | *Interpretation*<br>Settlement? | *Estimated size*<br>? | *Date range*<br>LC-EH | |
| *Topography*<br>Located in the vicinity of the village of Vodopojnoe | | | | |
| *Investigations*<br>Found by A.A. Ščepinskij, year unknown. Not verified in the DSP survey | | | | |
| *Site interpretation*<br>'Unfortified settlement' | | | | |
| *Bibliography*<br>Ščeglov 1978, fig. 8.42; Ščeglov 2002, pl. 4.26; Užencev 2006, fig. 2.73 | | | | |

| **Groty 1** | | *Coordinates*<br>491354/5052569 | | *Fig.*<br>4.6 |
|---|---|---|---|---|
| | *Interpretation*<br>Unfortified settlement? | *Estimated size*<br>max. 50 x 100 m | *Date range*<br>LC-EH; LH | |
| *Topography*<br>Located on E slope of high ravine close to the sea coast | | | | |
| *Investigations*<br>Found and partly excavated by V.A. Latyševa in the 1970s. Visited during the DSP **Extensive survey** 17.05.07; GS 3031: this sample is the largest grab sample collected during the DSP survey | | | | |
| *Site interpretation*<br>Unfortified settlement? | | | | |
| *Bibliography*<br>Ščeglov 1978, fig. 8.3; Lancov 2004, fig. 1.74 | | | | |

| **Groty 2** | | *Coordinates*<br>490021/5051391 | | |
|---|---|---|---|---|
| | *Interpretation*<br>Farmhouse? | *Estimated size*<br>50 x 50 m | *Date range*<br>LC-EH | |
| *Topography*<br>Located on cliff facing the sea | | | | |
| *Investigations*<br>Found in the DSP **Extensive survey** 17.05.07; GS 3032 | | | | |
| *Site interpretation*<br>Farmhouse? | | | | |
| *Bibliography*<br>– | | | | |

| **Jarylgač** | | *Coordinates*<br>? | | |
|---|---|---|---|---|
| | *Interpretation*<br>Settlement? | *Estimated size*<br>? | *Date range*<br>LC-EH | |
| *Topography*<br>Situated in the pediment S of Lake Džarylgač. The more precise location is unknown | | | | |
| *Investigations*<br>Found by members of the Tarchankut Expedition, year unknown. Not verified in the DSP survey | | | | |
| *Site interpretation*<br>'Unfortified settlement' | | | | |
| *Bibliography*<br>Ščeglov 1978, fig. 8.40; Užencev 2006, fig. 2.71 | | | | |

<table>
<tr><td rowspan="2">Jarylgač Severnoe</td><td></td><td colspan="2">Coordinates (approx.)<br>486151/5044498</td><td></td></tr>
<tr><td>Interpretation<br>Settlement?</td><td>Estimated size<br>?</td><td>Date range<br>Neolithic</td><td></td></tr>
<tr><td colspan="5">Topography<br>Located on flat ground close to sea on tip of promontory</td></tr>
<tr><td colspan="5">Investigations<br>Found by the Tarchankut Expedition, year unknown<br>Additional information It is likely that the three pieces of flint collected in GS 3001 in fact derive from this settlement</td></tr>
<tr><td colspan="5">Site interpretation<br>Settlement?</td></tr>
<tr><td colspan="5">Bibliography<br>Unpublished; mentioned in Blagovolin & Ščeglov 2002, 286, pl. 179.VI</td></tr>
</table>

<table>
<tr><td rowspan="2">Jarylgač Vostočnoe</td><td></td><td colspan="2">Coordinates<br>486151/5044498</td><td></td></tr>
<tr><td>Interpretation<br>Settlement?</td><td>Estimated size<br>?</td><td>Date range<br>Late BA; LC-EH;<br>Early Middle Ages<br>(Saltovo)</td><td></td></tr>
<tr><td colspan="5">Topography<br>Situated on W bank of present-day Lake Maloe Solenoe, partly submerged</td></tr>
<tr><td colspan="5">Investigations<br>Found by the Tarchankut Expedition in 1975. Three test trenches were excavated; most of the finds consisted of Late Bronze Age handmade pottery and flints</td></tr>
<tr><td colspan="5">Site interpretation<br>Multi-period settlement?</td></tr>
<tr><td colspan="5">Bibliography<br>Ščeglov et al. 1976; Blagovolin & Ščeglov 2002, 288-290, pl. 179.VII</td></tr>
</table>

| **Kula Šeich** | | *Coordinates*<br>492785/5053695 | | |
|---|---|---|---|---|
| | *Interpretation*<br>? | *Estimated size*<br>? | *Date range*<br>EM? | |

*Topography*
Situated on W slope of shallow ravine

*Investigations*
Found in the DSP **Extensive survey** 17.05.07. Part of an architectural element cut in local limestone was recorded. It may be part of a pilaster. The stone was left where it was found

*Site interpretation*
The Tatarian toponym Kula Šeich (the tower or castle of the Sheik) may allude to a stone construction no longer preserved (a tomb?). The stone may have pertained to such a structure

*Bibliography*
–

| **Kurama Kostel'** | | *Coordinates*<br>494473/5044892 | | |
|---|---|---|---|---|
| | *Interpretation*<br>Settlement?; ploughed-over kurgan? | *Estimated size*<br>300 x 300 m or more | *Date range*<br>LC-EH; EM | |

*Topography*
Located on flat ground slightly inland from the SE side of Džarylgač Lake

*Investigations*
Found during the DSP **Extensive survey** 05.05.07; GS 1003. Several patches of grey and powdery soil due to high content of ash

*Site interpretation*
Settlement or kurgan during the Hellenistic phase. Settlement during the Early Modern period

*Bibliography*
–

<table>
<tr><td rowspan="2">Masliny 1 (Vladimirovka)</td><td></td><td colspan="2">Coordinates<br>502309/5059394</td><td>Figs.<br>4.4-5</td></tr>
<tr><td>Interpretation<br>Fortified settlement</td><td>Estimated size<br>200 x 100 m</td><td>Date range<br>LC-EH; LH</td><td></td></tr>
<tr><td colspan="5">Topography<br>Situated on flat cliff immediately to the shore of a small bay; to the W a shallow ravine. Partly eroded away</td></tr>
<tr><td colspan="5">Investigations<br>Excavated by V.A. Latyševa between 1972 and 1986; visited during the DSP Extensive survey 17.05.07; GS 3007<br>Additional information The earliest constructions are in all likelihood three towers constructed in the usual Chersonesean style with drafted margin ashlars (Fig. Q, R) (mid-4th century BC?). These were later incorporated into farmhouses (second half of the 4th century BC?). The houses as well as the lanes inside the settlement are paved with slabs. The settlement perished in a fire in the second half of the 2nd century</td></tr>
<tr><td colspan="5">Site interpretation<br>Fortified settlement</td></tr>
<tr><td colspan="5">Bibliography<br>Latyševa 1978; Latyševa 1997; Ščeglov 1978, fig. 8.45 [Vladimirovka]; Lancov 2004, fig. 1.80</td></tr>
</table>

<table>
<tr><td rowspan="2">Masliny 2</td><td></td><td colspan="2">Coordinates<br>502394/5059420</td><td></td></tr>
<tr><td>Interpretation<br>Settlement</td><td>Estimated size<br>not mentioned</td><td>Date range<br>LC-EH</td><td></td></tr>
<tr><td colspan="5">Topography<br>See Masliny 1</td></tr>
<tr><td colspan="5">Investigations<br>See Masliny 1<br>Additional information The localisation of a pit house excavated in 1965, a number of household pits and an ash hill outside (to the E of) the Masliny Settlement have induced A.N. Ščeglov to suggest that barbarians lived in a separate settlement outside the main fortification (Ščeglov 1978, 82-83)</td></tr>
<tr><td colspan="5">Site interpretation<br>Dugout, household pits, ash hill</td></tr>
<tr><td colspan="5">Bibliography<br>Ščeglov 1978, 82-83</td></tr>
</table>

| **Masliny 3** | | *Coordinates*<br>501413/5059108 | | |
|---|---|---|---|---|
| | *Interpretation*<br>Farmhouse? | *Estimated size*<br>ca. 150 x 50 m | *Date range*<br>LC-EH | |
| *Topography*<br>Located on cliff facing the sea | | | | |
| *Investigations*<br>Found during the DSP **Extensive survey** 12.05.07; GS 3008 | | | | |
| *Site interpretation*<br>Farmhouse? | | | | |
| *Bibliography*<br>– | | | | |

| **Mežvodnoe** | | *Coordinates*<br>? | | |
|---|---|---|---|---|
| | *Interpretation*<br>Unfortified(?) farmhouse | *Estimated size*<br>? | *Date range*<br>LC-EH | |
| *Topography*<br>Located at the Jarylgač Bay; the more precise location is unknown | | | | |
| *Investigations*<br>Excavated by T.N. Vysotskaja in 1967. A test trench was occasioned by the occasional find of a limestone relief with a reclining Herakles. Of the farmhouse was only excavated a single room and a courtyard | | | | |
| *Site interpretation*<br>Unfortified(?) farmhouse | | | | |
| *Bibliography*<br>Vysotskaja 1968; 1969; Ščeglov 1978, fig. 8.41; Lancov 2004, fig. 1.72 | | | | |

<table>
<tr><td rowspan="2">Panskoe I<br>See also<br>DSP07-F01-01</td><td></td><td colspan="2">Coordinates<br>485536/5044250</td><td>Figs.<br>4.31, 4.33-37</td></tr>
<tr><td>Interpretation<br>Settlement</td><td>Estimated size<br>ca. 4 ha + ca. 2 ha submerged[100]</td><td>Date range<br>LC-EH; LH</td><td>Geomagnetic survey<br>T. Smekalova</td></tr>
<tr><td colspan="5">Topography<br>Situated on extended promontory between Lake Sasyk (now Panskoe) and Jarylgač Bay immediately at the coast of the lake</td></tr>
<tr><td colspan="5">Investigations<br>Excavated by the Tarchankut Expedition (1969-1994); GS 1004. In 2007 and 2008 the settlement of Panskoe I was surveyed with magnetometer by T.N. Smekalova assisted by A. Čudin. The purpose was to supplement the existing sketch plan, which has been published in several versions in order to determine the extension and composition of its layout (e.g. Ščeglov 2002, pl. 6). In 2007 were measured two large tracts S and W of U6 as well as a large tract E of U7 and a smaller one at the E outskirts of the settlement (Figs. 4.33, 35-36). In 2008 was measured a tract of land 35 m x 85 m E of U6. The results of these measurements allow us to complement the known picture of the settlement history of the site (Chapter 4.3.2.1)<br>Interpretation Unlike the majority of the other rural settlements of the region, which are represented by isolated fortified and unfortified farmhouses, the layout of Panskoe is rather unique, as it represents an agglomeration of both well constructed, rectilinear farmhouses and other larger building structures</td></tr>
<tr><td colspan="5">Site interpretation<br>Settlement</td></tr>
<tr><td colspan="5">Bibliography<br>Ščeglov 1978, fig. 6.5, 8.36; Panskoe I 2002; Lancov 2004, fig. 1.67; comprehensive bibliography available at http://www.pontos.dk/research/ra_3/publications-devoted-to-panskoye</td></tr>
</table>

<table>
<tr><td rowspan="2">Panskoe II<br>/DSP06-F03-01</td><td></td><td colspan="2">Coordinates (centre)<br>485810/5042440</td><td>Fig.<br>4.31</td></tr>
<tr><td>Interpretation<br>Settlement</td><td>Estimated size<br>ca. 40 m across (EM)</td><td>Date range<br>LC-EH; LH; EM</td><td></td></tr>
<tr><td colspan="5">Topography<br>Located close to the present S shore of the Sasyk Lake at junction between E-W road and the Panskoe promontory in an unploughed area in the NE corner of Field 3</td></tr>
<tr><td colspan="5">Investigations<br>Unexcavated; soundings at the settlement were made by T.N. Vysotskaja in the late 1960s<br>Extensive survey A scatter of ca. 40 m across in a patch of greyish soil containing many stones and Early Modern pottery among which pottery with a brown glaze. The halo of finds related to this site continue for only 60 to 80 m and at places much less. This means that there seems hardly any archaeological effect of the site on the surrounding terrain. Grab sample (GS 3002) collected in 2007<br>Additional information Scattered in the terrain are a number of ashlar blocks that, provided they have not been transported from elsewhere, indicate that the building was intended as a permanent one. A square tower 10 x 10 m is mentioned by A.N. Ščeglov (Chtcheglov 1992, 151). An isolated farm is marked on the half-verst map of the 1890s in a location corresponding to this site. It may be the material remains thereof we have located</td></tr>
<tr><td colspan="5">Site interpretation<br>Multi-period site. Hellenistic fort or fortified farmhouse; Early Modern scatter of pottery, stones and ash, probably a farmhouse</td></tr>
<tr><td colspan="5">Bibliography<br>Ščeglov 1978, fig. 6.2, 8.37; Chtcheglov 1992, 151; Lancov 2004, fig. 1.68</td></tr>
</table>

102 Ščeglov 1987, 241.

| **Panskoe III** | | *Coordinates (approx.)*<br>482117/5042790 | | *Figs.*<br>4.31-32 |
|---|---|---|---|---|
| | *Interpretation*<br>Farmhouse | *Estimated size*<br>? | *Date range*<br>LC-EH | |
| *Topography*<br>Located on gentle slope W of the Sasyk Ravine | | | | |
| *Investigations*<br>Located in 1972 (Ščeglov et al. 1973, 354) and excavated by Ju.A. Vinogradov, E.Ja. Rogov, D. Naman in 1978. The building, which lacks its SE side, because it has been eroded away by the lake, may have been square in plan. It consists of a number of rooms grouped around a central courtyard | | | | |
| *Site interpretation*<br>Unfortified, isolated farmstead | | | | |
| *Bibliography*<br>Ščeglov 1978, fig. 6.3, 8.38; Ščeglov 1985, 5; plan in Ščeglov 1987, 259, fig. 2; Lancov 2004, fig. 1.69 | | | | |

| **Panskoe IV** | | *Coordinates (approx.)*<br>484337/5043652 | | *Fig.*<br>4.31 |
|---|---|---|---|---|
| | *Interpretation*<br>Farmhouse | *Estimated size*<br>Unknown | *Date range*<br>LC-EH | |
| *Topography*<br>Originally located at side of ravine; today submerged | | | | |
| *Investigations*<br>Non-systematically investigated | | | | |
| *Site interpretation*<br>Isolated farmhouse | | | | |
| *Bibliography*<br>Ščeglov 1978, fig. 6.4, 8.39; Ščeglov 1987, 240; Lancov 2004, fig. 1.70 | | | | |

| **Panskoe V/ DSP07-F01-02** | *Core units*<br>1059-61, 1064, 1068-70, 2046, 2055, 2057-58, 2062 | *Sample density*<br>20%, 24% and 28% | *Coordinates (centre)*<br>485956/5043257 | *Figs.*<br>4.44-45, 4.47-49, |
|---|---|---|---|---|
| | *Adjacent units* | *Sample density* | | |
| | *Interpretation*<br>'Settlement' or ploughed-out kurgans? | *Estimated size*<br>? | *Date range*<br>LC-EH; LH | |
| **Feature 1** | *Core units*<br>2055 | *Sample density*<br>20% | *Coordinates (centre)*<br>485863/5043261 | |
| | *Interpretation*<br>Kurgan? | *Estimated size*<br>Unknown | *Date range*<br>LC-EH | |
| **Feature 2** | *Core units*<br>1064, 2046, 2057-58 | *Sample density*<br>20% | *Coordinates (centre)*<br>486017/5043323 | |
| | *Interpretation*<br>Kurgan? | *Estimated size*<br>Unknown | *Date range*<br>LC-EH | |
| **Feature 3** | *Core units*<br>1059-61, 1068-70, 2062 | *Sample density*<br>20% | *Coordinates (centre)*<br>485879/5043182 | |
| | *Interpretation*<br>Kurgan? | *Estimated size*<br>Unknown | *Date range*<br>LC-EH | |

*Topography*
**DSP07-F01-02** is located in the S part of the field 'between Panskoe I and II, but closest to II' (quoting Randsborg 1994, 179)

*Investigations*
The Tarchankut Expedition excavated a kurgan with a stone krepis wall in this general area in the 1980s, but this produced no finds at all. A "settlement" was identified in 1993 during a site visit (K. Randsborg, L. Hannestad and V. Stolba). In 1994, LC-EH pottery from ploughed-out ritual deposits associated with the kurgan cluster was collected and attributed by V. Stolba; a stone altar similar to the ones found in the necropolis of Panskoe I was found at the W edge of field DSP07-1
**Extensive survey** Inspection of the kurgan outside the surveyed field to the E. This kurgan is grown with grass and seems mostly undamaged. No archaeological finds were collected on the mound itself
**Intensive survey** 50 x 50 m units. High-density scatter of pottery most likely related to a large kurgan is located directly E of the survey units (Feature 1). Two small, non-discrete scatters of stones and pottery frs. (Features 2-3) approximately 230 m W of the kurgan but no clear elevations was observed in the field

*Site interpretation*
The location corresponds to a group of ploughed-out kurgans, which bordered the large Bronze Age kurgan located ca. 1.2 km SE of Panskoe I. It is likely that the material found derives from these kurgans

*Bibliography*
Randsborg 1994, 179, fig. 4.9

| **Severnoe 1** | | *Coordinates*<br>502749/5058057 | | |
|---|---|---|---|---|
| | *Interpretation*<br>Settlement | *Estimated size*<br>? | *Date range*<br>Late BA; EM | |
| *Topography*<br>Situated on gentle slope by shallow ravine | | | | |
| *Investigations*<br>V.A. Kolotuchin, year unknown | | | | |
| *Site interpretation*<br>Settlement | | | | |
| *Bibliography*<br>Kolotuchin 2003, 68 fig. 1.29 | | | | |

| **Severnoe 2** | | *Coordinates*<br>502748/5058316 | | |
|---|---|---|---|---|
| | *Interpretation*<br>Settlement? | *Estimated size*<br>? | *Date range*<br>LC-EH | |
| *Topography*<br>Located on gentle slope by shallow ravine | | | | |
| *Investigations*<br>**Razvedki** S.G. Koltuchov, year unknown; visited during the DSP **Extensive survey** 17.05.07; GS 3006 | | | | |
| *Site interpretation*<br>Settlement? | | | | |
| *Bibliography*<br>Lancov 2004, fig. 1.79 | | | | |

| **Skalistoe 1** | | *Coordinates*<br>493731/5054249 | | |
|---|---|---|---|---|
| | *Interpretation*<br>Settlement? | *Estimated size*<br>? | *Date range*<br>LC-EH | |
| *Topography*<br>Situated on low E slope of small ravine; built over with modern fishing station | | | | |
| *Investigations*<br>Partly excavated by V.A. Latyševa, year unknown | | | | |
| *Site interpretation*<br>Settlement? | | | | |
| *Bibliography*<br>– | | | | |

| **Skalistoe 2/ DSP08-F03-02** | *Core units*<br>6226, 6244-45, 6247 | *Sample density*<br>20%, 23% | *Coordinates (centre)*<br>493952/5053607 | *Figs.*<br>4.98-101, 4.105-108 |
|---|---|---|---|---|
| | *Adjacent units*<br>6227, 6243, 6246, 6248, 6265-67 | *Sample density*<br>20% | | |
| | *Interpretation*<br>Settlement | *Estimated size*<br>3,000 m², halo of ca. 12,500 m² | *Date range*<br>BA, LC-EH | |
| **Phase 1** | *Core units*<br>6244 | *Sample density*<br>20% | *Coordinates (centre)*<br>493983/5053614 | |
| | *Interpretation*<br>Settlement | *Estimated size*<br>110 x 130 m | *Date range*<br>BA | |
| **Phase 2** | *Core units*<br>6245, 6247 | *Sample density*<br>20% | *Coordinates (centre)*<br>494052/5053588 | |
| | *Interpretation*<br>Settlement | *Estimated size*<br>? | *Date range*<br>LC-EH | |

*Topography*
**DSP08-F03-02** is located in the SW corner of the surveyed field. The site is located shallow depression next to a ravine 650 m from the current coastline. Outside the field, on the other side of the ravine, an elevated plateau was visible which may be an extension of the site

*Investigations*
A small scale excavation by V.A. Kolotuchin (1973; 1987) revealed house foundations of field stones
**Intensive survey** 50 x 60 m units; TS 6243/02-06; 6246/02; 6247/02. A high-density scatter with two components: a strong Bronze Age presence, especially in Unit 6244, and a larger Hellenistic part. The Hellenistic finds include amphorai, table- and utility wares

| **Geomagnetic survey**<br>T. Smekalova<br>*Dimensions* 3,500 m² | *Description of measuring conditions*<br>Good |
|---|---|

*Interpretation of features* Six or seven weak positive anomalies, possibly pit houses

**Augering** Four augerings were conducted in a cross section over the site. The sediment seems very thin, except in auger 31 which has a 1,16 m layer of loam sediment. This core was taken at the eastern end of the site. No clear archaeological culture layer was detected.

| Auger S28: 493983/5053527 | |
|---|---|
| 0-0.20 m | Loam, some grit, 10YR 3/2 |
| 0.20-0.30 m | Loam, some small limestone inclusions, 10YR 3/2 |
| 0.30 m | End of augering: soil too hard |
| Auger S29: 494016/5053558 | |
| 0-0.20 m | Loam, topsoil, 10YR 3/2 |
| 0.20-0.40 m | Loam, 10YR 3/2 |
| 0.70 m | End of augering, bedrock |
| Auger S30: 494034/5053574 | |
| 0-0.10 m | Loam, 10YR 3/2 |
| 0.10-0.30 m | Loam, some limestone inclusions, 10YR4/2 |

<table>
<tr><td>0.30 m</td><td>End of augering: bedrock</td></tr>
<tr><td colspan="2">Auger S31: 494054 /5053594</td></tr>
<tr><td>0-0.49 m</td><td>Loam, 10YR 3/2, no effervescence</td></tr>
<tr><td>0.49-1.16 m</td><td>Loam, 10YR 5/4, effervescences</td></tr>
<tr><td>1.16 m</td><td>End of augering: bedrock</td></tr>
<tr><td colspan="2">Site interpretation<br>The location of the site and the combination of multiple functions suggests for both periods that this was a settlement</td></tr>
<tr><td colspan="2">Bibliography<br>Kolotuchin 2003, 39, 68 fig. 1.25[101]</td></tr>
</table>

<table>
<tr><td rowspan="2">Skalistoe 3 and 4</td><td></td><td>Coordinates<br>?</td><td></td><td></td></tr>
<tr><td>Interpretation<br>Settlement?</td><td>Estimated size<br>?</td><td>Date range<br>LC-EH</td><td></td></tr>
<tr><td colspan="5">Topography<br>The precise location is unknown; it may duplicate some of the following sites</td></tr>
<tr><td colspan="5">Investigations<br>Found by the North-Crimean Expedition in 1987</td></tr>
<tr><td colspan="5">Site interpretation<br>Settlement?</td></tr>
<tr><td colspan="5">Bibliography<br>Lancov 2004, fig. 1.75-76; p. 62; Lantsov & Uzhentzev 2007, 681</td></tr>
</table>

103 The site is mentioned as Skalistoe 1 in the figure, but as Skalistoe 2 in the text. Since the Hellenistic site under the fishing station was found first that one should be Skalistoe 1 and the Bronze Age site Skalistoe 2.

<table>
<tr><td rowspan="2">Skalistoe 5</td><td></td><td>Coordinates<br>493899/5054423</td><td></td><td>Figs.<br>4.12-20</td></tr>
<tr><td>Interpretation<br>Fortified farmhouse; ash hill</td><td>Estimated size<br>Ca. 1.5 ha</td><td>Date range<br>LC-EH; LH?</td><td></td></tr>
<tr><td colspan="5">Topography<br>Located on flat high cliff immediately to the sea shore; partly eroded away (wall in profile)</td></tr>
<tr><td colspan="5">Investigations<br>Partially excavated in 1973 by Z.Ch. Popandopulo; revisited by DSP Extensive survey 17.05.07; GS 3026-3027/2007 FL 10; sample of mudbrick from tower. Excavation: 2007 FL1-9.</td></tr>
<tr><td colspan="2">Geomagnetic survey<br>T. Smekalova<br>Dimensions ca. 90 x 40 m</td><td colspan="3">Description of measuring conditions<br>Good; NW part of the site eroded into the sea; at the tower, trenches left by stone looters</td></tr>
<tr><td colspan="5">Excavation Excavated and interpreted by S. Lancov, V. Stolba</td></tr>
<tr><td></td><td>Coordinates<br>493886/5054489</td><td>Trench Dimensions<br>2 x 2 m</td><td colspan="2">Features Exposed features include the remains of a rectangular tower measuring ca. 9 x 9 m and ash hill as well as E corner of a surrounding wall observable in the tracks of a dirt road<br>Trial trench Section of W wall of the building overlaid by ash hill</td></tr>
<tr><td colspan="5">Description of stratigraphy by layer:<br>Trial trench<br>1) 0-0.15 m topsoil<br>2) 0.15-0.30 m ashy layer<br>3) 0.30-0.40/0.45 m ashy layer with stone debris<br>4) 0.40/0.45 m top level of mudbrick debris<br>5) 0.45-0.60 m amorphous mudbrick debris<br>6) 0.60/0.75 m top level of buried topsoil<br>7) 0.75-1 m buried topsoil<br>8) 1 m -loam</td></tr>
<tr><td colspan="2">Finds<br>Pottery, shells, animal bones, tile</td><td>Date range<br>Late 4th-early 3rd c. BC</td><td colspan="2">Interpretation Among the transport amphorai and tableware the Chersonesean import is dominant. Pottery assamblage is characteristic of Chersonesean rural sites</td></tr>
<tr><td colspan="5">Site interpretation<br>Fortified farmhouse; ash hill</td></tr>
</table>

| **Skalistoe 6** | | *Coordinates*<br>4948812/5054794 | | |
|---|---|---|---|---|
| | *Interpretation*<br>Farmhouse?; settlement? | *Estimated size*<br>ca. 100 x 50 m | *Date range*<br>LC-EH | |
| *Topography*<br>Situated on small tongue between two small ravines close to sea | | | | |
| *Investigations*<br>Found during the DSP **Extensive survey** 17.05.07; GS 3023 | | | | |
| *Site interpretation*<br>Farmhouse?; settlement? | | | | |

| **Skalistoe 7** | | *Coordinates*<br>493890/5053886 | | |
|---|---|---|---|---|
| | *Interpretation*<br>Farmhouse? | *Estimated size*<br>50 x 50 m | *Date range*<br>LC-EH | |
| *Topography*<br>Located slightly inland at E side of small ravine on gentle slope facing the sea immediately S of Skalistoe *chutor*, which is depicted on the half-verst map | | | | |
| *Investigations*<br>Found during the DSP **Extensive survey** 12.05.07; GS 3028 | | | | |
| *Site interpretation*<br>Farmhouse? | | | | |

| **Skalistoe 8** | | *Coordinates*<br>493454/5054090 | | *Fig.*<br>4.11 |
|---|---|---|---|---|
| | *Interpretation*<br>Farmhouse? | *Estimated size*<br>ca. 100 x 50 m | *Date range*<br>LC-EH | |
| *Topography*<br>Situated on cliff facing the sea | | | | |
| *Investigations*<br>Found in DSP **Extensive survey** 17.05.07; GS 3029 | | | | |
| *Site interpretation*<br>Farmhouse? | | | | |

<table>
<tr><td rowspan="2">Skalistoe 9</td><td></td><td colspan="2">Coordinates<br>494744/50548723</td><td></td></tr>
<tr><td>Interpretation<br>Settlement? farmhouse?</td><td>Estimated size<br>ca. 200 x 100 m</td><td>Date range<br>LC-EH</td><td></td></tr>
<tr><td colspan="5">Topography<br>Situated on high cliff immediately along the present sea shore, very widely extended almost between two ravines</td></tr>
<tr><td colspan="5">Investigations<br>Found during the DSP Extensive survey 17.05.07; GS 3024</td></tr>
<tr><td colspan="5">Site interpretation<br>Settlement or farmhouse?</td></tr>
</table>

<table>
<tr><td rowspan="2">Skalistoe 10</td><td></td><td colspan="2">Coordinates<br>494334/5054685</td><td></td></tr>
<tr><td>Interpretation<br>Farmhouse?</td><td>Estimated size<br>ca. 50 x 50 m</td><td>Date range<br>LC-EH; LH</td><td></td></tr>
<tr><td colspan="5">Topography<br>Located on cliff facing the sea</td></tr>
<tr><td colspan="5">Investigations<br>Found during the DSP Extensive survey 17.05.07; GS 3025</td></tr>
<tr><td colspan="5">Site interpretation<br>Farmhouse?</td></tr>
</table>

<table>
<tr><td rowspan="2">Vetrenaja Bay</td><td></td><td colspan="2">Coordinates<br>478418/5041947</td><td>Figs.<br>4.29-30</td></tr>
<tr><td>Interpretation<br>Farmhouse</td><td>Estimated size<br>?</td><td>Date range<br>LC-EH</td><td></td></tr>
<tr><td colspan="5">Topography<br>Located slightly over 1.5 km NE of the Černomorskoe Settlement. It is positioned on a high cliff; today most of it is eroded away. Originally it was located in the SE corner of the plot as outlined by the cadastre</td></tr>
<tr><td colspan="5">Investigations<br>Partially excavated by the Tarchankut Expedition headed by A.N. Ščeglov between 1963 and 1966. The building consists of a series of rectangular rooms around a courtyard. 40 x 25-30 m of the building's extension is preserved. In one room measuring 24.6 m² 10 Sinopean pithoi were sitting side by side. In the building was also found tableware, e.g. black-gloss pottery (fish plate, kantharos: Ščeglov 1967, fig. 15). Interesting for the understanding of the relation to Chersonesos is the find of a terracotta representation of Herakles' club (Ščeglov 1967, fig. 16)</td></tr>
<tr><td colspan="5">Site interpretation<br>Farmhouse</td></tr>
<tr><td colspan="5">Bibliography<br>Ščeglov 1967; 1978</td></tr>
</table>

| **Vladimirovka** | | *Coordinates*<br>? | | |
|---|---|---|---|---|
| | *Interpretation*<br>Settlement? | *Estimated size*<br>? | *Date range*<br>Late BA | |
| *Topography*<br>Located inland; the more precise location is unknown | | | | |
| *Investigations*<br>Excavated by V.A. Kolotuchin; year unknown; the site could not be verified in the DSP survey | | | | |
| *Site interpretation*<br>Settlement? | | | | |
| *Bibliography*<br>Kolotuchin 2003, 68 fig. 1.28 | | | | |

| **Vodopojnoe 1 (see chapter 4.3.3)** | | *Coordinates*<br>493242/5047343 | | *Fig.*<br>4.38 |
|---|---|---|---|---|
| | *Interpretation*<br>Settlement | *Estimated size*<br>? | *Date range*<br>Late BA | |
| *Topography*<br>Located on W side of small ravine where it meets the Džarylgač Lake | | | | |
| *Investigations*<br>Found by V.A. Kolotuchin; year unknown<br>Systematic **Extensive survey** by DPS in 2007 | | | | |
| *Site interpretation*<br>Settlement | | | | |
| *Bibliography*<br>Kolotuchin 2003, 40, 68 fig. 1.22 | | | | |

| **Vodopojnoe 2** | | *Coordinates*<br>497552/5049033 | | |
|---|---|---|---|---|
| | *Interpretation*<br>Settlement, kurgan and farmhouse? | *Estimated size*<br>50 x 50 m | *Date range*<br>Late BA; LC-EH?; EM | |
| *Topography*<br>Situated in valley bottom on low elevation connected with soil bank to Bronze Age(?) kurgan | | | | |
| *Investigations*<br>*Razvedki* S.G. Koltuchov; year unknown; visited during the DSP **Extensive survey** 27.05.07 | | | | |
| *Site interpretation*<br>Multi-period settlement, kurgan and farmhouse? | | | | |
| *Bibliography*<br>Kolotuchin 2003, 68 fig. 1.23 | | | | |

| **Vodopojnoe 3A and B** | | *Coordinates*<br>4996923/5049333 (3A); 499872/5049211 (3B) | | |
|---|---|---|---|---|
| | *Interpretation*<br>Settlement | *Estimated size*<br>ca. 3 ha | *Date range*<br>Late BA; LC-EH; EM | |
| *Topography*<br>Extended area at W slope of low ravine 500-1000 m up a side branch of the main ravine | | | | |
| *Investigations*<br>*Razvedki* S.G. Koltuchov; year unknown; visited during the DSP **Extensive survey** 27.05.07; GS 3065 | | | | |
| *Site interpretation*<br>Multi-period settlement | | | | |
| *Bibliography*<br>– | | | | |

| **Vodopojnoe 4** | | *Coordinates*<br>500244/5048513 | | |
|---|---|---|---|---|
| | *Interpretation*<br>Settlement? | *Estimated size*<br>? | *Date range*<br>Late BA | |
| *Topography*<br>Located at side of ravine at confluence of two small streams | | | | |
| *Investigations*<br>*Razvedki* S.G. Koltuchov; year unknown. The site could not be verified in the DSP **Extensive survey** | | | | |
| *Site interpretation*<br>Settlement? | | | | |
| *Bibliography*<br>Kolotuchin 2003, 68 fig. 1.24 | | | | |

| **Vodopojnoe 5** | | *Coordinates*<br>498081/5046452 | | |
|---|---|---|---|---|
| | *Interpretation*<br>Settlement? | *Estimated size*<br>? | *Date range*<br>Late BA | |
| *Topography*<br>Situated at side of ravine at confluence of two small streams | | | | |
| *Investigations*<br>*Razvedki* S.G. Koltuchov; year unknown. The site could not be verified in the DSP **Extensive survey** | | | | |
| *Site interpretation*<br>Settlement? | | | | |
| *Bibliography*<br>– | | | | |

# Appendix 2A
# Finds per site

*S. Handberg*

This appendix lists the finds from each site, according to the contents of each of the bags of material collected, first in the central area of the site, i.e. 'core units', and where relevant, in the surrounding area, i.e. 'adjacent units'. The finds are presented in more detail in Appendix 2B, and they are discussed and analyzed in Chapter 5.2.

| AR | Architectural |
|---|---|
| BO | Bone |
| CW | Cookingware |
| HW | Handmade ware |
| IN | Indeterminable |
| IR | Iron |
| LI | Lithics |
| TE | Terracotta |
| TR | Transport amphorae |
| TW | Tabelware |
| UT | Utility ware |

Table appendix 2A.1: list of finds category abbreviations

| BagID | Function | Total | Body | Rim | Handle | Base | Fabric | Surface | Shape | Period | Cat. |
|---|---|---|---|---|---|---|---|---|---|---|---|
| 100601 | TW | 5 | 3 | 1 | 0 | 1 | Red | Slip | Jug | LC-EH | |
| 100601 | TW | 3 | 0 | 3 | 0 | 0 | Red | Slip | Bowl | LC-EH | 48, 50 |
| 100602 | TW | 2 | 0 | 2 | 0 | 0 | Red | Slip | Jug | LC-EH | 44 |
| 100602 | TW | 3 | 3 | 0 | 0 | 0 | Attic | Glazed | Cup | LC-EH | 52 |
| 100602 | TW | 1 | 0 | 0 | 1 | 0 | Orange | None | Jug | LC-EH | |
| 100602 | TW | 2 | 1 | 1 | 0 | 0 | Chersonesean | Slip | Jug | LC-EH | 36, 40 |
| 100602 | TW | 1 | 0 | 0 | 0 | 1 | Light | None | Lekythos | LC-EH | 54 |
| 100602 | TW | 1 | 0 | 1 | 0 | 0 | Red | Slip | Fish-plate | LC-EH | 55 |
| 100602 | TW | 5 | 5 | 0 | 0 | 0 | Grey | Slip | ? | LC-EH | |
| 100603 | TW | 1 | 0 | 1 | 0 | 0 | Light | None | Bowl | LC-EH | 45 |
| 100603 | TW | 2 | 1 | 1 | 0 | 0 | Red | Glazed | Kantharos | LC-EH | 58, 60 |
| 100603 | TW | 1 | 1 | 0 | 0 | 0 | Grey | None | Jug | LC-EH | 47 |
| 100603 | TW | 16 | 16 | 0 | 0 | 0 | Red | Decorated | ? | LC-EH | |
| 100603 | TW | 3 | 0 | 3 | 0 | 0 | Red | None | Bowl | LC-EH | 59 |
| 100603 | TW | 14 | 14 | 0 | 0 | 0 | Red | None | ? | LC-EH | |
| 100603 | TW | 1 | 0 | 0 | 0 | 1 | Grey | None | Fishplate | LC-EH | 56 |
| 100603 | TW | 7 | 7 | 0 | 0 | 0 | Grey | None | ? | LC-EH | |
| 100603 | TW | 2 | 2 | 0 | 0 | 0 | Attic | Glazed | Fishplate | LC-EH | 57 |
| 100603 | TW | 1 | 0 | 1 | 0 | 0 | Sinopean | None | Jug | LC-EH | 46 |
| 101101 | TW | 1 | 0 | 1 | 0 | 0 | Grey | Slip | Bowl | LC-EH | 61 |
| 101301 | TW | 2 | 2 | 0 | 0 | 0 | Red | None | Closed shape | LC-EH | |
| 101401 | TW | 2 | 1 | 0 | 1 | 0 | Red | None | Jug? | LC-EH | |
| 200201 | TW | 2 | 1 | 0 | 0 | 1 | Chersonesean | None | Jug | LC-EH | |
| 200301 | TW | 3 | 3 | 0 | 0 | 0 | Chersonesean | Slip | Closed shape | LC-EH | |
| 200301 | TW | 1 | 0 | 1 | 0 | 0 | Red | Glazed | lekthos | LC-EH | 53 |
| 200301 | TW | 1 | 0 | 1 | 0 | 0 | Red | None | Bowl | LC-EH | |
| 200501 | TW | 1 | 0 | 0 | 0 | 1 | Red | Slip | Jug | LC-EH | 62 |
| 200601 | TW | 1 | 1 | 0 | 0 | 0 | Grey | None | Jug? | LC-EH | |
| 200601 | TW | 1 | 0 | 1 | 0 | 0 | Chersonesean? | Slip | Jug | LC-EH | 37 |
| 200601 | TW | 2 | 2 | 0 | 0 | 0 | Red | None | ? | LC-EH | |
| 100401 | HW | 1 | 1 | 0 | 0 | 0 | Brown | None | Cooking vessel | LC_EH | 2 |
| 100401 | UT | 1 | 0 | 1 | 0 | 0 | Red | None | Louterion | LC-EH | 67 |
| 100401 | UT | 2 | 0 | 2 | 0 | 0 | Light | Slip | Louterion | LC-EH | 65-66 |
| 100601 | UT | 2 | 1 | 1 | 0 | 0 | Light | Slip? | Louterion | LC-EH | 64 |
| 100601 | UT | 2 | 2 | 0 | 0 | 0 | Red | None | ? | LC-EH | |
| 100601 | UT | 1 | 0 | 1 | 0 | 0 | Red | Slip | Basin | LC-EH | 68 |
| 100601 | UT | 2 | 2 | 0 | 0 | 0 | Orange | None | ? | LC-EH | |
| 100601 | UT | 5 | 5 | 0 | 0 | 0 | Brown | None | ? | LC-EH | |
| 100601 | UT | 3 | 0 | 3 | 0 | 0 | Red | None | Cooking vessel | LC-EH | 71-73 |
| 100602 | UT | 4 | 4 | 0 | 0 | 0 | Red | None | ? | LC-EH | |
| 100602 | UT | 1 | 0 | 1 | 0 | 0 | UtRed | None | Cooking vessel | LC-EH | 74 |
| 100602 | UT | 1 | 1 | 0 | 0 | 0 | Light | Smoothed | ? | LC-EH | |
| 100603 | UT | 4 | 4 | 0 | 0 | 0 | Red | None | ? | LC-EH | |

| BagID | Function | Total | Body | Rim | Handle | Base | Fabric | Surface | Shape | Period | Cat. |
|---|---|---|---|---|---|---|---|---|---|---|---|
| 100603 | UT | 1 | 0 | 1 | 0 | 0 | Brown | None | Basin | LC-EC | 77 |
| 100603 | UT | 2 | 0 | 2 | 0 | 0 | Light | None | Cooking vessel | LC-EH | 75-76 |
| 200301 | UT | 1 | 0 | 1 | 0 | 0 | Sinopean | None | Pithos | LC-EH | |
| 200301 | UT | 1 | 0 | 0 | 0 | 1 | Light | Slip | Louterion | LC-EH | 79 |
| 100603 | HW | 61 | 57 | 3 | 1 | 0 | Grey | None | Pot | LC-EH? | 3-8 |
| 100801 | HW | 1 | 1 | 0 | 0 | 0 | Grey | None | Pot | LC-EH? | |
| 100501 | IN | 1 | 1 | 0 | 0 | 0 | Red | None | ? | LC-EH? | |
| 100602 | IN | 23 | 23 | 0 | 0 | 0 | | None | ? | LC-EH? | |
| 100603 | IN | 26 | 26 | 0 | 0 | 0 | | | ? | LC-EH? | |
| 101101 | IN | 1 | 1 | 0 | 0 | 0 | Red | None | ? | LC-EH? | |
| 100401 | TR | 1 | 1 | 0 | 0 | 0 | Red mica | None | Amphora | LC-EH? | |
| 100601 | TR | 1 | 1 | 0 | 0 | 0 | Brown | None | Amphora | LC-EH? | |
| 100601 | TR | 3 | 2 | 0 | 1 | 0 | Red | Slip | Amphora | LC-EH? | 26 |
| 100701 | TR | 1 | 1 | 0 | 0 | 0 | Red mica | None | Amphora | LC-EH? | |
| 100401 | TW | 1 | 1 | 0 | 0 | 0 | Red mica | None | ? | LC-EH? | |
| 100602 | TW | 9 | 0 | 0 | 0 | 0 | Red | None | ? | LC-EH? | |
| 100701 | TW | 1 | 1 | 0 | 0 | 0 | Red mica | None | ? | LC-EH? | |
| 200101 | TW | 1 | 1 | 0 | 0 | 0 | Red mica | none | ? | LC-EH? | |
| 200201 | TW | 4 | 4 | 0 | 0 | 0 | Red | None | ? | LC-EH? | |
| 200301 | TW | 1 | 1 | 0 | 0 | 0 | Grey | None | ? | LC-EH? | |
| 200301 | TW | 2 | 2 | 0 | 0 | 0 | Red mica | None | ? | LC-EH? | |
| 200401 | TW | 3 | 3 | 0 | 0 | 0 | Red | None | ? | LC-EH? | |
| 200401 | TW | 1 | 1 | 0 | 0 | 0 | Grey | None | ? | LC-EH? | |
| 200601 | TW | 1 | 1 | 0 | 0 | 0 | Red mica | None | Closed shape | LC-EH? | |
| 100401 | UT | 1 | 0 | 1 | 0 | 0 | Sinopean | None | Pithos | LC-EH? | 69 |
| 100601 | UT | 1 | 0 | 1 | 0 | 0 | Sinopean | None | Pithos | LC-EH? | 70 |
| 200301 | UT | 1 | 0 | 1 | 0 | 0 | Red | None | Cooking vessel | LC-EH | 78 |
| 200301 | UT | 2 | 2 | 0 | 0 | 0 | Red | None | Closed shape | LC-EH? | |
| 200401 | UT | 1 | 0 | 1 | 0 | 0 | Red | None | Pithos | LC-EH? | 80 |
| 200401 | TW | 1 | 1 | 0 | 0 | 0 | Tatarian red | Glazed | ? | EM | |
| 200101 | AR | 1 | 0 | 1 | 0 | 0 | Light | None | Tile | | |
| 100401 | HW | 2 | 2 | 0 | 0 | 0 | Grey | Burnished | Pot | EH | 1 |
| 200301 | HW | 1 | 1 | 0 | 0 | 0 | Red | None | Pot | | |
| 200301 | HW | 1 | 1 | 0 | 0 | 0 | Grey | None | ? | | |
| 200701 | HW | 1 | 1 | 0 | 0 | 0 | Grey | None | ? | | |
| 100401 | IN | 1 | 1 | 0 | 0 | 0 | Red? | none | ? | | |
| 101301 | IN | 1 | 1 | 0 | 0 | 0 | Red mica | None | ? | | |
| 101301 | IN | 2 | 2 | 0 | 0 | 0 | Red coarse | None | ? | | |
| 101301 | IN | 1 | 1 | 0 | 0 | 0 | Red | None | ? | | |
| 200301 | IN | 1 | 1 | 0 | 0 | 0 | Red | None | ? | | |
| 200401 | IN | 13 | 13 | 0 | 0 | 0 | Red | None | ? | | |
| 200401 | IN | 1 | 1 | 0 | 0 | 0 | Red mica | None | ? | | |
| 200501 | IN | 5 | 5 | 0 | 0 | 0 | Red | None | ? | | |

| BagID | Function | Total | Body | Rim | Handle | Base | Fabric | Surface | Shape | Period | Cat. |
|---|---|---|---|---|---|---|---|---|---|---|---|
| 200901 | IN | 1 | 1 | 0 | 0 | 0 | Red mica | None | ? | | |
| 100401 | LI | 1 | 0 | 0 | 0 | 0 | | | | | |
| 200301 | LI | 3 | 0 | 0 | 0 | 0 | | | ? | | |
| 200601 | LI | 1 | 0 | 0 | 0 | 0 | | | | | |
| 200601 | LI | 1 | 0 | 0 | 0 | 0 | | | | | |
| 100401 | SH | 1 | 0 | 0 | 0 | 0 | | | | | |
| 101301 | SH | 1 | 0 | 0 | 0 | 0 | | | | | |
| 101201 | TW | 2 | 1 | 0 | 0 | 1 | | | | | |
| 200401 | UT | 1 | 1 | 0 | 0 | 0 | Red mica | None | ? | | |
| 200401 | UT | 1 | 1 | 0 | 0 | 0 | Light brown | None | ? | | |
| 200401 | UT | 1 | 1 | 0 | 0 | 0 | Brown | None | ? | | |

## DSP07-F02-01

Finds table core units

| BagID | Function | Total | Body | Rim | Handle | Base | Fabric | Surface | Shape | Period | Cat. |
|---|---|---|---|---|---|---|---|---|---|---|---|
| 216401 | HW | 1 | 1 | 0 | 0 | 0 | Red | None | Pot | LC-EH | |
| 216401 | HW | 3 | 3 | 0 | 0 | 0 | Grey | None | Pot | LC-EH | |
| 216402 | HW | 4 | 4 | 0 | 0 | 0 | Red | None | Pot | LC-EH | |
| 216402 | HW | 38 | 37 | 0 | 0 | 1 | Grey | None | Pot | LC-EH | 81 |
| 216501 | HW | 4 | 4 | 0 | 0 | 0 | Grey | None | Pot | LC-EH | |
| 220501 | HW | 2 | 2 | 0 | 0 | 0 | Grey | None | Pot | LC-EH | |
| 220502 | HW | 3 | 3 | 0 | 0 | 0 | Grey | None | Pot | LC-EH | |
| 220601 | HW | 2 | 2 | 0 | 0 | 0 | Grey | None | Pot | LC-EH | |
| 220602 | HW | 8 | 8 | 0 | 0 | 0 | Grey | None | Pot | LC-EH | |
| 216401 | IN | 3 | 3 | 0 | 0 | 0 | Red | None | ? | LC-EH | |
| 216402 | IN | 7 | 7 | 0 | 0 | 0 | Red | None | ? | LC-EH | |
| 216501 | IN | 2 | 2 | 0 | 0 | 0 | Red | None | ? | LC-EH | |
| 216501 | IN | 2 | 2 | 0 | 0 | 0 | Red mica | None | ? | LC-EH | |
| 220501 | IN | 2 | 2 | 0 | 0 | 0 | Red | None | ? | LC-EH | |
| 220502 | IN | 1 | 1 | 0 | 0 | 0 | Red | None | Closed shape? | LC-EH | |
| 220601 | IN | 1 | 1 | 0 | 0 | 0 | Red | None | ? | LC-EH | |
| 220601 | IN | 3 | 3 | 0 | 0 | 0 | Red | None | ? | LC-EH | |
| 216401 | TR | 1 | 1 | 0 | 0 | 0 | Sinopean | None | Amphora | LC-EH | |
| 216401 | TR | 2 | 2 | 0 | 0 | 0 | Herakleian | None | Amphora | LC-EH | |
| 216402 | TR | 2 | 1 | 0 | 1 | 0 | Red | None | Amphora | LC-EH | 83 |
| 216402 | TR | 1 | 1 | 0 | 0 | 0 | Sinopean | None | Amphora | LC-EH | |
| 216402 | TR | 1 | 1 | 0 | 0 | 0 | Chersonesean | None | Amphora | LC-EH | |
| 216501 | TR | 2 | 2 | 0 | 0 | 0 | Herakleian | None | Amphora | LC-EH | |
| 216501 | TR | 3 | 3 | 0 | 0 | 0 | Chersonesean | None | Amphora | LC-EH | |
| 220501 | TR | 1 | 1 | 0 | 0 | 0 | Herakleian | None | Amphora | LC-EH | |

| BagID | Function | Total | Body | Rim | Handle | Base | Fabric | Surface | Shape | Period | Cat. |
|---|---|---|---|---|---|---|---|---|---|---|---|
| 220501 | TR | 2 | 0 | 0 | 2 | 0 | Chian? | None | Amphora | LC-EH | **84** |
| 220502 | TR | 1 | 0 | 0 | 1 | 0 | Chersonesean | None | Amphora | LC-EH | |
| 220502 | TR | 2 | 2 | 0 | 0 | 0 | Herakleian | None | Amphora | LC-EH | |
| 220601 | TR | 3 | 3 | 0 | 0 | 0 | Herakleian | None | Amphora | LC-EH | |
| 220602 | TR | 6 | 6 | 0 | 0 | 0 | Chersonesean | None | Amphora | LC-EH | |
| 220602 | TR | 5 | 4 | 1 | 0 | 0 | Herakleian? | None | Amphora | LC-EH | **82** |
| 216401 | TW | 1 | 1 | 0 | 0 | 0 | Red | None | Jug | LC-EH | |
| 216402 | TW | 1 | 0 | 1 | 0 | 0 | Red | Slipped? | Jug | LC-EH | |
| 216402 | TW | 3 | 3 | 0 | 0 | 0 | Red | Slip | ? | LC-EH | |
| 220501 | TW | 1 | 1 | 0 | 0 | 0 | Red | Decorated | Jug? | LC-EH | **86** |
| 220502 | TW | 1 | 0 | 1 | 0 | 0 | Light | None | Jug? | LC-EH | **85** |
| 220502 | TW | 1 | 1 | 0 | 0 | 0 | Red | None | Closed shape | LC-EH | |
| 220602 | TW | 1 | 0 | 1 | 0 | 0 | Light | None | Open shape? | LC-EH | |
| 220602 | TW | 2 | 2 | 0 | 0 | 0 | Red | Slip | Open shape | LC-EH | |
| 216401 | UT | 1 | 0 | 1 | 0 | 0 | Red | Slip | Open shape | LC-EH | |
| 216501 | UT | 2 | 2 | 0 | 0 | 0 | Brown? | Smoothed? | Open shape? | LC-EH | |
| 220601 | UT | 1 | 0 | 1 | 0 | 0 | Brown? | None | Pithos? | LC-EH | |
| 216402 | IN | 3 | 3 | 0 | 0 | 0 | Light | None | ? | LC-EH? | |
| 216402 | IN | 2 | 2 | 0 | 0 | 0 | Brown | None | ? | LC-EH? | |

Finds table adjacent units

| BagID | Function | Total | Body | Rim | Handle | Base | Fabric | Surface | Shape | Period | Cat. |
|---|---|---|---|---|---|---|---|---|---|---|---|
| 216501 | HW | 4 | 4 | 0 | 0 | 0 | Grey | None | Pot | LC-EH | |
| 216501 | IN | 2 | 2 | 0 | 0 | 0 | Red mica | None | ? | LC-EH | |
| 216501 | IN | 2 | 2 | 0 | 0 | 0 | Red | None | ? | LC-EH | |
| 216501 | TR | 2 | 2 | 0 | 0 | 0 | Herakleian | None | Amphora | LC-EH | |
| 216501 | TR | 3 | 3 | 0 | 0 | 0 | Chersonesean | None | Amphora | LC-EH | |
| 216601 | TR | 1 | 1 | 0 | 0 | 0 | Chersonesean | None | Amphora | LC-EH | |
| 219401 | TR | 1 | 1 | 0 | 0 | 0 | Sinopean | None | Amphora | LC-EH | |
| 216501 | UT | 2 | 2 | 0 | 0 | 0 | Brown? | Smoothed? | Open shape? | LC-EH | |
| 219501 | UT | 1 | 1 | 0 | 0 | 0 | Red | None | ? | LC-EH? | |

## DSP07-F02-02

Finds table core units

| BagID | Function | Total | Body | Rim | Handle | Base | Fabric | Surface | Shape | Period | Cat. |
|---|---|---|---|---|---|---|---|---|---|---|---|
| 210201 | HW | 1 | 1 | 0 | 0 | 0 | Grey | None | Pot | LC-EH | |
| 214601 | HW | 1 | 1 | 0 | 0 | 0 | Grey | None | Pot | LC-EH | |
| 214701 | IN | 1 | 1 | 0 | 0 | 0 | Red | None | ? | LC-EH | |
| 210001 | TR | 1 | 1 | 0 | 0 | 0 | Herakleian | None | Amphora | LC-EH | |
| 210101 | TR | 1 | 1 | 0 | 0 | 0 | Herakleian | None | Amphora | LC-EH | |
| 210201 | TR | 1 | 1 | 0 | 0 | 0 | Chersonesean | None | Amphora | LC-EH | |
| 210202 | TR | 1 | 1 | 0 | 0 | 0 | Chersonesean | None | Amphora | LC-EH | |
| 211701 | TR | 2 | 1 | 0 | 1 | 0 | Chersonesean | None | Amphora | LC-EH | |
| 214701 | TR | 1 | 0 | 0 | 1 | 0 | Chersonesean | None | Amphora | LC-EH | |
| 210001 | TW | 1 | 1 | 0 | 0 | 0 | Red | Glazed | Open shape | LC-EH | |
| 211901 | TW | 1 | 1 | 0 | 0 | 0 | Red | None | Closed shape | LC-EH | |
| 210001 | UT | 1 | 0 | 1 | 0 | 0 | Red | None | Pithos | LC-EH | |
| 100201 | HW | 1 | 1 | 0 | 0 | 0 | Grey | None | Closed shape | LC-EH? | |
| 100201 | IN | 1 | 1 | 0 | 0 | 0 | Red | None | Closed shape | LC-EH? | |
| 100201 | TR | 1 | 1 | 0 | 0 | 0 | Red | None | Amphora | LC-EH? | |
| 100201 | TR | 1 | 0 | 0 | 1 | 0 | Herakleian | None | Amphora | LC-EH? | **87** |
| 100201 | UT | 1 | 0 | 0 | 0 | 1 | Red | None | Closed shape | EM | **89** |
| 100201 | TW | 1 | 0 | 0 | 0 | 1 | Red | None | Jug | EM? | **88** |
| 210202 | TW | 1 | 0 | 0 | 1 | 0 | Tatarian | None | Jug | PM | |
| 211801 | IN | 1 | 1 | 0 | 0 | 0 | Red | None | ? | | |
| 100201 | LI | 1 | 0 | 0 | 0 | 0 | | | | | |

Finds table adjacent units

| BagID | Function | Total | Body | Rim | Handle | Base | Fabric | Surface | Shape | Period | Cat. |
|---|---|---|---|---|---|---|---|---|---|---|---|
| 210401 | TR | 1 | 1 | 0 | 0 | 0 | Chersonesean | None | Amphora | LC-EH | |

## DSP07-F02-03

Finds table core units

| BagID | Function | Total | Body | Rim | Handle | Base | Fabric | Surface | Shape | Period | Cat. |
|---|---|---|---|---|---|---|---|---|---|---|---|
| 123301 | HW | 2 | 2 | 0 | 0 | 0 | Grey | None | Pot | LC-EH | |
| 123302 | HW | 7 | 7 | 0 | 0 | 0 | Grey | None | Pot | LC-EH | |
| 123302 | HW | 3 | 3 | 0 | 0 | 0 | Grey coarse | None | ? | LC-EH | |
| 123303 | HW | 1 | 1 | 0 | 0 | 0 | Red | None | Pot | LC-EH | |
| 123303 | HW | 5 | 5 | 0 | 0 | 0 | Grey | None | Pot | LC-EH | |
| 123401 | HW | 3 | 3 | 0 | 0 | 0 | Grey coarse | None | ? | LC-EH | |
| 123401 | HW | 3 | 3 | 0 | 0 | 0 | Grey | None | Pot | LC-EH | |
| 123402 | HW | 6 | 6 | 1 | 0 | 0 | Grey | None | Pot | LC-EH | |
| 123402 | HW | 2 | 2 | 0 | 0 | 0 | Red | None | Pot | LC-EH | |
| 123402 | HW | 1 | 1 | 0 | 0 | 0 | Grey coarse | None | ? | LC-EH | |
| 123403 | HW | 8 | 8 | 0 | 0 | 0 | Grey | None | Pot | LC-EH | |
| 214601 | HW | 1 | 1 | 0 | 0 | 0 | Grey | None | Pot | LC-EH | |
| 123301 | IN | 2 | 2 | 0 | 0 | 0 | Red mica | None | ? | LC-EH | |
| 123302 | IN | 2 | 2 | 0 | 0 | 0 | Red mica | None | ? | LC-EH | |
| 123303 | IN | 4 | 4 | 0 | 0 | 0 | Red mica | None | ? | LC-EH | |
| 123401 | IN | 1 | 1 | 0 | 0 | 0 | Red | None | Closed shape? | LC-EH | |
| 214701 | IN | 1 | 1 | 0 | 0 | 0 | Red | None | ? | LC-EH | |
| 123301 | TR | 2 | 2 | 0 | 0 | 0 | Herakleian | None | Amphora | LC-EH | |
| 123301 | TR | 1 | 1 | 0 | 0 | 0 | Chersonesean | None | Amphora | LC-EH | |
| 123302 | TR | 2 | 2 | 0 | 0 | 0 | Herakleian | None | Amphora | LC-EH | |
| 123302 | TR | 1 | 0 | 0 | 1 | 0 | Chersonesean | None | Amphora | LC-EH | |
| 123401 | TR | 1 | 0 | 0 | 1 | 0 | Chersonesean | None | Amphora | LC-EH | **90** |
| 123402 | TR | 1 | 0 | 0 | 1 | 0 | Chersonesean | None | Amphora | LC-EH | **91** |
| 123402 | TR | 3 | 3 | 0 | 0 | 0 | Herakleian | None | Amphora | LC-EH | |
| 211701 | TR | 2 | 1 | 0 | 1 | 0 | Chersonesean | None | Amphora | LC-EH | |
| 214701 | TR | 1 | 0 | 0 | 1 | 0 | Chersonesean | None | Amphora | LC-EH | |
| 123302 | TW | 1 | 1 | 0 | 0 | 0 | Red | None | ? | LC-EH | |
| 123402 | TW | 2 | 2 | 0 | 0 | 0 | Red | None | Closed shape | LC-EH | |
| 123403 | TW | 1 | 1 | 0 | 0 | 0 | Light | None | Closed shape | LC-EH | |
| 123403 | TW | 1 | 1 | 0 | 0 | 0 | Red mica | None | Closed shape | LC-EH | |
| 211901 | TW | 1 | 1 | 0 | 0 | 0 | Red | None | Closed shape | LC-EH | |
| 123403 | IN | 1 | 1 | 0 | 0 | 0 | South Pontic | None | ? | LC-EH? | |
| 123302 | UT | 1 | 1 | 0 | 0 | 0 | Brown | None | Open shape | LC-EH? | |
| 211801 | IN | 1 | 1 | 0 | 0 | 0 | Red | None | ? | ? | |
| 123301 | LI | 1 | 0 | 0 | 0 | 0 | | | | | |
| 123302 | LI | 1 | 0 | 0 | 0 | 0 | | | | | |
| 123302 | LI | 1 | 0 | 0 | 0 | 0 | | | | | |
| 123303 | LI | 1 | 0 | 0 | 0 | 0 | | | | | |
| 123401 | LI | 1 | 0 | 0 | 0 | 0 | | | | | |

| BagID | Function | Total | Body | Rim | Handle | Base | Fabric | Surface | Shape | Period | Cat. |
|---|---|---|---|---|---|---|---|---|---|---|---|
| 123403 | SH | 1 | 0 | 0 | 0 | 0 | | | | | |

Finds table adjacent units

| BagID | Function | Total | Body | Rim | Handle | Base | Fabric | Surface | Shape | Period | Cat. |
|---|---|---|---|---|---|---|---|---|---|---|---|
| 123001 | TW | 1 | 1 | 0 | 0 | 0 | Red | Slipped? | Closed shape | LC-EH | |
| 125901 | IN | 1 | 1 | 0 | 0 | 0 | Brown | None | ? | LC-EH? | |
| 122801 | UT | 1 | 1 | 0 | 0 | 0 | Brown | None | Closed shape? | LC-EH? | |

## DSP07-F02-04

Finds table core units

| BagID | Function | Total | Body | Rim | Handle | Base | Fabric | Surface | Shape | Period | Cat. |
|---|---|---|---|---|---|---|---|---|---|---|---|
| 113401 | HW | 4 | 3 | 1 | 0 | 0 | Grey | None | Pot | LC-EH | |
| 113402 | HW | 8 | 8 | 0 | 0 | 0 | Grey | None | Pot | LC-EH | |
| 113402 | HW | 4 | 4 | 0 | 0 | 0 | Red | None | Pot | LC-EH | |
| 113403 | HW | 6 | 6 | 0 | 0 | 0 | Grey | None | Pot | LC-EH | |
| 118101 | HW | 6 | 6 | 0 | 0 | 0 | Grey | None | Pot | LC-EH | |
| 118102 | HW | 1 | 1 | 0 | 0 | 0 | Grey coarse | None | ? | LC-EH | |
| 118102 | HW | 4 | 4 | 0 | 0 | 0 | Grey | None | Pot | LC-EH | |
| 118103 | HW | 1 | 1 | 0 | 0 | 0 | Red | None | Pot | LC-EH | |
| 118103 | HW | 4 | 4 | 0 | 0 | 0 | Grey | None | Pot | LC-EH | |
| 113401 | TR | 1 | 0 | 1 | 0 | 0 | Chersonesean? | None | Amphora | LC-EH | 94 |
| 113403 | TR | 1 | 1 | 0 | 0 | 0 | Herakleian | None | Amphora | LC-EH | |
| 113401 | TW | 1 | 1 | 0 | 0 | 0 | Red Tatarian | Glazed | Open shape | LC-EH | |
| 113401 | TW | 1 | 1 | 0 | 0 | 0 | Red coarse | None | Closed shape | LC-EH | |
| 113401 | TW | 1 | 1 | 0 | 0 | 0 | Red | Slip | Closed shape | LC-EH | |
| 118102 | TW | 1 | 1 | 0 | 0 | 0 | Red mica | None | ? | LC-EH | |
| 118103 | TW | 1 | 1 | 0 | 0 | 0 | Red mica | None | Closed shape | LC-EH | |
| 113401 | HW | 1 | 1 | 0 | 0 | 0 | Grey coarse | None | ? | LC-EH? | |
| 113401 | IN | 1 | 1 | 0 | 0 | 0 | Red | None | ? | LC-EH? | |
| 113402 | IN | 2 | 2 | 0 | 0 | 0 | Brown | None | ? | LC-EH? | |
| 118101 | IN | 1 | 1 | 0 | 0 | 0 | Brown | None | ? | LC-EH? | |
| 118101 | UT | 1 | 1 | 0 | 0 | 0 | Brown | Slip | ? | LC-EH? | |
| 118101 | UT | 1 | 1 | 0 | 0 | 0 | South Pontic | None | ? | LC-EH? | |
| 113403 | TW | 1 | 1 | 0 | 0 | 0 | China | None | ? | Modern | |
| 118103 | TW | 1 | 0 | 0 | 0 | 1 | China | Glazed | ? | Modern | |
| 118101 | SH | 1 | 0 | 0 | 0 | 0 | | | | | |
| 118102 | SH | 1 | 0 | 0 | 0 | 0 | | | | | |

Finds table adjacent units

| BagID | Function | Total | Body | Rim | Handle | Base | Fabric | Surface | Shape | Period | Cat. |
|---|---|---|---|---|---|---|---|---|---|---|---|
| 108901 | HW | 2 | 2 | 0 | 0 | 0 | Grey | None | Pot | LC-EH | |
| 108901 | HW | 1 | 1 | 0 | 0 | 0 | Red | None | Pot | LC-EH | 92 |
| 110301 | HW | 2 | 1 | 1 | 0 | 0 | Grey | None | Pot | LC-EH | 93 |
| 110301 | HW | 2 | 2 | 0 | 0 | 0 | Red | None | Pot | LC-EH | |
| 110401 | HW | 2 | 2 | 0 | 0 | 0 | Red | None | Pot | LC-EH | |
| 112001 | HW | 1 | 1 | 0 | 0 | 0 | Grey | None | Pot | LC-EH | |
| 112201 | HW | 1 | 1 | 0 | 0 | 0 | Red | None | Pot | LC-EH | |
| 113201 | HW | 1 | 1 | 0 | 0 | 0 | Red | None | Pot | LC-EH | |
| 112001 | IN | 2 | 2 | 0 | 0 | 0 | Red | None | ? | LC-EH | |
| 115001 | IN | 1 | 1 | 0 | 0 | 0 | South Pontic | None | Closed shape | LC-EH | |
| 110401 | TR | 1 | 1 | 0 | 0 | 0 | Chersonesean? | None | Amphora | LC-EH | |
| 112001 | TR | 1 | 0 | 0 | 1 | 0 | Chersonesean? | None | Amphora | LC-EH | 95 |
| 112001 | TR | 1 | 1 | 0 | 0 | 0 | Herakleian | None | Amphora | LC-EH | |
| 112201 | TR | 1 | 0 | 1 | 0 | 0 | Red mica | None | Amphora | LC-EH | 96 |
| 113201 | TR | 1 | 1 | 0 | 0 | 0 | Herakleian | None | Amphora | LC-EH | |
| 113201 | TR | 1 | 1 | 0 | 0 | 0 | Chersonesean | None | Amphora | LC-EH | |
| 115001 | TR | 2 | 1 | 0 | 1 | 0 | Herakleian | None | Amphora | LC-EH | |
| 115101 | TR | 1 | 1 | 0 | 0 | 0 | Sinopean | None | Amphora | LC-EH | |
| 115101 | TR | 1 | 0 | 0 | 0 | 1 | Chersonesean | None | Amphora | LC-EH | |
| 115101 | TR | 2 | 2 | 0 | 0 | 0 | Herakleian | Slip | Amphora | LC-EH | |
| 116201 | TR | 1 | 1 | 0 | 0 | 0 | Herakleian | None | Amphora | LC-EH | |
| 116201 | TR | 1 | 1 | 0 | 0 | 0 | Sinopean | None | Amphora | LC-EH | |
| 118001 | TR | 1 | 1 | 0 | 0 | 0 | Herakleian | None | Amphora | LC-EH | |
| 118002 | TR | 1 | 1 | 0 | 0 | 0 | Herakleian | None | Amphora | LC-EH | |
| 118002 | TR | 1 | 1 | 0 | 0 | 0 | Red mica? | None | Amphora? | LC-EH | |
| 115101 | TW | 1 | 1 | 0 | 0 | 0 | Light | Slip | Jug | LC-EH | |
| 118001 | UT | 1 | 0 | 1 | 0 | 0 | South Pontic | None | Pithos? | LC-EH | |
| 110401 | IN | 1 | 1 | 0 | 0 | 0 | Red mica | None | ? | LC-EH? | |
| 113201 | IN | 1 | 1 | 0 | 0 | 0 | Red | None | ? | LC-EH? | |
| 113301 | IN | 1 | 1 | 0 | 0 | 0 | Red | None | ? | LC-EH? | |
| 119201 | IN | 1 | 1 | 0 | 0 | 0 | Red | None | ? | LC-EH? | |
| 119201 | IN | 2 | 2 | 0 | 0 | 0 | Brown | None | ? | LC-EH? | |
| 116201 | TR? | 1 | 1 | 0 | 0 | 0 | Brown | None | Amphora | LC-EH? | |
| 112001 | TW | 2 | 2 | 0 | 0 | 0 | Red | None | Closed shape | LC-EH? | |
| 113301 | TW | 1 | 1 | 0 | 0 | 0 | Brown | None | ? | LC-EH? | |
| 119201 | UT | 1 | 1 | 0 | 0 | 0 | Brown | None | Open shape? | LC-EH? | |
| 119201 | TW | 1 | 0 | 0 | 0 | 1 | Red | None | Jug | Medieval? | |
| 113201 | TW | 1 | 1 | 0 | 0 | 0 | Red Tatarian | Glazed | Bowl/plate | EM | |
| 110401 | AR | 1 | 1 | 0 | 0 | 0 | Red modern | Smoothed | Tile? | Modern | |
| 108901 | AR? | 1 | 1 | 0 | 0 | 0 | Red modern | None | Tile? | Modern? | |
| 116201 | IN | 1 | 1 | 0 | 0 | 0 | - | None | ? | ? | |

| BagID | Function | Total | Body | Rim | Handle | Base | Fabric | Surface | Shape | Period | Cat. |
|---|---|---|---|---|---|---|---|---|---|---|---|
| 112201 | LI | 1 | 0 | 0 | 0 | 0 | | | | | |
| 119201 | LI | 1 | 0 | 0 | 0 | 0 | | | | | |

## DSP07-F02-05

Finds table adjacent units

| BagID | Function | Total | Body | Rim | Handle | Base | Fabric | Surface | Shape | Period | Cat. |
|---|---|---|---|---|---|---|---|---|---|---|---|
| 112501 | TR | 1 | 1 | 0 | 0 | 0 | Chersonesean | None | Amphora | LC-EH | |
| 112501 | AR | 1 | 0 | 1 | 0 | 0 | Red modern | None | Brick | Modern | |
| 109001 | TW | 1 | 1 | 0 | 0 | 0 | Brown mica | None | Closed shape | ? | |

## DSP07-F03-01

Finds table core units

| BagID | Function | Total | Body | Rim | Handle | Base | Fabric | Surface | Shape | Period | Cat. |
|---|---|---|---|---|---|---|---|---|---|---|---|
| 230101 | HW | 1 | 1 | 0 | 0 | 0 | Grey | None | Pot | LC-EH | |
| 230101 | HW | 1 | 1 | 0 | 0 | 0 | Grey | None | Pot | LC-EH | |
| 230401 | HW | 1 | 1 | 0 | 0 | 0 | Grey | None | Pot | LC-EH | |
| 136201 | IN | 1 | 1 | 0 | 0 | 0 | Red | None | ? | LC-EH | |
| 230101 | IN | 1 | 1 | 0 | 0 | 0 | Red | None | ? | LC-EH | |
| 230101 | IN | 1 | 1 | 0 | 0 | 0 | Red | None | ? | LC-EH | |
| 230401 | IN | 1 | 1 | 0 | 0 | 0 | Light | None | ? | LC-EH | |
| 230401 | IN | 1 | 1 | 0 | 0 | 0 | Red | None | ? | LC-EH | |
| 230401 | IN | 4 | 4 | 0 | 0 | 0 | Red | None | ? | LC-EH | |
| 230101 | TR | 2 | 2 | 0 | 0 | 0 | Herakleian? | Slipped? | Amphora | LC-EH | |
| 230101 | TR | 2 | 2 | 0 | 0 | 0 | Herakleian | None | Amphora | LC-EH | |
| 230402 | TR | 1 | 0 | 1 | 0 | 0 | Herakleian? | None | Amphora | LC-EH | **98** |
| 230401 | TR | 1 | 0 | 0 | 1 | 0 | Red | Slip | Amphora | LC-EH | **99** |
| 230401 | TR | 3 | 3 | 0 | 0 | 0 | Sinopean | None | Amphora | LC-EH | |
| 230401 | TR | 2 | 2 | 0 | 0 | 1 | Chersonesean | None | Amphora | LC-EH | **97** |
| 230401 | TR | 3 | 2 | 0 | 1 | 0 | Herakleian | None | Amphora | LC-EH | |
| 230402 | TR | 1 | 0 | 0 | 1 | 0 | Red | None | Amphora | LC-EH | **100** |
| 230402 | TR | 1 | 1 | 0 | 0 | 0 | Herakleian? | None | Amphora | LC-EH | |
| 230101 | TW | 1 | 0 | 1 | 0 | 0 | Red | Glazed | Cup | LC-EH | |
| 230101 | TW | 1 | 0 | 1 | 0 | 0 | Red | Glazed | Cup? | LC-EH | |
| 136201 | IN | 1 | 1 | 0 | 0 | 0 | Brown | None | ? | LC-EH? | |
| 230401 | IN | 1 | 1 | 0 | 0 | 0 | Brown | None | ? | LC-EH? | |
| 230101 | TR | 1 | 1 | 0 | 0 | 0 | Chian? | None | Amphora | LC-EH? | |
| 136201 | AR | 1 | 1 | 0 | 0 | 0 | Brown modern | None | Tile | Modern | |
| 230101 | IN | 1 | 1 | 0 | 0 | 0 | Red modern? | None | ? | Modern? | |
| 230101 | BO | 1 | 0 | 0 | 0 | 0 | | | | | |
| 230101 | BO | 1 | 0 | 0 | 0 | 0 | | | | | |
| 230401 | LI | 1 | 0 | 0 | 0 | 0 | | | | | |

Finds table adjacent units

| BagID | Function | Total | Body | Rim | Handle | Base | Fabric | Surface | Shape | Period | Cat. |
|---|---|---|---|---|---|---|---|---|---|---|---|
| 230001 | HW | 1 | 1 | 0 | 0 | 0 | Grey | None | Pot | LC-EH | |
| 230001 | TR | 2 | 2 | 0 | 0 | 0 | Herakleian | None | Amphora | LC-EH | |
| 230201 | IN | 1 | 1 | 0 | 0 | 0 | South Pontic | None | ? | LC-EH? | |
| 230201 | IR | 1 | 0 | 0 | 0 | 0 | | | | | |

## DSP07-F04-01

Finds table core units

| BagID | Function | Total | Body | Rim | Handle | Base | Fabric | Surface | Shape | Period | Cat. |
|---|---|---|---|---|---|---|---|---|---|---|---|
| 140001 | HW | 1 | 1 | 0 | 0 | 0 | BaGrey | None | Pot | BA? | |
| 140002 | HW | 1 | 1 | 0 | 0 | 0 | Grey | None | Pot | LC-EH | |
| 140003 | HW | 4 | 4 | 0 | 0 | 0 | Grey | None | Pot | LC-EH | |
| 140001 | IN | 1 | 1 | 0 | 0 | 0 | Brown | None | ? | LC-EH? | |
| 140001 | BO | 1 | 0 | 0 | 0 | 0 | | | | | |
| 140001 | LI | 1 | 0 | 0 | 0 | 0 | | | | | |

Finds table adjacent units

| BagID | Function | Total | Body | Rim | Handle | Base | Fabric | Surface | Shape | Period | Cat. |
|---|---|---|---|---|---|---|---|---|---|---|---|
| 139901 | HW | 1 | 1 | 0 | 0 | 0 | Grey | None | Pot | LC-EH | |
| 139401 | TR | 1 | 1 | 0 | 0 | 0 | Sinopean | None | Amphora | LC-EH | |
| 140101 | TR | 1 | 1 | 0 | 0 | 0 | Herakleian | None | Amphora | LC-EH | |
| 236101 | TR | 1 | 1 | 0 | 0 | 0 | Herakleian | None | Amphora | LC-EH | |
| 236101 | TR | 1 | 0 | 0 | 1 | 0 | Chersonesean | None | Amphora | LC-EH | |
| 236301 | TR | 1 | 1 | 0 | 0 | 0 | Herakleian | None | Amphora | LC-EH | |
| 236201 | TR | 1 | 1 | 0 | 0 | 0 | Orange | None | Amphora? | LC-EH? | |
| 236101 | UT | 1 | 1 | 0 | 0 | 0 | Light | None | Open shape | LC-EH? | |
| 140101 | IN | 1 | 1 | 0 | 0 | 0 | Red | None | ? | ? | |
| 236301 | LI | 1 | 0 | 0 | 0 | 0 | | | | | |

## DSP07-F04-02

Finds table core units

| BagID | Function | Total | Body | Rim | Handle | Base | Fabric | Surface | Shape | Period | Cat. |
|---|---|---|---|---|---|---|---|---|---|---|---|
| 138802 | TW | 1 | 0 | 1 | 0 | 0 | Grey | Slip | Jug | LC-H | **101** |
| 138805 | HW | 1 | 1 | 0 | 0 | 0 | Grey | None | Pot | LC-EH | |
| 138804 | IN | 2 | 2 | 0 | 0 | 0 | Tatarian | None | Closed shape | EM | |
| 138804 | IN | 2 | 2 | 0 | 0 | 0 | Tatarian | Slip | ? | EM | |
| 138805 | IN | 8 | 8 | 0 | 0 | 0 | Tatarian | None | Closed shape? | EM | |
| 138806 | IN | 3 | 2 | 0 | 0 | 1 | Tatarian | Slip | ? | EM | |
| 138806 | IN | 2 | 2 | 0 | 0 | 0 | Tatarian | None | ? | EM | |
| 138807 | IN | 5 | 5 | 0 | 0 | 0 | Tatarian | None | Closed shape | EM | |
| 138802 | TW | 1 | 1 | 0 | 0 | 0 | Tatarian | Slip | Jug? | EM | |
| 138804 | TW | 2 | 2 | 0 | 0 | 0 | Tatarian | Slip | Jug? | EM | |
| 138804 | TW | 1 | 0 | 1 | 0 | 0 | Tatarian | None | Jug | EM | **105** |
| 138804 | TW | 1 | 0 | 1 | 0 | 0 | Tatarian | None | Jug | EM | **106** |
| 138805 | TW | 8 | 8 | 0 | 0 | 0 | Tatarian | Slip | Jug | EM | |

| BagID | Function | Total | Body | Rim | Handle | Base | Fabric | Surface | Shape | Period | Cat. |
|---|---|---|---|---|---|---|---|---|---|---|---|
| 138805 | TW | 1 | 1 | 0 | 0 | 0 | Tatarian | None | Jug | EM | |
| 138805 | TW | 1 | 0 | 1 | 0 | 0 | Tatarian | None | Bowl | EM | |
| 138806 | TW | 1 | 1 | 0 | 0 | 0 | Tatarian | Glazed | Open shape | EM | |
| 138806 | TW | 5 | 4 | 0 | 0 | 0 | Tatarian | Slip | Jug | EM | |
| 138806 | TW | 1 | 1 | 0 | 0 | 0 | Tatarian | Grooved | Jug | EM | |
| 138807 | TW | 2 | 2 | 0 | 0 | 0 | Tatarian | Glazed | Open shape | EM | |
| 138807 | TW | 1 | 1 | 0 | 0 | 0 | Tatarian | Slip | Jug | EM | **109** |
| 138801 | UT | 6 | 6 | 0 | 0 | 0 | Tatarian | None | Pot? | EM | |
| 138801 | UT | 1 | 0 | 0 | 0 | 1 | Tatarian | Slip | Pot | EM | **103** |
| 138802 | UT | 1 | 1 | 0 | 0 | 0 | Tatarian | None | Open shape? | EM | |
| 138803 | UT | 1 | 0 | 0 | 0 | 1 | Tatarian | None | Closed shape? | EM | **104** |
| 138804 | UT | 1 | 0 | 1 | 0 | 0 | Tatarian | None | Basin/ cooking pot | EM | **102** |
| 138804 | UT | 1 | 0 | 1 | 0 | 0 | Tatarian | None | Pot | EM | |
| 138806 | UT | 9 | 9 | 0 | 0 | 0 | Tatarian | None | Closed shape? | EM | **107** |
| 138806 | UT | 1 | 1 | 0 | 0 | 0 | Tatarian | None | Cooking pot? | EM | |
| 138807 | UT | 1 | 0 | 1 | 0 | 0 | Tatarian | Slip | Pot | EM | **108** |
| 138807 | UT | 1 | 1 | 0 | 0 | 0 | Tatarian | None | Closed shape | EM | |
| 138805 | BO | 3 | 0 | 0 | 0 | 0 | | | | | |
| 138805 | BO | 11 | 0 | 0 | 0 | 0 | | | | | |
| 138806 | BO | 11 | 11 | 0 | 0 | 0 | | | | | |
| 138807 | BO | 5 | 0 | 0 | 0 | 0 | | | | | |
| 138802 | IN | 1 | 1 | 0 | 0 | 0 | - | None | - | ? | |
| 138805 | IN | 1 | 1 | 0 | 0 | 0 | Brown | None | Closed shape | ? | |
| 138804 | LI | 1 | 0 | 0 | 0 | 0 | | | | | |
| 138805 | LI | 1 | 0 | 0 | 0 | 0 | | | | | |
| 138805 | SH | 1 | 0 | 0 | 0 | 0 | | | | | |

Finds table adjacent units

| BagID | Function | Total | Body | Rim | Handle | Base | Fabric | Surface | Shape | Period | Cat. |
|---|---|---|---|---|---|---|---|---|---|---|---|
| 139001 | IN | 1 | 1 | 0 | 0 | 0 | Red | None | ? | LC-EH | |
| 138601 | TR | 1 | 1 | 0 | 0 | 0 | Chersonesean | None | Amphora | LC-EH | |
| 138601 | TR | 1 | 1 | 0 | 0 | 0 | Chersonesean | None | Amphora | LC-EH | |
| 138701 | TR | 1 | 0 | 1 | 0 | 0 | Chersonesean? | Slip | Amphora | LC-EH | |
| 140501 | TR | 1 | 0 | 0 | 1 | 0 | Sinopean | None | Amphora | LC-EH | |
| 138701 | TW | 1 | 1 | 0 | 0 | 0 | Chersonesean | None | Jug? | LC-EH | |
| 138901 | IN | 1 | 1 | 0 | 0 | 0 | Red | None | Closed shape? | LC-EH? | |
| 235601 | TW | 1 | 0 | 0 | 1 | 0 | Red | Slip | Jug? | LC-EH? | |
| 138901 | UT | 1 | 1 | 0 | 0 | 0 | ? | None | Open shape? | LC-EH? | |
| 140401 | TW | 1 | 1 | 0 | 0 | 0 | Tatarian | Glazed | Plate | EM | |
| 140601 | TW | 1 | 1 | 0 | 0 | 0 | Tatarian? | Slip | Jug? | EM? | |

## DSP07-F05-01

Finds table core units

| BagID | Function | Total | Body | Rim | Handle | Base | Fabric | Surface | Shape | Period | Cat. |
|---|---|---|---|---|---|---|---|---|---|---|---|
| 150201 | HW | 1 | 1 | 0 | 0 | 0 | Grey | None | Pot | LC-EH | |
| 150301 | IN | 2 | 2 | 0 | 0 | 0 | Red | None | Closed shape | LC-EH | |
| 150401 | IN | 2 | 2 | 0 | 0 | 0 | South Pontic | None | ? | LC-EH | |
| 150401 | IN | 1 | 1 | 0 | 0 | 0 | Red | None | ? | LC-EH | |
| 150201 | TR | 1 | 1 | 0 | 0 | 0 | Herakleian | None | Amphora | LC-EH | |
| 150301 | TR | 1 | 1 | 0 | 0 | 0 | Herakleian | None | Amphora | LC-EH | |
| 150301 | TR | 1 | 1 | 0 | 0 | 0 | Chersonesean | Slip | Amphora | LC-EH | |
| 150401 | TR | 1 | 1 | 0 | 0 | 0 | Sinopean | None | Amphora? | LC-EH | |
| 150401 | TR | 2 | 2 | 0 | 0 | 0 | Herakleian? | None | Amphora? | LC-EH | |
| 150201 | IN | 1 | 1 | 0 | 0 | 0 | Red | Smoothed | ? | LC-EH? | |
| 150401 | IN | 1 | 1 | 0 | 0 | 0 | Red | Slip | Closed shape | LC-EH? | |
| 150201 | TW | 1 | 0 | 1 | 0 | 0 | Tatarian | Glazed | Bowl | EM | **112** |
| 150201 | TW | 1 | 0 | 1 | 0 | 0 | Tatarian | Glaze | Plate | EM | **110** |
| 150201 | TW | 2 | 0 | 2 | 0 | 0 | Tatarian | Glazed | Jug | EM | **111** |
| 150202 | TW | 4 | 4 | 0 | 0 | 0 | Tatarian | Glazed | ? | EM | |
| 150201 | UT | 1 | 0 | 0 | 0 | 1 | Tatarian | Glazed | Storage vessel | EM | **113** |
| 150202 | UT | 1 | 1 | 0 | 0 | 0 | Brown | None | Casserole | EM | **115** |
| 150201 | AR | 2 | 2 | 0 | 0 | 0 | Red modern | None | Tile | Modern | |
| 150202 | AR | 2 | 2 | 0 | 0 | 0 | Red modern | None | Tile | Modern | |
| 150202 | AR | 1 | 1 | 0 | 0 | 0 | Red modern | None | Tile | Modern | |
| 150301 | AR | 1 | 0 | 1 | 0 | 0 | China | Glazed | Tile | Modern | |
| 150301 | AR | 2 | 2 | 0 | 0 | 0 | Red modern | None | Tile | Modern | |
| 150201 | TW | 1 | 0 | 1 | 0 | 0 | China | Glazed | Bowl | Modern | |
| 150202 | TW | 3 | 3 | 0 | 0 | 0 | China | Glazed | Open shape | Modern | |
| 150301 | TW | 1 | 1 | 0 | 0 | 0 | China | Glazed | ? | Modern | |
| 150201 | IR | 1 | 0 | 0 | 0 | 0 | | | | | |
| 150301 | IR | 1 | 0 | 0 | 0 | 0 | | | | | |
| 150201 | TW | 1 | 0 | 1 | 0 | 0 | Tatarian | Glazed | Jug | EM | **114** |
| 150202 | TW | 5 | 5 | 0 | 0 | 0 | | | | | |
| 150202 | TW | 2 | 2 | 0 | 0 | 0 | | | | | |

Finds table adjacent units

| BagID | Function | Total | Body | Rim | Handle | Base | Fabric | Surface | Shape | Period | Cat. |
|---|---|---|---|---|---|---|---|---|---|---|---|
| 150101 | HW | 1 | 1 | 0 | 0 | 0 | Grey | None | Pot | LC-EH | |
| 150601 | HW | 1 | 1 | 0 | 0 | 0 | Grey | None | Pot | LC-EH | |
| 148201 | IN | 1 | 1 | 0 | 0 | 0 | Red | None | ? | LC-EH | |
| 148201 | IN | 2 | 2 | 0 | 0 | 0 | Red | None | Closed shape | LC-EH | |
| 148201 | TR | 1 | 1 | 0 | 0 | 0 | Herakleian? | None | Amphora? | LC-EH | |
| 148401 | TR | 1 | 1 | 0 | 0 | 0 | Herakleian | Slip | Amphora | LC-EH | |
| 149901 | TR | 1 | 0 | 0 | 1 | 0 | Herakleian | None | Amphora | LC-EH | |
| 150001 | TR | 1 | 1 | 0 | 0 | 0 | Herakleian | None | Amphora | LC-EH | |
| 148201 | HW | 1 | 1 | 0 | 0 | 0 | Grey | Smoothed | Pot | LC-EH? | |
| 148001 | IN | 1 | 1 | 0 | 0 | 0 | Brown | None | Closed shape? | LC-EH? | |
| 148401 | IN | 1 | 1 | 0 | 0 | 0 | Red | None | ? | LC-EH? | |
| 152201 | IN | 1 | 1 | 0 | 0 | 0 | Red | None | Closed shape | LC-EH? | |
| 148301 | IN | 1 | 1 | 0 | 0 | 0 | Orange | None | ? | Medieval? | |
| 149901 | UT | 1 | 0 | 0 | 0 | 1 | Brown | None | Storage vessel | Medieval? | |
| 150601 | IN | 1 | 1 | 0 | 0 | 0 | Tatarian | Slip | Closed shape | EM | |
| 148001 | TW | 1 | 0 | 1 | 0 | 0 | Tatarian? | Smoothed | Bowl | EM | |
| 148001 | AR | 1 | 1 | 0 | 0 | 0 | Red modern | None | Tile | Modern | |
| 148201 | AR | 1 | 1 | 0 | 0 | 0 | China | Glazed | Tile | Modern | |
| 148201 | AR | 1 | 1 | 0 | 0 | 0 | Red modern | Smoothed | Tile | Modern | |
| 148301 | AR | 1 | 1 | 0 | 0 | 0 | Red modern | Smoothed | Tile | Modern | |
| 148301 | AR | 5 | 4 | 1 | 0 | 0 | Red modern | None | Tile | Modern | |
| 148401 | AR | 4 | 3 | 1 | 0 | 0 | Red modern | Slip | Tile | Modern | |
| 149901 | AR | 3 | 3 | 0 | 0 | 0 | Red modern | None | Tile | Modern | |
| 150001 | AR | 1 | 1 | 0 | 0 | 0 | Red modern | None | Tile | Modern | |
| 148001 | TW | 1 | 1 | 0 | 0 | 0 | China | Glazed | ? | Modern | |
| 148301 | IN | 1 | 1 | 0 | 0 | 0 | Red | Smoothed | ? | ? | |
| 150001 | LI | 1 | 0 | 0 | 0 | 0 | | | | | |

## DSP07-F05-02

Finds table adjacent units

| BagID | Function | Total | Body | Rim | Handle | Base | Fabric | Surface | Shape | Period | Cat. |
|---|---|---|---|---|---|---|---|---|---|---|---|
| 242201 | HW | 1 | 1 | 0 | 0 | 0 | BaGrey | Smoothed | Pot | Late BA? | |
| 242701 | HW | 1 | 1 | 0 | 0 | 0 | BaGrey | None | Pot | Late BA? | |
| 240301 | TR | 1 | 1 | 0 | 0 | 0 | Herakleian | None | Amphora | LC-EH | |
| 242202 | TR | 1 | 1 | 0 | 0 | 0 | Sinopean | None | Amphora | LC-EH | |
| 242701 | TR | 1 | 1 | 0 | 0 | 0 | Herakleian? | None | Amphora | LC-EH | |

| BagID | Function | Total | Body | Rim | Handle | Base | Fabric | Surface | Shape | Period | Cat. |
|---|---|---|---|---|---|---|---|---|---|---|---|
| 247702 | TR | 2 | 2 | 0 | 0 | 0 | Red | None? | Amphora? | LC-EH? | |
| 247702 | TW | 1 | 0 | 1 | 0 | 0 | Brown | None | Bowl | LC-EH? | **125** |
| 252002 | TW | 1 | 1 | 0 | 0 | 0 | Brown | None | Closed shape | LC-EH? | |
| 247702 | UT | 2 | 2 | 0 | 0 | 0 | ? | None | ? | LC-EH? | |
| 247702 | UT | 4 | 4 | 0 | 0 | 0 | Red? | None | ? | LC-EH? | |
| 251902 | UT | 1 | 1 | 0 | 0 | 0 | Grey | None | Closed shape | LC-EH? | |
| 247702 | IN | 2 | 0 | 0 | 0 | 0 | Light | None | ? | Medieval | |
| 247701 | IN | 1 | 1 | 0 | 0 | 0 | Brown | None | ? | Medieval? | |
| 251902 | TW | 1 | 0 | 1 | 0 | 0 | Tatarian | Slip | Jug | EM | |
| 247701 | AR | 3 | 3 | 0 | 0 | 0 | Red modern | None | Tile | Modern | |
| 251902 | AR | 5 | 4 | 1 | 0 | 0 | Red modern | None | Tile | Modern | |
| 247702 | UT | 2 | 2 | 0 | 0 | 0 | China | Glazed | ? | Modern | |
| 251902 | UT | 1 | 1 | 0 | 0 | 0 | China | Glazed | Porcelain | Modern | |
| 252002 | UT | 1 | 0 | 1 | 0 | 0 | China | Glazed | Porcelain | Modern | |
| 251901 | LI | 1 | 0 | 0 | 0 | 0 | | | | | |

## DSP07-F07-01

Finds table core units

| BagID | Function | Total | Body | Rim | Handle | Base | Fabric | Surface | Shape | Period | Cat. |
|---|---|---|---|---|---|---|---|---|---|---|---|
| 261301 | LI | 1 | 0 | 0 | 0 | 0 | | | | Late BA | |
| 258602 | AR | 1 | 0 | 1 | 0 | 0 | Sinopean | Smoothed | Tile | LC-EH | |
| 258602 | HW | 1 | 1 | 0 | 0 | 0 | Grey | None | Closed shape | LC-EH | |
| 258601 | IN | 2 | 2 | 0 | 0 | 0 | Red | None | ? | LC-EH | |
| 261002 | IN | 1 | 1 | 0 | 0 | 0 | Red | None | ? | LC-EH | |
| 261102 | IN | 2 | 2 | 0 | 0 | 0 | Red | None | ? | LC-EH | |
| 258601 | TR | 2 | 2 | 0 | 0 | 0 | Herakleian | None | Amphora | LC-EH | |
| 258602 | TR | 5 | 4 | 0 | 1 | 0 | Herakleian | None | Amphora | LC-EH | |
| 261103 | TR | 1 | 1 | 0 | 0 | 0 | Red mica | None | Amphora | LC-EH | |
| 261103 | TR | 1 | 1 | 0 | 0 | 0 | Red | None | Amphora | LC-EH | |
| 261202 | TR | 1 | 1 | 0 | 0 | 0 | Sinopean | Slip | Amphora | LC-EH | |
| 261202 | TR | 2 | 1 | 0 | 1 | 0 | Herakleian | Slip | Amphora | LC-EH | |
| 263601 | TR | 1 | 1 | 0 | 0 | 0 | Herakleian | None | Amphora | LC-EH | |
| 263901 | TR | 2 | 2 | 0 | 0 | 0 | Herakleian | None | Amphora | LC-EH | |
| 263902 | TR | 1 | 1 | 0 | 0 | 0 | Chersonesean | None | Amphora | LC-EH | |
| 263902 | TR | 3 | 3 | 0 | 0 | 0 | Herakleian | Slip | Amphora | LC-EH | |
| 263901 | UT | 1 | 1 | 0 | 0 | 0 | Brown | Slip | Pithos? | LC-EH | |
| 258602 | IN | 1 | 1 | 0 | 0 | 0 | Red | None | ? | LC-EH? | |
| 261001 | IN | 1 | 1 | 0 | 0 | 0 | Red | None | ? | LC-EH? | |
| 261202 | IN | 1 | 0 | 0 | 0 | 0 | Red | none | Closed shape | LC-EH? | |
| 263702 | IN | 1 | 1 | 0 | 0 | 0 | Brown | None | Closed shape | LC-EH? | |
| 263801 | IN | 1 | 1 | 0 | 0 | 0 | Red | None | ? | LC-EH? | |
| 263902 | IN | 1 | 1 | 0 | 0 | 0 | Brown | None | ? | LC-EH? | |
| 261202 | TR | 3 | 3 | 0 | 0 | 0 | Brown | None | Amphora | LC-EH? | |
| 263701 | TR | 1 | 1 | 0 | 0 | 0 | Herakleian | None | Amphora | LC-EH? | |
| 263902 | TR | 1 | 1 | 0 | 0 | 0 | Brown | Slip | Amphora | LC-EH? | |
| 261201 | ? | 1 | 1 | 0 | 0 | 0 | - | - | - | Modern? | |
| 261001 | IN | 1 | 1 | 0 | 0 | 0 | Red | None | Closed shape | ? | |
| 263702 | IN | 1 | 1 | 0 | 0 | 0 | | | | | |
| 263702 | IN | 1 | 1 | 0 | 0 | 0 | Red | None | ? | ? | |
| 263901 | LI | 1 | 0 | 0 | 0 | 0 | | | | | |

Finds table adjacent units

| BagID | Function | Total | Body | Rim | Handle | Base | Fabric | Surface | Shape | Period | Cat. |
|---|---|---|---|---|---|---|---|---|---|---|---|
| 258701 | IN | 1 | 1 | 0 | 0 | 0 | Red | None | Closed shape | LC-EH | |
| 266201 | TW | 1 | 1 | 0 | 0 | 0 | Red | Smoothed | Closed shape | LC-EH | |
| 264001 | IN | 1 | 1 | 0 | 0 | 0 | Red | None | ? | LC-EH? | |
| 266101 | IN | 1 | 0 | 0 | 1 | 0 | Red mica | None | ? | LC-EH? | |

## DSP07-F07-02

Finds table core units

| BagID | Function | Total | Body | Rim | Handle | Base | Fabric | Surface | Shape | Period | Cat. |
|---|---|---|---|---|---|---|---|---|---|---|---|
| 257301 | HW | 1 | 0 | 1 | 0 | 0 | Grey | None | Pot | LC-EH | |
| 257302 | HW | 1 | 1 | 0 | 0 | 0 | Grey | None | Pot | LC-EH | |
| 257302 | IN | 1 | 1 | 0 | 0 | 0 | Red | None | ? | LC-EH | |
| 257201 | TR | 1 | 0 | 0 | 1 | 0 | Thasian | None | Amphora | LC-EH | |
| 257302 | TR | 1 | 1 | 0 | 0 | 0 | Red | None | Amphora | LC-EH | |
| 257301 | UT | 1 | 1 | 0 | 0 | 0 | Brown | None | Cooking pot? | LC-EH | |
| 257302 | UT | 1 | 1 | 0 | 0 | 0 | Red | None | Closed shape | LC-EH? | |

Finds table adjacent units

| BagID | Function | Total | Body | Rim | Handle | Base | Fabric | Surface | Shape | Period | Cat. |
|---|---|---|---|---|---|---|---|---|---|---|---|
| 255201 | TR | 1 | 1 | 0 | 0 | 0 | Herakleian | None | Amphora | LC-EH | |

## DSP07-F08-01

Finds table core units

| BagID | Function | Total | Body | Rim | Handle | Base | Fabric | Surface | Shape | Period | Cat. |
|---|---|---|---|---|---|---|---|---|---|---|---|
| 899101 | HW | 1 | 1 | 0 | 0 | 0 | BaGrey | None | Cup? | Middle BA | |
| 899101 | HW | 1 | 1 | 0 | 0 | 0 | BaCoarse | ImpressedCord | Pot | Middle/ Late BA | |
| 271902 | HW | 1 | 1 | 0 | 0 | 0 | BaGrey | None | Closed shape? | Middle/ Late BA? | **126** |
| 271601 | AR | 1 | 0 | 1 | 0 | 0 | Sinopean? | None | Tile? | LC-EH | |
| 270601 | HW | 1 | 1 | 0 | 0 | 0 | Coarse | None | Pot | LC-EH | |
| 270602 | HW | 2 | 2 | 0 | 0 | 0 | Coarse | None | Pot | LC-EH | |
| 270602 | HW | 11 | 11 | 0 | 0 | 0 | Grey | None | Closed shape | LC-EH | |
| 270602 | HW | 1 | 1 | 0 | 0 | 0 | Grey | Smoothed | Open shape | LC-EH | |
| 270701 | HW | 6 | 6 | 0 | 0 | 0 | Grey | None | Closed shape | LC-EH | |
| 270703 | HW | 2 | 2 | 0 | 0 | 0 | Coarse | None | Pot | LC-EH | |
| 270703 | HW | 42 | 41 | 1 | 0 | 0 | Grey | None | Closed shape | LC-EH | **127** |
| 270703 | HW | 1 | 1 | 0 | 0 | 0 | Coarse | Smoothed | Open shape | LC-EH | |
| 270703 | HW | 1 | 0 | 1 | 0 | 0 | HwGrey | None | Pot? | LC-EH | **128** |
| 270703 | HW | 1 | 0 | 1 | 0 | 0 | HwGrey | None | Cup | LC-EH | **129** |
| 270801 | HW | 5 | 5 | 0 | 0 | 0 | Grey | None | Closed shape | LC-EH | |
| 270901 | HW | 1 | 0 | 1 | 0 | 0 | HwGrey | None | Bowl? | LC-EH | **130** |
| 270901 | HW | 6 | 5 | 1 | 0 | 0 | Grey | None | Pot | LC-EH | **131** |
| 270901 | HW | 1 | 1 | 0 | 0 | 0 | Grey | None | Open shape | LC-EH | |
| 270902 | HW | 12 | 12 | 0 | 0 | 0 | Coarse | None | Pot | LC-EH | |

| BagID | Function | Total | Body | Rim | Handle | Base | Fabric | Surface | Shape | Period | Cat. |
|---|---|---|---|---|---|---|---|---|---|---|---|
| 270902 | HW | 32 | 0 | 1 | 0 | 0 | Grey | None | Closed shape | LC-EH | |
| 270902 | HW | 1 | 1 | 0 | 0 | 0 | Grey | None | Open shape | LC-EH | |
| 270902 | HW | 3 | 2 | 0 | 0 | 0 | Coarse | Smoothed | Open shape | LC-EH | |
| 271001 | HW | 1 | 1 | 0 | 0 | 0 | Coarse | Smoothed | Open shape | LC-EH | |
| 271001 | HW | 8 | 8 | 0 | 0 | 0 | Grey | None | Closed shape | LC-EH | |
| 271001 | HW | 8 | 8 | 0 | 0 | 0 | Coarse | None | Pot | LC-EH | |
| 271002 | HW | 5 | 5 | 0 | 0 | 0 | Grey | None | Closed shape | LC-EH | |
| 271501 | HW | 1 | 1 | 0 | 0 | 0 | Grey | None | Closed shape | LC-EH | |
| 271501 | HW | 1 | 1 | 0 | 0 | 0 | Coarse | None | Pot | LC-EH | |
| 271601 | HW | 3 | 3 | 0 | 0 | 0 | Grey | None | Closed shape | LC-EH | |
| 271701 | HW | 1 | 1 | 0 | 0 | 0 | Coarse | None | Pot | LC-EH | |
| 271701 | HW | 1 | 1 | 0 | 0 | 0 | Grey | None | Closed shape | LC-EH | |
| 271701 | HW | 1 | 1 | 0 | 0 | 0 | Red | None | Closed shape | LC-EH | |
| 271801 | HW | 1 | 1 | 0 | 0 | 0 | Grey | None | Closed shape | LC-EH | |
| 271901 | HW | 1 | 1 | 0 | 0 | 0 | Coarse | None | Pot | LC-EH | |
| 271901 | HW | 1 | 0 | 1 | 0 | 0 | HwGrey | None | Bowl | LC-EH | **132** |
| 271901 | HW | 8 | 8 | 0 | 0 | 0 | Grey | None | Closed shape | LC-EH | |
| 271902 | HW | 6 | 5 | 1 | 0 | 0 | Coarse | None | Pot | LC-EH | **133** |
| 271902 | HW | 10 | 8 | 2 | 0 | 0 | Grey | None | Closed shape | LC-EH | **134-135** |
| 272001 | HW | 11 | 11 | 0 | 0 | 0 | Grey | None | Closed shape | LC-EH | |
| 272001 | HW | 2 | 2 | 0 | 0 | 0 | Coarse | None | Pot | LC-EH | |
| 272101 | HW | 1 | 1 | 0 | 0 | 0 | Grey | None | Closed shape | LC-EH | |
| 272401 | HW | 2 | 2 | 0 | 0 | 0 | Grey | None | Closed shape | LC-EH | |
| 272901 | HW | 1 | 1 | 0 | 0 | 0 | Coarse | Releif | Pot | LC-EH | |
| 899101 | HW | 10 | 10 | 0 | 0 | 0 | Grey | None | Pot | LC-EH | |
| 899101 | HW | 1 | 0 | 1 | 0 | 0 | HwCoarse | None | Bowl | LC-EH | |
| 899102 | HW | 1 | 1 | 0 | 0 | 0 | Black | Polished | Pot | LC-EH | |
| 899102 | HW | 2 | 2 | 0 | 0 | 0 | Grey | None | Pot | LC-EH | |
| 899104 | HW | 8 | 8 | 0 | 0 | 0 | Grey | None | Pot | LC-EH | |
| 899105 | HW | 1 | 1 | 0 | 0 | 0 | Grey | Impressed | Pot | LC-EH | |
| 899105 | HW | 1 | 0 | 1 | 0 | 0 | Grey | None | Pot | LC-EH | |
| 899105 | HW | 10 | 10 | 0 | 0 | 0 | Grey | None | Pot | LC-EH | |
| 899106 | HW | 6 | 6 | 0 | 0 | 0 | Grey | None | Pot | LC-EH | |
| 899201 | HW | 4 | 4 | 0 | 0 | 0 | Red | None | Pot | LC-EH | |
| 270602 | IN | 2 | 2 | 0 | 0 | 0 | Red | None | Closed shape? | LC-EH | |
| 270701 | IN | 1 | 1 | 0 | 0 | 0 | South Pontic | None | ? | LC-EH | |
| 270701 | IN | 1 | 1 | 0 | 0 | 0 | Red | None | Closed shape | LC-EH | |
| 270701 | IN | 1 | 1 | 0 | 0 | 0 | Red mica | None | Closed shape | LC-EH | |
| 270703 | IN | 8 | 8 | 0 | 0 | 0 | Red | None | ? | LC-EH | |
| 270902 | IN | 1 | 1 | 0 | 0 | 0 | Grey | None | ? | LC-EH | |
| 270902 | IN | 6 | 6 | 0 | 0 | 0 | Red | None | Closed shape? | LC-EH | |
| 271001 | IN | 5 | 5 | 0 | 0 | 0 | Red | None | Closed shape | LC-EH | |
| 271002 | IN | 1 | 1 | 0 | 0 | 0 | Red | None | Closed shape | LC-EH | |

| BagID | Function | Total | Body | Rim | Handle | Base | Fabric | Surface | Shape | Period | Cat. |
|---|---|---|---|---|---|---|---|---|---|---|---|
| 271002 | IN | 2 | 2 | 0 | 0 | 0 | Red | none | ? | LC-EH | |
| 271501 | IN | 1 | 1 | 0 | 0 | 0 | Red | None | ? | LC-EH | |
| 271601 | IN | 2 | 2 | 0 | 0 | 0 | Red mica | None | Closed shape? | LC-EH | |
| 271601 | IN | 2 | 2 | 0 | 0 | 0 | Red | None | Closed shape | LC-EH | |
| 271601 | IN | 2 | 2 | 0 | 0 | 0 | Light | None | Closed shape? | LC-EH | |
| 271902 | IN | 1 | 1 | 0 | 0 | 0 | Brown mica | None | Closed shape | LC-EH | |
| 271902 | IN | 4 | 4 | 0 | 0 | 0 | Red | None | Closed shape | LC-EH | |
| 272501 | IN | 1 | 1 | 0 | 0 | 0 | Red | None | Closed shape? | LC-EH | |
| 272901 | IN | 1 | 1 | 0 | 0 | 0 | Red | None | ? | LC-EH | |
| 270501 | TR | 1 | 1 | 0 | 0 | 0 | Herakleian | None | Amphora | LC-EH | |
| 270701 | TR | 1 | 1 | 0 | 0 | 0 | Red mica | None | Amphora | LC-EH | |
| 270801 | TR | 1 | 0 | 1 | 0 | 0 | Brown | None | Amphora | LC-EH | 137 |
| 271001 | TR | 2 | 2 | 0 | 0 | 0 | Herakleian | None | Amphora | LC-EH | |
| 271001 | TR | 1 | 1 | 0 | 0 | 0 | Chersonesean? | None | Amphora | LC-EH | |
| 271002 | TR | 3 | 2 | 1 | 0 | 0 | Herakleian | None | Amphora | LC-EH | |
| 271002 | TR | 1 | 1 | 0 | 0 | 0 | Red mica | None | Amphora | LC-EH | |
| 271501 | TR | 1 | 1 | 0 | 0 | 0 | Red | None | Amphora? | LC-EH | |
| 271901 | TR | 1 | 1 | 0 | 0 | 0 | Red | None | Amphora | LC-EH | |
| 271902 | TR | 1 | 0 | 1 | 0 | 0 | Chersonesean | None | Amphora | LC-EH | |
| 272001 | TR | 2 | 1 | 0 | 1 | 0 | Chersonesean | Slip | Amphora | LC-EH | |
| 272101 | TR | 1 | 1 | 0 | 0 | 0 | Sinopean | None | Amphora | LC-EH | |
| 272801 | TR | 1 | 1 | 0 | 0 | 0 | Herakleian | None | Amphora | LC-EH | |
| 899201 | TW | 1 | 0 | 0 | 0 | 1 | Grey | None | Jug | LC-EH | |
| 271601 | UT | 1 | 1 | 0 | 0 | 0 | Light | None | Open shape? | LC-EH | |
| 272401 | UT | 1 | 1 | 0 | 0 | 0 | Light | Slip | Louterion | LC-EH | |
| 899104 | HW | 2 | 2 | 0 | 0 | 0 | Coarse | None | Pot | LC-EH? | |
| 270703 | IN | 2 | 2 | 0 | 0 | 0 | Brown | None | ? | LC-EH? | |
| 270902 | IN | 1 | 1 | 0 | 0 | 0 | Brown | None | Closed shape | LC-EH? | |
| 271001 | IN | 2 | 2 | 0 | 0 | 0 | Brown | None | Closed shape | LC-EH? | |
| 270903 | TR | 1 | 1 | 0 | 0 | 0 | Herakleian | None | Amphora | LC-EH? | |
| 899102 | TR | 2 | 2 | 0 | 0 | 0 | Thasian | None | Amphora | LC-EH? | |
| 899102 | TR | 2 | 2 | 0 | 0 | 0 | Herakleian | None | Amphora | LC-EH? | |
| 899104 | TR | 1 | 1 | 0 | 0 | 0 | Red | None | Amphora? | LC-EH? | |
| 899105 | TR | 1 | 1 | 0 | 0 | 0 | Thasian | None | Amphora | LC-EH? | |
| 270902 | UT | 1 | 1 | 0 | 0 | 0 | Brown | Smoothed | Open shape | LC-EH? | |
| 271801 | UT | 1 | 1 | 0 | 0 | 0 | Brown | None | Closed shape | LC-EH? | |
| 270501 | TW | 1 | 1 | 0 | 0 | 0 | Tatarian? | None | Jug? | EM | |
| 270703 | TW | 1 | 1 | 0 | 0 | 0 | Tatarian | Slip | Lid? | EM | |
| 270801 | TW | 1 | 1 | 0 | 0 | 0 | Tatarian | Glazed | Closed shape | EM | |
| 270701 | AR | 1 | 0 | 1 | 0 | 0 | Red modern | Slip | Tile | Modern | |
| 272001 | AR | 1 | 1 | 0 | 0 | 0 | Red modern | None | Tile | Modern | |
| 272101 | AR | 1 | 1 | 0 | 0 | 0 | Red modern | None | Tile | Modern | |
| 272201 | AR | 3 | 3 | 0 | 0 | 0 | Red modern | None | Tile | Modern | |

| BagID | Function | Total | Body | Rim | Handle | Base | Fabric | Surface | Shape | Period | Cat. |
|---|---|---|---|---|---|---|---|---|---|---|---|
| 272301 | AR | 1 | 1 | 0 | 0 | 0 | Red modern | Slip | Tile | Modern | |
| 272401 | AR | 1 | 1 | 0 | 0 | 0 | Red modern | None | Tile | Modern | |
| 272801 | AR | 1 | 1 | 0 | 0 | 0 | Red modern | None | Tile | Modern | |
| 271002 | TW | 1 | 1 | 0 | 0 | 0 | China | Glazed | Open shape | Modern | |
| 270602 | AR | 1 | 1 | 0 | 0 | 0 | Red modern | None | Tile | Modern? | |
| 270601 | IN | 1 | 1 | 0 | 0 | 0 | Red | None | ? | ? | |
| 272001 | IN | 1 | 0 | 0 | 0 | 0 | Light | Impressed | ? | ? | |
| 899101 | IN | 4 | 4 | 0 | 0 | 0 | Red | None | ? | ? | |
| 899102 | IN | 3 | 3 | 0 | 0 | 0 | Red | None | ? | ? | |
| 270602 | LI | 1 | 0 | 0 | 0 | 0 | | | | | |
| 270702 | LI | 1 | 0 | 0 | 0 | 0 | | | | | |
| 270703 | LI | 1 | 0 | 0 | 0 | 0 | | | | | |
| 270901 | LI | 1 | 0 | 0 | 0 | 0 | | | | | |
| 270902 | LI | 2 | 0 | 0 | 0 | 0 | | | | | |
| 271001 | LI | 2 | 2 | 0 | 0 | 0 | | | | | |
| 271601 | LI | 1 | 0 | 0 | 0 | 0 | | | | | |
| 271801 | LI | 1 | 0 | 0 | 0 | 0 | | | | | |
| 271901 | LI | 1 | 0 | 0 | 0 | 0 | | | | | |
| 271902 | LI | 2 | 0 | 0 | 0 | 0 | | | | | |
| 272001 | LI | 1 | 0 | 0 | 0 | 0 | | | | | |
| 899101 | LI | 1 | 1 | 0 | 0 | 0 | | | | | |
| 901901 | HW | 2 | 2 | 0 | 0 | 0 | Grey | None | Pot | LC-EH | |
| 901901 | IN | 1 | 1 | 0 | 0 | 0 | Red mica | None | ? | ? | |
| 901901 | TR | 1 | 0 | 0 | 1 | 0 | Red | None | Amphora | LC-EH | |
| 901901 | TR | 1 | 0 | 0 | 1 | 0 | Sinopean | None | Amphora | LC-EH | **136** |

## DSP08-F01-01

Finds table core units

| BagID | Function | Total | Body | Rim | Handle | Base | Fabric | Surface | Shape | Period | Cat. |
|---|---|---|---|---|---|---|---|---|---|---|---|
| 515401 | TR | 1 | 0 | 0 | 0 | 1 | Herakleian | None | Amphora | Late Classical | **140** |
| 600901 | IN | 1 | 1 | 0 | 0 | 0 | Red mica | None | Closed shape | LC-EH | |
| 515301 | TR | 1 | 1 | 0 | 0 | 0 | Herakleian | None | Amphora | LC-EH | |
| 600901 | TR | 2 | 2 | 0 | 0 | 0 | Thasian | None | Amphora | LC-EH | |
| 600901 | TR | 1 | 1 | 0 | 0 | 0 | Sinopean | Slip | Amphora | LC-EH | |
| 600901 | TR | 4 | 4 | 0 | 0 | 0 | Herakleian | None | Amphora | LC-EH | |
| 601601 | TR | 1 | 0 | 0 | 1 | 0 | Herakleian | None | Amphora | LC-EH | **141** |
| 515401 | IN | 1 | 1 | 0 | 0 | 0 | Red mica | None | Closed shape? | LC-EH? | |
| 601601 | IN | 1 | 1 | 0 | 0 | 0 | Red coarse | Polished | Closed shape | Medieval? | |
| 601602 | TW | 1 | 0 | 1 | 0 | 0 | Red | Decorated | Bowl | Medieval? | |
| 601601 | IN | 1 | 1 | 0 | 0 | 0 | Red | None | Closed shape | EM | |

## DSP08-F02-01

Finds table core units

| BagID | Function | Total | Body | Rim | Handle | Base | Fabric | Surface | Shape | Period | Cat. |
|---|---|---|---|---|---|---|---|---|---|---|---|
| 817701 | TR | 5 | 4 | 1 | 0 | 0 | Red | None | Amphora | Late Classical | **153** |
| 817802 | HW | 3 | 3 | 0 | 0 | 0 | Grey | None | Pot | LC-EH | |
| 817802 | HW | 2 | 2 | 0 | 0 | 0 | Red | None | Pot | LC-EH | |
| 817701 | TR | 4 | 3 | 0 | 1 | 0 | Red | None | Amphora | LC-EH | **154** |
| 817701 | TR | 1 | 0 | 1 | 0 | 0 | Knidian | None | Amphora | LC-EH | **152** |
| 817701 | HW | 3 | 1 | 1 | 0 | 1 | HwRed | None | Pot | LC-EH? | **138-139** |
| 817702 | HW | 7 | 7 | 0 | 0 | 0 | HwRed | None | Pot | LC-EH? | |
| 818101 | HW | 2 | 2 | 0 | 0 | 0 | Red | None | Pot | LC-EH? | |
| 817702 | IN | 4 | 4 | 0 | 0 | 0 | Red | None | Closed shape? | LC-EH? | |
| 817702 | IN | 1 | 1 | 0 | 0 | 0 | Red sandy | None | Closed shape | LC-EH? | |
| 817802 | IN | 3 | 3 | 0 | 0 | 0 | Red mica | None | ? | LC-EH? | |
| 817802 | IN | 1 | 1 | 0 | 0 | 0 | Red | None | Closed shape | LC-EH? | |
| 817802 | IN | 2 | 2 | 0 | 0 | 0 | Red pyro | None | ? | LC-EH? | |
| 817802 | IN | 6 | 6 | 0 | 0 | 0 | Red | None | ? | LC-EH? | |
| 818101 | IN | 1 | 1 | 0 | 0 | 0 | Red mica | None | ? | LC-EH? | |
| 818101 | IN | 6 | 6 | 0 | 0 | 0 | Red | None | Closed shape | LC-EH? | |
| 817701 | TR | 2 | 2 | 0 | 0 | 0 | Sinopean | None | Amphora | LC-EH? | |
| 817702 | TR | 1 | 1 | 0 | 0 | 0 | Red | Slip | Amphora | LC-EH? | |
| 817702 | TR | 8 | 7 | 0 | 1 | 0 | Herakleian | None | Amphora | LC-EH? | **144** |
| 817702 | TR | 1 | 1 | 0 | 0 | 0 | Sinopean | None | Amphora | LC-EH? | **148** |
| 817702 | TR | 3 | 2 | 0 | 1 | 0 | Red | None | Amphora | LC-EH? | **156** |
| 817701 | TR | 1 | 0 | 0 | 1 | 0 | Red | None | Amphora | LC-EH? | **155** |
| 817702 | TR | 2 | 2 | 0 | 0 | 0 | Brown | None | Amphora | LC-EH? | |
| 817702 | TR | 3 | 3 | 0 | 0 | 0 | Thasian | None | Amphora | LC-EH? | |
| 817801 | TR | 1 | 0 | 0 | 1 | 0 | Red | None | Amphora | LC-EH? | |
| 817801 | TR | 6 | 5 | 0 | 1 | 0 | Herakleian | None | Amphora | LC-EH? | |
| 817801 | TR | 6 | 5 | 0 | 1 | 0 | Thasian | None | Amphora | LC-EH? | **151** |
| 817801 | TR | 1 | 1 | 0 | 0 | 0 | Red mica | None | Amphora | LC-EH? | |
| 817801 | TR | 1 | 1 | 0 | 0 | 0 | Sinopean | None | Amphora | LC-EH? | |
| 817802 | TR | 1 | 0 | 0 | 1 | 0 | Thasian | None | Amphora? | LC-EH? | **149-150** |
| 817802 | TR | 1 | 0 | 0 | 1 | 0 | Red | slip | Amphora | LC-EH? | **157** |
| 817802 | TR | 12 | 10 | 1 | 1 | 0 | Herakleian | None | Amphora | LC-EH? | **147** |
| 817802 | TR | 2 | 2 | 0 | 0 | 0 | Red mica | None | Amphora | LC-EH? | |
| 817802 | TR | 1 | 1 | 0 | 0 | 0 | Red coarse | None | Amphora | LC-EH? | |
| 817802 | TR | 4 | 2 | 0 | 2 | 0 | Red golden mica | None | Amphora | LC-EH? | **158-159** |
| 818101 | TR | 1 | 0 | 0 | 1 | 0 | Sinopean | None | Amphora | LC-EH? | |
| 818101 | TR | 6 | 4 | 0 | 2 | 0 | Herakleian | None | Amphora | LC-EH? | **145-146** |
| 818101 | TR | 1 | 0 | 0 | 1 | 0 | Thasian | None | Amphora | LC-EH? | |

| BagID | Function | Total | Body | Rim | Handle | Base | Fabric | Surface | Shape | Period | Cat. |
|---|---|---|---|---|---|---|---|---|---|---|---|
| 818101 | TR | 1 | 1 | 0 | 0 | 0 | Chersonesean | None | Amphora | LC-EH? | **142** |
| 817702 | TW | 1 | 1 | 0 | 0 | 0 | Red | Slip | Closed shape | LC-EH? | |
| 817801 | TW | 1 | 0 | 1 | 0 | 0 | Red | None | Bowl | LC-EH? | **161** |
| 818101 | TW | 1 | 1 | 0 | 0 | 0 | Grey | None | Closed shape | LC-EH? | |
| 818101 | UT | 1 | 1 | 0 | 0 | 0 | Red | None | Pithos? | LC-EH? | |
| 817701 | IN | 1 | 1 | 0 | 0 | 0 | Tatarian | Glaze | Open shape | EM | |
| 817702 | TW | 1 | 0 | 1 | 0 | 0 | Tatarian | Glazed | ? | EM | |
| 817801 | TW | 1 | 1 | 0 | 0 | 0 | Tatarian | Glazed | Open shape | EM | |
| 817702 | AR | 1 | 1 | 0 | 0 | 0 | Red modern | Slip | Tile | Modern? | |
| 817802 | AR | 1 | 1 | 0 | 0 | 0 | Red | Smoothed | Brick | Modern? | |
| 817702 | - | 1 | 0 | 0 | 0 | 0 | | | | | |
| 817801 | IN | 6 | 6 | 0 | 0 | 0 | Red | None | ? | ? | |
| 817801 | IN | 1 | 1 | 0 | 0 | 0 | Red | None | Closed shape | ? | |
| 817801 | IN | 1 | 1 | 0 | 0 | 0 | Red coarse | None | Closed shape | ? | |
| 817702 | LI | 1 | 0 | 0 | 0 | 0 | | | | | |
| 817801 | LI | 1 | 0 | 0 | 0 | 0 | | | | | |
| 817802 | LI | 1 | 0 | 0 | 0 | 0 | | | | | |
| 818101 | LI | 1 | 0 | 0 | 0 | 0 | | | | | |

Finds table adjacent units

| BagID | Function | Total | Body | Rim | Handle | Base | Fabric | Surface | Shape | Period | Cat. |
|---|---|---|---|---|---|---|---|---|---|---|---|
| 818301 | HW | 2 | 1 | 1 | 0 | 0 | HwCoarse | impressed | Pot | LC-EH | |
| 818201 | HW | 1 | 1 | 0 | 0 | 0 | HwLight | None | Pot | LC-EH? | |
| 818401 | HW | 2 | 2 | 0 | 0 | 0 | HwLight | None | Pot | LC-EH? | |
| 817501 | IN | 5 | 5 | 0 | 0 | 0 | Red | None | ? | LC-EH? | |
| 818301 | IN | 1 | 1 | 0 | 0 | 0 | Red | None | ? | LC-EH? | |
| 818401 | IN | 4 | 4 | 0 | 0 | 0 | Red | None | ? | LC-EH? | |
| 817501 | TR | 6 | 5 | 0 | 1 | 0 | Herakleian | None | Amphora | LC-EH? | |
| 817501 | TR | 1 | 1 | 0 | 0 | 0 | Red mica | None | Amphora | LC-EH? | |
| 817501 | TR | 2 | 2 | 0 | 0 | 0 | Sinopean | None | Amphora | LC-EH? | |
| 817501 | TR | 5 | 4 | 0 | 1 | 0 | Thasian | None | Amphora | LC-EH? | |
| 817501 | TR | 1 | 0 | 0 | 1 | 0 | Red | None | Amphora? | LC-EH? | |
| 817501 | TR | 3 | 3 | 0 | 0 | 0 | Red mica | None | Amphora | LC-EH? | |
| 818201 | TR | 4 | 4 | 0 | 0 | 0 | Herakleian | None | Amphora | LC-EH? | |
| 818201 | TR | 1 | 1 | 0 | 0 | 0 | Sinopean | None | Amphora | LC-EH? | |
| 818201 | TR | 3 | 3 | 0 | 0 | 0 | Red mica | None | Amphora | LC-EH? | |
| 818201 | TR | 1 | 1 | 0 | 0 | 0 | Chersonesean? | None | Amphora | LC-EH? | **143** |
| 818201 | TR | 1 | 0 | 0 | 1 | 0 | Red | None | Amphora | LC-EH? | **160** |
| 818301 | TR | 2 | 2 | 0 | 0 | 0 | Herakleian | None | Amphora | LC-EH? | |
| 818301 | TR | 5 | 5 | 0 | 0 | 0 | Thasian | None | Amphora | LC-EH? | |
| 818401 | TR | 1 | 1 | 0 | 0 | 0 | Chersonesean | Slip | Amphora | LC-EH? | |
| 818401 | TR | 4 | 4 | 0 | 0 | 0 | Herakleian | None | Amphora | LC-EH? | |

| BagID | Function | Total | Body | Rim | Handle | Base | Fabric | Surface | Shape | Period | Cat. |
|---|---|---|---|---|---|---|---|---|---|---|---|
| 818401 | TR | 1 | 1 | 0 | 0 | 0 | Red mica | None | Amphora | LC-EH? | |
| 817501 | TW | 1 | 0 | 1 | 0 | 0 | Light | Slip | Bowl? | Medieval? | |
| 818201 | UT | 1 | 1 | 0 | 0 | 0 | Red coarse | None | Closed shape | Medieval? | |
| 818401 | TW | 1 | 0 | 0 | 0 | 1 | Tatarian | Glazed | Plate? | EM | |
| 818301 | AR | 1 | 1 | 0 | 0 | 0 | Red Tatarian | None | Brick | EM? | |
| 818201 | TE | 1 | 1 | 0 | 0 | 0 | Tatarian | None | Pibe | EM? | **162** |
| 817501 | TW | 1 | 1 | 0 | 0 | 0 | Tatarian | Glazed | Open shape | EM? | |
| 818301 | | 1 | 0 | 0 | 0 | 0 | | | | | |

## DSP08-F03-01

Finds table core units

| BagID | Function | Total | Body | Rim | Handle | Base | Fabric | Surface | Shape | Period | Cat. |
|---|---|---|---|---|---|---|---|---|---|---|---|
| 623401 | HW | 2 | 2 | 0 | 0 | 0 | Red | None | Pot | LC-EH? | |
| 623402 | HW | 4 | 4 | 0 | 0 | 0 | Red | None | Pot | LC-EH? | |
| 623402 | HW | 4 | 4 | 0 | 0 | 0 | Red | None | Pot | LC-EH? | |
| 623401 | TR | 1 | 0 | 0 | 1 | 0 | Red | None | Amphora | LC-EH? | **163** |
| 623402 | TR | 1 | 0 | 0 | 1 | 0 | Red | None | Amphora | LC-EH? | **164** |
| 623402 | UT | 1 | 0 | 1 | 0 | 0 | Grey | Slip | Cooking ware | EM | **175** |
| 623401 | TW | 8 | 7 | 1 | 0 | 0 | Tatarian | Glazed | Bowl? | EM | **171** |
| 623402 | TW | 8 | 4 | 3 | 1 | 0 | Tatarian | Glazed | ? | EM | |
| 623401 | TW | 22 | 22 | 0 | 0 | 0 | Tatarian 2 | Slip | ? | EM? | |
| 623401 | UT | 2 | 0 | 2 | 0 | 0 | Tartarian | Slip | Casserole | EM | **170, 174** |
| 623402 | TW | 34 | 29 | 4 | 0 | 1 | Tatarian 2 | Slip | ? | EM? | **179, 183, 188** |
| 623401 | UT | 8 | 4 | 3 | 0 | 1 | Tatarian 2 | None | Cooking ware | EM? | **167-169, 172-173** |
| 623402 | UT | 14 | 10 | 4 | 0 | 0 | Tatarian 2 | None | Cooking ware | EM? | **176-178, 180-182, 184-187** |
| 623401 | - | 1 | 0 | 0 | 0 | 0 | | | | modern | |
| 623401 | AR | 10 | 9 | 1 | 0 | 0 | Red modern | None | Brick | Modern | |
| 623402 | AR | 5 | 4 | 1 | 0 | 0 | Red modern | Smoothed | Brick | Modern | |
| 623402 | IR | 1 | 0 | 0 | 0 | 0 | | | | Modern | |
| 623402 | TW | 1 | 1 | 0 | 0 | 0 | China | Glazed | Open shape | Modern | |
| 623401 | IR | 1 | 1 | 0 | 0 | 0 | | | | Modern? | |
| 623401 | IR | 1 | 0 | 0 | 0 | 0 | | | | | |
| 623402 | IR | 3 | 0 | 0 | 0 | 0 | | | | | |

Finds table adjacent units

| BagID | Function | Total | Body | Rim | Handle | Base | Fabric | Surface | Shape | Period | Cat. |
|---|---|---|---|---|---|---|---|---|---|---|---|
| 623301 | HW | 1 | 1 | 0 | 0 | 0 | Red | None | Pot | LC-EH? | |
| 623601 | HW | 3 | 3 | 0 | 0 | 0 | Red | None | Pot | LC-EH? | |
| 624001 | HW | 1 | 1 | 0 | 0 | 0 | Red | None | Pot | LC-EH? | |
| 625201 | HW | 2 | 2 | 0 | 0 | 0 | Grey | None | Pot | LC-EH? | |
| 623001 | IN | 7 | 6 | 0 | 0 | 1 | Red | None | ? | LC-EH? | |
| 623201 | IN | 1 | 1 | 0 | 0 | 0 | Red | None | ? | LC-EH? | |
| 623301 | IN | 1 | 1 | 0 | 0 | 0 | Red | None | ? | LC-EH? | |
| 625201 | IN | 1 | 1 | 0 | 0 | 0 | Red | None | ? | LC-EH? | |
| 623001 | TR | 2 | 2 | 0 | 0 | 0 | Red coarse | None | Amphora? | LC-EH? | |
| 623101 | TR | 1 | 1 | 0 | 0 | 0 | Herakleian | None | Amphora | LC-EH? | |
| 623301 | TR | 1 | 1 | 0 | 0 | 0 | Herakleian? | None | Amphora? | LC-EH? | |
| 623801 | TR | 1 | 1 | 0 | 0 | 0 | Herakleian | None | Amphora? | LC-EH? | |
| 623901 | TR | 1 | 1 | 0 | 0 | 0 | Thasian | None | Amphora | LC-EH? | |
| 624001 | TR | 1 | 1 | 0 | 0 | 0 | Herakleian | None | Amphora | LC-EH? | |
| 625201 | TR | 1 | 1 | 0 | 0 | 0 | Herakleian | None | Amphora | LC-EH? | |
| 625601 | TR | 1 | 0 | 0 | 1 | 0 | Red | None | Amphora | LC-EH? | **165** |
| 623901 | IN | 3 | 2 | 0 | 0 | 1 | Red | None | Jug? | Medieval? | |
| 623601 | IN | 1 | 1 | 0 | 0 | 0 | Tatarian 2 | None | ? | EM | |
| 623001 | TW | 1 | 1 | 0 | 0 | 0 | Tatarian | Glazed | Open shape | EM | |
| 623301 | TW | 1 | 0 | 0 | 1 | 0 | Tatarian | Slip | Jug? | EM | **166** |
| 623601 | TW | 1 | 0 | 1 | 0 | 0 | Tatarian 2 | glaze | Jug | EM | **189** |
| 623801 | TW | 2 | 2 | 0 | 0 | 0 | Tatarian 2 | Glazed | Open shape | EM | |
| 624001 | TW | 1 | 0 | 1 | 0 | 0 | Red | None | Jug | EM | **190** |
| 625301 | TW | 1 | 1 | 0 | 0 | 0 | Tatarian 2 | Glazed | Open shape | EM | |
| 625601 | TW | 2 | 1 | 1 | 0 | 0 | Tatarian 2 | Glazed | Bowl | EM | |
| 623501 | AR | 1 | 1 | 0 | 0 | 0 | Red | None | Tile? | EM? | |
| 623001 | UT | 3 | 3 | 0 | 0 | 0 | Tatarian 2 | Slip | ? | EM? | |
| 623101 | AR | 3 | 3 | 0 | 0 | 0 | Red modern | None | Brick | Modern | |
| 623301 | AR | 1 | 0 | 1 | 0 | 0 | Red modern | Slip | Tile | Modern | |
| 623601 | IN | 5 | 4 | 1 | 0 | 0 | Red | None | ? | ? | |
| 624001 | IN | 2 | 2 | 0 | 0 | 0 | Red | None | ? | ? | |
| 623101 | LI | 1 | 0 | 0 | 0 | 0 | | | | | |
| 623501 | LI | 1 | 0 | 0 | 0 | 0 | | | | | |
| 625201 | LI | 1 | 0 | 0 | 0 | 0 | | | | | |
| 625601 | TR | 1 | 0 | 0 | 0 | 1 | LC-EH | None | None | LC-EH | |
| 625601 | TW | 2 | 1 | 1 | 0 | 0 | PM | Glazed | Bowl? | Modern | **191** |

## DSP08-F04-01

Finds table core units

| BagID | Function | Total | Body | Rim | Handle | Base | Fabric | Surface | Shape | Period | Cat. |
|---|---|---|---|---|---|---|---|---|---|---|---|
| 851003 | UT | 4 | 4 | 0 | 0 | 0 | Sinopean | Slip | Louterion | Classical | |
| 851003 | TR | 1 | 0 | 0 | 0 | 1 | Red | Slip | Amphora | LC | |
| 851001 | HW | 9 | 9 | 0 | 0 | 0 | Grey | None | Pot | LC-EH | |
| 851002 | HW | 13 | 13 | 0 | 0 | 0 | Grey | None | Pot | LC-EH | |
| 851002 | HW | 3 | 3 | 0 | 0 | 0 | Red | None | Pot | LC-EH | |
| 851002 | HW | 2 | 2 | 0 | 0 | 0 | Coarse | None | Pot | LC-EH | |
| 851003 | HW | 12 | 12 | 0 | 0 | 0 | Grey | None | Pot | LC-EH | |
| 851003 | HW | 1 | 1 | 0 | 0 | 0 | Red | None | Pot | LC-EH | |
| 851101 | HW | 2 | 2 | 0 | 0 | 0 | Red | None | Pot | LC-EH | |
| 851101 | HW | 3 | 3 | 0 | 0 | 0 | Grey | None | Pot | LC-EH | |
| 851001 | IN | 1 | 1 | 0 | 0 | 0 | Red pyro | None | ? | LC-EH | |
| 851001 | IN | 3 | 3 | 0 | 0 | 0 | Red porous | Slip | Closed shape | LC-EH | |
| 851003 | IN | 1 | 1 | 0 | 0 | 0 | Red pyro | None | ? | LC-EH | |
| 851003 | IN | 1 | 1 | 0 | 0 | 0 | Red | None | Open shape | LC-EH | |
| 851003 | IN | 3 | 3 | 0 | 0 | 0 | Red | None | Closed shape | LC-EH | |
| 851001 | TR | 3 | 1 | 2 | 0 | 0 | Red | None | Amphora | LC-EH | **193-194** |
| 851002 | TR | 1 | 0 | 0 | 1 | 0 | Chersonesean? | None | Amphora | LC-EH | |
| 851002 | TR | 1 | 1 | 0 | 0 | 0 | Sinopean | None | Amphora | LC-EH | |
| 851101 | TR | 2 | 2 | 0 | 0 | 0 | Chersonesean? | None | Amphora? | LC-EH | |
| 847101 | TW | 1 | 0 | 1 | 0 | 0 | Chersonesean? | Smoothed | Bowl | LC-EH | **195** |
| 851001 | TW | 2 | 1 | 0 | 0 | 1 | Sinopean | Slip | Jug? | LC-EH | **196** |
| 851002 | TW | 1 | 1 | 0 | 0 | 0 | Red | None | Open shape | LC-EH | |
| 847001 | HW | 3 | 3 | 0 | 0 | 0 | Grey | None | Closed shape | LC-EH? | |
| 846901 | IN | 2 | 2 | 0 | 0 | 0 | Red | None | Closed shape | LC-EH? | |
| 851001 | IN | 4 | 4 | 0 | 0 | 0 | Red | None | ? | LC-EH? | |
| 851002 | IN | 3 | 3 | 0 | 0 | 0 | Red | None | ? | LC-EH? | |
| 851002 | IN | 1 | 1 | 0 | 0 | 0 | Light | Slip | Closed shape | LC-EH? | |
| 851101 | IN | 3 | 3 | 0 | 0 | 0 | Red | None | ? | LC-EH? | |
| 851101 | IN | 1 | 1 | 0 | 0 | 0 | Red | Slip | Open shape | LC-EH? | |
| 847001 | TR | 1 | 1 | 0 | 0 | 0 | Red pyro | None | Amphora | LC-EH? | |
| 851001 | TR | 1 | 1 | 0 | 0 | 0 | Brown sand | Smoothed | Amphora? | LC-EH? | |
| 851003 | TR | 1 | 0 | 1 | 0 | 0 | Herakleian | none | Amphora | LC-EH? | **192** |
| 851001 | UT | 1 | 1 | 0 | 0 | 0 | Brown mica | Smoothed | ? | LC-EH? | |
| 851001 | UT | 1 | 1 | 0 | 0 | 0 | Brown | Smoothed | ? | LC-EH? | |
| 851003 | UT | 1 | 1 | 0 | 0 | 0 | Red | Smoothed | ? | LC-EH? | **197** |
| 846901 | IN | 1 | 1 | 0 | 0 | 0 | Light | None | Closed shape | ? | |
| 851001 | IN | 1 | 1 | 0 | 0 | 0 | Light | None | ? | ? | |
| 846901 | LI | 1 | 1 | 0 | 0 | 0 | | | | | |
| 851003 | LI | 1 | 0 | 0 | 0 | 0 | | | | | |

| BagID | Function | Total | Body | Rim | Handle | Base | Fabric | Surface | Shape | Period | Cat. |
|---|---|---|---|---|---|---|---|---|---|---|---|
| 851101 | LI | 1 | 0 | 0 | 0 | 0 | | | | | |

Finds table adjacent units

| BagID | Function | Total | Body | Rim | Handle | Base | Fabric | Surface | Shape | Period | Cat. |
|---|---|---|---|---|---|---|---|---|---|---|---|
| 851601 | HW | 1 | 1 | 0 | 0 | 0 | Grey | None | Pot | LC-EH | |
| 847001 | HW | 3 | 3 | 0 | 0 | 0 | Grey | None | Closed shape | LC-EH? | |
| 850901 | IN | 2 | 2 | 0 | 0 | 0 | Red | None | Closed shape | LC-EH? | |
| 847001 | TR | 1 | 1 | 0 | 0 | 0 | Red pyro | None | Amphora | LC-EH? | |
| 850901 | TR | 1 | 1 | 0 | 0 | 0 | Chersonesean | Slip | Amphora | LC-EH? | |
| 851601 | IN | 2 | 2 | 0 | 0 | 0 | Red | None | Closed shape? | ? | |

## DSP08-F04-02

Finds table core units

| BagID | Function | Total | Body | Rim | Handle | Base | Fabric | Surface | Shape | Period | Cat. |
|---|---|---|---|---|---|---|---|---|---|---|---|
| 850003 | UT | 1 | 1 | 0 | 0 | 0 | Chersonesean | Slip | Louterion? | LC | |
| 849901 | HW | 2 | 2 | 0 | 0 | 0 | Grey | None | Pot | LC-EH | |
| 849901 | HW | 1 | 1 | 0 | 0 | 0 | Coarse | None | Pot | LC-EH | |
| 850001 | HW | 3 | 2 | 1 | 0 | 0 | Grey | None | Pot | LC-EH | |
| 850002 | HW | 27 | 27 | 0 | 0 | 0 | Grey | None | Pot | LC-EH | |
| 850002 | HW | 1 | 1 | 0 | 0 | 0 | Coarse | None | Pot | LC-EH | |
| 850003 | HW | 1 | 0 | 1 | 0 | 0 | Red | Smoothed | Open shape | LC-EH | |
| 850003 | HW | 1 | 1 | 0 | 0 | 0 | Grey | polished | Pot | LC-EH | |
| 850003 | HW | 40 | 40 | 1 | 0 | 0 | Grey | None | Pot | LC-EH | **198** |
| 850004 | HW | 9 | 9 | 0 | 0 | 0 | Red | none | Pot | LC-EH | |
| 850004 | HW | 3 | 3 | 0 | 0 | 0 | Coarse | None | Pot | LC-EH | |
| 850004 | HW | 23 | 23 | 0 | 0 | 0 | Grey | None | Pot | LC-EH | |
| 850004 | HW | 2 | 2 | 0 | 0 | 0 | Light | None | Pot | LC-EH | |
| 850004 | HW | 8 | 8 | 0 | 0 | 0 | Black | None | Pot | LC-EH | |
| 850005 | HW | 6 | 5 | 1 | 0 | 0 | Red | None | Bowl? | LC-EH | **199** |
| 850005 | HW | 11 | 11 | 0 | 0 | 0 | Grey | None | Pot | LC-EH | |
| 850005 | HW | 1 | 1 | 0 | 0 | 0 | Black | none | Pot | LC-EH | |
| 850006 | HW | 10 | 7 | 3 | 0 | 0 | Grey | Smoothed | Pot | LC-EH | **200-202** |
| 850007 | HW | 1 | 0 | 1 | 0 | 0 | HwRed | None | Pot | LC-EH | **203** |
| 852001 | HW | 1 | 1 | 0 | 0 | 0 | Black | None | Pot | LC-EH | |
| 852001 | HW | 4 | 4 | 0 | 0 | 0 | Grey | None | Pot | LC-EH | |
| 852001 | HW | 1 | 1 | 0 | 0 | 0 | Red | None | Pot | LC-EH | |
| 852101 | HW | 1 | 0 | 1 | 0 | 0 | Grey | Impressed | Open shape | LC-EH | |
| 852101 | HW | 10 | 10 | 0 | 0 | 0 | Grey | None | Pot | LC-EH | |
| 852101 | HW | 2 | 2 | 0 | 0 | 0 | Red | None | Pot | LC-EH | |

| BagID | Function | Total | Body | Rim | Handle | Base | Fabric | Surface | Shape | Period | Cat. |
|---|---|---|---|---|---|---|---|---|---|---|---|
| 852101 | HW | 2 | 2 | 0 | 0 | 0 | Grey | None | Pot | LC-EH | |
| 852101 | HW | 3 | 3 | 0 | 0 | 0 | Red | None | Pot | LC-EH | |
| 850003 | IN | 1 | 1 | 0 | 0 | 0 | Red pyro | None | ? | LC-EH | |
| 849901 | TR | 2 | 2 | 0 | 0 | 0 | Chersonesean? | None | Amphora | LC-EH | |
| 850002 | TR | 1 | 1 | 0 | 0 | 0 | Sandwich | None | Amphora | LC-EH | |
| 850002 | TR | 5 | 4 | 0 | 1 | 0 | Red | None | Amphora | LC-EH | |
| 850002 | TR | 1 | 1 | 0 | 0 | 0 | Sinopean | Slip | Amphora | LC-EH | |
| 850002 | TR | 1 | 1 | 0 | 0 | 0 | Grey | None | Amphora | LC-EH | |
| 850004 | TR | 2 | 2 | 0 | 0 | 0 | Sinopean | Smoothed | Amphora | LC-EH | |
| 850006 | TR | 1 | 1 | 0 | 0 | 0 | Sinopean | None | Amphora | LC-EH | |
| 850001 | HW | 11 | 10 | 0 | 0 | 1 | Grey | None | Pot | LC-EH? | |
| 850001 | HW | 7 | 6 | 1 | 0 | 0 | Red | None | Pot | LC-EH? | |
| 850001 | HW | 2 | 2 | 0 | 0 | 0 | Coarse | None | Pot | LC-EH? | |
| 849901 | IN | 3 | 3 | 0 | 0 | 0 | Red | None | ? | LC-EH? | |
| 850001 | IN | 4 | 4 | 0 | 0 | 0 | Red | None | ? | LC-EH? | |
| 850002 | IN | 1 | 1 | 0 | 0 | 0 | Red | Smoothed | Open shape? | LC-EH? | |
| 850002 | IN | 7 | 7 | 0 | 0 | 0 | Red | None | ? | LC-EH? | |
| 850003 | IN | 1 | 1 | 0 | 0 | 0 | Red coarse | None | Closed shape | LC-EH? | |
| 850004 | IN | 4 | 4 | 0 | 0 | 0 | Red | None | ? | LC-EH? | |
| 850005 | IN | 6 | 5 | 0 | 1 | 0 | Red | None | ? | LC-EH? | |
| 850006 | IN | 1 | 1 | 0 | 0 | 0 | Red mica | None | ? | LC-EH? | |
| 852001 | IN | 2 | 2 | 0 | 0 | 0 | Red | None | ? | LC-EH? | |
| 852101 | IN | 1 | 1 | 0 | 0 | 0 | Red | moulding | ? | LC-EH? | |
| 852101 | IN | 1 | 1 | 0 | 0 | 0 | Red pyro | Slip | Closed shape | LC-EH? | |
| 852101 | IN | 1 | 1 | 0 | 0 | 0 | Red | None | ? | LC-EH? | |
| 852101 | IN | 6 | 6 | 0 | 0 | 0 | Red | None | ? | LC-EH? | |
| 852101 | IN | 1 | 1 | 0 | 0 | 0 | Light | None | Closed shape | LC-EH? | |
| 850001 | TR | 1 | 1 | 0 | 0 | 0 | Red | Smoothed | Amphora | LC-EH? | |
| 850001 | TR | 1 | 0 | 1 | 0 | 0 | Chersonesean? | None | Amphora | LC-EH? | |
| 850001 | TR | 1 | 1 | 0 | 0 | 0 | Herakleian? | None | Amphora | LC-EH? | |
| 850001 | TR | 1 | 1 | 0 | 0 | 0 | Brown | Grooved | Amphora | LC-EH? | 207 |
| 850001 | TR | 1 | 1 | 0 | 0 | 0 | Red | None | Amphora | LC-EH? | |
| 850001 | TR | 5 | 4 | 0 | 0 | 0 | Herakleian | None | Amphora | LC-EH? | |
| 850001 | TR | 1 | 1 | 0 | 0 | 0 | Light | Slip | Amphora | LC-EH? | |
| 850001 | TR | 1 | 0 | 1 | 0 | 0 | Sinopean | None | Amphora | LC-EH? | 205 |
| 850001 | TR | 1 | 1 | 0 | 0 | 0 | Red | None | Amphora | LC-EH? | |
| 850003 | TR | 2 | 2 | 0 | 0 | 0 | Thasian | None | Amphora | LC-EH? | |
| 850003 | TR | 1 | 1 | 0 | 0 | 0 | Red | None | Amphora | LC-EH? | |
| 850003 | TR | 2 | 2 | 0 | 0 | 0 | Herakleian | None | Amphora | LC-EH? | |
| 850003 | TR | 2 | 2 | 0 | 0 | 0 | Chersonesean? | Smoothed | Amphora | LC-EH? | |
| 850004 | TR | 2 | 2 | 0 | 0 | 0 | Mendean? | None | Amphora | LC-EH? | |
| 850004 | TR | 2 | 2 | 0 | 0 | 0 | Chersonesean | Slip | Amphora | LC-EH? | |
| 850004 | TR | 1 | 1 | 0 | 0 | 0 | Red pyro | Smoothed | Amphora | LC-EH? | |

| BagID | Function | Total | Body | Rim | Handle | Base | Fabric | Surface | Shape | Period | Cat. |
|---|---|---|---|---|---|---|---|---|---|---|---|
| 850004 | TR | 1 | 1 | 0 | 0 | 0 | Red | None | Amphora | LC-EH? | **208** |
| 850005 | TR | 1 | 1 | 0 | 0 | 0 | Thasian | None | Amphora | LC-EH? | |
| 850005 | TR | 2 | 2 | 0 | 0 | 0 | Light | Slip | Amphora | LC-EH? | |
| 850005 | TR | 3 | 3 | 0 | 0 | 0 | Sinopean | Smoothed | Amphora | LC-EH? | |
| 852001 | TR | 1 | 1 | 0 | 0 | 0 | Light | None | Amphora? | LC-EH? | |
| 852001 | TR | 2 | 2 | 0 | 0 | 0 | Red | None | Amphora? | LC-EH? | |
| 852101 | TR | 1 | 1 | 0 | 0 | 0 | Herakleian? | None | Amphora | LC-EH? | |
| 852101 | TR | 1 | 1 | 0 | 0 | 0 | Red pyro | None | Amphora? | LC-EH? | |
| 852101 | TR | 3 | 2 | 0 | 1 | 0 | Chersonesean | Slip | Amphora | LC-EH? | |
| 852101 | TR | 2 | 2 | 0 | 0 | 0 | Herakleian | None | Amphora | LC-EH? | |
| 852103 | TR | 4 | 1 | 0 | 2 | 1 | Red | Slip | Amphora | LC-EH? | **209-210** |
| 852103 | TR | 1 | 0 | 0 | 0 | 1 | Chersonesean | Slip | Amphora | LC-EH? | **204** |
| 850001 | TW | 1 | 1 | 0 | 0 | 0 | Grey | Slip | Jug | LC-EH? | **211** |
| 850003 | TW | 1 | 1 | 0 | 0 | 0 | Chersonesean | Slip | Jug? | LC-EH? | |
| 852101 | UT | 5 | 5 | 0 | 0 | 0 | Light | Slip | Louterion? | LC-EH? | |
| 852101 | UT | 2 | 2 | 0 | 0 | 0 | Red | Wash | ? | LC-EH? | |
| 852101 | LI | 2 | 0 | 0 | 0 | 0 | | | | Modern | |
| 850005 | - | 1 | 1 | 0 | 0 | 0 | | | | | |
| 850003 | BO | 1 | 0 | 0 | 0 | 0 | | | | | |
| 850003 | IN | 5 | 5 | 0 | 0 | 0 | Red | None | ? | ? | |
| 850004 | IN | 1 | 1 | 0 | 0 | 0 | Light | None | Closed shape | ? | |
| 850003 | LI | 1 | 1 | 0 | 0 | 0 | | | | | |
| 852101 | LI | 1 | 0 | 0 | 0 | 0 | | | | | |

Finds table adjacent units

| BagID | Function | Total | Body | Rim | Handle | Base | Fabric | Surface | Shape | Period | Cat. |
|---|---|---|---|---|---|---|---|---|---|---|---|
| 851901 | HW | 1 | 1 | 0 | 0 | 0 | Grey | none | Pot | LC-EH | |
| 849801 | HW | 1 | 1 | 0 | 0 | 0 | Grey | None | Pot | LC-EH? | |
| 850101 | HW | 1 | 1 | 0 | 0 | 0 | Grey | None | Pot | LC-EH? | |
| 848001 | TR | 1 | 0 | 0 | 1 | 0 | Red | None | Amphora | LC-EH? | **206** |

## DSP08-F04-03

Finds table core units

| BagID | Function | Total | Body | Rim | Handle | Base | Fabric | Surface | Shape | Period | Cat. |
|---|---|---|---|---|---|---|---|---|---|---|---|
| 856802 | TW | 1 | 1 | 0 | 0 | 0 | Attic? | Glazed | Cup | Classical | **247** |
| 300501 | HW | 6 | 0 | 0 | 0 | 0 | | | | LC-EH | |
| 853301 | HW | 2 | 2 | 0 | 0 | 0 | Grey | None | Pot | LC-EH | |
| 853301 | HW | 2 | 2 | 0 | 0 | 0 | Red | None | Pot | LC-EH | |
| 853401 | HW | 3 | 3 | 0 | 0 | 0 | Grey | None | Pot | LC-EH | |
| 853401 | HW | 1 | 1 | 0 | 0 | 0 | Red | None | Pot | LC-EH | |
| 853401 | HW | 1 | 1 | 0 | 0 | 0 | Black | None | Pot | LC-EH | |
| 853402 | HW | 3 | 3 | 0 | 0 | 0 | Grey | None | Pot | LC-EH | |
| 853402 | HW | 3 | 1 | 2 | 0 | 0 | Red | None | Pot | LC-EH | |
| 856801 | HW | 1 | 1 | 0 | 0 | 0 | Red | None | Pot | LC-EH | |
| 856801 | HW | 1 | 1 | 0 | 0 | 0 | Black | polished | Pot | LC-EH | |
| 856802 | HW | 2 | 2 | 0 | 0 | 0 | Grey coarse | None | ? | LC-EH | |
| 856802 | HW | 4 | 4 | 0 | 0 | 0 | Red | None | Pot | LC-EH | |
| 856901 | HW | 2 | 2 | 0 | 0 | 0 | Grey | None | Pot | LC-EH | |
| 857001 | HW | 1 | 0 | 1 | 0 | 0 | Red | None | Pot | LC-EH | **213** |
| 857001 | HW | 1 | 0 | 1 | 0 | 0 | HwCoarse | None | Bowl | LC-EH | **212** |
| 857001 | HW | 15 | 15 | 0 | 0 | 0 | Grey | None | Pot | LC-EH | |
| 857001 | HW | 2 | 2 | 0 | 0 | 0 | Red coarse | None | ? | LC-EH | |
| 857003 | HW | 16 | 16 | 0 | 0 | 0 | Grey | None | Pot | LC-EH | |
| 857003 | HW | 1 | 0 | 1 | 0 | 0 | HwCoarse | None | Pot | LC-EH | **214** |
| 857004 | HW | 1 | 0 | 1 | 0 | 0 | Red | Impressed | Pot | LC-EH | |
| 857004 | HW | 2 | 1 | 1 | 0 | 0 | Red | None | Bowl | LC-EH | **215** |
| 857004 | HW | 8 | 8 | 0 | 0 | 0 | Grey | None | Pot | LC-EH | |
| 859101 | HW | 3 | 3 | 0 | 0 | 0 | Grey | None | Pot | LC-EH | |
| 859201 | HW | 1 | 1 | 0 | 0 | 0 | Red coarse | None | ? | LC-EH | |
| 859201 | HW | 5 | 5 | 0 | 0 | 0 | Grey | None | Pot | LC-EH | |
| 859201 | HW | 4 | 4 | 0 | 0 | 0 | Red | None | Pot | LC-EH | |
| 859201 | HW | 2 | 2 | 0 | 0 | 0 | Light | None | Pot | LC-EH | |
| 859202 | HW | 6 | 6 | 0 | 0 | 0 | Red | None | Pot | LC-EH | |
| 859202 | HW | 8 | 8 | 0 | 0 | 0 | Grey | None | Pot | LC-EH | |
| 859203 | HW | 5 | 5 | 0 | 0 | 0 | Grey | None | Pot | LC-EH | |
| 859203 | HW | 1 | 1 | 0 | 0 | 0 | Red coarse | None | Pot | LC-EH | |
| 859204 | HW | 3 | 3 | 0 | 0 | 0 | Red | None | Pot | LC-EH | |
| 859204 | HW | 13 | 12 | 1 | 0 | 0 | Grey | None | Pot | LC-EH | |
| 859205 | HW | 1 | 1 | 0 | 0 | 0 | Coarse | None | Pot | LC-EH | |
| 859205 | HW | 9 | 9 | 0 | 0 | 0 | Grey | None | Pot | LC-EH | |
| 859301 | HW | 3 | 3 | 0 | 0 | 0 | Grey | None | Pot | LC-EH | |
| 859301 | HW | 1 | 0 | 1 | 0 | 0 | HwCoarse | None | Bowl | LC-EH | |
| 859301 | HW | 1 | 1 | 0 | 0 | 0 | HwCoarse | Impressed | Pot | LC-EH | |

| BagID | Function | Total | Body | Rim | Handle | Base | Fabric | Surface | Shape | Period | Cat. |
|---|---|---|---|---|---|---|---|---|---|---|---|
| 859401 | HW | 3 | 3 | 0 | 0 | 0 | Red | None | Pot | LC-EH | |
| 300501 | IN | 19 | 0 | 0 | 0 | 0 | | | | LC-EH | |
| 300501 | TR | 3 | 0 | 0 | 0 | 0 | Red | None | Amphora | LC-EH | |
| 853301 | TR | 1 | 0 | 0 | 0 | 1 | Red | Slip | Amphora | LC-EH | |
| 853401 | TR | 1 | 1 | 0 | 0 | 0 | Sinopean | None | Amphora | LC-EH | |
| 856801 | TR | 1 | 0 | 0 | 1 | 0 | Chersonesean | None | Amphora | LC-EH | **216** |
| 856801 | TR | 1 | 0 | 0 | 1 | 0 | Sinopean | None | Amphora | LC-EH | **221** |
| 857001 | TR | 17 | 15 | 0 | 2 | 0 | Herakleian | None | Amphora | LC-EH | **229** |
| 859402 | TR | 1 | 1 | 0 | 0 | 0 | Light | relief | Amphora? | LC-EH | |
| 857001 | TW | 3 | 3 | 0 | 0 | 0 | Chersonesean | Slip | Jug? | LC-EH | |
| 857001 | TW | 1 | 0 | 1 | 0 | 0 | Red | Slip | Lid? | LC-EH | **248** |
| 857002 | TW | 1 | 1 | 0 | 0 | 0 | Chersonesean | Slip | Jug | LC-EH | |
| 857003 | TW | 1 | 0 | 0 | 1 | 0 | Red | Slip | Jug | LC-EH | |
| 857001 | UT | 1 | 1 | 0 | 0 | 0 | Red | Slip | Open shape | LC-EH | |
| 857001 | HW | 1 | 1 | 0 | 0 | 0 | Black | None | Pot | LC-EH? | |
| 857002 | HW | 10 | 10 | 0 | 0 | 0 | Grey | None | Pot | LC-EH? | |
| 857005 | HW | 2 | 2 | 0 | 0 | 0 | Grey | None | Pot | LC-EH? | |
| 859102 | HW | 6 | 6 | 0 | 0 | 0 | Grey | None | Pot | LC-EH? | |
| 859102 | HW | 5 | 3 | 1 | 0 | 1 | Red coarse | None | Pot | LC-EH? | |
| 853301 | IN | 1 | 1 | 0 | 0 | 0 | Red | None | ? | LC-EH? | |
| 853401 | IN | 1 | 1 | 0 | 0 | 0 | Red | slip | Closed shape | LC-EH? | |
| 853401 | IN | 2 | 2 | 0 | 0 | 0 | Brown | None | ? | LC-EH? | |
| 853401 | IN | 3 | 3 | 0 | 0 | 0 | Red | None | ? | LC-EH? | |
| 853402 | IN | 5 | 5 | 0 | 0 | 0 | Red | None | ? | LC-EH? | |
| 853402 | IN | 2 | 2 | 0 | 0 | 0 | Chersonesean | Slip | Closed shape? | LC-EH? | |
| 856801 | IN | 6 | 6 | 0 | 0 | 0 | Red | None | ? | LC-EH? | |
| 856802 | IN | 1 | 1 | 0 | 0 | 0 | Red pyro | Slip | Open shape | LC-EH? | |
| 856802 | IN | 2 | 2 | 0 | 0 | 0 | Red | Slip | ? | LC-EH? | |
| 856802 | IN | 6 | 6 | 0 | 0 | 0 | Red | None | ? | LC-EH? | |
| 856802 | IN | 1 | 1 | 0 | 0 | 0 | Light | None | ? | LC-EH? | |
| 856901 | IN | 1 | 1 | 0 | 0 | 0 | Brown | None | Closed vessel | LC-EH? | |
| 856901 | IN | 4 | 4 | 0 | 0 | 0 | Red | None | ? | LC-EH? | |
| 857001 | IN | 3 | 3 | 0 | 0 | 0 | Red mica | None | ? | LC-EH? | |
| 857001 | IN | 1 | 1 | 0 | 0 | 0 | Brown | Slip | Closed shape | LC-EH? | |
| 857001 | IN | 7 | 7 | 0 | 0 | 0 | Red | None | ? | LC-EH? | |
| 857001 | IN | 4 | 4 | 0 | 0 | 0 | Red | None | Closed shape | LC-EH? | |
| 857001 | IN | 2 | 2 | 0 | 0 | 0 | Red | Slip | Closed shape | LC-EH? | |
| 857001 | IN | 1 | 1 | 0 | 0 | 0 | Red | Slip | ? | LC-EH? | |
| 857002 | IN | 2 | 2 | 0 | 0 | 0 | Red coarse | None | Closed shape | LC-EH? | |
| 857002 | IN | 4 | 4 | 0 | 0 | 0 | Red | Smoothed | ? | LC-EH? | |
| 857003 | IN | 2 | 2 | 0 | 0 | 0 | Red mica | None | Closed shape | LC-EH? | |
| 857003 | IN | 1 | 1 | 0 | 0 | 0 | Red | None | Closed shape | LC-EH? | |
| 857003 | IN | 1 | 1 | 0 | 0 | 0 | Sinopean | None | ? | LC-EH? | |

| BagID | Function | Total | Body | Rim | Handle | Base | Fabric | Surface | Shape | Period | Cat. |
|---|---|---|---|---|---|---|---|---|---|---|---|
| 857003 | IN | 1 | 1 | 0 | 0 | 0 | Red mica | None | ? | LC-EH? | |
| 857004 | IN | 13 | 13 | 0 | 0 | 0 | Red | None | ? | LC-EH? | |
| 857004 | IN | 2 | 2 | 0 | 0 | 0 | Red pyro | None | ? | LC-EH? | |
| 857004 | TW | 2 | 2 | 0 | 0 | 0 | Chersonesean | Slip | Closed shape | LC-EH? | **244** |
| 857005 | IN | 5 | 5 | 0 | 0 | 0 | Red | None | ? | LC-EH? | |
| 859101 | IN | 4 | 4 | 0 | 0 | 0 | Red mica | None | Closed shape | LC-EH? | |
| 859102 | IN | 2 | 2 | 0 | 0 | 0 | Red pyro | None | ? | LC-EH? | |
| 859102 | IN | 1 | 1 | 0 | 0 | 0 | Red | Slip | Amphora | LC-EH? | |
| 859102 | IN | 2 | 2 | 0 | 0 | 0 | Grey | None | Closed shape | LC-EH? | |
| 859201 | IN | 1 | 1 | 0 | 0 | 0 | Red mica | None | ? | LC-EH? | |
| 859201 | IN | 2 | 2 | 0 | 0 | 0 | Red | None | ? | LC-EH? | |
| 859202 | IN | 3 | 3 | 0 | 0 | 0 | Red mica | None | Closed shape | LC-EH? | |
| 859203 | IN | 3 | 3 | 0 | 0 | 0 | Red | None | Closed shape | LC-EH? | |
| 859203 | IN | 14 | 14 | 0 | 0 | 0 | Red mica | None | ? | LC-EH? | |
| 859204 | IN | 13 | 13 | 0 | 0 | 0 | Red | None | ? | LC-EH? | |
| 859204 | IN | 2 | 2 | 0 | 0 | 0 | Red pyro | None | ? | LC-EH? | |
| 859204 | IN | 9 | 9 | 0 | 0 | 0 | Red mica | None | Closed shape | LC-EH? | |
| 859204 | IN | 1 | 1 | 0 | 0 | 0 | Red mIca | None | ? | LC-EH? | |
| 859205 | IN | 1 | 1 | 0 | 0 | 0 | Red pyro | None | ? | LC-EH? | |
| 859205 | IN | 1 | 1 | 0 | 0 | 0 | Red | Smoothed | Closed shape | LC-EH? | |
| 859205 | IN | 6 | 6 | 0 | 0 | 0 | Red | None | ? | LC-EH? | |
| 859301 | IN | 2 | 2 | 0 | 0 | 0 | Red | None | ? | LC-EH? | |
| 859401 | IN | 2 | 2 | 0 | 0 | 0 | Red | None | ? | LC-EH? | |
| 859401 | IN | 1 | 1 | 0 | 0 | 0 | Red mica | None | Closed shape | LC-EH? | |
| 853301 | TR | 2 | 2 | 0 | 0 | 0 | Chersonesean? | Slip | Amphora | LC-EH? | |
| 853401 | TR | 1 | 1 | 0 | 0 | 0 | Red | None | Amphora | LC-EH? | |
| 853401 | TR | 1 | 1 | 0 | 0 | 0 | Red coarse | None | Amphora | LC-EH? | |
| 853401 | TR | 1 | 1 | 0 | 0 | 0 | Red pyro | None | Amphora? | LC-EH? | |
| 853401 | TR | 5 | 4 | 0 | 1 | 0 | Chersonesean | Slip | Amphora | LC-EH? | **217** |
| 853401 | TR | 2 | 2 | 0 | 0 | 0 | Red | None | Amphora | LC-EH? | |
| 853402 | TR | 1 | 1 | 0 | 0 | 0 | Thasian | None | Amphora | LC-EH? | |
| 853402 | TR | 1 | 1 | 0 | 0 | 0 | Brown coarse | None | Amphora? | LC-EH? | |
| 853402 | TR | 1 | 1 | 0 | 0 | 0 | Chersonesean | Slip | Amphora | LC-EH? | |
| 853402 | TR | 4 | 3 | 0 | 1 | 0 | Light | Slip | Amphora | LC-EH? | |
| 853402 | TR | 1 | 1 | 0 | 0 | 0 | Red coarse | None | Amphora | LC-EH? | |
| 853402 | TR | 2 | 2 | 0 | 0 | 0 | Sinopean | None | Amphora | LC-EH? | |
| 856801 | TR | 2 | 1 | 1 | 0 | 0 | Light | Slip | Amphora | LC-EH? | **225** |
| 856801 | TR | 5 | 4 | 1 | 0 | 0 | Thasian? | None | Amphora | LC-EH? | **223** |
| 856802 | TR | 2 | 2 | 0 | 0 | 0 | Chersonesean | Slip | Amphora | LC-EH? | |
| 856802 | TR | 1 | 1 | 0 | 0 | 0 | Red | None | Amphora? | LC-EH? | |
| 856901 | TR | 3 | 3 | 0 | 0 | 0 | Light | Slip | Amphora | LC-EH? | |
| 856901 | TR | 1 | 0 | 0 | 1 | 0 | Red | Slip | Amphora | LC-EH? | |
| 856901 | TR | 3 | 3 | 0 | 0 | 0 | Chersonesean | Slip | Amphora | LC-EH? | |

| BagID | Function | Total | Body | Rim | Handle | Base | Fabric | Surface | Shape | Period | Cat. |
|---|---|---|---|---|---|---|---|---|---|---|---|
| 856901 | TR | 1 | 1 | 0 | 0 | 0 | Red | None | Amphora? | LC-EH? | |
| 856901 | TR | 3 | 3 | 0 | 0 | 0 | Herakleian | None | Amphora | LC-EH? | |
| 857001 | TR | 4 | 4 | 0 | 0 | 0 | Thasian | None | Amphora | LC-EH? | |
| 857001 | TR | 1 | 1 | 0 | 0 | 0 | Red coarse | Slip | Amphora | LC-EH? | 230 |
| 857001 | TR | 2 | 2 | 0 | 0 | 0 | Sinopean | None | Amphora | LC-EH? | |
| 857001 | TR | 1 | 0 | 0 | 1 | 0 | Red | None | Amphora? | LC-EH? | 226 |
| 857001 | TR | 2 | 2 | 0 | 0 | 0 | Red mica | Smoothed | Amphora? | LC-EH? | |
| 857001 | TR | 2 | 1 | 0 | 1 | 0 | Light | Slip | Amphora | LC-EH? | 227 |
| 857001 | TR | 1 | 1 | 0 | 0 | 0 | Red coarse | Slip | Amphora | LC-EH? | |
| 857001 | TR | 3 | 3 | 0 | 0 | 0 | Red | None | Amphora | LC-EH? | |
| 857001 | TR | 2 | 2 | 0 | 0 | 0 | Chersonesean | Slip | Amphora | LC-EH? | |
| 857001 | TR | 1 | 0 | 0 | 1 | 0 | Red | Slip | Amphora | LC-EH? | 228 |
| 857001 | TR | 1 | 0 | 1 | 0 | 0 | Knidian? | None | Amphora | LC-EH? | 224 |
| 857002 | TR | 3 | 3 | 0 | 0 | 0 | Herakleian | None | Amphora | LC-EH? | |
| 857002 | TR | 1 | 1 | 0 | 0 | 0 | Red pyro | Slip | Amphora? | LC-EH? | |
| 857002 | TR | 3 | 3 | 0 | 0 | 0 | Thasian | None | Amphora | LC-EH? | |
| 857002 | TR | 1 | 0 | 0 | 1 | 0 | Red | None | Amphora? | LC-EH? | 231 |
| 857003 | TR | 2 | 2 | 0 | 0 | 0 | Red | Slip | Amphora | LC-EH? | |
| 857003 | TR | 2 | 2 | 0 | 0 | 0 | Chersonesean | None | Amphora | LC-EH? | |
| 857003 | TR | 1 | 0 | 0 | 1 | 0 | Light | Slip | Amphora | LC-EH? | |
| 857003 | TR | 1 | 1 | 0 | 0 | 0 | Red | Slip | Amphora | LC-EH? | |
| 857003 | TR | 1 | 1 | 0 | 0 | 0 | Red coarse | Slip | Amphora | LC-EH? | |
| 857004 | TR | 4 | 4 | 0 | 0 | 0 | Herakleian | None | Amphora | LC-EH? | |
| 857004 | TR | 1 | 0 | 0 | 1 | 0 | Red | None | Amphora? | LC-EH? | |
| 857005 | TR | 1 | 0 | 0 | 1 | 0 | Thasian | None | Amphora | LC-EH? | |
| 857005 | TR | 3 | 3 | 0 | 0 | 0 | Sinopean | None | Amphora | LC-EH? | |
| 857005 | TR | 3 | 3 | 0 | 0 | 0 | Red lime | None | Closed shape | LC-EH? | |
| 857005 | TR | 6 | 4 | 1 | 0 | 1 | Chersonesean | Slip | Amphora | LC-EH? | |
| 857005 | TR | 1 | 1 | 0 | 0 | 0 | Red | Smoothed | Amphora? | LC-EH? | |
| 857005 | TR | 4 | 4 | 0 | 0 | 0 | Orange coarse | None | Amphora | LC-EH? | |
| 859101 | TR | 1 | 1 | 0 | 0 | 0 | Chersonesean | Slip | Amphora | LC-EH? | |
| 859101 | TR | 1 | 1 | 0 | 0 | 0 | Sinopean | None | Amphora | LC-EH? | |
| 859101 | TR | 1 | 1 | 0 | 0 | 0 | Red | Slip | Amphora | LC-EH? | |
| 859101 | TR | 1 | 1 | 0 | 0 | 0 | Light | None | Amphora | LC-EH? | |
| 859101 | TR | 5 | 5 | 0 | 0 | 0 | Herakleian | None | Amphora | LC-EH? | |
| 859101 | TR | 1 | 0 | 1 | 0 | 0 | Thasian | None | Amphora | LC-EH? | |
| 859102 | TR | 1 | 1 | 0 | 0 | 0 | Orange pyro | Smoothed | Amphora | LC-EH? | |
| 859102 | TR | 1 | 1 | 0 | 0 | 0 | Light | None | Amphora | LC-EH? | |
| 859102 | TR | 1 | 1 | 0 | 0 | 0 | Red coarse | Slip | Amphora | LC-EH? | |
| 859102 | TR | 4 | 4 | 0 | 0 | 0 | Thasian | None | Amphora | LC-EH? | |
| 859102 | TR | 9 | 9 | 0 | 0 | 0 | Herakleian | None | Amphora | LC-EH? | 219 |
| 859201 | TR | 9 | 8 | 0 | 1 | 0 | Herakleian | None | Amphora | LC-EH? | |
| 859201 | TR | 2 | 2 | 0 | 0 | 0 | Sinopean | None | Amphora | LC-EH? | |

| BagID | Function | Total | Body | Rim | Handle | Base | Fabric | Surface | Shape | Period | Cat. |
|---|---|---|---|---|---|---|---|---|---|---|---|
| 859201 | TR | 2 | 1 | 0 | 1 | 0 | Red | None | Amphora | LC-EH? | |
| 859201 | TR | 3 | 3 | 0 | 0 | 0 | Red | Slip | Amphora | LC-EH? | |
| 859201 | TR | 4 | 4 | 0 | 0 | 0 | Thasian | None | Amphora | LC-EH? | |
| 859202 | TR | 1 | 1 | 0 | 0 | 0 | Red | None | Amphora | LC-EH? | |
| 859202 | TR | 7 | 4 | 1 | 2 | 0 | Herakleian | none | Amphora | LC-EH? | **233-234** |
| 859203 | TR | 1 | 1 | 0 | 0 | 0 | Thasian | None | Amphora | LC-EH? | |
| 859203 | TR | 1 | 1 | 0 | 0 | 0 | Red porous | Slip | Amphora? | LC-EH? | |
| 859203 | TR | 1 | 1 | 0 | 0 | 0 | Red mica | Slip | Amphora | LC-EH? | |
| 859203 | TR | 1 | 1 | 0 | 0 | 0 | Red | Slip | Amphora? | LC-EH? | |
| 859203 | TR | 2 | 2 | 0 | 0 | 0 | Orange pyro | None | Amphora | LC-EH? | |
| 859203 | TR | 13 | 13 | 0 | 0 | 0 | Herakleian | None | Amphora | LC-EH? | |
| 859203 | TR | 1 | 0 | 0 | 1 | 0 | Red | None | Amphora | LC-EH? | |
| 859204 | TR | 1 | 1 | 0 | 0 | 0 | Red coarse | None | Amphora | LC-EH? | **235** |
| 859204 | TR | 2 | 2 | 0 | 0 | 0 | Sinopean | None | Amphora | LC-EH? | |
| 859204 | TR | 3 | 3 | 0 | 0 | 0 | Brown | None | Amphora | LC-EH? | |
| 859204 | TR | 1 | 0 | 0 | 0 | 1 | Thasian | None | Amphora | LC-EH? | **222** |
| 859204 | TR | 1 | 0 | 0 | 1 | 0 | Red | None | Amphora | LC-EH? | |
| 859204 | TR | 2 | 2 | 0 | 0 | 0 | Red pyro | None | Amphora | LC-EH? | |
| 859204 | TR | 1 | 1 | 0 | 0 | 0 | Red mica | Grooved | Amphora | LC-EH? | |
| 859204 | TR | 13 | 11 | 0 | 2 | 0 | Herakleian | None | Amphora | LC-EH? | |
| 859205 | TR | 7 | 7 | 0 | 0 | 0 | Herakleian | None | Amphora | LC-EH? | **220** |
| 859205 | TR | 1 | 1 | 0 | 0 | 0 | Thasian | None | Amphora | LC-EH? | |
| 859205 | TR | 1 | 1 | 0 | 0 | 0 | Chersonesean? | None | Amphora | LC-EH? | |
| 859301 | TR | 5 | 4 | 0 | 2 | 0 | Thasian | None | Amphora | LC-EH? | |
| 859301 | TR | 1 | 1 | 0 | 0 | 0 | Red | None | Amphora | LC-EH? | |
| 859301 | TR | 7 | 6 | 0 | 0 | 1 | Herakleian | None | Amphora | LC-EH? | |
| 859301 | TR | 1 | 0 | 0 | 1 | 0 | Red | None | Amphora | LC-EH? | |
| 859401 | TR | 3 | 3 | 0 | 0 | 0 | Chersonesean | Slip | Amphora | LC-EH? | |
| 859401 | TR | 2 | 2 | 0 | 0 | 0 | Herakleian | None | Amphora | LC-EH? | |
| 853301 | TW | 1 | 1 | 0 | 0 | 0 | Red | None | Jug? | LC-EH? | |
| 853401 | TW | 1 | 0 | 0 | 1 | 0 | Chersonesean | Slip | Jug | LC-EH? | **237** |
| 853402 | TW | 1 | 0 | 1 | 0 | 0 | Red | Slip | Bowl | LC-EH? | **246** |
| 853402 | TW | 2 | 0 | 0 | 1 | 1 | Chersonesean? | Slip | Jug? | LC-EH? | **238-239** |
| 857001 | TW | 2 | 1 | 0 | 0 | 1 | Chersonesean | Slip | Jug | LC-EH? | **240** |
| 857003 | TW | 2 | 0 | 0 | 1 | 1 | Light | Slip | Jug | LC-EH? | **242-243** |
| 857003 | TW | 1 | 0 | 0 | 0 | 1 | Chersonesean | Slip | Jug | LC-EH? | **241** |
| 859102 | TW | 1 | 0 | 0 | 1 | 0 | Red | None | Jug? | LC-EH? | **245** |
| 859203 | TW | 1 | 1 | 0 | 0 | 0 | Red | Decorated | Open shape | LC-EH? | |
| 859204 | TW | 1 | 1 | 0 | 0 | 0 | Red | None | ? | LC-EH? | |
| 859203 | UT | 1 | 0 | 1 | 0 | 0 | Brown pyro | None | Pithos? | LC-EH? | |
| 859205 | UT | 3 | 3 | 0 | 0 | 0 | Red | None | ? | LC-EH? | |
| 857001 | IN | 2 | 2 | 0 | 0 | 0 | Red | Slip | Closed shape | Byzantine? | |
| 857001 | TW | 1 | 0 | 1 | 0 | 0 | Tatarian | Slip | Jug | EM | |

| BagID | Function | Total | Body | Rim | Handle | Base | Fabric | Surface | Shape | Period | Cat. |
|---|---|---|---|---|---|---|---|---|---|---|---|
| 857002 | TE | 1 | 0 | 1 | 0 | 0 | Red | None | Pibe? | EM? | |
| 857003 | AR | 1 | 1 | 0 | 0 | 0 | Red modern | Glaze | Tile | Modern | |
| 859102 | | 1 | 0 | 0 | 0 | 0 | | | | | |
| 857003 | IN | 19 | 19 | 0 | 0 | 0 | Red | None | ? | | |
| 859102 | IN | 5 | 5 | 0 | 0 | 0 | Red | None | ? | | |
| 859201 | IN | 6 | 6 | 0 | 0 | 0 | Red | None | Closed shape | | |
| 859201 | IN | 1 | 1 | 0 | 0 | 0 | Red pyro | Smoothed | ? | | |
| 859201 | IN | 5 | 5 | 0 | 0 | 0 | Red | None | ? | | |
| 859202 | IN | 14 | 14 | 0 | 0 | 0 | Red | None | ? | | |
| 859204 | IN | 1 | 1 | 0 | 0 | 0 | Red | Smoothed | Closed shape | | |
| 856901 | LI | 1 | 0 | 0 | 0 | 0 | | | | | |
| 859203 | LI | 1 | 1 | 0 | 0 | 0 | | | | | |
| 859201 | TR | 2 | 2 | 0 | 0 | 0 | Orange coarse | None | Amphora? | | |
| 859202 | TR | 8 | 6 | 0 | 2 | 0 | Red | Slip | Amphora | | **232** |

Finds table adjacent units

| BagID | Function | Total | Body | Rim | Handle | Base | Fabric | Surface | Shape | Period | Cat. |
|---|---|---|---|---|---|---|---|---|---|---|---|
| 859501 | HW | 2 | 2 | 0 | 0 | 0 | Grey | None | Pot | LC-EH | |
| 863801 | HW | 6 | 6 | 0 | 0 | 0 | Grey | None | Pot | LC-EH | |
| 856701 | IN | 3 | 3 | 0 | 0 | 0 | Red | None | ? | LC-EH? | |
| 863701 | IN | 1 | 1 | 0 | 0 | 0 | Red | Slip | Closed shape | LC-EH? | |
| 856701 | TR | 1 | 0 | 0 | 1 | 0 | Herakleian | none | Amphora | LC-EH? | **218** |
| 856701 | TR | 1 | 1 | 0 | 0 | 0 | Red coarse | None | Amphora? | LC-EH? | |
| 859501 | TR | 1 | 1 | 0 | 0 | 0 | Herakleian | None | Amphora | LC-EH? | |
| 863701 | TR | 2 | 2 | 0 | 0 | 0 | Herakleian? | None | Amphora? | LC-EH? | |
| 863801 | TR | 1 | 0 | 0 | 1 | 0 | Herakleian? | None | Amphora? | LC-EH? | **236** |
| 863701 | TW | 1 | 1 | 0 | 0 | 0 | Light | Ribbed | Jug | LC-EH | **249** |
| 863701 | IN | 1 | 0 | 0 | 0 | 0 | Light red | None | ? | ? | |
| 863701 | IN | 2 | 2 | 0 | 0 | 0 | Red | None | ? | ? | |
| 863801 | IN | 1 | 1 | 0 | 0 | 0 | Light | Slip | ? | ? | |
| 863801 | IN | 1 | 1 | 0 | 0 | 0 | Light pyro | None | Closed shape | ? | |
| 863801 | IN | 2 | 2 | 0 | 0 | 0 | Red | None | ? | ? | |
| 856701 | LI | 1 | 0 | 0 | 0 | 0 | | | | | |
| 863801 | TR | 1 | 0 | 0 | 1 | 0 | Light | Slip | Amphora | ? | |

## DSP08-F04-04

Finds table core units

| BagID | Function | Total | Body | Rim | Handle | Base | Fabric | Surface | Shape | Period | Cat. |
|---|---|---|---|---|---|---|---|---|---|---|---|
| 877101 | HW | 1 | 0 | 1 | 0 | 0 | BaBlack | Coomed | Cup | BA | **250** |
| 874501 | HW | 1 | 1 | 0 | 0 | 0 | Grey | None | Pot | LC-EH | |
| 873802 | TR | 2 | 1 | 0 | 1 | 0 | Brown coarse | None | Amphora | LC-EH? | |
| 873802 | TR | 1 | 0 | 0 | 1 | 0 | Red | None | Amphora | LC-EH? | **251** |
| 873701 | TW | 1 | 0 | 1 | 0 | 0 | Tatarian | Glazed | Closed shape | EM | |
| 873801 | TW | 1 | 1 | 0 | 0 | 0 | Tatarian 2 | Slip | Closed shape | EM | |
| 885901 | IN | 1 | 1 | 0 | 0 | 0 | Red | Slip | Closed shape | ? | |
| 885901 | IN | 2 | 2 | 0 | 0 | 0 | Red | None | ? | ? | |
| 873802 | LI | 0 | 0 | 0 | 0 | 0 | | | | | |

## DSP08-F04-05

Finds table core units

| BagID | Function | Total | Body | Rim | Handle | Base | Fabric | Surface | Shape | Period | Cat. |
|---|---|---|---|---|---|---|---|---|---|---|---|
| 658101 | AR | 2 | 1 | 1 | 0 | 0 | Light Red | Slip | Tile | Modern | |
| 658101 | TR | 1 | 0 | 0 | 1 | 0 | Brown | Slip | Amphora | LC-EH | **254** |
| 661701 | TR | 1 | 1 | 0 | 0 | 0 | Red | Slip | Amphora | LC-EH? | |
| 661701 | TR | 1 | 1 | 0 | 0 | 0 | Brown | Slip | Amphora | LC-EH? | |
| 880501 | UT | 1 | 1 | 0 | 0 | 0 | Tatarian | Ribbed | Closed shape | EM | **256** |

Finds table adjacent units

| BagID | Function | Total | Body | Rim | Handle | Base | Fabric | Surface | Shape | Period | Cat. |
|---|---|---|---|---|---|---|---|---|---|---|---|
| 661601 | HW | 1 | 0 | 0 | 0 | 1 | HwRed | None | Pot | LC-EH | **252** |
| 661601 | IN | 1 | 1 | 0 | 0 | 0 | Light pyro | None | Closed shape | LC-EH? | |
| 661801 | IN | 1 | 1 | 0 | 0 | 0 | Red pyro | None | Closed shape | LC-EH? | |
| 661601 | TR | 1 | 0 | 0 | 1 | 0 | Red | None | Amphora | LC-EH? | |
| 661801 | TR | 1 | 0 | 1 | 0 | 0 | Herakleian | None | Amphora | LC-EH? | **253** |
| 661601 | TW | 1 | 0 | 0 | 0 | 1 | Red | None | Jug | LC-EH? | **255** |
| 880401 | IN | 1 | 1 | 0 | 0 | 0 | Tatarian? | None | Closed shape | EM? | |

## DSP08-F04-06

Finds table core units

| BagID | Function | Total | Body | Rim | Handle | Base | Fabric | Surface | Shape | Period | Cat. |
|---|---|---|---|---|---|---|---|---|---|---|---|
| 652604 | HW | 1 | 0 | 1 | 0 | 0 | BaRed | None | Pot | Late BA | **259** |

| BagID | Function | Total | Body | Rim | Handle | Base | Fabric | Surface | Shape | Period | Cat. |
|---|---|---|---|---|---|---|---|---|---|---|---|
| 652604 | HW | 2 | 1 | 1 | 0 | 0 | BaCoarse | none | Pot | Late BA | **257-258** |
| 652604 | HW | 1 | 0 | 1 | 0 | 0 | BaCoarse | None | Bowl | Late BA | **260** |
| 652201 | TW | 2 | 0 | 0 | 2 | 0 | Red | None | Jug | LC-H | **262-263** |
| 652601 | HW | 6 | 6 | 0 | 0 | 0 | Grey | None | Pot | LC-EH | |
| 652602 | HW | 1 | 1 | 0 | 0 | 0 | Grey | None | Pot | LC-EH | |
| 652603 | HW | 2 | 2 | 0 | 0 | 0 | Grey | None | Pot | LC-EH | |
| 652604 | HW | 11 | 11 | 0 | 0 | 0 | Grey | None | Pot | LC-EH | |
| 652604 | HW | 1 | 1 | 0 | 0 | 0 | Grey | Polished | Open shape? | LC-EH | |
| 652605 | HW | 6 | 6 | 0 | 0 | 0 | Grey | None | Pot | LC-EH | |
| 652606 | HW | 1 | 1 | 0 | 0 | 0 | Coarse | None | Pot | LC-EH | |
| 652601 | TR | 1 | 1 | 0 | 0 | 0 | Herakleian | None | Amphora | LC-EH | |
| 652201 | IN | 1 | 1 | 0 | 0 | 0 | Red pyro | None | ? | LC-EH? | |
| 652604 | IN | 1 | 1 | 0 | 0 | 0 | Red pyro | None | ? | LC-EH? | |
| 652604 | IN | 1 | 1 | 0 | 0 | 0 | Red | None | Closed shape | LC-EH? | |
| 652602 | TR | 1 | 0 | 1 | 0 | 0 | Red | None | Jug | LC-EH? | **264** |
| 652604 | TR | 2 | 2 | 0 | 0 | 0 | Red | Slip | Amphora? | LC-EH? | |
| 652604 | TR | 6 | 6 | 0 | 0 | 0 | Red | Slip | Amphora | LC-EH? | |
| 652604 | TW | 1 | 0 | 1 | 0 | 0 | Red pyro | Slip | Bowl | LC-EH? | **265** |
| 652605 | IN | 1 | 1 | 0 | 0 | 0 | Red lime | Slip | Closed shape | Medieval? | |
| 652605 | IN | 1 | 1 | 0 | 0 | 0 | Red | Slip | Closed shape | Medieval? | |
| 652201 | TR | 4 | 4 | 0 | 0 | 0 | Red | None | Amphora? | Medieval? | |
| 652604 | TR | 1 | 0 | 0 | 1 | 0 | Red | Slip | Amphora | Medieval? | **261** |
| 652101 | TW | 1 | 0 | 0 | 1 | 0 | Red | Slip | Jug? | EM | **266** |
| 652501 | TW | 1 | 1 | 0 | 0 | 0 | Tatarian | Glazed | Open shape | EM | |
| 652602 | TW | 1 | 1 | 0 | 0 | 0 | Tatarian | Glazed | Open shape | EM | |
| 652101 | IN | 2 | 2 | 0 | 0 | 0 | Red | None | Closed shape | ? | |
| 652501 | IN | 1 | 1 | 0 | 0 | 0 | Light | None | ? | ? | |
| 652501 | IN | 3 | 3 | 0 | 0 | 0 | Red | None | ? | ? | |
| 652601 | IN | 2 | 2 | 0 | 0 | 0 | Red | Slip | Closed shape | ? | |
| 652601 | IN | 2 | 2 | 0 | 0 | 0 | Red | None | Closed shape | ? | |
| 652602 | IN | 2 | 2 | 0 | 0 | 0 | Red | None | Closed shape | ? | |
| 652603 | IN | 1 | 1 | 0 | 0 | 0 | Red | None | Closed shape | ? | |
| 652604 | IN | 1 | 1 | 0 | 0 | 0 | Red mica | None | ? | ? | |
| 652604 | IN | 1 | 1 | 0 | 0 | 0 | Red | None | ? | ? | |
| 652604 | IN | 1 | 1 | 0 | 0 | 0 | Red coarse | None | ? | ? | |
| 652605 | IN | 1 | 1 | 0 | 0 | 0 | Red | Slip | Closed shape | ? | |
| 652605 | IN | 1 | 0 | 0 | 1 | 0 | Red | None | ? | ? | |
| 652605 | IN | 3 | 3 | 0 | 0 | 0 | Red | None | ? | ? | |
| 652605 | IN | 1 | 1 | 0 | 0 | 0 | Red lime | None | Closed shape | ? | |
| 652501 | TR | 1 | 1 | 0 | 0 | 0 | Red | None | Amphora | ? | |

Finds table adjacent units

| BagID | Function | Total | Body | Rim | Handle | Base | Fabric | Surface | Shape | Period | Cat. |
|---|---|---|---|---|---|---|---|---|---|---|---|
| 652701 | TW | 1 | 0 | 0 | 0 | 1 | Chersonesean | Slip | Jug | LC-EH? | |
| 652301 | IN | 1 | 1 | 0 | 0 | 0 | Red | None | Closed shape? | ? | |
| 652701 | IN | 1 | 1 | 0 | 0 | 0 | Red coarse | None | Closed shape | ? | |

## DSP08-F04-07

Finds table core units

| BagID | Function | Total | Body | Rim | Handle | Base | Fabric | Surface | Shape | Period | Cat. |
|---|---|---|---|---|---|---|---|---|---|---|---|
| 649201 | HW | 3 | 3 | 0 | 0 | 0 | Coarse | None | Pot | LC-EH | |
| 649202 | HW | 1 | 0 | 1 | 0 | 0 | Red | impressed | Pot | LC-EH | |
| 649202 | HW | 6 | 6 | 0 | 0 | 0 | Red | None | Pot | LC-EH | |
| 649202 | HW | 15 | 15 | 0 | 0 | 0 | Grey | None | Pot | LC-EH | |
| 649202 | HW | 2 | 2 | 0 | 0 | 0 | Coarse | None | Pot | LC-EH | |
| 649203 | HW | 7 | 6 | 1 | 0 | 0 | Grey | None | Pot | LC-EH | |
| 649205 | HW | 2 | 2 | 0 | 0 | 0 | Grey | None | Pot | LC-EH | |
| 649302 | HW | 17 | 16 | 1 | 0 | 0 | Black | None | Pot | LC-EH | **268** |
| 649302 | HW | 1 | 1 | 0 | 0 | 0 | Coarse | None | Pot | LC-EH | |
| 649302 | HW | 2 | 2 | 0 | 0 | 0 | Red | None | Pot | LC-EH | |
| 649303 | HW | 21 | 20 | 1 | 0 | 0 | Grey | None | Pot | LC-EH | **269** |
| 649401 | HW | 1 | 1 | 0 | 0 | 0 | Red coarse | None | ? | LC-EH | |
| 649401 | HW | 2 | 2 | 0 | 0 | 0 | Red | None | Pot | LC-EH | |
| 649401 | HW | 8 | 8 | 0 | 0 | 0 | Grey | None | Pot | LC-EH | |
| 649402 | HW | 15 | 15 | 0 | 0 | 0 | Grey | None | Pot | LC-EH | |
| 649402 | HW | 1 | 1 | 0 | 0 | 0 | Coarse | None | Pot | LC-EH | |
| 649201 | HW | 10 | 9 | 1 | 0 | 0 | Grey | None | Pot | LC-EH? | |
| 649204 | HW | 5 | 4 | 1 | 0 | 0 | Grey | None | Pot | LC-EH? | |
| 649204 | HW | 2 | 2 | 0 | 0 | 0 | Black | None | Pot | LC-EH? | |
| 649204 | HW | 1 | 1 | 0 | 0 | 0 | Red coarse | None | Closed shape? | LC-EH? | |
| 649301 | HW | 9 | 9 | 0 | 0 | 0 | Red | None | Pot | LC-EH? | |
| 649301 | HW | 1 | 0 | 1 | 0 | 0 | Red | None | Open shape | LC-EH? | **267** |
| 649303 | HW | 6 | 6 | 0 | 0 | 0 | Black | None | Pot | LC-EH? | |
| 649391 | HW | 3 | 3 | 0 | 0 | 0 | Black | Grooved | Pot | LC-EH? | |
| 649391 | HW | 1 | 1 | 0 | 0 | 0 | Light | None | Pot | LC-EH? | |
| 649201 | IN | 1 | 1 | 0 | 0 | 0 | Red | Smoothed | Closed shape | LC-EH? | |
| 649201 | IN | 2 | 2 | 0 | 0 | 0 | Red | None | Closed shape | LC-EH? | |
| 649201 | IN | 1 | 1 | 0 | 0 | 0 | Red | none | ? | LC-EH? | |
| 649201 | IN | 1 | 1 | 0 | 0 | 0 | Chersonesean | Slip | Closed shape | LC-EH? | |
| 649202 | IN | 6 | 6 | 0 | 0 | 0 | Red | None | ? | LC-EH? | |
| 649202 | IN | 1 | 1 | 0 | 0 | 0 | Red | None | Closed shape | LC-EH? | |
| 649202 | IN | 1 | 1 | 0 | 0 | 0 | Red mica | None | ? | LC-EH? | |

| BagID | Function | Total | Body | Rim | Handle | Base | Fabric | Surface | Shape | Period | Cat. |
|---|---|---|---|---|---|---|---|---|---|---|---|
| 649203 | IN | 2 | 2 | 0 | 0 | 0 | Red | None | Closed shape | LC-EH? | |
| 649204 | IN | 16 | 16 | 0 | 0 | 0 | Red | None | ? | LC-EH? | |
| 649204 | IN | 1 | 1 | 0 | 0 | 0 | Red pyro | Slip | Closed shape | LC-EH? | |
| 649204 | IN | 3 | 3 | 0 | 0 | 0 | Red | None | Closed shape | LC-EH? | |
| 649204 | IN | 5 | 5 | 0 | 0 | 0 | Brown | None | ? | LC-EH? | |
| 649204 | IN | 4 | 4 | 0 | 0 | 0 | Red mica | None | ? | LC-EH? | |
| 649205 | IN | 1 | 1 | 0 | 0 | 0 | Red mica | None | ? | LC-EH? | |
| 649301 | IN | 6 | 6 | 0 | 0 | 0 | Red | None | ? | LC-EH? | |
| 649302 | IN | 1 | 1 | 0 | 0 | 0 | Red pyro | None | ? | LC-EH? | |
| 649303 | IN | 9 | 9 | 0 | 0 | 0 | Red | None | ? | LC-EH? | |
| 649401 | IN | 5 | 5 | 0 | 0 | 0 | Red | None | ? | LC-EH? | |
| 649402 | IN | 1 | 1 | 0 | 0 | 0 | Red pyro | Slip | Closed shape | LC-EH? | |
| 649201 | TR | 1 | 0 | 1 | 0 | 0 | Red | Slip | Amphora | LC-EH? | |
| 649201 | TR | 5 | 5 | 0 | 0 | 0 | Thasian | None | Amphora | LC-EH? | |
| 649201 | TR | 1 | 1 | 0 | 0 | 0 | Red pyro | Slip | Amphora? | LC-EH? | |
| 649201 | TR | 6 | 5 | 0 | 1 | 0 | Red | None | Amphora | LC-EH? | |
| 649201 | TR | 11 | 9 | 1 | 0 | 1 | Herakleian | None | Amphora | LC-EH? | |
| 649202 | TR | 9 | 8 | 0 | 1 | 0 | Red | Slip | Amphora | LC-EH? | 277 |
| 649202 | TR | 1 | 1 | 0 | 0 | 0 | Red coarse | Slip | Amphora | LC-EH? | |
| 649202 | TR | 2 | 2 | 0 | 0 | 0 | Red | Slip | Amphora | LC-EH? | |
| 649202 | TR | 1 | 1 | 0 | 0 | 0 | Red | Slip | Amphora | LC-EH? | |
| 649202 | TR | 11 | 11 | 0 | 0 | 0 | Herakleian | None | Amphora | LC-EH? | |
| 649202 | TR | 1 | 0 | 0 | 1 | 0 | Red | Slip | Amphora | LC-EH? | 276 |
| 649203 | TR | 1 | 1 | 0 | 0 | 0 | Thasian | None | Amphora | LC-EH? | |
| 649203 | TR | 2 | 2 | 0 | 0 | 0 | Red pyro | Slip | Amphora? | LC-EH? | |
| 649203 | TR | 5 | 4 | 1 | 0 | 0 | Herakleian | None | Amphora | LC-EH? | 273 |
| 649204 | TR | 4 | 4 | 0 | 0 | 0 | Red mica | Smoothed | Amphora | LC-EH? | |
| 649204 | TR | 1 | 1 | 0 | 0 | 0 | Brown coarse | Slip | Amphora | LC-EH? | |
| 649204 | TR | 12 | 12 | 0 | 0 | 0 | Herakleian | None | Amphora | LC-EH? | |
| 649204 | TR | 4 | 4 | 0 | 0 | 0 | Sinopean | None | Amphora | LC-EH? | |
| 649204 | TR | 2 | 2 | 0 | 0 | 0 | Thasian | None | Amphora | LC-EH? | |
| 649205 | TR | 4 | 4 | 0 | 0 | 0 | Herakleian | None | Amphora | LC-EH? | |
| 649301 | TR | 1 | 1 | 0 | 0 | 0 | Red | Slip | Amphora | LC-EH? | |
| 649301 | TR | 1 | 0 | 1 | 0 | 0 | Herakleian? | None | Amphora | LC-EH? | 274 |
| 649301 | TR | 3 | 3 | 0 | 0 | 0 | Red | None | Amphora | LC-EH? | |
| 649302 | TR | 1 | 1 | 0 | 0 | 0 | Herakleian? | None | Amphora | LC-EH? | |
| 649302 | TR | 1 | 1 | 0 | 0 | 0 | Red | Slip | Amphora | LC-EH? | |
| 649302 | TR | 5 | 4 | 0 | 1 | 0 | Thasian | None | Amphora? | LC-EH? | 275 |
| 649302 | TR | 1 | 1 | 0 | 0 | 0 | Sinopean | None | Amphora | LC-EH? | |
| 649302 | TR | 1 | 0 | 0 | 1 | 0 | Red | None | Amphora | LC-EH? | 278 |
| 649303 | TR | 1 | 1 | 0 | 0 | 0 | Grey | Smoothed | Amphora? | LC-EH? | |
| 649303 | TR | 1 | 1 | 0 | 0 | 0 | Red | None | Amphora | LC-EH? | |
| 649303 | TR | 2 | 2 | 0 | 0 | 0 | Red mica | Smoothed | Amphora | LC-EH? | |

| BagID | Function | Total | Body | Rim | Handle | Base | Fabric | Surface | Shape | Period | Cat. |
|---|---|---|---|---|---|---|---|---|---|---|---|
| 649303 | TR | 3 | 2 | 0 | 1 | 0 | Red | slip | Amphora | LC-EH? | **280** |
| 649303 | TR | 3 | 3 | 0 | 0 | 0 | Herakleian | Smoothed | Amphora | LC-EH? | |
| 649391 | TR | 1 | 1 | 0 | 0 | 0 | Thasian | None | Amphora | LC-EH? | |
| 649391 | TR | 1 | 1 | 0 | 0 | 0 | Red mica | None | Amphora | LC-EH? | |
| 649391 | TR | 2 | 2 | 0 | 0 | 0 | Herakleian | None | Amphora | LC-EH? | |
| 649391 | TR | 1 | 1 | 0 | 0 | 0 | Red coarse | Slip | Amphora | LC-EH? | |
| 649401 | TR | 8 | 8 | 0 | 0 | 0 | Herakleian | None | Amphora | LC-EH? | |
| 649401 | TR | 1 | 1 | 0 | 0 | 0 | Red | Slip | Amphora? | LC-EH? | |
| 649401 | TR | 1 | 1 | 0 | 0 | 0 | Red mica | None | Amphora | LC-EH? | |
| 649401 | TR | 1 | 1 | 0 | 0 | 0 | Chersonesean? | Smoothed | Amphora | LC-EH? | |
| 649402 | TR | 2 | 1 | 1 | 0 | 0 | Herakleian | None | Amphora | LC-EH? | **272** |
| 649402 | TR | 2 | 2 | 0 | 0 | 0 | Red pyro | None | Amphora | LC-EH? | |
| 649205 | TW | 1 | 0 | 1 | 0 | 0 | Red | None | Flask | LC-EH? | **282** |
| 649201 | UT | 1 | 1 | 0 | 0 | 0 | Red | Slip | Open shape | LC-EH? | |
| 649302 | IN | 1 | 1 | 0 | 0 | 0 | Red | None | Closed shape | Byzantine? | |
| 649204 | TW | 2 | 2 | 0 | 0 | 0 | Red | Slip | Jug? | Byzantine? | |
| 649302 | BO | 1 | 0 | 0 | 0 | 0 | | | | | |
| 649201 | HW | 1 | 1 | 0 | 0 | 0 | Black | Grooved | Closed shape? | ? | |
| 649204 | IN | 1 | 1 | 0 | 0 | 0 | Red coarse | Slip | ? | ? | |
| 649205 | IN | 1 | 1 | 0 | 0 | 0 | Red | None | Closed shape | ? | |
| 649302 | IN | 9 | 9 | 0 | 0 | 0 | Red | None | ? | ? | |
| 649302 | IN | 4 | 4 | 0 | 0 | 0 | Red mica | None | ? | ? | |
| 649402 | IN | 2 | 2 | 0 | 0 | 0 | Red | None | ? | ? | |
| 649201 | LI | 1 | 1 | 0 | 0 | 0 | | | | | |
| 649202 | LI | 1 | 1 | 0 | 0 | 0 | | | | | |
| 649202 | LI | 3 | 0 | 0 | 0 | 0 | | | | | |
| 649204 | LI | 1 | 1 | 0 | 0 | 0 | | | | | |
| 649204 | LI | 1 | 0 | 0 | 0 | 0 | | | | | |
| 649205 | LI | 1 | 1 | 0 | 0 | 0 | | | | | |
| 649302 | LI | 1 | 0 | 0 | 0 | 0 | | | | | |
| 649303 | LI | 1 | 1 | 0 | 0 | 0 | | | | | |
| 649303 | LI | 2 | 2 | 0 | 0 | 0 | | | | | |
| 649391 | LI | 1 | 1 | 0 | 0 | 0 | | | | | |
| 649401 | LI | 1 | 0 | 0 | 0 | 0 | | | | | |
| 649204 | TR | 2 | 2 | 0 | 0 | 0 | Brown sand | Smoothed | Amphora? | ? | |
| 649302 | TR | 2 | 1 | 0 | 1 | 0 | Red | None | Amphora | ? | **279** |
| 649302 | TR | 1 | 1 | 0 | 0 | 0 | Light | None | Amphora | ? | |
| 649302 | TR | 1 | 0 | 0 | 1 | 0 | Red | Smoothed | Amphora | ? | |
| 649204 | TW | 1 | 1 | 0 | 0 | 0 | Red mica | Slip | Bowl? | ? | |

Finds table adjacent units

| BagID | Function | Total | Body | Rim | Handle | Base | Fabric | Surface | Shape | Period | Cat. |
|---|---|---|---|---|---|---|---|---|---|---|---|
| 647601 | HW | 1 | 1 | 0 | 0 | 0 | Coarse | None | Pot | LC-EH | |
| 647801 | HW | 1 | 1 | 0 | 0 | 0 | Black | None | Pot | LC-EH | |
| 649001 | HW | 2 | 2 | 0 | 0 | 0 | Grey | None | Pot | LC-EH | |
| 649101 | HW | 1 | 0 | 1 | 0 | 0 | Light | None | Pot | LC-EH | |
| 649102 | HW | 1 | 1 | 0 | 0 | 0 | Red | None | Pot | LC-EH | |
| 649102 | HW | 4 | 4 | 0 | 0 | 0 | Grey | None | Pot | LC-EH | |
| 649103 | HW | 2 | 2 | 0 | 0 | 0 | Grey | None | Pot | LC-EH | |
| 649103 | HW | 2 | 2 | 0 | 0 | 0 | Coarse | None | Pot | LC-EH | |
| 649103 | HW | 2 | 2 | 0 | 0 | 0 | Red | None | Pot | LC-EH | |
| 649501 | HW | 1 | 1 | 0 | 0 | 0 | Grey | None | Pot | LC-EH | |
| 660701 | TW | 1 | 0 | 0 | 0 | 1 | Red | Slip | Jug | LC-EH | |
| 649001 | IN | 1 | 1 | 0 | 0 | 0 | Brown | None | Closed shape | LC-EH? | |
| 649001 | IN | 1 | 1 | 0 | 0 | 0 | Red mica | None | ? | LC-EH? | |
| 649102 | IN | 6 | 6 | 0 | 0 | 0 | Red | None | ? | LC-EH? | |
| 649103 | IN | 1 | 1 | 0 | 0 | 0 | Red coarse | None | Closed shape | LC-EH? | |
| 649103 | IN | 1 | 1 | 0 | 0 | 0 | Brown | None | Closed shape | LC-EH? | |
| 649103 | IN | 1 | 1 | 0 | 0 | 0 | Red coarse slip | None | Closed shape | LC-EH? | |
| 649501 | IN | 1 | 1 | 0 | 0 | 0 | Red mica | None | ? | LC-EH? | |
| 659701 | IN | 1 | 1 | 0 | 0 | 0 | Red | None | Closed shape | LC-EH? | |
| 647801 | TR | 1 | 0 | 0 | 0 | 1 | Chersonesean | None | Amphora | LC-EH? | 271 |
| 647801 | TR | 1 | 1 | 0 | 0 | 0 | Red mica | Smoothed | Amphora | LC-EH? | |
| 648001 | TR | 1 | 1 | 0 | 0 | 0 | Thasian | none | Amphora | LC-EH? | |
| 648001 | TR | 1 | 1 | 0 | 0 | 0 | Light mica | None | Amphora? | LC-EH? | |
| 649001 | TR | 1 | 1 | 0 | 0 | 0 | Herakleian | None | Amphora | LC-EH? | |
| 649101 | TR | 1 | 1 | 0 | 0 | 0 | Sinopean | None | Amphora | LC-EH? | |
| 649101 | TR | 2 | 2 | 0 | 0 | 0 | Red | Smoothed | Amphora | LC-EH? | |
| 649101 | TR | 3 | 3 | 0 | 0 | 0 | Herakleian | None | Amphora | LC-EH? | |
| 649102 | TR | 4 | 4 | 0 | 0 | 0 | Herakleian | None | Amphora | LC-EH? | |
| 660601 | TW | 1 | 0 | 1 | 0 | 0 | Red | Slip | Bowl | LC-EH? | |
| 647701 | UT | 1 | 0 | 1 | 0 | 0 | Red pyro | None | Open shape? | LC-EH? | |
| 647701 | AR | 1 | 0 | 1 | 0 | 0 | Red coarse modern | None | Tile | Medieval? | |
| 647601 | IN | 1 | 1 | 0 | 0 | 0 | Red | Grooved | Closed shape? | Medieval? | |
| 647701 | IN | 1 | 1 | 0 | 0 | 0 | Red mica | impressed | Closed shape? | Medieval? | |
| 647501 | IN | 1 | 1 | 0 | 0 | 0 | Red coarse mica | None | ? | ? | |
| 647601 | IN | 1 | 1 | 0 | 0 | 0 | Red | None | ? | ? | |
| 647701 | IN | 1 | 1 | 0 | 0 | 0 | Brown coarse | None | Closed shape | ? | |
| 647801 | IN | 1 | 1 | 0 | 0 | 0 | Red | None | ? | ? | |
| 648001 | IN | 1 | 1 | 0 | 0 | 0 | Red | None | ? | ? | |
| 649001 | IN | 2 | 2 | 0 | 0 | 0 | Red | None | Closed shape | ? | |
| 649101 | IN | 1 | 1 | 0 | 0 | 0 | Red mica | None | Closed shape | ? | |
| 649101 | IN | 4 | 4 | 0 | 0 | 0 | Red | None | ? | ? | |

| BagID | Function | Total | Body | Rim | Handle | Base | Fabric | Surface | Shape | Period | Cat. |
|---|---|---|---|---|---|---|---|---|---|---|---|
| 649103 | IN | 3 | 3 | 0 | 0 | 0 | Red | None | ? | ? | |
| 649103 | IN | 1 | 1 | 0 | 0 | 0 | Red | Slip | ? | ? | |
| 649501 | IN | 1 | 1 | 0 | 0 | 0 | Red | None | ? | ? | |
| 650601 | TR | 1 | 0 | 0 | 1 | 0 | Red | Smoothed | Amphora | | |
| 650601 | TR | 1 | 1 | 0 | 0 | 0 | Brown Gold Mica | None | Amphora? | LC-EH? | |
| 650601 | TR | 1 | 0 | 0 | 1 | 0 | Light | None | Amphora? | LC-EH | **281** |
| 650601 | HW | 9 | 9 | 0 | 0 | 0 | Grey | None | | | |
| 650601 | UT | 1 | 0 | 1 | 0 | 0 | Sinopean | Slip | Louterion | LC-EH? | |
| 650601 | IN | 1 | 1 | 0 | 0 | 0 | Red | Slip | Closed shape | | |
| 650601 | IN | 1 | 1 | 0 | 0 | 0 | Red Pyro | None | Closed shape | | |
| 650601 | IN | 1 | 1 | 0 | 0 | 0 | Red | None | ? | | |
| 650602 | BO | 1 | 1 | 0 | 0 | 0 | | | | | |
| 650602 | TR | 2 | 1 | 0 | 1 | 0 | Herakleian? | None | Amphora | LC-EH | |
| 650602 | TR | 1 | 0 | 0 | 1 | 0 | Sinopean | None | Amphora | LC-EH | |
| 650602 | TR | 1 | 1 | 0 | 0 | 0 | Red Pyro | Wash | Amphora | LC-EH | |
| 650602 | HW | 1 | 1 | 0 | 0 | 0 | Coarse | None | Pot | LC-EH | |
| 650602 | HW | 5 | 5 | 0 | 0 | 0 | Grey | None | Pot | LC-EH | |
| 650602 | HW | 4 | 4 | 0 | 0 | 0 | Red | Relief | Pot | LC-EH | |
| 650602 | IN | 2 | 2 | 0 | 0 | 0 | Red | None | Closed shape | LC-EH | |
| 650602 | IN | 2 | 2 | 0 | 0 | 0 | Red Mica | None | Closed shape | LC-EH | |
| 650602 | IN | 7 | 7 | 0 | 0 | 0 | Red | None | ? | LC-EH | |
| 650701 | TR | 2 | 2 | 0 | 0 | 0 | Herakleian | None | Amphora | LC-EH | |
| 650701 | TR | 1 | 1 | 0 | 0 | 0 | Red | None | Amphora | LC-EH | |
| 650701 | TR | 1 | 1 | 0 | 0 | 0 | Red | Slip | Amphora | LC-EH | |
| 650701 | HW | 8 | 7 | 1 | 0 | 0 | Grey | Impressed | Pot | LC-EH | |
| 650701 | IN | 1 | 1 | 0 | 0 | 0 | Red Mica | None | Closed shape | LC-EH | |
| 650701 | IN | 5 | 5 | 0 | 0 | 0 | Red | None | ? | LC-EH | |
| 650702 | TR | 1 | 1 | 0 | 0 | 0 | Herakleian | None | Amphora | LC-EH | |
| 650702 | TR | 2 | 2 | 0 | 0 | 0 | Thasian | None | Amphora | LC-EH | |
| 650702 | TR | 2 | 2 | 0 | 0 | 0 | Red | Slip | Amphora? | LC-EH | |
| 650702 | TR | 1 | 0 | 0 | 1 | 0 | Red | None | Amphora | LC-EH | |
| 650702 | TR | 1 | 1 | 0 | 0 | 0 | Red | Slip | Amphora? | LC-EH | |
| 650702 | HW | 6 | 6 | 0 | 0 | 0 | Grey | None | Pot | LC-EH | |
| 650702 | HW | 4 | 4 | 0 | 0 | 0 | Grey | None | Pot | LC-EH | |
| 650702 | HW | 1 | 0 | 1 | 0 | 0 | Coarse | None | Pot | LC-EH | |
| 650702 | IN | 1 | 1 | 0 | 0 | 0 | Red | Slip | Closed shape | Byzantine? | |
| 650702 | IN | 2 | 2 | 0 | 0 | 0 | Red | Slip | Closed shape | LC-EH | |
| 650702 | IN | 2 | 2 | 0 | 0 | 0 | Red | None | Closed shape | LC-EH | |
| 650702 | IN | 1 | 1 | 0 | 0 | 0 | Red Mica | None | Closed shape | LC-EH | |
| 650702 | IN | 1 | 1 | 0 | 0 | 0 | Red Mica | None | ? | LC-EH | |
| 650702 | IN | 1 | 1 | 0 | 0 | 0 | Red | None | ? | LC-EH | |
| 650703 | TR | 2 | 0 | 0 | 1 | 1 | Chersonesean | Slip | Amphora | LC-EH | **270** |

| BagID | Function | Total | Body | Rim | Handle | Base | Fabric | Surface | Shape | Period | Cat. |
|---|---|---|---|---|---|---|---|---|---|---|---|
| 660601 | IN | 1 | 1 | 0 | 0 | 0 | Red | None | ? | ? | |
| 660701 | IN | 1 | 1 | 0 | 0 | 0 | Red coarse | None | Closed shape | ? | |
| 647501 | LI | 1 | 0 | 0 | 0 | 0 | | | | | |
| 647601 | LI | 1 | 0 | 0 | 0 | 0 | | | | | |
| 648001 | LI | 1 | 0 | 0 | 0 | 0 | | | | | |

## DSP08-F05-01

Finds table core units

| BagID | Function | Total | Body | Rim | Handle | Base | Fabric | Surface | Shape | Period | Cat. |
|---|---|---|---|---|---|---|---|---|---|---|---|
| 902401 | HW | 1 | 1 | 0 | 0 | 0 | Grey | None | Pot | LC-EH | |
| 902501 | HW | 1 | 1 | 0 | 0 | 0 | Red | None | Pot | LC-EH | |
| 902401 | IN | 1 | 1 | 0 | 0 | 0 | Brown | None | Closed shape | LC-EH? | |
| 902401 | TR | 1 | 1 | 0 | 0 | 0 | Red golden mica | None | Amphora | LC-EH? | |
| 902401 | TR | 1 | 0 | 0 | 1 | 0 | Red | None | Amphora | LC-EH? | |
| 902501 | TR | 1 | 0 | 0 | 1 | 0 | Red | None | Amphora | LC-EH? | |
| 902501 | TR | 1 | 1 | 0 | 0 | 0 | Herakleian | None | Amphora | LC-EH? | |
| 902501 | TW | 1 | 0 | 0 | 0 | 1 | Tatarian | Glazed | Plate | EM | **286** |
| 902501 | IN | 5 | 4 | 1 | 0 | 0 | Tatarian | None | Closed shape | EM? | |
| 902401 | TW | 1 | 1 | 0 | 0 | 0 | Tatarian | Slip | Closed shape | EM? | |
| 902401 | IN | 1 | 1 | 0 | 0 | 0 | Red | None | ? | ? | |
| 902401 | IN | 1 | 1 | 0 | 0 | 0 | | None | ? | ? | |
| 902501 | IN | 2 | 2 | 0 | 0 | 0 | Red | None | Closed shape | ? | |

Finds table adjacent units

| BagID | Function | Total | Body | Rim | Handle | Base | Fabric | Surface | Shape | Period | Cat. |
|---|---|---|---|---|---|---|---|---|---|---|---|
| 901301 | IN | 1 | 1 | 0 | 0 | 0 | Red mica | None | ? | LC-EH? | |
| 902601 | TR | 1 | 0 | 1 | 0 | 0 | Knidian? | Slip | Amphora | LC-EH? | **283** |
| 902601 | TR | 1 | 0 | 0 | 1 | 0 | Red | None | Amphora | LC-EH? | **284** |
| 902801 | IN | 1 | 1 | 0 | 0 | 0 | Tatarian | None | Closed shape | EM | |
| 902601 | TW | 1 | 1 | 0 | 0 | 0 | Tatarian | Glazed | Closed shape | EM | |
| 902801 | TW | 1 | 0 | 1 | 0 | 0 | Tatarian | Slip | Jug | EM | **285** |
| 902801 | IN | 1 | 1 | 0 | 0 | 0 | Red | None | ? | ? | |
| 902601 | LI | 1 | 1 | 0 | 0 | 0 | | | | | |

## DSP08-F06-01

Finds table core units

| BagID | Function | Total | Body | Rim | Handle | Base | Fabric | Surface | Shape | Period | Cat. |
|---|---|---|---|---|---|---|---|---|---|---|---|
| 920211 | HW | 18 | 18 | 0 | 0 | 0 | BaGrey | Grooved | Pot | BA | |
| 920001 | HW | 1 | 1 | 0 | 0 | 0 | BaRed | None | Pot | BA? | |
| 920001 | HW | 2 | 2 | 0 | 0 | 0 | BaBlack | None | Pot | BA? | |
| 920003 | HW | 1 | 1 | 0 | 0 | 0 | BaGrey | burnished | | BA? | |
| 920003 | HW | 9 | 9 | 0 | 0 | 0 | BaGrey | None | Pot | BA? | |
| 920003 | HW | 1 | 1 | 0 | 0 | 0 | BaCoarse | None | Pot | BA? | |
| 920007 | HW | 4 | 4 | 0 | 0 | 0 | BaGrey | None | Pot | BA? | |
| 920007 | HW | 1 | 1 | 0 | 0 | 0 | BaRed | None | Pot | BA? | |
| 920007 | HW | 11 | 11 | 0 | 0 | 0 | BaBlack | None | Pot | BA? | |
| 920103 | HW | 1 | 1 | 0 | 0 | 0 | BaBlack | Grooved | Pot | BA? | |
| 920103 | HW | 10 | 10 | 0 | 0 | 0 | BaBlack | None | Pot | BA? | |
| 920103 | HW | 1 | 0 | 0 | 1 | 0 | BaGrey | None | Pot | BA? | **320** |
| 920103 | HW | 2 | 1 | 1 | 0 | 0 | BaGrey | None | Pot | BA? | |
| 920103 | HW | 1 | 0 | 1 | 0 | 0 | BaRed | Incised | Pot | BA? | |
| 920104 | HW | 1 | 1 | 0 | 0 | 0 | BaRed | Incised | Pot | BA? | |
| 920104 | HW | 2 | 2 | 0 | 0 | 0 | BaBlack | None | Pot | BA? | |
| 920105 | HW | 27 | 27 | 0 | 0 | 0 | BaGrey | None | Pot | BA? | |
| 920106 | HW | 6 | 5 | 1 | 0 | 0 | BaBlack | None | Pot | LBA? | **323** |
| 920106 | HW | 1 | 1 | 0 | 0 | 0 | BaRed | None | Pot | BA? | |
| 920106 | HW | 17 | 17 | 0 | 0 | 0 | BaGrey | None | Pot | BA? | |
| 920106 | HW | 1 | 1 | 0 | 0 | 0 | BaBlack | Smoothed | ? | BA? | |
| 920107 | HW | 3 | 2 | 1 | 0 | 0 | BaBlack | None | Pot | BA? | |
| 920107 | HW | 0 | 0 | 0 | 0 | 0 | BaRed | None | Pot | BA? | |
| 920108 | HW | 4 | 4 | 0 | 0 | 0 | BaGrey | None | Pot | BA? | |
| 920108 | HW | 1 | 1 | 0 | 0 | 0 | BaBlack | Polished | Pot | BA? | |
| 920109 | HW | 12 | 12 | 0 | 0 | 0 | BaBlack | None | Pot | BA? | |
| 920109 | HW | 12 | 12 | 0 | 0 | 0 | BaGrey | None | Pot | BA? | |
| 920109 | HW | 1 | 1 | 0 | 0 | 0 | BaGrey | Impressed | Pot | BA? | **301** |
| 920109 | HW | 2 | 2 | 0 | 0 | 0 | BaBlack | Grooved | Pot | BA? | |
| 920109 | HW | 15 | 15 | 0 | 0 | 0 | BaBlack | None | Pot | BA? | |
| 920110 | HW | 13 | 13 | 0 | 0 | 0 | BaGrey | None | Pot | BA? | |
| 920111 | HW | 3 | 3 | 0 | 0 | 0 | BaBlack | None | Pot | BA? | |
| 920111 | HW | 5 | 5 | 0 | 0 | 0 | BaGrey | None | Pot | BA? | |
| 920201 | HW | 1 | 1 | 0 | 0 | 0 | BaBlack | Incised | Pot | BA? | |
| 920203 | HW | 1 | 1 | 0 | 0 | 0 | BaRed | Polished | Pot | BA? | |
| 920202 | HW | 6 | 5 | 1 | 0 | 0 | BaGrey | None | Pot | BA? | **334** |
| 920202 | HW | 1 | 0 | 1 | 0 | 0 | BaGrey | Incised | Pot | BA | **333** |
| 920205 | HW | 11 | 10 | 1 | 0 | 0 | BaGrey | None | Pot | BA? | **335** |
| 920206 | HW | 3 | 3 | 0 | 0 | 0 | BaGrey | None | Pot | BA? | |

| BagID | Function | Total | Body | Rim | Handle | Base | Fabric | Surface | Shape | Period | Cat. |
|---|---|---|---|---|---|---|---|---|---|---|---|
| 920207 | HW | 12 | 11 | 1 | 0 | 0 | BaGrey | impressed | Pot | BA? | **289** |
| 920209 | HW | 18 | 18 | 0 | 0 | 0 | BaGrey | None | Pot | BA? | |
| 920210 | HW | 1 | 1 | 0 | 0 | 0 | BaRed | None | Pot | BA? | |
| 920210 | HW | 15 | 13 | 2 | 0 | 0 | BaGrey | None | Pot | BA? | **290, 303** |
| 920210 | HW | 3 | 3 | 0 | 0 | 0 | BaBlack | None | Pot | BA? | |
| 920211 | HW | 10 | 10 | 0 | 0 | 0 | BaGrey | None | Pot | BA? | |
| 920211 | HW | 15 | 13 | 2 | 0 | 0 | BaBlack | None | Pot | LBA? | **337, 340** |
| 920211 | HW | 1 | 0 | 1 | 0 | 0 | BaBlack | None | Cup | LBA? | **339** |
| 920211 | HW | 1 | 1 | 1 | 0 | 0 | BaBlack | None | Frying pan | LBA | **338** |
| 920211 | HW | 16 | 13 | 3 | 0 | 0 | BaBlack | None | Pot | LBA? | |
| 920301 | HW | 6 | 6 | 0 | 0 | 0 | BaGrey | None | Pot | BA? | |
| 920301 | HW | 1 | 1 | 0 | 0 | 0 | BaGrey | Incised | Pot | BA? | |
| 920302 | HW | 1 | 1 | 0 | 0 | 0 | BaRed | None | Pot | BA? | |
| 920302 | HW | 1 | 1 | 0 | 0 | 0 | BaCoarse | None | Pot | BA? | |
| 920302 | HW | 9 | 9 | 0 | 0 | 0 | BaBlack | None | Pot | BA? | |
| 920302 | HW | 13 | 13 | 0 | 0 | 0 | BaGrey | None | Pot | BA? | |
| 920701 | HW | 2 | 2 | 0 | 0 | 0 | BaGrey | None | Pot | BA? | |
| 920701 | HW | 1 | 1 | 0 | 0 | 0 | BaBlack | None | Pot | BA? | |
| 920701 | HW | 1 | 1 | 0 | 0 | 0 | BaGrey | Incised | Pot | BA? | |
| 920701 | HW | 2 | 2 | 0 | 0 | 0 | BaBlack | None | Pot | BA? | |
| 920701 | HW | 1 | 1 | 0 | 0 | 0 | BaBlack | None | Pot | BA? | |
| 920701 | HW | 1 | 1 | 0 | 0 | 0 | BaRed | None | Pot | BA? | |
| 920702 | HW | 4 | 4 | 0 | 0 | 0 | BaBlack | None | Pot | BA? | |
| 920703 | HW | 13 | 13 | 0 | 0 | 0 | BaGrey | None | Pot | BA? | |
| 920703 | HW | 2 | 2 | 0 | 0 | 0 | BaBlack | None | Pot | BA? | |
| 920801 | HW | 16 | 14 | 1 | 0 | 1 | BaGrey | None | Pot | LBA? | **347, 364** |
| 920801 | HW | 1 | 1 | 0 | 0 | 0 | BaGrey | Polished | Pot | BA? | **365** |
| 920801 | HW | 1 | 1 | 0 | 0 | 0 | BaGrey | Relief | Pot | BA? | **344** |
| 920802 | HW | 29 | 28 | 1 | 0 | 0 | BaRed | None | Pot | Middle/Late BA? | **308** |
| 920802 | HW | 2 | 2 | 0 | 0 | 0 | BaRed | Burnished | | BA? | |
| 920802 | HW | 17 | 14 | 3 | 0 | 0 | BaBlack | None | Pot | BA? | **293, 348, 350** |
| 920802 | HW | 1 | 1 | 0 | 0 | 0 | BaCoarse | None | Pot | BA? | |
| 920802 | HW | 213 | 207 | 6 | 0 | 0 | BaGrey | None | Pot | BA | **351-355, 366** |
| 920802 | HW | 1 | | 1 | | | BaGrey | Impressed | Pot | BA | **311** |
| 920901 | HW | 13 | 13 | 0 | 0 | 0 | BaGrey | None | Pot | BA? | |
| 920902 | HW | 7 | 7 | 0 | 0 | 0 | BaRed | none | Pot | BA? | |
| 920902 | HW | 1 | 1 | 0 | 0 | 0 | BaRed | Nail | Pot | BA? | **359** |
| 920902 | HW | 1 | 1 | 0 | 0 | 0 | BaCoarse | None | Pot | BA? | |
| 921901 | HW | 2 | 1 | 1 | 0 | 0 | BaGrey | None | Pot | BA? | **368** |
| 921901 | HW | 5 | 5 | 0 | 0 | 0 | BaGrey | None | Pot | BA? | |
| 920202 | HW | 1 | 0 | 1 | 0 | 0 | BaBlack | None | Bowl? | BA? | **363** |

| BagID | Function | Total | Body | Rim | Handle | Base | Fabric | Surface | Shape | Period | Cat. |
|---|---|---|---|---|---|---|---|---|---|---|---|
| 920101 | HW | 2 | 2 | 0 | 0 | 0 | BaBlack | Coombed | Pot | Middle BA | **287-288** |
| 920101 | HW | 1 | 1 | 0 | 0 | 0 | BaBlack | Cord | Pot | Middle BA | **298** |
| 920102 | HW | 1 | 1 | 0 | 0 | 0 | BaGrey | Impressed | Pot | Middle BA | |
| 920702 | HW | 1 | 0 | 1 | 0 | 0 | BaBlack | None | Kitchen Vessel | Middle BA | **291** |
| 920802 | HW | 1 | 1 | 0 | 0 | 0 | BaBlack | Impressed | Pot | Middle BA | **294** |
| 920902 | HW | 1 | 1 | 0 | 0 | 0 | BaBlack | Incised | Pot | Middle BA | **295** |
| 920101 | HW | 3 | 3 | 0 | 0 | 0 | BaBlack | Incised | Pot | Middle BA? | |
| 920101 | HW | 9 | 9 | 0 | 0 | 0 | BaRed | None | Pot | Middle BA? | |
| 920101 | HW | 8 | 7 | 0 | 0 | 1 | BaBlack | Grooved | Pot | BA | **362** |
| 920211 | HW | 1 | 1 | 0 | 0 | 0 | BaBlack | Impressed | Pot | Middle BA? | **304** |
| 920802 | HW | 7 | 7 | 0 | 0 | 0 | BaBlack | Incised | Pot | Middle BA? | |
| 920002 | HW | 1 | 1 | 0 | 0 | 0 | BaGrey | Incised | Pot | Middle/ Late BA | **296** |
| 920101 | HW | 1 | 1 | 0 | 0 | 0 | BaRed | Cord | ? | Middle/ Late BA | **297** |
| 920101 | HW | 10 | 9 | 1 | 0 | 0 | BaBlack | Impressed Cord | Pot | Middle/ Late BA | **316, 318** |
| 920301 | HW | 1 | 1 | 0 | 0 | 0 | BaGrey | impressed | Pot | Middle/ Late BA | **306** |
| 920801 | HW | 1 | 0 | 1 | 0 | 0 | BaBlack | None | Pot | Middle/ Late BA | **307** |
| 920902 | HW | 39 | 36 | 2 | 0 | 1 | BaGrey | None | Pot | Middle/ Late BA | **312, 367** |
| 921902 | HW | 1 | 1 | 0 | 0 | 0 | BaBlack | Impressed Cord | Pot | Middle/ Late BA | **313** |
| 921902 | HW | 1 | 0 | 1 | 0 | 0 | BaGrey | None | Pot | Middle/ Late BA | **314** |
| 920001 | HW | 1 | 1 | 0 | 0 | 0 | BaGrey | Impressed | Pot | Late BA | |
| 920001 | HW | 3 | 3 | 0 | 0 | 0 | BaGrey | Grooved | Pot | Late BA | |
| 920001 | HW | 3 | 3 | 0 | 0 | 0 | BaGrey | Grooved | Pot | Late BA | |
| 920001 | HW | 2 | 2 | 0 | 0 | 0 | BaBlack | Grooved | Pot | Late BA | |
| 920003 | HW | 7 | 7 | 0 | 0 | 0 | BaGrey | Grooved | Pot | Late BA | |
| 920004 | HW | 1 | 1 | 0 | 0 | 0 | BaGrey | Frooved | Pot | Late BA | |
| 920005 | HW | 6 | 6 | 0 | 0 | 0 | BaGrey | Grooved | Pot | Late BA | |
| 920007 | HW | 3 | 3 | 0 | 0 | 0 | BaGrey | Grooved | Pot | Late BA | |
| 920007 | HW | 5 | 5 | 0 | 0 | 0 | BaGrey | Grooved | Pot | Late BA | |
| 920102 | HW | 3 | 3 | 0 | 0 | 0 | BaBlack | Grooved | Pot | Late BA | |
| 920102 | HW | 10 | 10 | 0 | 0 | 0 | BaGrey | None | Pot | Late BA | |
| 920102 | HW | 1 | 1 | 0 | 0 | 0 | BaBlack | Incised | Pot | Late BA | **319** |
| 920102 | HW | 17 | 17 | 0 | 0 | 0 | BaGrey | Grooved | Pot | Late BA | |
| 920103 | HW | 1 | 1 | 0 | 0 | 0 | BaGrey | Cord Decoration | Pot | Late BA | **299** |
| 920103 | HW | 3 | 2 | 1 | 0 | 0 | BaBlack | Grooved | Pot | Late BA | |
| 920103 | HW | 1 | 1 | 0 | 0 | 0 | BaGrey | Incised | Pot | Late BA | |
| 920104 | HW | 2 | 2 | 0 | 0 | 0 | BaGrey | Grooved | Pot | Late BA | |
| 920104 | HW | 4 | 4 | 0 | 0 | 0 | BaBlack | Grooved | Pot | Late BA | |

| BagID | Function | Total | Body | Rim | Handle | Base | Fabric | Surface | Shape | Period | Cat. |
|---|---|---|---|---|---|---|---|---|---|---|---|
| 920105 | HW | 14 | 14 | 0 | 0 | 0 | BaGrey | Grooved | Pot | Late BA | |
| 920105 | HW | 14 | 14 | 0 | 0 | 0 | BaGrey | Grooved | Pot | Late BA | |
| 920106 | HW | 1 | 1 | 0 | 0 | 0 | BaBlack | Impressed | Pot | Late BA | **300** |
| 920106 | HW | 2 | 2 | 0 | 0 | 0 | BaRed | Impressed | Pot | Late BA | **321-322** |
| 920106 | HW | 1 | 0 | 1 | 0 | 0 | BaRed | Incised | Pot | Late BA | **324** |
| 920106 | HW | 9 | 9 | 0 | 0 | 0 | BaGrey | Grooved | Pot | Late BA | |
| 920107 | HW | 3 | 3 | 0 | 0 | 0 | BaRed | Grooved | Pot | Late BA | |
| 920108 | HW | 1 | 1 | 0 | 0 | 0 | BaGrey | Grooved | Pot | Late BA | |
| 920109 | HW | 6 | 5 | 1 | 0 | 0 | BaBlack | Grooved | Pot | Late BA | **326** |
| 920109 | HW | 14 | 10 | 4 | 0 | 0 | BaGrey | Grooved | Pot | Late BA | **327-330** |
| 920109 | HW | 3 | 3 | 0 | 0 | 0 | BaGrey | Impressed | Pot | Late BA | **325** |
| 920110 | HW | 11 | 11 | 0 | 0 | 0 | BaBlack | Grooved | Pot | Late BA | |
| 920110 | HW | 16 | 15 | 1 | 0 | 0 | BaGrey | Grooved | | Late BA | |
| 920111 | HW | 7 | 7 | 0 | 0 | 0 | BaGrey | Grooved | Pot | Late BA | |
| 920111 | HW | 4 | 4 | 0 | 0 | 0 | BaBlack | Grooved | Pot | Late BA | |
| 920201 | HW | 1 | 1 | 0 | 0 | 0 | BaGrey | Impressed | Pot | Late BA | |
| 920201 | HW | 1 | 1 | 0 | 0 | 0 | BaBlack | Impressed | Pot | Late BA | |
| 920201 | HW | 24 | 24 | 0 | 0 | 0 | BaGrey | Grooved | Pot | Late BA | |
| 920201 | HW | 24 | 24 | 0 | 0 | 0 | BaGrey | None | Pot | Late BA | |
| 920202 | HW | 2 | 2 | 0 | 0 | 0 | BaGrey | Grooved | Pot | Late BA | |
| 920203 | HW | 6 | 6 | 0 | 0 | 0 | BaGrey | Grooved | Pot | Late BA | |
| 920205 | HW | 2 | 2 | 0 | 0 | 0 | BaGrey | Incised | Pot | Late BA | |
| 920205 | HW | 1 | 1 | 0 | 0 | 0 | BaGrey | Grooved | Pot | Late BA | |
| 920206 | HW | 4 | 4 | 0 | 0 | 0 | BaGrey | Grooved | Pot | Late BA | |
| 920207 | HW | 9 | 9 | 0 | 0 | 0 | BaGrey | Grooved | Pot | Late BA | |
| 920207 | HW | 2 | 2 | 0 | 0 | 0 | BaGrey | Impressed | Pot | Late BA | **302, 336** |
| 920208 | HW | 3 | 3 | 0 | 0 | 0 | BaGrey | Grooved | Pot | Late BA | |
| 920210 | HW | 25 | 25 | 0 | 0 | 0 | BaGrey | Grooved | Pot | Late BA | |
| 920210 | HW | 8 | 8 | 0 | 0 | 0 | BaBlack | Grooved | Pot | Late BA | |
| 920210 | HW | 1 | 1 | 0 | 0 | 0 | BaBlack | Incised | Pot | Late BA | |
| 920210 | HW | 1 | 0 | 1 | 0 | 0 | Brown | None | Bowl | Hellenisic | **374** |
| 920211 | HW | 2 | 2 | 0 | 0 | 0 | BaBlack | Grooved | Pot | Late BA | |
| 920211 | HW | 1 | 1 | 0 | 0 | 0 | BaBlack | Relief | Pot | Late BA | **341** |
| 920211 | HW | 2 | 2 | 0 | 0 | 0 | BaBlack | Incised | Pot | Middle/Late BA | **305** |
| 920302 | HW | 1 | 1 | 0 | 0 | 0 | BaBlack | Grooved | Pot | Late BA | |
| 920302 | HW | 2 | 2 | 0 | 0 | 0 | BaGrey | Grooved | Pot | Late BA | |
| 920302 | HW | 5 | 5 | 0 | 0 | 0 | BaGrey | None | Pot | Late BA | |
| 920302 | HW | 2 | 2 | 0 | 0 | 0 | BaBlack | Grooved | Pot | Late BA | |
| 920701 | HW | 8 | 7 | 1 | 0 | 0 | BaGrey | Grooved | Bowl | Late BA | **342** |
| 920701 | HW | 6 | 6 | 0 | 0 | 0 | BaGrey | Grooved | Pot | Late BA | |
| 920701 | HW | 1 | 1 | 0 | 0 | 0 | BaGrey | Grooved | Pot | Late BA | |
| 920702 | HW | 4 | 4 | 0 | 0 | 0 | BaGrey | Grooved | Pot | Late BA | |

| BagID | Function | Total | Body | Rim | Handle | Base | Fabric | Surface | Shape | Period | Cat. |
|---|---|---|---|---|---|---|---|---|---|---|---|
| 920703 | HW | 5 | 5 | 0 | 0 | 0 | BaGrey | Grooved | Pot | Late BA | |
| 920801 | HW | 15 | 14 | 1 | 0 | 0 | BaGrey | Grooved | Pot | Late BA | **345** |
| 920801 | HW | 4 | 1 | 3 | 0 | 0 | BaBlack | Impressed | Pot | Late BA | **346** |
| 920801 | HW | 1 | 1 | 0 | 0 | 0 | BaBlack | Incised | Pot | Late BA | |
| 920801 | HW | 1 | 1 | 1 | 0 | 0 | BaGrey | incised | Pot | Late BA | |
| 920802 | HW | 121 | 121 | 0 | 0 | 0 | BaGrey | Grooved | Pot | Late BA | |
| 920802 | HW | 17 | 17 | 0 | 0 | 0 | BaRed | Grooved | Pot | Late BA | |
| 920802 | HW | 13 | 12 | 1 | 0 | 0 | BaGrey | Incised | Pot | Late BA | **292** |
| 920902 | HW | 17 | 17 | 0 | 0 | 0 | BaGrey | Grooved | Pot | Late BA | |
| 920902 | HW | 2 | 0 | 2 | 0 | 0 | BaGrey | None | Bowl | Late BA | **356, 358** |
| 921801 | HW | 4 | 3 | 1 | 0 | 0 | BaBlack | None | Pot | Late BA | **360** |
| 921901 | HW | 2 | 2 | 0 | 0 | 0 | BaGrey | Grooved | Pot | Late BA | |
| 921901 | HW | 1 | 1 | 0 | 0 | 0 | BaBlack | Grooved | Pot | Late BA | |
| 921901 | HW | 17 | 16 | 1 | 0 | 0 | BaGrey | None | Pot | Late BA | **361** |
| 920001 | HW | 13 | 13 | 0 | 0 | 0 | BaGrey | None | Pot | Late BA? | |
| 920005 | HW | 1 | 1 | 0 | 0 | 0 | BaGrey | Burnished | | Late BA? | **315** |
| 920101 | HW | 1 | 0 | 1 | 0 | 0 | BaRed | None | Bowl | Late BA? | **317** |
| 920110 | HW | 1 | 0 | 1 | 0 | 0 | BaGrey | None | Frying pan? | Late BA? | **331** |
| 920110 | HW | 1 | 1 | 0 | 0 | 0 | BaGrey | Incised | Pot | Late BA? | |
| 920201 | HW | 1 | 0 | 1 | 0 | 0 | BaCoarse | None | Pot | Late BA? | **332** |
| 920201 | HW | 1 | 1 | 0 | 0 | 0 | BaBlack | Coombed | Pot | Late BA? | |
| 920201 | HW | 2 | 1 | 0 | 0 | 1 | BaBlack | None | Pot | Late BA? | |
| 920201 | HW | 1 | 1 | 0 | 0 | 0 | BaBlack | Relief | Open shape | Late BA? | |
| 920701 | HW | 2 | 2 | 0 | 0 | 0 | BaGrey | Grooved | Pot | Late BA? | |
| 920702 | HW | 1 | 0 | 1 | 0 | 0 | BaBlack | None | Bowl | Late BA? | **343** |
| 920802 | HW | 4 | 4 | 0 | 0 | 0 | BaGrey | Impressed | Pot | Late BA? | **349** |
| 920802 | HW | 2 | 2 | 0 | 0 | 0 | BaBlack | Impressed | Cup? | Late BA? | **309-310** |
| 920902 | HW | 1 | 0 | 1 | 0 | 0 | BaBlack | None | Pot | Late BA? | **357** |
| 920002 | HW | 10 | 9 | 1 | 0 | 0 | Red | None | Pot | LC-EH | **371** |
| 920005 | HW | 15 | 14 | 1 | 0 | 0 | Brown | None | Pot | LC-EH | **372** |
| 920702 | HW | 14 | 13 | 1 | 0 | 0 | Brown | None | Pot | LC-EH | **375** |
| 920702 | HW | 1 | 0 | 1 | 0 | 0 | Brown | None | Pot | LC-EH | **376** |
| 920801 | HW | 1 | 0 | 1 | 0 | 0 | Grey | none | Pot | LC-EH | **378** |
| 920801 | HW | 1 | 1 | 0 | 0 | 0 | Brown | Impressed | Pot | LC-EH | **377** |
| 920802 | TW | 1 | 0 | 0 | 1 | 0 | Red | None | Jug? | LC-EH | **383** |
| 920001 | HW | 9 | 8 | 0 | 0 | 1 | Grey | None | Pot | LC-EH? | **370** |
| 920201 | HW | 4 | 3 | 1 | 0 | 0 | Red | None | Pot | LC-EH? | |
| 920703 | HW | 1 | 1 | 0 | 0 | 0 | Grey | Relief | Pot | LC-EH? | |
| 920802 | HW | 1 | 0 | 1 | 0 | 0 | Grey | None | Jug | LC-EH | **379** |
| 920802 | HW | 1 | 1 | 0 | 0 | 0 | Brown | None | Pot | LC-EH | **381** |
| 920802 | HW | 1 | 0 | 1 | 0 | 0 | Grey | None | Pot | LC-EH | **380** |
| 920004 | IN | 1 | 1 | 0 | 0 | 0 | Red | None | | LC-EH? | |
| 920007 | IN | 1 | 1 | 0 | 0 | 0 | Red | None | ? | LC-EH? | |

| BagID | Function | Total | Body | Rim | Handle | Base | Fabric | Surface | Shape | Period | Cat. |
|---|---|---|---|---|---|---|---|---|---|---|---|
| 920107 | IN | 1 | 1 | 0 | 0 | 0 | Red | None | | LC-EH? | |
| 920109 | IN | 1 | 1 | 0 | 0 | 0 | Red mica | None | ? | LC-EH? | |
| 920203 | IN | 1 | 1 | 0 | 0 | 0 | Red mIca | None | | LC-EH? | |
| 920207 | IN | 1 | 1 | 0 | 0 | 0 | Red | None | | LC-EH? | |
| 920211 | IN | 1 | 1 | 0 | 0 | 0 | Red | None | | LC-EH? | |
| 920701 | IN | 3 | 3 | 0 | 0 | 0 | Red | None | | LC-EH? | |
| 920802 | IN | 1 | 1 | 0 | 0 | 0 | Red | Slip | Closed shape | LC-EH? | |
| 920902 | IN | 3 | 3 | 0 | 0 | 0 | Red | None | Pot | LC-EH? | |
| 920007 | TR | 1 | 1 | 0 | 0 | 0 | Thasian | None | Amphora | LC-EH? | |
| 920007 | TR | 2 | 2 | 0 | 0 | 0 | Herakleian | None | Amphora | LC-EH? | |
| 920104 | TR | 1 | 1 | 0 | 0 | 0 | Red | None | Amphora | LC-EH? | |
| 920201 | TR | 1 | 1 | 0 | 0 | 0 | Thasian | None | Amphora? | LC-EH? | |
| 920302 | TR | 2 | 1 | 0 | 1 | 0 | Red | None | Amphora | LC-EH? | |
| 920902 | TR | 1 | 1 | 0 | 0 | 0 | Red mica | None | Amphora | LC-EH? | |
| 920902 | TR | 1 | 1 | 0 | 0 | 0 | Red | None | Amphora | LC-EH? | |
| 921901 | TR | 1 | 0 | 0 | 1 | 0 | Red | Slip | Amphora | LC-EH? | 382 |
| 920001 | TW | 1 | 0 | 1 | 0 | 0 | Red | None | Bowl? | LC-EH? | |
| 920101 | TW | 1 | 1 | 0 | 0 | 0 | Red | Grooved | | Byzantine? | |
| 920003 | AR | 1 | 1 | 0 | 0 | 0 | Red | None | Tile | EM | |
| 920109 | TW | 1 | 0 | 0 | 1 | 0 | Tatarian | None | Jug? | EM | |
| 920201 | TW | 1 | 1 | 0 | 0 | 0 | Tatarian | Ribbed | Jug | EM? | |
| 920101 | AR | 2 | 2 | 0 | 0 | 0 | Red | None | | Modern | |
| 920103 | AR | 1 | 1 | 0 | 0 | 0 | Red modern | None | Tile | Modern | |
| 920103 | AR | 1 | 1 | 0 | 0 | 0 | | | | Modern | |
| 920211 | AR | 1 | 1 | 0 | 0 | 0 | Red | None | Brick | Modern | |
| 920206 | AR | 1 | 1 | 0 | 0 | 0 | | | | | |
| 920703 | AR | 1 | 1 | 0 | 0 | 0 | | | | | |
| 920102 | BO | 1 | 0 | 0 | 0 | 0 | | | | | |
| 920111 | BO | 1 | 1 | 0 | 0 | 0 | | | | | |
| 920701 | BO | 1 | 1 | 0 | 0 | 0 | | | | | |
| 920001 | HW | 1 | 1 | 0 | 0 | 0 | Red | None | Pot | ? | |
| 920001 | HW | 2 | 2 | 0 | 0 | 0 | Black | None | Pot | ? | |
| 920004 | HW | 8 | 8 | 0 | 0 | 0 | Grey | None | Pot | ? | |
| 920005 | HW | 2 | 2 | 0 | 0 | 0 | Coarse | None | Pot | ? | |
| 920006 | HW | 4 | 4 | 0 | 0 | 0 | Grey | None | Pot | ? | |
| 920101 | HW | 21 | 21 | 0 | 0 | 0 | Grey | None | Pot | ? | |
| 920101 | HW | 3 | 3 | 0 | 0 | 0 | Coarse | None | Pot | ? | |
| 920104 | HW | 2 | 2 | 0 | 0 | 0 | Grey | None | Pot | ? | |
| 920105 | HW | 3 | 3 | 0 | 0 | 0 | Coarse | None | Pot | ? | |
| 920107 | HW | 1 | 1 | 0 | 0 | 0 | Coarse | None | Pot | ? | |
| 920107 | HW | 1 | 1 | 0 | 0 | 0 | Grey | None | Pot | ? | |
| 920109 | HW | 2 | 1 | 1 | 0 | 0 | Brown | None | Pot | ? | 373 |
| 920109 | HW | 1 | 1 | 0 | 0 | 0 | Coarse | None | Pot | ? | |

| BagID | Function | Total | Body | Rim | Handle | Base | Fabric | Surface | Shape | Period | Cat. |
|---|---|---|---|---|---|---|---|---|---|---|---|
| 920110 | HW | 2 | 2 | 0 | 0 | 0 | Red | Grooved | Pot | ? | |
| 920110 | HW | 1 | 1 | 0 | 0 | 0 | Coarse | None | Pot | ? | |
| 920111 | HW | 2 | 2 | 0 | 0 | 0 | Coarse | None | Pot | ? | |
| 920111 | HW | 2 | 2 | 0 | 0 | 0 | Red | None | Pot | ? | |
| 920208 | HW | 2 | 2 | 0 | 0 | 0 | Black | None | Pot | ? | |
| 920208 | HW | 10 | 10 | 0 | 0 | 0 | Grey | None | Pot | ? | |
| 920211 | HW | 1 | 1 | 0 | 0 | 0 | Coarse | None | Pot | ? | |
| 920701 | HW | 6 | 6 | 0 | 0 | 0 | Grey | None | Pot | ? | |
| 920702 | HW | 1 | 1 | 0 | 0 | 0 | Red | None | Pot | ? | |
| 920001 | IN | 1 | 1 | 0 | 0 | 0 | Red | None | Closed shape | ? | |
| 920003 | IN | 3 | 3 | 0 | 0 | 0 | Red | None | | ? | |
| 920005 | IN | 3 | 3 | 0 | 0 | 0 | Red mica | None | | ? | |
| 920103 | IN | 2 | 2 | 0 | 0 | 0 | Red | None | | ? | |
| 920105 | IN | 1 | 1 | 0 | 0 | 0 | Red | None | | ? | |
| 920107 | IN | 3 | 3 | 0 | 0 | 0 | Red | None | | ? | |
| 920109 | IN | 1 | 1 | 0 | 0 | 0 | Red | None | ? | ? | |
| 920110 | IN | 3 | 3 | 0 | 0 | 0 | Red | None | | ? | |
| 920201 | IN | 1 | 1 | 0 | 0 | 0 | Red | None | ? | ? | |
| 920802 | IN | 3 | 3 | 0 | 0 | 0 | Red | None | | ? | |
| 920001 | LI | 1 | 1 | 0 | 0 | 0 | | | | | |
| 920002 | LI | 3 | 0 | 0 | 0 | 0 | | | | | |
| 920005 | LI | 2 | 0 | 0 | 0 | 0 | | | | | |
| 920007 | LI | 2 | 0 | 0 | 0 | 0 | | | | | |
| 920101 | LI | 1 | 0 | 0 | 0 | 0 | | | | | |
| 920103 | LI | 1 | 1 | 0 | 0 | 0 | | | | | |
| 920104 | LI | 5 | 0 | 0 | 0 | 0 | | | | | |
| 920105 | LI | 1 | 0 | 0 | 0 | 0 | | | | | |
| 920106 | LI | 2 | 0 | 0 | 0 | 0 | | | | | |
| 920108 | LI | 3 | 0 | 0 | 0 | 0 | | | | | |
| 920109 | LI | 3 | 0 | 0 | 0 | 0 | | | | | |
| 920110 | LI | 5 | 0 | 0 | 0 | 0 | | | | | |
| 920201 | LI | 3 | 0 | 0 | 0 | 0 | | | | | |
| 920204 | LI | 3 | 0 | 0 | 0 | 0 | | | | | |
| 920208 | LI | 2 | 2 | 0 | 0 | 0 | | | | | |
| 920209 | LI | 2 | 0 | 0 | 0 | 0 | | | | | |
| 920701 | LI | 5 | 0 | 0 | 0 | 0 | | | | | |
| 920702 | LI | 3 | 0 | 0 | 0 | 0 | | | | | |
| 920801 | LI | 2 | 2 | 0 | 0 | 0 | | | | | |
| 920802 | LI | 51 | 2 | 0 | 0 | 0 | | | | | |
| 920902 | LI | 4 | 0 | 0 | 0 | 0 | | | | | |
| 921901 | LI | 4 | 0 | 0 | 0 | 0 | | | | | |

Finds table adjacent units

| BagID | Function | Total | Body | Rim | Handle | Base | Fabric | Surface | Shape | Period | Cat. |
|---|---|---|---|---|---|---|---|---|---|---|---|
| 921701 | HW | 1 | 0 | 0 | 0 | 0 | Grey | None | Pot | LC-EH? | |
| 921001 | TR | 1 | 0 | 0 | 1 | 0 | Red | None | Amphora | LC-EH? | |
| 921001 | TR | 1 | 1 | 0 | 0 | 0 | Herakleian | None | Amphora | LC-EH? | |

## DSP08-F06-02

Finds table core units

| BagID | Function | Total | Body | Rim | Handle | Base | Fabric | Surface | Shape | Period | Cat. |
|---|---|---|---|---|---|---|---|---|---|---|---|
| 922601 | HW | 1 | 1 | 0 | 0 | 0 | Grey | None | Pot | LC-EH? | |
| 922701 | TR | 1 | 1 | 0 | 0 | 0 | Grey mica | None | Amphora? | LC-EH? | |

## DSP08-F06-03

Finds table core units

| BagID | Function | Total | Body | Rim | Handle | Base | Fabric | Surface | Shape | Period | Cat. |
|---|---|---|---|---|---|---|---|---|---|---|---|
| 904501 | HW | 3 | 3 | 0 | 0 | 0 | Grey | None | Pot | LC-EH | |
| 904501 | IN | 4 | 4 | 0 | 0 | 0 | Red | None | ? | LC-EH? | |
| 904501 | TR | 1 | 0 | 0 | 1 | 0 | Herakleian | Slip | Amphora | LC-EH? | **385** |

Finds table adjacent units

| BagID | Function | Total | Body | Rim | Handle | Base | Fabric | Surface | Shape | Period | Cat. |
|---|---|---|---|---|---|---|---|---|---|---|---|
| 905001 | HW | 1 | 0 | 1 | 0 | 0 | Coarse | Impressed | Pot | LC-EH | **384** |
| 904401 | TR | 1 | 0 | 0 | 1 | 0 | Sinopean | None | Amphora | LC-EH? | **386** |
| 904401 | TR | 1 | 1 | 0 | 0 | 0 | Herakleian | None | Amphora | LC-EH? | |
| 905001 | TR | 1 | 1 | 0 | 0 | 0 | Sinopean | None | Amphora | LC-EH? | |
| 905001 | LI | 1 | 1 | 0 | 0 | 0 | | | | | |

## DSP07-H01-01

Finds table core units

| BagID | Function | Total | Body | Rim | Handle | Base | Fabric | Surface | Shape | Period | Cat. |
|---|---|---|---|---|---|---|---|---|---|---|---|
| 306001 | HW | 1 | 0 | 0 | 0 | 0 | HwGrey2 | | | LC-EH | |
| 305001 | IN | 1 | 0 | 0 | 0 | 0 | | | | LC-EH | |
| 305201 | IN | 1 | 0 | 0 | 0 | 0 | | | | LC-EH | |
| 305401 | IN | 1 | 0 | 0 | 0 | 0 | | | | LC-EH | |
| 305601 | IN | 1 | 0 | 0 | 0 | 0 | | | | LC-EH | |
| 305901 | IN | 1 | 0 | 0 | 0 | 0 | | | | LC-EH | |
| 305301 | TR | 1 | 0 | 0 | 0 | 0 | Sinopean | | Amphora | LC-EH | |
| 305801 | TR | 1 | 0 | 0 | 0 | 0 | Sinopean | | Amphora | LC-EH | |
| 304301 | IN | 1 | 0 | 0 | 0 | 0 | | | | LC-EH? | |
| 304601 | IN | 3 | 0 | 0 | 0 | 0 | | | | LC-EH? | |
| 304901 | IN | 1 | 0 | 0 | 0 | 0 | | | | Modern? | |
| 305701 | IN | 1 | 0 | 0 | 0 | 0 | | | | Modern? | |
| 304501 | IN | 1 | 0 | 0 | 0 | 0 | | | | | |
| 305101 | IN | 1 | 0 | 0 | 0 | 0 | | | | | |
| 305501 | IN | 1 | 0 | 0 | 0 | 0 | | | | | |

## DSP07-H01-01.

**Feature 5**. Excavation finds

| Findslist 6. In a stone premise on a site of ancient settlement | | | | |
|---|---|---|---|---|
| No. | Description | Quantity | Cat. | Note |
| 1 | Reddish-brown clay body fr. | 4 | | |
| 2a | Handmade rim | 1 | **400** | BA |
| 2b | Handmade grey polished body frs. | 2 | **397-398** | BA |
| 2c | Handmade ribbed and smoothed body frs. | 2 | | BA |
| 2d | Handmade ribbed body frs. | 2 | **399** | |
| 3 | Animal bones | 14 | | |
| 4 | Grape snails (*Helix vulgaris*) | | | |
| Material from a layer of a stone blockage above floor (Layer 2), 0.5-0.7 m | | | | |
| 5 | Body fr. of Heraclean amphora | 1 | | |
| 6 | Body fr. of Mediterranean amphora with fine mica inclusions | 1 | | |
| 7 | Body fr. of amphora? | 1 | | |
| 8 | Handmade smoothed moulded body fr. | 1 | | BA |
| 9 | Bones of animals | 31 | | |
| 10 | Grape snails (*Helix vulgaris*) | 2 | | |
| Material from a friable layer of yellow clay with ashes (layer 3) Above a floor of a room, below a stone blockage | | | | |
| No. | Description | Quantity | Cat. | Note |
| 11 | Body fr. of brown-clay amphora | 1 | | |
| 12 | Handmade rim with dark interior | 1 | | BA |
| 13 | Handmade ribbed body fr. of dark clay | 1 | | |
| 14 | Bones of animals | 33 | | |
| 15 | Grape snails (*Helix vulgaris*) | 12 | | |
| Dense yellow clay (layer 4) above continental clay. Depth 0.7-0.8 m | | | | |
| 16 | Handmade red-clay body fr. | 1 | | |
| 17 | Bones of animals | 3 | | |
| 18 | Grape snails (*Helix vulgaris*) | 2 | | |

**Feature 18.** Excavation finds

| Finds list 1. Material from filling of pit 1 | | | | |
|---|---|---|---|---|
| No. | Description | Quantity | Cat. | Note |
| 1a | Rim of Chersonesean amphora | 1 | **395** | Traces of secondary firing |
| 1b | Body frs. of Chersonesean amphora | 7 | | Traces of secondary firing |
| 2a | Rim of Thasian amphora | 1 | **396** | Traces of secondary firing |
| 2b | Body of Thasian amphora | 1 | | |
| 3 | Body frs. of red-clay jugs with light slip | 2 | | |
| 4a | Handmade rim with rough surface. Brown clay with grog and grits | 1 | **391** | |
| 4b | Handmade base with rough surface. Brown clay with grog and grits | 1 | | |
| 4c | Handmade body frs. with rough surface. Brown clay with grog and grits | 11 | | |
| 5a | Handmade rim of yellow clay with pinched decoration | 1 | **389** | |
| 5b | Handmade base of yellow clay | 1 | **392** | |
| 5c | Handmade body frs. of yellow clay | 19 | | |
| 6 | Handmade rim with impressed finger decoration. Coarse clay with shell inclusions | 1 | **393** | |
| 7 | Handmade rims of brown clay with grit | 3 | **394** | Two joining |
| 8 | Handmade rim of bowl | 1 | | |
| 9 | Handmade body fr. with smoothed exterior | 1 | | |
| 10 | Fired clay from kiln? | 2 | | |
| 11 | Animal bones | 8 | | |
| 12 | Grape snails (*Helix vulgaris*) | 1 | | |
| Finds list 2. material from turf above yellow clay and above Pit 1 | | | | |
| No. | Description | Quantity | Cat. | Note |
| 1a | Handmade rim with impressed finger decoration | 1 | **388** | |
| 1b | Handmade body frs. | 4 | | |
| Finds list 3. The Same. In east extension to trench 1 above Pit 1 (or the top of the hole) | | | | |
| No. | Description | Quantity | Cat. | Note |
| 1 | Body fr. of Thasian amphora | 1 | | |
| 2 | Body fr. of red-clay jug | 1 | | |
| 3 | Handmade body frs. | 4 | | |
| Findslist 4. material from turf above yellow clay layer over entire trench. Depth up to 0.3 m | | | | |
| No. | Description | Quantity | Cat. | Note |
| 1 | Body fr. of Heraclean amphora | 1 | | |
| 2 | Body frs. of Chersonesean? amphora | 6 | | |
| 3 | Amphora/jug? of pale brown clay with grey temper | 2 | | |
| 4a | Handmade body frs. from different vessels | 40 | | |
| 4b | Handmade bases of pale brown clay with chrushed shells | 2 | **387** | Possibly, from one vessel |
| 5 | Bones of animals | 2 | | |
| 6 | Grape snails (*Helix vulgaris*) | 16 | | |
| Finds list 5. Material from filling of Pit 1 | | | | |
| No. | Description | Quantity | Cat. | Note |

| | | | | |
|---|---|---|---|---|
| 1 | Body fr. of Mediterranean amphora of red clay with yellow mica and pyrite and dark slip | 1 | | |
| 2 | Body frs. of Sinopean amphora with pale slip | 2 | | |
| 3 | Body fr. of thick-walled Mediterranean? Amphora of brown clay with benthonic | 1 | | |
| 4a | Handmade body frs. from two-three vessels tempered with grog | 7 | | |
| 4b | Handmade base frs. | 2 | **390** | Joining |
| 5 | Animal bones | 1 | | |

## DSP07-H02-01

Finds table core and adjacent units

| BagID | Function | Total | Body | Rim | Handle | Base | Fabric | Surface | Shape | Period | Cat. |
|---|---|---|---|---|---|---|---|---|---|---|---|
| 307701 | IN | 0 | 0 | 0 | 0 | 0 | | | | BA | |
| 308001 | HW | 5 | 5 | 0 | 0 | 0 | BaGrey | None | Pot | BA? | |
| 308201 | ? | 0 | 0 | 0 | 0 | 0 | | | | LC-EH | |
| 307701 | IN | 20 | 0 | 0 | 0 | 0 | | | | LC-EH | |
| 307801 | IN | 3 | 0 | 0 | 0 | 0 | | | | LC-EH | |
| 308201 | HW | 1 | 1 | 0 | 0 | 0 | Black | None | Pot | LC-EH? | |
| 308201 | IN | 3 | 3 | 0 | 0 | 0 | Red mica | None | ? | LC-EH? | |
| 308201 | IN | 1 | 1 | 0 | 0 | 0 | Red | Slip | Closed shape | LC-EH? | |
| 308201 | IN | 20 | 20 | 0 | 0 | 0 | Red | None | ? | LC-EH? | |
| 308201 | IN | 1 | 1 | 0 | 0 | 0 | Orange | None | Amphora | LC-EH? | |
| 308201 | TR | 13 | 12 | 1 | 0 | 0 | Herakleian | None | Amphora | LC-EH? | |
| 308201 | TR | 1 | 0 | 0 | 1 | 0 | Red | None | Amphora | LC-EH? | |
| 308201 | TR | 3 | 3 | 0 | 0 | 0 | Red coarse | None | Amphora | LC-EH? | |
| 308201 | TR | 2 | 2 | 0 | 0 | 0 | Brown | Wash | Amphora | LC-EH? | |
| 307901 | IN | 1 | 0 | 0 | 0 | 0 | | | | Medieval? | |

## DSP07-H04-01

Finds table core units

| BagID | Function | Total | Body | Rim | Handle | Base | Fabric | Surface | Shape | Period | Cat. |
|---|---|---|---|---|---|---|---|---|---|---|---|
| 306801 | IN | 1 | 0 | 0 | 0 | 0 | | | | LC-EH | |
| 306901 | IN | 1 | 0 | 0 | 0 | 0 | | | | LC-EH | |
| 307001 | IN | 1 | 0 | 0 | 0 | 0 | | | | LC-EH | |
| 307101 | IN | 1 | 0 | 0 | 0 | 0 | | | | LC-EH | |
| 307201 | IN | 1 | 0 | 0 | 0 | 0 | | | | LC-EH | |

## DSP07-H04-02

Finds table core units

| BagID | Function | Total | Body | Rim | Handle | Base | Fabric | Surface | Shape | Period | Cat. |
|---|---|---|---|---|---|---|---|---|---|---|---|
| 307501 | IN | 1 | 0 | 0 | 1 | 0 | | | | EM? | **401** |

## DSP07-H04-03

Finds table core units

| BagID | Function | Total | Body | Rim | Handle | Base | Fabric | Surface | Shape | Period | Cat. |
|---|---|---|---|---|---|---|---|---|---|---|---|
| 307401 | IN | 1 | 0 | 0 | 0 | 0 | | | | | |

## DSP07-H06-01

Finds table core units (2007 FL8, 3085, 20002/01-02, 20007/02-03, 3201-03)

| BagID | Function | Total | Body | Rim | Handle | Base | Fabric | Surface | Shape | Period | Cat. |
|---|---|---|---|---|---|---|---|---|---|---|---|
| 308501 | HW | 1 | 1 | 0 | 0 | 0 | Grey | polished | Closed shape? | LC-EH | |
| 320101 | HW | 1 | 1 | 0 | 0 | 0 | Grey | None | Pot | LC-EH | |
| 2000201 | HW | 1 | 1 | 0 | 0 | 0 | Grey | None | Pot | LC-EH | |
| 320201 | HW | 1 | 1 | 0 | 0 | 0 | Black | Polished | Closed shape? | LC-EH? | |
| 320301 | HW | 2 | 2 | 0 | 0 | 0 | Grey | None | Pot | LC-EH? | |
| 2000202 | IN | 1 | 1 | 0 | 0 | 0 | Red | None | ? | LC-EH? | |
| 2000703 | IN | 2 | 2 | 0 | 0 | 0 | Red | None | ? | Medieval? | |
| 308501 | BO | 2 | 2 | 0 | 0 | 0 | | | | | |
| 2000702 | LI | 1 | 1 | 0 | 0 | 0 | | | | | |
| 2007 FL8 | HW | 6 | 6 | 0 | 0 | 0 | | poorly expressed horizontal layer | Pot | BA | |

## DSP07-H07-01

Finds table core units

| BagID | Function | Total | Body | Rim | Handle | Base | Fabric | Surface | Shape | Period | Cat. |
|---|---|---|---|---|---|---|---|---|---|---|---|
| 306201 | IN | 2 | 0 | 0 | 0 | 0 | | | | LC-EH | |
| 306301 | IN | 3 | 0 | 0 | 0 | 0 | | | | LC-EH | |
| 2007 FL9 | IN | 2 | 2 | 0 | 0 | 0 | Red | | Jug | ? | |
| 2007 FL9 | IN | 2 | 2 | 0 | 0 | 0 | Reddish | | Jug | + | |

## DSP07-H07-02

Finds table core units

| BagID | Function | Total | Body | Rim | Handle | Base | Fabric | Surface | Shape | Period | Cat. |
|---|---|---|---|---|---|---|---|---|---|---|---|
| 2007 FL10 | TR | 5 | 5 | 0 | 0 | 0 | Red | | Amphora? | ? | |
| 2007 FL10 | TR | 1 | 1 | 0 | 0 | 0 | Light red | | Amphora? | ? | |
| 2007 FL10 | TR | 1 | 1 | 0 | 0 | 0 | Light | | Amphora? | ? | |
| 2007 FL10 | TW | 1 | 1 | 0 | 0 | 0 | Red | | Jug | ? | |

## DSP08-H07-03

Finds table core units

| BagID | Function | Total | Body | Rim | Handle | Base | Fabric | Surface | Shape | Period | Cat. |
|---|---|---|---|---|---|---|---|---|---|---|---|
| 2006503 | IN | 1 | 1 | 0 | 0 | 0 | Red | None | ? | LC-EH? | |
| 322101 | IN | 1 | 0 | 1 | 0 | 0 | Tatarian | Combing | Jug or flask? | EM | **402** |
| 322101 | UT | 1 | 1 | 0 | 0 | 0 | Tatarian | Interior clear glaze | ? | EM | |
| 322101 | UT | 1 | 1 | 0 | 0 | 0 | Tatarian | Interior green glaze | ? | EM | |
| 322101 | IN | 1 | 1 | 0 | 0 | 0 | Tatarian | watery selfslip | ? | EM? | |
| 322101 | HW | 1 | 1 | 0 | 0 | 0 | | | ? | ? | |
| 322101 | IN | 1 | 1 | 0 | 0 | 0 | | | ? | ? | |
| 2006401 | IN | 1 | 1 | 0 | 0 | 0 | Red | None | ? | ? | |
| 2006501 | IN | 1 | 1 | 0 | 0 | 0 | Red | None | ? | ? | |
| 2006502 | IN | 1 | 1 | 0 | 0 | 0 | Red | None | | ? | |
| 2006504 | IN | 1 | 1 | 0 | 0 | 0 | Red | None | ? | ? | |

## DSP07-H08-01

Finds table core units

| BagID | Function | Total | Body | Rim | Handle | Base | Fabric | Surface | Shape | Period | Cat. |
|---|---|---|---|---|---|---|---|---|---|---|---|
| 2001302 | HW | 1 | 1 | 0 | 0 | 0 | BaBlack | Grooved | Pot | BA | |
| 2001302 | HW | 1 | 1 | 0 | 0 | 0 | BaGrey | Grooved | Pot | BA | |
| 2001402 | HW | 1 | 1 | 0 | 0 | 0 | BaGrey | Grooved | Pot | BA | |
| 309501 | HW | 4 | 4 | 0 | 0 | 0 | | Incised | Pot | BA | |
| 2001301 | HW | 1 | 1 | 0 | 0 | 0 | Grey | None | Pot | LC-EH | |
| 2001501 | HW | 1 | 1 | 0 | 0 | 0 | Grey | None | Pot | LC-EH | |
| 2001503 | HW | 1 | 1 | 0 | 0 | 0 | Grey | None | Pot | LC-EH | |
| 2001504 | HW | 1 | 1 | 0 | 0 | 0 | Red | none | Pot | LC-EH | |
| 2001506 | HW | 3 | 3 | 0 | 0 | 0 | Grey | None | Pot | LC-EH | |
| 2001507 | HW | 3 | 3 | 0 | 0 | 0 | Grey | None | Pot | LC-EH | |
| 2001510 | HW | 1 | 1 | 0 | 0 | 0 | Red | Smoothed | Pot | LC-EH | |
| 2001601 | HW | 1 | 1 | 0 | 0 | 0 | Grey | None | Pot | LC-EH | |
| 2001602 | HW | 1 | 1 | 0 | 0 | 0 | Grey | None | Pot | LC-EH | |
| 2001604 | HW | 1 | 1 | 0 | 0 | 0 | Red | None | Pot | LC-EH | |
| 2001502 | HW | 1 | 1 | 0 | 0 | 0 | Red | None | Pot | LC-EH? | |
| 2001603 | IN | 1 | 1 | 0 | 0 | 0 | Tatarian | None | Closed shape? | EM? | |
| 2001401 | BO | 1 | 1 | 0 | 0 | 0 | | | | | |
| 2001509 | IN | 1 | 1 | 0 | 0 | 0 | Black | None | ? | ? | |
| 2001508 | LI | 1 | 0 | 0 | 0 | 0 | | | | | |
| 2007 FL11 | HW | 15 | 13 | 2 | 0 | 0 | | | | | |
| 2007 FL12 | HW | 13 | 13 | 0 | 0 | 0 | | | | BA | |

**Feature 2**. Excavations finds

| Bag 18: Top soil: 0-0.2 m | | | | |
|---|---|---|---|---|
| No. | Description | Quantity | Cat. | Notes |
| # | Animal bones, smaller frs. | App 10 | | |
| # | Snail | 1 | | |
| Bag 19: Level: 0.2-0.3 m | | | | |
| No. | Description | Quantity | Cat. | |
| 1 | Handmade body fr. with two rows of slanting dots stamped with a hollow pointy instrument | 1 | **404** | |
| 2 | Handmade body fr. decorated with two parallel lines of slanting grooves, primitive herring-bone pattern | 1 | **405** | |
| 3 | Thick-walled Handmade body fr. with grooved/shaved interior | 1 | **406** | |
| 4 | Handmade body frs., coarser exterior surface | 3 | | |
| 5 | Handmade body fr., finer exterior surface | 1 | | |
| # | Animal bones, smaller frs. dominate, several burnt | App 30 | | |

| # | Snails | 13 | | |
|---|---|---|---|---|
| Bag 20: Extension on the SE side, level 0.2-0.3 m | | | | |
| No. | Description | Quantity | Cat. | |
| 1 | Handmade body fr. with decoration stamped with instrument with rounded tip | 1 | **407** | |
| 2 | Handmade body frs. | 2 | | |
| # | Animal bones | App 5 | | |
| # | Snail | 4 | | |
| Bag 21: Level 0.35-0.45 m | | | | |
| No. | Description | Quantity | Cat. | |
| # | Animal bones, some burnt | App 40 | | |
| # | Snail | 1 | | |
| Bag 22: Level 0.45-0.5 m (1) | | | | |
| No. | Description | Quantity | Cat. | |
| # | Animal bones, some burnt | App 40 | | |
| Bag 23: Level 0.4 m | | | | |
| No. | Description | Quantity | Cat. | |
| # | Animal bones | 2 | | |
| Bag 24: Level 0.45-0.5 m (2) | | | | |
| No. | Description | Quantity | Cat. | |
| 1 | Handmade body fr. with row of triangular stamped decoration | 1 | **408** | |
| 2 | Undecorated Handmade body frs. from same vessel | 3 | | |
| 3 | Assorted undecorated Handmade body frs. | 4 | | |
| # | Animal bones, some burnt | App 15 | | |
| Bag 25: Level 0.45-0.6 m | | | | |
| No. | Description | Quantity | Cat. | |
| # | Animal bones | App 10 | | |
| Bag 26: Level 0.6-0.65 m | | | | |
| No. | Description | Quantity | Cat. | |
| 1 | Tiny Handmade body fr. | 1 | | |
| # | Animal bones | App 5 | | |
| Bag 32: Level 0.3-0.4 m | | | | |
| No. | Description | Quantity | Cat. | |
| 1 | Handmade body fr. with grooved/shaved interior and exterior decoration in the form of row of vertical grooves framed with horizontal groove | 1 | **409** | |
| # | Animal bones | App 7 | | |

**Feature** 7. Excavations finds

| Bag 27: Top soil 0-0.25 m | | | | |
|---|---|---|---|---|
| No. | Description | Quantity | Cat. | Notes |
| 1 | Handmade rim, open vessel | 1 | **410** | |
| 2 | Handmade rim, closed vessel | 1 | **411** | |

| 3 | Handmade base, very thick | 1 | **412** | |
|---|---|---|---|---|
| 4 | Handmade base | 1 | **413** | |
| 5 | Handmade body fr. with repair hole | 1 | **414** | |
| 6 | Assorted undecorated handmade body frs. | 13 | | |
| # | Animal bones | App 20 | | |
| Bag 28: Level 0.2 m, stone debris | | | | |
| No. | Description | Quantity | Cat. | Notes |
| 1 | Handmade base, joining | 2 | **415** | |
| 2 | Handmade rim | 1 | **416** | |
| 3 | Handmade body fr. with row of slanting stamped lines | 1 | **417** | |
| 4 | Assorted undecorated handmade body frs. | 7 | | |
| # | Animal bones | App 10 | | |
| Bag 29: Level 0.25 m (1) | | | | |
| No. | Description | Quantity | Cat. | Notes |
| 1 | Handmade base | 1 | **418** | |
| 2 | Handmade rim, closed vessel, joining | 3 | **419** | |
| 3 | Handmade rim, large closed vessel | 1 | **420** | |
| 4 | Handmade rim, open vessel | 2 | **421** | |
| 5 | Handmade body fr. possible gossam | 1 | **422** | |
| 6 | Assorted undecorated handmade body frs. | 18 | | |
| # | Animal bones | App 20 | | |
| Bag 30: Level 0.25 m (2) | | | | |
| No. | Description | Quantity | Cat. | Notes |
| 1 | Assorted undecorated handmade body frs. | 8 | | |
| # | Animal bones | App 30 | | |
| Bag 31: Floor level, 0.3 m | | | | |
| No. | Description | Quantity | Cat. | Notes |
| 1 | Handmade rim, closed vessel | 1 | **423** | |
| 2 | Handmade rim, closed vessel | 1 | **424** | |
| 3 | Handmade rim, open vessel with stamped decoration on the rim | 1 | **427** | |
| 4 | Handmade rim, open vessel | 1 | **425** | |
| 5 | Handmade rim, open vessel | 1 | **426** | |
| 6 | Handmade rim, open vessel | 1 | **428** | |
| 7 | Handmade rim, closed vessel, thin-walled | 1 | **429** | |
| 8 | Handmade rim, open vessel, possibly lid | 1 | **430** | |
| 9 | Handmade body fr. with burnished interior | 1 | | |
| 10 | Fr. of stone axe | 1 | **431** | |
| 11 | Micro-lith | 1 | **432** | |
| 12 | Assorted undecorated handmade body frs. | 26 | | |
| # | Animal bones | App 30 | | |
| # | Snail | 1 | | |

## DSP07-H08-02

Finds table core units

| BagID | Function | Total | Body | Rim | Handle | Base | Fabric | Surface | Shape | Period | Cat. |
|---|---|---|---|---|---|---|---|---|---|---|---|
| 2001802 | UT | 1 | 0 | 0 | 0 | 1 | Tatarian | None | ? | EM | **403** |
| 2001801 | IN | 1 | 1 | 0 | 0 | 0 | Tatarian? | None | Closed shape | EM? | |

## DSP07-H09-01

Finds table core units

| BagID | Function | Total | Body | Rim | Handle | Base | Fabric | Surface | Shape | Period | Cat. |
|---|---|---|---|---|---|---|---|---|---|---|---|
| 308901 | HW | 2 | 2 | 0 | 0 | 0 | Grey | Grooved | Pot | LC-EH? | |
| 308901 | IN | 9 | 9 | 0 | 0 | 0 | Red | None | Closed shape | Medieval? | |
| 308801 | UT | 1 | 1 | 0 | 0 | 0 | Tatarian 2 | Slip | ? | EM | |
| 308801 | BO | 2 | 2 | 0 | 0 | 0 | | | | | |
| 308801 | IN | 2 | 2 | 0 | 0 | 0 | Red | None | ? | ? | |
| 309001 | IN | 1 | 1 | 0 | 0 | 0 | Red | None | ? | ? | |
| 2007 FL14 | HW | 1 | 1 | 0 | 0 | 0 | | | | BA | |

## DSP07-H10-01 AND DSP07-H18-01

Finds table core units

| BagID | Function | Total | Body | Rim | Handle | Base | Fabric | Surface | Shape | Period | Cat. |
|---|---|---|---|---|---|---|---|---|---|---|---|
| 308601 | HW | 1 | 1 | 0 | 0 | 0 | Red | None | Pot | LC-EH | |
| 2007 FL13 | HW | 1 | 0 | 1 | 0 | 0 | Lime | | Pot | LC-EH | **444** |
| 2007 FL13 | HW | 4 | 4 | 0 | 0 | 0 | Red grog | | | LC-EH | |
| 2007 FL13 | TR | 6 | 6 | 0 | 0 | 0 | Herakleian | | Amphora | LC-EH | |
| 2007 FL13 | TR | 1 | 1 | 0 | 0 | 0 | Sinopean | | Amphora | LC-EH | |
| 2007 FL13 | TR | 1 | 1 | 0 | 0 | 0 | Thasian | | Amphora | LC-EH | |
| 2007 FL13 | TR | 1 | 1 | 0 | 0 | 0 | Chersonesean? | | Amphora | LC-EH | |
| 308401 | TR | 1 | 0 | 0 | 1 | 0 | Light | None | Amphora | LC-EH | **435** |
| 308601 | TR | 1 | 1 | 0 | 0 | 0 | Grey mica | None | Amphora | LC-EH | |
| 2007 FL13 | TR | 4 | 4 | 0 | 0 | 0 | | | Amphora | LC-EH | |
| 2007 FL13 | TW | 1 | 1 | 0 | 0 | 0 | Red | | Jug | LC-EH | |
| 308401 | HW | 1 | 1 | 0 | 0 | 0 | Grey | Burnished | Closed shape | BA? | **433** |
| 308701 | HW | 1 | 0 | 1 | 0 | 0 | Brown | None | Closed shape | LC-EH? | |
| 308401 | TR | 1 | 0 | 0 | 1 | 0 | Herakleian? | None | Amphora | LC-EH? | **434** |
| 308401 | TW | 1 | 0 | 0 | 0 | 1 | Brown coarse | None | Jug | EM | **436** |
| 308601 | LI | 1 | 1 | 0 | 0 | 0 | | | | | |

**Feature 2**. Excavation finds

| Find list 12: Layer 2: stone debris W of the wall, depth 0.2-0.4 m | | | | |
|---|---|---|---|---|
| No. | Description | Number | Cat.no. | Notes |
| 1 | Frs. from toe, lower handle attachment and body of Chersonesean amphora. Clay reddish-brown with small inclusions of lime. Without engobe | 5 | **447-448** | One vessel |
| 2 | Large body frs. of Chersonesean amphora. Clay light-brown, with small and abundant inclusions of lime. Typical Chersonesean. Light-brown engobe (lighter than in the break of sherd) | 4 | | One vessel |
| 3 | Shells of *Helix* | 4 | | |

**Feature 5.** Excavation finds

| Find list 1: Top-soil layer, 0-0.2 m | | | | |
|---|---|---|---|---|
| No. | Description | Quantity | Cat. | Notes |
| 1 | Body fr. of Sinopean amphora. | 1 | | Broken in two parts |
| 2 | Body fr. of amphora with traces of repair hole. Clay gritty, with small inclusions of quartz, lime and brown particles | 1 | | |
| 3 | Small body frs. of red-clay jugs | 2 | | |
| 4 | Fr. of handmade(?) earthenware of unclear shape. Clay well fired, terracotta-coloured | 1 | | |
| 5 | Small body frs. of handmade vessels | 2 | | |
| 6 | Frs. of animal bones | 4 | | |
| 7 | Small bone frs. | 2 | | |
| Find list 2: Below top-soil layer, 0.2-0.4 m | | | | |
| № | Description | Quantity | Cat. | Notes |
| 1 | Rim fr. of Chersonesean amphora, rounded on top. Engobe yellowish-green | 1 | 445 | |
| 2 | Body fr. of Chersonesean amphora | 1 | | Same vessel as no. 1(?) |
| 3 | Body frs. of Chersonesean amphorai | 3 | | |
| 4 | Small handle fr. of Chersonesean amphora (split) | 1 | | |
| 5 | Body fr. of amphora. Fabric similar to FL 1/2. With quartz | | | Same vessel as FL 1/2 (?) |
| 6 | Body frs. of red-clay amphora. Clay reddish-brown with inclusions of pyroxene | 2 | | |
| 7 | Small body frs. of amphora. Clay brown with inclusions of golden mica | 1 | | |
| 8 | Body fr. of amphora | 1 | | |
| 9 | Fr. from base with part of lower body of handmade pot | 1 | | |
| 10 | Body frs. of handmade vessels | 3 | | |
| 11 | Fr. of whetstone | 1 | | |
| 12 | Frs. of animal bones | 5 | | |
| 13 | Shells of *Helix* | 8 | | |
| Find list 3: Extension of the trench at its E side (above Pit 1). Humus layer, 0-0.2 m | | | | |
| № | Description | Quantity | Cat. | Notes |
| 1 | Body fr. of red-clay jug | 1 | | |
| Find list 4: Extension of the trench at its E side (above Pit 1). Below top-soil layer, 0.2-0.4 m | | | | |
| № | Description | Quantity | Cat. | Notes |
| 1 | Lower-body fr. of Chersonesean amphora | 1 | | |
| 2 | Body fr. of jug. Fabric similar to FL 3/1 | 1 | | Same vessel as FL 3/1 (?) |
| 3 | Body fr. of thin-walled jug. Clay dark-brown | 1 | | |
| 4 | Rim fr. of handmade pot, rounded on top. Oblique cuts on top, made by knife back or stick | 1 | 438 | |
| 5 | Body frs. of handmade vessel | 2 | | Two different vessels; one is thick-walled |
| 6 | Animal tooth | 1 | | |
| 7 | Fr. of animal bone | 1 | | |
| 8 | Shell of small snail | 1 | | |
| 9 | Shell of *Helix* | 1 | | |

| Find list 5: Pit 1, depth 0.4-0.6 m | | | | |
|---|---|---|---|---|
| № | Description | Quantity | Cat. | Notes |
| 1 | Rim and body frs. of Chersonesean amphora | 2 | **446** | Same vessel as FL 2/1, 2/2 |
| 2 | Body frs. of Chersonesean amphorai. Clay with inclusions of pyroxene | 2 | | |
| 3 | Body frs. of Sinopean amphorai | 2 | | |
| 4 | Body frs. of amphorai | 3 | | |
| 5 | Body frs. of handmade vessels | 4 | | |
| 5a | Fr. of burned mud brick | 1 | | |
| 6 | Animal bones | 16 | | |
| Find list 6: Pit 1, depth 0.6-0.97 m | | | | |
| № | Description | Quantity | Cat. | Notes |
| 1 | Rim fr. of handmade pot, outer edge ornamented with finger impressions | 1 | **439** | |
| 2 | Body frs. of handmade vessels | 16 | | |
| 3 | Animal bones | 28 | | |
| 4 | Grains of cereals | 28 | | |
| Find list 7: Control dig at the S edge of trench 1, depth 0.6-0.7 m | | | | |
| № | Description | Quantity | Cat. | Notes |
| 1 | Body fr. of Chersonesean (?) amphora | 1 | | |
| Find list 8: Extension of the trench at its N side. Top-soil layer, depth 0-0.2 m | | | | |
| № | Description | Quantity | Cat. | Notes |
| 1 | Body fr. of amphora. Clay reddish-brown with inclusions of lime, quartz and feldspar. Herakleian (?) | 1 | | |
| 2 | Body frs. of handmade vessels. Clay greyish-black with inclusions of lime and grog (little) | 2 | | |
| Find list 9: Pit 2. Extension of the trench at its N side (above Pit 1). Below top-soil layer, depth 0.2-0.6 m (= Upper fill) | | | | |
| № | Description | Quantity | Cat. | Notes |
| 1 | Body fr. of Chersonesean amphora | 1 | | 50 x 45 mm<br>Depth: 0.2-0.4 m |
| 2 | Body fr. of amphora. Clay brown. Sparse inclusions of lime. Chersonesos | 1 | | Depth: 0.2-0.4 m |
| 3 | Small body fr. of amphorai. Clay reddish-brown | 1 | | Depth: 0.2-0.4 m |
| 4 | Splits from body of Chersonesean (?) amphora | 2 | | Depth: 0.2-0.4 m |
| 5 | Base fr. with ring-foot of black-glazed bowl or plate. Glaze black and dense beneath, brown on upper surface. Shallow glazed groove on resting surface | 1 | **449** | Depth: 0.2-0.4 m Another fr. of this vessel was founded in the lower fill of the pit. Drawing |
| 6 | Small body fr. of handmade vessel. Clay light-brown on the outer surface, grey on interior | 1 | | Depth: 0.2-0.4 m |
| 7 | Rim fr. of handmade thick-walled bowl. Clay two-layered in cross section, light-brown to reddish-brown on outer surface, dark-grey on interior. With inclusions of limestone | 1 | | Depth: 0.4-0.6 m |
| 8 | Body fr. of handmade pot | 1 | | From handmade pot with applied decoration found in lower fill (joining).<br>Depth: 0.4-0.6 m |
| Find list 10: Pit 2. Fill, 0.6-1.15 m | | | | |
| № | Description | Quantity | Cat. | Notes |

| | | | | |
|---|---|---|---|---|
| 1 | Body fr. of Chersonesean amphora. Typical Chersonesean clay with inclusions of lime | 1 | | Depth: 0.75-0.85 m |
| 2 | Body fr. of Chersonesean amphora. Clay brown from secondary exposure to fire. Traces of burning at the edge of sherd | 1 | | Depth: 0.75-0.85 m |
| 3 | Splits from body of amphora | 1 | | Depth: 0.9 m |
| 3a | Fr. from floor of black-glazed bowl or plate. Glaze brown to black | 1 | 449 | Depth: 1.05-1.07 m<br>Same vessel as FL 9/5 |
| 4 | Rim fr. of handmade bowl, rounded on top. Clay porous due to burned organic inclusions. Light-brown on the surface, grey on interior | 1 | 437 | Depth: 1.07 m |
| 5 | Body frs. of handmade pot | 5 | | Depth: 1.05-1.07; 0.9 m<br>From handmade pot with applied ornament found in lower fill |
| 6 | Body fr. of thick-walled handmade pot | 1 | | Depth: 0.6-0.75 m |
| 7 | Small body frs. of handmade vessels | 3 | | Depth: 0.6-0.75 m |
| 8 | Fragmentary whetstone | 1 | 450 | |
| 9 | Animal bones | 11 | | Depth: 0.6-0.75 m;<br>0.75-0.85 m; 1.05-1.07 m |
| Find list 11: Pit 2. Fill below stone blocking, 115-136 cm | | | | |
| № | Description | Quantity | Cat. | Notes |
| 1 | Neck frs. of Chersonesean amphora | 2 | | Depth: 1.25-1.3 m |
| 2 | Body frs. of Chersonesean amphorai. One with traces of burning | 2 | | Depth: 1.25-1.3 m |
| 3 | Body fr. of amphora. Fabric gritty, similar to Herakleian. Grey. Centre? | 1 | | Depth: 1.15-1.17 m |
| 4 | Body fr. of amphora. Clay brown, gritty with sparkles of quartz. Traces of burning on the outer surface and in break | 1 | | Depth: 1.25-1.3 m |
| 5 | Fragmentary handmade pot. Whole profile. Preservation about 60-70%. Rounded, gently out-turned rim. Slightly raised base. Ornamented on rim by finger-nail impressions. On shoulder, four cone-like attachments | 1 | 442 | Depth: 1.25-1.3 m<br>48 frs.,<br>36 of them joining frs.<br>Frs. of same pot- FL 9/8 and 10/5 |
| 6 | Base fr. with part of body of thick-walled handmade vessel (pot?) | 1 | 443 | Depth: 1.25-1.3 m |
| 7 | Body frs. of handmade pot (?) | 13 | | Depth: 1.25-1.3; 1.35 m |
| 8 | Small body frs. of handmade vessels | 2 | | Depth: 1.25-1.3 m<br>Same vessel as FL 11/5 (?) |
| 9 | Animal bones and bones frs. | 26 | | On the whole from depth 1.25-1.36 m |
| 10 | Shells of *Helix* | 2 | | Depth: 1.15-1.25 m;<br>1.35-1.36 m |
| 11 | Grains of cereals | | | From lower fill |

## SP07-H11-01

Finds table core units

| BagID | Function | Total | Body | Rim | Handle | Base | Fabric | Surface | Shape | Period | Cat. |
|---|---|---|---|---|---|---|---|---|---|---|---|
| 308301 | HW | 3 | 3 | 0 | 0 | 0 | BaBlack | Grooved | Pot | Late BA | 451-453 |
| 308302 | HW | 2 | 1 | 1 | 0 | 0 | BaBlack | Grooved | Pot | Late BA | 454 |
| 308302 | TR | 1 | 1 | 0 | 0 | 0 | Chian? | None | Amphora | Classical? | 468 |
| 308301 | HW | 2 | 2 | 0 | 0 | 0 | HwCoarse | none | Pot | LC-EH | |
| 308301 | HW | 1 | 0 | 1 | 0 | 0 | HwRed | Decoration | Open shape | LC-EH | 489 |
| 309301 | HW | 4 | 4 | 0 | 0 | 0 | Grey | None | Pot | LC-EH | |
| 308301 | TR | 3 | 0 | 0 | 3 | 0 | Herakleian | None | Amphora | LC-EH | 460-461, 476 |
| 308301 | TR | 3 | 0 | 0 | 3 | 0 | Red | None | Amphora | LC-EH | 473, 475, 478 |
| 308301 | TR | 1 | 0 | 0 | 0 | 1 | Herakleian | None | Amphora | LC-EH | 459, 462 |
| 308301 | TR | 2 | 0 | 2 | 0 | 0 | Herakleian | None | Amphora | LC-EH | 458, 474 |
| 308301 | TR | 1 | 0 | 1 | 0 | 0 | Light brown | None | Amphora | LC-EH | 477 |
| 308301 | TR | 2 | 0 | 1 | 0 | 1 | Chersonesean? | None | Amphora | LC-EH | 456 |
| 308301 | TR | 3 | 0 | 0 | 1 | 2 | Sinopean | Slip | Amphora | LC-EH | |
| 308302 | TR | 3 | 3 | 0 | 0 | 0 | Herakleian? | None | Amphora | LC-EH | |
| 308301 | TR | 4 | 1 | 0 | 2 | 2 | Red | Slip | Amphora | LC-EH | 469, 470-472 |
| 308302 | TR | 1 | 0 | 0 | 1 | 0 | Herakleian? | slip? | Amphora | LC-EH | 480 |
| 308302 | TR | 9 | 6 | 0 | 3 | 0 | Herakleian | None | Amphora | LC-EH | 463-465 |
| 308302 | TR | 2 | 0 | 0 | 2 | 0 | Red | None | Amphora | LC-EH | 483-484 |
| 308302 | TR | 2 | 2 | 0 | 0 | 0 | Sinopean | Smoothed | Amphora | LC-EH | |
| 308301 | TW | 1 | 0 | 0 | 0 | 1 | Chersonesean | Slip | Jug | LC-EH | 485 |
| 308302 | TW | 2 | 2 | 0 | 0 | 0 | Grey ware | Glazed | Jug | LC-EH | 486 |
| 309301 | TW | 1 | 1 | 0 | 0 | 0 | Red | None | Jug? | LC-EH | 487 |
| 308302 | UT | 1 | 1 | 0 | 0 | 0 | Sinopean | None | Louterion | LC-EH | 488 |
| 308301 | AR | 1 | 0 | 1 | 0 | 0 | Sinopean | Slip | Tile | LC-EH? | 455 |
| 308302 | HW | 9 | 7 | 0 | 0 | 2 | HwGrey | None | Pot | LC-EH? | 490-491 |
| 308302 | IN | 1 | 1 | 0 | 0 | 0 | Thasian? | None | ? | LC-EH? | 467 |
| 308302 | IN | 1 | 1 | 0 | 0 | 0 | Brown | none | Closed shape | LC-EH? | |
| 308302 | IN | 1 | 1 | 0 | 0 | 0 | Brown | Slip | Closed shape | LC-EH? | |
| 308302 | IN | 1 | 1 | 0 | 0 | 0 | Brown gold mica | None | Closed shape | LC-EH? | |
| 308302 | IN | 2 | 2 | 0 | 0 | 0 | Red | Smoothed | Amphora? | LC-EH? | |
| 308302 | IN | 3 | 3 | 0 | 0 | 0 | Red | Smoothed | Closed shape | LC-EH? | |
| 309301 | IN | 5 | 5 | 0 | 0 | 0 | Red | None | ? | LC-EH? | |
| 309301 | IN | 3 | 3 | 0 | 0 | 0 | Red | none | ? | LC-EH? | |
| 308301 | TR | 1 | 0 | 0 | 1 | 0 | Chersonesean? | Slip | Amphora | LC-EH? | 457 |
| 308301 | TR | 2 | 0 | 0 | 2 | 0 | Red | None | Amphora | LC-EH? | 479 |
| 308301 | TR | 1 | 0 | 0 | 0 | 1 | Thasian | None | Amphora | LC-EH? | 466 |
| 308302 | TR | 1 | 0 | 0 | 1 | 0 | Red coarse mica | none | Amphora | LC-EH? | 481 |

| BagID | Function | Total | Body | Rim | Handle | Base | Fabric | Surface | Shape | Period | Cat. |
|---|---|---|---|---|---|---|---|---|---|---|---|
| 308302 | TR | 2 | 2 | 0 | 0 | 0 | Red lime | Gloss | Amphora? | LC-EH? | |
| 308302 | TR | 2 | 2 | 0 | 0 | 0 | Red | Smoothed | Amphora? | LC-EH? | |
| 308302 | TR | 1 | 0 | 0 | 1 | 0 | Red | slip | Amphora | LC-EH? | **482** |
| 308302 | TR | 1 | 1 | 0 | 0 | 0 | Light | none | Amphora? | LC-EH? | |
| 309301 | TR | 1 | 1 | 0 | 0 | 0 | Red | None | Amphora | LC-EH? | |
| 309301 | UT | 1 | 1 | 0 | 0 | 0 | Red | None | Basin? | LC-EH? | |
| 308302 | TW | 4 | 3 | 1 | 0 | 0 | China | Decorated | Plate | Modern | |
| 308302 | BO | 1 | 0 | 0 | 0 | 0 | | | | | |
| 309301 | BO | 1 | 1 | 0 | 0 | 0 | | | | | |
| 309301 | IR | 3 | 3 | 0 | 0 | 0 | | | | | |
| 308301 | LI | 4 | 0 | 0 | 0 | 0 | | | | | |
| 308301 | LI | 1 | 0 | 0 | 0 | 0 | | | | | |
| 308302 | LI | 1 | 0 | 0 | 0 | 0 | | | | | |
| 309301 | LI | 1 | 1 | 0 | 0 | 0 | | | | | |

## DSP07-H11-02

Finds table core units

| BagID | Function | Total | Body | Rim | Handle | Base | Fabric | Surface | Shape | Period | Cat. |
|---|---|---|---|---|---|---|---|---|---|---|---|
| 309401 | HW | 1 | 1 | 0 | 0 | 0 | Light | None | Pot | LC-EH | |
| 321801 | HW | 1 | 1 | 0 | 0 | 0 | Red | Polished | Pot | LC-EH | |
| 309401 | IN | 1 | 1 | 0 | 0 | 0 | Red mica | None | Closed shape | LC-EH | |
| 309401 | HW | 1 | 1 | 0 | 0 | 0 | Coarse | Relief | ? | BA? | **492** |
| 309401 | HW | 1 | 1 | 0 | 0 | 0 | Grey | Grooved | Pot | LC-EH? | |
| 2005009 | HW | 1 | 1 | 0 | 0 | 0 | Grey | None | Pot | LC-EH? | |
| 2005201 | HW | 1 | 1 | 0 | 0 | 0 | Grey | None | Pot | LC-EH? | |
| 2005402 | HW | 1 | 1 | 0 | 0 | 0 | Grey | None | Pot | LC-EH? | |
| 2005404 | HW | 1 | 1 | 0 | 0 | 0 | Red | None | Pot | LC-EH? | |
| 309401 | IN | 2 | 2 | 0 | 0 | 0 | Red | None | ? | LC-EH? | |
| 321801 | IN | 1 | 1 | 0 | 0 | 0 | Red | None | Closed shape | LC-EH? | |
| 2005313 | IN | 1 | 1 | 0 | 0 | 0 | Red | None | ? | LC-EH? | |
| 309401 | TR | 1 | 0 | 0 | 0 | 1 | Sinopean? | None | Amphora | LC-EH? | **495** |
| 309401 | TR | 1 | 0 | 0 | 1 | 0 | Red | Slip | Amphora | 1st c. AD | **500** |
| 309401 | TR | 1 | 0 | 0 | 1 | 0 | Red | None | Amphora | LC-EH? | |
| 309401 | TR | 1 | 1 | 0 | 0 | 0 | Chian? | slip | Amphora | LC-EH? | |
| 309401 | TR | 1 | 0 | 0 | 1 | 0 | Chersonesean? | Slip | Amphora | LC-EH? | **501** |
| 309401 | TR | 7 | 5 | 0 | 2 | 0 | Herakleian? | None | Amphora | LC-EH? | **502-504** |
| 321801 | TR | 2 | 0 | 0 | 2 | 0 | Thasian? | None | Amphora | LC-EH? | **505** |
| 321801 | TR | 2 | 1 | 0 | 1 | 0 | Herakleian | None | Amphora | LC-EH? | **493** |
| 321801 | TR | 1 | 1 | 0 | 0 | 0 | Light | None | Amphora | LC-EH? | |
| 321801 | TR | 1 | 0 | 0 | 1 | 0 | Red | Slip | Amphora | Medieval? | **507** |

| BagID | Function | Total | Body | Rim | Handle | Base | Fabric | Surface | Shape | Period | Cat. |
|---|---|---|---|---|---|---|---|---|---|---|---|
| 321801 | TR | 1 | 1 | 0 | 0 | 0 | Red | None | Amphora | LC-EH? | |
| 2005012 | TR | 1 | 1 | 0 | 0 | 0 | Red | None | Amphora | LC-EH? | |
| 2005103 | TR | 2 | 2 | 0 | 0 | 0 | Red | None | Amphora? | LC-EH? | |
| 2005204 | TR | 1 | 1 | 0 | 0 | 0 | Red | Slip | Amphora | LC-EH? | |
| 2005304 | TR | 1 | 1 | 0 | 0 | 0 | Thasian | None | Amphora | LC-EH? | |
| 2005305 | TR | 1 | 1 | 0 | 0 | 0 | Herakleian | None | Amphora | LC-EH? | |
| 2005307 | TR | 1 | 0 | 0 | 1 | 0 | Sinopean | None | Amphora | LC-EH? | **494** |
| 2005309 | TR | 2 | 2 | 0 | 0 | 0 | Herakleian | None | Amphora | LC-EH? | |
| 2005311 | TR | 1 | 1 | 0 | 0 | 0 | Herakleian | None | Amphora | LC-EH? | |
| 2005405 | TR | 1 | 1 | 0 | 0 | 0 | Sinopean | None | Amphora | LC-EH? | |
| 2005406 | TR | 1 | 0 | 0 | 1 | 0 | Red | None | Amphora | LC-EH? | **498** |
| 2005407 | TR | 1 | 1 | 0 | 0 | 0 | Sinopean | None | Amphora | LC-EH? | |
| 2005501 | TR | 1 | 1 | 0 | 0 | 0 | Thasian | None | Amphora | LC-EH? | |
| 2005502 | TR | 1 | 1 | 0 | 0 | 0 | Red | None | Amphora | LC-EH? | |
| 321801 | UT | 1 | 0 | 1 | 0 | 0 | Sinopean | None | Louterion | LC-EH? | **506** |
| 2005005 | IN | 1 | 1 | 0 | 0 | 0 | Tatarian | None | | EM | |
| 2005206 | IN | 1 | 1 | 0 | 0 | 0 | Tatarian? | None | | EM | |
| 2005303 | IN | 8 | 8 | 0 | 0 | 0 | Tatarian | None | ? | EM | |
| 309401 | AR | 1 | 0 | 1 | 0 | 0 | Red | None | Cover tile | EM? | |
| 2005007 | IN | 2 | 2 | 0 | 0 | 0 | Tatarian? | None | | EM? | |
| 309401 | IN | 1 | 1 | 0 | 0 | 0 | | | | Modern | |
| 2005108 | IN | 1 | 1 | 0 | 0 | 0 | Red modern | None | | Modern | |
| 2005004 | IN | 1 | 1 | 0 | 0 | 0 | Red | None | | Modern? | |
| 2005110 | IN | 1 | 1 | 0 | 0 | 0 | Red | None | | Modern? | |
| 309401 | BO | 1 | 0 | 0 | 0 | 0 | | | | | |
| 2005106 | HW | 3 | 3 | 0 | 0 | 0 | Grey | None | Pot | | |
| 2005312 | HW | 1 | 1 | 0 | 0 | 0 | Grey | None | Pot | | |
| 309401 | IN | 1 | 1 | 0 | 0 | 0 | Red | Slip | Jug? | | |
| 2005001 | IN | 1 | 1 | 0 | 0 | 0 | Red | None | ? | | |
| 2005003 | IN | 2 | 2 | 0 | 0 | 0 | Grey | None | ? | | |
| 2005006 | IN | 1 | 1 | 0 | 0 | 0 | Red mica | None | | | |
| 2005006 | IN | 1 | 1 | 0 | 0 | 0 | Red | None | ? | | |
| 2005007 | IN | 1 | 1 | 0 | 0 | 0 | Red | None | | | |
| 2005008 | IN | 1 | 1 | 0 | 0 | 0 | Red | None | ? | | |
| 2005010 | IN | 1 | 1 | 0 | 0 | 0 | Red | None | | | |
| 2005012 | IN | 1 | 1 | 0 | 0 | 0 | Red | None | | | |
| 2005013 | IN | 1 | 1 | 0 | 0 | 0 | Red | None | ? | | |
| 2005014 | IN | 1 | 1 | 0 | 0 | 0 | Red | None | | | |
| 2005015 | IN | 2 | 2 | 0 | 0 | 0 | Red | None | ? | | |
| 2005016 | IN | 1 | 1 | 0 | 0 | 0 | Red | Slip | Closed shape | | |
| 2005017 | IN | 1 | 1 | 0 | 0 | 0 | Red | None | | | |
| 2005101 | IN | 1 | 1 | 0 | 0 | 0 | Red | None | ? | | |
| 2005104 | IN | 1 | 1 | 0 | 0 | 0 | Red | None | ? | | |

| BagID | Function | Total | Body | Rim | Handle | Base | Fabric | Surface | Shape | Period | Cat. |
|---|---|---|---|---|---|---|---|---|---|---|---|
| 2005107 | IN | 2 | 2 | 0 | 0 | 0 | Red | None | | | |
| 2005202 | IN | 1 | 1 | 0 | 0 | 0 | Red | None | ? | | |
| 2005203 | IN | 1 | 1 | 0 | 0 | 0 | Red | Slip | Closed shape | | |
| 2005205 | IN | 1 | 1 | 0 | 0 | 0 | Red | None | ? | | |
| 2005302 | IN | 1 | 1 | 0 | 0 | 0 | Red | None | ? | | |
| 2005303 | IN | 1 | 1 | 0 | 0 | 0 | Light | None | ? | | |
| 2005306 | IN | 1 | 1 | 0 | 0 | 0 | Red | None | ? | | |
| 2005308 | IN | 1 | 1 | 0 | 0 | 0 | Red | None | ? | | |
| 2005309 | IN | 1 | 1 | 0 | 0 | 0 | Red | None | ? | | |
| 2005310 | IN | 1 | 1 | 0 | 0 | 0 | Red | None | Closed shape | | |
| 2005312 | IN | 4 | 4 | 0 | 0 | 0 | Red | None | Closed shape | | |
| 2005312 | IN | 1 | 1 | 0 | 0 | 0 | Light | None | Closed shape | | |
| 2005401 | IN | 1 | 1 | 0 | 0 | 0 | Red | None | Closed shape | | |
| 2005403 | IN | 1 | 1 | 0 | 0 | 0 | Red | None | Amphora | | |
| 2005504 | IN | 1 | 1 | 0 | 0 | 0 | Red | None | ? | | |
| 2005102 | LI | 1 | 1 | 0 | 0 | 0 | | | | | |
| 2005105 | LI | 1 | 0 | 0 | 0 | 0 | | | | | |
| 2005201 | LI | 1 | 1 | 0 | 0 | 0 | | | | | |
| 2005002 | TR | 1 | 1 | 0 | 0 | 0 | Red | None | Amphora? | | |
| 2005011 | TR | 1 | 0 | 1 | 0 | 0 | Red | None | Amphora | | |
| 2005109 | TR | 1 | 0 | 0 | 1 | 0 | Red | None | Amphora | | 496 |
| 2005201 | TR | 2 | 2 | 0 | 0 | 0 | Herakleian | None | Amphora? | | |
| 2005301 | TR | 1 | 1 | 0 | 0 | 0 | Red | Slip | Amphora? | | |
| 2005303 | TR | 1 | 1 | 0 | 0 | 0 | Red | None | Amphora? | | 497 |
| 2005503 | TR | 1 | 0 | 0 | 1 | 0 | Red | None | Amphora | | 499 |

**Feature 9**. Excavation finds

| Layer 2: Grey ashy layer, 0.2-0.45 m. All frs. have strong lime incrustations | | | | |
|---|---|---|---|---|
| No. | Description | Quantity | Cat. | Note |
| 7 | Body frs. of Chersonesean amphorai | 5 | | |
| 8 | Body frs. of Sinopean amphorai/jugs | 2 | 522 | One with hole from repair |
| 9 | Body frs. of jug of pale brown sandwich clay with mica | 15 | | Samos? |
| 10 | Glossed Grey Ware rim of cup of poor quality | 1 | 525 | |
| 11a | Handmade rim | 2 | 509-510 | |
| 11b | Handmade rim with circular impressions | 1 | 511 | |
| 11c | Handmade flat bases | 2 | 512-513 | |
| 11d | Handmade body frs. | 17 | | |
| 12 | Rim of a handmade jug? | 1 | 508 | |
| 13 | Grape snails | 16 | | |
| 14 | Bones of animals | 22 | | |
| Layer 3: Loamy sand, 0.4-1.47 m. All frs. have strong lime incrustations | | | | |
| No. | Description | Quantity | Cat. | Note |

| | | | | |
|---|---|---|---|---|
| 15 | Body frs. of Sinopean amphorai | 19 | | From two amphoras |
| 16 | Body frs. of Thasian amphorai | 8 | | From two amphoras |
| 17 | Body frs. of Chersonesean amphorai | 7 | 527 | |
| 18 | Body fr. of Heraclean amphorai | 1 | 526 | |
| 19a | Handle of Chersonesean jug | 2 | 523 | |
| 19b | Body frs. of Chersonesean jugs | 3 | | |
| 20 | Body frs. of light brown jug | 4 | | |
| 21 | Large handmade rims of light brown clay with impressed finger decoration | 2 | 514 | Joining |
| 22 | Out-turned handmade rim of light brown clay with frequent impressed finger decoration | 1 | 519 | |
| 23 | Handmade rim of light brown clay with impressed finger decoration, black interior | 1 | 518 | |
| 24 | Handmade rims of light brown clay with impressed finger decoration, black exterior, light interior. Slashed decoration on the shoulder | 2 | 516-517 | |
| 25 | Rim fr. of handmade bowl of light brown clay, lightly polished | 1 | 515 | |
| 26 | Handmade body frs. | 23 | | |
| 27 | Frs. of Stone slabs, up to 12 × 10 × 1.7 cm | 2 | | Imported, could be from southeastern Crimea |
| 28 | Small fr. of kiln | 1 | | |
| 29 | Bones of animals | 42 | | |
| 30 | Polished handle fr. of Chersonesean amphora | 1 | 527 | |
| 31 | Body frs. of Chersonesean amphorai | 6 | | |
| 32a | Toe fr. of Heraclean amphorai | 1 | | |
| 32b | Body frs. of Heraclean amphorai | 2 | | |
| 33 | Body frs. of Sinopean amphorai | 6 | | From two amphorai |
| 34 | Body frs. of Thasian amphorai | 2 | | |
| 35 | Body frs. of red clay amphorai | 3 | | |
| 36a | Mushroom-shaped rim fr. of light brown clay with mica and with greyish brown slip | 1 | 528 | |
| 36b | Body fr. of light brown clay with mica and with greyish brown slip | 1 | 524 | Mediterranean? |
| 37 | Body frs. of red clay jugs | 2 | | |
| 38a | Handmade slashed rim fr. | 1 | 520 | |
| 38b | Handmade body frs. | 6 | | |
| 38c | Handmade flat base | 1 | 521 | |
| 39 | Bones of animals | 4 | | |

## DSP08-H12-01

Finds table core units

| BagID | Function | Total | Body | Rim | Handle | Base | Fabric | Surface | Shape | Period | Cat. |
|---|---|---|---|---|---|---|---|---|---|---|---|
| 2006308 | HW | 1 | 1 | 0 | 0 | 0 | BaGrey | Grooved | Pot | Late BA? | |
| 321001 | HW | 1 | 1 | 0 | 0 | 0 | Grey | None | Pot | LC-EH | |
| 321201 | HW | 1 | 1 | 0 | 0 | 0 | Grey | None | Pot | LC-EH | |
| 321501 | HW | 1 | 1 | 0 | 0 | 0 | Grey | None | Pot | LC-EH | |
| 2003510 | HW | 1 | 0 | 0 | 0 | 1 | Red | None | Pot | LC-EH | **529** |
| 2003602 | HW | 1 | 1 | 0 | 0 | 0 | Grey | None | Pot | LC-EH | |
| 320901 | TR | 1 | 0 | 0 | 1 | 0 | Red | None | Amphora | LC-EH | **536** |
| 2006102 | HW | 2 | 2 | 0 | 0 | 0 | Red | None | Pot | LC-EH? | |
| 2006202 | HW | 1 | 1 | 0 | 0 | 0 | Grey | None | Pot | LC-EH? | |
| 2006204 | HW | 1 | 1 | 0 | 0 | 0 | Black | None | Pot | LC-EH? | |
| 2006305 | HW | 1 | 1 | 0 | 0 | 0 | Grey | None | Pot | LC-EH? | |
| 2006309 | HW | 1 | 1 | 0 | 0 | 0 | Grey | None | Pot | LC-EH? | |
| 2006312 | HW | 1 | 0 | 1 | 0 | 0 | Red | None | Pot | LC-EH? | **530** |
| 321101 | IN | 1 | 1 | 0 | 0 | 0 | Red mica | None | ? | LC-EH? | |
| 321201 | IN | 1 | 1 | 0 | 0 | 0 | Red | None | ? | LC-EH? | |
| 321501 | IN | 1 | 1 | 0 | 0 | 0 | Red mica | None | ? | LC-EH? | |
| 2006101 | IN | 1 | 1 | 0 | 0 | 0 | Red | None | | LC-EH? | |
| 2006102 | IN | 1 | 1 | 0 | 0 | 0 | Red | None | | LC-EH? | |
| 2006103 | IN | 1 | 1 | 0 | 0 | 0 | Red | None | ? | LC-EH? | |
| 2006104 | IN | 1 | 1 | 0 | 0 | 0 | Red | None | | LC-EH? | |
| 2006205 | IN | 1 | 1 | 0 | 0 | 0 | Red | None | | LC-EH? | |
| 2006207 | IN | 1 | 1 | 0 | 0 | 0 | Red | None | ? | LC-EH? | |
| 2006304 | IN | 1 | 1 | 0 | 0 | 0 | Red | None | | LC-EH? | |
| 2006306 | IN | 1 | 1 | 0 | 0 | 0 | Red | None | | LC-EH? | |
| 321401 | TR | 1 | 1 | 0 | 0 | 0 | Chian? | None | Amphora | LC-EH | **531** |
| 2003603 | TR | 2 | 2 | 0 | 0 | 0 | Brown coarse gold mica | None | Amphora | LC-EH? | |
| 2003604 | TR | 1 | 1 | 0 | 0 | 0 | Red mica | None | Amphora | LC-EH? | |
| 2003605 | TR | 1 | 1 | 0 | 0 | 0 | Thasian? | None | Amphora | LC-EH? | |
| 2003607 | TR | 1 | 1 | 0 | 0 | 0 | Sinopean | None | Amphora | LC-EH? | |
| 2003608 | TR | 1 | 1 | 0 | 0 | 0 | Herakleian | None | Amphora | LC-EH? | |
| 2003609 | TR | 1 | 1 | 0 | 0 | 0 | Red | None | Amphora | LC-EH? | |
| 2003611 | TR | 1 | 1 | 0 | 0 | 0 | Red | None | Amphora | LC-EH? | |
| 2006105 | TR | 1 | 0 | 0 | 1 | 0 | Red | None | Amphora | LC-EH? | **532** |
| 2006201 | TR | 1 | 1 | 0 | 0 | 0 | Herakleian? | None | Amphora | LC-EH? | |
| 2006202 | TR | 1 | 1 | 0 | 0 | 0 | Herakleian | None | Amphora | LC-EH? | |
| 2006206 | TR | 1 | 1 | 0 | 0 | 0 | Sinopean | None | Amphora | LC-EH? | |
| 2006207 | TR | 1 | 1 | 0 | 0 | 0 | Red | None | Amphora | LC-EH? | |
| 2006208 | TR | 1 | 0 | 1 | 0 | 0 | Red | None | Amphora | LC-EH? | **533** |
| 2006301 | TR | 1 | 0 | 0 | 1 | 0 | Red | None | Amphora | LC-EH? | **534** |

| BagID | Function | Total | Body | Rim | Handle | Base | Fabric | Surface | Shape | Period | Cat. |
|---|---|---|---|---|---|---|---|---|---|---|---|
| 2006303 | TR | 1 | 1 | 0 | 0 | 0 | Sinopean | None | Amphora | LC-EH? | |
| 2003508 | TR | 1 | 1 | 0 | 0 | 0 | Red coarse | Grooved | Amphora? | Byzantine? | |
| 2003612 | TR | 1 | 0 | 0 | 1 | 0 | Red | None | Amphora | Medieval? | 537 |
| 2006106 | IN | 1 | 1 | 0 | 0 | 0 | Tatarian | None | | EM | |
| 2006203 | IN | 1 | 1 | 0 | 0 | 0 | Red | None | Brick? | Modern | |
| 321601 | ? | 1 | 0 | 0 | 0 | 0 | | | | | |
| 321301 | HW | 1 | 1 | 0 | 0 | 0 | | None | ? | | |
| 2006206 | HW | 1 | 1 | 0 | 0 | 0 | Grey | None | Pot | | |
| 2003509 | IN | 1 | 1 | 0 | 0 | 0 | Red | None | ? | | |
| 2003601 | IN | 1 | 1 | 0 | 0 | 0 | Red mica | None | ? | | |
| 2003606 | IN | 1 | 1 | 0 | 0 | 0 | Red | None | ? | | |
| 2006307 | IN | 1 | 1 | 0 | 0 | 0 | | | | | |
| 320801 | LI | 1 | 0 | 0 | 0 | 0 | | | | | |
| 2003610 | LI | 1 | 0 | 0 | 0 | 0 | | | | | |
| 2006302 | LI | 1 | 0 | 0 | 0 | 0 | | | | | |
| 2006310 | LI | 3 | 0 | 0 | 0 | 0 | | | | | |
| 2006311 | TR | 1 | 0 | 0 | 1 | 0 | Red | None | Amphora? | | 535 |

Finds table adjacent units

| BagID | Function | Total | Body | Rim | Handle | Base | Fabric | Surface | Shape | Period | Cat. |
|---|---|---|---|---|---|---|---|---|---|---|---|
| 2003613 | TW | 23 | 21 | 1 | 0 | 1 | Tatarian | None | Pot | EM? | |
| 2003507 | AR | 1 | 1 | 0 | 0 | 0 | Red modern | None | Brick | Modern | |

## DSP08-H13-01

Finds table core units

| BagID | Function | Total | Body | Rim | Handle | Base | Fabric | Surface | Shape | Period | Cat. |
|---|---|---|---|---|---|---|---|---|---|---|---|
| 2004902 | AR | 1 | 1 | 0 | 0 | 0 | Herakleian? | None | | LC-EH? | |
| 2005902 | TR | 1 | 1 | 0 | 0 | 0 | Red mica | None | Amphora | LC-EH? | |
| 2005903 | TR | 1 | 0 | 0 | 1 | 0 | Thasian? | None | Amphora | LC-EH? | 538 |
| 2004901 | IN | 1 | 0 | 0 | 0 | 0 | Tatarian? | None | | EM? | |
| 2006002 | IN | 1 | 1 | 0 | 0 | 0 | Red modern | None | Closed shape | Modern | |
| 2006002 | HW | 1 | 1 | 0 | 0 | 0 | Grey | None | ? | | |
| 2004903 | IN | 1 | 1 | 0 | 0 | 0 | Red | None | ? | | |
| 2005901 | IN | 1 | 1 | 0 | 0 | 0 | Red | None | Closed shape | | |
| 2006001 | IN | 1 | 1 | 0 | 0 | 0 | Red mica | None | Closed shape | | |
| 2006003 | IN | 1 | 1 | 0 | 0 | 0 | Red | None | Closed shape | | |
| 2006004 | IN | 1 | 1 | 0 | 0 | 0 | Red mica | None | ? | | |

## DSP08-H14-01

Finds table core and adjacent units

| BagID | Function | Total | Body | Rim | Handle | Base | Fabric | Surface | Shape | Period | Cat. |
|---|---|---|---|---|---|---|---|---|---|---|---|
| 2004702 | HW | 2 | 2 | 0 | 0 | 0 | BaGrey | Incised | Pot | BA? | |
| 2004702 | HW | 1 | 1 | 0 | 0 | 0 | BaBlack | Incised | Pot | BA? | |
| 2004702 | HW | 1 | 1 | 0 | 0 | 0 | Grey | None | Pot | LC-EH? | |
| 2004802 | HW | 1 | 1 | 0 | 0 | 0 | Grey | None | Pot | LC-EH? | |
| 2004803 | HW | 1 | 1 | 0 | 0 | 0 | Red | None | Pot | LC-EH? | |
| 2004301 | TR | 1 | 0 | 0 | 1 | 0 | Red | None | Amphora | LC-EH? | **539** |
| 2004501 | TR | 1 | 1 | 0 | 0 | 0 | Red mica | None | Amphora | LC-EH? | |
| 2004502 | TR | 2 | 2 | 0 | 0 | 0 | Herakleian | None | Amphora | LC-EH? | |
| 2004801 | TR | 1 | 1 | 0 | 0 | 0 | Red | Ribbed | Amphora? | Byzantine? | |
| 2004601 | IN | 1 | 1 | 0 | 0 | 0 | Tatarian | None | ? | EM | |
| 2004701 | IN | 1 | 1 | 0 | 0 | 0 | Orange | Slip | Closed shape | Modern? | |
| 2004702 | HW | 1 | 1 | 0 | 0 | 0 | Red | None | Pot | | |
| 2004702 | HW | 1 | 1 | 0 | 0 | 0 | | | | | |
| 2004702 | HW | 4 | 4 | 0 | 0 | 0 | Grey | None | Pot | | |
| 2004703 | HW | 3 | 3 | 0 | 0 | 0 | Grey | None | Pot | | |
| 2004702 | IN | 1 | 1 | 0 | 0 | 0 | Red | None | ? | | |
| 2004804 | LI | 1 | 1 | 0 | 0 | 0 | | | | | |

**Feature 2.** Excavation finds

| Layer 4: grey ashy layer, 0.4-0.65 m | | | | |
|---|---|---|---|---|
| No. | Description | Quantity | Cat. | Note |
| 1a | Handmade black polished rim | 44 | **547-550, 556-557** | BA |
| 1b | Handmade light polished rim | 4 | **552, 555, 558** | BA |
| 1c | Handmade light polished rim with horizontal moulding with impressed dots | 2 | **554, 564** | BA |
| 1d | Handmade black polished body fr. with impressed nail decoration | 1 | **553** | BA |
| 1e | Handmade black polished body frs. | 48 | **559-560** | BA |
| 1f | Handmade light polished body frs. | 4 | | BA |
| 1g | Handmade base frs. | 4 | **543-545** | BA |
| 2 | Bones of animals | 23 | | |
| Layer 3: Material from the lower part of the stone debris, 0.65-0.7 m, 0-0.1 m above the bedrock. | | | | |
| 3a | Handmade lightly polished rims of light brown clay | 2 | **551** | |
| 3b | Handmade body frs. | 4 | | |
| 3c | Handmade flat base | 1 | **563** | |
| 4a | Handmade rim of light polished clay | 1 | | |
| 4b | Handmade shoulder frs. with horizontal moulding | 3 | **540-541, 561** | |
| 4c | Handmade ribbed body fr. | 1 | **542** | |
| 4d | Handmade body frs. | 13 | **562** | |
| 4e | Handmade flat base | 1 | **546** | |
| 5a | Handmade grey polished rim | 1 | | |
| 5b | Handmade grey polished body frs. | 4 | | |
| 6 | Bones of animals | 13 | | One lime encrusted |

## DSP08-H17-01

Finds table core and adjacent units

| BagID | Function | Total | Body | Rim | Handle | Base | Fabric | Surface | Shape | Period | Cat. |
|---|---|---|---|---|---|---|---|---|---|---|---|
| 320401 | IN | 1 | 1 | 0 | 0 | 0 | Red | None | ? | | |

## DSP08-H17-02

Finds table core units

| BagID | Function | Total | Body | Rim | Handle | Base | Fabric | Surface | Shape | Period | Cat. |
|---|---|---|---|---|---|---|---|---|---|---|---|
| 2000304 | IN | 1 | 1 | 0 | 0 | 0 | Red | None | ? | LC-EH? | |
| 320501 | TR | 1 | 0 | 0 | 1 | 0 | Thasian | None | Amphora | LC-EH? | |
| 2000301 | TW | 1 | 0 | 1 | 0 | 0 | Tatarian red | None | Bowl? | EM? | **565** |
| 320601 | IN | 1 | 1 | 0 | 0 | 0 | Red | None | ? | | |
| 2000303 | LI | 1 | 0 | 0 | 0 | 0 | | | | | |
| 2000302 | TR | 1 | 0 | 0 | 1 | 0 | Red porous | None | Amphora | | |

## DSP07-H19-01

**SW of Kurgan 3**. Excavation finds

| Finds list 7: Top soil, 0-0.55 m | | | | |
|---|---|---|---|---|
| No. | Description | Quantity | Cat. | Notes |
| 1 | Body frs. of red-clay amphora, Chersonesean? | 2 | **567** | |
| 2 | Body frs. of red-clay jugs | 4 | | |
| 3 | Handmade body fr. | 1 | | |
| On top of layer 2, 0.55 m | | | | |
| 4 | Body fr. of Thasian amphora | 1 | | |
| On the bedrock, 0.85 m | | | | |
| 5 | Toe of Chersonesea amphora | 1 | **566** | Ca. 350-325 BC |

## DSP07-H19-03

Finds table core units

| BagID | Function | Total | Body | Rim | Handle | Base | Fabric | Surface | Shape | Period | Cat. |
|---|---|---|---|---|---|---|---|---|---|---|---|
| 2003618 | IN | 1 | 1 | 0 | 0 | 0 | Red pyro | None | ? | LC-EH? | |
| 2003604 | TR | 1 | 1 | 0 | 0 | 0 | Red mica | None | Amphora? | LC-EH? | |
| 2003605 | TR | 1 | 1 | 0 | 0 | 0 | Thasian? | None | Amphora? | LC-EH? | |
| 2003607 | TR | 1 | 1 | 0 | 0 | 0 | Sinopean | None | Amphora | LC-EH? | |
| 2003608 | TR | 1 | 1 | 0 | 0 | 0 | Herakleian | None | Amphora | LC-EH? | |
| 2003609 | TR | 1 | 1 | 0 | 0 | 0 | Red | None | Amphora | LC-EH? | |
| 2003611 | TR | 1 | 1 | 0 | 0 | 0 | Red | None | Amphora? | LC-EH? | |
| 2003615 | TR | 1 | 1 | 0 | 0 | 0 | Red | None | Amphora? | LC-EH? | |
| 2003616 | TR | 1 | 1 | 0 | 0 | 0 | Red | None | Amphora | LC-EH? | |
| 2003617 | TR | 1 | 1 | 0 | 0 | 0 | Herakleian | None | Amphora? | LC-EH? | |
| 2003612 | TR | 1 | 0 | 0 | 1 | 0 | Red | None | Amphora | Medieval? | |
| 2003506 | IN | 6 | 6 | 0 | 0 | 0 | Tatarian? | None | Closed shape | EM? | |
| 2003613 | TW | 23 | 21 | 1 | 0 | 1 | Tatarian | None | Pot | EM? | **568-569** |
| 2003505 | IN | 1 | 0 | 0 | 1 | 0 | Red | None | ? | | |
| 2003606 | IN | 1 | 1 | 0 | 0 | 0 | Red | None | ? | | |
| 2003614 | IN | 1 | 1 | 0 | 0 | 0 | Red | None | ? | | |
| 2003610 | LI | 1 | 0 | 0 | 0 | 0 | | | | | |

## DSP08-H19-09

Finds table core units

| BagID | Function | Total | Body | Rim | Handle | Base | Fabric | Surface | Shape | Period | Cat. |
|---|---|---|---|---|---|---|---|---|---|---|---|
| 2003501 | TW | 1 | 0 | 1 | 0 | 0 | Tatarian? | None | Flask | EM | |

## DSP08-H20-01

Finds table core units

| BagID | Function | Total | Body | Rim | Handle | Base | Fabric | Surface | Shape | Period | Cat. |
|---|---|---|---|---|---|---|---|---|---|---|---|
| 2000902 | HW | 1 | 1 | 0 | 0 | 0 | Grey | None | Pot | LC-EH | |
| 2000904 | HW | 1 | 1 | 0 | 0 | 0 | Grey | None | | LC-EH | |
| 2000901 | IN | 2 | 2 | 0 | 0 | 0 | Red | None | ? | LC-EH? | |
| 2000903 | TR | 1 | 0 | 0 | 1 | 0 | Red | None | Amphora | LC-EH? | 571 |
| 2000904 | LI | 1 | 0 | 0 | 0 | 0 | | | | | |
| 2001103 | LI | 1 | 0 | 0 | 0 | 0 | | | | | |

## DSP08-H22-01

Finds table core units

| BagID | Function | Total | Body | Rim | Handle | Base | Fabric | Surface | Shape | Period | Cat. |
|---|---|---|---|---|---|---|---|---|---|---|---|
| 2005702 | IN | 1 | 1 | 0 | 0 | 0 | Red | None | ? | ? | |

## DSP08-H22-02

Finds table adjacent units

| BagID | Function | Total | Body | Rim | Handle | Base | Fabric | Surface | Shape | Period | Cat. |
|---|---|---|---|---|---|---|---|---|---|---|---|
| 2005701 | UT | 6 | 3 | 3 | 0 | 0 | Tatarian? | Slip | Pot | EM | 572 |

## DSP08-H23-01

Finds table core units

| BagID | Function | Total | Body | Rim | Handle | Base | Fabric | Surface | Shape | Period | Cat. |
|---|---|---|---|---|---|---|---|---|---|---|---|
| 2004202 | TR | 1 | 0 | 0 | 1 | 0 | Red | None | Amphora? | | 573 |

## DSP08-H23-02

Finds table core units:

| BagID | Function | Total | Body | Rim | Handle | Base | Fabric | Surface | Shape | Period | Cat. |
|---|---|---|---|---|---|---|---|---|---|---|---|
| 2004106 | IN | 1 | 1 | 0 | 0 | 0 | Red | None | | | |
| 2004201 | IN | 1 | 1 | 0 | 0 | 0 | Red | None | ? | LC-EH? | |

## DSP08-H24-01

Finds table core units

| BagID | Function | Total | Body | Rim | Handle | Base | Fabric | Surface | Shape | Period | Cat. |
|---|---|---|---|---|---|---|---|---|---|---|---|
| 321701 | TR | 1 | 1 | 0 | 0 | 0 | Light | None | Amphora | LC-EH? | |

Finds table adjacent units

| BagID | Function | Total | Body | Rim | Handle | Base | Fabric | Surface | Shape | Period | Cat. |
|---|---|---|---|---|---|---|---|---|---|---|---|
| 2004004 | IN | 1 | 1 | 0 | 0 | 0 | Red pyro | None | ? | LC-EH? | |
| 2004003 | TR | 1 | 1 | 0 | 0 | 0 | Red | Paint | Amphora | LC-EH? | 574 |
| 2003901 | LI | 1 | 0 | 0 | 0 | 0 | | | | | |

## DSP08-H24-02

Finds table core and adjacent units

| BagID | Function | Total | Body | Rim | Handle | Base | Fabric | Surface | Shape | Period | Cat. |
|---|---|---|---|---|---|---|---|---|---|---|---|
| 2003903 | TR | 1 | 1 | 0 | 0 | 0 | Red | None | Amphora? | LC-EH? | |
| 2003904 | IN | 1 | 1 | 0 | 0 | 0 | Red | None | ? | | |
| 2003905 | IN | 1 | 1 | 0 | 0 | 0 | Orange | None | Closed shape | | |
| 2003906 | IN | 1 | 1 | 0 | 0 | 0 | Red | None | ? | | |
| 2003907 | IN | 1 | 1 | 0 | 0 | 0 | Red | None | Closed shape | | |
| 2004001 | IN | 1 | 1 | 0 | 0 | 0 | Red | None | ? | | |

## DSP08-H24-03

Finds table core units

| BagID | Function | Total | Body | Rim | Handle | Base | Fabric | Surface | Shape | Period | Cat. |
|---|---|---|---|---|---|---|---|---|---|---|---|
| 2003910 | TR | 1 | 0 | 1 | 0 | 0 | Chersonesean? | None | Amphora | LC-EH? | 575 |

## DSP08-H25-01

Finds table core units

| BagID | Function | Total | Body | Rim | Handle | Base | Fabric | Surface | Shape | Period | Cat. |
|---|---|---|---|---|---|---|---|---|---|---|---|
| 2003101 | IN | 1 | 1 | 0 | 0 | 0 | Red | None | ? | | |

## DSP08-H26-01

Finds table core units

| BagID | Function | Total | Body | Rim | Handle | Base | Fabric | Surface | Shape | Period | Cat. |
|---|---|---|---|---|---|---|---|---|---|---|---|
| 320001 | TR | 1 | 0 | 0 | 1 | 0 | Red | None | Amphora | LC-EH? | 576 |

## DSP08-H26-02

Finds table core units

| BagID | Function | Total | Body | Rim | Handle | Base | Fabric | Surface | Shape | Period | Cat. |
|---|---|---|---|---|---|---|---|---|---|---|---|
| 2000101 | TR | 2 | 2 | 0 | 0 | 0 | Herakleian? | None | Amphora? | LC-EH? | |
| 2003001 | TR | 3 | 3 | 0 | 0 | 0 | Herakleian | None | Amphora | LC-EH? | |
| 2002901 | IN | 1 | 1 | 0 | 0 | 0 | Red | None | Closed shape | | |

## Baj Kijat 1 (and 2)

Finds table core units

| BagID | Function | Total | Body | Rim | Handle | Base | Fabric | Surface | Shape | Period | Cat. |
|---|---|---|---|---|---|---|---|---|---|---|---|
| 300902 | HW | 72 | 72 | 0 | 0 | 0 | BaBlack | None | Pot | Middle BA | **581** |
| 300902 | HW | 3 | 0 | 3 | 0 | 0 | BaGrey | Impressed | bowl | MBA-LBA | **578, 580, 583** |
| 300902 | HW | 1 | 0 | 1 | 0 | 0 | BaGrey | None | cup | Late BA? | **579** |
| 300901 | HW | 9 | 8 | 1 | 0 | 0 | BaGrey | None | Pot | Late BA? | **577, 582** |
| 300901 | IN | 1 | 1 | 0 | 0 | 0 | Red | None | | Late BA? | |
| 300901 | IN | 6 | 5 | 0 | 0 | 0 | Red | None | | LC-EH | |
| 300902 | TR | 1 | 0 | 0 | 1 | 0 | Red | None | Amphora | LC-EH | **584** |
| 300901 | BO | 10 | 0 | 0 | 0 | 0 | | | | | |
| 300902 | BO | 2 | 0 | 0 | 0 | 0 | | | | | |
| 300902 | LI | 3 | 0 | 0 | 0 | 0 | | | | | |

## Baj Kijat 3

Finds table *Core units*

| BagID | Function | Total | Body | Rim | Handle | Base | Fabric | Surface | Shape | Period | Cat. |
|---|---|---|---|---|---|---|---|---|---|---|---|
| 301001 | TR | 1 | 1 | 0 | 0 | 0 | Chersonesean | | Amphora | LC-EH | |
| 301001 | HW | 28 | 23 | 5 | 0 | 0 | Brown | None | Pot | LC-EH? | **585-589** |
| 301001 | IN | 1 | 0 | 0 | 0 | 0 | Tatarian | | | EM | |
| 301001 | BO | 2 | 0 | 0 | 0 | 0 | | | | | |

## Burun Eli 2

Finds table core units

| BagID | Function | Total | Body | Rim | Handle | Base | Fabric | Surface | Shape | Period | Cat. |
|---|---|---|---|---|---|---|---|---|---|---|---|
| 301101 | TW | 2 | 1 | 0 | 0 | 1 | Chersonesan | None | Jug | LC-EH | **590** |
| 301101 | IN | 8 | 8 | 0 | 0 | 0 | | | | LC-EH? | |
| 301101 | BO | 1 | 1 | 0 | 0 | 0 | | | | | |

## Burun Eli 4

Finds table core units

| BagID | Function | Total | Body | Rim | Handle | Base | Fabric | Surface | Shape | Period | Cat. |
|---|---|---|---|---|---|---|---|---|---|---|---|
| 302101 | HW | 1 | 1 | 0 | 0 | 0 | - | None | pot | LC-EH | |
| 302101 | IN | 28 | 28 | 0 | 0 | 0 | | | | LC-EH | |
| 302101 | IN | 1 | 1 | 0 | 0 | 0 | | | | Medieval? | |

## Burun Eli 6

Finds table core units

| BagID | Function | Total | Body | Rim | Handle | Base | Fabric | Surface | Shape | Period | Cat. |
|---|---|---|---|---|---|---|---|---|---|---|---|
| 302201 | TR | 1 | 0 | 0 | 1 | 0 | Chersonesean | None | Amphora | LC-EH | |
| 302201 | TW | 23 | 23 | 0 | 0 | 0 | | | | LC-EH | |
| 302201 | TW | 4 | 4 | 0 | 0 | 0 | | | | EM | |
| 302201 | BO | 2 | 0 | 0 | 0 | 0 | | | | | |

## Groty 1

Finds table core units

| BagID | Function | Total | Body | Rim | Handle | Base | Fabric | Surface | Shape | Period | Cat. |
|---|---|---|---|---|---|---|---|---|---|---|---|
| 303101 | BO | 3 | 0 | 0 | 0 | 0 | - | - | | | |
| 303101 | Slag | 4 | 0 | 0 | 0 | 0 | - | - | | | |
| 303101 | AR | 4 | 0 | 0 | 0 | 0 | Sinopean | None | Tile | LC-EH | |
| 303101 | AR | 27 | 0 | 0 | 0 | 0 | Red | None | Tile | LC-EH | |
| 303101 | HW | 56 | 51 | 4 | 0 | 1 | - | - | ? | LC-EH | |
| 303101 | IN | 103 | 103 | 0 | 0 | 0 | - | None | ? | LC-EH | |
| 303101 | TR | 1 | 0 | 1 | 0 | 0 | Red | None | Amphora | LC-EH | **599** |
| 303101 | TR | 1 | 0 | 1 | 0 | 0 | Brown | None | Amphora | LC-EH | **600** |
| 303101 | TR | 1 | 0 | 1 | 0 | 0 | Greyish Brown | None | Amphora | LC-EH | **601** |
| 303101 | TR | 1 | 0 | 0 | 1 | 0 | Herakleian? | None | Amphora | LC-EH | **597** |
| 303101 | TR | 3 | 0 | 2 | 1 | 0 | Herakleian? | None | Amphora | LC-EH | **598, 602-603** |
| 303101 | TW | 1 | 0 | 1 | 0 | 0 | Grey | None | Jug | LC-EH | **604** |
| 303101 | TW | 1 | 0 | 1 | 0 | 0 | Red | None | Jug | LC-EH | **605** |
| 303101 | TW | 1 | 0 | 1 | 0 | 0 | Chersonesan | None | Jug | LC-EH | **606** |
| 303101 | TW | 1 | 0 | 0 | 0 | 1 | Attic | Glazed | Bowl | LC-EH | **607** |
| 303101 | TW | 1 | 1 | 0 | 0 | 0 | Grey | Banded | Jug | LC-EH | **608** |
| 303101 | TW | 1 | 0 | 1 | 0 | 0 | Grey | Banded | Bowl | LC-EH | **609** |
| 303101 | TR | 1 | 1 | 0 | 0 | 0 | Colchian? | None | Amphora | LC-EH? | |

## Groty 2

Finds table core units

| BagID | Function | Total | Body | Rim | Handle | Base | Fabric | Surface | Shape | Period | Cat. |
|---|---|---|---|---|---|---|---|---|---|---|---|
| 303201 | HW | 1 | 1 | 0 | 0 | 0 | - | None | ? | LC-EH | |
| 303201 | IN | 5 | 5 | 0 | 0 | 0 | | | | LC-EH | |

## Kurama Kostel

Finds table core units

| BagID | Function | Total | Body | Rim | Handle | Base | Fabric | Surface | Shape | Period | Cat. |
|---|---|---|---|---|---|---|---|---|---|---|---|
| 100301 | TR | 1 | 0 | 1 | 0 | 0 | Thasian | None | Amphora | LC-EH | **615** |
| 100301 | TR | 1 | 1 | 0 | 0 | 0 | Chersonesean | Slip | Amphora | LC-EH | |
| 100301 | TR | 1 | 0 | 0 | 0 | 1 | Herakleian | None | Amphora | LC-EH | **613** |
| 100301 | TR | 1 | 0 | 0 | 1 | 0 | Herakleian | None | Amphora | LC-EH | **614** |
| 100301 | IN | 2 | 2 | 0 | 0 | 0 | Light | None | ? | LC-EH? | |
| 100301 | IN | 44 | 39 | 5 | 0 | 0 | Tatarian | None | ? | EM | **622, 631, 635** |
| 100301 | TW | 1 | 0 | 1 | 0 | 0 | Tatarian | None | Pibe | EM | **634** |
| 100301 | TW | 4 | 3 | 1 | 0 | 0 | Tatarian | Glazed | ? | EM | |
| 100301 | TW | 1 | 0 | 1 | 0 | 0 | Tatarian | Glazed | Cup | EM | |
| 100301 | TW | 7 | 3 | 2 | 0 | 3 | Tatarian | Glazed | Plate | EM | **619-621, 630** |
| 100301 | TW | 15 | 2 | 9 | 1 | 2 | Tatarian | None | Jug | EM | **617, 626-628, 636** |
| 100301 | TW | 1 | 0 | 1 | 0 | 0 | Tatarian | None | Bowl | EM | **624** |
| 100301 | TW | 4 | 4 | 0 | 0 | 0 | Tatarian | Grooved | Jug? | EM | **637-639** |
| 100301 | UT | 5 | 1 | 0 | 0 | 4 | Tatarian | None | Pot | EM | **625, 629, 632-633** |
| 100301 | UT | 1 | 1 | 0 | 0 | 0 | Tatarian | Glazed | Jug | EM | **618** |
| 100301 | UT | 3 | 0 | 0 | 0 | 0 | Tatarian | None | Cooking pot | EM | **623** |
| 100301 | AR | 7 | 0 | 0 | 0 | 0 | Red modern | None | Tile | Modern | **610-612** |
| 100301 | IN | 1 | 0 | 0 | 1 | 0 | Red | Slip | Jug | LC-EH? | **616** |

## Masliny 1

Finds table core units

| BagID | Function | Total | Body | Rim | Handle | Base | Fabric | Surface | Shape | Period | Cat. |
|---|---|---|---|---|---|---|---|---|---|---|---|
| 300701 | HW | 4 | 0 | 0 | 0 | 0 | | None | | LC-EH | 640 |
| 300701 | IN | 20 | 0 | 0 | 0 | 0 | | None | | LC-EH | |
| 300701 | UT | 2 | 1 | 1 | 0 | 0 | | None | Louterion | EH | 641 |
| 300701 | BO | 1 | 0 | 0 | 0 | 0 | | | | | |

## Masliny 3

Finds table core units

| BagID | Function | Total | Body | Rim | Handle | Base | Fabric | Surface | Shape | Period | Cat. |
|---|---|---|---|---|---|---|---|---|---|---|---|
| 300801 | TR | 1 | 1 | 0 | 0 | 0 | Herakleian | None | Amphora | LC-EH | |
| 300801 | IN | 9 | 0 | 0 | 0 | 0 | | None | | LC-EH? | |
| 300801 | IN | 1 | 0 | 0 | 0 | 0 | | None | | | |
| 300801 | BO | 3 | 0 | 0 | 0 | 0 | | | | | |
| 300801 | LI | 1 | 0 | 0 | 0 | 0 | | | | | |

## Panskoe II/DSP06-F03-01

Finds table core units

| BagID | Function | Total | Body | Rim | Handle | Base | Fabric | Surface | Shape | Period | Cat. |
|---|---|---|---|---|---|---|---|---|---|---|---|
| 300201 | HW | 1 | 0 | 1 | 0 | 0 | HwBrown | None | cup | LC-EH | **643** |
| 300201 | HW | 1 | 1 | 0 | 0 | 0 | Black | None | Pot | LC-EH? | |
| 300201 | HW | 1 | 0 | 1 | 0 | 0 | HwGrey | Cut | Pot | LC-EH | **642** |
| 300201 | HW | 1 | 1 | 0 | 0 | 0 | Red | Decorated | Pot | LC-EH | |
| 300201 | HW | 3 | 3 | 0 | 0 | 0 | Grey | None | cooking pot? | LC-EH | |
| 300201 | TR | 1 | 0 | 0 | 0 | 1 | Mendean | None | Amphora | Hellenistic | **647** |
| 300201 | TR | 1 | 0 | 0 | 1 | 0 | Light | None | Amphora | Hellenistic | **655** |
| 300201 | TR | 5 | 0 | 0 | 5 | 1 | Red | None | Amphora | LC-ER | **648-652** |
| 300201 | TR | 4 | 1 | 0 | 0 | 2 | Chersonesean | Slip | Amphora | EH | **644** |
| 300201 | TR | 1 | 1 | 0 | 0 | 0 | Brown | None | Amphora | Hellenistic? | **653** |
| 300201 | TR | 1 | 1 | 0 | 0 | 0 | Brown | Slip | Amphora | ? | |
| 300201 | TR | 4 | 3 | 1 | 0 | 0 | Brown | None | Amphora | Hellenistic? | |
| 300201 | TR | 1 | 0 | 0 | 0 | 1 | Sinopean | Slip | Amphora | Hellenistic | **645** |
| 300201 | TR | 1 | 1 | 0 | 0 | 0 | Brown | Slip | Amphora | Hellenistic | **654** |
| 300201 | TR | 2 | 2 | 0 | 0 | 0 | Red | None | Amphora | Hellenistic | |
| 300201 | TR | 3 | 3 | 0 | 0 | 0 | Sinopean | None | Amphora | Hellenistic | **646** |
| 300201 | TR | 2 | 2 | 0 | 0 | 0 | Herakleian | None | Amphora | Hellenistic | |
| 300201 | TW | 1 | 0 | 0 | 1 | 0 | Chersonesean | None | Jug | LC-H | **657** |
| 300201 | AR | 3 | 3 | 0 | 0 | 0 | Red | None | Tile | LC-EH | |
| 300201 | IN | 1 | 1 | 0 | 0 | 0 | South Pontic | Slip | Closed shape | LC-EH | |
| 300201 | IN | 18 | 18 | 0 | 0 | 0 | Red | None | ? | LC-EH | |
| 300201 | IN | 6 | 6 | 0 | 0 | 0 | Red | Slip | Closed shape | LC-EH | |
| 300201 | IN | 6 | 5 | 0 | 1 | 0 | South Pontic | None | Closed shape? | LC-EH | |
| 300201 | TW | 1 | 1 | 0 | 0 | 0 | Chersonesean | Slip | Plate? | LC-EH | |
| 300201 | TW | 1 | 0 | 0 | 0 | 1 | Chersonesean | Slip | Jug | LC-EH | **658** |
| 300201 | TW | 2 | 1 | 0 | 0 | 0 | Chersonesean? | Slipped/ decorated | Jug | LC-EH | **656** |
| 300201 | UT | 1 | 1 | 0 | 0 | 0 | Red | Slip | Louterion? | LC-EH | |

| BagID | Function | Total | Body | Rim | Handle | Base | Fabric | Surface | Shape | Period | Cat. |
|---|---|---|---|---|---|---|---|---|---|---|---|
| 300201 | IN | 5 | 5 | 0 | 0 | 0 | Light | None | ? | LC-EH? | |
| 300201 | IN | 5 | 5 | 0 | 0 | 0 | Brown | None | ? | LC-EH? | |
| 300201 | LI | 1 | 0 | 0 | 0 | 0 | | | | | |
| 300201 | LI | 1 | 0 | 0 | 0 | 0 | | | | | |
| 300201 | SH | 1 | 0 | 0 | 0 | 0 | | | | | |

## Panskoe V/ DSP07-F01-02

Finds table core units

| BagID | Function | Total | Body | Rim | Handle | Base | Fabric | Surface | Shape | Period | Cat. |
|---|---|---|---|---|---|---|---|---|---|---|---|
| 106801 | HW | 1 | 1 | 0 | 0 | 0 | Grey | Polished Int. | Open shape | LC-EH | |
| 106801 | HW | 1 | 0 | 1 | 0 | 0 | Grey | None | Large open | LC-EH | **661** |
| 106401 | IN | 2 | 2 | 0 | 0 | 0 | Sinopean | None | ? | LC-EH | |
| 206201 | IN | 1 | 1 | 0 | 0 | 0 | Red mica | None | ? | LC-EH | |
| 206201 | IN | 1 | 1 | 0 | 0 | 0 | Red | None | ? | LC-EH | |
| 105901 | TR | 1 | 0 | 0 | 1 | 0 | Herakleian | None | Amphora | LC-EH | **659** |
| 106001 | TR | 1 | 1 | 0 | 0 | 0 | Chersonesean | Slip | Amphora | LC-EH | |
| 106901 | TR | 2 | 2 | 0 | 0 | 0 | Herakleian | None | Amphora | LC-EH | |
| 106901 | TR | 2 | 2 | 0 | 0 | 0 | Chersonesean | None | Amphora | LC-EH | |
| 204601 | TR | 1 | 1 | 0 | 0 | 0 | Red | None | Amphora | LC-EH | |
| 205501 | TR | 1 | 1 | 0 | 0 | 0 | Herakleian | None | Amphora | LC-EH | |
| 205701 | TR | 2 | 2 | 0 | 0 | 0 | Herakleian | None | Amphora | LC-EH | |
| 205701 | TR | 2 | 2 | 0 | 0 | 0 | Chersonesean | None | Amphora | LC-EH | |
| 107001 | HW | 1 | 1 | 0 | 0 | 0 | Red | None | Pot | LC-EH? | |
| 107001 | HW | 1 | 1 | 0 | 0 | 0 | Grey | None | Pot | LC-EH? | |
| 106401 | IN | 1 | 1 | 0 | 0 | 0 | Red | None | ? | LC-EH? | |
| 205501 | IN | 1 | 1 | 0 | 0 | 0 | Red | None | ? | LC-EH? | |
| 205801 | IN | 1 | 1 | 0 | 0 | 0 | Red | None | ? | LC-EH? | |
| 106401 | TW | 1 | 1 | 0 | 0 | 0 | Red | None | ? | LC-EH? | |
| 106401 | TW | 1 | 0 | 1 | 0 | 0 | Red mica | None | ? | LC-EH? | |
| 205501 | TW | 1 | 0 | 0 | 0 | 1 | Red | Glazed | Cup | 500-450 BC | **662** |
| 206201 | TW | 1 | 1 | 0 | 0 | 0 | Red | None | Closed shape | LC-EH? | |
| 106401 | TW | 1 | 0 | 1 | 0 | 0 | Tatarian | None | Cup? | EM | |
| 105901 | AR | 1 | 0 | 1 | 0 | 0 | Red modern | Smoothed | Tile | Modern | |
| 205801 | AR | 1 | 1 | 0 | 0 | 0 | Grey? | Smoothed | Tile | ? | |
| 106001 | HW | 1 | 0 | 1 | 0 | 0 | Grey | Smoothed | ? | ? | **660** |
| 106101 | HW | 1 | 1 | 0 | 0 | 0 | Grey | None | ? | ? | |
| 205801 | HW | 1 | 1 | 0 | 0 | 0 | Red | None | Pot | ? | |
| 105901 | IN | 1 | 1 | 0 | 0 | 0 | Red mica | None | ? | ? | |
| 106101 | IN | 1 | 1 | 0 | 0 | 0 | Light | Uneven | ? | ? | |

| BagID | Function | Total | Body | Rim | Handle | Base | Fabric | Surface | Shape | Period | Cat. |
|---|---|---|---|---|---|---|---|---|---|---|---|
| 107001 | IN | 1 | 1 | 0 | 0 | 0 | Light | None | ? | ? | |
| 106901 | LI | 1 | 0 | 0 | 0 | 0 | | | | | |
| 107001 | LI | 1 | 0 | 0 | 0 | 0 | | | | | |
| 204601 | LI | 1 | 0 | 0 | 0 | 0 | | | | | |
| 205801 | LI | 2 | 0 | 0 | 0 | 0 | | | | | |
| 206201 | LI | 1 | 1 | 0 | 0 | 0 | | | | | |
| 106901 | SH | 1 | 0 | 0 | 0 | 0 | | | | | |
| 107001 | SH | 1 | 0 | 0 | 0 | 0 | | | | | |
| 106801 | TW | 1 | 0 | 1 | 0 | 0 | Light | Slip | ? | ? | |
| 106801 | TW | 1 | 1 | 0 | 0 | 0 | Light | Slip | Jug | ? | |

## Severnoe 2

Finds table core units

| BagID | Function | Total | Body | Rim | Handle | Base | Fabric | Surface | Shape | Period | Cat. |
|---|---|---|---|---|---|---|---|---|---|---|---|
| 300601 | IN | 1 | 1 | 0 | 0 | 0 | | | | LC-EH | |
| 300601 | TW | 1 | 0 | 0 | 1 | 0 | Chersonesean | | Jug | LC-EH | **663** |

## Skalistoe 2/ DSP08-F03-02

Finds table core units

| BagID | Function | Total | Body | Rim | Handle | Base | Fabric | Surface | Shape | Period | Cat. |
|---|---|---|---|---|---|---|---|---|---|---|---|
| 624401 | HW | 11 | 11 | 0 | 0 | 0 | BaRed | Grooved | Pot | BA | |
| 624401 | HW | 3 | 3 | 0 | 0 | 0 | BaLight | Grooved | Pot | BA | |
| 624401 | HW | 5 | 5 | 0 | 0 | 0 | BaBlack | Grooved | Pot | BA | |
| 624402 | HW | 18 | 16 | 2 | 0 | 0 | BaBlacK | Grooved | Pot | BA | **664,** |
| 624702 | HW | 1 | 1 | 0 | 0 | 0 | BaBlack | Impressed | Pot | BA | |
| 622602 | HW | 6 | 6 | 0 | 0 | 0 | BaGrey | Grooved | Pot | BA? | **665** |
| 624702 | HW | 68 | 68 | 0 | 0 | 0 | BaGrey | None | Pot | BA? | |
| 624702 | HW | 1 | 1 | 0 | 0 | 0 | BaRed | Incised | Pot | BA? | |
| 624702 | HW | 3 | 3 | 0 | 0 | 0 | BaBlack | Incised | Pot | BA? | |
| 624402 | HW | 1 | 1 | 1 | 0 | 0 | BaBlacK | Incised | Pot | Late BA | **678** |
| 624702 | HW | 1 | 1 | 0 | 0 | 0 | BaBlack | Combed | Pot | Middle BA | **666** |
| 624702 | HW | 1 | 1 | 0 | 0 | 0 | BaBlack | Incised | Pot | Middle BA | **686** |
| 624401 | HW | 1 | 0 | 1 | 0 | 0 | BaCoarse | None | Closed shape | Middle/ Late BA | **674** |
| 624701 | HW | 1 | 1 | 0 | 0 | 0 | BaGrey | Grooved | Pot | Middle/ Late BA | **673** |
| 624401 | HW | 1 | 1 | 0 | 0 | 0 | BaBrown | Decorated | Pot | Late BA | **669** |
| 624401 | HW | 1 | 1 | 0 | 0 | 0 | BaLight | Incised | Pot | Late BA | **676** |
| 624401 | HW | 3 | 1 | 2 | 0 | 0 | BaRed | Smoothed | Closed shape | Late BA | 675, 677 |

| BagID | Function | Total | Body | Rim | Handle | Base | Fabric | Surface | Shape | Period | Cat. |
|---|---|---|---|---|---|---|---|---|---|---|---|
| 624402 | HW | 31 | 30 | 1 | 0 | 0 | BaGrey | Grooved | Pot | Late BA | |
| 624402 | HW | 2 | 2 | 0 | 0 | 0 | BaBlacK | Incised | Pot | Late BA | **670-671** |
| 624701 | HW | 7 | 6 | 1 | 0 | 0 | BaBlack | Grooved | Pot | Late BA | **680** |
| 624702 | HW | 4 | 2 | 2 | 0 | 0 | BaCoarse | None | Pot | Late BA | **683-684** |
| 624702 | HW | 9 | 8 | 1 | 0 | 0 | BaBlack | Grooved | Pot | Late BA | |
| 624702 | HW | 1 | 0 | 1 | 0 | 0 | BaGrey | Impressed | Kitchen vessel | Late BA | **682** |
| 624702 | HW | 1 | 0 | 0 | 0 | 1 | BaBlack | None | Pot | Late BA | **687** |
| 624702 | HW | 7 | 6 | 1 | 0 | 0 | BaGrey | None | Pot | Late BA | **685** |
| 622602 | TW | 1 | 1 | 0 | 0 | 0 | Brown mica | Decorated | Jug? | Classical? | |
| 624401 | HW | 1 | 0 | 1 | 0 | 0 | BaCoarse | None | Open shape | LC-H | **690** |
| 624402 | HW | 1 | 0 | 1 | 0 | 1 | Brown | None | Bowl | LC-EH | **691** |
| 622601 | HW | 4 | 4 | 0 | 0 | 0 | Grey | None | Pot | LC-EH? | |
| 622601 | HW | 1 | 1 | 0 | 0 | 0 | Light | None | Pot | LC-EH? | |
| 622601 | HW | 2 | 2 | 0 | 0 | 0 | Red | None | Pot | LC-EH? | |
| 622602 | HW | 31 | 30 | 1 | 0 | 0 | Grey | None | Pot | LC-EH? | |
| 622602 | HW | 10 | 10 | 0 | 0 | 0 | Red | None | Pot | LC-EH? | |
| 622602 | HW | 5 | 5 | 0 | 0 | 0 | Coarse | None | Pot | LC-EH? | |
| 624401 | HW | 1 | 1 | 0 | 0 | 0 | Grey | None | Pot | LC-EH? | |
| 624401 | HW | 1 | 1 | 0 | 0 | 0 | Brown | None | Pot | LC-EH? | |
| 624401 | HW | 2 | 2 | 0 | 0 | 0 | Brown mica | None | Pot | LC-EH? | |
| 624402 | HW | 10 | 10 | 0 | 0 | 0 | Coarse | None | Pot | LC-EH? | |
| 624501 | HW | 6 | 6 | 0 | 0 | 0 | Red | None | Pot | LC-EH? | |
| 624501 | HW | 5 | 5 | 0 | 0 | 0 | Grey | None | Pot | LC-EH? | |
| 624701 | HW | 15 | 5 | 0 | 0 | 0 | Grey | None | Pot | LC-EH? | |
| 624701 | HW | 13 | 13 | 0 | 0 | 0 | Black | None | Pot | LC-EH? | |
| 624702 | HW | 16 | 15 | 0 | 0 | 1 | Red | None | Pot | LC-EH? | |
| 622601 | IN | 1 | 1 | 0 | 0 | 0 | Red | None | Closed shape | LC-EH? | |
| 622602 | IN | 9 | 9 | 0 | 0 | 0 | Red | None | ? | LC-EH? | |
| 624401 | IN | 2 | 2 | 0 | 0 | 0 | Brown | None | ? | LC-EH? | |
| 624401 | IN | 2 | 2 | 0 | 0 | 0 | Red | None | Closed shape | LC-EH? | |
| 624401 | IN | 4 | 4 | 0 | 0 | 0 | Red | None | ? | LC-EH? | |
| 624402 | IN | 17 | 17 | 0 | 0 | 0 | Red | None | ? | LC-EH? | |
| 624402 | IN | 1 | 1 | 0 | 0 | 0 | Red mica | None | Closed shape | LC-EH? | |
| 624402 | IN | 3 | 3 | 0 | 0 | 0 | Red | None | Closed shape | LC-EH? | |
| 624501 | IN | 2 | 2 | 0 | 0 | 0 | Red mica | None | ? | LC-EH? | |
| 624501 | IN | 1 | 1 | 0 | 0 | 0 | Light | None | ? | LC-EH? | |
| 624501 | IN | 1 | 1 | 0 | 0 | 0 | Red mica | None | ? | LC-EH? | |
| 624702 | IN | 1 | 1 | 0 | 0 | 0 | Red | None | Closed shape | LC-EH? | |
| 622601 | TR | 2 | 2 | 0 | 0 | 0 | Herakleian | None | Amphora | LC-EH? | |
| 622601 | TR | 1 | 0 | 0 | 1 | 0 | Thasian | None | Amphora | LC-EH? | |
| 622602 | TR | 2 | 2 | 0 | 0 | 0 | Thasian | None | Amphora | LC-EH? | |
| 622602 | TR | 1 | 0 | 0 | 1 | 0 | Red | None | Amphora | LC-EH? | **705** |
| 624401 | TR | 1 | 1 | 0 | 0 | 0 | Orange | None | Amphora? | LC-EH? | |

| BagID | Function | Total | Body | Rim | Handle | Base | Fabric | Surface | Shape | Period | Cat. |
|---|---|---|---|---|---|---|---|---|---|---|---|
| 624401 | TR | 3 | 3 | 0 | 0 | 0 | Red mica | None | Amphora? | LC-EH? | |
| 624402 | TR | 2 | 2 | 0 | 0 | 0 | Thasian | None | Amphora | LC-EH? | |
| 624402 | TR | 1 | 1 | 0 | 0 | 0 | Herakleian? | None | Amphora? | LC-EH? | |
| 624402 | TR | 1 | 0 | 0 | 1 | 0 | Red | None | Amphora? | LC-EH? | |
| 624501 | TR | 1 | 1 | 0 | 0 | 0 | Herakleian | None | Amphora | LC-EH? | |
| 624501 | TR | 2 | 2 | 0 | 0 | 0 | Red | None | Amphora | LC-EH? | |
| 624501 | TR | 2 | 2 | 0 | 0 | 0 | Thasian | None | Amphora | LC-EH? | |
| 624702 | TR | 1 | 1 | 0 | 0 | 0 | Herakleian | None | Amphora | LC-EH? | |
| 624702 | TR | 3 | 3 | 0 | 0 | 0 | Thasian | None | Amphora | LC-EH? | |
| 624402 | UT | 1 | 1 | 0 | 0 | 0 | Red | None | ? | LC-EH? | |
| 624402 | HW | 68 | 65 | 3 | 0 | 0 | Red | None | Pot | Early Hellenistic | **695-696, 698** |
| 624402 | HW | 1 | 0 | 1 | 0 | 0 | Brown | None | Pot | Early Hellenistc | |
| 624701 | IN | 1 | 1 | 0 | 0 | 0 | Red | Grooved | Closed shape | Byzantine? | |
| 622602 | IN | 3 | 3 | 0 | 0 | 0 | Red | None | Closed shape | Medieval? | |
| 624402 | IN | 3 | 3 | 0 | 0 | 0 | Red | None | Closed shape | Medieval? | |
| 624702 | IN | 2 | 2 | 0 | 0 | 0 | Red | None | Closed shape | Medieval? | |
| 624702 | UT | 2 | 0 | 0 | 1 | 1 | Tatarian | None | Jug? | EM | |
| 622602 | TW | 1 | 1 | 0 | 0 | 0 | Tatarian | Glazed | Jug | EM? | |
| 624702 | AR | 1 | 1 | 0 | 0 | 0 | Red modern | Slip | Tile | Modern | |
| 622601 | IN | 1 | 1 | 0 | 0 | 0 | Red modern | None | ? | Modern | |
| 622602 | TW | 1 | 1 | 0 | 0 | 0 | China | Glazed | Open shape | Modern | |
| 622602 | BO | 5 | 0 | 0 | 0 | 0 | | | | | |
| 622602 | BO | 1 | 0 | 0 | 0 | 0 | | | | | |
| 624401 | BO | 7 | 0 | 0 | 0 | 0 | | | | | |
| 624402 | BO | 23 | 23 | 0 | 0 | 0 | | | | | |
| 624702 | BO | 1 | 0 | 0 | 0 | 0 | | | | | |
| 624401 | HW | 20 | 20 | 0 | 0 | 0 | Red | None | Pot | ? | |
| 624401 | HW | 4 | 2 | 0 | 0 | 0 | Coarse | None | Pot | ? | |
| 624401 | HW | 2 | 0 | 2 | 0 | 0 | Grey | None | ? | ? | **692-693** |
| 624401 | HW | 4 | 4 | 0 | 0 | 0 | Black | None | Pot | ? | |
| 624402 | HW | 114 | 114 | 0 | 0 | 0 | Grey | None | ? | ? | |
| 624402 | HW | 42 | 42 | 0 | 0 | 0 | Black | None | Pot | ? | |
| 624701 | IN | 3 | 3 | 0 | 0 | 0 | Red | None | ? | ? | |
| 624702 | IN | 13 | 13 | 0 | 0 | 0 | Red | None | ? | ? | |
| 622602 | IR | 1 | 0 | 0 | 0 | 0 | | | | | |
| 622601 | LI | 3 | 0 | 0 | 0 | 0 | | | | | |
| 622602 | LI | 2 | 0 | 0 | 0 | 0 | | | | | |
| 624401 | LI | 2 | 0 | 0 | 0 | 0 | | | | | |
| 624401 | LI | 2 | 0 | 0 | 0 | 0 | | | | | |
| 624401 | LI | 1 | 0 | 0 | 0 | 0 | | | | | |
| 624401 | LI | 12 | 12 | 0 | 0 | 0 | | | | | |
| 624401 | LI | 2 | 0 | 0 | 0 | 0 | | | | | |

| BagID | Function | Total | Body | Rim | Handle | Base | Fabric | Surface | Shape | Period | Cat. |
|---|---|---|---|---|---|---|---|---|---|---|---|
| 624402 | LI | 1 | 1 | 0 | 0 | 0 | | | | | |
| 624501 | LI | 1 | 0 | 0 | 0 | 0 | | | | | |
| 624701 | LI | 1 | 0 | 0 | 0 | 0 | | | | | |
| 624701 | LI | 1 | 0 | 0 | 0 | 0 | | | | | |
| 624702 | LI | 1 | 0 | 0 | 0 | 0 | | | | | |
| 624702 | LI | 3 | 0 | 0 | 0 | 0 | | | | | |
| 624702 | LI | 2 | 0 | 0 | 0 | 0 | | | | | |

Finds table adjacent units

| BagID | Function | Total | Body | Rim | Handle | Base | Fabric | Surface | Shape | Period | Cat. |
|---|---|---|---|---|---|---|---|---|---|---|---|
| 622701 | HW | 3 | 3 | 0 | 0 | 0 | BaGrey | Grooved | Pot | BA? | |
| 622702 | - | 1 | 0 | 0 | 0 | 0 | | | | LC-EH? | |
| 622701 | HW | 7 | 7 | 0 | 0 | 0 | Grey | None | Pot | LC-EH? | |
| 622701 | HW | 6 | 6 | 0 | 0 | 0 | Red | None | Pot | LC-EH? | |
| 622701 | HW | 1 | 1 | 0 | 0 | 0 | Coarse | None | Pot | LC-EH? | |
| 622702 | HW | 6 | 5 | 0 | 0 | 1 | Light | Smoothed | Pot | LC-EH? | |
| 622702 | HW | 23 | 23 | 0 | 0 | 0 | Red | None | Pot | LC-EH? | |
| 622702 | HW | 18 | 18 | 0 | 0 | 0 | Grey | None | Pot | LC-EH? | |
| 624801 | HW | 5 | 4 | 1 | 0 | 0 | Grey | None | Pot | LC-EH? | |
| 626501 | HW | 1 | 1 | 0 | 0 | 0 | Black | None | Pot | LC-EH? | |
| 622702 | IN | 1 | 1 | 0 | 0 | 0 | Red | None | ? | LC-EH? | |
| 622702 | IN | 1 | 1 | 0 | 0 | 0 | Orange | None | Closed shape | LC-EH? | |
| 626501 | IN | 2 | 2 | 0 | 0 | 0 | Red | None | Closed shape | LC-EH? | |
| 626601 | IN | 1 | 1 | 0 | 0 | 0 | Red | None | ? | LC-EH? | |
| 622702 | TR | 1 | 1 | 0 | 0 | 0 | Sinopean | None | Amphora | LC-EH? | |
| 626601 | TR | 1 | 1 | 0 | 0 | 0 | Herakleian | None | Amphora | LC-EH? | |
| 622702 | TR | 1 | 0 | 0 | 1 | 0 | Red | None | Amphora? | Byzantine? | |
| 622701 | TW | 1 | 1 | 0 | 0 | 0 | Tatarian | Glazed | Open shape | EM | |
| 622702 | TW | 1 | 1 | 0 | 0 | 0 | Tatarian | Glazed | Open shape | EM? | |
| 622701 | IN | 2 | 2 | 0 | 0 | 0 | Red | None | Closed shape | ? | |
| 622701 | LI | 1 | 1 | 0 | 0 | 0 | | | | | |
| 622702 | LI | 1 | 0 | 0 | 0 | 0 | | | | | |
| 626501 | LI | 1 | 0 | 0 | 0 | 0 | | | | | |
| 626501 | TW | 1 | 0 | 0 | 0 | 0 | | | | | |

## Skalistoe 5[104]

Finds table core units

| BagID | Function | Total | Body | Rim | Handle | Base | Fabric | Surface | Shape | Period | Cat. |
|---|---|---|---|---|---|---|---|---|---|---|---|
| 302601 | IN | 32 | 32 | 0 | 0 | 0 | | | | LC-EH? | |
| 302601 | TW | 2 | 0 | 2 | 0 | 0 | Red | None | Jug | LC-EH | **709-710** |

Trial trench

| **Findslist 1: 0- 0.15 m. Top soil** | | | | |
|---|---|---|---|---|
| No. | Description | Quantity | Cat. | Note |
| | **Amphorai.** *Unknown production* | | | |
| 1 | Wall fr. Clay: Brown with spares pyroxene inclusions. Slip in colour of clay | 1 | | 2x3 cm |
| 2 | Wall fr. Clay: Reddish brown with spares pyroxene. Slip in colour of clay | 1 | | |
| | **Plain red ware** | | | |
| 3 | Wall fr. of Herakleian jug. Clay: Brown with fine pyroxene and spares lime and feldspar inclusions. Slip in colour of clay | 1 | | 3.5x2.5 cm |
| | **Handmade ware** | | | |
| 4 | Fr. of outturned rim. Clay: Grey on int. Brown on ext. | 1 | | 1.3x1.7 cm |
| 5 | Thick wall fr. Clay: Grey on int., light brown on ext. lime inclusions | 1 | | 1.5x1.5 cm |
| 6 | Wall fr. Clay: Grey on int. and ext. lime inclusions | 1 | | 2.5x1.3 cm |
| | **Organic material** | | | |
| 7 | Oyster shell | 1 | | |
| 8 | Mussel shell | 2 | | |
| **Findslist: 2. 0.15-0.30 m. Ash layer undet the top soil** | | | | |
| | **Amphorai** | | | |
| | *Herakleia* | | | |
| 1 | Wall fr. Clay: brown | 1 | | 8x6.5 cm |
| | *Chersonesos* | | | |
| 2 | Wall frs. Clay: Reddish brown with spares pyroxene and lime inclusions. Brown slip | 2 | | Join. Traces of burning |
| 3 | Wall fr. Clay: Reddish brown with spares pyroxene and lime inclusions | 1 | | 6 × 3.5 cm |
| 4 | Wall fr. Clay: lime inclusions. Greenish white slip | 1 | | Traces of soot |
| | *Unknow production* | | | |
| 5 | Wall frs. with grooves | 4 | | No joins. One fr. possibly from shape vessel as FL3/11 |
| 6 | Wall frs. | 6+1 | 743 | |
| 7 | Fine chips of wall | 9+1 | | |

[104] Descriptions of the excavation finds in the finds catalogue by S. Handberg (App. 2B) are based on autosy. In some cases there is not complete congruency with the findslists recorded by Lancov & Stolba for the annual reports of the Department of Antiquities.

| | **Table ware** | | | |
|---|---|---|---|---|
| 8 | Rim fr. of a jug with triangular lip. Part of handle preserved. Clay: Light green with fine pyroxene inclusions. Slip in colour of clay | 2 | 741 | Join. Similar clay as FL6/3,possibly from the samevessel |
| 9 | Wall. Frs. of a Hereakleian jug. Clay: reddish brown with fine pyroxene and feldspar inclusions. Cream slip | 2 | | From same vessel |
| 10 | Wall fr. of Chersonesean jug. Clay: reddish brown with lime inclusions | 2 | | From same vessel |
| 11 | Wall frs. of Chersonesean jug. Clay: Reddish brown with fine pyroxene and quartz inclusions. Light brown slip. Two horizontal red bands | 1 | 759 | |
| 12 | Wall frs. of jugs | 8 | | From same vessel |
| 13 | Wall fr. of Chersonesean louterion(?) with repair hole | 1 | | |
| | Cooking ware | | | |
| 14 | Rim fr. with ledge for a lid. Herakleian? | 1 | | |
| 15 | Wall frs. of cooking vessels | 2 | | |
| | **Glossed ware** | | | |
| 16 | Rim fr. of glazed kantharos | 1 | | |
| 17 | Fr. of bowl with brown gloss. Repair hole | | 761 | |
| 18 | Wall fr. of a Grey Ware jug. Slip in colour of clay | 1 | | |
| | **Handmade ware** | | | |
| 19 | Rim fr. | 1 | 712 | 1.3x1.4 cm |
| 20 | Rim fr. of pot with rounded lip | 1 | 713 | |
| 21 | Shoulder fr. of a pot with a reflief decoration | 1 | 714 | |
| 22 | Rim and base frs. of a pot | 3 | 715-716 | 2 joining frs. From same vessel |
| 23 | Rim fr. | 1 | 717 | |
| 24 | Wall frs. | 36 | | |
| 25 | Wall frs. | 4 | | |
| 26 | Kiln frs. | 3 | | |
| | **Organic material** | | | |
| 27 | Animal bones | 10 | | |
| 28 | Oyster shells | 11 | | |
| 29 | Mussel shells | 10 | | |
| 20 | Molluscs | 1 | | |
| **Findslist 3: Ash layer with a stone blocking** | | | | |
| | **Amphorai** | | | |
| | *Chersonesos* | | | |
| 1 | Handle with stamp | 1 | 735 | |
| 2 | Rim fr. | 1 | 736 | |
| 3 | Base fr. | | 737 | |
| 4 | Neck frs. with horizontal grooves | 2 | | |
| 5 | Neck fr. | 1 | | |
| 6 | Wall frs. | 25 | | One fr. with traces of burning. One fr. eroded by the sea |
| | *Sinope* | | | |

| | | | | |
|---|---|---|---|---|
| 7 | Handle fr. | 1 | 742 | |
| 8 | Wall fr. | 1 | | Traces of burning |
| 9 | Chips of wall frs. | 12 | | From same vessel |
| | *Thasos* | | | |
| 9a | Wall frs. | 2 | | |
| | *Unknown production* | | | |
| 10 | Base fr. | 2 | | 4 joining frs. |
| 11 | Wall fr. | 1 | | The same vessels as FL5/5 |
| 12 | Wall frs. | 6 | | |
| | **Red clay ware** | | | |
| 13 | Rim fr. of Chersonesean jug | 1 | | |
| 14 | Handled fr. | 1 | 745 | |
| 15 | Handled fr. | 1 | 746 | |
| 16 | Large base fr. of a Chersonesean jug | 1 | 748 | |
| 17 | Wall frs. of a Chersonesean jug with horizontal bands | 2 | | |
| 18 | Wall frs. | 11 | | |
| 19 | Wall frs. of jugs | 2 | | |
| 20 | Wall frs. of a Sinopean jug | 1 | | |
| 21 | Wall frs. of jugs | 8 | | |
| 22 | Handle fr. with contour of Chersonesean louterion. Repair hole | 1 | | |
| 23 | Fr. of Chersonesean louterion | 1 | 722 | |
| 24 | Rim fr. of Chersonesean bowl with inturned rim and yellowish slip | 1 | 762 | |
| 25 | Rim fr. of Chersonesean bowl with inturned rim and light slip | 1 | 763 | |
| 25a | Rim fr. of a bowl | 1 | 764 | |
| | **Kitchen ware** | | | |
| 26 | Rim fr. of pot with everted rim | 1 | | |
| 27 | Neck and shoulder fr. | 1 | | |
| 28 | Wall frs. | 5 | | |
| 29 | Wall fr. of thin walled saucepan | 1 | | |
| 30 | Rim fr. of bowl. Slip in colour of clay | | 765 | |
| | **Black Glazed** | | | |
| 31 | Neck of kantharos. Matt gloss | 1 | 747 | |
| 32 | Frs. of an echinus bowl | 6 | 766 | |
| 33 | Base fr. of a Grey Ware fish-plate | 1 | 767 | |
| | **Glossed pottery** | | | |
| 34 | Rim fr. of fish-plate with grooved edge | ? | 768 | |
| | **Handmade ware** | | | |
| 35 | Rim fr. wall with outturned rim | 1 | 718 | |
| 36 | Rim fr. wall with outturned rim | 2 | 719 | |
| 37 | Rim fr. of a large pot with outturned rim | 1 | 720 | |
| 38 | Rim fr. of a small pot | 1 | 723 | |
| 39 | Base fr. | 1 | 724 | |

| | | | | |
|---|---|---|---|---|
| 40 | Base fr. | 1 | | |
| 41 | Rim fr. of a pot with outturned rim. Part of handle attachment | 1 | **721** | |
| 42 | Handle fr. | 1 | **726** | |
| 43 | Rim fr. of a cup with inturned rim and handle attachment | 3 | **725, 727** | Not joining |
| 44 | Frs. of a perforated wall | 4 | **728** | |
| 45 | Thick walled frs. | 3 | | |
| 46 | Wall frs. with burnishing | 72 | | |
| 47 | Wall fr. | 20 | | |
| 48 | Wall frs. with polishing | 3 | | |
| 49 | Kiln frs. | 16 | | |
| | **Organic material** | | | |
| 50 | Animal bones | 16 | | |
| 51 | Oyster shells | 19 | | |
| 52 | Mussel shells | 29 | | |
| 52 | Shells | 1 | | |
| **Findslist 4: Upper part of the debris. 0.40-0.45 m** | | | | |
| | **Amphorai** | | | |
| | *Chersonesos* | | | |
| 1 | Base fr. | 1 | | |
| 2 | Wall frs. | 2 | | Not joining |
| 3 | Wall frs. | 5 | | |
| 4 | Wall frs. | 2 | | |
| | **Table ware** | | | |
| 5 | Wall fr. of a red clay jug | 1 | | |
| 6 | Wall fr. of a Grey Ware jug | 1 | | |
| 7 | Rim fr. of a Chersonesean bowl | 1 | **770** | |
| 8 | Rim fr. of a bowl. External part has chipped off | 1 | | |
| | **Black Glossed** | | | |
| 9 | Wall fr. of a cup | 1 | **769** | |
| | **Handmade ware** | | | |
| 10 | Rim frs. | 2 | **730-731** | |
| 11 | Wall frs. | 12 | | |
| | **Organic material** | | | |
| 12 | Animal bones | 3 | | |
| 13 | Mussel shells | 7 | | |
| **Findslist 5: Debris 0.45-0.60/65 m** | | | | |
| | **Tile** | | | |
| 1 | *Sinope* | 1 | **734** | |
| | **Amphorai** | | | |
| | *Chersonesos* | | | |
| 2 | Wall frs. | | | |
| | *Sinope* | | | |
| 3 | Wall frs. | 1 | | |

| | | | | |
|---|---|---|---|---|
| | *Unidentified production* | | | |
| 4 | Wall. frs. | 2 | | |
| 5 | Wall frs. chips | 1 | | |
| | **Red clay ware** | | | |
| 6 | Fr. of a Chersonesean mortar with handle attachment | 1 | | Composed from 10 frs. |
| 7 | Shoulder fr. of a Chersonesean jug with horizontal bands | 1 | 771 | |
| | **Handmade ware** | | | |
| 8 | Wall frs. | 3 | 732 | |
| 9 | Wall frs. with polishing | 1 | | |
| 10 | Kiln frs. | 1 | | |
| | **Metals** | | | |
| 11 | Frs. of an iron hammer | 1 | | |
| | **Organic material** | | | |
| 12 | Animal bones | 1 | | |
| 13 | Oyster shells | 1 | | |
| 14 | Mussel shells | 1 | | |
| 15 | Eggs | 3 | | |
| **Findslist 6: Level of the ancient surface 0.60 m** | | | | |
| | **Amphorai** | | | |
| | *Chersonesos* | | | |
| 1 | Wall frs. | 6 | | |
| | **Table ware** | | | |
| 2 | Wall fr. of a Red clay jug | 1 | | |
| 3 | Wall fr. of a jug. Slip in colour of clay | | | |
| | **Handmade ware** | | | |
| 4 | Rim fr. of a pot | 1 | 729 | |
| 5 | Base fr. of a pot | 1 | | |
| 6 | Wall frs. | 3 | | |
| | **Metals** | | | |
| 7 | Iron nails | 1 | | |
| 8 | Bones of animals | 6 | | |
| 9 | Oyster shells | 1 | | |
| 10 | Mussel shells | 1 | | |
| **Findslist 7: Foundation layer 1 m** | | | | |
| 1 | Chip of the base of a red clay jug | 1 | | |
| 2 | Kiln frs. | 1 | | |
| 3 | Mussel shells | 1 | | |
| **Findslist 8: Pit of filling to the north-east from wall no. 1** | | | | |
| 1 | Chip of wall fr. of an amphora | 1 | | |
| 2 | Base fr. of a Chersonesean jug | 1 | | |
| 3 | Wall fr. | 5 | | |
| 4 | Animal bones | 2 | | |
| **Findslist 9: Room 1 (to the north-east of wall no. 1), floor level** | | | | |
| 1 | Wall fr. of a red clay amphora | 1 | | |

| | | | | |
|---|---|---|---|---|
| 2 | Wall fr. of a handmade pot | 1 | | |
| 3 | Flat sandstone | 1 | | |
| **Findslist: 10: Finds exposed in the cliff** | | | | |
| No. | Description | Quantity | Cat. | Note |
| | **Tile** | | | |
| 1 | Edge fr. of unknown production | 1 | 733 | 8x8 cm |
| 2 | Fr. of a Chersonesean tile | 1 | | 6.5x3 cm |
| 3 | Fr. of a Sinopean tile | 1 | | 9.5x7.5 cm |
| | **Amphorai** | | | |
| | *Chersonesos* | | | |
| 4 | Frs. of rim and neck with part of handle and shoulder preserved. White horizontal band on neck | 7 | 738 | Four joining frs. |
| 5 | Rim fr. | | 740 | |
| 6 | Neck fr. with white horizontal band | 1 | | |
| 7 | Neck frs. | 4 | | One fr. with horizontal groove |
| 8 | Wall fr. | 20 | | |
| 9 | Fr. of handle | 1 | 739 | |
| | *Unknown production* | | | |
| 10 | Wall fr. | 2 | 744 | |
| | **Red clay ware** | | | |
| 11 | Fragmented Chersonesean jug with three bands of brown paint on the shoulder. Frs. of the neck and shoulder + the lowest part of a flat handle. Clay: light brown with an impurity of limestone | 10 + 1 | 749 | 7 + 2 frs. join |
| 12 | Fr. of the neck and upper part of handle from a Chersonesean jug | 1 | | |
| 13 | Handle fr. | 1 | 750 | Possibly from same vessel as FL10/15 and 10/17 |
| 14 | Handle fr. | 1 | | Possibly from same vessel as FL10/15 and 10/17 |
| 15 | Wall frs. of a Chersonesean jug with bands of a red paint on a shoulder. Yellow-greenish slip | 5 | | Two joining frs. |
| 16 | Wall frs. of a Chersonesean jug with bands of a red paint on a shoulder. Light brown slip | 3 | | Probably, from the same vessel, as FL10/14-15 |
| 17 | Wall frs. of Chersonesean jugs | 2 | | |
| 18 | Base fr. of a Chersonesean jug. Yellow-greenish slip | 1 | 751 | |
| 19 | Base fr. of a Chersonesean jug with false ring foot | 1 | 752 | |
| 20 | Base fr. of a Chersonesean jug with part of wall preserved | 7 | 753-754 | From one vessel |
| 21 | Base fr. of a Chersonesean jug with a ring foot | 1 | 755 | |
| 22 | Wall frs. | 20 | | one fr. with red paint |
| 23 | Wall fr. of a Herakleian jug | 1 | | |
| 24 | Base fr. of a jug from an unknown production. Clay: red-brown with sparse pyroxene and lime inclusions. Chersonesos? | 1 | 756 | |
| | **Plain red ware** | | | |
| 25 | Rim fr. of a deep Chersonesean bowl with inturned rim | 3 | 757 | Two joining frs. with traces of burning |
| | **Black glossed** | | | |

| 26 | Deep Chersonesean bowl with an outturned rim. Matt gloss | 2 | 758 | Joining |
|---|---|---|---|---|
| | **Handmade ware** | | | |
| 27 | Base fr. of a pot | 1 | 711 | |
| 28 | Wall frs. | 3 | | |
| | **Organic material** | | | |
| 4 | Animal bones | 12 | | |

## Skalistoe 6

Finds table core units

| BagID | Function | Total | Body | Rim | Handle | Base | Fabric | Surface | Shape | Period | Cat. |
|---|---|---|---|---|---|---|---|---|---|---|---|
| 302301 | IN | 4 | 0 | 0 | 0 | 0 | | | | LC-EH | |
| 302301 | BO | 1 | 0 | 0 | 0 | 0 | | | | | |
| 302301 | TR | 1 | 0 | 0 | 1 | 0 | Chersonesean? | | Amphora | LC-EH | 773 |

## Skalistoe 7

Finds table core units

| BagID | Function | Total | Body | Rim | Handle | Base | Fabric | Surface | Shape | Period | Cat. |
|---|---|---|---|---|---|---|---|---|---|---|---|
| 302801 | HW | 7 | 0 | 0 | 0 | 0 | | | | LC-EH | |
| 302801 | IN | 16 | 0 | 0 | 0 | 0 | | | | LC-EH | |

## Skalistoe 8

Finds table core units

| BagID | Function | Total | Body | Rim | Handle | Base | Fabric | Surface | Shape | Period | Cat. |
|---|---|---|---|---|---|---|---|---|---|---|---|
| 302901 | HW | 4 | 0 | 0 | 0 | 0 | HwGrey2 | | | LC-EH | |
| 302901 | IN | 25 | 0 | 0 | 0 | 0 | | | | LC-EH | |

## Skalistoe 9

Finds table core units

| BagID | Function | Total | Body | Rim | Handle | Base | Fabric | Surface | Shape | Period | Cat. |
|---|---|---|---|---|---|---|---|---|---|---|---|
| 302401 | HW | 8 | 0 | 0 | 0 | 0 | | | | LC-EH | |
| 302401 | IN | 49 | 0 | 0 | 0 | 0 | | | | LC-EH | |

## Skalistoe 10

Finds table core units

| BagID | Function | Total | Body | Rim | Handle | Base | Fabric | Surface | Shape | Period | Cat. |
|---|---|---|---|---|---|---|---|---|---|---|---|
| 302501 | IN | 3 | 0 | 0 | 0 | 0 | | | | LC-EH | |

## Vodopojnoe 1

Finds table core units

| BagID | Function | Total | Body | Rim | Handle | Base | Fabric | Surface | Shape | Period | Cat. |
|---|---|---|---|---|---|---|---|---|---|---|---|
| 306401 | IN | 7 | 0 | 0 | 0 | 0 | | | | LC-EH | |

## Vodopojnoe 3A and B

Finds table core units

| BagID | Function | Total | Body | Rim | Handle | Base | Fabric | Surface | Shape | Period | Cat. |
|---|---|---|---|---|---|---|---|---|---|---|---|
| 306501 | IN | 6 | 0 | 0 | 0 | 0 | | | | LC-EH | |
| 306501 | | 0 | 0 | 0 | 0 | 0 | | | | Medieval | |
| 306501 | BO | 1 | 0 | 0 | 0 | 0 | | | | | |

# Appendix 2B
# Finds catalogue

*S. Handberg*

## DSP06-F03-01, see Panskoe II

## DSP07-F01-01

Core units: 1005-1014, 2001-2009
Grab samples: 1004
Adjacent units: none

### Handmade ware

**Cat. 1** **1004/01/10**

Wall fr. of pot with smooth ext. and int. surface. Soot marks on the ext. H: 3; W: 4.9; T: 0.8. Sandwich fabric. Core: 5YR 4/? reddish brown. Ext.: 5YR 3/1 very dark grey. Frequent tiny lime inclusions; medium small to tiny quartz.

Early Hellenistic.

**Cat. 2** **1004/01/9**

Wall fr. of pot with smooth ext. and int. surface. Soot marks on the ext. H: 4.2; W: 2; T: 0.8. Sandwich fabric. Core: 5YR 4/? reddish brown. Ext.: 5YR 3/1 very dark grey. Frequent tiny lime inclusions; medium small to tiny quartz.

Early Hellenistic.

**Cat. 3** **1006/03/05**

Rim fr. of vessel with concave, outturned rim. H: 3.1; W: 3.1; T: 0.7; Ø? Clay: 7.5YR 5/4 brown. Core: 7.5YR 2.5/1 black. Sparse tiny lime inclusions; very sparse very tiny mica; very sparse sand.

Cf. Stolba 2002, type 1 Cat. D1-D68. Early Hellenistic.

**Cat. 4** **1006/03/08**

Body fr. of pot. H: 2.1; W: 2.9; T: 0.7. Core: 7.5YR 3/2 dark brown. Surface: 7.5YR 2.5/1 black. Sparse sand; sparse tiny to large quarz; sparse tiny lime.

Cf. Stolba 2002, type 1 Cat. D1-D68. Early Hellenistic.

**Cat. 5** **1006/03/09**

Rim fr. of vessel with concave, outturned rim. H: 1.7; W: 3.2; T: 0.8; Ø? Clay: 2.5YR 5/8 red. Sparse very tiny lime inclusions; medium sand.

Cf. Stolba 2002, type 1 Cat. D1-D68. Early Hellenistic.

**Cat. 6** **1006/03/10**

Body fr. of pot. Soot? marks on the ext. H: 3.6; W: 3.2; T: 0.5. Clay: 7.5YR 2.5/1 black. Sparse sand, sparse very tiny lime inclusions.

Hellenistic.

**Cat. 7** **1006/03/14**

Rim fr. of pot with outturned rim. H: 3.6; W: 3.6; T: 0.8; Ø? Clay: 2.5YR 5/6 red. Core: 7.5YR 2.5/1 black. Sparse tiny lime inclusions; medium sand.

Cf. Stolba 2002, type 1 Cat. D1-D68. Early Hellenistic.

**Cat. 8** **1006/03/20**

Wall fr. of jug? with small part of handle preserved. H: 3.9; W: 2.9; T: 0.5. Core: 7.5YR 3/2 dark brown. Surface: 7.5YR 2.5/1 black. Sparse very tiny lime inclusions; sparse sand.

Early Hellenistic.

**Cat. 9** **1006/03/23**

Wall fr. of Kizil-Koba pot with combed and impressed dot decoration. H: 4.5; W: 6.9; T: 0.6. Clay: 7.5YR 4/1 dark grey to 7.5YR 2.5/1 black. Medium tiny to small lime inclusions; sparse medium pebbles; medium sand; very sparse small quartz.

Cf. Stolba 2002, 185 Cat. D104. Early Hellenistic.

### Architectural elements

**Cat. 10** **1006/02/4**

Edge of roof tile. H: 4.2; W: 4.6; T: 1.7. Clay: 10R 7/4 pale red. Sparse small lime inclusions; sparse tiny to small feldspar; very sparse tiny pyroxene.

Hellenistic.

### Amphorai

*Chersonesos*

**Cat. 11** **2005/01/3**

Rim fr. H: 3; W: 4.7; T: 1.4; Ø: 14; D: 35. Clay: 2.5YR 6/6 light red. Sparse tiny lime; sparse tiny feldspar.

Early Hellenistic.

**Cat. 12** **1006/03/22**

Handle fr. H: 4.9; W: 4.3; T: 2.4. Clay: 2.5YR 5/8 red. Medium tiny lime inclusions; very sparse tiny pyroxene.

Late Classical-Early Hellenistic.

## *Herakleia*

**Cat. 13 1004/01/3**

Handle fr. H: 5.3; W: 4.5; T: 3.1. Core: 2.5YR 6/6 light red. Surface: 10R 6/8 light red. Sparse small pyroxene; frequent tiny to small feldspar; tiny to large lime.

4th century BC.

**Cat. 14 1004/01/4**

Handle fr. H: 8.6; W: 4.1; T: 2.2. Clay: 2.5YR 6/6 light red. Medium tiny to small lime inclusions; sparse small quartz; frequent tiny pyroxene; tiny feldspar.

4th century BC.

**Cat. 15 1006/01/2**

Handle fr. H: 6.1; W: 3.9; T: 2.6. Clay: 2.5YR 6/6 light red. Slip: 2.5YR 7/6 light red. Frequent tiny to medium pyroxene; medium tiny lime inclusions; sparse tiny to small feldspar.

4th century BC.

**Cat. 16 1006/02/10**

Handle fr. H: 3.3; W: 3.2; T: 1.1. Clay: 5YR 6/4 light reddish brown. Sparse tiny lime; medium tiny feldspar.

4th century BC.

**Cat. 17 1006/02/3**

Rim fr. H: 2.5; W: 3.3; T: 1.2; Ø: 7; D: 30. Clay: 10R 6/8 light red. Medium tiny to small lime inclusions; sparse tiny pyroxene; sparse small feldspar.

4th century BC.

**Cat. 18 2003/01/1**

Handle fr. H: 4.9; W: 3.7; T: 2.4. Clay: close to 10R 6/6 light red. Surface: 10R 6/8 light red. Sparse small pyroxene. Medium tiny to small lime inclusions; sparse tiny feldspar; sparse small shells.

Conical or biconical type. 4th century BC.

**Cat. 19 2005/01/1**

Handle fr. (badly preserved). H: 9.1; W: 4.2; T: 2.3? Clay: 2.5YR 6/8 light red. Medium tiny pyroxene; medium small to tiny lime inclusions; sparse tiny shells; sparse tiny feldspar.

4th century BC.

## *Sinope*

**Cat. 20 1006/01/3**

Handle fr. with finger impression at attachment. H: 10.8; W: 6.5; T: 2. Clay: 2.5YR 5/8 red. Slip: 10YR 7/4 very pale brown. Medium tiny to large pyroxene; medium tiny to large lime inclusions.

Late Classical-Hellenistic.

**Cat. 21 1006/02/2**

Fr. of lower body? Badly preserved. H: 5.2; W: 5.8; T: 0.9. Clay: 10YR 6/2 light brownish grey. Frequent tiny to small pyroxene; very sparse small golden mica; very sparse tiny lime inclusions.

4th century BC.

**Cat. 22 2005/01/2**

Handle fr. H: 4.5; W: 4.4; T: 2.5. Clay: 7.5YR 6/4 light brown. Large to small pyroxene; medium tiny to small lime inclusions; sparse sand.

Probably pithoid type. Ca. 350-250 BC.

## *Thasos*

**Cat. 23 1006/02/1**

Handle fr. H: 6.5; W: 5.3; T: 2.7. Clay: 10R 4/8 red. Slip: close to 10R 4/6 red. Frequent tiny mica; medium tiny to small lime inclusions; sparse large feldspar.

Conical type II-C-2? Cf. Monachov 2003, pl. 48. First half of 4th century BC.

**Cat. 24 1013/01/1**

Base fr. Slightly flaring with cavity on underside. H: 9; Ø: 6. Core: 2.5YR 5/8 red. Surface: 2.5YR 6/4 light reddish brown. Frequent very tiny to small mica; very sparse medium quartz; sparse tiny to medium lime inclusions; sparse sand.

Biconical type II-B-1. Cf. Monachov 2003, pls. 42-43. First half of 4th century BC.

## *Unidentified production*

**Cat. 25 1004/01/1**

Handle fr. H: 2.7; W: 3.8; T: 1.6. Clay: 5YR 6/6 reddish yellow. Very sparse very tiny quartz; sparse very tiny to tiny lime inclusions.

Early Hellenistic?

**Cat. 26 1006/01/1**

Handle fr. H: 4.9; W: 3.4; T: 1.7. Clay: 5YR 5/8 yellowish red. Slip: 7.5YR 7/4 pink. Tiny to large lime inclusions; sparse tiny feldspar; very sparse tiny pyroxene.

Late Classical-Early Hellenistic.

**Cat. 27 1006/03/04**

Handle fr. H: 4.9; W: 2.7; T: 1.9. Clay: 10R 6/4 pale red. Sparse tiny lime inclusions; very sparse golden and silver mica; very sparse tiny feldspar.

Late Classical-Early Hellenistic.

**Cat. 28 2002/01/1**

Handle fr. H: 11.3; W: 4; T: 2.3. Clay: 5YR 5/8 yellowish red. Medium tiny to small lime inclusions; frequent tiny to medium feldspar; very sparse small pyroxene.

Classical-Early Hellenistic.

**Cat. 29 2004/01/1**

Body fr. with part of the handle preserved. H: 4.7; W: 7.2; T: 0.9. Clay: 5YR 5/6 yellowish red. Core: 5Y 5/1 grey. Frequent tiny to very tiny mica; very sparse small quartz; very sparse small feldspar.

Classical-Hellenistic.

**Cat. 30 2004/01/3**

Handle fr. H: 2.9; W: 3.4; T: 1.4. Core: 2.5Y 5/2 greyish brown to 2.5YR 5/6 red. Surface: 10YR 6/4 light yellowish brown. Frequent tiny to medium lime inclusions; sparse small quartz.

Classical-Hellenistic.

**Cat. 31 2004/01/4**

Handle fr. H: 6.3; W: 4.2; T: 2.2. Clay: 7.5YR 7/4 pink. Frequent tiny to medium pyroxene; sparse tiny lime inclusions; medium tiny to small feldspar.

Sinope?, Classical-Hellenistic.

**Cat. 32 2004/01/5**

Handle fr. H: 6.6; W: 4.2; T: 2.8. Clay: 2.5Y 6/3 light yellowish brown. Slip: 2.5Y 7/4 pale yellow. Sparse medium to small brown peebles; sparse tiny lime inclusions; sparse small to tiny feldspar.

Chersonesos?, Classical-Hellenistic.

## Tabel ware

### *Chersonesean jugs*

**Cat. 33 1004/01/12**

Wall fr. of jug. H: 1.7; W: 2.8; T: 0.8. Core: 2.5YR red. Slip: 5YR 7/4 reddish yellow. Medium tiny lime inclusions; sparse very tiny shells; very sparse small quartz.

Cf. Kašaev 2002, 151-152, Cat. C7-23. Late 4th-3rd century BC.

**Cat. 34 1004/01/2**

Handle fr. of jug. H: 4.2; W: 3.4; T: 1.2. Clay: 5YR 7/6 reddish yellow. Sparse medium lime inclusions; sparse medium pyroxene; medium small to tiny feldspar.

Cf. Kašaev 2002, 151-152, Cat. C7-23. Late 4th-3rd century BC.

**Cat. 35 1004/01/5**

Base fr. of jug with flat underside and slightly moulded foot. H: 2.2; W: 7.3; T: 1-1.3; Ø: 12; D: 7.5. Clay: 10R 7/8 light red. Very sparse tiny lime inclusions; sparse small pyroxene; sparse medium to small quartz; sparse small to tiny feldspar.

The shape of the foot is similar to the jugs from Panskoe I, cf., e.g., Kašaev 2002, Cat. C13, C16. Late 4th-3rd century BC.

**Cat. 36 1006/02/8**

Wall fr. of jug. H: 2.3; W: 3.2; T: 0.7. Clay: 5YR 7/6 reddish yellow. Slip: 10YR 8/2 very pale brown. Sparse tiny lime inclusions; sparse tiny pyroxene; sparse tiny feldspar.

Cf. Kašaev 2002, 151-152, Cat. C7-23. Late 4th-3rd century BC.

**Cat. 37 2006/01/1**

Fr. of projecting rim of jug. H: 3.4; W: 4.6; T: 0.7; Ø: ca. 10. Clay: 2.5YR 6/8 light red to 5YR 5/6 yellowish red. Small part of slip is preserved on neck. Glaze: 7.5YR 3/1 very dark grey. Medium small to tiny lime inclusions; very sparse sand; very sparse pyroxene; very sparse tiny feldspar.

Cf. Kašaev 2002, 151-152, Cat. C7-23. Late 4th-3rd century BC.

**Cat. 38 1004/01/6**

Fr. of grooved handle of jug? H: 3.4; W: 1.4; T: 1.2. Core: Gley 2 4/10 BG dark greenish grey. Slip: 2.5YR 4/8 red. Very sparse very tiny lime inclusions.

Chersonesos? Similar handles found in Panskoe I, cf.Kašaev 2002, Cat. C15. Late 4th-3rd century BC.

**Cat. 39 1004/01/7**

Base fr. of jug? with a false ring foot. H: 1.5; W: 5.7; T: 1.4; Ø: ca. 12; D: 60. Clay: 2.5YR 5/8 red. Frequent small to tiny lime inclusions; sparse medium quartz; medium small to tiny shells.

Chersonesos? Similar to some of the jugs in Panskoe I, cf. Kašaev 2002, C9, C13, C15. Late 4th-3rd century BC.

**Cat. 40 1006/02/9**

Rim fr. of jug? with a groove on the upper part of the lip. H: 1.6; W: 3.9; T: 0.5; Ø: 8-10. Clay: 10R 5/6 red. Slip: 10YR 7/3 very pale brown. Frequent tiny to small lime inclusions.

Chersonesos? Cf. Kašaev 2002, 151-152, Cat. C16. Late 4th-3rd century BC.

### *Other*

**Cat. 41 1004/01/8**

Rim fr. of small jug with ledge on the upper lip possibly for lid. H: 2.2; W: 3; T: 0.4; Ø: ca. 4; D: 60. Core: 2.5YR 6/8 light red. Slip: 2.5YR 5/8 red. Sparse tiny lime inclusions.

Late Classical-Hellenistic.

**Cat. 42 1005/01/01**

Wall fr. of closed shape, presumably a jug. Small horizontal relief line on the ext. Part of repair hole preserved. H: 2.6; W: 2.5; T: 0.3-0.4. Core: 7.5YR 5/3 brown. Surface: 2.5Y 6/2 light brownish grey. Smoothed on the ext. Sparse tiny lime inclusions; medium small to tiny feldspar.

Late Classical-Hellenistic.

**Cat. 43 1006/01/11**

Base fr. of jug with flat underside and false ring foot. H: 1.6; W: 4.3; T: 0.9. Clay: 5YR 5/8 yellowish red. Medium tiny lime inclusions.

Similar to some of the Chersonesean jugs in Panskoe I, cf. Kašaev 2002, C9, C13, C15. Late 4th-3rd century BC.

**Cat. 44 1006/02/7**

Rim fr. of jug with projecting rim. H: 3.1; W: 4.8; T: 0.7; Ø: ca. 12. Core: 5YR 4/3 reddish brown to 5YR 4/2 dark reddish grey. Slip: 2.5Y 6/2 light brownish grey. Sparse small to tiny lime inclusions; sparse tiny feldspar.

Cf., e.g., Kašaev 2002, Cat. C8. Late 4th-3rd century BC.

**Cat. 45 1006/03/16**

Rim fr. of jug? with a flat lip. H: 1.6; W: 3; T: 0.4-0.8; Ø: ca. 9. Clay: 5YR 8/4 pink. Sparse tiny lime inclusions; sparse tiny feldspar.

Late Classical-Hellenistic.

**Cat. 46 1006/03/18**

Two joining rim frs. of jug with flaring rim and flat lip. H: 2.8; W: 5.2; T: 0.5; Ø: 11. Clay: 2.5YR 5/8 red. Surface: 2.5YR 6/6 light red. Medium small pyroxene; medium small to tiny lime inclusions; sparse small quartz.

Sinope, Late Classical-3rd century BC.

**Cat. 47 1006/03/19**

Fr. of neck from a Grey Ware jug with ribbed neck. H: 4.4; W: 3.2; T: 0.6. Clay: 7.5YR 5/1 grey. Slip: 10YR 7/2 light grey. Sparse small to tiny lime inclusions.

Late 4th-Early 3rd century BC.

**Cat. 48 1006/01/10**

Rim fr. of bowl with inturned rim. Possibly from a one-handler. H: 2.9; W: 3.9; T: 0.9; Ø: ca. 19. Core: 2.5YR 5/8 red. Slip: 7.5YR 7/4 pink. Very sparse very tiny lime inclusions; very sparse tiny mica.

For a comparison to one-handlers, cf. Kašaev 2002, Cat. 37. Early Hellenistic.

**Cat. 49 1006/01/4**

Base fr. with a ring foot. H: 3; W: 5.7; T: 0.5; Ø: ca. 8. Clay: 5YR 5/8 yellowish red. Slip: 7.5YR 8/3 pink. Sparse tiny lime inclusions; medium small to tiny pyroxene; very sparse sand.

Chersonesos, cf., e.g., Kašaev 2002, Cat. C17. Late 4th-3rd century BC.

**Cat. 50 1006/01/5**

Rim fr. of small bowl or salt-cellar. H: 2.3; W: 2.1; T: 0.2; Ø: ca. 5. Clay: 7.5YR 5/3 brown. Glaze: Gley 2 2.5/5 B bluish black. No visible inclusions.

Late 4th-3rd century BC.

**Cat. 51** 1006/01/6

Wall fr. of closed shape, glazed. H: 1.4; W: 1.7; T: 0.4. Clay: 7.5YR 5/4 Brown. Glaze: Gley 2 2.5/5 PB bluish black. Very sparse very tiny lime inclusions.

Late Classical-Hellenistic.

**Cat. 52** 1006/02/11

Body fr. of Attic cup, possibly a cup-skyphos. Badly preserved. H: 2.8; W: 3.6; T: 0.8; Ø? Clay: 5YR 6/6 reddish yellow. Glaze: 7.5YR 2.5/1 black. Sparse very tiny light-reflecting particles; very sparse tiny lime inclusions.

Cf. Sparkes-Talcott 1970, Cat. 612-623. Late 5th-Early 4th century BC.

**Cat. 53** 2003/01/3

Rim fr. of lekythos. H: 2.6; W: 2.9; T: 0.4; Ø: ca. 5. Clay: 10R 4/3 dusky red. Glaze: 7.5YR 2.5/1 black. Sparse tiny to small lime inclusions.

Cf. Kašaev 2002, Cat. C88-C93. Late Classical-Early Hellenistic.

**Cat. 54** 1006/02/6

Base fr. of small closed vessel, possibly a lekythos, with a ring foot. H: 1.2; W: 2.3; T: 0.6; Ø? Clay: 5Y 8/2 pale yellow. Sparse tiny lime inclusions; sparse tiny feldspar.

Cf. e.g., Kašaev 2002, Cat. C89-C93. Late 4th-3rd century BC.

**Cat. 55** 1006/02/5

Downturned rim fr. of fish-plate? H: 1.2; W: 3.8; T: 0.4; Ø: ca. 20. Clay: 10R 6/8 light red. Glaze: 2.5YR 2.5/1 reddish black. Sparse tiny lime inclusions; very sparse tiny feldspar.

Cf. Rotroff 1997, 148-152. 3rd century BC.

**Cat. 56** 1006/03/02

Fr. of ring foot of Grey Ware fish-plate. H: 2.9; W: 2.6; T: 0.4-0.8; Ø? Clay: 7.5YR 5/1 grey. Slip: Gley 2 4/10 BG dark greenish grey. Sparse tiny lime inclusions.

Cf. Stolba 2002, Cat. B222-B235. Late 4th-early 3rd century BC.

**Cat. 57** 1006/03/15

Small rim fr. of fish-plate with a reserved band and a groove at the edge. Badly preserved. H: 2.3; W: 3; Ø? Clay: 5YR 6/6 reddish yellow. Glaze: 7.5YR 2.5/1 black. Very sparse light-reflecting particles.

Cf. Rotroff 1997, 148-152. Early Hellenistic.

**Cat. 58** 1006/03/07

Fr. of lower part of kantharos with stem attachment preserved. H: 3.7; W: 6.1; T: 0.7; Ø: ca. 7. Clay: 2.5YR 5/8 red. Glaze ext.: 2.5YR 3/2 dusky red to 5YR 4/4 reddish brown. Glaze int.: 2.5YR 4/8 red. Sparse very tiny light-reflecting particles.

Cf. Rotroff 1997, 82-85 for the type. Ca. 325-250 BC.

**Cat. 59** 1006/03/21

Rim fr. of bowl with incurved rim. H: 2.9; W: 5; T: 0.7; Ø: ca. 20. Clay: 5YR 5/6 yellowish red. Frequent very tiny lime inclusions.

Cf. Rotroff 1997, Cat. 984-989. Ca. 310-250 BC.

**Cat. 60** 1006/03/3

Rim fr. from a classical type kantharos with plain rim. Part of one handle attachment preserved. H: 3.3; W: 5.1; T: 0.4; Ø: 7. Clay: 10R 5/6 red. Glaze: 7.5YR 2.5/1 black, semi-lustrous. No visible inclusions.

Cf. Rotroff 1997, 82-85 for the type. Ca. 325-250 BC.

**Cat. 61** 1011/01

Rim fr. from a Grey Ware bowl with moulded lip. Part of repair hole preserved. H: 2.5; W: 4.6; T: 0.5-0.8; Ø: 16. Clay: 10YR 6/2 light brownish grey. Slip: 5Y 3/1 very dark grey. Very sparse tiny lime inclusions; very sparse tiny feldspar.

Ca. 400-350 BC.

**Cat. 62** 2005/01/5

Base fr. of jug with a flat underside. H: 3.1; W: 4.4; T: 0.8-1.8; Ø? Clay: 2.5YR 5/6 red. Slip: 5Y 8/3 pale yellow. Sparse tiny lime inclusions; very sparse tiny feldspar.

Cf. e.g., Kašaev 2002, Cat. C254. Late 4th-3rd century BC.

**Cat. 63** 2006/01/2

Rim fr. of cup? Badly preserved. H: 1.5; W: 2.2; T: 0.6; Ø? Clay: 2.5YR 6/8 light red. Tiny lime inclusions; tiny feldspar.

Late Classical-Early Hellenistic?

## Utility ware

**Cat. 64** 1006/01/12

Rim fr. of deep louterion with downturned rim. H: 2.4; W: 5.5; T: 1.1; Ø: >30. Clay: 2.5YR 6/6 light red. Sparse small to tiny lime inclusions; medium sand; very sparse small pebbles; very sparse medium to small quartz; frequent small to medium voids.

Numerous examples were found in Panskoe I U6, but no exact parallels for the lip can be found among the published material, cf. Kašaev 2002, 158-160, Cat. C188-C201. Early Hellenistic.

**Cat. 65** 1004/01/13

Rim fr. of louterion with a part of applied 'finger' handle. Somewhat rough surface. H: 2.8; W: 4.1; T: 1.2; Ø: Ca. 19. Clay: 10R 7/6 light red. Slip: 7.5YR 8/4 pink. Sparse tiny lime inclusions; sparse sand inclusion.

Numerous examples of the type in Panskoe I U6, cf. Kašaev 2002, 158-160, Cat. C202-C214. Early Hellenistic.

**Cat. 66** 1004/01/14

Rim fr. of deep louterion/mortar with an applied broad handle that overlaps the rim. Lip is square with a projecting moulding towards the int. H: 5.4; W: 13.4; T: 2.8; Ø? Clay: between 7.5YR 5/3 brown to 7.5YR 5/4 brown. Slip: close to 5YR 8/4 pink. Sparse medium to tiny lime inclusions; sparse medium to tiny pyroxene.

Numerous examples were found in Panskoe I U6, but no exact parallels for the lip can be found among the published material. Kašaev 2002, 158-160-Cat. C188-C201. Early Hellenistic.

**Cat. 67** 1004/01/17

Rim fr. of louterion/basin with outturned rim. Badly preserved. H: 3.7; W: 5.8; T: 1.7; Ø? Clay: 2.5YR 7/6 light red. Medium tiny lime inclusions; medium tiny to small feldspar.

Cf. Kašaev 2002, 159, cf. Cat. C215-C220. Early Hellenistic.

**Cat. 68** 1006/01/14

Fr. of projecting rim of small basin. H: 1.9; W: 3.9; T: 0.9; Ø? Clay: 2.5YR 5/8 red. Slip: 7.5YR 8/4 pink. Sparse very tiny lime particle; very sparse sand; very sparse tiny to small quartz.

Resembles the larger louteria in Panskoe I, cf. Kašaev 2002, Cat. C215-C220. Early Hellenistic.

**Cat. 69** 1004/01/19

Fr. of broad flat projecting rim. H: 4.5; W: 6.9; T: 1-3.6; Ø: ca. 30; D: 40. Clay: 2.5YR 6/8 red. Frequent medium to tiny pyroxene; sparse small to tiny feldspar.

Sinopean pithos. Similar, although with a straight rim, was found in Panskoe I U6, cf. Kac et al. 2002, 103-104, Cat. Ac3. Early Hellenistic.

**Cat. 70** **1006/01/13**

Fr. of straight rim. H: 2.3; W: 6.2; T: 2.5; Ø? Clay: 2.5YR 6/6 light red. Very sparse tiny lime inclusions; frequent large to tiny pyroxene; sparse small feldspar.

Sinopean pithos, cf. Kac et al. 2002, 103-104, Cat. Ac3. Early Hellenistic.

**Cat. 71** **1006/01/7**

Rim fr. of cooking pot. H: 1.7; W: 1.7; T: 0.6; Ø? Clay: 5YR 6/6 reddish yellow. Very sparse tiny lime inclusions.

Cf. Kašaev 2002, Cat. C122-127, especially C126. Late 4th-3rd century BC.

**Cat. 72** **1006/01/8**

Rim fr. of cooking pot with an int. ledge for a lid. H: 2.9; W: 3.5; T: 0.5; Ø? Clay: 2.5YR 5/8 red. Slip: 10R 4/6 red. Medium sand; very sparse small to tiny quartz.

Cf. Kašaev 2002, Cat. C122-127. Late 4th-3rd century BC.

**Cat. 73** **1006/01/9**

Rim fr. of cooking pot with an int. ledge for a lid. H: 2.5; W: 2.8; T: 0.6; Ø? Clay: 5YR 5/8 yellowish red. Slip: 2.5YR 5/6 red. Very sparse tiny to very tiny lime inclusions; very sparse tiny mica.

Cf. Kašaev 2002, Cat. C122-127. Late 4th-3rd century BC.

**Cat. 74** **1006/02/12**

Rim fr. of cooking pot. H: 1.4; W: 1.7; T: 0.6; Ø? Clay: 10R 4/8 red. Sparse tiny lime inclusions; sparse tiny pyroxene.

Cf. Kašaev 2002, Cat. C122-127. Late 4th-3rd century BC.

**Cat. 75** **1006/03/01**

Rim fr. of basin/louterion with projecting rim. H: 2.6; W: 6.1; T: 0.9; Ø? Clay: 10R 5/8 red. Frequent tiny to medium lime inclusions; very sparse tiny pyroxene. Similar to 1006/01/14.

Hellenistic.

**Cat. 76** **1006/03/06**

Rim fr. of cooking pot with small ledge int. for a lid. H: 2.7; W: 2.9; T: 0.5; Ø? Clay: 5YR 5/8 yellowish red. Slip: 2.5YR 5/6 red. Medium tiny lime inclusions; medium tiny mica.

Cf. e.g., Kašaev 2002, Cat. C132. Late 4th-3rd century BC.

**Cat. 77** **1006/03/17**

Rim fr. of small basin with projecting rim. Badly preserved. H: 3.9; W: 3.2; T: 2.2; Ø? Clay: 7.5YR 6/4 brown. Sparse tiny lime inclusions; sparse sand; medium small pyroxene. Similar to 1006/01/14.

Early Hellenistic.

**Cat. 78** **2003/01/2**

Rim fr. of cooking pot with an int. ledge for a lid. Flat lip. H: 2; W: 2.7; T: ca. 0.7; Ø: 18. Clay: 7.5YR 5/4 brown. Sparse tiny lime inclusions; sparse tiny feldspar.

Cf. Kašaev 2002, Cat. C122-127. Late 4th-3rd century BC.

**Cat. 79** **2003/01/4**

Base fr. of base or louterion with a straight base and a flat underside. H: 6.8; W: 7.9; T: 0.8-1.8; Ø? Clay: 10R 6/8 light red. Slip: 2.5YR 7/6 light red. Sparse tiny to small lime inclusions; sparse tiny feldspar.

Kašaev 2002, e.g. Cat. C192-C193, C197-C198, although all these examples have a moulded base. Early Hellenistic.

**Cat. 80** **2004/01/2**

Fr. of broad projecting rim. H: 3.1; W: 6.1; T: 3.5; Ø? Clay: 5YR 6/6 reddish yellow. Frequent tiny to medium pyroxene; sparse tiny lime inclusions; sparse small to medium quartz.

Sinopean pithos, similar to 1004/01/19. Early Hellenistic.

# DSP07-F01-02, see Panskoe V

# DSP07-F02-01

Core units: 2164-65, 2205-206
Adjacent units: 2166, 2194-95

## Handmade ware

**Cat. 81** **2164/02/1**

Base fr. of pot with moulded base and flat underside. H: 1.9; W: 1.8; T: 0.6-0.8; Ø?; D: 5. Core: 5YR 2.5/1 black. Surface: 5YR 6/3 light reddish brown. Very sparse tiny lime inclusions; medium organic inclusions.

Cf. e.g., Stolba 2002, Cat. D67. Early Hellenistic.

## Amphorai

### *Herakleia*

**Cat. 82** **2206/02/1**

Rim fr. H: 1.5; W: 4; T: 1.2; Ø: 11.8. Clay: 10YR 7/4 very pale brown. Very sparse medium quartz; sparse medium to tiny lime inclusions; sparse small to tiny feldspar.

Possibly type II-3. Cf. Monachov 2003, Pl. 94. Ca. 360-330 BC.

### *Unidentified production*

**Cat. 83** **2164/02/2**

Handle fr. H: 6.6; W: 3.5; T: 1.9. Clay: 5YR 5/6 yellowish red. Surface: 10R 7/6 light red. Sparse tiny lime particle; very sparse small pyroxene; sparse tiny feldspar.

Classical-Hellenistic.

**Cat. 84** **2205/01/1a**

Handle fr., badly preserved. H: 8; W: 4.6; T: 4.3? Clay: 5YR 6/6 reddish yellow. Core: 5YR 6/4 light reddish brown. Small to tiny lime inclusions; very sparse tiny feldspar.

Chian? Late Classical-Hellenistic.

## Tableware

**Cat. 85** **2205/02/1**

Rim fr. of jug with protruding lip. H: 1.8; W: 2.2; T: 0.6; Ø? Clay: 10YR 7/4 very pale brown. Sparse small lime inclusions; very sparse small mica.

Late 4th-3rd century BC.

**Cat. 86** **2205/01/1b**

Body fr. of jug with wavy band on the ext. H: 3.5; W: 4.4; T: 0.7. Clay: 2.5YR 6/6 light red. Slip: 10YR 7/3 very pale brown. Decoration: 2.5YR 6/4 light reddish brown. Sparse small to tiny lime inclusions.

Chersonesos, cf. Kašaev 2002, 151-152, Cat. C7-23. Late 4th-3rd century BC.

# DSP07-F02-02

Core units: 2100-02, 2117-19, 2146-47
Adjacent units: 2104
Grab samples: 1002

## Amphorai

### *Herakleia*

**Cat. 87 1002/2**

Handle fr. H: 7.9; W: 4.1; T: 2.5. Core: 5YR 6/6 reddish yellow. Slip: 7.5YR 6/6 reddish yellow. Medium tiny to medium lime inclusions; sparse small to medium pyroxene; sparse tiny to small feldspar.
4th-early 3rd century BC.

## Tableware

**Cat. 88 1002/4**

Base fr. of jug? with straight walls and a flat underside. H: 2.9; W: 3.6; T: 1.1; Ø? Clay: 2.5YR 5/6 red. Medium small to medium pebbles; sparse tiny to medium quartz; sparse tiny to medium lime inclusions.
Early Modern?

## Utility ware

**Cat. 89 1002/3**

Fr. of flat base of closed shape. H: 2; W: 4.5; T: 1.2; Ø? Core: 7.5YR 6/4 light brown. Sparse small to medium lime.
Early Modern.

# DSP07-F02-03

Core units: 1233-34
Adjacent units: 1228, 1230, 1259, 1261

## Amphorai

### *Chersonesos*

**Cat. 90 1234/01/1**

Handle fr. H: 4.2; W: 4; T: 2.3. Clay: 2.5YR 5/6 red. Medium tiny to small lime inclusions; sparse tiny to small feldspar.
Early Hellenistic.

**Cat. 91 1234/02/1**

Handle fr. H: 4.6; W: 5.6; T: 2.2. Clay: 2.5YR 4/8 red. Slip: 2.5YR 5/6 red. Medium small pyroxene; medium tiny to large lime inclusions; sparse small feldspar.
Early Hellenistic.

# DSP07-F02-04

Core units: 1134, 1181
Adjacent units: 1089, 1103-04, 1120, 1122, 1132-33, 1150-51, 1162, 1180, 1192

## Handmade ware

**Cat. 92 1089/01/1**

Wall fr. of pot with a horizontal moulding on the ext. H: 2.8; W: 2.7; T: 0.7. Clay: 2.5Y 3/1 very dark grey. Surface: 10YR 5/3 brown. Sparse very tiny lime inclusions; sparse smal quartz.
Cf. Stolba 2002, Cat. D69-D70. Early Hellenistic.

**Cat. 93 1103/01/2**

Rim fr. with slightly concave, outturned rim and flat lip. H: 3.2; W: 4.2; T: 0.9; Ø: 19. Clay: 2.5Y 2.5/1 black. Surface: 2.5Y 6/2 light brownish grey. Sparse tiny lime inclusions; sparse sand.
Cf. Stolba 2002, type 1 Cat. D1-D68. Early Hellenistic.

## Amphorai

### *Chersonesos*

**Cat. 94 1134/01/1**

Rim fr. H: 3.2; W: 5.8; T: 2; Ø: 15; D: 50. Clay: 2.5YR 6/8 light red. Frequent tiny to large lime inclusions.
The flat upper lip is most common in Monachov's rim group 3. Cf. Monachov 1989, 71. Chersonesos? Early Hellenistic.

**Cat. 95 1120/01/1**

Handle fr. H: 5.1; W: 4.6; T: 1.9. Clay: 10R 5/6 red. Slip: 10YR 8/2 very pale brown. Very sparse very tiny mica; sparse tiny quartz; medium tiny to small lime inclusions; sparse tiny to small feldspar.
Chersonesos? Early Hellenistic.

### *'Samos'*

**Cat. 96 1122/01/1**

Fr. of slightly undercut rim. H: 3.5; W: 3.4; T: 1.8; Ø: 12; D. 35. Clay: 2.5YR red. 2.5YR 6/6 light red. Frequent very tiny light-reflecting particles; sparse very tiny to tiny lime inclusions; sparse very tiny feldspar.
Zeest, 'Samian' type, variant, cf. Cook & Dupont 1998 fig. 23.12b. Late Archaic.

# DSP07-F02-05

Core units: 1096, 1100
Adjacent units: 1090, 1125
No diagnostic material.

# DSP07-F03-01

Core units: 1361-1362, 2301, 2304
Adjacent units: 1353, 2300, 2302

### Amphorai

#### *Chersonesos*

**Cat. 97 2304/01/2**

Fr. of base with slightly flaring foot and cavity on underside. Dividing horizontal groove on upper part. H: 4.5; W: 4.8; Ø? Clay: 5YR 6/4 light reddish brown. Frequent very tiny to medium lime; medium tiny to small feldspar.

Hellenistic.

#### *Herakleia?*

**Cat. 98 2304/02/2**

Rim fr. H: 4.3; W: 3.5; T: 1.2; Ø: 11. Clay: 5YR 6/6 reddish yellow. Sparse tiny lime inclusions; sparse sand; sparse tiny feldspar.

Somewhat resembles Herakleian type II-A-2, cf. Monachov 2003, pl. 96.2. Late 4th-early 3rd century BC?

#### *Unknown productions*

**Cat. 99 2304/01/1**

Handle fr. H: 5.5; W: 2.3; T: 2. Clay: 2.5YR 6/6 light red. Sparse tiny lime inclusions; medium tiny to small feldspar.

Classical-Hellenistic.

**Cat. 100 2304/02/1**

Handle fr. H: 6.5; W: 5.2; T: 2.2. Clay: 5YR 5/6 yellowish red. Sparse sand; medium tiny lime inclusions; very sparse tiny to small quartz; medium tiny to small feldspar.

Classical-Hellenistic.

## DSP07-F04-01

Core units: 1400
Adjacent units: 1394, 1399, 1401, 2361-2363
No diagnostic material.

## DSP07-F04-02

Core units: 1388
Adjacent units: 1386-1387, 1389-1390, 1404-1406, 2356.
No diagnostic material.

### Tableware

**Cat. 101 1388/02/1**

Rim fr. from a jug with offset rim. H: 2.6; W: 3.9; T: 0.5; Ø? Clay: 10YR 5/2 greyish brown. Slip: 10YR 5/1 grey. Sparse tiny lime inclusions.

### Utility ware

**Cat. 102 1388/04/3**

Rim fr. of a bowl with part of the rim preserved. Badly preserved. H: 2.9; W: 4.3; T: 0.8; Ø? Clay: 2.5Y 3/1 very dark grey. Surface: Gley 1 4/N dark grey. Very sparse very tiny to tiny lime inclusions; very sparse tiny feldspar.

Late Classical-Hellenistic.

### Post-Antique pottery

**Cat. 103 1388/01/01**

Fr. of flat base of utility vessel. H: 4.6; W: 7.1; T: 1.2; Ø: 10. Clay: 7.5YR 6/6 reddish yellow. Very sparse very tiny lime.

Early Modern.

**Cat. 104 1388/03/01**

Fr. of flat base of utility vessel. H: 4.6; W: 7.1; T: 1.2; Ø: 10. Core: 5YR 6/6 reddish yellow. Surface: 2.5YR 3/2 dusky red. Sparse tiny to small lime; sparse tiny feldspar.

Early Modern.

**Cat. 105 1388/04/01**

Fr. of jar with outflaring rim H: 5.2; W: 4; T: 0.7; Ø? Clay: 5YR 6/6 reddish yellow. Sparse sand; sparse tiny lime.

Cf. Hayes 1992, 342-343, pl. 46.a-d. 18th-20th century AD

**Cat. 106 1388/04/02**

Fr. from a jug with everted plain rim. H: 2.7; W: 4.1; T: 0.8; Ø: 12. Clay: 5YR 5/8 yellowish red. Sparse very tiny to tiny lime; single large grog.

Cf. Bakirtzis 1980, fig. 10 right; Hayes 1992, 342-342, pl. 46.f-j, fig. 106. 18th-20th century AD

**Cat. 107 1388/06/01**

Fr. of flat base. No measurements. Large. Fine sandy fabric with few tiny lime inclusions and voids. Occasional larger voids. Self-slipped exterior.

Early Modern

**Cat. 108 1388/07/01**

Fr. of jug/jar with flaring, slightly thickened rim. Horizontal rib below the exterior lip and combing below the rib H: 5.2; W: 4.6; T: 1; Ø: 12. Core: 5YR 6/6 reddish yellow. Surface: 5YR 5/4 reddish brown. Very sparse tiny feldspar.

Cf. Hayes 1992, 342-343, pl. 46.a-d. 18th-20th century AD

**Cat. 109 1388/07/02**

Spout from jug. H: 8.6; W: 7.5; T: 0.7; Ø? Clay: 5YR 7/8 reddish yellow. Surface: 2.5YR 5/4 reddish brown. Very sparse tiny lime; very sparse tiny feldspar.

Cf. Bakirtzis 1980, fig. 10 right; Hayes 1992, 342-342, pl. 46.f-j, fig. 106. 18th-20th century AD

## DSP07-F05-01

Core units: 1502-1505
Adjacent units: 1480, 1482-1484, 1499-1501, 1506, 1519, 1522

### Post-Antique pottery

**Cat. 110 1502/01/01**

Fr. of plate with flat rim with upturned lip. Glazed interior and exterior. H: 2.7; W: 2.6; T: 0.9; Ø? Clay: 2.5YR 6/8 light red. Glaze: clear. Sparse tiny to medium lime; sparse tiny to small feldspar?

Cf. Hayes 1992, fig. 144:10. Late 19th century AD

**Cat. 111 1502/01/02**

Rim fr. of jug with overhanging rim with glazed interior and exterior. H: 2.4; W: 4.7; T: 0.5; Ø: 14. Clay: 2.5YR 6/8 light red. Glaze: yellowish. Sparse tiny to medium lime; sparse tiny to small feldspar?

Early Modern

**Cat. 112 1502/01/03**

Rim fr. of bowl with rounded lip with glazed interior and exterior. H: 3.3; W: 4.1; T: 1.2; Ø: 16. Clay: 2.5YR 6/8 light red. Glaze: yellowish brown. Sparse tiny to medium lime; sparse tiny to small feldspar?

Early Modern.

**Cat. 113 1502/01/04**

Fr. of flat flaring base of jar/jug. H: 4.9; W: 7.7; T: 1.1; Ø: 15. Clay: 2.5YR 6/8 light red. Glaze: green. Sparse tiny to medium lime; sparse tiny to small feldspar?

Possibly similar to Korre-Zougraphou 2000, 280 no. A138 or 348 no. B94a. 19th century AD

**Cat. 114 1502/01/05**

Fr. of exterior off-set flat base of jar/jug. H. 1.9; W: 5.8; T: 0.6; Ø: 7. Clay: 2.5YR 6/8 light red. Glaze: clear yellowish. Sparse tiny to medium lime; sparse tiny to small feldspar?

Early Modern

**Cat. 115 1502/02/01**

Body fr. from casserole with a hole in the middle. H: 4.1; W: 3.7; T: 0.8; Ø? Clay: 10YR 6/3 pale brown. Sparse sand; tiny lime.

Cf. Hayes 1992, pl. 12 no. e; Dmimrienko & Maslovski 2006, fig. 10.8. Probably Early Modern.

# DSP07-F05-02

Core units: 2404
Adjacent units: 2403, 2422, 2427
No diagnostic material.

# DSP07-F06-01

Core units: 1607
Adjacent units: 1587, 1608-1609, 2525

## Amphorai

### *Chersonesos*

**Cat. 116 1607/01/1**

Rim fr. H: 3.2; W: 6.2; T: 0.7-1.4; Ø: 12; D: 20. Clay: 2.5YR 5/8 red. Medium tiny lime inclusions; sparse tiny feldspar.

Closest to Monachov rim type 1. Cf. Monachov 1989, 69-70. Ca. 350-300 BC.

### *'Samian'*

**Cat. 117 1607/02/1**

Handle fr. H: 6.9; W: 5.3; T: 2.2. Clay: 7.5YR 6/4 light brown. Tiny to medium lime inclusions; very sparse very tiny to tiny mica.

Short and rather thick handle. Resembles early amphorai of the 6th and 5th century BC. Cf. e.g. Grace 1971, 71 Fig. 2.4. Late Archaic?

## Tableware

**Cat. 118 1607/01/2**

Flat base fr. of jug. H: 2.1; W: 4.1; T: 0.7-1.8; Ø?; D: <15. Clay: 2.5YR 6/6 light red. Slip: 5YR 7/4 pink. Sparse tiny lime inclusions; sparse tiny feldspar.

Chersonesos, cf. e.g., Kašaev 2002, Cat. C254. Late 4th-3rd century BC.

# DSP07-F06-02

Core units: 1579
Adjacent units: 1558, 1578, 1580

## Amphorai

### *Chersonesos*

**Cat. 119 1579/01/1**

Handle fr. H: 7.2; W: 4.1; T: 2.5. Clay: 2.5YR 6/8 light red. Slip: 7.5YR 6/6 reddish yellow. Medium tiny to medium lime inclusions; medium tiny to medium feldspar.

Early Hellenistic.

### *Herakleia?*

**Cat. 120 1579/01/3**

Handle fr. H: 8.6; W: 4.5; T: 2.4. Clay: 2.5YR 6/6 light red. Medium tiny to medium mica; medium tiny lime inclusions; very sparse tiny quartz; sparse tiny to medium feldspar.

4th century BC?

## Tableware

**Cat. 121 1579/01/2**

Small part of body and handle from a jug. H: 2.5; W: 3.6; T: 1.6. Clay: 10R 5/8 red. Surface: close to 10R 7/8 light red. Sparse tiny lime inclusions; sparse tiny feldspar.

Late Classical-Hellenistic.

# DSP07-F06-03

Core units: 2476-2477, 2519-2520
Adjacent units: none

## Amphorai

### *Herakleia*

**Cat. 122 2477/02/2**

Rim fr. with handle attachment preserved. H: 5.4; W: 5.4; T: 1.1-1.3; Ø: 11; D: 45. Clay: 2.5YR 6/8 light red. Very sparse tiny to medium pyroxene; tiny to medium lime inclusions; medium tiny to medium feldspar.

Close to type II-A-2. Cf. Monachov 2003, pl. 95.6-8 & pl. 96. Late 4th-Early 3rd century BC.

### *Unknown production*

**Cat. 123** **2519/02/1**

Rim fr. H: 2.6; W: 5; T: 1.3; Ø: 13. Core: 2.5YR 5/8 red. Medium tiny to small lime; medium tiny to medium feldspar.
Hellenistic.

**Cat. 124** **2519/02/3**

Fr. of base. H: 5; W: 3; T: 2.6; Ø? Core: 2.5YR 6/6 light red. Surface: 5YR 5/8 yellowish red. Medium tiny to small lime inclusions; medium tiny to medium feldspar.
Hellenistic.

### Tableware

**Cat. 125** **2477/02/1**

Rim fr. of bowl with incurved rim. H: 2.5; W: 2.9; T: 0.9; Ø: 22. Clay: 10YR 6/3 pale brown. Sparse tiny to small lime inclusions; very sparse sand.
Cf. Kašaev 2002, Cat. C85-C86. Early Hellenistic.

## DSP07-F07-01

Core units: 2586, 2610-2613, 2636-2639
Adjacent units: 2587, 2640, 2661-2662
No diagnostic material.

## DSP07-F07-02

Core units: 2572-2573
Adjacent units: 2552
No diagnostic material.

## DSP07-F08-01

Core units: 2705-2710, 2715-2720, 8991-8992, 9019
Adjacent units: none

### Handmade ware

### *Middle/Late Bronze Age*

**Cat. 126** **2719/02/5**

Body fr. of closed vessel. Several horizontal furrows on the ext. H: 2.6; W: 3.1; T: 0.9. Core: 5Y 3/1 very dark grey. Surface: 10YR 6/3 pale brown. Medium tiny to small lime.; very sparse small pebbles.
Middle Bronze Age/Late Bronze Age.

### *Hellenistic*

**Cat. 127** **2707/03/1**

Rim fr. of bowl? with slightly outcurved rim and a round projecting lip. H: 1.9; W: 2; T: 0.8; Ø? Core: 5Y 2.5/1 black. Surface: 10YR 6/3 pale brown. Sparse very tiny to tiny lime inclusions; sparse sand.
Hellenistic.

**Cat. 128** **2707/03/2**

Rim fr. of pot with straight neck and rounded lip. H: 1.7; W: 2.1; T: 1; Ø? Core: 10YR 6/3 pale brown. Very sparse very tiny to tiny lime inclusions.
Hellenistic.

**Cat. 129** **2707/03/3**

Rim fr. of small pot with flat lip, possibly a cup. H: 1.1; W: 2; T: 0.7; Ø: ca. 15. Core: 5Y 3/1 very dark grey. Surface: 2.5Y 5/3 light olive brown. Very sparse very tiny lime inclusions; very sparse very tiny black inclusions.
Hellenistic.

**Cat. 130** **2709/01/1**

Rim fr. of bowl? with rounded projecting rim. Badly preserved. H: 1.8; W: 2.3; T: 0.8; Ø? Clay: 10YR 6/2 light brownish grey. Surface: 10YR 6/4 light yellowish brown. Sparse tiny lime inclusions; sparse tiny to small feldspar.
Hellenistic.

**Cat. 131** **2709/01/2**

Rim fr. of pot with straight neck. Diagonally cut decoration on rim. H: 3.1; W: 3.4; T: 1; Ø: ca. 12. Clay: 5YR 2.5/1 black. Very sparse very tiny mica; very sparse small quartz; sparse tiny to small lime.
Cf. Stolba 2002, type 1B Cat. D74. Early Hellenistic.

**Cat. 132** **2719/01/1**

Rim fr. possibly of bowl with everted rim. H: 2.4; W: 2.2; T: 0.7; Ø? Core: 5YR 2.5/1 black. Surface: 5YR 3/1 very dark grey. Medium tiny to small lime inclusions; sparse tiny to small quartz.
Possibly similar to bowls from farmhouse 338 at Chersonesos, cf. Kovalevskaja-Suchnova 2005, fig. 4.7. Early Hellenistic.

**Cat. 133** **2719/02/1**

Rim fr. of pot with outturned rim. H: 2.2; W: 2.6; T: 0.7; Ø? Gley 2 4/5 BG dark greenish grey. Surface: 7.5YR 6/3 light brown. Medium very tiny to tiny lime.
Cf. Stolba 2002, type 1 Cat. D1-D68. Early Hellenistic.

**Cat. 134** **2719/02/2**

Rim fr. of large pot with straight outturned rim. A layer of clay seems to have been pulled over the lip from the interior towards the exterior. H: 2.3; W: 4.1; T: 1.2; Ø? Core: 7.5YR 2.5/1 black. Surface: 7.5YR 6/4 light brown. Frequent tiny to medium lime inclusions; medium tiny to large shells; sparse tiny to small quartz.
Cf. Stolba 2002, type 1 Cat. D1-D68. Early Hellenistic.

**Cat. 135** **2719/02/3**

Rim fr. of pot with a straight outturned rim and rounded lip. H: 3.5; W: 2.5; T: 1; Ø? Core: 10YR 4/1 dark grey. Surface: 2.5Y 6/2 light brownish grey. Sparse very tiny to small lime inclusions; medium organic inclusions.
Cf. Stolba 2002, type 1 Cat. D1-D68. Early Hellenistic.

### Amphorai

#### *Sinope*

**Cat. 136 9019/01/1**

Handle fr. H: 5.2; W: 4.24; T: 2.74. Core: 2.5YR 7/6 light red. Surface: 10YR 8/4 very pale brown. Sparse medium quartz; sparse medium pyroxene; single tiny lime particle.

Late Classical-Hellenistic.

#### *Unknown production*

**Cat. 137 2708/01/1**

Rim fr. H: 2; W: 3.9; T: 1.7; Ø? 10YR 5/3 brown. Medium tiny lime inclusions; very sparse tiny to small quartz.

Classical-Hellenistic?

## DSP08-F01-01

Core units: 5153-5154, 6009-10, 6016
Adjacent units: none

### Handmade ware

**Cat. 138 8177/01/4**

Base fr. of pot with flat underside. H: 2.6; W: 3.5; T: 0.8-1.4; Ø: ca. 10. Core: 2.5YR 6/8 light red to 7.5YR 5/3 brown. Surface: 2.5Y 5/1 grey. Medium tiny to medium quartz; frequent tiny to small feldspar; frequent small to medium stones.

Cf., e.g., Stolba 2002, Cat. D43. Early Hellenistic.

**Cat. 139 8177/01/6**

Rim fr. of pot with straight rim. H: 3.2; W: 3.5; T: 0.7-0.8; Ø? Ext. surface: 10YR 7/6 yellow, int. surface: Gley 1 3/5 B very dark bluish grey. Frequent tiny to medium lime inclusions; single large lime particle; medium large stones.

Cf. Stolba 2002, Cat. D86-D87. Early Hellenistic.

### Amphorai

#### *Herakleia*

**Cat. 140 5154/01/1**

Fr. of toe. H: 8.4; W: 5-5.2.; Ø: 5.0. Clay: 10R 7/8 light red. Sparse small pyroxene; medium tiny to small lime inclusions; medium small feldspar.

Type I-A-2 to I-A-4. Ca. 375-350 BC.

**Cat. 141 6016/01/1**

Handle fr. Re-used as a tool at both ends. H: 10.2; W: 3.6-3.8; T: 2.3. Clay: 10R 7/8 light red to 2.5YR 7/8 light red. Medium small to medium quartz; sparse tiny mica; medium small to medium pyroxene; sparse small feldspar.

4th century BC.

## DSP08-F02-01

Core units: 8177-8178, 8181
Adjacent units: 8175-8176, 8179, 8182-8184

### Amphorai

#### *Chersonesos*

**Cat. 142 8181/01/3**

Fr. of lower part. H: 5.1; W: 6; T: 1.5. Clay: 2.5YR 6/6 light red. Surface: 5YR 6/6 reddish yellow. Thin slip. Medium tiny lime inclusions; sparse tiny quartz; sparse small feldspar.

Early Hellenistic.

**Cat. 143 8182/01/1**

Fr. of lower body. H: 4.7; W: 4.4-5.8; T: 1.1. Clay: 2.5YR 7/8 light red. Sparse tiny to small quartz; medium tiny lime inclusions; medium sand; medium tiny feldspar.

Early Hellenistic?

#### *Herakleia*

**Cat. 144 8177/02/3**

Wall fr. with a hole Ø 0.52-0.97. L: 4.74; W: 3.1; T: 0.96. Core: 2.5YR 7/8 light red. Surface: 10YR 8/4 very pale brown. Frequent small to medium pyroxene; frequent small to medium quartz; frequent small to medium feldspar; sparse small shells.

4th century BC.

**Cat. 145 8181/01/1**

Fr. of upper part of handle with part of the wall preserved. H: 6.9; W: 5.8; T: 5.2. Clay: 10R 6/8 light red. Surface: 10R 7/6 light red. Sparse tiny to small pyroxene; medium tiny to small lime inclusions; frequent tiny feldspar.

4th century BC.

**Cat. 146 8181/01/2**

Handle fr. H: 5.1; W: 3.7; T: 2.4. Clay: 2.5YR 6/8 light red. Medium tiny lime inclusions; medium tiny to small quartz; sparse small red stones; frequent small to medium feldspar.

4th century BC.

**Cat. 147 8178/02/4**

Handle fr. H: 6.3; W: 4.4; T: 2.6. Clay: 5YR 6/8 reddish yellow to 5YR 5/8 yellowish red. Frequent tiny to small pyroxene; frequent tiny to small lime inclusions; medium small quartz; single large stone; frequent tiny feldspar.

4th century BC.

#### *Sinope*

**Cat. 148 8177/01/2**

Shoulder fr. H: 3.5; W: 2.9; T: 0.8-1. Clay: 10R 7/4 pale red to 2.5YR 7/4 light reddish brown. Sparse light-reflecting particles; medium small to medium pyroxene; single small quartz mineral.

Classical-Hellenistic.

### *Thasos*

**Cat. 149 8178/02/2**

Handle fr. Re-used as a tool in one end. H: 8.2; W: 5.3; T: 2.45. Clay: 10R 6/8 light red to 2.5YR 6/8 light red. Frequent tiny mica; medium tiny to smal lime inclusions; sparse smal quartz; medium tiny feldspar.

Possibly Monachov 2003, pl. 48.2, type II-C-2. Early Hellenistic.

**Cat. 150 8178/02/3**

Handle fr. with small part of the inner wall preserved. H: 3.2; W: 6.3; T: 2.7. Clay: 10R 6/8 light red to 2.5YR 6/8 light red. Frequent tiny mica; medium tiny to smal lime inclusions; sparse smal quartz; medium tiny feldspar.

Possibly Monachov 2003, pl. 48.2, type II-C-2. Early Hellenistic.

**Cat. 151 8178/01/3**

Handle fr. H: 11; W: 4.7; T: 2.2. Clay: 2.5YR 7/8 light red. Medium tiny lime inclusions; frequent tiny mica; frequent tiny quartz; frequent tiny feldspar.

Late conical type II-C-3? Ca. 325-250 BC?

### *Knidos*

**Cat. 152 8177/01/1a**

Fr. of rounded projecting rim, beveled on the underside. H: 3.8; W: 5.4; T: 0.5-1.4. Core: 10R 4/4 dusky red. Surface: 2.5YR 5/6 red. Sparse tiny lime inclusions; sparse small quartz; frequent mica.

Type II-B-1. Cf. Monachov 2003, pl. 73.6. Ca. 350-early 3rd century BC.

### *Solocha type?*

**Cat. 153 8177/01/3**

Rim fr. H: 1.8; W: 3.9; T: 1; Ø: ca. 13. Clay: 5YR 6/6 light red. Sparse tiny lime inclusions; sparse tiny feldspar; sparse tiny stones.

Close to Solocha type I-A, cf. Monachov 2003, pl. 68.3. 400-350 BC.

### *Unknown production*

**Cat. 154 8177/01/1b**

Badly worn handle fr. Re-used as a tool. H: 5.6; W: 2.4-2.7; T: 1.3-1.9. Clay: 10R 6/6 light red. Sparse light-reflecting particles; sparse tiny lime inclusions; sparse tiny feldspar; single small golden mica.

Classical-Hellenistic.

**Cat. 155 8177/01/5**

Handle fr. H: 5.7; W: 3.8; T: 2.4. Clay: 2.5YR 6/6 light red. Surface: 2.5YR 7/8 light red. Sparse tiny lime inclusions; sparse tiny quartz; sparse small to medium stones; medium tiny to small feldspar.

Classical-Hellenistic.

**Cat. 156 8177/02/1**

Handle fr. H: 4.15; W: 4-4.5; T: 2.56. Clay: 5YR 6/6 reddish yellow. Sparse tiny to small pyroxene; single tiny shell and quartz mineral.

Classical-Hellenistic.

**Cat. 157 8178/01/4**

Handle fr. H: 5.75; W: 2.9; T: 2. Clay: 2.5YR light red. Slip: 2.5YR 6/6 light red. Sparse tiny lime inclusions; sparse single tiny quartz mineral; sparse small feldspar.

Classical-Hellenistic.

**Cat. 158 8178/02/5**

Handle fr. Pierced through by an irregular-shaped hole. H: 2.8; W: 2.9; T: 2.1. Clay: 2.5YR 5/6 red. Sparse tiny light-reflecting particles; medium tiny to medium lime inclusions.

Classical-Hellenistic.

**Cat. 159 8178/02/6**

Handle fr. Incompletely preserved. H: 3.25; W: 2.8; T: 2.5. Clay: 2.5YR 5/8 red. Slip: 10R 5/2 weak red. Medium tiny lime inclusions.

Classical-Hellenistic.

**Cat. 160 8182/01/2**

Handle fr. with smoothed surface. H: 7.4; W: 3.95; T: 2.6. Clay: 5YR 6/8 reddish yellow. Surface: 2.5YR 6/6 light red. Frequent tiny to medium quartz; frequent tiny to medium lime inclusions; frequent small to medium red stones; sparse small pyroxene inclusions; frequent tiny to medium feldspar.

Classical-Hellenistic.

## Tableware

**Cat. 161 8178/01/2**

Rim fr. of shallow bowl with thickened rim. Ribbed below the rim. H: 2.2; W: 2.05; T: 0.6-1.1; Ø: ca. 15. Clay: 7.5YR 7/6 reddish yellow to 2.5YR 7/6 light red. Light slip on both int. and ext. Sparse tiny mica; medium tiny to small quartz; sparse tiny red stones.

Late Classical-Hellenistic.

## Post-Antique pottery

**Cat. 162 8182/01/03**

Fr. of pibe with a rounded lower body and a moulding at the junction with the upper body. Tiny stamped dots in the shape of rays on the underside. H: 2.4; W: 2.05; T: 0.35-0.6. Clay: 10R 5/6 red. Surface: 10R 4/1 dark reddish grey. Sparse tiny lime; medium sand.

Early Modern.

# DSP08-F03-01/Skalistoe chutor

Core units: 6234
Adjacent units: 6230-6233, 6235-6240, 6252-6256

## Amphorai

### *Unknown production*

**Cat. 163 6234/01/5**

Handle fr. H: 4.3; W: 4.1; T: 2-2.1. Clay: 10R 7/8 light red. Sparse small quartz; medium tiny to small lime inclusions; single gold mica inclusion; frequent tiny to small feldspar.

Probably not Herakleian due to the presence of golden mica. 4th century BC?

**Cat. 164 6234/02/6**

Handle fr. H: 6.7; W: 4.15-4.4; T: 2.5-2.7. Clay: 5YR 7/6 reddish yellow. Sparse tiny to medium lime inclusions; medium tiny to small feldspar; sparse sand inclusion.

Classical-Hellenistic.

**Cat. 165** 6256/01/1

Handle fr. with part of the wall preserved. H: 6.6; W: 8; T: 1.15-2.3. Clay: 5YR 5/6 yellowish red. Surface: close to 5YR 7/6 reddish yellow. Medium small quartz; medium tiny to large lime inclusions; medium small to large stones; medium smal feldspar.

Classical-Hellenistic.

## Post-Antique pottery

**Cat. 166** 6233/01/01

Handle fr. of jug. H: 2.41; W: 3.07-3.32; T: 1.25-1.3. Core: 2.5YR 6/6, 2.5YR 7/4. Surface: 5YR 7/6.

Early Modern.

**Cat. 167** 6234/01/01

Fr. of cooking pot with high collared rim. Corrugated exterior. Gritty surface. H: 4; W: 4.6; T: 0-7-1; Ø: 22. Clay: 10YR 8/3 pale brown. Slip: Gley 2 5/5BG greenish grey. Sparse tiny quartz; sparse grog.

Early Modern.

**Cat. 168** 6234/01/02

Fr. of cooking pot with high collared rim. Softly corrugated exterior. Gritty surface. H: 2.65; W: 5.2; T: 0.4-0.8; Ø? Clay: 7.5YR 5/1 grey. Slip: Gley 2 3/10BG very dark greenish grey. Sparse tiny lime.

Early Modern.

**Cat. 169** 6234/01/03

Fr. of cooking pot with high collared rim. Softly corrugated exterior. Gritty surface. H: 2.3; W: 2.9; T: 0.4-0.6; Ø: 16? Clay: 10YR 8/3 pale brown. Slip: Gley 2 5/5BG green to 10R6/8 light red. Sparse tiny quartz; sparse grog.

Early Modern.

**Cat. 170** 6234/01/04

Fr. of casserole with almost flat rim and interior ledge for lid. Gritty surface. H: 2.4; W: 2.9; T: 0.4-0.8; Ø: 24? Clay: 10YR 8/3 pale brown. Slip: Gley 2 5/5BG greenish grey. Sparse tiny quartz; sparse grog.

Early Modern.

**Cat. 171** 6234/01/06

Rim fr. from deep open vessel with rounded slightly incurved lip and glazed interior and exterior. Horisontal combing below exterior lip and circular combing below. H: 3.7; W: 5.3; T: 0.7-0.9; Ø? Clay: 2.5YR 6/8 light red. Interior glaze: bright green lip, yellow below. Exterior glaze: bright green. Sparse small feldspar; sparse tiny lime; sparse sand.

Early Modern.

**Cat. 172** 6234/01/07

Shoulder of cooking pot, probably from vessel similar to 6234/01/2. H: 3.6; W: 5.5; T: 0.45; Ø? Core and interior: 2.5Y 7/4 pale yellow. Exterior: 2.5Y 4/1 dark grey. Coarse sandy fabric with numerous grits mostly clear rounded quartz of tiny to small size. Also some tiny to small grog.

Early Modern.

**Cat. 173** 6234/01/08

Fr. of flat base of cooking pot H: 1.87; W: 3.09; T: 0.5; Ø? Core: 2.5Y 7/4 pale yellow. Interior slip: 10YR 5/1, exterior: 2.5Y 4/1 dark grey, possibly soot. Coarse sandy fabric with numerous grits mostly clear rounded quartz of tiny to small size. Also some tiny to small grog.

Early Modern.

**Cat. 174** 6234/01/09

Fr. of casserole rim with everted plain rim with interior ledge for lid. H: 2.7; W: 3.7; T: 0.5; Ø? Clay: 2.5Y 4/1 dark grey from secondary firing. Coarse sandy fabric with numerous grits, mostly clear rounded quartz of tiny to small size and some tiny to small grog.

Early Modern.

**Cat. 175** 6234/02/01

Fr. of cooking pot with high collared rim. Gritty surface. H: 2.5; W: 3.6; T: 0.6-0.7; Ø: 18. Clay: 10YR 7/4 yellow. Slip: Gley 2 4/5 BG dark greenish grey. Single large lime; medium sand.

Early Modern.

**Cat. 176** 6234/02/02

Fr. of slightly everted, thickened rim. Lip is not preserved. Cooking vessel? H: 2.4; W: 1.9; T: 0.4-0.6; Ø: 23. Core: 10R 2.5/1 reddish black, possibly from secondary firing. Single tiny lime.

Early Modern.

**Cat. 177** 6234/02/03

Fr. of flat rim of cooking vessel possibly a casserole. Interior ledge for lid. Gritty surface. H: 2.3; W: 2.9; T: 0.4-0.6; Ø: 16? Clay: 10YR 8/3 pale brown. Core: Gley 2 4/5 BG dark greenish grey. Slip: Gley 2 5/5BG green to 10R 6/8. Sparse tiny quartz; sparse grog.

Early Modern.

**Cat. 178** 6234/02/04

Fr. of cooking pot with high collared rim. Gritty surface. H: 2.6; W: 2.7; T: 0.5-0.7; Ø: 15. Clay: 10YR 6/4 light yellowish brown. Slip: 10R 6/8 light red. Sooted lip: 5YR 2.5/1 black. Single small quartz; sparse tiny lime

Early Modern.

**Cat. 179** 6234/02/05

Fr. of jug with thickened rim and glazed interior and exterior. H: 2.4; W: 1.8; T: 0.4-0.85; Ø: ca. 10. Clay: 7.5YR 6/6 reddish yellow. Interior glaze: yellowish. Exterior glaze: dark brown. Sparse tiny lime; sparse tiny to small feldspar; sparse small quartz.

Early Modern.

**Cat. 180** 6234/02/07

Fr. of cooking pot with high collared rim. Gritty surface. H: 3.3; W: 4.2; T: 0.6-0.7; Ø: 20. Clay: 7.5YR 6/4 light brown. Slip:2.5YR 5/8 red. Medium sand; medium medium quartz; sparse tiny lime.

Early Modern.

**Cat. 181** 6234/02/08

Fr. of cooking pot with high collared rim. Gritty surface. H: 1.8; W: 3.4; T: 0.6-0.65; Ø? Clay: 7.5YR 4/3 brown. Slip: 7.5YR 3/2.5 very dark grey. Sparse small to medium lime; single small quartz.

Early Modern.

**Cat. 182** 6234/02/09

Fr. of casserole rim with everted rim and interior ledge for lid. Gritty surface. H: 2; W: 2.4; T:0.7; Ø: 22. Clay: 2.5YR 6/6 light red. Slip: 2.5YR 5/6 red. Medium tiny to small feldspar.

Early Modern.

**Cat. 183** 6234/02/10

Fr. of jug with overfolded, exterior rim and interior glazed. H: 2.2; W: 4.2; T: 0.4-1; Ø: 16. Clay: 7.5YR 7/4 pink. Glaze: clear. Medium tiny to medium lime; sparse small feldspar.

Early Modern.

**Cat. 184** **6234/02/11**

Fr. of cooking pot with high collared rim. Gritty surface. Ø: 23. Clay: 7.5YR 6/4 light brown. Slip: 2.5YR 5/8 red. Medium sand; medium medium quartz; sparse tiny lime.
Early Modern.

**Cat. 185** **6234/02/12**

Fr. of flat base of utility vessel. H: 2.95; W: 6.65; T: 0.63-0.7; Ø: 12. Core: 10YR 8/4 very pale brown. Exterior: 10YR 4/1 dark grey. Sparse small quartz; single large quartz.
Early Modern.

**Cat. 186** **6234/02/13**

Fr. of cooking pot with everted rim for lid. H: 3.7; W: 5.7; T: 0.5-0.8; Ø: ca. 26. Clay: 7.5YR 4/2 brown. Slip: 7.5YR 4/1 dark grey to Gley 2 3/5 BG very dark greenish grey. Single small quartz; sparse tiny lime.
Early Modern.

**Cat. 187** **6234/02/14**

Fr. of cooking pot with everted rim with rounded lip. Glazed interior. H: 3.1; W: 3.7; T: 0.6; Ø: 24. Clay: 10YR 6/4 light yellowish brown. Glaze: clear. Medium sand; sparse small quartz; sparse tiny lime.
Early Modern.

**Cat. 188** **6234/02/15**

Base fr. of bowl with interior glaze with painted, concentric circles. H: 4.1; W: 6.5; T: 0.6-0.7. Clay: 2.5YR 5/8 red. Glaze: 2.5Y 8/6 yellow and dark orange. Sparse light-reflecting particles; sparse tiny lime; sparse tiny feldspar.
Early Modern.

**Cat. 189** **6236/01/01**

Fr. of jug with high overfolded rim. H: 4.2; W: 5.3; T: 0.4-0.9; Ø: 27. Clay: 7.5YR 7/6 reddish yellow. Slipped and glazed int. Slip: 7.5YR 5/2 brown. Glaze: light brown. Sparse tiny lime; sparse tiny feldspar.
Early Modern.

**Cat. 190** **6240/01/01**

Rim fr. of jug with carinated upper part. Concave groove on upper part of lip. H: 1.8; W: 2.5; T: 0.3-0.9; Ø: 13. Clay: 5YR 5/8 yellowish red. Slip: 2.5YR 6/8 light red. Sparse tiny to small lime inclusions.
Early Modern.

**Cat. 191** **6256/01/02**

Fr. of outturned rim with glazed interior. H: 3.9; W: 2.6; T: 0.6-0.9; Ø? Clay: 5YR 7/8 reddish yellow. Glaze: clear yellowish. Medium tiny feldspar.
Early Modern.

## DSP08-F03-02, see Skalistoe 2

## DSP08-F04-01

Core units: 8468-8471, 8510-8511
Adjacent units: 8470, 8509, 8516

### Amphorai

#### *Herakleia*

**Cat. 192** **8510/03/2**

Rim fr. with beveled lip. H: 2.41; W: 3.42; T: 0.76 (wall); Ø: 12. Clay: 10YR 5/3 brown. Medium tiny to small pyroxene; single tiny quartz mineral; medium tiny to small lime inclusions; medium tiny to small feldspar; sparse small voids.
Possibly type I-4. Cf. Monachov 2003, pl. 88.2. Ca. 400-375 BC.

#### *Unknown production*

**Cat. 193** **8510/01/1**

Fr. of protruding rim with beveled lip. H: 2.2; W: 2.9; T: 0.6-1; Ø: 13. Clay: 2.5YR 6/8 light red. Sparse tiny lime; sparse tiny feldspar.
Hellenistic?

**Cat. 194** **8510/03/1**

Fr. of lower body with smoothed edges caused by re-use. H: 8.9; Ø 4.3; W: 7.9; T: 1.38-1.9. Core: 10R 6/8 light red. Surface: 2.5Y 8/6 yellow. Sparse small to medium quartz; frequent medium-large voids.
Classical-Hellenistic.

### Tableware

**Cat. 195** **8471/01/1**

Fr. of bowl with inturned rim. H: 3.4; W: 5.65; T: 0.8-1.2; Ø: 23. Clay: 10R 6/8 light red. Surface: 5YR 7/6 reddish yellow. Sparse tiny to small lime inclusions; single medium pyroxene inclusion; sparse tiny to medium feldspar.
Cf. Kašaev 2002, Cat. C82, C86. Early Hellenistic.

**Cat. 196** **8510/02/1**

Base fr. jug with ring foot. H: 1.7; W: 2.9; T: 0.7-1.2; Ø: ca. 10. Clay: 10R 6/6 light red. Slip: 2.5YR 8/4 pink. Sparse tiny to small lime inclusions; medium small to medium pyroxene.
Sinope? Late Classical-Early Hellenistic.

### Utility ware

**Cat. 197** **8510/03/3**

Wall fr. with part of applied handle preserved. H: 2.3; W: 3.4; T: 0.75-1.32. Core: 7.5YR 6/4 light brown. Surface: 10YR 8/4 very pale brown. Sparse large pyroxene; frequent medium pyroxene; single large void.
Sinopean louterion, Early Hellenistic.

## DSP08-F04-02

Core units: 8499-8500, 8520-8521
Adjacent units: 8480, 8498, 8501, 8519

## Handmade ware

**Cat. 198 8500/03/1**

Rim fr. of bowl with smoothed surface and flat lip. H: 3.08; W: 2.87; T: 0.87-0.92; Ø: ca. 12. Core: 5YR 5/4 reddish brown. Surface: 10YR 8/4 very pale brown and 10YR 6/1 grey. Medium quartz; medium grog; medium small to medium feldspar.

Cf. Stolba 2002, Cat. D118-D119. Early Hellenistic.

**Cat. 199 8500/05/1**

Rim fr. of bowl? Oblique cut decoration on the lip. H: 1.4; W: 2.6; T: 0.65; Ø: 13-14. Core: 5YR 5/2 reddish grey. Surface: 5YR 6/6 reddish yellow. Sparse small to medium lime inclusions; single light-reflecting particle; medium small stones (quartz?).

Cf. Stolba 2002, Cat. D118-D119. Early Hellenistic.

**Cat. 200 8500/06/1**

Rim fr. of pot with outturned rim. H: 2.8; W: 1.9; T: 0.8-0.9; Ø: 14. Clay: 5YR 6/3 light reddish brown. Medium tiny to small fedspar; single large stone; sparse tiny quartz; sparse light-reflecting particles.

Cf. Stolba 2002, type 1 Cat. D1-D68. Early Hellenistic.

**Cat. 201 8500/06/2**

Rim fr. of small open vessel with straight rim and rounded lip. H: 2.5; W: 1.5; T: 0.7-0.8; Ø? Core: 7.5YR 5/3 brown. Surface: 7.5YR 6/4 light brown. Slip: 10R 5/6 red. Medium small quartz; medium tiny to small feldspar; sparse light-reflecting particles. Probably from the same vessel as 8500/06/3.

Cf. Stolba 2002, type 1, variant c, Cat. D86-D87. Early Hellenistic.

**Cat. 202 8500/06/3**

Rim fr. of small open vessel with straight rim and rounded lip. H: 1.8; W: 1.9; T: 0.8; Ø: ca. 11. Clay: close to 7.5YR 4/3 brown. Surface: 5YR 5/6 yellowish red. Medium small quartz; medium tiny to small feldspar; sparse light-reflecting particles. Probably from the same vessel as 8500/06/2.

Cf. Stolba 2002, type 1, variant c Cat. D86-D87. Early Hellenistic.

**Cat. 203 8500/07/1**

Rim fr. of pot with straight outturned rim and rounded lip. Possibly with impressed decoration on the lip. H: 1.8; W: 1.6; T: 0.5; Ø? Core: 10YR 3/1 very dark grey. Surface: 7.5YR 7/4 pink. Sparse small lime inclusions; sparse medium pebbles.

Cf. Stolba 2002, type 1 Cat. D1-D68. Early Hellenistic.

## Amphorai

### *Chersoneos*

**Cat. 204 8521/03/2**

Fr. of toe with a shallow concave underside. H: 5.9; W: 2.5; T: 1.6. Clay: 10R 6/8 light red. Slip: 5YR 7/4 pink. Medium tiny to small lime inclusions; sparse small to medium red inclusions; sparse tiny quartz.

Early Hellenistic.

### *Sinope*

**Cat. 205 8500/01/3**

Rim fr. with rounded protruding rim. H: 3; W: 2.9; T: 0.6-1.2; Ø: 12. Clay: 2.5YR 8/4 pink. Slip: close to 10R 8/1 white. Frequent tiny to small pyroxene; sparse small red stones.

Cf. Monachov 2003, pl. 101.1, type I-D, possibly also type II-C or II-E. Cf. Monachov 2003. Ca. 375-early 3rd century BC.

### *Unknown production*

**Cat. 206 8480/01/1**

Handle fr. H: 5.4; W: 4; T: 2.2. 2.5YR 6/6 light red. Surface: close to 5YR 6/6 reddish yellow. Frequent tiny to small lime inclusions; sparse tiny shells; sparse small pyroxene; frequent tiny to small feldspar.

Classical-Hellenistic.

**Cat. 207 8500/01/1**

Shoulder fr. with part of three incised horizontal grooves. H: 5.8; W: 5.9; T: 0.9-1.1. Clay: 5YR 5/4 reddish brown. Surface: 7.5YR 5/2 brown to 7.5YR 6/1 grey. Frequent tiny to small pyroxene; frequent tiny lime inclusions; sparse small quartz; sparse tiny light-reflecting particles.

Classical-Hellenistic?

**Cat. 208 8500/04/1**

Body fr. with two horizontal fillets and accidental scratches. Fillets width 0.17, 0.06, depth 0.05, 0.02. H: 6.84; W: 7.8; T: 1.28-1.46. Core: 5YR 7/6 reddish yellow and 5YR 7/8 reddish yellow. Surface: 10YR 6/3 pale brown. Sparse tiny shells.

Classical-Hellenistic.

**Cat. 209 8521/03/1**

Fr. of lower body. H: 4.5; W: 3.6; T: 0.9-1.8. Clay: 10R 5/8 light red. Frequent tiny to small lime particle; singel large quartz mineral.

Classical-Hellenistic.

**Cat. 210 8521/03/3**

Handle fr. H: 6.9; W: 4.6-4.7; T: 2.8. Clay: 10R 6/6 light red. Slip: 5YR 7/4 pink 7/4 pink. Sparse sand inclusion; medium tiny to medium lime inclusions; sparse small red inclusions; medium small to medium feldspar.

Classical-Hellenistic.

## Tableware

**Cat. 211 8500/01/2**

Shoulder fr. of small Grey Ware jug. H: 3.3; W: 3; T: 0.4. Clay: 10R 6/1 reddish grey. Slip: 2.5YR 5/2 weak red. Frequent light-reflecting particles; single small quartz mineral; sparse small feldspar.

4th-early 3rd century BC.

# DSP08-F04-03

Core units: 8531, 8533-8534, 8568-8570, 8591-8594
Adjacent units: 8530, 8567, 8595, 8637-38

## Handmade ware

**Cat. 212 8570/01/5**

Rim fr. of bowl? with projecting uneven rim. H: 2.4; W: 2.3; T: 0.8; Ø: 13-14. Core: 10R 7/3 pale red. Surface: 2.5YR 8/3 pink. Medium large lime inclusions; sparse small feldspar.

Somewhat similar to a handmade bowl from a grave on the Kerč Peninsula, cf. Maslenikov 1995, fig. 20.2. 3rd-2nd century BC?

**Cat. 213 8570/01/9**

Rim fr. of pot with irregular rim with one impressed indentation. H: 3.6; W: 3.4; T: 0.7; Ø? Core: 5YR 2.5/1 black. Surface: 2.5YR 7/4,

YR 6/4 light reddish brown. Medium tiny to medium large shells; single large stone.

Late Classical-Hellenistic.

**Cat. 214 8570/03/4**

Rim fr. of pot with slightly concave, outturned rim and flat lip. H: 1.53; W: 2.08; T: 0.8; Ø? Core: Gley 1 3/N very dark grey. Surface: Gley 1 5/10Y greenish grey. Sparse tiny shells.

Cf. Stolba 2002, type 1 Cat. D1-D68. Early Hellenistic.

**Cat. 215 8570/04/1**

Rim fr. of bowl with slightly inturned rim. H: 2.5; W: 2.8; T: 0.7. Ext. surface: 5YR 7/6 reddish yellow, int. surface: 7.5YR 4/1 dark grey. Medium tiny to large lime inclusions; sparse small feldspar; sparse sand; single medium black inclusions; single very large pebble.

Cf. Stolba 2002, Cat. D122-D123. Early Hellenistic.

## Amphorai

### *Chersonesos*

**Cat. 216 8568/01/3**

Handle fr. with ridge on the ext. H: 6.76; W: 3.52-3.82; T: 2.03. Core: 10YR 6/2 light brownish grey. Surface: 5YR 7/6 reddish yellow. Single tiny mica; sparse small feldspar; sparse shells.

Early Hellenistic.

**Cat. 217 8534/01/1**

Handle attachment. H: 3.6; W: 6.8, handle; W: ca. 5.2; T: 2.2. Clay: 2.5YR 6/6 light red. Slip 7.5YR 8/4 pink. Frequent tiny to small lime inclusions; sparse tiny pyroxene; sparse tiny quartz; medium tiny to small feldspar.

Chersonesos? Early Hellenistic.

### *Herakleia*

**Cat. 218 8567/01/1**

Handle fr. H: 10.3; W: 4.2; T: 2.6. Clay: 5YR 7/4 pink. Frequent tiny to medium lime inclusions; medium small to medium pyroxene; sparse small to large grog; medium tiny to small feldspar.

4th century BC.

**Cat. 219 8592/01/2**

Fr. of protruding rim. H: 2; W: 3.25; T: 0.8-1.3; Ø? Clay: 7.5YR 7/4 pink. Sparse tiny to small lime inclusions; sparse small red inclusions; sparse small feldspar.

4th century BC.

**Cat. 220 8592/04/2**

Handle fr. L: 4.2; W: 4.35-4.7; T: 2.52. Core: 7.5YR 8/6 reddish yellow. Surface: 2.5YR 6/8 light red. Frequent small to medium pyroxene; sparse medium quartz; single tiny shell.

4th century BC.

### *Sinope*

**Cat. 221 8568/01/4**

Handle fr. H: 7.42; W: 3.9; T: 1.75-2.1. Core: 2.5Y 8/4 pale yellow. Reddish layer near the surface. 5YR 7/6 reddish yellow. Sparse medium to large pyroxene; sparse medium quartz; single small shell; sparse small feldspar.

Late 4th-3rd century BC.

### *Thasos*

**Cat. 222 8592/04/1**

Fr. of toe. H: 6.8; W: 4.43; Ø: 7. Clay: 5YR 7/8 reddish yellow. Sparse tiny mica; single tiny void.

Late Classical-Early Hellenistic.

**Cat. 223 8568/01/2**

Rim fr. with flat upper lip. H: 3.06; W: 4.4; T: 0.54 (wall)-1.08; Ø: 12. Core: 2.5YR 4/1 dark reddish grey. Surface: 10YR 8/4 very pale brown. Medium small to medium pyroxene; medium small to medium quartz; single tiny shell.

Profile resembles Thasian type II-A-3, cf. Monachov 2003, pl. 40.5. First third of 5th century BC.

### *Knidos?*

**Cat. 224 8570/01/1**

Fr. of undercut protruding rim. Handle attachment preserved below rim. H: 2.9; W: 4.7; T: 0.8-1.3; Ø: 13; D: 45. Clay: 2.5YR 6/8 light red. Frequent tiny to small lime inclusions; medium light-reflecting particles; medium tiny to small feldspar.

Possibly type II-F, cf. Monachov 2003, pl. 77.3. Late 3rd century BC.

### *Unknown production*

**Cat. 225 8568/01/1**

Rim fr. with beveled lip. H: 2.87; W: 2.94; T: 0.87 (wall)-1.28; Ø: 12. Core: 10R 6/6 light red. Surface: 2.5YR 7/8 light red. Sparse tiny feldspar; sparse tiny voids.

Classical-Hellenistic.

**Cat. 226 8570/01/3**

Wall fr. Cutting marks made by a knife. H: 6.2; W: 5.6; T: 1.1-1.6. Clay: 2.5YR 6/8 light red. Slip ext.: 10YR 6/4 very pale brown. Medium tiny to small lime inclusions; sparse small to medium red inclusions; medium sand inclusion; medium tiny to small feldspar.

Classical-Hellenistic.

**Cat. 227 8570/01/4**

Handle fr. H: 5.4; W: 4.1; T: 2.6. Clay: 10R 7/8 light red. Slip: 10YR 8/4 very pale brown. Sparse tiny lime inclusions; medium sand inclusion; sparse tiny feldspar.

Classical-Hellenistic.

**Cat. 228 8570/01/6**

Handle fr. H: 5.9; W: 2.9-3.2; T: 2.3. Clay: 10YR 5/4 yellowish brown. Surface: 10YR 6/4 light yellowish brown. Sparse tiny to medium pyroxene; medium tiny to small lime inclusions; medium small red inclusions.

Classical-Hellenistic.

**Cat. 229 8570/01/7**

Handle fr. H: 7.5; W: 3.8-4.1; T: 2.7-2.8. Clay: 2.5YR 7/6 light red. Frequent small to medium red inclusions; medium tiny to medium lime inclusions; sparse small to medium quartz; sparse small pyroxene; medium tiny to small feldspar.

Herakleia, 4th century BC.

**Cat. 230 8570/01/10**

Wall fr.? Repair hole at the edge. H: 3.8; W: 6.3; T: 0.8-1. Core: 2.5YR 5/6 red. Slip: 5YR 8/4 pink. Sparse tiny to small lime inclusions; sparse tiny stones.

Classical-Hellenistic.

**Cat. 231** 8570/02/1

Handle fr. Re-used as a grinding tool with a hole (Ø 0.47). H: 2.9; W: 2.8 T. 2.05. Clay: 2.5YR 6/8 light red. Medium tiny to small mica; single tiny quartz mineral.

Classical-Hellenistic.

**Cat. 232** 8592/02/2

Handle fr. H: 6.2; W: 4; T: 2.3. Clay: 2.5YR 6/8 light red. Medium tiny to small lime inclusions; sparse small to medium black inclusions; medium tiny to small feldspar.

Classical-Hellenistic.

**Cat. 233** 8592/02/1

Wall fr. with part of handle preserved. H: 5.7; W: 4; T: 0.7. Clay: 2.5 5/8 red. Slip: 2.5YR 6/8 light red. Sparse tiny to medium lime inclusions; sparse small pyroxene; sparse small to medium red inclusions; sparse small feldspar.

Classical-Hellenistic.

**Cat. 234** 8592/02/3

Handle fr. Slight ridge on int. H: 5.2; W: 4.65; T: 2.9. Clay: 10R 6/6 light red. Surface: 5YR 7/6 reddish yellow. Sparse small pyroxene; sparse small to medium red inclusions; sparse tiny to medium lime; sparse small quartz.

Herakleia, 4th century BC.

**Cat. 235** 8592/04/3

Fr. of lower body. H: 4.2; W: 5.03; Ø: 5. Clay: 5YR 8/4 pink. Sparse small pyroxene; sparse small quartz; sparse shells.

Classical-Hellenistic.

**Cat. 236** 8638/01/1

Handle fr. H: 6.75; W: 2.3-2.8; T: 2. Clay: 10YR 7/9 light red. Surface: 10R 7/6 light red. Sparse sand inclusion; sparse tiny to medium lime inclusions; sparse tiny feldspar.

Classical-Hellenistic.

## Tableware

**Cat. 237** 8534/01/2

Large handle fr. of jug? H: 10.9; W: 2.8; T: 1.3. Clay: 2.5YR 5/6 red. Slip 5YR 7/4 pink. Frequent tiny to medium lime inclusions; sparse small shells; sparse tiny to medium feldspar; sparse small quartz.

Chersonesos? Late 4th-3rd century BC.

**Cat. 238** 8534/02/2

Base fr. of jug?, badly preserved. H: 2.3; W: 3.42; T: 0.52-1.26; Ø? Core: 2.5YR 6/8 light red. Surface: 5YR 7/8 reddish yellow. Single small lime particle; medium small to large pyroxene; single small quartz.

Chersonesos? Late 4th-3rd century BC.

**Cat. 239** 8534/02/3

Fr. of red-clay jug handle. H: 9.27; W: 3.48; T: 1.32. Core: 5YR 8/4 pink to 5YR 8/2. Slip: 2.5Y 8/3 pale yellow. Sparse tiny to small lime inclusions; medium quartz; medium pyroxene; medium feldspar; single tiny shell.

Cf. Kašaev 2002, 151-152, Cat. C7-23. Late 4th-3rd century BC.

**Cat. 240** 8570/01/8

Base fr. with a low conical ring foot of jug. H: 2.2; W: 8.5; T: 0.8-1.3; Ø: 12. Clay: 2.5YR 6/8 light red. Slip: 10YR 6/4 very pale brown. Medium tiny to small lime inclusions; medium tiny to small feldspar; medium small to medium red inclusions; medium sand inclusion.

Chersonesos, cf. Kašaev 2002, Cat. C231. Late 4th-3rd century BC.

**Cat. 241** 8570/03/1

Base fr. with a low ring foot of jug. H: 2.25; W: 2.95; T: 1.06 (wall)-0.83 (bottom); Ø: ca. 12. Core: 2.5YR 6/8 light red. Surface: 5YR 7/8 reddish yellow. Sparse small to medium quartz; sparse small shells; sparse tiny voids.

Chersonesos, cf. Kašaev 2002, Cat. C231. Late 4th-3rd century BC.

**Cat. 242** 8570/03/2

Flat broad handle fr. of jug. Low ridge on the ext. of the handle. H: 6.75; W: 4.35; T: 1.53. Clay: 2.5YR 7/8 light red to 7.5YR 8/4 pink. Slip: 5Y 8/2 pale yellow. Medium tiny shells; medium tiny quartz; medium tiny voids.

Chersonesos, cf. Kašaev 2002, 151-152, Cat. C7-23. Late 4th-3rd century BC.

**Cat. 243** 8570/03/3

Base fr. of jug with ring foot. H: 2.15; W: 5.04; T: 0.6; Ø? Core: 5Y 8/2 pale yellow. Surface: 5Y 3/1 very dark grey. Sparse small to medium quartz; sparse small to medium shells; medium tiny voids.

Late 4th-3rd century BC.

**Cat. 244** 8570/04/2

Wall fr. of jug with part of repair hole. H: 1.9; W: 3.1; T: 0.6. Clay: 2.5YR 6/8 light red. Slip ext.: 2.5Y 8/3 pale yellow. Frequent small to medium lime inclusions; sparse small to medium quartz; medium small feldspar.

Chersonesos, cf. Kašaev 2002, 151-152, Cat. C7-23. Late 4th-3rd century BC.

**Cat. 245** 8591/02/1

Handle fr. of small jug or flask. Badly preserved. H: 2.67; W: 1.3; T: 0.51. Clay: 2.5YR 7/6 light red to 7.5YR 7/4 pink. Sparse tiny lime and mica inclusions.

Late Classical-Hellenistic.

**Cat. 246** 8534/02/1

Rim fr. of bowl with inturned rim. H: 4.1; W: 3.56; T: 0.78-1.06; Ø: ca. 25. Clay: 5YR 7/6 reddish yellow. Sparse tiny lime inclusions; medium tiny to medium voids.

Cf. Kašaev 2002, Cat. C48-C49. Early Hellenistic.

**Cat. 247** 8568/02/1

Floor fr. of Attic? black-glossed cup. Slightly lustrous gloss; flaking on the int. Central circle surrounded by two concentric circles on the underside. W: 2.7 x 2; T: 0.4-0.5. Clay: 5YR 5/4 reddish brown. Sparse tiny lime inclusions.

Similar treatment of the underside is found on various Attic cups like stemless cups, cup-skyphoi, bolsals and one-handlers, primarily dating to the 5th century BC, but continue into the 4th century BC. cf., e.g., Sparkes-Talcott 1970, Cat. 464, 481, 500, 539, 578. 5th century BC.

**Cat. 248** 8570/01/11

Rim fr. of lid. H: 2.6; W: 3.6; T: 0.5-0.6; Ø? Clay: 5YR 5/6 yellowish red. Slip: 5YR7/6 reddish yellow. Sparse tiny lime inclusions; sparse light-reflecting particles; sparse tiny feldspar.

Late Classical-Hellenistic.

**Cat. 249** 8637/01/1

Fr. of the neck of a jug with ribbings H: 4.8 W: 3.85 T: 0.8. Clay: 7.5YR 7/4 pink. Spares tiny-small lime inclusions; single. medium quartz; sparse tiny feldspar.

Cf. Bylkova 2005, fig. 6.3. Early Hellenistic.

## DSP08-F04-04

Core units: 8707, 8737-8738, 8745, 8771, 8784, 8859
Adjacent units: none

### Handmade ware

*Bronze Age*

**Cat. 250** 8771/01/1

Cup with flaring wall, rounded lip and a thick base with slightly concave underside. Deep vertical striations on the exterior. Composed of four frs. Complete profile preserved. H: 10.2 Ø base: 9.2 T: 0.6-0.1.7. Clay: greenish-grey to greyish black. Frequent small-large lime particles.

Cf. Toščev 2007, 183, fig. 94.4 (from Vinogradov). For the striations see also Toščev 2007, 154 fig. 82.18 (from Planerskoe I). Middle Bronze Age

### Amphorai

*Unknown production*

**Cat. 251** 8738/02/1

Handle fr. H: 3.9; W: 3.42; T: 2.3. Core: 2.5YR 7/8 light red. Surface: 7.5YR 8/6 reddish yellow. Sparse tiny mica; sparse shells; single medium shell.

Classical-Hellenistic.

## DSP08-F04-05

Core units: 6581, 6602, 6617, 8805-8806
Adjacent units: 6582, 6616, 6618, 8804

### Handmade ware

**Cat. 252** 6616/01/1

Base fr. of pot with moulded base and flat underside. H: 3.75; W: 4.4; T: 0.9-1.6; Ø? Clay: 2.5YR 6/6 light red to 2.5YR 5/6 red. Sparse tiny to small lime inclusions; medium tiny to small feldspar; single small to medium quartz mineral.

Cf., e.g., Stolba 2002, Cat. D67. Early Hellenistic.

### Amphorai

*Herakleia*

**Cat. 253** 6618/01/1

Rim fr. with slightly beveled lip. H: 3.5; W: 3.29; T: 0.94 (wall)-1.2 (rim); Ø: 7. Core: 5YR 8/2 pinkish white. Surface: 5YR 8/3 pink to 2.5YR 7/8 light red. Frequent medium pyroxene; frequent medium quartz; frequent medium feldspar; single small shell.

Possibly type II-A-3 or type IV, cf. Monachov 2003, pl. 97. Early Hellenistic.

*Unknown production*

**Cat. 254** 6581/01/1

Handle fr. with flaky coloured inner structure. H: 5.33; W: 4.1-4.6; T: 2.18-2.27. Core: 10YR 7/3 very pale brown to 10R 6/8 light red. Clay: 7.5YR 7/4 pink to 2.5YR 6/6 light red. Covered by dark grey slip. Sparse tiny shells; sparse tiny lime inclusions; sparse small oval voids.

Classical-Hellenistic.

### Tableware

**Cat. 255** 6616/01/2

Base fr. of closed shape, possibly a jug, with a flat underside. H: 3; W: 5; T: 1-1.7; Ø: ca. 9. Clay: 10R 5/2 weak red. Surface: 2.5YR 5/4 reddish brown. Sparse tiny lime inclusions; frequent tiny feldspar; sparse tiny to small pyroxene.

Late Classical-Hellenistic.

### Post-Antique pottery

**Cat. 256** 8805/01/1

Wall fr. of pot. H: 2.13; W: 3.32; T: 0.46-0.64. Core: Gley 1 6/10Y greenish grey. Surface: 2.5Y 8/2 pale yellow and 5YR 7/4 pink. Slip: 5YR 5/3 reddish brown. Single small lime inclusion; sparse tiny to small mica; single tiny quartz.

Early Modern

## DSP08-F04-06

Core units: 6521-6522, 6525-6526
Adjacent units: 6523, 6527

### Handmade ware

*Bronze Age*

**Cat. 257** 6526/04/2

Rim fr. of cooking vessel with flaring rim. H: 2.87; W: 4.62; T: 0.75-0.79; Ø? Core: Gley 1 3/N very dark grey. Surface: Gley 1 6/10Y greenish grey and 7.5YR 8/6 reddish yellow. Sparse small to medium shells and quartz.

Cf. Gerškovič 1999 type T1II2, cf. Fig. 12.10. Late Bronze Age.

**Cat. 258** 6526/04/3

Wall fr. of cooking vessel? with a repair hole. H: 4.45; W: 3; T: 0.78-1.11. Core: Gley 2 4/10B dark bluish grey. Surface: 10R 4/1 dark reddish grey to 5YR 6/4 light reddish brown. Sparse small to medium shells and quartz.

Late Bronze Age.

**Cat. 259** 6526/04/4

Rim fr. of pot with smoothed surface. H: 3.4; W: 2.3; T: 6.4-7.2; Ø? Clay: 7.5YR 4.1 dark grey. Medium small to medium lime inclusions; medium small to medium shells.

Similar to some of the vessels from Baj Kijat, cf. Kolotuchin 2003, e.g. pl. 52.8-9. Late Bronze Age.

**Cat. 260** **6526/04/5**

Outturned rim fr. of bowl? H: 2.05; W: 1.74; T: 0.49-0.69; Ø? Clay: 7.5YR 7/4 pink. Frequent medium lime inclusions; frequent medium quartz; frequent medium feldspar.

Possibly similar to Gerškovič 1999 type S1I1, cf. Fig. 12.1. Late Bronze Age?

## Amphorai

### *Unknown production*

**Cat. 261** **6526/04/6**

Handle fr. of a large jug. H: 6.94; W: 3.85; T: 2.03-2.06. Core: 10R 7/6 light red. Surface: 5YR 7/6 reddish yellow. Single tiny shell; medium tiny to medium voids.

Medieval?

## Tableware

**Cat. 262** **6522/01/1**

Handle fr. of jug? H: 4.34; W: 3; T: 1.18. Clay: 10R 5/6 red to 2.5YR 6/8 light red. Single medium lime inclusions; sparse medium quartz; sparse medium feldspar; sparse medium voids.

Late Classical-Hellenistic.

**Cat. 263** **6522/01/2**

Handle fr. of jug H: 2.42; W: 3.1; T: 1.2. Clay: 2.5YR 6/8 light red. Single medium lime particle; sparse medium quartz; single medium void.

Late Classical-Hellenistic.

**Cat. 264** **6526/02/1**

Rim fr. of jug with outturned, slightly beveled rim. H: 3.51; W: 3.77; T: 0.53 (wall), 0.93 (rim); Ø? Core: 2.5YR 7/8 light red. Surface: 7.5YR 8/6 reddish yellow. Frequent small to medium shells; sparse small to medium quartz; single tiny pyroxene inclusion; sparse tiny voids.

For the shape, cf., e.g., Zolotarev 2005, fig. 10.4. Late 4th-3rd century BC.

**Cat. 265** **6526/04/1**

Rim fr. of bowl with inturned rim. H: 2.9; W: 5.52; T: 0.96-1.16; Ø: ca. 25. Core: 10YR 6/8 light red. Surface: 7.5YR 7/6 reddish yellow and 7.5YR 6/2 pinkish grey. Single tiny lime particle; medium to large pyroxene; sparse medium to large quartz; sparse medium to large feldspar.

Cf. Kašaev 2002, Cat. C45. Early Hellenistic.

## Post-Antique pottery

**Cat. 266** **6521/01/01**

Fr. of jug with flat, broad handle. H: 6.3; W: 3.8; T: 1.1. Core: 5YR 5/8 yellowish red. Surface: 2.5 6/6 light red. Sparse tiny lime inclusions; single large pebble; sparse medium dark inclusions.

Early Modern.

# DSP08-F04-07

Core units: 6492-6494
Adjacent units: 6475-6480, 6490-6491, 6495, 6506-6507

## Handmade ware

**Cat. 267** **6493/01/2**

Rim fr. of pot with straight rim and impressed finger decoration on the lip. H: 3.1; W: 3.7; T: 1.2-1.4; Ø: 13-16. Core: 2.5YR 5/1 reddish grey. Surface: 10R 7/8-6/8 light red. Frequent tiny to large lime inclusions; sparse small to medium voids.

Cf. Stolba 2002, type 1C, Cat. D86-D87. Early Hellenistic.

**Cat. 268** **6493/02/3**

Rim fr. of vessel with concave, outturned rim and rounded lip. H: 3.8; W: 2.3; T: 1-1.2; Ø? Core: 10R 5/1 reddish grey. Surface: 10R 3/1 dark reddish grey to 10R 6/6 light red. Sparse tiny to small lime inclusions; sparse small dark inclusions; sparse sand.

Cf. Stolba 2002, type 1 Cat. D1-D68. Early Hellenistic.

**Cat. 269** **6493/03/2**

Rim fr. of pot with straight outturned rim and a flat lip. Horizontal moulding below the lip on the ext. H: 3.4; W: 3; T: 0.9-1.1; Ø: 16. Core: Gley 2 3/10 BG to 2.5YR 7/4 light reddish brown. Surface: Gley 2 3/10 BG. Medium small to large grog; sparse tiny mica.

Cf. Stolba 2002, type 1 with relief decoration, Cat. D69-D70. Early Hellenistic.

## Amphorai

### *Chersonesos*

**Cat. 270** **6507/03/1**

Handle fr. H: 10.85; W: 4.4; T: 2.7. Core: 2.5YR 4/8 red. Near surface: 10YR 5/2 greyish brown. Slip: 10YR 7/3 very pale. Frequent tiny to medium lime inclusions; sparse tiny to medium shells; sparse medium quartz; medium tiny silver mica; sparse sand inclusions; sparse medium voids

Classical-Hellenistic.

**Cat. 271** **6478/01/1**

Fr. of toe. H: 3.3; Ø: 7.3. Core: 2.5YR 7/8 light red. Surface: 2.5YR 7/4 light reddish brown. Single tiny mica and shell.

Chersonesos? Type I-A-1 or I-A-2 (Monachov 2003 type 1 base). Ca. 350-300 BC.

### *Herakleia*

**Cat. 272** **6494/02/1**

Fr. of protruding rim. Lip is not preserved. H: 3.3; W: 3.4; T: 0.75-0.8; Ø: ca. 13. Clay: 10R 6/8 light red. Sparse small pyroxene; sparse tiny to small quartz; sparse tiny lime inclusions; sparse tiny to small feldspar.

4th century BC.

**Cat. 273** **6492/03/1**

Rim fr. H: 3.3; W: 2.8; T: 0.9-1.1; Ø: 10. Clay: 7.5YR 7/4 pink. Sparse small pyroxene; sparse tiny to small feldspar.

Herakleia? Type I-4 or II-1, cf. Monachov 2003 e.g., pl. 90.3 and 93.6. Ca. 400-325 BC.

**Cat. 274** **6493/01/1**

Rim fr. H: 2.4; W: 3.3; T: 0.9-1; Ø: 10. Clay: 2.5YR 6/8 light red. Sparse tiny pyroxene; sparse tiny lime inclusions; sparse tiny feldspar.
Herakleia? Monachov 2003, pl. 89.4 and 95.3. 4th century BC.

### *Thasos*

**Cat. 275** **6493/02/4**

Handle fr. H: 6.65; W: 4.6; T: 2-2.2. Clay: 5YR 6/6 reddish yellow. Slip: 10R 5/4 weak red. Frequent tiny silver mica; sparse tiny to small golden mica; sparse tiny lime inclusions; sparse tiny to small feldspar.
Late conical type, II-C-3. Cf. Monachov 2003, pl. 49-50. Ca. 325-250 BC.

### *Unknown production*

**Cat. 276** **6492/02/1**

Handle attachment. Body H: 3.8; T: 1.2 handle; W: 4.7; T: 2. Core: 5YR 7/6 reddish yellow. Surface: 7.5YR 8/4 pink. Frequent tiny to large golden mica; medium tiny to small lime inclusions; sparse small shells; sparse small feldspar?; sparse small quartz?
Hellenistic?

**Cat. 277** **6492/02/2**

Handle fr. with three broad shallow grooves on the ext. H: 6.7; W: 4.7; T: 2.8. Clay: 5YR 6/6 reddish yellow. Surface: 10YR 6/4 light yellowish brown. Frequent tiny to large golden mica; medium small lime inclusions; sparse small feldspar; sparse sand; sparse large black inclusions.
Classical-Hellenistic.

**Cat. 278** **6493/02/2**

Handle fr. H: 3.3; W: 3; T: 1.95. 2.5YR 7/6 light red. Surface: 2.5YR 6/4 light reddish brown. Medium tiny mica; sparse sand; medium tiny lime inclusions; sparse tiny to medium feldspar.
Classical-Hellenistic.

**Cat. 279** **6493/02/5**

Handle fr. H: 5.8; W: 3.2-3.9; T: handle 2.6; wall: 0.8. Clay: 2.5YR 5/8 red. Surface: 6/6 light red. Single tiny mica; sparse tiny lime inclusions; sparse tiny feldspar; sparse small pebbles.
Classical-Hellenistic.

**Cat. 280** **6493/03/1**

Handle fr. H: 7.1; W: 3.6-4; T: 2.6-2.8. Clay: 2.5YR 5/6 red. Slip: 2.5YR 6/3 light reddish brown. Frequent tiny to small mica; sparse small quartz; sparse small feldspar.
Classical-Hellenistic.

**Cat. 281** **6506/01/2**

Handle fr. Incompletely preserved. H: 6.3; W: 3.4; T: 2. Clay: 10R 6/8 light red. Medium tiny to medium lime inclusions; medium tiny feldspar.
Classical-Hellenistic.

## Tableware

**Cat. 282** **6492/05/1**

Rim fr. of small flask with shallow neck and a rounded lip. H: 1.8; W: 2.05; T: 0.9; Ø: 5. Core: 10R 6/1 reddish grey. Surface: 10R 7/4-6/4 pale red. Sparse small lime inclusions; sparse tiny mica; sparse small quartz; sparse tiny to small feldspar.
Late Classical-Hellenistic.

# DSP08-F05-01

Core units: 9024-9025
Adjacent units: 9013, 9026, 9028

## Amphorai

### *Knidos?*

**Cat. 283** **9026/01/2**

Rim fr. with protruding rim. H: 3.5; W: 6.4; T: 0.75-3.5; Ø: ca. 27. Clay: 7.5YR 7/4 pink. Slip: 5YR 8/4 pink. Sparse tiny golden mica; sparse small quartz; sparse tiny feldspar.
Type I-A to I-C. Cf. Monachov 2003, tav 71-72. Ca. 400-350 BC.

### *Unknown production*

**Cat. 284** **9026/01/01**

Handle fr. H:4.5; W: 5.4; T: handle: 1.9, wall: 1. Clay: 2.5YR 7/8 light red. Sparse small to medium grog; medium tiny to medium lime inclusions; medium small to medium pebbles; sparse small quartz.
Medieval?

## Post-Antique pottery

**Cat. 285** **9028/01/1**

Rim fr. of jug with thickened rim. Concave groove on the upper part of the lip. H: 1.86; W: 2.76; T: 0.54 (wall)-1.05 (rim); Ø: 7. Core: 5YR 7/6 reddish yellow. Surface: 10YR 6/1 grey. Sparse very tiny to tiny lime inclusions.

**Cat. 286** **9025/01/01**

Fr. of ring-base glazed on interior and exterior. H: 2.2; W: 3.13; T: 1.06 (wall). Core: light red 2.5YR 7/6 and 5YR 8/4 pink. Glaze: 10YR 5/8 yellowish brown and 2.5Y 6/8 olive yellow. Sparse very tiny lime.
Early Modern.

# DSP08-F06-01

Core units: 9200-9203, 9207-9209, 9218-9219
Adjacent units: 9210, 9217

## Handmade ware

### *Bronze Age*

**Cat. 287** **9201/01/1**

Wall fr. with combed ext. surface. H: 4.4; W: 2.68; T: 0.66. Core: 5YR 2.5/1 black. Surface: 2.5Y 5/2 greyish brown. Single medium grog?; medium tiny to small feldspar; sparse sand.
Cf. Toščev 2007, fig. 82.18 (Planerskoe).

**Cat. 288** **9201/01/3**

Wall fr. of vessel with grooved ext. H: 3.3; W: 4.38; T: 0.87. Core: Gley 1 2.5/N black. Int. surface: Gley 1 5/N grey. Ext. surface: 10YR 7/4 very pale brown. Sparse tiny lime particles; sparse sand.
Cf. Toščev 2007, fig. 82.18 (Planerskoe). Middle Bronze Age?

**Cat. 289 9202/07/2**

Rim fr. of vessel with three horizontal rows of impressed cord decoration. H: 2.13; W: 1.5; T: 0.71; Ø? Core: Gley 1 2.5/N black. Surface: 2.5Y 2/1 grey. Sparse sand; sparse tiny lime inclusions.

Cf. Leskov 1970, fig. 3.7 (Kirovo).

**Cat. 290 9202/10/1**

Wall fr. of vessel with smoothed ext. surface and horizontal furrow (width: 0.27, depth: 0.16). H: 3.64; W: 2.15; T: 1-1.12. Core: Gley 1 3/N very dark grey. Int. surface: Gley 1 4/N dark grey. Ext. surface: Gley 1 7/N light grey and 2.5 Y 7/4 pale yellow. Sparse tiny lime inclusions.

Cf. Leskov 1970, fig. 3.5.

**Cat. 291 9207/02/1**

Rim fr. of vessel with a short concave rim. H: 1.7; W: 2.1; T: 0.72; Ø? Core: Gley 1 2.5/N black. Sparse light-reflecting particles; sparse tiny lime inclusions.

Cf. Leskov 1970, fig. 3.1.

**Cat. 292 9208/02/19**

Rim fr. of vessel with two incised horizontal furrows below the rim on the ext. H: 1.6; W: 2.2; T: 0.9; Ø? Core: Gley 1 2.5/N black. Surface: 2.5Y 7/1 light grey. Sparse tiny lime inclusions; single small quartz; sparse small red inclusions.

Cf. Leskov 1970, fig. 3.5-6 (Kirovo). Middle Bronze Age?

**Cat. 293 9208/02/2**

Rim fr. of vessel with a short very concave rim. H: 1.9; W: 2.36 T: 1; Ø: 16-17. Clay: Gley 1 3/N very dark grey. Sparse tiny lime inclusions; sparse sand.

Cf. Leskov 1970, fig. 3.1.

**Cat. 294 9208/02/4**

Wall fr. of smaller vessel with impressed 'herringbone' decoration and a horizontal moulding. H: 1.84; W: 1.6; T: 0.5. Core: Gley 1 2.5/N black. Surface: 10YR 7/3 very pale brown. Single tiny lime inclusions; single small quartz.

Cf. Toščev 2007, fig. 80.2,9 (Kimmerik).

**Cat. 295 9209/02/1**

Wall fr. of vessel with incised squares. H: 3; W: 3.2; T: 1.1. Core: 5YR 2.5/1 black. Int. surface: Gley 2 5/5 B bluish grey. Ext. surface: 5YR 8/4 pink. Sparse tiny lime inclusions; medium sand.

Cf. Toščev 2007, fig. 82.10 (Planerskoe I); Rybalova 1974, fig. 14.2,4 (Kamenka).

**Cat. 296 9200/02/1**

Wall fr. of vessel with deeply incised furrows on the ext. H: 3.35; W: 2.51; T: 0.87. Core: Gley 1 2.5/N black to 10YR 5/3 brown. Ext. surface: 10YR 8/2 very pale brown. Int. surface: sparse tiny lime inclusions; sparse small quartz; sparse sand.

Cf. Rybalova 1974, fig. 14.2,4 (Kamenka).

**Cat. 297 9201/01/4**

Rim fr. of vessel with straight rim and flat lip. Six horizontal lines of impressed cord decoration on the exterior below the rim. H: 3.1; W: 3; T: 1.15; Ø ca. 20. Core: Gley 1 3/N very dark grey. Ext. surface: Gley 1 6/N grey, int. surface: Gley 1 4/N dark grey. Sparse tiny to medium lime inclusions; sparse tiny feldspar.

For the docoration, see Toščev 2007, fig. 80.2 (Kimerik) and Gerškovič 1999, pl. 14.8 (Uskalka).

**Cat. 298 9201/01/5**

Shoulder fr. of vessel with impressed cord decoration. H: 6.4; W: 3.5; T: 0.9. Core: Gley 1 2.5/N black. Int. surface: 10YR 5/1 grey to 10YR 6/2 light brownish grey. Frequent tiny to medium lime inclusions; sparse tiny feldspar.

Cf. Leskov 1970, fig. 3.8 (Kirovo).

**Cat. 299 9201/03/2**

Wall fr. of vessel with three horizontal lines of impressed cord decoration. H: 1.9; W: 2.55; T: 0.72. Clay: Gley 1 2.5/N black. Sparse tiny lime inclusions; single medium quartz;

For the decoration, see, e.g., Leskov 1970, fig. 3.8.

**Cat. 300 9201/06/3**

Shoulder fr. of vessel with obligue impressed decoration. H: 2.5; W: 3.2; T: 0.7-0.8. Core: Gley 1 2.5/N black to 10YR 4/2 dark greyish brown. Ext. surface: 10YR 8/1 white. Sparse tiny to small lime inclusions; tiny to small quartz; sparse sand inclusions.

**Cat. 301 9201/09/4**

Wall fr with two rows of impression. H: 2.64; W: 2.64; T: 1. Core: Gley 1 2.5/N black. Surface: 10YR 7/3 very pale brown. Sparse tiny lime inclusions; sparse sand; single small shell.

Cf. Rybalova 1974, fig. 14.12 (Kamenka).

**Cat. 302 9202/07/1**

Wall fr. of vessel with impressed 'herringbone' decoration. H: 2.1; W: 1.75; T: 0.73. Core: Gley 1 2.5N black. Int. surface: 10YR 5/1 grey, ext. surface: 10YR 8/2 very pale brown. Sparse tiny lime inclusions; single light-reflecting particle.

For the decoration, see Toščev 2007, fig. 80.9 (Kimmerik).

**Cat. 303 9202/10/2**

Rim fr. of vessel with flat lip with obligue cuts. H: 2.06; W: 1.95; T: 0.8-1.02; Ø: ca. 19. Core: Gley 1 4/N dark grey. Int. surface: Gley 1 6/N grey, ext. surface: 10YR 7/4 very pale brown. Sparse very tiny lime particles.

**Cat. 304 9202/11/1**

Wall fr. of vessel with two horizontal rows of impressed cord decoration. H: 3.2; W: 3.4; T: 0.8-1. Core: Gley 2 2.5/5 BG greenish black. Ext. surface: 10R 4/2 pale red, int. surface: Gley 2 5/10B. Single tiny quartz; single light-reflecting particle; sparse sand inclusions.

Cf. Toščev 2007, fig. 79.21 (Kamenka).

**Cat. 305 9202/11/2**

Wall fr. of vessel with incised furrows on the ext. H: 2.1; W: 3.1; T: 0.7-1. Core: Gley 2 3/10 BG. Surface: Gley 2 6/5 BG greenish grey. Sparse light-reflecting particles; single small quartz.

Cf. Rybalova 1974, fig. 14.2,4 (Kamenka).

**Cat. 306 9203/01/1**

Wall fr. of vessel with impressed decoration. H: 3.3; W: 3.84; T: 1. Core: Gley 1 2.5/N black. Surface: 10YR 5/1 grey. Frequent tiny lime inclusions; sparse tiny feldspar.

**Cat. 307 9208/01/9**

Rim fr. of vessel with straight walls and a rounded lip. H: 1.77; W: 1.9; T: 0.85; Ø: 19. Core and int. surface: Gley 1 2.5/N black, ext. surface: 2.5Y 8/2 pale yellow. Medium tiny lime inclusions; sparse small quartz; sparse sand.

Middle Bronze Age/Late Bronze Age?

**Cat. 308 9208/02/13**

Rim fr. of vessel with flat and slightly thickend rim. H: 1.34; W: 1.56; T: 0.96; Ø? Core: Gley 1 2.5/N black. Surface: 10YR 7/4 very pale brown. Single tiny lime particle; sparse tiny light-reflecting particles.

Cf. e.g. Rybalova 1974, fig. 5.8. Middle Bronze Age/Late Bronze Age?

**Cat. 309 9208/02/3**

Wall fr. of cup? with two rows of small impressed squares. H: 1.66; W: 1.25; T: 0.46. Core: Gley 1 2.5/N black. Surface: 10YR 5/1 grey. Sparse small lime inclusions.

**Cat. 310 9208/02/5**

Rim fr. of larger storage vessel? with slightly flaring rim. Three horizontal rows of impressed double cord decoration. H: 3.23; W: 3.7; T: 1.07; Ø: 30-35. Clay: Gley 1 2.5/N black. Medium tiny to small lime inclusions; sparse light-reflecting particles.

Cf. e.g., Toščev 2007, fig. 79.21 (Kamenka).

**Cat. 311 9208/02/8**

Rim fr. of vessel with slightly outturned rim. Three rows of small impressed squares on the ext. below the rim. (Not well preserved) H: 2.59; W: 1.96; T: 1; Ø? Core: Gley 1 2.5/N black. Surface: 2.5Y 7/1 light grey. Sparse tiny lime inclusions; sparse sand.

Cf. Rybalova 1974, fig. 14.12 (Kamenka).

**Cat. 312 9209/02/5**

Base fr. of pot with impressed cord? decoration near the bottom. H: 3.35; W: 3.6; T: 0.8-1.6; Ø: ca. 10. Core: Gley 2 2.5/10 BG greenish grey. Int. surface: Gley 2 3/10 B very dark bluish grey. Single tiny lime particle; sparse tiny feldspar.

Middle Bronze Age/Late Bronze Age?

**Cat. 313 9219/02/1**

Wall fr. of vessel with two horizontal rows of impressed cord decoration in the form of 'herringbone' pattern. H: 4.33; W: 2.84; T: 1.52. Core: 10YR 2/1 black. Surface: Gley 1 4/N dark grey. Frequent tiny to medium lime inclusions; sparse small to medium quartz; sparse tiny feldspar; sparse light-reflecting particles.

For horizontal rows of impressed cord decoration, see, e.g., Leskov 1970, fig. 3.7.

**Cat. 314 9219/02/3**

Rim fr. of vessel with a short concave rim. H: 4.54; W: 5.1; T: 1.27; Ø: ca. 25. Core: Gley 1 3/N very dark grey. Int. surface: 2.5Y 2/5 grey. Ext. surface: 10YR 6/3 pale brown. Sparse tiny to medium pyroxene; sparse tiny lime inclusions; sparse light-reflecting particles.

Cf. Leskov 1970, fig. 3.1.

**Cat. 315 9200/05/1**

Wall fr. of burnished vessel. H: 3.93; W: 4.32; T: 0.7. Core: 10YR 4/1 dark grey. Surface: 10YR 6/4 light yellowish brown. Sparse medium grog; sparse light-reflecting particles.

Late Bronze Age?

**Cat. 316 9201/01/6**

Wall fr. with impressed decoration. H: 3.5; W: 2.8; T: 0.7. Core: Gley 1 2.5/N black. Int. surface: Gley 1 4/N dark grey, ext. surface 2.5Y 4/1 dark grey. Medium tiny to small lime inclusions.

Late Bronze Age?

**Cat. 317 9201/01/7**

Rim fr. of vessels with straigt walls and rounded lip, possibly a bowl. H: 1.8: W: 1.86; T: 0.85; Ø ca. 13. Core: 2.5Y 6/1-3/1 grey to very dark grey. Surface: 10YR 7/4 very pale brown. Single medium grog.

Late Bronze Age?

**Cat. 318 9201/01/8**

Rim fr. of open vessel with outturned rim, possibly frying pan. H: 2.93; W: 3.25; T: 0.8; Ø: 18-21. Core: Gley 1 3/N very dark grey. Surface: 10YR 6/3 pale brown. Frequent tiny to large lime particles.

Somewhat resembles Gerškovič 1999, pl. 14.9. Late Bronze Age?

**Cat. 319 9201/02/1**

Wall fr. of vessel with horizontal moulding on the ext. with two obligue cuts. H: 2.5; W: 2.1; T: 0.5-1. Core: Gley 2 2.5/10 BG greenish black. Surface: 5YR 8/3 pink. Sparse tiny lime inclusions; sparse tiny mica.

Cf. Toščev 2007, fig. 97.8 (Sazonovka).

**Cat. 320 9201/03/1**

Rim fr. of vessel with slightly concave rim and flat lip. H: 2.75; W: 2.66; T: 0.8 Ø:? Clay: Gley 1 2.5/N black. Surface: 10YR 6/1 grey. Sparse tiny lime inclusions; single small quartz; sparse sand.

Somewhat similar to Gerškovič 1999, pl. 13.5 (Uškalka).

**Cat. 321 9201/06/1**

Wall fr. of vessel with impressed decoration. H: 1.7; W: 2; T: 0.9. Core: Gley 1 2.5/N black to Gley 1 5/N grey. Int. surface: Gley 1 2.5/N black, ext. surface: 10YR? Sparse tiny lime inclusions; sparse tiny feldspar.

Late Bronze Age?

**Cat. 322 9201/06/2**

Wall fr. of pot with impressed decoration. H: 3.2; W: 3.9; T: 1-1.2. Core: Gley 1 3/N very dark grey. Int. surface: 10YR 7/4 very pale brown. Ext. surface: 7.5YR 6/4. Medium tiny to large lime; medium tiny to medium feldspar.

**Cat. 323 9201/06/4**

Rim fr. of vessel with slightly flaring rim and a flat lip. H: 2.34; W: 1.87; T: 0.71; Ø: 8-12. Core: Gley 1 3/N very dark grey. Surface: 10YR 5/1-6/2 grey to light brownish grey. Single light-reflecting particle; single tiny lime particle.

**Cat. 324 9201/06/5**

Rim fr. of vessel with protruding rim with an undercut moulding below the rim. H: 2.9; W: 2.55; T: 1.21; Ø: >20. Core: Gley 1 2.5/N black to 10YR 7/3 very pale brown. Surface: 7.5YR 6/4 light brown. Medium tiny to small lime inclusions; medium small to medium quartz; medium sand inclusions.

**Cat. 325 9201/09/1**

Wall fr. of cooking vessel with a horizontal moulding with impressed decoration. H: 3.04; W: 3.3; T: 1. Core: Gley 1 2.5/N black. Ext. of core and surface: 2.5YR 7/1 light grey. Sparse tiny lime particle; single tiny mica particle.

Cf. Kolotuchin 2003, pl. 58.9 (Družnoe).

**Cat. 326 9201/09/2**

Rim fr. of vessel with a out flaring rim and a beveled lip. H: 3.4; W: 4.6; T: 0.65; Ø: 24-26. Core: Gley 1 2.5/N black. Surface: 7.5YR 7/3 very pale brown. Sparse tiny lime inclusions; sparse tiny light-reflecting particles; single large dark inclusion.

Cf. Gerškovič 1999, type T1II2, fig. 11.10.

**Cat. 327 9201/09/3**

Fr. of straight rim with obligue cuts on top of the lip. H: 1.9; W: 1.9; T: 0.86; Ø? Core: Gley 1 2.5/N black. Int. surface: Gley 1 6/N grey, ext. surface: Gley 1 4/N dark grey to 5YR 7/4 pink. Sparse tiny lime inclusions; sparse tiny feldspar; single medium quartz.

Late Bronze Age?

**Cat. 328 9201/09/6**

Rim fr. of vessel with protruding lip. Obligue impressions on lip and round impressions below the rim on the ext. H: 3.36; W: 2.49 T. 0.94; Ø: ca. 30. Core: 10YR 5/1 grey. Surface: 10YR 5/2 greyish brown to 7.5YR 6/6 reddish yellow. Frequent light-reflecting particles; sparse tiny lime inclusions; sparse tiny feldspar.

**Cat. 329 9201/09/7**

Rim fr. of storage vessel? with flaring rim. H: 3.7; W: 3.89; T: 0.78; Ø: 28-30. Sandwich clay. Int. core: 10YR 8/2 very pale brown, ext. core: Gley 1 2.5/N black. Single tiny lime particle; medium small to medium quartz; sparse tiny feldspar.

Cf. Gerškovič 1999, type VT1I2 or VT2I2, fig. 13.2,4.

**Cat. 330 9201/09/8**

Rim fr. of storage vessel? with flaring rim. Some grooves on the ext. H: 3.6; W: 3.3; T: 1.15; Ø: ca. 25. Core: Gley 1 7/N light grey. Sparse tiny lime inclusions; single large quartz; sparse sand.

Cf. Gerškovič 1999, type VT1I2 or VT2I2, fig. 13.2,4.

**Cat. 331 9201/10/2**

Rim fr. of frying pan? with straight wall and a protruding lip. Few grooves on the ext. and int. H: 1.9; W: 2.2; T: 0.9; Ø? Core: Gley 1 2.5/N black. Surface: 10YR 6/3 pale brown. Single medium quartz; sparse light-reflecting particles; sparse sand.

Possibly same shape as Gerškovič 1999, pl. 8.27 (Bol'šie Kopani).

**Cat. 332 9202/01/1**

Rim fr. of vessel with slightly flaring rim. H: 2.22; W: 2.78; T: 0.9 Ø: 19. Core: Gley 1 2.5/N black. Sparse small quartz; sparse sand.

Possibly belonging to Gerškovič 1999, type T1II2, fig. 11.10.

**Cat. 333 9202/02/1**

Rim fr. of cooking vessel with incised triangles on the ext. H: 1.83; W: 1.72; T: 0.4; Ø: ca. 13. Core: 10YR 3/1 very dark grey. Surface: 10YR 7/2 light grey. Sparse tiny lime inclusions; sparse small to medium quartz; sparse small to medium dark inclusions.

Possibly belonging to Gerškovič 1999, type T1II2, fig. 11.10.

**Cat. 334 9202/02/3**

Rim fr. of vessel with slightly flaring rim. H: 2; W: 3.01; T: 0.66; Ø: 13. Core: Gley 1 2.5/N black. Int. surface: 2.5Y 6/1 grey. Ext. surface: 5YR 7/6 reddish yellow. Sparse light-reflecting particles; sparse tiny lime inclusions.

Possibly belonging to Gerškovič 1999, type T1II2, fig. 11.10.

**Cat. 335 9202/05/1**

Rim fr. of frying pan? with widely flaring walls. H: 1.85; W: 2.3; T: 0.94; Ø: ca. 28. Core: Gley 1 3/N very dark grey. Surface: 10YR 6/3 pale brown. Sparse medium quartz; sparse tiny lime inclusions.

Cf. Kolotuchin 2003, pl. 53.13 (Baj Kijat).

**Cat. 336 9202/07/3**

Wall fr. of vessel with two rows of impressed decoration on the ext. H: 1.98; W: 1.99; T: 0.86. Clay: Gley 1 2.5/N black. Ext. surface: 10YR 6/2 light brownish grey. Int. surface: 10YR 5/3 brown. Single small quartz; sparse tiny to small lime inclusions.

For the decoration, see e.g. Gerškovič 1999, pl. 14.1 (Uškalka).

**Cat. 337 9202/11/3**

Rim fr. of vessel with outturned rim. H: 1.65; W: 2.6; T: 0.9-1.2; Ø: 12-14. Core: Gley 2 2.5/5 BG greenish black. Surface: Gley 2 3/5 B very dark bluish grey. Medium sand inclusion; single tiny lime inclusions; single tiny light-reflecting particle.

Late Bronze Age?

**Cat. 338 9202/11/4**

Rim fr. of small frying pan with slightly protruding rim. H: 2.3; W: 2.2; T: 0.7-0.9; Ø: ca. 10. Core: 5YR 2.5/1 black. Surface: Gley 2 5/10 BG greenish grey. Single tiny lime particle; sparse sand.

Cf. e.g. Kolotuchin 2003, pl. 49.16-17.

**Cat. 339 9202/11/5**

Wall fr. of cup? with part of the handle? preserved. H: 3.2; W: 3.5; T: 0.7-1.35. Core: 5YR 2.5/1 black. Ext. surface: Gley 2 3/5 BG very dark greenish grey. Int. surface: 10R 4/1. Sparse small lime inclusions; sparse sand.

Late Bronze Age?

**Cat. 340 9202/11/6**

Rim fr. of vessel with slightly concave rim. H: 1.6; W: 2.1; T: 0.8-0.9; Ø: ca. 16. Core: 5YR 2.5/1 black. Surface: Gley 2 5/10 BG greenish grey. Single tiny lime particle; sparse sand.

Possibly belonging to Gerškovič 1999, type T1II2, fig. 11.10.

**Cat. 341 9202/11/7**

Wall fr. of vessel with incised and impressed decoration. H: 1.9; W: 2.15; T: 0.6. Core: 10R 2.5/1 reddish black. Ext. surface: 10R 6/3 pale red to 10R 6/6 light red. Sparse tiny lime inclusions.

Late Bronze Age?

**Cat. 342 9207/01/1**

Rim fr. of bowl with inturned rim. H: 3.2; W: 4.9; T: 1.1; Ø: ca. 19. Core: Gley 1 2.5/N black. Surface: Gley 1 4/N dark grey. Sparse tiny lime inclusions; sparse tiny to small feldspar.

Cf. Gerškovič 1999, type N2I2, fig. 11.6.

**Cat. 343 9207/02/2**

Fr. of shallow bowl with inturned rim. H: 3.45; W: 2.07; T: 1.09; Ø: 13. Core: Gley 1 2.5/N black. Surface: Gley 1 2.5/N black to 10YR 8/2 very pale brown. Sparse light-reflecting particles; single small lime inclusions; sparse sand.

Late Bronze Age?

**Cat. 344 9208/01/1**

Wall fr. with part of horizontal moulding with two obligue cuts. H: 1.83; W: 2.27; T: 1. Core: Gley 1 2.5/N black. Surface: 5YR 7/6 reddish yellow. Sparse tiny lime inclusions; sparse tiny feldspar.

**Cat. 345 9208/01/2**

Rim fr. of vessel with projecting flat rim. H: 2.15; W: 3.05; T: 0.93; Ø: 13-15. Core: 10YR 4/1 dark grey. Int. surface: 10YR 6/3 pale brown. Ext. surface: 5YR 7/4 pink. Sparse tiny to small red inclusions; sparse tiny lime inclusions; sparse sand.

Cf. Gerškovič 1999, pl. 5.4 (Berislav); Rybalova 1974, fig. 6.1-2, 4.

**Cat. 346 9208/01/3**

Rim fr. of large vessel with a high rim. Impressed 'herringbone' motive with an incised furrow below. H: 4.7; W: 2.17; T: 1.1; Ø: ca. 24. Core: Gley 1 2.5/N black. Ext. core: Gley 1 2.5/N black to Gley 1 6/N grey. Sparse light-reflecting particles; sparse sand; single tiny lime particle; sparse tiny dark inclusions.

Resembles some of the higher rims from Kamenka, cf. e.g. Rybalova 1974, figs. 5.1, 14.3.

**Cat. 347 9208/01/8**

Rim fr. of storage vessel? with slightly flaring rim. H: 1.88; W: 2.4; T: 0.9; Ø: 19-20. Core and int. surface: Gley 1 2.5/N black. Ext. surface: Gley 1 6/N grey. Frequent tiny lime inclusions; sparse sand.

Cf. Gerškovič 1999, type VT1I2, fig. 13.2.

**Cat. 348 9208/02/10**

Rim fr. of vessel with slightly concave rim. H: 2.1; W: 1.46; T: 0.92; Ø? Core: Gley 1 2.5/N black. Int. surface: Gley 1 5/N grey. Ext. surface: 2.5Y 8/2 pale yellow. Sparse tiny lime inclusions; sparse tiny to small quartz; sparse sand.

Late Bronze Age?

**Cat. 349 9208/02/11**

Wall fr. of vessel with impressed V-shaped decoration. H: 2.45; W: 1.73; T: 1. Core: Gley 1 2.5/N black. Surface: 10YR 8/3 very pale brown. Medium tiny lime inclusions; sparse tiny to small quartz; single tiny light-reflecting particles.

Late Bronze Age?

**Cat. 350 9208/02/12**

Rim fr. of vessel with rounded lip. Badly preserved. H: 0.9; W: 2.11; T: 1.33; Ø? Core: Gley 1 2.5/N black. Surface: 10YR 7/2 light grey. Sparse tiny lime inclusions.

Late Bronze Age?

**Cat. 351 9208/02/14**

Rim fr. of vessel with a high slightly flaring rim. H: 3.34; W: 2.45; T: 0.74; Ø? Core: Gley 1 2.5/N black. Surface: 10YR 7/2 light grey. Sparse tiny lime inclusions; medium tiny to small quartz; sparse sand.

Resembles some of the higher rims from Kamenka, cf. e.g. Rybalova 1974, figs. 5.1, 14.3.

**Cat. 352 9208/02/15**

Rim fr. of cooking vessel with outturned rim. H: 2.25; W: 3.55; T: 1.08; Ø: ca. 16. Clay: Gley 1 2.5/N black. Sparse tiny lime inclusions; sparse tiny light-reflecting particles.

Cf. Gerškovič 1999, type T1II2, fig. 11.10.

**Cat. 353 9208/02/16**

Rim fr. of vessel with flat rim. Badly preserved. H: 1.5; W: 1.35; T: 1; Ø? Core: Gley 1 2.5/N black. Surface: 7.5YR 7/4 pink. Sparse tiny lime inclusions; sparse tiny shells; sparse tiny dark inclusions.

Late Bronze Age?

**Cat. 354 9208/02/17**

Rim fr. of vessel with flaring rim. H: 2.84; W: 3.52; T: 0.99; Ø: 20-25. Core: Gley 1 2.5/N black. Sparse tiny lime inclusions; sparse small quartz.

**Cat. 355 9208/02/20**

Rim fr. of cooking vessel with slightly flaring rim. H: 2.27; W: 2.21; T: 0.78; Ø? Core: Gley 1 3/1 very dark grey. Surface: 10YR 7/2 light grey to 10YR 7/4 very pale brown. Sparse tiny lime inclusions; sparse tiny quartz; sparse sand.

Cf. Gerškovič 1999, type T1II2, fig. 11.10.

**Cat. 356 9209/02/2**

Rim fr. of bowl with straight walls. Badly preserved. H: 3.3; W: 2.5; T: 0.6?; Ø: 16-18. Core: Gley 2 3/10 BG very dark bluish grey. Surface: 10R 7/8 light red. Sparse tiny to medium lime inclusions; medium tiny feldspar.

Possibly of Gerškovič 1999, type N1I2 or N1II2, fig. 11.2,4.

**Cat. 357 9209/02/3**

Very small rim fr. of vessel with rounded rim. Badly preserved. H: 1.5; W: 2.5; T: 0.65-1; Ø? Core: Gley 1 2.5/N black. Surface: 10R 7/4 pale red. Sparse tiny to medium lime inclusions; sparse small to medium quartz; sparse sand; single small dark inclusions.

Late Bronze Age?

**Cat. 358 9209/02/4**

Rim fr. of cooking vessel with slightly flaring rim. H: 3.4; W: 3.2; T: 0.4-1; Ø? Core: Gley 2 2.5/10 BG greenish grey. Int. surface: Gley 2 6/5 BG greenish grey. Ext. surface: sparse tiny to large lime inclusions.

Cf. Gerškovič 1999, type T1II2, fig. 11.10.

**Cat. 359 9209/02/6**

Wall fr. of vessel with impressed nail decoration. H: 3; W: 2.5; T: 0.9. Core: Gley 2 2.5/10 B bluish black. Int. surface: 2.5YR 6/1 reddish grey. Ext. surface: 2.5YR 7/6. Single tiny lime particle; single small quartz.

Late Bronze Age?

**Cat. 360 9218/01/1**

Rim fr. of cooking vessel with concave rim. H: 3.4; W: 3.6; T: 1.1; Ø? Core: Gley 1 2.5/N black. Surface: Gley 1 3/N very dark grey. Sparse tiny lime inclusions; frequent light-reflecting particles.

Cf. Gerškovič 1999, type T1II2, fig. 11.10.

**Cat. 361 9219/01/1**

Rim fr. of cooking vessel with concave rim. H: 2.87; W: 2.7; T: 0.86; Ø: 12-15. Core: Gley 1 3/N very dark grey. Surface: 2.5Y 6/2 light brown grey. Single tiny lime particle; sparse sand.

Cf. Gerškovič 1999, type T1II2, fig. 11.10.

**Cat. 362 9201/01/2**

Fr. of base with flat underside. H: 1.58; W: 3.82; T: 1.6.; Ø: <10. Core: Gley 1 2.5/N black. Surface: 10YR 4/1 dark grey. Sparse tiny light-reflecting particles; sparse medium quartz; sparse sand; sparse tiny lime inclusions.

Bronze Age?

**Cat. 363 9202/02/2**

Rim fr. of vessel with flat lip. H. 1.83; W: 2.03; T: 0.93; Ø: 15-16. Core: Gley 1 2.5/N Black. Surface: Gley 1 4/N dark grey. Sparse light-reflecting particles; sparse tiny lime inclusions.

**Cat. 364 9208/01/5**

Base fr. of vessel with flat base. H: 2.2; W: 3.45; T: 1.3-0.87; Ø? Core and int. surface: Gley 1 2.5/N black. Ext. surface: 10YR 7/3 very pale brown. Single light-reflecting particle; sparse tiny voids.

Bronze Age?

**Cat. 365 9208/01/6**

Wall fr. of vessel with burnished interior. H: 1.95; W: 2.59; T: 0.7. Core and int. surface: Gley 1 2.5/N black to 2.5Y 8/2 pale yellow. Ext. surface: 8/4 pale yellow. Sparse tiny lime inclusions; single large quartz; sparse sand.

Bronze Age?

**Cat. 366 9208/02/9**

Rim fr. of thin-walled vessel with thickened, outturned lip. H: 2.6; W: 2.7; T: 1.06; Ø? Core: Gley 1 2.5/N black. Surface: 2.5Y 7/4 pale yellow. Sparse tiny lime inclusions; single small quartz; sparse tiny feldspar.

Bronze Age?

**Cat. 367 9209/02/7**

Rim fr. of vessel with straight walls and rounded rim. H: 2.4; W: 3.35; T: 0.7-1-1; Ø: ca. 14. Core and int. surface: 5YR 2.5/1 black. Ext. surface: 10R 6/3 pale red. Sparse tiny lime inclusions; sparse sand.

Bronze Age?

**Cat. 368 9219/01/2**

Small rounded rim fr. Badly preserved. H: 1.8; W: 1.45; T: 0.79; Ø? Core: Gley 1 3/N very dark grey. Surface: 10YR 7/3 very pale brown. Sparse tiny to medium lime inclusions; sparse small quartz.

Bronze Age?

**Cat. 369 9219/02/2**

Rim fr. of large bowl? with a projecting flat rim. H: 2.15; W: 2.45; T: 1.02; Ø: 28-30. Core: 10YR 3/1 very dark grey. Surface: 10YR 5/2 greyish brown. Sparse tiny lime inclusions; sparse light-reflecting particles.

Bronze Age?

*Hellenistic*

**Cat. 370 9200/01/1**

Base fr. of pot with flat underside. H: 1.68; W: 2.76; T: 0.95; Ø: 12-14. Core and int. surface: Gley 1 2.5/N black. Ext. surface: 10YR 7/3 very pale brown. Sparse tiny lime inclusions; sparse sand.

Cf., e.g., Stolba 2002, Cat. D43. Early Hellenistic.

**Cat. 371 9200/02/2**

Rim fr. of pot with outturned rim. H: 1.95; W: 1.9; T: 0.53; Ø: ca. 15. Core: 5YR 6/6 reddish yellow to 2.5/1 black. Surface: 10YR 6/3 pale brown. Sparse tiny lime inclusions; sparse light-reflecting particles.

Cf. Stolba 2002, type 1 Cat. D1-D68. Early Early Hellenistic.

**Cat. 372 9200/05/2**

Rim fr. of pot with slightly concave rim. H: 1.72; W: 2.13; T: 0.76; Ø: ca. 22. Core: 10YR 2/1 black. Surface: 10YR 7/4 very pale brown. Sparse tiny lime inclusions; sparse sand.

Cf. Stolba 2002, type 1 Cat. D1-D68. Early Early Hellenistic.

**Cat. 373 9201/09/5**

Rim fr. of pot with straight rim. H: 1.83; W: 2.36; T: 0.72; Ø: 14-16. Core: 10YR 4/1 dark grey. Surface: 10YR7/4 very pale brown. Sparse tiny lime inclusions; frequent tiny feldspar; single tiny mica.

Cf. Stolba 2002, type 1 Cat. D1-D68. Early Early Hellenistic.

**Cat. 374 9202/10/3**

Rim fr. of bowl with flat projecting and undercut rim. H: 1.16; W: 2.05; T: 0.82 (wall), 1.26 (rim). Core: Gley 1 3/N very dark grey. Surface: 10YR 7/4 very pale brown. Sparse very tiny lime inclusions.

Somewhat similar to a handmade bowl from a grave on the Kerč Peninsula, cf. Maslenikov 1995, fig. 20.2. 3rd-2nd century BC.

**Cat. 375 9207/02/3**

Rim fr. of pot with straight rim. H: 1.75; W: 1.86; T: 0.62; Ø: ca. 16. Core: 10YR 3/1 very dark grey. Int. surface: 10YR 5/3 brown. Ext. surface: 7.5YR 6/4 light brown. Frequent tiny lime inclusions; sparse sand.

Cf. Stolba 2002, type 1, variant c Cat. D86-D87. Early Hellenistic.

**Cat. 376 9207/02/4**

Rim fr. of pot with a straight rim. H: 1.7; W: 1.83; T: 0.75; Ø? Core: Gley 1 2.5/n black. Surface: 10YR 7/3 very pale brown. Sparse tiny lime inclusions.

Cf. Stolba 2002, type 1, variant c Cat. D86-D87. Early Hellenistic.

**Cat. 377 9208/01/4**

Wall fr. of pot with impressed decorarion on the ext. H: 2.7; W: 3.7; T: 0.76. Core: 10YR 4/1 dark grey. Int. surface: 10YR 5/2 greyish brown. Ext. surface: 7.5YR 6/4 light brown. Sparse tiny lime inclusions; single light-reflecting particle; sparse small to medium quartz; single medium grog.

Early Hellenistic.

**Cat. 378 9208/01/7**

Small rim fr. of pot. H: 1.2; W: 1.55; T: 0.79; Ø? Core and int. surface: Gley 1 2.5/N black. Ext. surface: 2.5Y 7/2 light grey. Single medium lime inclusions; sparse sand.

Cf. Stolba 2002, type 1 Cat. D1-D68. Early Hellenistic.

**Cat. 379 9208/02/18**

Rim fr. of vessel with straight rim. H: 1.45; W: 1.42; T: 0.49; Ø? Core: Gley 1 3/N very dark grey. Surface: 10YR 6/1 grey. Single small lime particle.

Cf. Stolba 2002, type 1 Cat. D1-D68. Early Hellenistic.

**Cat. 380 9208/02/6**

Fr. of pot with slightly concave, outturned rim. H: 1.7; W: 1.23; T: 0.69; Ø? Core and ext. surface: Gley 1 2.5/N black. Int. surface: 10YR 7/2 light grey. Sparse tiny to medium lime inclusions; sparse sand.

Cf. Stolba 2002, type 1 Cat. D1-D68. Early Early Hellenistic.

**Cat. 381 9208/02/7**

Wall fr. of pot. H: 1.74.; W: 1.75; T: 0.66. Core: 10YR 3/1 very dark grey, 7.5YR 6/6 reddish yellow to 10YR 3/2 very dark greyish brown. Interior: 10YR 5/1 grey. Single tiny lime particle; sparse small quartz.

## Amphorai

*Unknown production*

**Cat. 382 9219/01/3**

Handle fr. H: 6.46; W: 4.57; T: 2.98. Core: 2.5YR 5/8 red. Slip: 10YR 8/4 very pale brown. Sparse tiny mica; sparse light-reflecting inclusions; sparse tiny feldspar.

Classical-Hellenistic.

## Tableware

**Cat. 383 9208/02/1**

Handle fr. of small jug. H: 2.76; W: 2.26; T: 0.8. Clay: 7.5YR 4/2 brown. Sparse tiny lime inclusions.

Late Classical-Hellenistic.

# DSP08-F06-02

Core units: 9226-9227
Adjacent units: 9218-19, 9225
See DSP08-F06-01 for 9218-19.

# DSP08-F06-03

Core units: 9045
Adjacent units: 9044, 9049-50

## Handmade ware

**Cat. 384 9050/01/1**

Rim fr. of pot with impressed finger decoration on the lip. H: 2.15; W: 2.76; T: 0.68-0.79; Ø: 18. Core: Gley 1 3/N very dark grey. Surface: Gley 1 6/N grey and 10YR 7/4 very pale brown. Frequent small to medium lime inclusions.

Cf. Stolba 2002, type 1 Cat. D1-D68. Early Hellenistic.

## Amphorai

### *Herakleia*

**Cat. 385 9045/01/1**

Handle fr. H: 8.3; W: 4.4-5.1; T: 2.7-2.9. Clay: 10R 6/8 light red. Slip: 5YR 6/6 reddish yellow. Medium tiny to medium pyroxene; sparse tiny to small lime inclusions; sparse tiny feldspar.

4th century BC.

### *Sinope*

**Cat. 386 9044/01/1**

Handle fr. H: 8.2; W: 6.2; T: 2.8-2.9. Clay: 10R 7/6 light red. Frequent tiny to large pyroxene; sparse tiny to medium lime inclusions; single tiny mica.

Late Classical-Early Hellenistic.

# DSP07-H01-01

Grab samples: 3043, 3045-46, 3049-3060
No diagnostic material.

# DSP07-H01-01

Trial trench Feature 18

## Handmade ware

**Cat. 387 H01-1/B1/5**

Two non-joining frs. of vessel with flat offset base. H: 3.1; W: 4.3; T: 0.6-0.7; Ø: ca. 8. Core: 7.5YR 5/1 to 7/6 grey to reddish yellow. Surface: 7.5YR 7/6 reddish yellow. Poorly sorted sandy clay with some large grits; large brown, possibly lime inclusions visible on the surface.

Cf., e.g., Stolba 2002, Cat. D12. Early Hellenistic.

**Cat. 388 H01-1/B2/2**

Rim fr. of vessel with concave, outturned rim. Impressed decoration on lip. H: 3.35; W: 3.4; T: 0.65-0.85 (wall), 1.05 (rim); Ø: ca. 13. Clay: Gley 1 5/N grey. Small lime inclusions and lime implosion reaction rims.

Cf. Stolba 2002, type 1 Cat. D1-D86. Early Hellenistic.

**Cat. 389 H01-1/B3/2**

Rim fr. of vessel with concave, outturned rim. Impressed decoration on flat lip. H: 3.35; W: 3.2; T: 0.6-0.7 (wall), 0.8 (rim); Ø? Core: Gley 1 4/N dark grey. Surface: 7.5YR 7/6 reddish yellow. Poorly sorted clay with tiny lime quartz and occasional large to very large reddish inclusions, possibly grog.

Cf. Stolba 2002, type 1 Cat. D1-D86. Early Hellenistic.

**Cat. 390 H01-1/B4/4**

Two joining base frs. of vessel with offset base and flat underside. H: 4.3; W: 6.5; T: 0.5-0.6 (wall), 0.8 (bottom); Ø: 7. Core: 2.5Y 5/1 to 10YR 7/3 grey to very pale brown. Int. surface: 10YR 7/1 light grey. Ext. surface: 2.5Y 5/1 grey. Very coarse clay with large to very large grits.

Cf. Stolba 2002, type 1 Cat. D1-D86. Early Hellenistic.

**Cat. 391 H01-1/B5/2**

Rim fr. of pot with concave rim and rounded lip. Stamped impressions on upper part of lip. H: 2.9; W: 4; T: 0.7-0.85 (wall); Ø? Core: Gley 1 3/N very dark grey. Surface: 10YR 3/1 very dark grey. Sandy clay with large grits and quartz.

Cf. Stolba 2002, type 1 Cat. D1-D86. Early Hellenistic.

**Cat. 392 H01-1/B5/3**

Base fr. of vessel with offset base and flat underside. H: 3.65; W: 2.8; T: 0.6-0.9 (wall); Ø: ca. 8. Core: 7.5YR 5/1 grey. Surface: 7.5YR 6/4 light brown. Sandy clay, poorly sorted with some large pale brown inclusions, possibly grog and some tiny red inclusions.

Cf. Stolba 2002, type 1 Cat. D1-D86. Early Hellenistic.

**Cat. 393 H01-1/B5/4**

Rim fr. of pot with concave rim and rounded lip. Impressions on upper part of lip. H: 3.45; W: 4.3; T: 0.7 (wall), 0.8 (rim); Ø? Core: 10YR 4/1 to 2 dark grey to dark greyish brown. Int. surface: 10YR 6/3 pale brown. Ext. surface: 7.5YR 7/6 reddish yellow. Sandy clay with quartz and shell and some large grits.

Cf. Stolba 2002, type 1 Cat. D1-D86. Early Hellenistic.

**Cat. 394 H01-1/B5/5**

Fr. of bowl(?) with plain, everted, angular rim. H: 2.7; W: 2.45; T: 0.55 (rim), 0.9 (wall); Ø? Core: Gley 1 5/N grey. Int. surface: 7.5YR 3/2 dark brown to 10YR 7/2 light grey. Ext. surface: 5YR 6/6 reddish yellow. Occasional large red inclusions; some shells and sand.

Possibly similar to Kavalevskaja-Suchanova 2005, fig. 4.4-7. Early Hellenistic.

## Amphorai

### *Chersonesos*

**Cat. 395 H01-1/B3/1**

Fr. of thickened rim. H: 5.25; W: 3.5; T: 0.5-0.9 (wall), 1.4 (rim); Ø: ca. 14. Clay: 2.5YR 6/8 light red. Slip ext.: 7.5YR 7/6 reddish yellow. Sandy clay with many tiny lime inclusions occasional medium size as well.

Cf. Monachov 1989, 60-61, pl. XIX.88, rim type II. Ca. 300-275 BC.

### *Thasos*

**Cat. 396 H01-1/B5/1**

Rim fr. with triangular lip. H: 2.9; W: 6.35; T: 1.55; Ø: ca. 9. Clay: 7.5YR 6/4 to 4/2 light brown to brown. Sandy clay with some tiny to small quartz and few red inclusions, possibly grog and numerous silver mica/clear quartz.

Most likely as Monachov 2003, pl. 45.2,4, type II-B-2. Ca. 375-325 BC.

# DSP07-H01-01

Trial trench Feature 5

## Handmade ware

**Cat. 397 H01-2/B3/1**

Body fr. with interior grooves. H: 5.1; W: 4.1; T: 0.9-1.1. Clay: Gley 1 3/N very dark grey. Coarse poorly sorted clay with large quartz and grits.
Cf. Točev 2007, fig. 94.4. Late Bronze Age.

**Cat. 398 H01-2-B1/1**

Body fr. with interior grooves. H: 2.38; W: 1.8; T: 0.75-0.85. Core: Gley 1 3/N. Surface: 10YR 4/1 to 7/4 dark grey to very pale brown. Some grits, all burnt black.
Točev 2007, fig. 94.4. Late Bronze Age.

**Cat. 399 H01-2-B1/2**

Body fr. with interior and exterior grooves. H: 3.3; W: 2.95; T: 0.55-0.7. Core and int.: Gley 1 2.5/N black. Ext. 7.5YR 6/3 light brown. Sandy clay with tiny quartz.
Točev 2007, fig. 94.4. Late Bronze Age.

**Cat. 400 H01-2-B1/3**

Fr. of vessel with straight rim, flat lip and flaring shoulders with two grooves. H: 6.4; T: 0.84-1.4; Ø: ca. 25. Sandwich clay. Core: Gley 2 4/5B dark bluish grey. Ext. surface: 2.5YR 7/6 light red to 2.5YR 7/4 light reddish brown. Sparse tiny to medium lime inclusions; sparse tiny feldspar.
Late Bronze Age.

# DSP07-H02-01

Grab samples: 3077-3082
No diagnostic material.

# DSP07-H02-02

See DSP07-H02/01

# DSP07-H04-01

Grab samples: 3068-3072
No diagnostic material.

# DSP07-H04-02

Grab samples: 3075

## Post-Antique pottery

**Cat. 401 3075/01**

Handle fr. of jug. H: 5.3; W: 4; T: 1.6. Clay: 5YR 6/8 reddish yellow. Medium very tiny to small lime particle; sparse tiny to medium feldspar.
Medieval?

# DSP07-H06-01

Standard samples: 20002/01-02, 20007/02-03
Grab samples: 3085, 2007 FL8, 3201-03
No diagnostic material.

# DSP07-H07-01

Grab samples: 3062-3063, 2007 FL 9
No diagnostic material.

# DSP07-H07-02

Grab samples: 2007 FL 10
No diagnostic material.

# DSP08-H07-03

Standard samples: 20064-65
Grab samples: 3221

## Post-Antique pottery

**Cat. 402 3221/01/01**

Fr. of plain rim with horizontal combing under the exterior lip. H: 2.5; W: 1.7; T: 0.8; Ø: 9. Clay: 5YR 6/8 reddish-yellow. Sandy fabric with numerous tiny lime and quartz and occasional small rounded red inclusions.
Early Modern.

# DSP07-H08-01

Standard samples: 20013/01-02, 20014/01-02, 20015/01-10, 20016/01-04
Grab samples: 3095, 2007 FL 11-12
No diagnostic material.

# DSP07-H08-01

Trial trench Feature 2

## Handmade ware

**Cat. 404 H08-3/B19/1**

Body fr. decorated with two rows of slanting dots made with a hollow pointed instrument. H: 2.59; W: 2.13; T: 0.83. Clay: Gley 1 3/N very dark grey. Sandy clay with some tiny quartz and lime inclusions.

Točev 2007, fig. 80.9. Middle Bronze Age.

**Cat. 405 H08-3/B19/2**

Body fr. decorated with two parallel lines of slanting grooves, primitive herringbone pattern. H: 2.94; W: 2; T: 0.94. Clay: Gley 1 2.5/N black. Sandy clay with some tiny quartz inclusions.

Decoration possibly similar to Točev 2007, fig. 81.11. Middle Bronze Age.

**Cat. 406 H08-3/B19/3**

Thick-walled body fr. with striations on the interior. H: 3.48; W: 3.5; T: 1.36. Clay: Gley 1 3/N very dark grey. Poorly sorted sandy clay with some large rounded quartz visible in completely black burnt fabric.

Middle-Late Bronze Age.

**Cat. 407 H08-3/B20/1**

Body fr. with decoration stamped with an instrument with a rounded tip. H: 3.25; W: 2.35; T: 0.92. Core: Gley 1 3/N very dark grey. Interior: 10YR 6/1 grey. Ext.: 7.5YR 7/4 pink. Poorly sorted clay with large grits and tiny lime inclusions.

Middle-Late Bronze Age.

**Cat. 408 H08-3/B24/1**

Body fr. with a row of stamped triangular decoration. H: 2.26; W: 2.64; T? Clay: 7.5YR 6/6 reddish yellow. Sandy clay with many voids possibly after burnt-out organic material, some tiny quartz inclusions.

Middle-Late Bronze Age.

**Cat. 409 H08-3/B32/1**

Body fr. with striations on the interior. Exterior decoration in the form of row of vertical grooves framed by a horizontal groove. H: 3.22; W: 3.11; T: 0.69. Core: 10YR 7/2 light grey to Gley 1 3/N very dark grey. Surface: Gley 1 3/N very dark grey. Relatively fine sandy clay with tiny quartz and lime inclusions.

Točev 2007, fig. 94.6. Late Bronze Age?

# DSP07-H08-01

Trial trench Feature 7

## Handmade ware

### *Middle-Late Bronze Age*

**Cat. 410 H08-4/B27/1**

Rim fr. of vessel with concave, outturned rim. Impressions on the lip. H: 5.9; W: 5; T: 0.75-0.9; Ø? Clay: 7.5YR 6-5/2 light brownish grey to brownish grey. Sandy clay with small grits and many pale red inclusions, probably lime.

Gerškovič 1999, type T1II2, fig. 11.10. Late Bronze Age.

**Cat. 411 H08-4/B27/2**

Closed vessel with flaring shoulders. H: 4.3; W: 3.8; T: 0.8; Ø: ca. 18. Clay: Gley 1 2.5/N black. Sandy clay with some tiny quartz and lime inclusions.

Close to a vessel from Burun Eli, cf. Točev 2007, fig. 99.1. Late Bronze Age.

**Cat. 412 H08-4/B27/3**

Base fr. of vessel with offset base and flat underside. Badly preserved. H: 2.75; W: 3.28; T: 1.1(bottom); Ø? Clay: 10YR 5/2 greyish brown. Sandy clay with tiny grits and lime inclusions.

Late Bronze Age?

**Cat. 413 H08-4/B27/4**

Base fr. of vessel with flat exterior offset base. Badly preserved. H: 2; W: 3.4; T: 0.65-0.8; Ø? Clay: Gley 1 4/N dark grey. Sandy clay with some tiny lime and large grits.

Late Bronze Age?

**Cat. 414 H08-4/B27/5**

Body fr. with repair hole? Striations on the interior surface. H: 4.55; W: 4.35; T: 0.7-0.75. Core and int. surface: Gley 1 2.5/N black. Ext. surface: 7.5YR 6/4 light brown. Very sandy fabric with multiple rounded transparent quartz inclusions.

Late Bronze Age?

**Cat. 415 H08-4/B28/1**

Two joining base frs. of vessel with offset base and flat underside. H: 3.05; W: 7.75; T: 0.65-0.95 (wall), 0.95 (bottom); Ø: ca. 8. Core and int. surface: Gley 1 3/N very dark grey. Ext. surface: 7.5YR 7/4 pink. Sandy clay with some large grits and lime inclusions.

Gerškovič 1999, type T1II2, fig. 11.10. Late Bronze Age.

**Cat. 416 H08-4/B28/2**

Fr. of vessel with concave, outturned rim. H: 1.88; W: 2.37; T: 0.8. Core: Gley 1 3/N very dark grey. Surface: 7.5YR 7/6 to 5YR 7/6 reddish yellow. Sandy clay with some quartz and large grits.

Gerškovič 1999, type T1II2, fig. 11.10. Late Bronze Age.

**Cat. 417 H08-4/B28/3**

Body fr. decorated with a row of slanting stamped lines. Burnished on the exterior. H: 8.07; W: 4.66; T: 0.62-0.75. Core and ext. surface: Gley 1 3/N very dark grey. Int. surface: 10YR 6/2 light brownish grey. Sandy clay with large grits and some small orange inclusions, possibly burned lime.

Late Bronze Age.

**Cat. 418 H08-4/B29/1**

Base fr. of vessel with offset base and flat underside. H: 2.9; W: 5.5; T: 0.75; Ø: ca. 10. Core and int. surface: Gley 1 3/N very dark grey. Ext. surface: 10YR 7/3 very pale brown. Coarse clay with medium to large grits, seems sandy.

Gerškovič 1999, type T1II2, fig. 11.10. Late Bronze Age.

**Cat. 419 H08-4/B29/2**

Three joining frs. of closed vessel with flaring shoulders. H: 3.05; W: 5.5; T: 0.6-0.75; Ø: ca. 14. Core and int. surface: Gley 1 3/N very dark grey. Ext. surface: 7.5YR 7/6 reddish yellow. Sandy clay with tiny to large grits and some lime inclusions.

Close to a vessel from Burun Eli, cf. Točev 2007, fig. 99.1. Late Bronze Age.

**Cat. 420 H08-4/B29/3**

Large closed vessel (wall has only one surface). H: 3; W: 4.4; T: 1.2 (rim); Ø: ca. 18. Core: Gley 1 4/N dark grey. Surface: 10YR 7/2 light grey. Multiple tiny to small angular lime and grits.

Late Bronze Age?

**Cat. 421 H08-4/B29/4a**

Rim fr. of open vessel. Badly preserved. H: 1.45; W: 1.7; T: 1.05; Ø? Core: Gley 1 3/N very dark grey. Surface: 10YR 7/3 very pale brown. Sandy clay with some tiny lime inclusions and large grits.

Possibly belonging to Gerškovič 1999, type T1II2, fig. 11.10. Late Bronze Age?

**Cat. 422 H08-4/B29/4b**

Open vessel with a slightly overfolded rim. H: 2.3; W: 3.45; T: 0.85 (wall), 1 (rim); Ø? Core: Gley 1 3/N very dark grey. Surface: 7.5YR 6/6 reddish yellow. Sandy clay with tiny to medium grits and quartz inclusions.

Possibly belonging to Gerškovič 1999, type T1II2, fig. 11.10. Late Bronze Age.

**Cat. 423 H08-4/B31/1**

Rim fr. of vessel with concave, outturned rim. H: 4.45; W: 4.65; T: 0.65-0.85; Ø: ca. 18. Core and int. surface: Gley 1 3/N very dark grey. Ext. surface: 10YR 6/2 light brownish grey. Sandy clay with some lime and occasional large porous inclusions of light grey colour.

Cf. Gerškovič 1999, type T1II2, fig. 11.10. Late Bronze Age.

**Cat. 424 H08-4/B31/2**

Rim fr. of vessel with outturned rim. H: 3.67; W: 4.18; T: 0.75-0.85; Ø: ca. 22. Clay: Gley 1 3/N very dark grey. Sandy clay with rounded quartz inclusions and medium-large grits.

Cf. Gerškovič 1999, type T1II2, fig. 11.10. Late Bronze Age.

**Cat. 425 H08-4/B31/4**

Rim fr. of vessel with outturned rim. H: 6.32; W: 5.5; T: 0.9-1.1; Ø: ca. 27. Clay: 10YR 6/3 pale brown. Sandy clay with tiny to large grits.

Cf. Gerškovič 1999, type T1II2, fig. 11.10. Late Bronze Age.

**Cat. 426 H08-4/B31/6**

Rim fr. of vessel with a plain rim. H: 2.1; W: 2.7; T: 0.95-1.05; Ø? Clay: Gley 1 3/N very dark grey. Sandy clay, all burned black.

Late Bronze Age?

**Cat. 427 H08-4/B31/3**

Open vessel, possibly a pan, with stamped decoration on the rim. H: 3.5; W: 4.31; T: 0.8-0.9 (wall), 1.27 (rim); Ø: ca. 32. Clay: 10YR 6/4 light yellowish brown. Sandy clay with numerous tiny to medium grits and some lime and pink (possibly lime) inclusions.

Possibly similar to a pan from Burun Eli, cf. Točev 2007, fig. 99.17. Late Bronze Age.

**Cat. 428 H08-4/B31/5**

Open vessel, possibly a pan, with stamped decoration on the rim. H: 3.05; W: 4.7; T: 1.05-1.3; Ø: ca. 33. Clay: 10YR 4/1 dark grey. Sandy clay with rounded quartz and grits.

Possibly similar to a pan from Burun Eli, cf. Točev 2007, fig. 99.17. Late Bronze Age.

**Cat. 429 H08-4/B31/7**

Thin-walled cup. H: 2.65; W: 3.5; T: 0.4-0.77; Ø: ca. 9. Clay: 10YR 5/1-2 grey to greyish brown. Some shell and occasional very large grits.

Cf. Točev 2007, figs. 94.2, 99.16. Late Bronze Age.

**Cat. 430 H08-4/B31/8**

Open vessel with an overfolded rim, possibly lid. H: 5.3; W: 3.45; T: 1.05-1.45; Ø? Clay: 10YR 6/3-4 pale brown to light yellowish brown. Sandy clay with grits and lime inclusions.

Late Bronze Age?

### Lithics

**Cat. 431 H08-4/B31/10**

Fr. of stone axe.

**Cat. 432 H08-4/B31/11**

Micro-lith.

## DSP07-H08-02

Standard samples: 20018/01-02

### Post-Antique pottery

**Cat. 403 20018/02/1**

Fr. of flat base of utility vessel. No measurements available. No description of fabric.

Early Modern.

## DSP07-H09-01

Grab Samples: 3088-3090, 2007 FL 14
No diagnostic material.

## DSP07-H10-01 and DSP07-H18-01

Grab samples: 3084, 3086-87 and 2007 FL 13

### Handmade ware

**Cat. 433 3084/01/3**

Body fr. of vessel. Burnished on the ext. H: 3.1; W: 2.4; T: 0.8. Core: 5YR 3/3 dark reddish brown. Surface: Gley 1 3/N very dark grey. Sparse medium shells and voids.

Bronze Age?

### Amphorai

#### *Herakleia?*

**Cat. 434 3084/01/4**

Handle fr. H: 6.1; W: 3.9; T: 2.4. Clay: 2.5YR 5/6 red. Frequent tiny to medium lime inclusions; frequent small to medium shells; sparse medium pyroxene.

4th century BC?

## Unknown production

**Cat. 435 3084/01/1**

Handle fr. H: 12.1; W: 4; T: 2.5. Clay: 7.5YR 7/6 reddish yellow. Sparse medium lime inclusions; sparse medium to large shells; sparse medium to large quartz; sparse medium grog; medium sand; frequent small to large feldspar.

Classical-Hellenistic?

## Post-Antique pottery

**Cat. 436 3084/01/2**

Base fr. of jug with a ring foot. H: 2.7; W: 7.2; T: 4.8; Ø: 8. Core: 2.5YR 5/6 red to 5YR 6/1 grey. Surface: 10YR 5/2 greyish brown to 5/3 brown. Frequent tiny to small lime inclusions; sparse small shells; medium very large stones; sparse large pyroxene; frequent large voids.

Medieval?

# DSP07-H10-01

Trial trench Feature 2

## Handmade ware

**Cat. 437 H10-1/Fl01/4**

Rim(?) fr. of very fine clay, which may be unfired. Possibly a secondarily fired sealing. H: 3.25; W: 6.2; T: 0.71; Ø? Core: 10YR 7/6 yellow. Surface: 5YR 7-6/6 reddish yellow. Sandy clay with tiny to small lime.

Early Hellenistic.

**Cat. 438 H10-1/Fl04/4**

Rim fr. of vessel with rounded lip with oblique strokes. H: 1.86; W: 2.85; T: 0.52-0.71. Clay: Gley 1 3/N very dark grey and 10R 4/1 dark reddish grey. Sparse, very tiny shells.

Early Hellenistic.

**Cat. 439 H10-1/Fl06/1**

Fr. of vessel with outturned rim. Finger decoration on lip. H: 3.27; W: 3.65; T: 0.65; Ø: ca. 26. Core: Gley 1 5/10Y greenish grey. Surface: 2.5Y 7/3 pale yellow and 2.5Y 3/1 very dark grey. Medium and large quartz and grog; sparse large oval voids.

Stolba 2002, type 1 Cat. D1-D86. Early Hellenistic.

**Cat. 440 H10-1/Fl09/7**

Fr. of vessel with concave, outturned rim. H: 3.4; W: 3.87; T: 0.95-1; Ø? Core: Gley 1 4/N dark grey and 5YR 8/4 pink. Surface: 5YR 7/8 reddish yellow. Sparse medium quartz and lime inclusions.

Stolba 2002, type 1 Cat. D1-D86. Early Hellenistic.

**Cat. 441 H10-1/Fl10/4**

Fr. of vessel with concave, outturned rim. H: 3.46; W: 2.94; T: 0.9-1.11; Ø: ca. 20. Core: Gley 1 3/N very dark grey. Surface: 10YR 7/4 very pale brown. Sparse tiny quartz and lime.

Stolba 2002, type 1 Cat. D1-D86. Early Hellenistic.

**Cat. 442 H10-1/Fl11/5**

An almost complete vessel. Reconstructed from many frs. Concave, outturned rim with impressed decoration on the lip. Ovoid body with four protruding knobs on the shoulder. Concave underside. H: 19.4; Ø max: 16.3 Ø rim: 10.5 Ø base: 7.9. No information on clay.

Similar to Panskoe I type 1, but with possible Taurian influence. Cf. Stolba 2002, type 1 Cat. D1-D68 Early Hellenistic.

**Cat. 443 H10-1/Fl11/6**

Base fr. of vessel with flat underside. H: 4.57; W: 5.5; T: 1.1; Ø base: ca. 10. Core: 10YR 6/2 light brownish grey. Int. surface: 2.5Y 5/1 grey. Ext. surface: 2.5Y 7/2 light grey. Coarse sandy clay with large red grogs and grits; long black lines indicating burned organic material. Some tiny quartz.

Stolba 2002, type 1 Cat. D1-D86. Early Hellenistic.

**Cat. 444 H10-1/Fl13/7**

Tiny rim fr. of vessel with impressions on the lip. H: 2.39; W: 1.93; T: 0.61; Ø? Core: Gley 3/N very dark grey. Surface: 10YR 7/4 very pale brown. Poorly sorted clay with large angular grey and some tiny lime.

Possibly belonging to Stolba 2002, type 1 Cat. D1-D87. Early Hellenistic.

## Amphorai

### *Chersonesos*

**Cat. 445 H10-1/Fl02/1**

Rim fr. with oval lip. Possibly from same amphora as H10-1/Fl05/1. H: 5; W: 4.08; T: 0.7-0.83 (wall), 1.26 (rim); Ø: ca. 12. Core: 5YR 6/2 pinkish grey. Surface: 5YR 7/3 pink and 2.5Y 8/2 pale yellow. Yellowish slip. Sparse small lime and quartz; medium small oval voids.

Cf. Monachov 1989, pl. 19 and 138, pl. 11, rim type II. Early Hellenistic.

**Cat. 446 H10-1/Fl05/1**

Rim fr. with oval lip. Possibly from same amphora as H10-1/Fl02/1. H: 2.15; W: 3.54; T: 0.73 (wall), 1.29 (rim); Ø: ca. 12. Core: 5YR 6/2 pinkish grey. Slip: 5YR 7/3 pink and 2.5Y 8/2 pale yellow. Sparse small lime and quartz inclusions; medium small oval voids.

Cf. Monachov 1989, pls. 19 and 138, pl. 11, rim type II Early Hellenistic.

**Cat. 447 H10-2/Fl12/1a**

Fr. of lower part of handle with part of the body preserved. H: 5.86; W: 5.97; T: 2.53. Clay: 5YR 6/8 reddish yellow. Large shells; large rounded sand(?) and occasional large black inclusions. Probably from same amphora as H10-2/Fl12/1b.

Late Classical-Early Hellenistic.

**Cat. 448 H10-2/Fl12/1b**

Fr. of foot with depression on underside. H: 7; W: 5.24; Ø? Clay: 5YR 6/6 reddish yellow. Tiny to large lime; some tiny black and shell inclusions. Probably from same amphora as H10-2/Fl12/1a.

Cf. Monachov 1989, tav. XXI-XXIV, foot type I or II. Late Classical-Early Hellenistic.

## Tableware

**Cat. 449 H10-1/Fl09/5**

Base fr. of Attic black-glazed cup with ring foot. A badly preserved body fr. (H10-1-Fl10-3) probably belongs to the same cup. H: 1.65; W: 3.96; T: 0.63 (bottom); Ø: 6. Core: 10YR 7/4 very pale brown. Surface glaze: 5Y 2.5/1 black, 2.5YR 3/4 dark reddish brown and 2.5YR 3/6 dark red. Single very tiny voids.

4th century BC.

## Lithics

**Cat. 450 H10-1/Fl10/8**

Fr. of grinding stone. L: 7.4; W: 4.39-4.55; T: 1.5-1.82. Clay: Gley 1 3/N very dark grey.

# DSP07-H11-01

Grab samples: 3083, 3093

## Handmade ware

**Cat. 451 3083/01/24**

Wall fr. of cooking vessel. Ribbed on the ext. H: 4.3; W: 4.9; T: 0.8. Sandwich clay. Int. core: Gley 2 4/5B dark bluish grey. Ext. core: 2.5YR 7/6 light red to 2.5YR 7/4 light reddish brown. Sparse tiny to medium lime inclusions; sparse tiny feldspar.

Middle Bronze Age/Late Bronze Age.

**Cat. 452 3083/01/25**

Wall fr. of cooking vessel. Ribbed on both int. and ext. H: 3.7; W: 4.2; T: 0.7-0.8. Core: Gley 2 2.5/5 BG greenish black. Int. surface: Gley 2 6/5BG greenish grey. Ext. surface: 2.5YR 7/6 light red. Medium small to medium lime inclusions; medium small dark stones.

Middle Bronze Age/Late Bronze Age.

**Cat. 453 3083/01/26**

Wall fr. of cooking vessel. H: 2.55; W: 3.3; T: 0.9-1. Core: Gley 2 2.5/5 BG greenish black. Int. surface. 2.5YR 7/6 light red. Ext. surface: Gley 1 5/N. Medium tiny lime inclusions; medium tiny feldspar; sparse small quartz.

Middle Bronze Age/Late Bronze Age.

**Cat. 454 3083/02/12**

Rim fr. of cooking vessels with outcurved rim and flat lip. Convex shoulder. Ribbed on the int. H: 3.4; W: 3.9; T: 1.1-1.2; Ø: ca. 16. Core: Gley 2 5/5 BG greenish grey. Surface: 2.5YR 7/6 light red. Sparse tiny lime inclusions; sparse tiny light-reflecting particles; sparse tiny feldspar.

Cf. Gerškovič 1999, type T 1I2, cf. 53, abb. 11.8. Late Bronze Age.

## Architectural elements

**Cat. 455 3083/01/1a**

Fr. of the edge of Sinopean roof tile. H: 17.7; W: 9.2; T: 2.7-2.9. Clay: 2.5YR 7/6 light red. Sparse small to medium lime inclusions; frequent tiny to medium pyroxene; single small golden mica.

Hellenistic.

## Amphorai

### *Chersonesos*

**Cat. 456 3083/01/21**

Rim fr. with part of handle attachment preserved on neck. H: 6.45; W: 4.4; T: 0.65-1.3; Ø? Clay: 2.5YR 5/8 red. Slip: 7.5YR 8/4 pink. Frequent tiny to small shells; frequent tiny lime inclusions; medium small feldspar; sparse small stones.

Hellenistic.

**Cat. 457 3083/02/3**

Handle fr. H: 4.9; W: 4.3. Clay: 10R 7/8 light red. Slight traces of light slip. Sparse tiny lime inclusions; sparse tiny mica; sparse tiny feldspar.

Early Hellenistic.

### *Herakleia*

**Cat. 458 3083/01/1b**

Rim fr. H: 3.75; W: 4.1; T: 0.5-0.9; Ø: 9. Clay: 2.5YR 6/8 light red. Frequent tiny to small lime; sparse small pyroxene; single small quartz; sparse tiny to small feldspar.

Hellenistic.

**Cat. 459 3083/01/9**

Fr. of lower part. H: 8.7; W: 6.3; T: 1.1-1.4. Clay: 10R 7/8 light red. Medium small lime; sparse small to medium pyroxene; sparse tiny to small feldspar; medium small brown stones.

Hellenistic.

**Cat. 460 3083/01/10**

Handle fr. H: 7.1; W: 4.25; T: 2.7. Clay: 2.5YR 6/8 light red. Medium tiny to small pyroxene; medium small to medium stones; sparse medium red stones; sparse small lime inclusions; medium small to medium quartz.

4th century BC.

**Cat. 461 3083/01/11**

Handle fr. Slightly smoothed on ext. H: 9.15; W: 4.6; T: 2.4. Clay: 7.5YR 6/6 reddish yellow. Frequent tiny to medium lime inclusions; frequent tiny feldspar; sparse tiny pyroxene; sparse tiny quartz; sparse tiny red inclusions.

4th century BC?

**Cat. 462 3083/01/19**

Slightly flaring toe with conical cavity in underside. Incision on ext. which could be an inscribed Λ or Δ. H: 5.7; W: 5.4. Clay: 10R 7/6-7/8 light red. Medium tiny to small pyroxene; frequent tiny feldspar; sparse tiny to small red stones.

4th century BC.

**Cat. 463 3083/02/2**

Lower handle fr. H: 9.9; W: 5.7. Clay: 10R 6/8 light red. Sparse tiny pyroxene; medium tiny to small lime inclusions; medium tiny feldspar.

4th century BC.

**Cat. 464 3083/02/8**

Handle fr. Re-used as a tool on one side. H: 5.3; W: 3.9; T: 2.5. Clay: 10R 7/8 light red. Sparse tiny pyroxene; sparse tiny quartz; medium tiny to small lime inclusions; medium tiny feldspar.

4th century BC.

**Cat. 465 3083/02/15**

Handle fr. with part of the wall preserved. One finger impression preserved on the lower part of the handle. H: 3.7; W: 5.7; T: 0.9-2.2. Clay: 2.5YR 6/8 light red. Sparse, tiny lime inclusions; medium sand; medium tiny feldspar.

4th century BC.

### *Thasos*

**Cat. 466 3083/01/7**

Fr. of toe. Lower part slightly flaring. Underside worn caused by re-using as a tool. H: 7.2; W: 5.65. Core: Gley 1 5/5G greenish grey. Surface: between 5YR 6/6 reddish yellow and 5YR 5/8. Frequent tiny to medium lime inclusions; frequent tiny to small mica; sparse medium quartz; sparse tiny feldspar.

Hellenistic.

**Cat. 467 3083/02/9**

Fr. of lower wall. H: 3.7; W: 5.7; T: 1.1-1.3. Clay: 2.5YR 6/8 light red. Frequent tiny mica; sparse tiny feldspar; frequent tiny to small lime; medium sand.

Hellenistic.

### *Chios*

**Cat. 468 3083/02/5**

Wall fr. with part of repair hole. H: 7.6; W: 6.5; T: 1.3. Clay: 10R 6/8 light red. Sparse light-reflecting inclusions; sparse tiny lime.

Hellenistic.

### *Unknown production*

**Cat. 469 3083/01/8**

Toe with conical cavity on underside. Smoothed surfave. Part of int. bottom preserved. H: 9.3; W: 5.6; Ø: 4.7. Clay: 5YR 5/6 yellowish red. Medium tiny to small lime; medium tiny to small quartz; medium small feldspar.

Hellenistic.

**Cat. 470 3083/01/12**

Handle fr. H: 8.9; W: 4.9; T: 3. Clay: 10R 7/6 light red. Sparse tiny lime inclusions; sparse small stones; sparse tiny to small shells; medium tiny to small feldspar.

Classical-Hellenistic.

**Cat. 471 3083/01/13**

Handle fr. H: 5.5; W: 3.1; T: 2.1. Clay: 2.5YR 6/8 light red. Medium tiny mica; frequent tiny lime inclusions; medium tiny quartz; sparse tiny feldspar.

Classical-Hellenistic.

**Cat. 472 3083/01/14**

Fr. of lower part. Part of int. bottom preserved. H: 6.4; W: 5.5-4.3. Clay: 10R 7/8 light red. Sparse small quartz; medium tiny to small lime; sparse tiny red inclusions; sparse small pyroxene; sparse tiny feldspar

Hellenistic.

**Cat. 473 3083/01/15**

Handle fr. H: 8.1; W: 4.4; T: 2.6-2.8. Clay: 2.5YR 7/6 light red. Sparse tiny to medium lime inclusions; sparse small stones; medium tiny to small feldspar.

Classical-Hellenistic.

**Cat. 474 3083/01/16**

Lower part. Bottom part worn of due to re-using as a grinding tool. H: 8.7; W: 7; T: 1.1-1.7. Clay: 5YR 5/6 yellowish red. Sparse tiny mica; medium tiny to medium lime; sparse tiny to small feldspar.

Hellenistic.

**Cat. 475 3083/01/17**

Handle fr. Low ridge on ext. H: 4.9; W: 2.75; T: 1.8. Clay: 5YR 6/6 reddish yellow. Slip: 7.5YR 8/3 pink. Medium large brown inclusions; medium tiny to small lime inclusions; medium small feldspar.

Classical-Hellenistic.

**Cat. 476 3083/01/18**

Handle fr. Two grooves on ext. H: 6.1; W: 3.8; T: 2.3-2.5. Clay: 10YR 7/4 very pale brown. Medium small brown inclusions; sparse small pyroxene; sparse tiny lime inclusions; sparse tiny feldspar.

Classical-Hellenistic?

**Cat. 477 3083/01/20**

Rim fr. with protruding lip. H: 3.1; W: 3.2; T: 1.11.5; Ø? Clay: 10YR 6/4 light yellowish brown. Medium tiny to medium lime; sparse small pyroxene; sparse small quartz.

Hellenistic.

**Cat. 478 3083/01/22**

Handle fr. Incompletely preserved. H: 5.1; W: 3.6; T: 2.75. Clay: 7.5YR 6/6 reddish yellow. Surface: 5YR 6/6 reddish yellow. Sparse tiny lime inclusions; sparse small stones (quartz?); medium sand.

Classical-Hellenistic.

**Cat. 479 3083/01/27**

Fr. of upper part of handle. H: 5.5; W: 6; T: 2.4. Clay: 7.5YR 7/6 reddish yellow. Medium tiny to small lime inclusions; frequent tiny feldspar; single large quartz.

Classical-Hellenistic?

**Cat. 480 3083/02/4**

Handle fr. H: 5.7; W: 5. Clay: 2.5YR 7/8 light red. Medium tiny to small lime inclusions; sparse tiny quartz; sparse tiny feldspar.

Classical-Hellenistic.

**Cat. 481 3083/02/6**

Handle fr. H: 5.8; W: 6.1. Clay: 10YR 7/6 light red, 10R 7/8 light red to 2.5YR 5/6 red. Medium small to medium quartz; frequent very tiny light-reflecting particles; medium tiny to small lime.

Thasos? Hellenistic.

**Cat. 482 3083/02/7**

Upper part of handle. H: 6.6; W: 6. Clay: 2.5YR light red. Slip: 5YR 8/3 pink. Medium tiny lime inclusions; sparse tiny feldspar; medium tiny voids.

Early Hellenistic.

**Cat. 483 3083/02/10**

Handle fr. H: 6.2; W: 3.1. Clay: 2.5YR 5/8 red. Wash: 2.5YR 5/6 reddish brown. Medium tiny to large lime inclusions; medium small to large quartz; sparse tiny feldspar; sparse tiny mica.

Classical-Hellenistic?

**Cat. 484 3083/02/11**

Round handle fr. H: 6.2; W: 3.2; T? Clay: 10R 7/8 light red. Slip: 10R 6/3 pale red. Tiny light-reflecting particles; sparse tiny lime inclusions; sparse tiny feldspar.

Classical-Hellenistic.

## Tableware

**Cat. 485 3083/01/6**

Base fr. of jug with moulded foot and flat underside. H: 3; W: 6.2; T: 1-2.4; Ø? Clay: 10R 7/6 light red. Slip on ext. and partly on int.: 7.5YR 8/4 pink. Sparse tiny lime inclusions; sparse small quartz; frequent tiny stones; sparse tiny feldspar.

Chersonesos, cf. e.g. Kašaev 2002, Cat. C14-C15. Late 4th-3rd century BC.

**Cat. 486** **3083/02/18**

Wall fr. of closed vessel, presumably part of the shoulder of small jug. Faint traces of painted horizontal band. H: 2.4; W: 2.1; T: 0.4-0.6. Clay: 10R 4/1 dark reddish grey. Surface: 10R 6/4 reddish grey. Sparse tiny lime inclusions; sparse tiny feldspar.

Late Classical-Early Hellenistic.

**Cat. 487** **3083/02/19**

Wall fr. of closed vessel, presumably a lekythos or a small jug. Unevenly glossed on the ext. H: 2.2; W: 2.8; T: 0.6. Clay: 10R 6/1 reddish grey. Slip: 5YR 4/4 reddish brown to 7.5YR 3/1 very dark grey. Sparse tiny lime inclusions; single light-reflecting particle; single tiny feldspar.

Late Classical-Early Hellenistic.

## Utility ware

**Cat. 488** **3083/02/16**

Wall fr. H: 7.8; W: 6.4; T: 1-1.1. Clay: 2.5YR 6/8 light red. Surface: 10R 7/8 light red. Sparse tiny lime inclusions; medium tiny to small pyroxene; single light-reflecting particles; sparse tiny to small feldspar.

Sinopean louterion? Early Hellenistic.

## Handmade ware

**Cat. 489** **3083/01/23**

Rim fr. of pot with slightly outturned, rounded rim. Impressed decoration on shoulder just below rim. H: 2.6; W: 3.75; T: 0.9-1.1; Ø? Core: Gley 2 2.5/10 G greenish black. Surface: 10YR 7/3-7/4 very pale brown. Sparse small to medium quartz; medium small red inclusions; sparse tiny to medium lime; sparse tiny feldspar.

Cf. Stolba 2002, type 1 Cat. D1-D68. Early Hellenistic.

**Cat. 490** **3083/02/14**

Base fr. of open pot. Slightly polished on int. H: 5.6; W: 6.8; T: 1.2-1.3; Ø? Clay: Gley 1 2.5/N black. Surface: Gley 2 3/10 G very dark greenish grey. Frequent tiny to small lime inclusions; medium sand; sparse small quartz; sparse small to medium grog.

Late Classical-Hellenistic.

**Cat. 491** **3083/02/17**

Base fr. of pot with flat underside. H: 2; W: 4.2; T: 1.1-1.4; Ø? Clay: Gley 2 4/5 B dark bluish grey and 1 Gley 8/10 GY light greenish grey. Medium tiny to small lime inclusions; sparse small quartz; single tiny mica; medium tiny to large voids.

Early Hellenistic.

# DSP08-H11-02

Standard samples: 20050-20055
Grab samples: 3094, 3218

## Handmade ware

**Cat. 492** **3094/01/6**

Body fr. of vessel decorated with relief ornament on the ext. H: 2.8; W: 3.1; T: 0.55-0.74. Core: Gley 2 3/10BG very dark greenish grey. Surface: 7.5YR 7/4 pink. Frequent medium and large shells.

Bronze Age?

## Amphorai

### *Herakleia*

**Cat. 493** **3218/01/4**

Handle fr. H: 7.86; W: 4.02; T: 2.85. Clay: 5YR 7/8 reddish yellow. Frequent medium pyroxene and quartz; single tiny lime particle.

4th century BC.

### *Sinope*

**Cat. 494** **20053/07/1**

Handle fr. (badly preserved). H: 8.7; W: 4.9; T: 2.8? Clay: 10YR 8/3 very pale brown. Surface: 10YR 7/4 very pale brown. Medium small to large pyroxene; sparse small to medium lime inclusions.

Probably of pithoid type. Ca. 350-250 BC.

**Cat. 495** **3094/01/2**

Fr. of toe. H: 6.9; Ø 4.45-5. Core: 7.5YR 8/4 pink. Surface: 10YR 8/4. Frequent small and medium pyroxene; single little quartz.

Sinope?, 3rd century BC?

### *Unknown production*

**Cat. 496** **20051/09/1**

Handle fr. H: 8.2; W: 4.77; T: 3.52. Core: 5YR 5/6 yellowish red. Slip: 10YR 7/3 very pale brown. Sparse sand inclusion; single tiny lime particle; sparse light-reflecting particles; medium small to medium dark inclusions.

**Cat. 497** **20053/03/1**

Wall fr. with a hole. Re-used as weight. H: 5.88; W: 2.9; T: 1.6. Clay: 2.5YR 5/8 red. Sparse tiny to small lime inclusions.

Classical-Hellenistic.

**Cat. 498** **20054/06/1**

Handle fr. H: 4.74; W: 4.2; T: 2.27. Clay: 2.5YR 5/6 red. Single small green olivin; sparse small red inclusions; sparse small quartz; sparse tiny to small lime inclusions; medium small to medium quartz.

Classical-Hellenistic.

**Cat. 499** **20055/03/1**

Handle fr. H: 7.3; W: 3.49; T: 2.25. Core: 2.5YR 7/4 light red brown. Slip. 2YR 8/2 pale yellow. Frequent tiny lime inclusions; medium medium shells; sparse small quartz.

Classical-Hellenistic.

**Cat. 500** **3094/01/1**

Handle fr. with part of the wall of narrow neck. H: 5.05; W: 4 (handle), 5 (wall); T: 2.5 (handle),0.9-1.5 (wall) Core: 2.5YR 6/6 light red. Surface: 7.5YR 8/4. Single tiny to small lime particle; sparse small pyroxene and feldspar inclusions.

**Cat. 501** **3094/01/3**

Handle fr. with part of the wall. H: 5.75; W: 3.89 (handle), 5.36 (wall); T: 2.72 (handle), 0.9 (wall). Core: 2.5YR 5/6 red. Surface: 7.5YR 8/4 pink. Medium tiny to small lime inclusions; sparse tiny voids.

Classical-Hellenistic.

**Cat. 502** **3094/01/4**

Handle fr. H: 8.56; W: 3.42-3-56; T: 2.17-2.23. Core: 7.5YR 7/6 reddish yellow. Surface: 10YR 8/4 very pale brown. Medium tiny to small light-reflecting particles.

Classical-Hellenistic.

**Cat. 503 3094/01/5**

Handle fr. H: 7; W: 3.96; T: 2.57. Clay: 5YR 7/6 reddish yellow. Frequent small and medium quartz; shells; single tiny pyroxene and feldspar.

Hellenistic.

**Cat. 504 3094/01/7**

Wall fr. with smoothed ext. surface. Int. surface has clear throwing marks. H: 8.55; W: 7.63; T: 0.8-1.4. Int. surface and core: 7.5YR 5/3 brown. Ext. surface: 7.5 6/6 reddish yellow. Medium small light-reflecting particles; sparse tiny shells.

Classical-Hellenistic.

**Cat. 505 3218/01/1**

Handle fr. H: 9.32; W: 4.85; T: 2.12-2.33. Core: 5YR 7/4 pink. Surface: 5YR 7/8 reddish yellow. Sparse medium quartz; sparse small to tiny lime inclusions.

Classical-Hellenistic.

### Utility ware

**Cat. 506 3218/01/2**

Rim fr. of louterion. H: 3.4; W: 8.5; T: 1.73 (wall), 3.16 (rim); Ø: 38. Core: 5YR 7/4 pink. Surface: 5YR 7/8 reddish yellow. Sparse medium lime inclusions; frequent medium to large pyroxene; sparse medium feldspar; sparse medium quartz.

Sinopean louterion, Early Hellenistic.

### Post-Antique pottery

**Cat. 507 3218/01/3**

Handle fr. H: 6; W: 3.48-3.6; T: 1.95-2.21. Core: 2.5YR 6/8 light red and 2.5YR 7/6 light red. Surface: 10YR 8/4 very pale brown; yellowish slip. Sparse tiny lime inclusions; sparse tiny voids.

Medieval?

## DSP08-H11-02

Trial trench Feature 9

### Handmade ware

**Cat. 508 H11-6/B1/2**

Rim fr. of vessel with preserved handle attachment, possibly a jug. Plain thickened and slightly outturned rim with flat lip. H: 3.89; W: 3.22; T: 0.6 (wall), 0.88 (rim); Ø: ca. 12. Clay: 5YR 6/6 reddish yellow. Some grits and many dark inclusions, probably pyroxene.

A distant parallel could be Stolba 2002, 197 Cat. D106. Early Hellenistic.

**Cat. 509 H11-6/B1/3**

Rim fr. of closed vessel with a plain rim and flaring shoulders. H: 5.8; W: 4.38; T: 0.45-0.7; Ø: ca. 20. Clay: 10YR 6/3 pale brown. Very coarse fabric with large milky quartz/lime(?) and dark inclusions.

Similar to Stolba 2002, type 1C, Cat. D87 in shape, but much larger. Early Hellenistic.

**Cat. 510 H11-6/B1/4**

Rim fr. of vessel with concave, outturned rim. H: 3.57; W: 5.45; T: 0.5 (rim), 0.8 (wall); Ø: ca. 15. Core: Gley 1 3/N very dark grey. Surface: 10YR 7/3 very pale brown. Some grits and shells.

Stolba 2002, type 1 Cat. D1-D86. Early Hellenistic.

**Cat. 511 H11-6/B1/5**

Rim fr. of vessel with slightly concave, outturned rim. Impressed decoration on the exterior lip. H: 3.55; W: 3.9; T: 0.55-0.7; Ø: ca. 10. Clay: 10YR 4/1 to 6/3 dark grey to pale brown. Some lime and dark red iron oxid inclusions.

Stolba 2002, type 1 Cat. D1-D86. Early Hellenistic.

**Cat. 512 H11-6/B1/6**

Small completely preserved base of closed vessel with flat underside. H: 2.05; W: 3.48; T: 0.45 (wall), 0.6 (bottom); Ø: ca. 4.5. Core: 10YR 4/1 dark grey. Surface: 5YR 7/4 pink. Sandy fabric with some large grits all burned black.

Early Hellenistic.

**Cat. 513 H11-6/B1/7**

Base fr. H: 5.4; W: 4.11; T: 0.85-0.98; Ø: ca. 8. Clay: Gley 1 3/N very dark grey. Poorly sorted with large grits and large pieces of grog.

Early Hellenistic.

**Cat. 514 H11-6/B2/2**

Two joining frs. of vessel with concave, outturned rim. Impressed decoration on lip. H: 6.5; W: 16; T: 0.7; Ø: ca. 23. Core: 10YR 7/2 light grey. Surface: 10YR 7/3 very pale brown. Very coarse with tiny to large quartz inclusions; large rounded reddish-brown inclusions and grits.

Stolba 2002, type 1 Cat. D1-D86. Early Hellenistic.

**Cat. 515 H11-6/B2/3**

Rim fr. of bowl with inturned rim and flat horizontal lip. H: 3.8; W: 3.8; T: 0.9; Ø? Core and int.: Gley 1 3/N very dark grey. Ext.: 10YR 7/3 very pale brown. Very poorly sorted, many large to very large grey grits; some lime.

Stolba 2002, Cat. D116-D119. Early Hellenistic.

**Cat. 516 H11-6/B2/4**

Rim fr. of vessel with concave, outturned rim. Possibly from same vessel as H11-6/B2/5. H: 5.1; W: 7.6; T: 0.9; Ø: ca. 16. Core and int. surface: 10YR 4/1 dark grey. Ext. surface. 7.5YR 4/1 to 7/4 dark grey to pink. Sandy clay with rounded grits; shells; reddish-beige inclusions and lime inclusions.

Stolba 2002, type 1 Cat. D1-D86. Early Hellenistic.

**Cat. 517 H11-6/B2/5**

Rim fr. of vessel with concave, outturned rim. Possibly from same vessel as H11-6/B2/4. H: 6.4; W: 9.2; T: 0.8; Ø: ca. 16. Core and int. surface: 10YR 7/4 very pale brown. Ext. surface: Gley 1 3/N very dark grey. Some small grits and lime inclusions.

Stolba 2002, type 1 Cat. D1-D86. Early Hellenistic.

**Cat. 518 H11-6/B2/6**

Rim fr. of vessel with concave, outturned rim. Impressed decoration on the exterior lip. H: 5.6; W: 6.7; T: 0.8; Ø? Core and int. surface: 10YR 4/1 dark grey. Ext. surface: 10YR 7/3 very pale brown. Sandy clay with small to large grits and tiny to large reddish-brown inclusions.

Stolba 2002, type 1 Cat. D1-D86. Early Hellenistic.

**Cat. 519 H11-6/B2/7**

Rim fr. of vessel with concave, outturned rim. Impressed decoration on the exterior lip. H: 6.1; W: 7.7; T: 0.96; Ø: ca. 17. Core and int.

surface: 10YR 5/1-2 grey to greyish brown. Ext. surface: 10YR 7/3 very pale brown. Sandy clay with some shell and reddish brown inclusions.
Stolba 2002, type 1 Cat. D1-D86. Early Hellenistic.

**Cat. 520 H11-6/B3/3**

Two joining frs. of vessel with concave, outturned rim. Impressed decoration on lip. H: 3.6; W: 5; T: 0.65-0.9; Ø: ca. 14. Core: 10YR 5/1 grey. Surface: 10YR 7/2 light grey. Poorly sorted clay with tiny to very large rounded grits; quartz and lime inclusions.
Stolba 2002, type 1 Cat. D1-D86. Early Hellenistic.

**Cat. 521 H11-6/B3/4**

Base fr. of pot with flat underside and burnished exterior. H: 2.6; W: 5.2; T: 0.6-0.8 (wall), 1.1(bottom); Ø: ca. 15. Clay: 10YR 4/1 to 5YR 6/6 dark grey to reddish yellow. Sandy clay full of large grits; quartz and lime inclusions.
Stolba 2002, type 1 Cat. D1-D86. Early Hellenistic.

## Tableware

**Cat. 522 H11-6/B1/8**

Body fr. of jug(?) with repair hole. H: 5.4; W: 2.8; T: 0.75-0.85. Clay: 2.5YR 6/6 light red. Many dark shiny inclusions, probably pyroxene.
Hellenistic.

**Cat. 523 H11-6/B2/1**

Lower handle attachment of jug or small table amphora. H: 3.2; W: 4.7; T: 1.7. Clay: 5YR 6/6 reddish yellow. Many medium dark inclusions, probably pyroxene, and shell.
Hellenistic.

**Cat. 524 H11-6/B2/8**

Body fr. of jug(?) with repair hole. H: 2.9; W: 4.3; T: 0.6. Clay: 10YR 6/3 pale brown. Very fine sandy clay with multiple quartz/silver mica(?).
Hellenistic.

**Cat. 525 H11-6/B1/1**

Rim fr. of Grey Ware bowl with thickened rim. H: 1.57; W: 2.9; T:0.45 (wall), 0.7 (rim); Ø: ca. 15. Clay: 2.5Y 7/1 light grey. Very fine clay; single medium greenish lime(?) inclusion.
The same type is known from Panskoe I, cf. Handberg, Stolba & Ušakov 2009, fig. 7b. Early Hellenistic.

## Amphorai

### *Herakleia*

**Cat. 526 H11-6/B3/5**

Fr. of the lower part. H: 2.7; W: 5.55; T: ca. 5. Clay: 7.5YR 7/4 pink. Numerous medium-sized grits; milky quartz; lime and dark and reddish-brown rounded inclusions.
Early Hellenistic.

### *Unknown Production*

**Cat. 527 H11-6/B3/2**

Handle fr. Reused as a tool. H: 4.2; W: 4; T: 1.8-2.3. Core: 5YR 6/4 light reddish brown. Surface: 10YR 5/1 grey. Sandy fabric with numerous grits; rounded quartz; lime and some dark inclusions.
Late Classical-Early Hellenistic.

**Cat. 528 H11-6/B3/1**

Fr. of mushroom rim with two exterior grooves. H: 5.65; W: 8.77; T: 0.5-0.75 (wall), 1.76 (rim); Ø: 14. Clay: 5YR 6/6 to 10YR 5/2 reddish yellow to greyish brown. Fine clay with tiny quartz and lime inclusions; slight sparkle on the surface possibly from quartz.

# DSP08-H12-01

Standard samples: 20035/08-10, 20036/01-12, 20061-20063
Grab samples: 3208-3216

## Handmade ware

**Cat. 529 20035/10/1**

Fr. of base of a pot. H: 2.55; W: 2.5; T: 0.5-0.9; Ø: 10. Clay: 2.5YR 6/8 light red. Sparse tiny lime particles; single small pyroxene and single tiny mica inclusions; sparse small-medium dark inclusions.
Hellenistic.

**Cat. 530 20063/12/1**

Rim fr. of small pot with slightly concave, outturned rim and rounded lip. H: 2.2; W: 2.6; T: 0.5-0.6; Ø: 11. Clay: 10R 7/6 light red. Sparse tiny lime inclusions.
Cf. Stolba 2002, type 1 Cat. D1-D68. Early Hellenistic.

## Amphorai

### *Chios?*

**Cat. 531 3214/01/1**

Lower wall fr. H: 11.3; W: 8.4; T: 2-2.4. Clay: 5YR 5/8 yellowish red. Sparse tiny mica; sparse tiny lime; sparse tiny feldspar.
Hellenistic.

### *Unknown production*

**Cat. 532 20061/05/1**

Handle fr. H: 4.96; W: 3.96; T: 1.76. Clay: 5YR 6/8 red yellow. Slip: as clay. Sparse tiny lime inclusions; sparse tiny mica.
Classical-Hellenistic.

**Cat. 533 20062/08/1**

Rim fr. H: 2.6; W: 4.15; T: 0.63; Ø: 8.5; D: 65. Core: 2.5YR 6/8 light red. Medium small to medium quartz; medium tiny to small mica; sparse tiny lime inclusions; sparse small red inclusions.
Hellenistic.

**Cat. 534 20063/01/1**

Handle fr. H: 5.4; W: 4.38; T: 2.84. Clay: 2.5YR 6/8 light red. Sparse small quartz; sparse tiny lime inclusions; sparse light-reflecting particles; sparse tiny to medium feldspar.
Classical-Hellenistic.

**Cat. 535 20063/11/1**

Handle fr. with grooved ext. H: 7.9; W: 5; T: 2.7. Clay: 2.5YR 6/8 light red. Sparse tiny mica; sparse tiny to small lime inclusions; sparse small quartz.

**Cat. 536** **3209/01/1**

Handle fr. H: 7.76; W: 3.91; T: 3. Clay: 5YR 7/8 reddish yellow and 2.5YR 6/8 light red. Sparse small quartz; sparse small lime inclusions; sparse very tiny mica; sparse small voids.

Classical-Hellenistic.

### Post-Antique pottery?

**Cat. 537** **20036/12/1**

Fr. of neck with part of handle preserved. H: 5.35; W: 5.2; T: handle: 2.1 x 5.1, wall: 0.8. Clay: 5YR 6/8 reddish yellow. Surface: 5YR 5/6 yellowish red. Sparse small to large lime inclusions; sparse tiny to medium lime inclusions; sparse medium dark inclusions; sparse small to medium grits.

Medieval?

## DSP08-H13-01

Standard samples: 20049, 20058-20060

### Amphorai

*Thasos?*

**Cat. 538** **20059/03/1**

Handle fr. H: 5.67; W: 5.94; T: 4.1 x 1.84. Clay: 7.5YR 6/6 reddish yellow. Frequent small to medium quartz; frequent tiny to small golden mica; sparse tiny to small lime inclusions; sparse small red inclusions.

Conical-biconical type II-C (on account of the flat handle). 4th century BC.

## DSP08-H14-01

Standard samples: 20043-20048

### Amphorai

*Unknown production*

**Cat. 539** **20043/01/1**

Upper part of handle fr. H: 9.8; W: 6.85; T: Wall: 0.8-1 Handle: 3.5 x 4.4. Clay: 10R 7/8 light red.

Classical-Hellenistic.

## DSP08-H14-01

Trial trench Feature 2

### Handmade ware

**Cat. 540** **H14-5/B1/1**

Body fr. decorated with a horizontal rib with impressed decoration. H: 2; W: 2; T: 0.5. Core: Gley 1 3/N very dark grey. Surface: 10YR 7/4 very pale brown. Sandy clay with grits and tiny black inclusions.

Bronze Age?

**Cat. 541** **H14-5/B1/10**

Body fr. of vessel decorated with a horizontal rib with impressed decoration. H: 2.4; W: 2.8; T: 0.53. Core and int. surface: Gley 1 3/N very dark grey. Ext. surface: 10YR 7/3 very pale brown. Sandy clay with small grits and quartz.

Late Bronze Age?

**Cat. 542** **H14-5/B1/12**

Body fr. of vessel decorated with impressed oblique strokes. H: 4.8; W: 4.1; T: 0.5. Clay: Gley 1 3/N very dark grey. Sandy clay with small to large grits, all burned black.

Similar decoration is known from Burun Eli, cf., e.g., Kolotuchin 2003, figs. 49.2 and 50.4. Late Bronze Age.

**Cat. 543** **H14-5/B1/13**

Base fr. of vessel with flat underside. H: 2.1; W: 4.9; T: 0.55; Ø: 15. Clay: Gley 1 3/N very dark grey. Sandy fabric with grits and quartz all fired black.

Possibly belonging to Gerškovič 1999, type T1II2, fig. 11.10. Late Bronze Age?

**Cat. 544** **H14-5/B1/14**

Base fr. of pot with flat underside. H: 1.56; W: 3.4; T: 0.8; Ø: 10. Clay: Gley 1 3/N very dark grey. Sandy clay with small grits and lime inclusions.

Possibly belonging to Gerškovič 1999, type T1II2, fig. 11.10. Late Bronze Age?

**Cat. 545** **H14-5/B1/15**

Base fr. of vessel with flat underside. Possibly belonging to the same vessel as H14-5/B1/16. L: 3.3; W: 2.5; T: 0.8; Ø: 14. Core and int. surface: 7.5YR 7/3 pink. Ext. surface: 7.5YR 4/1 dark grey. Sandy fabric with tiny to small grits and quartz inclusions; some tiny dark red inclusions.

Possibly belonging to Gerškovič 1999, type T1II2, fig. 11.10. Late Bronze Age?

**Cat. 546** **H14-5/B1/16**

Base fr. of vessel with flat underside. Possibly belonging to the same vessel as H14-5/B1/15. H: 1.6; W: 3.2; T: 0.9; Ø: 14. Core and int. surface: Gley 1 3/N very dark grey. Ext. surface: 10YR 7/3 very pale brown. Sandy clay with small rounded quartz and grits.

Possibly belonging to Gerškovič 1999, type T1II2, fig. 11.10. Late Bronze Age?

**Cat. 547** **H14-5/B1/2**

Rim fr. of vessel with concave, slightly outturned rim. Burnished on the interior and exterior. H: 3.7; W: 2; T: 0.9; Ø? Clay: Gley 1 3/N very dark grey. Sandy clay with tiny grits and quartz inclusions.

Cf. Gerškovič 1999, type T1II2, fig. 11.10. Late Bronze Age.

**Cat. 548** **H14-5/B1/3**

Rim fr. of small open vessel with impressed decoration on projecting rim. H: 2.3; W: 3; T: 0.6; Ø: ca. 11. Clay: Gley 1 3/N very dark grey. Sandy clay with small grits and rounded quartz.

Possibly of the same type as Leskov 1970, fig. 9.9. Late Bronze Age.

**Cat. 549** H14-5/B1/4

Rim fr. of vessel with concave, outturned rim. H: 3.4; W: 2.5; T: 0.66; Ø? Core and int. surface: 10YR 5/2 greyish brown. Ext. surface: Gley 1 3/N very dark grey. Sandy clay with small grits and rounded quartz inclusions.

Gerškovič 1999, type T1II2, fig. 11.10. Late Bronze Age.

**Cat. 550** H14-5/B1/5

Rim fr. of vessel with concave, outturned rim. Slightly burnished on the interior and exterior. H: 2.6; W: 2.5; T: 0.8; Ø: ca. 15. Clay: Gley 1 3/N very dark grey. Sandy clay with some small grits and tiny quartz inclusions.

Cf. Gerškovič 1999, type T1II2, fig. 11.10. Late Bronze Age.

**Cat. 551** H14-5/B1/6

Rim fr. of vessel with straight rim and thickened lip. H: 2.3; W: 3; T: 0.73; Ø? Clay: 7.5YR 7/6 reddish yellow. Sandy fabric with small grits; quartz and lime inclusions.

Possibly of the same type as Leskov 1970, fig. 9.9. Late Bronze Age.

**Cat. 552** H14-5/B1/7

Rim fr. of vessel with outturned rim. H: 2.94; W: 3.4; T: 0.87; Ø: ca. 14. Core and int. surface: Gley 1 3/N very dark grey. Ext. surface: 10YR 7/3 very pale brown. Fine clay with some small grits; quartz and lime inclusions.

Possibly belonging to Gerškovič 1999, type T1II2, fig. 11.10. Late Bronze Age.

**Cat. 553** H14-5/B1/8

Rim fr. of vessel with slightly outturned rim and impressed decoration on the lip. H: 2.67; W: 1.7; T: 0.8; Ø? Core: 10YR 5/2 greyish brown. Surface: 10YR 6/3 pale brown. Numerous matt grey inclusions and occasional rounded dark red ones.

Late Bronze Age?

**Cat. 554** H14-5/B1/9

Rim fr. of vessel with concave, outturned rim. H: 1.7; W: 0.96; T: 0.5; Ø? Core and int. surface: Gley 1 3/N very dark grey. Ext. surface: 7.5YR 7/4 pink. Sandy clay with tiny grits; quartz and occasional dark brown inclusions.

Cf. Gerškovič 1999, type T1II2, fig. 11.10 or Stolba 2002, type 1 Cat. D1-D86. Early Hellenistic/Late Bronze Age.

**Cat. 555** H14-5/B2/1

Rim fr. of vessel with concave, outturned rim. Slight striations on the exterior. H: 4.96; W: 3.74; T: 0.75-0.85; Ø? Core: 5YR 5/4-6 reddish brown to yellowish red. Surface: 10YR 6/4 light yellowish brown. Sandy fabric with grits; quartz and lime inclusions.

Cf. Gerškovič 1999, type T1II2, fig. 11.10. Late Bronze Age.

**Cat. 556** H14-5/B2/2

Rim fr. of vessel with concave, outturned rim. H: 2.6; W: 3.9; T: 0.55 (wall), 0.75 (rim); Ø? Core and int. surface: Gley 1 3/N very dark grey. Ext. surface: 7.5YR 6/3 light brown. Sandy fabric with some large, porous, grey grits, lime and quartz.

Possibly belonging to Gerškovič 1999, type T1II2, fig. 11.10. Late Bronze Age?

**Cat. 557** H14-5/B2/3

Rim fr. of vessel with concave, outturned rim. H: 3; W: 3.7; T: 0.55-0.6; Ø: ca. 13. Clay: Gley 1 3/N very dark grey. Sandy clay with small grits and quartz.

Gerškovič 1999, type T1II2, fig. 11.10. Late Bronze Age.

**Cat. 558** H14-5/B2/4

Rim fr. of vessel with concave, outturned rim. H: 2.9; W: 3.4; T: 0.65-0.75; Ø: ca. 11. Core: Gley 1 3/N very dark grey. Surface: 10YR 6/2 light brownish grey. Sandy clay with medium to large grits.

Cf. Gerškovič 1999, type T1II2, fig. 11.10. Late Bronze Age.

**Cat. 559** H14-5/B2/5

Body fr. decorated with a horizontal moulding. H: 3.7; W: 3.2; T: 0.5-0.6. Clay: Gley 1 3/N very dark grey. Sandy clay with medium to large grits; quartz and lime inclusions.

Late Bronze Age.

**Cat. 560** H14-5/B2/6

Body fr. decorated with a horizontal moulding. H: 4.3; W: 2.7; T: 0.5-0.6. Core and int. surface: 10YR 6/4 light yellowish brown. Ext. surface: Gley 1 3/N very dark grey. Sandy fabric with medium to large grits and lime, burned brown.

Late Bronze Age.

**Cat. 561** H14-5/B2/7

Body fr. decorated with rib. H: 5.4; W: 3.05; T: 0.5-0.6. Clay: Gley 1 3/N very dark grey. Sandy clay with small grits and quartz.

Late Bronze Age.

**Cat. 562** H14-5/B2/8

Body fr. decorated with shallow combing in multiple directions. H: 4.5; W: 3; T: 0.5-0.65. Clay: Gley 1 3/N very dark grey. Sandy clay with grits, all burned black, some quartz.

Late Bronze Age.

**Cat. 563** H14-5/B2/9

Base fr. of vessel with flat underside. H: 1.35; W: 5.3; T: 0.75 (wall), 0.8-0.95 (bottom); Ø? Clay: Gley 1 3/N very dark grey. Sandy fabric with small grits, burned all black.

Possibly belonging to Gerškovič 1999, type T1II2, fig. 11.10. Late Bronze Age?

**Cat. 564** H14-5/B1/11

Body fr. of cup decorated with horizontal rib with impressed decoration. H: 1.9; W: 1.8; T: 0.4. Core and int. surface: Gley 1 3/N very dark grey. Ext. surface: 7.5YR 7/4 pink. Fine clay with some lime and quartz.

Similar cups are known from Burun Eli, cf. Točev 2007, fig. 100.16 and Planerskoe, cf. Kolotuchin 2003, fig. 31.5. Late Bronze Age.

## DSP08-H17-01

Grab samples: 3204
No diagnostic material.

## DSP08-H17-02

Standard samples: 20003/01-04
Grab samples: 3205-3206

### Tableware

**Cat. 565 20003/01/1**

Fr. of rim of bowl with slightly incurved rim. Incised groove below lip on ext. H: 1.85; W: 1.8; T: 0.6-0.8; Ø: 13. Clay: 2.5YR 5/8 red. Sparse sand.

Early Modern.

## DSP07-H18-01

## See DSP07-H10-01

## DSP07-H19-01 Kurgan Group I

Trial Trench SW of Kurgan 3

### Amphorai

*Chersonesos*

**Cat. 566 KG1-1/3/2**

Fr. of toe from a Chersonesan amphora. H: 2.2; Ø: 7 7.5YR 7/6 reddish yellow. Medium tiny to small lime and feldspar inclusions; sparse small to medium quartz.

Monachov 1989, pl. XXI. 3-4, 9 and 73. Ca. 350-325 BC.

*Unknown production*

**Cat. 567 KG1-1/3/1**

Fr. of the lower part of amphora. H: 4; W: 7.2; T: 1.2-1.8. 7.5YR 7/6 reddish yellow. Fine sandy clay with tiny sand and some gold mica inclusions.

Late Classical-Early Hellenistic.

## DSP07-H19-03

Standard samples: 20035/05-06, 20036/14-18
Grab samples: 3205-3206

### Post-Antique pottery

**Cat. 568 20036/13/01**

Fr. of rim of jar with slightly outturned rim. H: 3.6; W: 4.5; T: 0.6-0.7; Ø: ca. 14. Clay: 10R 6/8 light red. Surface 2.5YR 5/4 reddish brown in places. Single tiny lime; single tiny to small feldspar.

Early Modern.

**Cat. 569 20036/13/02**

Fr. of flat base. H: 4.1; W: 4.45; T: 07-1; Ø: 13. Clay: 10R 6/8 light red. Surface 2.5YR 5/4 reddish brown in places. Single tiny lime; single tiny to small feldspar.

Early Modern.

## DSP07-H19-09

Standard samples: 20035/01

### Post-Medieval pottery

**Cat. 570 20035/1**

Rim of juglet. Not recorded in diagnostic database.

## DSP08-H20-01

Standard samples: 20009/01-04, 20011/03

### Amphorai

*Unknown production*

**Cat. 571 20009/03/1**

Handle fr. H: 2.2; W: 3.4; T: 2.1. Clay: 2.5YR 6/6 light red. Frequent tiny mica; sparse tiny to small lime inclusions; sparse small quartz; sparse tiny feldspar.

Classical-Early Hellenistic.

## DSP08-H22-01

Standard sample: 20057/02
No diagnostic material.

## DSP08-H22-02

Adjacent units: 20057/01

**Cat. 572 20057/01**

Six frs. of cooking pot with high collared rim. Softly corrugated exterior. Gritty surface. H: 2.9.; W: 3.69; T: 0.7; Ø: ca. 18. Core: 2.5Y 2/1 very dark grey. Slip int.: 10YR /76 yellow. Ext.: 7.5YR 4/2 brown to 2.5YR 5/4 reddish brown. Sparse small quartz; single medium grog.

Early Modern.

## DSP08-H23-01

Standard samples: 20042/02

### Amphorai

*Unknown production*

**Cat. 573 20042/02/1**

Handle fr. of amphora? (poorly preserved). H: 6; W: 2.86?; T: 1.94? Clay: 10YR 7/4 very pale brown. Sparse tiny lime inclusions; sparse medium dark inclusions; sparse tiny light-reflecting particles; frequent tiny to small red inclusions.

## DSP08-H23-02

Standard samples: 20041/06, 20042/01
No diagnostic material.

## DSP08-H24-01

Standard samples: 20039/01, 20040/03-04
Grab samples: 3217

### Amphorai

*Unknown production*

**Cat. 574 20040/03/1**

Wall fr. H: 5; W: 5.8; T: 0-95-1.45. Clay: 10R 6/8 light red. Slip: 7.5YR 5/2 brown. Sparse tiny mica; sparse tiny lime inclusions; sparse tiny to small voids.

Classical-Hellenistic.

## DSP08-H24-02

Standard samples: 20039/03-07, 20040/01
No diagnostic material.

## DSP08-H24-03

Standard samples: 20039/10

### Amphorai

*Unknown production*

**Cat. 575 20039/10/1**

Fr. of rounded protruding rim. H: 2.3; W: 4.1; T: 0.9-1.35; Ø: 11. Core: 2.5YR 7/6 light red. Surface: 2.5YR 5/6 red. Frequent tiny to small pyroxene; sparse tiny lime inclusions.

Similar to Monachov's rim type 2, cf. Monachov 1989, 70. Late 4th-3rd century BC.

## DSP08-H25-01

Standard samples: 20031/01
No diagnostic material.

## DSP08-H26-01

Grab samples: 3200

### Amphorai

*Unknown production*

**Cat. 576 3200/01/1**

Handle fr. H: 5.3; W: 3.8-3.9; T: 2.5. Clay: 10R 7/6 light red. Surface: 5YR 5/3 reddish brown. Frequent tiny to medium lime inclusions; medium tiny to small feldspar.

Classical-Hellenistic.

## DSP08-H26-02

Standard samples: 20001/01, 20029/01, 20030/01
No diagnostic material.

## Baj Kijat 1

Grab sample: 3009

### Handmade ware

**Cat. 577 3009/01/1**

Body fr. of vessel with outturned rim. H: 5.2; W: 3.8; T: 0.8-1.4. Core: 10YR 2/1 black. Surface: Gley 2 2.5/5B bluish black. Sparse sand; sparse tiny to small lime inclusions; very sparse tiny silver mica.

Late Bronze Age.

**Cat. 578 3009/02/1**

Rim fr. of bowl with flat rim with impressed decoration below the rim ext. H: 1.9; W: 2.4; T: 1.1; Ø? Core: 10YR 7/1 light grey. Surface: 10YR 6/4 light yellowish brown. Sparse sand; very sparse tiny lime inclusions.

Possibly similar to bowls from Kirovo, cf. Leskov 1970, fig. 9.6-7. Middle Bronze Age/Late Bronze Age.

**Cat. 579 3009/02/2**

Rim fr. of cup with projecting rim and round lip. H: 2.8; W: 3.8; T: 0.4-1.1; Ø: ca. 7. D. 60. Core: Gley 2 2.5/1 bluish grey. Surface: 10YR 6/4 light yellowish brown. Sparse sand; very sparse tiny lime inclusions; sparse dark inclusions.

Resembles cups from Kirovo, cf. Leskov 1970, fig. 15.1-7. Late Bronze Age.

**Cat. 580 3009/02/3**

Rim fr. of bowl with straight walls. H: 1.6; W: 3; T: 0.4; Ø: ca. 9. Core: 10R 4/1 dark reddish grey. Surface: 10YR 6/4 light yellowish brown. Sparse sand; very sparse tiny lime inclusions; sparse small dark inclusions.

Cf. Gerškovič 1999 type Sch 1I2, cf. Abb. 12.17. Late Bronze Age.

**Cat. 581 3009/02/4**

Body fr. of vessels with pot with ribbed ext. H: 2.1; W: 1.8; T: 0.8. Ext. surface: 10R 6/6 light red. Int. surface: Gley 2 2.5/1 bluish black. Sparse sand; sparse tiny lime inclusions; very sparse small quartz.

Middle Bronze Age/Late Bronze Age.

**Cat. 582 3009/02/5**

Rim fr. of cooking vessel with straight rim and rounded lip. H: 2.5; W: 1.7; T: 0.7; Ø? Clay: Gley 2 2.5/1 5B bluish black. Medium small lime inclusions; sparse sand.

Possibly belonging to Gerškovič 1999, 'Becher' type, cf. 54, abb. 12.5-8. Late Bronze Age.

**Cat. 583 3009/02/6**

Rim fr. of bowl with straight walls and rounded lip. Impressed decoration with incised horizontal furrows below. H: 2.7; W: 2.4; T: 1; Ø? Clay: Gley 2 2.5/1 5B bluish black. Medium small lime inclusions; sparse sand.

Middle Bronze Age/Late Bronze Age.

### Amphorai

*Unknown production*

**Cat. 584 3009/02/7**

Handle fr. H: 2.9; W: 3.1. Clay: close to 7.5YR 8/2 pinkish white. No visible inclusions.

Late Classical-Hellenistic.

## Baj Kijat 3

Grab sample: 3010

### Handmade ware

**Cat. 585 3010/1**

Rim fr. of pot with slightly outturned rim. H: 2.4; W: 2.9; T: 0.8; Ø: ca. 10-12. Core: 10R 2.5/1 reddish black. Surface: 7.5YR 5/3 brown. Sparse tiny lime inclusions; sparse sand; frequent very tiny silver mica.

Cf. Stolba 2002, type 1 Cat. D1-D68. Early Hellenistic.

**Cat. 586 3010/2**

Rim fr. of pot with outturned rim. H: 2.2; W: 2.6; T: 0.7; Ø?; D: <15. Core: 10R 2.5/1 reddish black. Surface: 7.5YR 5/3 brown. Sparse tiny lime inclusions; sparse sand.

Cf. Stolba 2002, type 1 Cat. D1-D68. Early Hellenistic.

**Cat. 587 3010/3**

Rim fr. of pot with straight outturned rim and slightly protruding lip. H: 3.9; W: 4.2; T: 0.9; Ø: 18; D: 28. Core: 10R 2.5/1 reddish black. Surface: 7.5YR 5/3 brown. Sparse tiny lime inclusions; sparse sand; frequent very tiny silver mica.

Cf. Stolba 2002, type 1 Cat. D1-D68. Early Hellenistic.

**Cat. 588 3010/4**

Rim fr. of pot with concave, outturned rim. H: 2.1; W: 3.5; T: 0.9; Ø: ca. 10; D: 20-25. Core: 10R 2.5/1 reddish black. Surface: 7.5YR 5/3 brown. Sparse tiny lime inclusions; sparse sand.

Cf. Stolba 2002, type 1 Cat. D1-D68. Early Hellenistic.

**Cat. 589 3010/5**

Rim fr. of pot with outturned rim. H: 3.1; W: 4.6; T: 0.9-1.1; Ø? Core: 7.5YR 2.5/1 black. Surface: 7.5YR 5/4 brown. Sparse medium lime inclusions; sparse tiny silver mica; sparse sand; medium dark inclusions.

Cf. Stolba 2002, type 1 Cat. D1-D68. Early Hellenistic.

## Burun Eli 2

Grab sample: 3011

### Tableware

**Cat. 590 3011/1**

Base fr. of jug or other closed shape with flat base. H: 4; T: 0.5-1.2 Ø: ca. 18-20. Clay: 2.5YR 5/8 red. Surface: close to 10R 6/8 light red. Tiny silver mica.

Late Classical-Hellenistic?

## Burun Eli 4

Grab sample: 3021
No diagnostic material.

## Burun Eli 6

Grab sample: 3022

### Handmade ware

**Cat. 591 3031/1**

Rim fr. of pot with concave, outturned rim. H: 4.8; W: 5; T: 0.6; Ø? Core: 7.5YR 5/3 brown. Surface: 10YR 3/1 very dark grey. Frequent shell; frequent small pebbles; medium tiny to medium quartz; sparse sand.

Cf. Stolba 2002, type 1 Cat. D1-D68. Early Hellenistic

**Cat. 592 3031/2**

Rim fr. of pot with very concave, outturned rim. H: 3.2; W: 4.7; T: 0.4-0.6; Ø: 17. Core: 5YR 4/1 dark grey. Surface: 5YR 5/8 yellowish red. Medium sand; medium small pebbles; sparse small shell; sparse small feldspar.

Cf. Stolba 2002, type 1 Cat. D1-D68. Early Hellenistic.

**Cat. 593 3031/3**

Base fr. of pot. H: 4.8; W: 5.1; T: 0.6-1.3; Ø? Clay: 5YR 6/4 light reddish brown. Int. core: 7.5YR 3/1 very dark grey. Frequent small and large pebbles; sparse small lime; sparse feldspar.

Early Hellenistic

**Cat. 594 3031/4**

Rim fr. of pot with slightly concave, outturned rim and flat lip. H: 4.3; W: 3.4; T: 0.7-0.9; Ø: >20. Hard fired, but friable fabric. Clay: 7.5YR 8/4 pink. Sparse tiny lime inclusions; sparse medium pebbles.

Cf. Stolba 2002, type 1 Cat. D1-D68. Hellenistic.

**Cat. 595 3031/5**

Rim fr. of pot with slightly concave, outturned rim and flat lip. H: 2.1; W: 3.1; T: 0.6; Ø: 12. Core: 2.5Y 3/1 very dark grey. Surface: 10YR 6/6 brownish yellow. Sparse tiny lime inclusions; medium sand; medium small pebbles; very sparse small quartz.

Cf. Stolba 2002, type 1 Cat. D1-D68. Early Hellenistic.

### Amphorai

#### *Chersonesos*

**Cat. 596 3022/1**

Handle fr. H: 8.4; W: 4.3. Clay: 2.5YR 5/8 red. Core: Gley 2 5/1 bluish grey. Slip: 2.YR 5/6 red. Medium tiny to small lime inclusions.
Hellenistic.

## Groty 1

Grab sample: 3031

### Amphorai

#### *Herakleia*

**Cat. 597 3031/7**

Handle fr. H: 3.7; W: 3.6. Clay: 10R 6/8 red. Slip: close to 10R 6/8 light red. Medium lime inclusions; sparse shells; very sparse quartz.
4th century BC.

**Cat. 598 3031/8**

Rim fr. with part of handle attachment. H: 3.1; W: 5.3; T: 0.9-1.2; Ø? Clay: 10R 6/8 red. Slip: close to 10R 6/8 light red. Medium lime; sparse shell; very sparse quartz; sparse tiny pyroxene.
Hellenistic.

#### *Unknown production*

**Cat. 599 3031/10**

Small rim fr. H: 1.7; W: 2.8; T: 1; Ø? Clay: 2.5YR 5/8 red. Medium small to tiny lime inclusions.
Classical-Hellenistic.

**Cat. 600 3031/11**

Rim fr. H: 2.5; W: 4.2; T: 1.1; Ø: 9. Clay: 10YR 4/3 brown. Medium sand; sparse tiny lime.
Hellenistic.

**Cat. 601 3031/12**

Rim fr. H: 3.9; W: 2.4; T: 1.5 Ø? Clay: 10YR 4/2 dark greyish brown. Sparse sand; sparse tiny lime; one large pebble; very sparse feldspar.
Hellenistic.

**Cat. 602 3031/6**

Handle fr. H: 8.3; W: 5.1. Clay: 5YR 4/6 yellowish red. Sparse lime inclusions; sparse shells; sparse feldspar.
Classical-Hellenistic.

**Cat. 603 3031/9**

Undercut rim fr. H: 2.4; W: 3.5; T: 1.2; Ø? Medium lime; sparse sand; sparse feldspar.
Hellenistic.

### Tableware

**Cat. 604 3031/13**

Rim fr. of jug with a projecting rim. H: 3.8; W: 2.4; T: 1.4; Ø? Clay: 5YR 4/1 dark grey. Sparse tiny lime inclusions; sparse sand; sparse tiny feldspar.
Late Classical-Hellenistic.

**Cat. 605 3031/14**

Rim fr. of jug with a projecting rim. H: 2.6; W: 3; T: 1.1; Ø? Clay: 2.5YR 5/8 red. Sparse tiny lime inclusions; frequent tiny silver mica; sparse sand.
Late Classical-Hellenistic.

**Cat. 606 3031/15**

Two joining frs. of jug with projecting rim. H: 3.9; W: 4; T: 0.3; Ø: 8. Clay: 10R 5/8 red. Slip: 10R 7/8 light red. Sparse tiny to medium lime inclusions.
Chersonesos, cf. Kašaev 2002, 151-152, Cat. C7-23. Late 4th-3rd century BC.

**Cat. 607 3031/16**

Two joining frs. of ring foot, presumably of bowl or cup. H: 1.7; W: 2.7; T: 0.7; Ø? Clay: 5YR 6/6 reddish yellow. Very sparse sand.
Late Classical-Early Hellenistic.

**Cat. 608 3031/17**

Wall fr. of small flask or jug. Two broad horizontal bands on shoulder. H: 2.6; W: 3; T: 0.5. Clay: 10R 4/1 dark reddish grey. Decoration: 10R 3/2 dusky red. Sparse very tiny lime inclusions; medium sand; sparse very tiny silver mica.
Late 4th-3rd century BC.

**Cat. 609 3031/18**

Two joining rim frs. of bowl. Three light concentric bands on int. H: 4.8; W: 8.6; Ø: 13. Core: Gley 2 4/10G dark greenish grey. Slip on both int. and ext. Sparse small to medium lime inclusions; sparse small to medium quartz; sparse small to medium feldspar; sparse small to medium pyroxene.
Hellenistic?

## Groty 2

Grab samples: 3032
No diagnostic material.

## Kurama Kostel'

Grab sample: 1003

### Architectural elements

**Cat. 610 1003/01/25**

Fr. of cover tile with scraped exterior. Pink sandy fabric with yellow interior surface with numerous lime and reaction rims, some fired reddish-brown.
Early Modern.

**Cat. 611 1003/01/22**

Corner fr. of cover tile? Fine orange sandy fabric with a few tiny voids Early Modern.

**Cat. 612 1003/01/27**

Fr. of cover tile rim. Gritty fabric with numerous tiny lime and reaction rims.

Early Modern.

## Amphorai

### *Herakleia*

**Cat. 613 1003/01/1**

Base fr. Slightly flaring with a cavity on underside. H: 7; W: 5.4; Ø? Clay: 5YR 6/4 light reddish brown. Medium tiny to small lime inclusions; sparse tiny feldspar.

4th century BC.

**Cat. 614 1003/01/4**

Fr. of the upper part of handle. H: 8.3; W: 6.3; T: 2.8. Clay: 7.5YR 7/6 reddish yellow. Medium tiny to medium pyroxene; medium tiny to small lime inclusions; medium tiny to small feldspar.

4th century BC.

### *Thasos*

**Cat. 615 1003/01/3**

Rim fr. H: 2.1; W: 3.9; T: 1; Ø: 9. Clay: 7.5YR 7/6 reddish yellow. Frequent very tiny to tiny mica; sparse small golden mica.

Conical-biconical type II-B 4th century BC.

## Tableware

**Cat. 616 1003/01/02**

Handle fr. of jug. H: 7.6; W: 2.7; T: 0.9; Ø? Clay: 2.5YR 5/8 red. Slip: 2.5YR 6/6 light red. Very sparse tiny lime.

## Post-Antique pottery

**Cat. 617 1003/01/02b**

Handle fr. of jug. H: 7.6; W: 2.7; T: 0.9; Ø? Clay: 2.5YR 5/8 red. Surface: 2.5YR 6/6 light red. Very sparse tiny lime.

Early Modern.

**Cat. 618 1003/01/05**

Fr. of spout of jug glazed on the exterior. H: 6.4; W: 8.1; T: 0.8; Ø? Clay: 5YR 5/8 yellowish red. Glaze: pale green. Sparse tiny to small lime.

Cf. Bakirtzis 1980, fig. 10 right; Hayes 1992, 342-342, pl. 46.f-j, fig. 106; Zographou-Korre 2000, 348 no. B94a. 18th-20th century AD

**Cat. 619 1003/01/06**

Fr. of interior glazed ring-base. H: 1.6; W: 1.6; T: 0.7; Ø? Clay: 5YR 5/8 yellowish red. Glaze: strong green. Very sparse very tiny lime; very sparse tiny feldspar.

Early Modern.

**Cat. 620 1003/01/07**

Fr. of interior and exterior glazed rim of jug or jar. H: 2.7; W: 2.2; T: 0.5; Ø? Clay: 5YR 5/8 yellowish red. Glaze: strong green. Very sparse tiny lime.

Early Modern.

**Cat. 621 1003/01/08**

Fr. of slightly downturned rim with glazed exterior lip. Jug? Two grooves below exterior rim. H: 3; W: 4.8; T: 0.7; Ø: 20. Clay: 7.5YR 7/6 reddish yellow. Glaze: clear. Sparse sand; very sparse very tiny to tiny lime; very sparse tiny feldspar.

Early Modern.

**Cat. 622 1003/01/09**

Fr. of shoulder of jug with two exterior horizontal bands of combing. H: 5.7; W: 6.5; T: 0.9; Ø? Clay: 2.5YR 5/6 red. Medium very tiny to tiny lime.

Cf. Bakirtzis 1980, fig. 10 right; Hayes 1992, 342-342, pl. 46.f-j, fig. 106. 18th-20th century AD

**Cat. 623 1003/01/10**

Fr. of cooking pot with everted plain rim with sooted exterior. H: 2.8; W: 4.9; T: 0.7; Ø: 16. Core: 7.5YR 7/4 pink. Surface: 10YR 5/2 greyish brown. Sparse very tiny lime.

Early Modern.

**Cat. 624 1003/01/11**

Two frs. of deep bowl with inverted rim and rounded lip. H: 2.8; W: 5.2; T: 1; Ø: 20. Clay: 5YR 6/8 reddish yellow. Sparse very tiny lime; one large grog.

Early Modern.

**Cat. 625 1003/01/12**

Fr. of flat base of utility vessel. H: 2.3; W: 3.9; T: 1.1; Ø: 10. Clay: 5YR 5/8 yellowish red. Very sparse very tiny light-reflecting particles; very sparse tiny lime.

Early Modern.

**Cat. 626 1003/01/13**

Fr. of slightly everted, high overfolded rim of jug. H: 4.5; W: 6.1; T: 0.5; Ø: 6. Clay: 5YR 6/8 reddish yellow. Very sparse very tiny lime

Cf. Bakirtzis 1980, fig. 10 right; Hayes 1992, 342-342, pl. 46.f-j, fig. 106. 18th-20th century AD

**Cat. 627 1003/01/14**

Fr. of everted rim of jug with rounded lip and interior ledge. H: 3; W: 4.4; T: 0.5; Ø: 7. Clay: 5YR 6/8 reddish yellow. Surface: 2.5YR 5/4 reddish brown. Tiny lime.

Cf. Bakirtzis 1980, fig. 10 right; Hayes 1992, 342-342, pl. 46.f-j, fig. 106. 18th-20th century AD

**Cat. 628 1003/01/15**

Fr. of thickened, exterior offset, angular rim of jug. H: 3.2; W: 3.3; T: 0.5; Ø: 6. Clay: 2.5YR 5/8 red. Medium sand; sparse tiny lime.

Cf. Bakirtzis 1980, fig. 10 right; Hayes 1992, 342-342, pl. 46.f-j, fig. 106. 18th-20th century AD

**Cat. 629 1003/01/16**

Fr. of flat base of utility vessel. H: 4.7; W: 11.2; T: 0.7-0.9; Ø: 12. Clay: 2.5YR 6/8 light red. Surface: 10R 4/4 weak red. Very sparse very tiny to tiny lime.

Early Modern.

**Cat. 630 1003/01/17**

Fr. of ring base of open vessel. Sparse traces of interior glaze preserved. H: 2.3; W: 8.1; T: 1.1; Ø: 8. Clay: 5YR 6/6 reddish yellow. Glaze: strong green. Sparse very tiny to tiny lime.

Early Modern.

**Cat. 631 1003/01/18**

Fr. of everted rim of open vessel, possibly basin. H: 2.8; W: 6.9; T: 1.2; Ø: 28. Clay: 2.5YR 5/8 red. Very sparse very tiny lime.

Cf. Hayes 1992, 342-343, pl. 46.a-d. 18th-20th century AD

**Cat. 632 1003/01/19**

Fr. of flat base of utility vessel. H: 2; W: 6.4; T: 0.9-1; Ø: 12. Clay: 10YR 4/2 dark greyish brown. Sparse sand; sparse very tiny to tiny lime.

Early Modern.

**Cat. 633 1003/01/20**

Fr. of flat base of utility vessel H: 5.6; W: 9.4; T: 1.3-2; Ø: 11. Clay: 2.5YR 6/6 light red. Tiny lime.

Early Modern.

**Cat. 634 1003/01/21**

Fr. head of pipe with a thin horizontal relief line on the exterior body. Lightly sooted inside. H: 3.8; W: 2.8; T: 0.3 Body; Ø: 2.5. Clay: 5YR 5/6 yellowish red. Very sparse very tiny lime.

Early Modern.

**Cat. 635 1003/01/23**

Plain rim fr. of jar? Ø: 17. Fine orange sandy fabric with a few tiny voids

Early Modern.

**Cat. 636 1003/01/24**

Fr. of band-handle. Very fine, powdery, orange, sandy fabric with some longish voids and tiny to small red inclusions.

Early Modern.

**Cat. 637 1003/01/26**

Fr. of ribbed shoulder of closed vessel. Fine sandy, exterior self-slipped fabric with common tiny to small lime and voids.

Cf. Hayes 1992, fig. 124: j1.2-3. End 15th to mid-16th century AD

**Cat. 638 1003/01/28**

Two fr. of closed vessel decorated with horizontal combing. Clay: 2.5YR 6/6 light red. Fine sandy fabric with many tiny to large pink inclusions, possible lime.

Cf. Bakirtzis 1980, fig. 10 right; Hayes 1992, 342-342, pl. 46.f-j, fig. 106. 18th-20th century AD

**Cat. 639 1003/01/29**

Fr. of closed vessel decorated with horizontal combing and exterior glaze. 2.5YR 6/8 light red. Fine sandy fabric with tiny lime.

Cf. Zographou-Korre 2000, 348 no. B94a. 18th-20th century AD

# Masliny 1-2

Grab samples: 3007

### Handmade ware

**Cat. 640 3007/2**

Shoulder fr. of pot. H: 3.3; W: 4.6; T: 0.8. Core: 7.5YR 5/1 grey. Surface: 7.5YR 6/4 light brown. Sparse small to large lime inclusions; sparse silver mica; sparse medium quartz; sparse medium. Organic inclusions.

Late Classical-Hellenistic.

### Utility ware

**Cat. 641 3007/1**

Joining rim and body fr. of louterion with outturned rim. H: 7.7; W: 6.9; T: 0.6-1.3; Ø: ca. 18. Clay: 2.5YR 6/4 light yellowish brown. Slip: as clay. Sparse small to large lime inclusions; very sparse pyroxene; medium medium pebbles.

Cf. Kašaev 2002, Cat. C215-C220. Early Hellenistic.

# Masliny 2

Grab sample: 3008
No diagnostic material.

# Panskoe II/DSP06-F03-01

Grab sample: 3002

### Handmade ware

**Cat. 642 3002/9**

Rim fr. of pot with straight neck. Oblique cut decoration on the lip. H: 2.9; W: 3.5; T: 1; Ø? Core: 10YR 6/2 light brownish grey. Surface: 7.5YR 6/6 reddish yellow. Medium very tiny to tiny lime inclusions.

Cf. Stolba 2002, type 1 Cat. D1-D68. Early Hellenistic.

**Cat. 643 3002/13**

Fr. of cup with part of the handle preserved. H: 3.3; W: 3.5; T: 1-1.6. Core: 5YR 5/3 reddish brown. Ext. surface: 10R 5/4 weak red and Gley 1 3/N very dark grey. Sparse tiny lime inclusions; sparse small to medium quartz; medium tiny to small mica.

Cf. Senatorov 2003, fig. 4.18. Late Classical-Hellenistic.

### Amphorai

*Chersonesos*

**Cat. 644 3002/10**

Base fr. with cavity on underside. H: 4.3; W: 3.8; Ø? Clay: 2.5YR 6/6 light red. Slip: 7.5YR 7/6 reddish yellow. Very sparse small to medium pyroxene; frequent very tiny to small lime; sparse tiny to small feldspar.

Early Hellenistic.

### *Sinope*

**Cat. 645 3002/14**

Base fr. H: 5.1; W: 4.4; Ø? Core: 10YR 7/3 very pale brown. Slip: 2.5YR 7/3 pale yellow. Medium small to medium pyroxene; sparse tiny to small lime inclusions.

3rd-Early 2nd century BC.

**Cat. 646 3002/17**

Body fr. with part of the handle preserved. H: 4.4; W: 8.2; T: 1.2. Clay: 2.5YR 5/6 red. Slip: 10YR 7/4 very pale brown. Medium tiny to small lime inclusions; sparse small to medium pyroxene; sparse tiny feldspar.

Sinope? Classical-Hellenistic.

### *Mende*

**Cat. 647 3002/2**

Base fr. with slightly flaring stem and angular transition to base with convex underside. H: 3.7; W: 6.1; Ø: 9.2. Clay: 2.5YR 5/4 reddish brown. Sparse sand; medium very tiny to tiny lime inclusions; very sparse tiny feldspar.

Type II-C, cf. Monachov 2003, pl. 63.4-6, 64-65. Ca. 375-325 BC.

### *Unknown production*

**Cat. 648 3002/1**

Handle fr. H: 4.9; W: 3.7; T: 2.5. Clay: 5YR 6/4 light reddish brown. Slip: 10YR 7/4 very pale brown. Sparse tiny to medium lime inclusions; sparse tiny feldspar.

Classical-Hellenistic.

**Cat. 649 3002/12**

Handle fr. H: 5.6; W: 3.4; T: 2.4. Clay: 2.5YR 5/8 red. Slip: 10YR 6/4 light yellowish brown. Sparse very tiny to tiny lime inclusions; sparse tiny feldspar.

Classical-Hellenistic.

**Cat. 650 3002/15**

Base fr. H: 5.6; W: 5.7. Clay: 10R 5/6 red. Slip: 2.5YR 7/6 light red. Sparse sand; sparse tiny lime inclusions; sparse tiny feldspar.

Classical-Hellenistic.

**Cat. 651 3002/16**

Handle fr. H: 5.1; W: 3.6; T: 2.2. Clay: 7.5YR 6/4 light brown. Slip: 2.5Y 7/3 pale yellow. Very sparse very tiny quartz; medium very tiny to tiny lime inclusions. Medium tiny feldspar.

Classical-Hellenistic.

**Cat. 652 3002/3**

Handle fr. H: 8.7; W: 4.3; T: 2. Clay: 2.5YR 5/8 red. Slip: 10R 6/4 pale red. Sparse tiny lime inclusions; frequent tiny to small feldspar.

Classical-Hellenistic.

**Cat. 653 3002/4**

Body fr. with an incised lambda. H: 2.9; W: 4.5; T: 0.8. Clay: 5YR 6/4 light reddish brown. Slip: 10YR 7/4 very pale brown. Very sparse tiny lime inclusions; very sparse tiny quartz.

Classical-Hellenistic.

**Cat. 654 3002/6**

Body fr. Repair hole. H: 4; W: 3.2; T: 1. Clay: 5YR 6/6 reddish yellow. Sparse sand; sparse medium to large quartz.

Classical-Hellenistic.

**Cat. 655 3002/8**

Fr. of double-barreled handle. H: 6.3; W: 4.2; T: 1.9. Clay: 2.5Y 8/2 pale yellow. Sparse tiny to small pebbles; sparse tiny feldspar; sparse sand.

Hellenistic?

## Tableware

**Cat. 656 3002/11**

Wall fr. of closed vessel with two horizontal bands on the ext. H: 2.4; W: 2.4; T: 0.4. Clay: 2.5YR 5/8 red. Decoration: 2.5YR 4/4 reddish brown. Sparse very tiny lime inclusions; medium tiny feldspar.

Possibly bulbous Chersonesean jug, cf. Zolotarev 2005, fig. 9. Ca. 300-225 BC.

**Cat. 657 3002/5**

Handle fr. of jug. H: 1.8; W: 2.1; T: 0.9. Clay: 7.5YR 6/6 reddish yellow. Sparse tiny lime inclusions; sparse tiny pyroxene; sparse very tiny to tiny feldspar.

Late Classical-Hellenistic?

**Cat. 658 3002/7**

Base fr. with moulded foot. H: 2.6; W: 6.4; T: 1-1.7; Ø: 9.5. Clay: 2.5YR 6/6 light red. Slip: 7.5YR 6/6 reddish yellow. Frequent tiny to large lime inclusions; medium tiny to small feldspar; medium tiny to small pyroxene.

Chersonesos, cf. Kašaev 2002, 151-152, Cat. C7-23. Late 4th-3rd century BC.

# Panskoe V/DSP07-F01-02

Core units: 1059-61, 1064, 1068-70, 2046, 2057-58, 2062

## Amphorai

### *Herakleia*

**Cat. 659 1059/01/1**

Handle fr. H: 9.7; W: 4.1; T: 2.5; Ø? Clay: 10YR 6/4 light yellowish brown. Frequent tiny to small pyroxene; medium tiny to small lime; sparse small quartz.

4th century BC.

## Handmade Ware

**Cat. 660 1060/01/1**

Straight rim fr. of pot. H: 1.8; W: 2.1; T: 1.2; Ø? Surface: clay: 7.5YR 5/3 brown. Core: 7.5YR 2.5/1 black. Very sparse silver mica. Sparse sand; very sparse tiny lime.

Hellenistic.

**Cat. 661 1068/01/2**

Rim fr. of open vessel with slightly flaring walls and a projecting lip. H: 2.4; W: 4; T: 1.1; Ø? Clay: 7.5YR 4/1 dark grey. Surface: 10YR 7/2 light grey. Medium small to tiny lime inclusions; sparse sand.

Somewhat similar to a handmade bowl from a grave on the Kerč Peninsula, cf. Maslenikov 1995, fig. 20.2. 3rd-2nd century BC?

### Tableware

**Cat. 662 2055/01/1**

Fr. of thin ring foot, possibly from a stemless cup. Glazed on both int. and ext. of the ring foot. H: 1.2; W: 2.9; T: 0.3; Ø: 9. Clay: 2.5YR 5/8 red. Glaze: 10YR 2/1 black. Sparse tiny silver mica.

Ca. 500-450 BC.

## Severnoe

Grab samples: 3006

### Tableware

**Cat. 663 3006/1**

Rim fr. of jug with a slightly flaring mouth. H: 2.1; T: 0.5; Ø: ca. 6. Clay: 10R 6/8 red. Sparse lime particle; sparse pyroxene or feldspar.

Chersonesos, cf. Kašaev 2002, 151-152, Cat. C7-23. Late 4th-3rd century BC.

## Skalistoe 2/DSP08-F03-02

Core units: 6226, 6244-45. 6247
Adjacent units: 6227, 6243, 6246

### Handmade ware

#### *Bronze Age*

**Cat. 664 6244/02/7**

Wall fr. from pot with grooves on int. and ext. H: 5.3; W: 2; T: 0.7. Core: 10YR 4/3 brown to Gley 2 3/5 B very dark bluish grey. Medium tiny to small lime inclusions; medium sand; single medium stone.

Cf. Toščev 2007, fig. 82.18 (Planerskoe I). Middle Bronze Age.

**Cat. 665 6246/02/5**

Wall fr. of vessel with combed decoration. H: 5.2; W: 6.7; T: 0.6-0.7. Core: Gley 2 3/10 BG very dark greenish grey. Surface: 10YR 3/1 very dark grey. Sparse tiny to small lime inclusions; single large lime particle; sparse sand.

Cf. Toščev 2007, fig. 82.18 (Planerskoe) Middle Bronze Age.

**Cat. 666 6247/02/7**

Wall fr. of vessel with combed decoration. H: 3.46; W: 5.9; T: 0.94. Core: Gley 1 3/N black. Medium to large shells and quartz.

Cf. Toščev 2007, fig. 82.18 (Planerskoe I). Middle Bronze Age.

**Cat. 667 6243/01/1**

Body fr. of cooking vessel. Moulding below lip with obliqely impressed decorations. Incised furrow between rim and moulding. H: 3.6; W: 3.15; T: 0.8. Core: Gley 2 3/10 BG very dark greenish grey. Int. surface: Gley 2 5/5B bluish grey. Ext. surface: 5YR 7/4 pink. Tiny lime inclusions; medium sand; sparse small quartz.

Similar type is found in Kirovo, cf. Leskov 1970, fig. 10.4. Middle Bronze Age/Late Bronze Age.

**Cat. 668 6243/02/1**

Wall fr. of vessel with four horizontal furrows (width 0.1, depth 0.07). H: 2.6; W: 2.26; T: 0.9. Core: Gley 1 4/N dark grey. Int. surface: 7.5YR 7/6 reddish yellow. Ext. surface: 10YR 7/4. Very sparse tiny shells.

Cf. Rybalova 1974, fig. 14.2,4 (Kamenka). Middle Bronze Age/Late Bronze Age.

**Cat. 669 6244/01/6**

Wall fr. of vessel with two broad furrows (width 0.2, deepth 0.1) and a row of impressed cord decoration. H: 2.1; W: 2.4; T: 0.88. Core: Gley 1 5/N grey. Surface: 10YR 8/2 very pale brown. Sparse small shells and quartz.

Middle Bronze Age/Late Bronze Age.

**Cat. 670 6244/02/6**

Wall fr. of pot with deep grooves on the ext. Probably from same vessels as 6244/02/8. H: 2.7; W: 2.2; T: 1.1. Core: Gley 2 3/5B very dark bluish grey. Int. surface: Gley 2 5/10 BG bluish grey. Ext. surface: Sparse tiny lime inclusions; sparse small to medium quartz.

Cf. Rybalova 1974, fig. 14.2,4 (Kamenka). Middle Bronze Age/Late Bronze Age.

**Cat. 671 6244/02/8**

Wall fr. of vessel with incised horizontal furrows on the ext. Probably from same vessels as 6244/02/6. H: 2.6; W: 2.2; T: 1. Core: Gley 2 3/5 B very dark bluish grey. Int. surface: Gley 2 5/10 BG bluish grey. Ext. surface: 7.5YR 8/3 pink. Sparse tiny lime inclusions; sparse small to medium quartz.

Cf. Rybalova 1974, fig. 14.2,4 (Kamenka). Middle Bronze Age/Late Bronze Age.

**Cat. 672 6246/02/1**

Rim fr. of thick-walled vessel with straight rim and flat lip. Two deep horizontal furrows on the ext. below rim. H: 3.4; W: 4.6; T: 1.6; Ø: 26-28. Core: 10R 4/2 dusky red to Gley 2 3/5 B. Int. surface: 10R 4/1 dark red. Ext. surface: 2.5YR 7/4. Medium tiny to small lime inclusions; medium tiny feldspar; sparse sand; sparse small voids.

Cf. Leskov 1970, fig. 3.5-6 (Kirovo). Middle Bronze Age/Late Bronze Age.

**Cat. 673 6247/01/2**

Wall fr. of vessel with broad grooves on the ext. H: 2.38; W: 2.47; T: 0.85. Core: Gley 1 4/N dark grey. Surface: 10YR 8/3 very pale brown. Sparse tiny quartz and shells.

Cf. Rybalova 1974, fig. 14.2,4 (Kamenka). Middle Bronze Age/Late Bronze Age

**Cat. 674 6244/01/01**

Rim fr. of cooking vessel with horizontal moulding with oblique cuts (width 0.95; height 0.65). H: 4.59; W: 7.37; T: 0.83-1.14. Core: Gley 1 4/N dark grey. Surface: Gley 1 6/N grey. Sparse small quartz; sparse small shells.

Cf. e.g. Kolotuchin 2003, fig. 49.3,8 (Burun Eli). Late Bronze Age.

**Cat. 675 6244/01/2**

Rim fr. of cooking vessel with horizontal moulding with oblique cuts (width 1.05; elevation 0.35). H: 4.36; W: 5.5; T: 0.83-1.14. Core: Gley 1 4/N dark grey. Surface: 7.5YR 8/6 reddish yellow and 7.5YR 5/2 brown. Sparse small quartz; sparse small shells.

Cf. e.g. Kolotuchin 2003, fig. 49.3,8 (Burun Eli). Late Bronze Age.

**Cat. 676 6244/01/3**

Wall fr. of vessel with two mouldings with impressed finger decoration. H: 3.44; W: 2.94; T: 1.1. Core: Gley 1 4/N dark grey. Surface: 10YR 7/2 light grey. Sparse small shells.

Cf. Kolotuchin 2003, fig. 58.6 (Petrovskaja Balka). Late Bronze Age.

**Cat. 677 6244/01/7**

Rim fr. of vessel with slightly outcurved rim and rounded lip. H: 1.85; W: 2.04; T: 0.52-0.66. Clay: Gley 1 2.5/N black. Sparse small shells.

Somewhat similar to Kolotuchin 2003, fig. 58.9 (Družnoe). Late Bronze Age.

**Cat. 678 6244/02/3**

Rim fr. of vessel with thickned rim. Impressed oblique strokes on the upper part of the lip. H: 2.6; W: 3.75; T: 0.9-1.3; Ø: ca. 16. Core: Gley 2 4/5 BG dark greenish grey. Outer edge of core and surface: Gley 2 8/5 BG light greenish grey. Sparse tiny lime.

Similar rims are found among Sabatinovka ceramic assemblages, cf. e.g. Gerškovič 1999, pl. 16.7, but could also be from a frying pan, cf. Gerškovič 1999, pl. 8.27. Late Bronze Age.

**Cat. 679 6246/02/2**

Rim fr. of cooking vessel with slightly flaring rim and flat lip. Upper part of lip is decorated with obligue cuts. Few grooves on the exterior below the lip. H: 2.9; W: 2.9; T: 0.9-1.1; Ø: 20. Core and ext. surface: 10R 4/1 dark reddish grey. Int. surface: Gley 2 3/10 BG very dark greenish grey. Sparse tiny lime inclusions; frequent tiny feldspar.

Cf. Gerškovič 1999 type T2I2 fig. 11.14. Late Bronze Age.

**Cat. 680 6247/01/1**

Rim fr. of cooking vessel with vertical walls. H: 3; W: 2.58; T: 0.65-0.9; Ø? Core: Gley 1 2.5/N black. Surface: Gley 1 3/N very dark grey to Gley 2 4/10 B dark bluish grey. Sparse tiny quartz and shells.

Cf. Gerškovič 1999, type T2I2 fig. 11.14. Late Bronze Age.

**Cat. 681 6247/01/3**

Wall fr. of cup with part of handle attachment preserved. H: 2.77; W: 2.7; T: 0.85 (wall). Core: Gley 1 3/N very dark grey. Surface: 5YR 7/6 reddish yellow. Sparse tiny lime inclusions.

Cf. Gerškovič 1999, Type H2, fig. 12.11-14. Late Bronze Age.

**Cat. 682 6247/02/1**

Rim fr. of cooking vessel with slightly flaring rim. H: 3.67; W: 5.65; T: 0.89-1.02; Ø: 20? Core: 7.5YR 2.5/1 black. Surface: 7.5YR 5/2 brown to 10YR 7/4 very pale brown. Frequent medium and large shells; frequent medium lime inclusions; frequent medium quartz.

Cf. Gerškovič 1999, type T2I2 fig. 11.14. Late Bronze Age.

**Cat. 683 6247/02/2**

Rim fr. of storage vessel with concave rim. H: 3.37; W: 4.20; T: 1.02-1.11; Ø: ca. 18. Core: Gley 1 3/N black. Surface: 10YR 8/2 very pale brown. Medium to large shells; medium quartz.

Cf. Gerškovič 1999, type VT 1I2, cf. Fig. 13.2. Late Bronze Age.

**Cat. 684 6247/02/3**

Rim fr. of frying pan? with thickened rim and oblique cuts on lip. Impressed decoration on the ext. below rim. H: 1.81; W: 2.91; T: 0.97-1.44; Ø: 23. Core: Gley 1 3/N black. Medium shells and quartz.

Similar rims are found among Sabatinovka ceramic assemblages, cf. e.g. Gerškovič 1999, pl. 16.7, but could also be from a frying pan, cf. Gerškovič 1999, pl. 8.27. Late Bronze Age.

**Cat. 685 6247/02/4**

Rim fr. of cooking vessel with flaring rim. H: 2.74; W: 3.41; T: 0.6-1.45 Ø? Core: Gley 2 4/5 PB dark bluish grey. Surface: 5YR 7/6 reddish yellow and 5YR 5/1 grey. Medium lime inclusions; medium to large shells; medium quartz; frequent medium to large oval voids.

Cf. Gerškovič 1999, type T1I2 or T2II2, fig. 11.8,10. Late Bronze Age.

**Cat. 686 6247/02/5**

Rim fr. of cooking vessel with flaring rim. Horizontal moulding with triangular impressions. H: 2.19; W: 3.04; T: 0.86-1.09. Core: Gley 1 2.5/N black. Surface: 10YR 8/4 very pale brown and 10YR 5/1 grey. Medium to large shells and quartz.

Cf. Gerškovič 1999, type T1I1 or T2II1, fig. 11.7,9. Late Bronze Age.

**Cat. 687 6247/02/6**

Base fr. H: 1.85; W: 2.53; T: 0.6-0.74. Core: Gley 1 3/N black. Surface: 10YR 8/2 very pale brown. Medium to large feldspar and quartz.

The flat base is typical of most types in the Sabatinovka ceramic assemblages. Late Bronze Age?

## *Hellenistic*

**Cat. 688 6243/05/3**

Rim fr. of small bowl with projecting rim. H: 2.2; W: 2.2; T: 0.7; Ø: 7. Clay: 10R 7/8 light red. Sparse tiny to small lime inclusions; medium small to medium quartz minerlas.

Somewhat similar to a handmade bowl from a grave on the Kerč Peninsula, cf. Maslenikov 1995, fig. 20.2. 3rd-2nd century BC?

**Cat. 689 6243/02/2**

Base fr. of cup with rounded base. H: 3.12; W: 2.85; T: 0.65 (wall), 1.02 (bottom). Core: Gley 1 3/N very dark grey. Int. surface: Gley 1 5/N grey. Ext. surface: 10YR 7/1 light grey and 10YR 8/4 very pale brown. Sparse tiny shells.

Possibly similar to a cup from Panskoe I U6, cf. Stolba 2002, Cat. D111. Early Hellenistic.

**Cat. 690 6244/01/4**

Rim fr. of pot with slightly concave, outturned rim. H: 2.71; W: 3.03; T: 0.83-1.16; Ø: ca. 24. Core: Gley 2 3/10 G very dark greenish grey. Surface: Gley 1 5/10Y greenish grey. Medium small to medium shells; single large shell; sparse tiny oval and round voids.

Cf. Stolba 2002, type 1 Cat. D1-D68. Early Hellenistic.

**Cat. 691 6244/02/1**

Rim fr. of pot with concave, outturned rim. Impressed finger decoration on the lip. H: 4.1; W: 5; T: 1.1; Ø: ca. 26. Sandwich fabric. Core: 10R 5/1 reddish grey. Edges: 10YR 6/4 light yellowish brown. Frequent small to large shells; sparse small quartz; medium tiny feldspar.

Cf. Stolba 2002, type 1 Cat. D1-D68. Early Hellenistic.

**Cat. 692 6243/04/1**

Rim fr. of vessel with tall straight rim and rounded lip. Irregularly fired. H: 4.6; W: 3.8; T: 0.8; Ø? Core: 7.5YR 2.5/1 black. Int. surface: 10R 6/8 light red. Ext. surface: 10R 6/6 red, 5YR 6/4 light red brown to Gley 2 6/5 B bluish grey. Slight traces of slip on ext.: 5YR 5/4 reddish brown. Sparse tiny to small lime inclusions; sparse tiny light-reflecting particles; sparse tiny feldspar.

Early Hellenistic.

**Cat. 693 6244/01/5**

Rim fr. of pot with outturned rim and beveled lip. H: 2.9; W: 3.4; T: 0.6-0.93; Ø? Clay: 7.5YR 7/6 reddish yellow. Medium small to large shells.

Hellenistic.

**Cat. 694 6244/01/8**

Rim fr. of pot with flat slightly undercut lip. H: 2.2; W: 2.4; T: 0.9-0.94; Ø: 19. Core: Gley 1 3/N very dark grey. Surface: 7.5YR 7/3 pink to 7.5YR 5/1 grey. Sparse small shells; sparse small quartz.

Late Classical-Hellenistic.

**Cat. 695** 6244/02/2

Small rim fr. of pot with rounded rim. H: 1.7; W: 2.4; T: 1; Ø: ca. 24. Core: Gley 2 4/5 BG dark greenish grey. Surface:10YR 8/ 2 very pale brown. Sparse tiny lime inclusions; sparse tiny feldspar.

Late Classical-Hellenistic.

**Cat. 696** 6244/02/4

Rim fr. of pot with straight walls and projecting lip. Oblique cut decoration on upper part of the lip. H: 1.45; W: 1.6; T: 0.9; Ø: ca. 19. Core: 10YR 4/1 dark grey. Surface: 2.5Y 7/3 pale yellow. Sparse small lime inclusions; sparse tiny feldspar.

Somewhat similar to a handmade bowl from a grave on the Kerč Peninsula, cf. Maslenikov 1995, fig. 20.2. 3rd-2nd century BC?

**Cat. 697** 6244/02/5

Base fr. of pot with widely flaring walls. Slightly polished on ext. H: 2.3; W: 5; T: 0.7-1.2; Ø: ca. 15. Core: 2.5YR 2.5/1 black. Surface: 10YR 6/3 pale brown. Single tiny mica; sparse tiny to small quartz; medium tiny feldspar.

Late Classical-Hellenistic.

**Cat. 698** 6244/02/9

Rim fr. of pot with concave, outturned rim. H: 2.4; W: 2.1; T: 0.9; Ø: ca. 20. Sandwich fabric: core: 10YR 5/1 grey. Edges: 2.5Y 7/3 pale yellow. Sparse tiny to small lime inclusions; sparse tiny quartz.

Cf. Stolba 2002, type 1 Cat. D1-D68. Early Hellenistic.

**Cat. 699** 6247/02/08

Fr. of handmade handle with smoothed surface. H: 3.46; W: 2.98; T: 0.65. Core: 10R 5/1 reddish grey. Outside surface: 5YR 7/8 reddish yellow. Medium medium shells; quartz.

## Amphorai

### *Herakleia*

**Cat. 700** 6246/01/3

Fr. of outturned, rounded rim. Incised groove at junction between neck and rim. H: 6.4; W: 4.4; T: 0-65-0.95; Ø: 10. Core: 5YR 6/8 reddish yellow. Slip: 10R 5/4 weak red. Medium tiny to small lime inclusions; sparse small to medium quartz; sparse tiny mica; sparse tiny to small feldspar.

4th century BC.

**Cat. 701** 6246/01/4

Fr. of toe. Deep conical cavity on underside. H: 3.55; T: 1.5-1.6; Ø: 4.95. Clay: 10R 6/8 light red. Medium tiny to small lime; frequent tiny feldspar; sparse small quartz; sparse small to medium pyroxene.

4th century BC.

### *Sinope*

**Cat. 702** 6246/01/1a

Handle fr. H: 8.4; W: 4.5-4.7; T: 3-3.1. Clay: 10R 7/8 light red. Frequent tiny to medium pyroxene; sparse small red inclusions; single tiny lime particle; single medium quartz.

Late Classical-Early Hellenistic.

### *Chios?*

**Cat. 703** 6243/05/2

Wall fr. of lower part. Thin horizontal band. H: 4.6; W: 5.65; T: 1.6-1.7. Core: 10R 6/6 light red. Slip: 2.5YR 5/6 red. Band: 10R 5/5 red. Single tiny mica; sparse tiny lime; single medium quartz; sparse tiny to small feldspar.

Classical.

### *'Samian'*

**Cat. 704** 6246/01/1b

Rim fr. H: 3.5; W: 4.4; T: 0.6-1.5; Ø: 14 D. 30. Core: 7.5YR 6/3 light brown. Slip: 7.5YR6/4 light brown. Frequent tiny mica; sparse medium to large lime inclusions; sparse tiny feldspar.

Zeest, 'Samian' type, variant. Cf. Cook & Dupont 1998 fig. 23.12b. Late Archaic

### *Unknown production*

**Cat. 705** 6226/02/1

Handle fr. H: 6.1; W: 4.22-4.47; T: 2.7. Clay: close to 5YR 5/8 red. Single small pyroxene inclusion; sparse tiny shells; medium tiny to small mica; sparse tiny to small feldspar.

Classical-Hellenistic.

**Cat. 706** 6243/05/1

Fr. of toe. H: 4.3; W: 2.5; T: 1.8. Core: 10R 5/6 red. Surface: 10R 6/6 light red. Sparse tiny to medium lime; sparse tiny feldspar.

Hellenistic.

**Cat. 707** 6246/01/2

Rim fr. with protruding lip. H: 4.4; W: 2.6; T: 0.8-1.2; Ø: 11. Clay: 2.5YR 5/8 red. Surface: 2.5YR 6/8 light red. Medium tiny lime inclusions; sparse tiny to small pyroxene; sparse small red inclusions.

Classical-Hellenistic.

## Tableware

**Cat. 708** 6226/02/2

Wall fr. of jug with painted horizontal band (width: 0.64). H: 3.46; W: 2.41; T: 0.47. Core: 7.5YR 6/2 pinkish grey. Surface: 7.5YR 5/4 brown. Decoration: 5YR 2.5/2 dark reddish brown. Medium tiny to medium lime inclusions; frequent tiny to small light-reflecting particles; frequent small voids on the int.

Chersonesos, cf., e.g., Zolotarev 2005, fig. 9-12. Late 4th-3rd century BC.

# Skalistoe 5[105]

Grab samples: 3026 and 3027/2007 FL 10

[105] Descriptions of the excavation finds in the finds catalogue by S. Handberg (Appendix 2B) are based on autosy. In some cases there is not complete congruency with the findslists recorded by Lancov and Stolba for the annual reports of the Department of Antiquities.

## Tableware

**Cat. 709 3026/1**

Rim fr. of bowl with inturned rim. H: 1.9; W: 6.7; T: 1; Ø: 23. Clay: 2.5YR 5/6 red. Slip: 7.5YR 6/6 reddish yellow. Tiny to small lime particle; sparse small pyroxene; sparse small feldspar; very sparse small quartz.

Cf. e.g. Kašaev 2002, Cat. C39-C51. Late 4th-early 3rd century BC.

**Cat. 710 3026/2**

Rim fr. of jug? with moulded rim. H: 2.1; W: 2.2; T: 0.6; Ø: 10; D: 25. Clay: 2.5YR 6/6 light red. Slip: 2.5Y 7/3 pale yellow. Sparse very tiny to tiny lime inclusions; sparse tiny feldspar.

Late Classical-Hellenistic.

# Skalistoe 5

Trial trench

## Handmade ware

**Cat. 711 Skal5/Fl10/22b**

Base fr. of vessel with flat underside. H: 7.8; W: 7.05; T: 8.5-1.2; Ø: 10. Core: Gley 1 5/N grey. Surface: 10YR 8/4 to 10YR 7/4 very pale brown. Sparse large quartz and grog; sparse medium shells.

Early Hellenistic.

**Cat. 712 Skal5/Fl2/10**

Rim fr. of vessel with rounded lip. H: 1.4; W: 1.7; T: 0.6; Ø? Clay: Gley 1 2.5/N black. Sparse small and medium quartz.

**Cat. 713 Skal5/Fl2/20**

Rim fr. of vessel with rounded lip. H: 3.8; W: 2.1; T: 0.65-0.7. Core: 5YR 7/6 reddish yellow. Surface: 10YR 8/4 very pale brown and 10YR 5/1 grey. Sparse small lime inclusions; single very large rounded inclusion.

Early Hellenistic.

**Cat. 714 Skal5/Fl2/21**

Shoulder fr. of closed vessel. Horizontal moulding on the ext. H: 4; W: 3.35; T: 0.8-0.86. Core: Gley 1 3/N very dark grey. Int. surface: 5YR 8/1 white. Ext. surface: 5YR 7/6 reddish yellow. Sparse medium shells and quartz inclusions.

Shape possibly similar to Stolba 2002, 197 Cat. D105. Early Hellenistic.

**Cat. 715 Skal5/Fl2/22**

Two joining frs. of bowl with incurved rim and flat underside. Complete profile preserved. H: 4.5; W: 6.8; T: 0.6-0.9 Ø rim: ca. 19; Ø base: ca. 9. Core: Gley 1 4/N dark grey. Surface: 10YR 8/3 very pale brown and 10YR 5/1 grey. Sparse medium quartz; grog and lime inclusions.

Similar bowls are known from Panskoe I, cf. Stolba 2002, 198 Cat. D121-D125. Early Hellenistic.

**Cat. 716 Skal5/Fl2/22a**

Rim fr. of bowl with incurved rim. H: 4.95; W: 11.; T: 0.7-0.87; Ø: ca. 25. Core: Gley 1 4/N dark grey. Surface: 10YR 8/3 very pale brown and 10YR 5/1 grey. Sparse medium quartz; grog and lime inclusions.

Similar bowls are known from Panskoe I, cf. Stolba 2002, 198, Cat. D121-D125. Early Hellenistic.

**Cat. 717 Skal5/Fl2/23**

Partly preserved rim fr. of pot. H: 1.7; W: 2.7; T: 1.35 (full T not preserved); Ø: 40-44. Clay: 10R 7/8 light red. Sparse small to medium quartz; sparse small feldspar inclusions; single tiny lime inclusion.

Early Hellenistic.

**Cat. 718 Skal5/Fl3/26**

Rim fr. of pot with flaring neck. H: 4.75; W: 4.45; T: 0.6; Ø: ca. 20. Core: 2.5YR 4/3 reddish brown. Surface: 7.5YR 3/1 very dark grey. Sparse light-reflecting particles and tiny lime inclusions.

Hellenistic.

**Cat. 719 Skal5/Fl3/35**

Rim fr. of pot with slightly flaring neck. H: 5.35; W: 4.8; T: 0.8; Ø: 20-22. Int. surface: 10R 6/4 pale red. Ext. surface: 10R 7/8 light red. Medium tiny to medium lime inclusion; single large pebble; sparse small to large voids.

Hellenistic.

**Cat. 720 Skal5/Fl3/36a**

Rim fr. of pot with flat lip. Repair hole(?) near rim. H: 3.4; W: 3.95; T: 0.7-0.75; Ø: 22. Core: 2.5YR 6/4 light reddish brown and Gley 1 5/N grey. Int. surface: Gley 1 2.5/5GY greenish black. Ext. surface: 10YR 8/4 very pale brown. Sparse small and medium lime and quartz.

Cf. Stolba 2002, type 1 Cat. D1-D68. Early Hellenistic.

**Cat. 721 Skal5/Fl3/36b**

Rim fr. of pot with straight rim. H: 3; W: 3; T: 0.5-0.8; Ø: ca. 13. Clay: 2.5YR 6/4 light reddish brown. Frequent small to medium lime inclusions; single medium pebble.

Possibly similar to Kavalevskaja-Suchanova 2005, fig. 4.6. Early Hellenistic.

**Cat. 722 Skal5/Fl3/37**

Rim fr. of vessel with outturned rim. H: 2.7; W: 3; T: 0.8; Ø: 16. Core: Gley 1 5/N grey. Int. surface: 2.5YR 7/6 light red. Ext. surface: 2.5YR 4/1 dark reddish grey. Sparse small to medium lime; grog and quartz.

Possibly similar to Kavalevskaja-Suchanova 2005, fig. 4.4-6. Early Hellenistic.

**Cat. 723 Skal5/Fl3/38**

Rim fr. of jug? with smoothed surface. H: 1.95; W: 2.92; T: 0.55-0.6; Ø: ca. 9. Core: Gley 1 3/N very dark grey. Ext. surface: 10YR 8/4 very pale brown. Sparse small to tiny lime and quartz.

Possibly similar to Stolba 2002, 197 Cat. D105. Early Hellenistic.

**Cat. 724 Skal5/Fl3/39**

Base fr. of miniature vessel. H: 1.7; W: 4; T: 0.6-0.8; Ø: ca. 4. Core: 2.5YR 3/2 dusky red. Surface: 10R 6/8 light red. Single tiny mica inclusions; frequent tiny to medium lime inclusions; sparse medium shells; sparse small quartz inclusions.

Possibly similar to Stolba 2002, 199 Cat. D126. Early Hellenistic.

**Cat. 725 Skal5/Fl3/41**

Rim fr. of cup with handle attachment. Slightly concave neck and rounded lip. H: 3.3; W: 3.6; T: 0.5-1.1; Ø: ca. 13. Int. surface: 2.5YR 6/6 light red. Ext. surface: 2.5YR 5/3 reddish brown. Frequent tiny to large lime inclusions; sparse light-reflecting particles; sparse sand.

Possibly similar to Stolba 2002, 197, Cat. D111, but see also Senatarov 2003, fig. 4.16 for the shape. Late 4th-early 3rd century BC (according to the context of the cup from Panskoe I U6).

**Cat. 726 Skal5/Fl3/42**

Fr. of round handle. H: 1.91; Ø: 1.35-1.52. Core: 10YR 4/1 dark grey and 7.5YR 6/4 light brown. Surface: 10YR 7/4 very pale brown and

10YR 5/1 grey. Sparse small quartz and shell inclusions; sparse small round voids.

Early Hellenistic?

**Cat. 727 Skal5/Fl3/43**

Two rim frs. of bowl with bevelled lip. Combed decoration(?) on exterior surface. Part of handle attachment preserved. Largest fr. H: 4.3; W: 3.7; T: 0.5; Ø: ca. 21. Core: Gley 2 2.5/10G greenish black. Surface: 2.5YR 5/4 reddish brown to 10R 7/8 light red. Occasionally 7.5YR 4/1 dark grey. Sparse light-reflecting particles; medium tiny to medium lime inclusions; sparse medium pebbles.

Cf. Senatarov 2003 fig. 7; Stolba 2002, 197, Cat. D104 for a similar decoration. 4th century BC?

**Cat. 728 Skal5/Fl3/44**

Four non-joining frs. of strainer jug. Largest fr. H: 2.5; W: 3.8; T: 0.9-1. Clay: 2.5YR 7/6 light red. Frequent tiny to medium lime inclusions; medium small to large pebbles; single small quartz(?) inclusion.

Gavriljuk 2010, Cat. I-62 (Olbia). Early Hellenistic.

**Cat. 729 Skal5/Fl4/10a**

Rim fr. of vessel with concave, outturned rim with rounded lip. H: 2.8; W: 3.4; T: 0.6-0.7; Ø: ca. 22. Core: Gley 1 5/N grey. Int. surface: 10YR 8/3 very pale brown. Ext. surface: 2.5Y 8/4 pale yellow and 2.5Y 3/1 very dark grey. Sparse small quartz and shell inclusions.

Cf. Stolba 2002, type 1 Cat. D1-D68. Early Hellenistic.

**Cat. 730 Skal5/Fl4/10b**

Rim fr. of closed vessel with burnished exterior surface. H: 8.3; W: 8.8; T: 0.55-0.75. Core: Gley 1 4/N dark grey. Int. surface: Gley 1 3/N very dark grey. Ext. surface: 7.5YR 5/3 brown and Gley 1 3/N very dark grey. Frequent small to large quartz; shell and lime inclusions; sparse medium voids.

Early Hellenistic?

**Cat. 731 Skal5/Fl4/10c**

Rim fr. of vessel with rounded lip. H: 3.55; W: 3.9; T: 0.6-0.8; Ø: 32. Core: 2.5YR 7/6 light red. Surface: 10YR 6/4 light yellowish brown to 10YR 4/1 dark grey. Sparse small quartz and lime inclusions; single large lime inclusion.

Cf. Stolba 2002, type 1 Cat. D1-D68. Early Hellenistic.

**Cat. 732 Skal5/Fl5/8**

Rim fr. of vessel with rounded lip. H: 1.72; W: 2.36; T: 0.64; Ø? Core: 10R 4/1 dark reddish grey to 10R 5/6 red. Surface: Gley 1 3/N very dark grey. Sparse small lime; quartz and grog.

Cf. Stolba 2002, type 1 Cat. D1-D68? Early Hellenistic.

## Architectural elements

**Cat. 733 Skal5/Fl10/1**

Fr. of the edge of tile with part of the border preserved. L: 8.3; W: 8.18; T: 2.55-2.7 (tile), 3.9 (edge). H: of border: 3.6, tile 6.3 Core: 2.5YR 6/8 light red. Surface: 10YR 8.4 pink. Sparse medium shells and quartz.

Chersonesos, Hellenistic.

**Cat. 734 Skal5/Fl5/1**

Fr. of tile. L: 7.14; W: 4.96; T: 2.49-2.6. Core: 2.5YR 7/2 pale red. Surface: 2.5YR 7/8 light red to 7.5YR 8/3 pink. Frequent small to large pyroxene; single tiny lime inclusions.

Sinope, Early Hellenistic?

## Amphorai

### *Chersonesos*

**Cat. 735 Skal5/Fl3/1**

Fr. of handle with impressed stamp with the name of the *astynomos* in retrograde.

[ΑΣ]ΤΥΝΟΜΟΥΝ[ΤΟΣ] ←

[ΑΠ]ΟΛΛΑ[ΘΕΟΥ] ←

H: 6.4; W: 7.4; T: wall: 1, handle: 3.4 x 2.2. Core: 10R 7/6 light red. Surface: 2.5YR 8/3 pink. Frequent small to tiny pyroxene inclusions; medium tiny lime inclusions; sparse small quartz.

Cf. Kač 1985, 108, pl II 1Б; Stolba 2005a, Tab. 2. 315-300 BC (Kač), 321-304 BC (Stolba 2005a).

**Cat. 736 Skal5/Fl3/13**

Rim fr. with thickened lip. H: 3.6; W: 4.1; T: 0.5-1; Ø: 10. Clay: 5YR 5/6 yellowish red. Sparse tiny lime inclusions; single small quartz(?) inclusions; sparse tiny to small feldspar.

The shape of the rim is close to the rims of the Chersonesan amphorai, Monachov 1989, type IIA. Ca. 350-250 BC.

**Cat. 737 Skal5/Fl4/1**

Fr. of toe. H: 4.26; Ø: 6.91 (upper), 4.95 (bottom). Core: 7.5YR 6/4 light brown. Ext. surface 2.5YR 7/6 light red. Sparse small to large lime, shell and quartz. Single small round voids.

Cf. Monachov 1989, 138, tav. XXI-XXII, type 1. Ca. 350-250 BC.

**Cat. 738 Skal5/Fl10/14**

Handle fr. H: 11.55; W: 3.18-3.8; T: 2.21-2.54. Core: 10R 7/6 light red. Surface: 2.5YR 8/3 pink. Sparse medium and single large pyroxene inclusion; sparse tiny lime inclusions.

Early Hellenistic.

**Cat. 739 Skal5/Fl10/15**

Handle fr. H: 3.25; W: 3.49; T: 2.41. Core: 10R 7/6 light red. Surface: 2.5YR 8/3 pink. Sparse medium and single large pyroxene inclusion; sparse tiny lime inclusions.

Early Hellenistic.

**Cat. 740 Skal5/Fl10/6**

Fr. of undercut rim. H: 4.15; W: 7.03; T: 0.7-0.8 (wall), 1.2 (rim); Ø: 9.5. Core: 5YR 6/6 reddish yellow. Surface: 5YR 7/6 reddish yellow. Medium-sized and sparse large pyroxene inclusions; single tiny shells.

Close to Monachov 2003, pl. 96.1,6, type II-A-2. Late 4th-early 3rd century BC.

### *Sinope*

**Cat. 741 Skal5/Fl2/8**

Two joining frs. of triangular rim. Handle attachment preserved. H: 5.8; W: 6.2; T: 0.6-1.2; Ø: 11. Clay: 10R 7/4 pale red. Medium tiny to small pyroxene; sparse tiny lime inclusions; single small gold mica; sparse small to large quartz; sparse tiny to small feldspar.

Cf. Monachov 2003, pl. 100.6, type IC. Ca. 350-325 BC.

**Cat. 742 Skal5/Fl3/7**

Handle fr. H: 7.8; W: 3.9-4.1; T: 2.35. Clay: 2.5YR 8/4 pink. Frequent tiny to medium pyroxene; medium tiny to small grog; sparse tiny to small lime inclusions; single large pebble; medium tiny feldspar.

Late Classical-Early Hellenistic.

## *Unknown production*

**Cat. 743 Skal5/Fl2/6**

Fr. of rim from amphora with protruding and slightly undercut rim. H: 3.2; W: 5.4; T: 0.75-1.1; Ø: 13. Clay: 2.5YR 5/6 red. Sparse tiny to small lime; pyroxene and feldspar.

Hellenistic.

**Cat. 744 Skal5/Fl10/10**

Handle fr. Surface is much abraded. H: 3.5; W: 3.8; T: 2.44. Clay: 2.5YR 7/8 light red. Tiny mica; single large quartz; sparse tiny voids.

Early Hellenistic?

**Cat. 745 Skal5/Fl3/14**

Handle fr. H: 6.2; W: 3.5; T: 1.9-2.1. Clay: 5YR 6/6 reddish yellow. Frequent tiny to medium lime inclusions; sparse tiny to small pyroxene; sparse tiny feldspar.

Early Hellenistic.

**Cat. 746 Skal5/Fl3/15**

Handle fr. H: 4.4; W: 3.1; T: 2.25-2.3. Clay: 10R 7/8 light red. Single medium gold mica; sparse tiny to medium lime inclusions; sparse tiny to small feldspar.

Early Hellenistic.

## Tableware

**Cat. 747 Skal5/3/31**

Rim fr. of Attic classical kantharos with West-Slope decoration. Part of olive branch. Matt glaze. H: 3.1; W: 2.6; T: 0.35-0.45. Clay: 2.5YR 7/8 light red. Glaze: 5YR 2.5/1 black. Slightly granular fabric. Sparse tiny mica and lime inclusions.

Cf. e.g. Rotroff 1997, 244, Cat. 29, fig. 5 and pl. 3. Ca. 285-275 BC.

**Cat. 748 Skal5/3/16**

Base fr. H: 3; W: 9.65; T: 0.8-1; Ø: 11. Clay: 10R 6/8 light red. Slip: 5Y 8/3 pale yellow. Sparse small to medium pyroxene; sparse tiny to small lime; single small gold mica; sparse tiny feldspar; medium small to large voids.

Chersonesan jug. Similar to some of the Chersonesan jugs in Pankoe I, cf. Kašaev 2002, 151-152, 161-162, Cat. C7-C19. Early Hellenistic.

**Cat. 749 Skal5/Fl10/12**

Large body fr. and base with three horizontal bands on the upper body. H: 13.9; W: 17; T: 0.5-0.68. Core: 10R 6/1 reddish grey and 10R 6/6 light red. Int. surface: 5YR 8/4 pink. Ext. surface: 2.5Y 8.3 pale yellow. Bands: 10YR 4/3 brown. Frequent medium-large lime inclusions.

Chersonesan jug, Kašaev 2002, 151-152, 161-162, Cat. C7-C19. Early Hellenistic.

**Cat. 750 Skal5/Fl10/13**

Handle fr. with part of wall. H: 6; W: 2.75; T: 0.6-0.7 (wall), 1.5 (handle). Core: 2.5YR 6/8 light red. Slip: 7.5YR 8/4 pink. Sparse small pyroxene and lime inclusions. Single tiny quartz and large lime.

Chersonesan jug, cf. Kašaev 2002, 151-152, 161-162, Cat. C7-C19. Early Hellenistic.

**Cat. 751 Skal5/Fl10/19**

Base fr. of jug with flat underside. H: 6.7; W: 8.65; T: 0.5-0.7; Ø: 10. Core: 5YR 7/6 reddish yellow and 5YR 5/6 yellowish red. Surface: 2.5Y 8/4 pale yellow. Sparse small lime inclusions; single tiny quartz.

Cf. Kašaev 2002, 151-152, 161-162, Cat. C7-C19. Early Hellenistic.

**Cat. 752 Skal5/Fl10/20**

Base fr. of jug with flat underside. H: 3.66; W: 5.36; T: 0.6-0.82 (bottom), 0.8-1.19 (wall). Core and int. surface: 2.5YR 7/6 light red. Ext. surface: 7.5YR 8/4 pink. Sparse tiny pyroxene and lime inclusions.

Chersonesan jug, Cf. Kašaev 2002, 151-152, 161-162, Cat. C7-C19. Early Hellenistic.

**Cat. 753 Skal5/Fl10/21a**

Base fr. of jug with flat underside. H: 5.1; W: 8.93; T: 0.9-1.45; Ø: 10. Core: 2.5YR 6/8 light red. Ext. surface: 5YR 8/4 pink. Sparse small lime inclusions; single tiny quartz.

Chersonesan jug, Cf. Kašaev 2002, 151-152, 161-162, Cat. C7-C19. Early Hellenistic.

**Cat. 754 Skal5/Fl10/21b**

Base fr. of jug with flat underside. H: 6.6; W: 8.05; T: 0.8-1.5 (wall), 1.2 (bottom). Core and int. surface: 2.5YR 7/8 light red. Ext. surface: 2.5YR 8/3 pink. Sparse tiny and single medium pyroxene and lime inclusions.

Chersonesan jug, Cf. Kašaev 2002, 151-152, 161-162, Cat. C7-C19. Early Hellenistic.

**Cat. 755 Skal5/Fl10/22a**

Base fr. of jug with a ring foot. H: 5.57; W: 12; T: 0.9-1.1; Ø: 9. Core and int. surface: 2.5YR 7/6 light red. Ext. surface: 10YR 8/3 very pale brown. Frequent small pyroxene; single small lime inclusions; sparse large voids.

Sinope, Early Hellenistic.

**Cat. 756 Skal5/Fl10/25**

Base fr. of jug with flat underside. H: 4.5; W: 6.2; T: 0.55-0.8 (wall), 0.5-0.9 (bottom); Ø: 6. Core: 5YR 7/6 reddish yellow. Surface: 7.5YR 6/4 light brown. Sparse small mica and quartz; single medium pyroxene; medium tiny voids.

Cf. Kašaev 2002, 151-152, 161-162, Cat. C7-C19. Early Hellenistic.

**Cat. 757 Skal5/Fl10/26**

Rim fr. of bowl with inturned rim. H: 4.9; W: 12.8; T: 0.8-1.1; Ø: 18. Core: 7.5YR 7/6 reddish yellow. Surface: 10YR 8/3 very pale brown. Sparse small pyroxene; sparse small voids.

Cf. Kašaev 2002, 153-154, 164-165, Cat. C39-C59. Early Hellenistic.

**Cat. 758 Skal5/Fl10/3a**

Fr. of lower body of black-glazed cup. Traces of attachment of ring foot preserved. L: 2.65; W: 2.63; T: 0.15-0.26. Core: 10YR 7/4 very pale brown. Surface glaze: 5Y 2.5/1 black and 2.5YR 3/4 dark reddish brown and 2.5YR 3/6 dark red. Single very tiny voids.

4th century BC?

**Cat. 759 Skal5/Fl2/11**

Body fr. with horizontal bands. H: 2.95 W: 3.3; T: 0.45-0.55. Core and int. surface: 2.5YR 7/8 light red. Ext. surface: 7.5YR8/6 reddish yellow. Yellowish slip. Bands: 10R 5/8 red. Frequent small and medium pyroxene and quartz; single small lime inclusions; sparse small oval voids.

Chersonesan jug, cf. Kašaev 2002, 151-152, 161-162, Cat. C7-C19. Early Hellenistic.

**Cat. 760 Skal5/Fl2/15**

Fr. of cup. H: 1.95; W: 1.7; T: 0.3. Clay: 2.5YR 7/8 light red. Metallic slip ext.: 5YR 4/6 yellowish red. Single tiny lime inclusions.

Hellenistic.

**Cat. 761 Skal5/Fl2/17**

Fr. of glazed cup. Broken off on the ext. One repair hole. H: 2.8; W: 3.4; T: 0.02. Clay: 10R 7/8 light red. Slip: 5YR 4/2 dark reddish grey. Singel tiny lime and mica inclusions.

Hellenistic.

**Cat. 762 Skal5/Fl3/24**

Rim fr. of bowl with rounded slightly incurved rim. H: 3.8; W: 4.05; T: 0.65-0.9; Ø: 20. Core: 5YR 5/4 reddish brown. Slip: 7.5YR 8/4 pink. Frequent tiny to medium lime inclusions; sparse small to medium quartz; sparse tiny feldspar.

Cf., e.g., Kašaev 2002, Cat. C43, C44, C54, C63 and C66-C67. Early Hellenistic.

**Cat. 763 Skal5/Fl3/25**

Rim fr. of large bowl with slightly incurved rim and rounded lip. H: 3; W: 7.1; T: 0.7-1.1; Ø: 29. Clay: 5YR 5/6 yellowish red. Slip: 7.5YR 8/4 pink. Frequent tiny to large lime inclusions; sparse small pyroxene; single small gold mica and quartz; sparse small to medium dark inclusions.

Early Hellenistic?

**Cat. 764 Skal5/Fl3/25a**

Small rim fr. of small bowl. H: 1.25; W: 2.7; T: 0.6; Ø: 11. Clay: 10R 7/8 light red. Single tiny mica and lime inclusions.

Early Hellenistic.

**Cat. 765 Skal5/Fl3/30**

Rim fr. of bowl? H: 3.85; W: 5.2; T: 0.5; Ø: 20. Clay: 2.5YR 6/8 light red. Single tiny mica; sparse tiny to large lime inclusions; sparse tiny feldspar.

Hellenistic.

**Cat. 766 Skal5/Fl3/32**

Two joining rim frs. of Attic black-glazed echinus bowl with incurved rim. H: 3.35; W: 2.35; T: 0.4; Ø: ca. 20. Core: 5YR 7/6 reddish yellow. Glaze: 2.5YR 2.5/1 reddish black. Sparse tiny lime inclusions.

Cf. Rotroff 1997, Cat. 974-976. Ca. 325-275 BC.

**Cat. 767 Skal5/Fl3/33**

Base fr. of Grey Ware fish-plate. High ridge around the despression and low ring foot. H: 2.7 W: 15.8; Ø base: 9. Black slip.

Cf. e.g. Stolba 2002, 147-148, Cat. B222-235. Ca. 325-275 BC.

**Cat. 768 Skal5/Fl3/34**

Fr. of small fish-plate or plate with a rilled rim. Incompletely preserved. H: 4; W: 3.5; T: 0.3. Clay: 10R 6/8 light red. Slip: 2.5YR 4/8 red. Sparse small pyroxene; sparse tiny to small lime and feldspar.

Early Hellenistic.

**Cat. 769 Skal5/Fl4/9**

Body fr. of glazed cup(?). Ribbed exterior surface. H: 2.5; W: 1.4; T: 0.3-0.4-0.55. Core: 2.5YR 6/6 light red. Int. surface: 5YR 3/2 dark reddish brown. Ext. surface: 5YR 3/1 very dark grey and 2.5YR 3/6 dark red. Sparse light-reflecting particles.

Early Hellenistic.

**Cat. 770 Skal5/Fl4/7**

Rim fr. of bowl with rounded lip. H: 2.6; W: 1.95; T: 0.59-0.72; Ø: 30. Core: 2.5YR 5/6 red. Surface: 2.5YR 7/6 light red. Sparse small pyroxene and quartz; single small lime inclusions and small voids.

Early Hellenistic.

**Cat. 771 Skal5/Fl5/7**

Body fr. of closed shape decorated with four thin horizontal bands. Width of bands 0.15-0.19. H: 4.8; W: 4.68; T: 0.5-1.02. Core and int. surface: 5YR 7/6 reddish yellow and 10R 7/6 light red. Slip: 10YR 8/4 very pale brown. Bands: 10R 5/4 weak red. Sparse small pyroxene and lime inclusions; single large quartz(?).

Hellenistic.

### Utility ware

**Cat. 772 Skal5/Fl3/22**

Rim fr. of louterion. H: 5.2; W: 7.6; T: 0.8-2.1; Ø: 36-40. Clay: 10R 7/8 light red. Slip: 5YR 8/4 reddish yellow. Medium tiny to large pyroxene; medium tiny lime inclusions; sparse small to medium quartz; sparse small red inclusions (grog?); sparse tiny feldspar.

Sinope? cf. e.g Kašaev 2002, Cat. C189. Early Hellenistic.

## Skalistoe 6

Grab sample: 3023

### Amphorai

*Chersonesos?*

**Cat. 773 3023/1**

Handle. H: 5.6; W: 4.9. Clay: 5YR 5/8 yellowish red. Sparse lime (very tiny), sparse sand; sparse feldspar (small).

Hellenistic.

## Skalistoe 7

Grab sample: 3028
No diagnostic material.

## Skalistoe 8

Grab sample: 3029
No diagnostic material.

## Skalistoe 9

Grab sample: 3024
No diagnostic material.

## Skalistoe 10

Grab sample: 3025

### Amphorai

**Cat. 774 3025/2**

Small part of flat handle. H: 3.9; W: 4.3; T: 0.8. Clay: 5YR 5/8 yellowish red. Surface: 5YR 5/6 yellowish red. Very sparse medium lime; very sparse small pyroxene; very sparse sand.

Late Classical-Hellenistic.

### Tableware

**Cat. 775 3025/1**

Upper part of carinated plate with protruding flat lip. H: 3.4; T: 0.5-0.7; Ø: 22. Clay: 5YR 5/6 yellowish red. Slip: 7.5YR 4/3 brown. Sparse lime (tiny-small); sparse quartz (small).

Second half of 3rd-2nd century BC.

## Vodopojne 1

Grab sample: 3064

### *Tableware*

**Cat. 776 3064/1**

Rim fr. of cup with straight walls and rounded lip. Slightly grooved on the int. H: 1.4; W: 2.1; T: 0.7; Ø: 13-16. Clay: 2.5YR 4/6 red. Medium tiny lime inclusions; sparse tiny feldspar.

### *Utility ware*

**Cat. 777 3064/3**

Rim fr. of cooking pot with straight rim. H: 2.4; W: 3.2; T: 0.7; Ø: 24. Clay: 7.5YR 5/1 grey. Surface: 5YR 6/6 reddish yellow. Sparse tiny lime; sparse tiny feldspar.

Hellenistic.

## Vodopojne 3A and B

Grab sample: 3065
No diagnostic material.

## Off-Site material

### Terracotta

**Cat. 778 8129/01/1**

Almost intact votive egg with a small hole in the top. H: 3.4; W: 2.4. Clay: 2.5YR 6/6 light red. Slip: 2.5YR 5/6 red. Medium tiny to medium lime; sparse small pyroxene; sparse tiny to small feldspar; single small quartz.

Cf. Ščeglov 2002, 221 Cat. No. G20 and Rogov 2002, 274, Cat. M22 for a similar egg from Panskoe I. Early Hellenistic.

### Handmade ware

#### *Bronze Age*

**Cat. 779 1445/01/1**

Body fr. of vessel with parts of three deep furrows on the ext. H: 1.7; W: 1.6; T: 0.6. Clay: Gley 1 4/N dark grey. Sparse tiny lime.

Cf. Leskov 1970, fig. 3.11,13. Middle Bronze Age.

**Cat. 780 6219/01/1**

Rim fr. with small part of lip preserved. Impressed cord decoration on the ext. H: 2.9; W: 2.8; T: 1.4. Core: Gley 2 2.5/5 BG greenish black. Surface: 2.5YR 7/6 light red to 2.5YR 7/4 light reddish. Frequent small to medium lime inclusions; sparse small quartz.

Similar character as some of the finds from Kimmerik, cf. Toščev 2007, fig. 80. Middle Bronze Age/Late Bronze Age.

### Amphorai

#### *Chersonesos*

**Cat. 781 8174/01/1**

Handle fr. reused as tool.

No information available.

#### *Herakleia*

**Cat. 782 2017/01/1**

Flaring foot with grooved transition to the lower part. H: 8.6; W: 5.3; Ø: 5.4. Clay: between 2.5YR 6/8 light red and 2.4YR 5/6 red. Medium tiny to large lime inclusions; medium small pyroxene; sparse tiny quartz; sparse small to large feldspar.

Herakleia, pithoid type. First half of 4th century BC.

**Cat. 783 2344/01/1**

Foot. H: 7.1; W: 4.9; Ø: 4.8. Clay: 2.5YR 6/6 light red. Slip: 2.5YR 7/6 light red. Sparse tiny to small pyroxene; sparse tiny lime inclusions; very sparse large quartz; sparse small feldspar.

Herakleia, type I-4 to I-6 or type I-A. Cf. Monachoc 2003, pl. 87.3-8, 88-90. First half of 4th century BC.

**Cat. 784 3003/1**

Foot. H: 3.9; Ø: 4.6. Clay: 2.5YR 6/8 light red. Medium lime and medium black inclusions.

Herakleia, type I-A-2, type I-A-4 or type II-1 to II-3. Cf. Monachov 2003, pl. 91-92.1-4. Second quarter of 4th century BC-early 3rd century BC.

#### *Thasos*

**Cat. 785 8403/01/1**

Fr. of rounded rim with a slight groove at the junction between rim and neck. H: 3.4; W: 3.1; T: 0.7-1.4; Ø: ca. 12; D: ca. 25. Clay: 2.5YR 6/6 light red. Frequent tiny mica; frequent tiny to small lime inclusions; single medium quartz; sparse tiny feldspar.

Thasos, type II-A-3, cf. Monachov 2003, pl. 40.4. Ca. 375-350 BC.

**Cat. 786 6124/01/1**

Handle fr. H: 6.2; W: 4.7; T: 2.2. Core: 5YR 5/6 yellowish red to 10R 5/8. Surface: 5YR 6/6 reddish yellow. Frequent tiny to small lime inclusions; sparse small shells; frequent tiny to medium golden mica; sparse tiny to large quartz; sparse tiny to medium feldspar.

Thasos, probably type II-C-2. Cf. Monachov 2003, pl. 48. Ca. 400-350 BC.

### *Chios*

**Cat. 787 8536/01/1**

Rim fr. and upper part of neck. H: 4.95; W: 5.6; T: 0.7-1.7; Ø: 11; D: 39. Clay: 10R 7/8 light red. Sparse tiny lime inclusions; sparse tiny feldspar; single medium quartz; single medium dark stones.

Chios, type III-B and III-C with with bulging neck. Cf. Monachov 2003, pl. 5-6. 5th century BC.

### *Unknown production*

**Cat. 788 1063/01/1**

Body fr. Reworked to be used as a tool with a hole in the middle and smoothed edges. H: 2.9; W: 5.1; T: 0.8; Ø? Clay: 10YR 5/3 brown. Slip: 10YR 6/4 light yellowish brown. Sparse tiny lime inclusions.

Classical-Early Hellenistic.

## Tableware

**Cat. 789 3091/01/2**

Wall fr. of closed vessel. Parts of three painted horizontal bands on the ext. H: 2.6; W: 3.6; T: 0.3. Clay: 2.5YR 5/6 red. Surface. 2.5YR 7/6 light red to 5YR 7/4 pink. Band: 2.5YR 4/3 reddish brown. Frequent tiny to medium lime inclusions; sparse small feldspar.

Cf. Kašaev 2002, Cat. C14. Late 4th-3rd century BC.

## Utility ware

**Cat. 790 2100/01/1**

Rim fr. of larger shape, presumably a pithos with straight rim. H: 3; W: 3.8; T: 2.5; Ø? Clay: 2.5YR 6/8 light red. Sparse very tiny to tin lime inclusions; very sparse tiny pyroxene; very sparse small to medium quartz; medium tiny to large feldspar.

For a pithos with straight rim from Panskoe I U6, see Kac et al. 2002, 103-104, Cat. Ac3. Early Hellenistic.

**Cat. 791 2409/01/1**

Rim fr. of larger basin with pointed rim. H: 1.7; W: 3.3; T: 1.1; Ø: 8 (int.). Clay: 5YR 6/6 reddish yellow. Sparse tiny lime inclusions; very sparse small pyroxene; sparse small to medium quartz.

Resembles the larger louteria in Panskoe, cf. Kašaev 2002, Cat. C215-C220. Early Hellenistic.

## Handmade ware

**Cat. 792 8991/01/1**

Rim fr. of bowl with slightly concave body and projecting rim. H: 1.88; W: 1.56; T: 0.7-0.82. Core: Gley 1 3/N very dark grey. Surface: 10YR 8/4 very pale brown and 7.5YR 7/6 reddish yellow. Sparse tiny lime inclusions and quartz.

Somewhat similar to a bowl from a grave on the Kerč Peninsula, cf. Maslenikov 1995, fig. 20.2. 3rd-2nd century BC.

## Post-Antique pottery

**Cat. 793 1064/01/01**

Small rim fr. of cup? H: 1.6; W: 1.6; T: 0.5; Ø? Clay: 2.5YR 4/6 red. Sparse tiny lime sparse tiny feldspar.

Early Modern.

**Cat. 794 1192/01/01**

Fr. of flat base of utility vessel. H: 2.9; W: 4.9; T: 0.8-1.4; Ø: 12. Clay: 2.5YR 6/6 light red. Very sparse tiny to small lime; very sparse small pebbles.

Early Modern.

**Cat. 795 1353/01/01**

Fr. of everted rim with glazed interior. Plate? H: 3.4; W: 4.7; T: 0.8 Ø? Clay: 7.5YR 6/6 reddish yellow. Glaze: yellowish. Very sparse very tiny lime.

Early Modern.

**Cat. 796 1687/01/01**

Handle fr. of large amphora with a slight groove. H: 8.6; W: 5.6; T: 3.2. Clay: 5YR 4/6 yellowish red. Slip: 2.5YR 4/4 reddish brown on the ext. badly preserved. Frequent very tiny to tiny lime inclusions; very sparse very large lime inclusions; very sparse sand; very sparse tiny feldspar.

Byzantine?

**Cat. 797 1823/01/01**

Fr. of flat base of utility vessel. H: 3.7; W: 6.5; T: 1-1.9; Ø? Clay: 5YR 5/6 yellowish red. Self-slip: 10YR 7/4 very pale brown. Sparse tiny lime.

Early Modern.

**Cat. 798 1931/01/01**

Fr. of jug with flat everted rim and glazed exterior. H: 1.9; W: 3.9; T: 0.5; Ø: 14. Clay: 5YR 5/4 reddish brown. Glaze: strong green. Sparse very tiny to tiny lime.

Early Modern.

**Cat. 799 1945/01/01**

Rim fr. of slipped pibe with exterior band of tiny rouletting below rim. H: 1.4; W: 2.1; T: 0.4; Ø: 2.1 Clay: 10R 5/6 red. No visible incl.

Early Modern.

**Cat. 800 1999/01/01**

Fr. of flat base of utility vessel. H: 3.4; W: 4.6; T: 0.8-1.1; Ø? Clay: 5YR 5/8 yellowish red Sparse very tiny lime.

Early Modern.

**Cat. 801 2519/02/02**

Fr. of jug with everted rim with upturned lip. H: 3.4; W: 4.9; T: 0.6; Ø: ca. 9. Clay: 5YR 6/8 reddish yellow. Slip int.: dark red. Sparse small lime.

Early Modern.

**Cat. 802 2696/01/04**

Fr. of glazed rim with thin painted bands on the exterior lip and rim. H: 2.9; W: 3.7; T: 0.7; Ø? Clay: 5YR 5/8 yellowish red. Glaze: clear. Paint: Gley 1 2.5/10GY greenish black. Sparse sand; very sparse tiny lime.

Early Modern.

**Cat. 803 2739/01/01**

Fr. of glazed jug with everted rim. H: 4; W: 3.7; T: 0.6; Ø: 13. Clay: 2.5YR 5/2 greyish brown. Glaze: yellowish. Very sparse tiny lime.

Early Modern.

**Cat. 804 3073/01**

Fr. of thick glazed handle with oval section. H: 4.6; W: 3.2; T: 1.8. Clay: 2.5YR 5/6 red. Glaze: pale green. Medium tiny to small lime inclusions; sparse tiny feldspar.
Early Modern.

**Cat. 805 3220/01/01**

Vertical glazed handle of jug or amphora. Central exterior groove. H: 11.3; W: 3.4; T: 2. Core: 2.5YR 6/8 light red. Glaze: bright pale green. Sandy clay with numerous tiny lime and occasional small ones. Tiny to large internal fractures
Early Modern.

**Cat. 806 5013/01/01**

Fr. of flat base of closed utility vessel. Lower part of ext. retains manufacturing marks (cuts with an instrument that creates a smooth surface). H: 3.9; W: 7.8; T: 0.9-1.7; Ø: 13. Clay: 2.5YR 6/6 light red.
Early Modern.

**Cat. 807 5019/01/01**

Rim fr. of bowl with glazed exterior and interior upper part. H: 2.5; W: 2.4; T: 0.8; Ø? Clay: 10YR 6/3 pale brown. Glaze: close to 10YR 6/8 brownish yellow. Medium sand; sparse tiny lime; sparse tiny feldspar.
Early Modern.

**Cat. 808 5099/01/01**

Fr. of flat base of closed utility vessel. H: 4.2; W: 7.2; T: 0.9; Ø? Clay: 2.5YR 6/6 light red. Very sparse tiny lime.
Early Modern.

**Cat. 809 6164/01/01**

Handle fr. of amphora. Battered in places. H: 7.2; W: 3.3-4.1; T: 1.5-1.8. Clay: 5YR 6/4 light reddish brown. Slip: 10R 5/6 red. Medium to large stones; medium small feldspar.
Medieval?

**Cat. 810 6188/01/01**

Shoulder fr. with deep corrugation. H: 4.2; W: 4.05; T: 0.4-0.75. Interior surface: 2.5YR 5/6. Exterior: 2.5YR 4/2 weak red. Medium tiny lime; sparse medium red inclusions; sparse large pebbles; sparse tiny feldspar
Byzantine?

**Cat. 811 6331/01/01**

Body fr. open vessel with interior polychrome glaze. Ornament: greyish-blue petals surrounded by cream lines on a rose coloured background. H: 3.1; W: 2.72 0.61-0.74. Core: 7.5YR 7/8 reddish yellow. Paint: Gley 1 5/N grey, 2.5Y 8/2 pale yellow and 10R 5/6 red. Sparse medium quartz.
Early Modern.

**Cat. 812 6611/01/01**

Corrugated body fr. H: 8; W: 5; T: 1.2-1.3. Clay: 2.5YR 7/8-6/8 light red. Frequent tiny to small mica; sparse small lime inclusions; sparse sand.
Byzantine?

**Cat. 813 8277/01/01**

Body fr. of open glazed vessel. Moulding with impressed strokes. H: 2.1; W: 2.1; T: 0.5. Clay: 2.5YR 6/8 light red. Glaze: strong green. Very fine with sparse sand.
Early Modern.

**Cat. 814 8286/01/01**

Fr. of flat base of closed utility vessel. H: 2.9; W: 7.9; T: 1.2-1.3; Ø: 12. Clay: 5YR 6/8 reddish yellow. Medium large pebbles; single light-reflecting particle; sparse tiny lime.
Early Modern.

**Cat. 815 8570/01/02**

Fr. of small jug with overfolded rim. H: 2.7; W: 3.2; T: 0.6-0.8; Ø: 12. Clay: 5YR 6/6 reddish yellow. Slip: 2.5YR 7/8 light red. Sparse light-reflecting particles; frequent tiny to medium lime; medium tiny feldspar.
Early Modern.

**Cat. 816 8570/02/02**

Body fr. of smoking pipe. L: 4.36; W: 1.78; T: 1.34. Core: 2.5Y 7/1 light grey. Surface: 5YR 7/6 reddish yellow. Single tiny shell.
Early Modern.

**Cat. 817 8737/01/01**

Fr. of everted, rounded, glazed rim. Jug? H: 2; W: 2.9; T: 0.7; Ø: 11. Clay: 2.5YR 6/6 light red. Glaze: yellow to light brownish green. Sparse sand; single tiny lime.
Early Modern.

**Cat. 818 8767/01/01**

Body fr. of jar with two bands of ribbings on the ext. H: 7.2; W: 7; T: 0.8. Clay: 10R 7/8-6/8 light red. Sparse tiny to medium lime inclusions; sparse small to medium pebbles.
Early Modern?

**Cat. 819 8994/01/01**

Handle fr. with part of the neck preserved. H: 3.1; W: 4.3; T: 1.74-2.4. Core: 2.5YR 7/6 light red-5YR 7/6 reddish yellow. Surface: 5YR 7/6 reddish yellow, 7.5YR 8/6 reddish yellow and 10YR 8/4 very pale brown. Sparse tiny to small lime inclusions.
Medieval.

**Cat. 820 9027/01/01**

Fr. of glazed candlestick. H: 8.85; T: (walls) 0.57-0.8-1.1; Ø: 2.37-3.51-4.74. Core: 2.5YR 6/8 light red and 5YR 8/4 pink. Glaze: bright pale and dark green. Sparse tiny lime; single tiny grog.
Cf. Hayes 1992, fig. 143 no 6. Mid-19th century AD.

**Cat. 821 9032/01/01**

Rounded handle fr. of amphora? H: 4.13; W: 1.65; T: 1.98. Core: 5YR 6/1 grey and 2.5YR 6/8 light red. Self-slip: 2.5Y 8/3 pale yellow. Sparse tiny lime inclusions.
Medieval?

**Cat. 822 20009/06**

16 frs. of flat base of open vessel with glazed interior. H: 2.25; W: 4.1; T: 0.4-0.7; Ø: 7. Clay: 10R 7/8 light red. Ext.: 5YR 6/8 reddish yellow. Glaze: brown. Single tiny to small lime; single medium quartz.
Early Modern.

**Cat. 823 1088/01/01**

Shoulder fr. of an amphora with corrugations. H: 5.0; W: 7.1; T: 1.1. Clay: 2.5YR 5/6 red. Frequent very tiny light reflecting particles; sparse tiny-small lime inclusions.
Byzantine-Medieval.

# Appendix 3A
# Soil augering descriptions: land type Pediment

*W. de Neef*

| **Auger T1** | Coordinates<br>492260 / 5043110 | Surface<br>35% limestone coverage at surface. Sized stones at surface around 5 m, some larger up to 20 m | |
|---|---|---|---|

| | |
|---|---|
| 0-0.22 m | Loam, 10YR 4/3 (d), 25% stones in soil (limestone) |
| 0.22-0.45 m | Loam with some ground limestone, 2.5Y 7/4, Stoniness soil 50% |
| 0.45 m | End of auger, soil too hard |

| **Auger T2** | Coordinates<br>492454 / 5043160 | Surface | Slope angle 4% |
|---|---|---|---|

| | |
|---|---|
| 0-0.34 m | Loam, 10YR 2/3, |
| 0.34-0.60 m | Loam, 10YR 5/4, Small gravel 10% |
| 0.60-0.75 m | Loam with 50% gritty frs., 10YR 7/6 |
| 0.75 m | Solid bedrock |

| **Auger T3** | Coordinates<br>492156 / 5043349 | Surface<br>60% stone sized slightly larger than 10-25 m | Slope angle 3% |
|---|---|---|---|

| | |
|---|---|
| 0-0.24 m | Loam, 10YR 3/4, effervescences |
| 0.24 m | End of auger, soil too stony |

| **Auger T4** | Coordinates<br>492083 / 5043507 | Surface | |
|---|---|---|---|

| | |
|---|---|
| 0-0.30 m | Loam, 7.5YR 4/4 |
| 0.30-0.45 m | Loam mixed with pink gritty material, 2.5YR 5/6 |
| 0.45 m | End of auger, bedrock |

| **Auger T5** | Coordinates<br>492006 / 5043645 | Surface<br>Some large stones (20-30 m) + smaller stones | |
|---|---|---|---|

| | |
|---|---|
| 0-0.38 m | Loam, 7.5YR 4/4 |
| 0.38-0.54 m | Transitional layer, loam, 7.5YR 4/4 (M) |
| 0.54-0.65 m | Loam, gritty, 5YR 4/4, effervescences |

| **Auger T6** | Coordinates<br>491946 / 5043785 | Surface | |
|---|---|---|---|

| | |
|---|---|
| 0-0.34 m | Loam, no inclusions, 7.5YR 4/4 |
| 0.34-0.46 m | Transition from topsoil to rock, loam with light pink limestone grit, 10YR 3/3 |
| 0.46 m | End of auger: bedrock |

| Auger T7 | Coordinates<br>491872 / 5043927 | Surface | |
|---|---|---|---|

| | |
|---|---|
| 0-0.18 m | Loam, 10YR 2/2 (m), effervescences |
| 0.18-0.52 m | Loam, 10YR 3/2 |
| 0.52-0.68 m | Loam, 10YR 5/4, effervescences |
| 0.68-0.75 m | Loam with pinkish white lime inclusions, 10YR 5.4 |

| Auger T8 | Coordinates<br>491783 / 5044105 | Surface<br>No stones at surface, topsoil effervescences | |
|---|---|---|---|

| | |
|---|---|
| 0-0.40 m | Loam, 10YR 5/5 (d) |
| 0.40-0.68 m | Loam, 10YR 5/4 |
| 0.68-0.76 m | Loam mixed with limestone |
| 0.76 m | End of auger: bedrock |

| Auger T9 | Coordinates<br>491713 / 5044270 | Surface<br>An occasional stone at the surface | |
|---|---|---|---|

| | |
|---|---|
| 0-0.45 m | Loam, 10YR 5/3 |
| 0.45-0.75 m | Gritty loam, 10YR 4/3 (m), upper part clayey with limestone intrusions |
| 0.75-0.86 m | Limestone-bedrock transition |
| 0.86 m | End of auger: bedrock |

| Auger T10 | Coordinates<br>491622 / 5044465 | Surface<br>No stones at the surface, topsoil effervescences | |
|---|---|---|---|

| | |
|---|---|
| 0-0.45 m | Loam, 10YR 5/3 (d) |
| 0.45-1.12 m | Loam with lime concretions, looks like pure loess, 10YR 4/6 |
| 1.12-1.26 m | Loam, mixed with some limestone particles, 10YR 4/3. Layer of loess on limestone weathering layer? |
| 1.26 m | End of auger: bedrock |

| Auger T11 | Coordinates<br>491539 / 5044600 | Surface<br>Stones at the surface, max 10x10, very sparse. Topsoil effervescences | |
|---|---|---|---|

| | |
|---|---|
| 0-0.33 m | Loam, 10YR 3/2-4/2 |
| 0.33-0.49 m | Loam, 10YR 4/3, somewhat lighter and grittier than top soil |
| 0.49 m | End of auger: bedrock |

| Auger T12 | Coordinates<br>491462 / 5044746 | Surface<br>Large stones at the surface, max 30-20 m. Topsoil effervescences | |
|---|---|---|---|

| | |
|---|---|
| 0-0.50 m | Loam, 10YR 5/5, lower down getting somewhat lighter |
| 0.50-0.73 m | Loam with grit, 10YR 4/3 |
| 0.73 m | End of auger, bedrock |

| Auger T13 | Coordinates<br>491389 / 5044870 | Surface<br>15-20% stones at the surface, 15x15 m, some larger ones. Topsoil effervescences | |
|---|---|---|---|

| | |
|---|---|
| 0-0.50 m | Loam, grittier lower down, 10YR 5/3 |
| 0.50-0.78 m | Loam, with grit, 10YR 4/3 |
| 0.78 m | End of auger: bedrock |

| **Auger T14** | Coordinates<br>491 317 / 5045024 | Surface<br>10-15% stones at the surface 15x15 m, some larger ones.<br>Topsoil effervescences | |
|---|---|---|---|

0-0.40 m Loam, 10YR 4/3
0.40-0.77 m Loam with limestone grit, very compact, from 10YR 4/3 to 10YR 5/4
0.77 m End of auger: loam layer was too dense. Increase of grit indicates that bedrock is close

| **Auger T15** | Coordinates<br>491230 / 5045133 | Surface<br>No stones at the surface, grown with grass | |
|---|---|---|---|

0-0.40 m Loam, 10YR 5/3
0.40-1.54 m Pure loess, 10YR 5/4
1.54-1.87 m Sandy loam with limestone grit, 2.5Y 8/4
1.87 m End of auger, soil too hard. Bedrock probably near.

# Appendix 3B
# Soil augering descriptions: land type Lowland Ridge

*W. de Neef*

| Auger T16 | Coordinates<br>489942 / 5047888 | Surface<br>Partly grown with grass, 30 m from northern cliff of Lake Džarylgač | The lake cliff is approx. 6 m high. The soil is very thick here, bedrock was not reached |
|---|---|---|---|

| | |
|---|---|
| 0-0.45 m | Loam, effervescences, 10YR 4/3 |
| 0.45-1.12 m | Loam, very dry, effervescences, 10YR 5/4 |
| 1.12-1.78 m | Loam, effervescences, 10YR 5/4 |
| 1.78-1.92 m | Loam, effervescences, 10YR 6/4 |

| Auger T17 | Coordinates<br>489881 / 5048164 | Surface<br>In barley field surface not visible | |
|---|---|---|---|

| | |
|---|---|
| 0-0.45 m | Loam, effervescences lightly, 10YR 6/4-5/4 (pure loess), darker patches 10YR 4/4 |
| 0.45-0.95 m | Loam (loess), effervescences lightly, 10YR 5/4 |
| 0.95 m | End of auger: loam too compact to continue |

| Auger T18 | Coordinates<br>489768 / 5048164 | Surface<br>North of a dirt road, on top of a local ridge. Slope dips slightly to the north. Surface is sandy and lighter than previous locations | |
|---|---|---|---|

| | |
|---|---|
| 0-0.21 m | Loam (loess), 10YR 4/4 |
| 0.21-0.70 m | Loam, 10YR 5/4 |
| 0.70-0.80 m | Loam with lime concretions, 10YR 6/4 |
| 0.80 m | End of auger: loam too compact to continue |

| Auger T19 | Coordinates<br>489720 / 5048238 | Surface | |
|---|---|---|---|

| | |
|---|---|
| 0-0.54 m | Loam, 10YR 4/3 |
| 0.54-0.70 m | Loam (loess), 7.5YR 5/4 |
| 0.70 m | End of auger: bedrock |

| Auger T20 | Coordinates<br>489617 / 5048373 | Surface | |
|---|---|---|---|

| | |
|---|---|
| 0-0.55 m | Loam mixed with small lime stones (1.5 x 1.5 m), 10YR 3/3 |
| 0.55-0.67 m | Loam with whitish limestone grit, 10YR 4/3 and 10YR 5/4 |
| 0.67 m | End of auger: bedrock |

| Auger T21 | Coordinates<br>489527 / 5048614 | Surface | |
|---|---|---|---|

| | |
|---|---|
| 0-0.50 m | Loam mixed with lime stones, effervescences, 10YR 4/3 |
| 0.50 m | End of auger: bedrock |

| **Auger T22** | Coordinates<br>489400 / 5048795 | Surface<br>Limestone bedrock with grit and some grassy patches | Bedrock at surface |
|---|---|---|---|

0 m Bedrock and limestone grit

| **Auger T23** | Coordinates<br>489324 / 5048936 | Surface<br>40% of the surface covered with limestones | |
|---|---|---|---|

0-0.25 m Loam with limestone grit, 10YR 4/2
0.25 m End of auger: bedrock

| **Auger T24** | Coordinates<br>489222 / 5049091 | Surface<br>Surface covered with limestones | |
|---|---|---|---|

0-0.15 m Loam with limestone grit, effervescences, 10YR 4/3
0.15 m End of auger: bedrock

| **Auger T25** | Coordinates<br>489082 / 5049343 | Surface | |
|---|---|---|---|

0-0.20 m Loam (loess) without limestone inclusions, effervescences, 10YR 3/3
0.20 m End of auger: bedrock

| **Auger T26** | Coordinates<br>488940 / 5049543 | Surface<br>Almond trees on flat upper part of coastal ridge. Hardly any stones at surface | |
|---|---|---|---|

0-0.40 m Loam without limestone inclusions, effervescences, 10YR 3/3
0.40-0.80 m Loam (loess), 10YR 5/4
0.80 -1.12 m Loam with white calcite particles, 10YR 6/6
1.12 m End of auger: bedrock

| **Auger T27** | Coordinates<br>488832 / 5049752 | Surface<br>No stones at surface | |
|---|---|---|---|

0-0.33 m Loam, no inclusions, effervescences, 10YR 3/3
0.33-0.47 m Loam with limestone grit, effervescences, 10YR 5/4
0.47 m End of auger: bedrock

| **Auger T28** | Coordinates<br>488701 / 5050019 | Surface | |
|---|---|---|---|

0-0.33 m Loam, no inclusions, 10YR 3/3
0.33 m End of auger, bedrock

| **Auger T29** | Coordinates<br>488573 / 5050213 | Surface<br>40 m from Black Sea coastal cliff. No stones at surface. Soil at cliff is several meters thick | |
|---|---|---|---|

0-0.37 m Loam, no limestone inclusions, 10YR 3/2
0.37-0.70 m Loam with limestone grit, 10YR 5/4
0.70 m End of auger: bedrock

# Appendix 3C
# Augerings at locations with geological observations

*W. de Neef*

| **Auger G1**<br>(Hill 12) | Coordinates<br>489510 / 5039030 | Surface<br>No stones at surface | |
|---|---|---|---|

Auger conducted at an extension of the plateau, with valleys on both sides. The beginning of the valley has a saucer-shape with shallow overflows from various directions that merge, creating a convex break of slope in the length profile as the factual start of the valley. Interpretation: Chernozem soil in loess

0-0.55 m Loam, 10YR 3/1 (d). 0-40 m was decalcified
0.55-1 m Loam, 10YR 6/4 (d)
1-2 m Same, with spots and veins of lime
2 m End of auger because soil was hard and information was sufficient

| **Auger G2**<br>(Hill 7) | Coordinates<br>488688 / 5041385 | Surface<br>Grown hillside | |
|---|---|---|---|

Between Quarries 1 and 2, 14 m from a quarry with a loess deposit.
0-0.26 m Loam, 10YR 3/2
0.26-1.08 m Loam, 10YR 5/4
1.08 m End of auger: bedrock

| **Auger G3**<br>(Hill 7) | Coordinates<br>488684 / 5041380 | Surface<br>Grown hillside | |
|---|---|---|---|

0-0.25 m Loam, 10YR 4/3
0.25-0.65 m Loam with limestone particles, 10YR 5/4
0.65 m End of auger: bedrock

| **Auger G4**<br>(Hill 7) | Coordinates<br>488685 / 5041369 | Surface<br>Grown hillside | |
|---|---|---|---|

Auger conducted between G1 and G2.
0-0.50 m Loam, 10YR 3/2. Transition is very gradual via 10YR 5/4 to the next layer
0.50-1.35 m Loam, 10YR 5/6
1.35 m End of auger: limestone bedrock

| **Auger G5**<br>(Hill 7) | Coordinates<br>488679 / 5041363 | Surface<br>Grown hillside | |
|---|---|---|---|

Auger conducted in the loess quarry, 2.5 m below the original hillside surface.
0-0.35 m Loam, 10YR 4/4
0.35 m End of auger: bedrock

| **Auger G6** | Coordinates<br>481636 / 5040614 | Surface<br>Ploughed field, lot of limestone grit and bigger pieces, 7.5YR 6/6 | |
|---|---|---|---|

0-0.01 m Loam, 7.5YR 5/4 (d), effervescences
0.01-0.30 m Loam, 7.5YR 5/4
0.30 m End of auger: bedrock

| **Auger G7** | Coordinates<br>461587 / 5040626 | Surface<br>Ploughed field red soil, only few bigger lime stones (15 x 15 m) | |
|---|---|---|---|

| | |
|---|---|
| 0-0.47 m | Loam, 7.5YR 3/2 (m), effervescences, no lime inclusions |
| 0.47-0.59 m | Loam, 7.5YR 4/3 |
| 0.59-0.66 m | Loam mixed with limestone grit, 7.5YR 4/3 |
| 0.66 m | End of auger, close to bedrock |

| **Auger G8** | Coordinates<br>481715 / 5040636 | Surface<br>Ploughed field, dark soil | |
|---|---|---|---|

| | |
|---|---|
| 0-0.42 m | Loam with limestone grit, 10YR 4/3, effervescences |
| 0.42-0.68 m | Same, mixed with white limestone |
| 0.68-0.96 m | Loam with grit, 7.5-10YR 5/4 |
| 0.96-1.02 m | Loam with grit (more than previous), 10YR 5/4 |
| 1.02 m | End of auger: bedrock |

# Bibliography


Abramov, A.P. & A.A. Maslenikov 2005. Keramičeskij kompleks V v. do. n.e. iz rajona poselenija "General'skoe-Zapadnoe", *Drevnosti Bospora* 8, 12-24.

Ajbabin, A.I. et al. (eds.) 2003. Materialy po archeologii, istorii i etnografii Tavrii, *MAIET* X.

Alcock, S.E. & J.F. Cherry (eds.) 2004. *Side-by-Side Survey, Comparative Regional Studies in the Mediterranean World*. Oxford.

Alexeev, A.Ju. 2003. *Chronografija Evropejskoj Skifija*. St. Petersburg.

Alekseev, A.Ju., V.Ju. Murzin & R. Rolle 1991. Čertomlyk (Skifskij kurgan IV veka do n.e.). Kiev.

Andreeva, O. 2003. Medieval pottery from Insula XXXVI, *The Study of Ancient Territories Chersonesos & South Italy. 2003 Annual Report*, 53-56 (http://www.utexas.edu/research/ica/publications/ICA_publications/ICA_annual_reports/2003AnRpt_SCRN.pdf).

Applebaum, A. 2003. *Gulag: A History*. New York.

Arthur, P. 2000. Bezymyannaya Late Roman and Early Byzantine Pottery. Preliminary Notes, *The Study of Ancient Territories Chersonesos & South Italy. 2000 Annual Report*, 45-49 (http://www.utexas.edu/research/ica/publications/ICA_publications/ICA_annual_reports/2000AnRpt_SCRN.pdf).

Attema, P.A.J. 2008. Survey in de steppe, het Dzarylgaĉ Survey Project (DSP) (Noordwestelijke Krim, Oekraïne), *Tijdschrift voor Mediterrane Archeologie* 40, 1-12.

Attema, P.A.J., G.J. Burgers & P.M. van Leusen 2010. *Regional Pathways to Complexity, Landscape and Settlement Dynamics in ancient Italy*. Amsterdam.

Attema, P.A.J., P. Guldager Bilde & C. Williamson, 2010. Nederzettingsdynamiek in de Sibritide (Calabrië, Italië) en de noordwestelijke Krim (Oekraïne). Een vergelijking van de impact van de koloniale omgeving op nederzetting en maatschappij, Tijdschrift voor Mediterrane Archeologie 43, 20-31.

Attema, P. et al. 2009. Survey in de steppe: de tweede veldcampagne van het Džarylgaĉ project (Krim, Oekraïne). Landschaps – en siteclassificatie, *Paleoaktueel* 20, 62-68.

Attema, P. & T. de Haas 2008. Survey in de steppe, verslag van de eerste veldcampagne van het Džarylgaĉ project (Krim, Oekraïne), *Paleoaktueel* 19, 142-150.

Attema, P.A.J. & G.J.M. Oortmerssen 1997-1998. Ceramics of millennium BC from a survey at Lanuvium in the Alban Hills, Central Italy: Method, aims and first result of regional fabric classification, *Palaeohistoria* 39/40, 413-440.

Attema, P.A.J., E. van Joolen & P.M. van Leusen 2001. A Marginal Landscape: Fieldwork on the beach ridge complex near Fogliano (South Lazio), *Palaeohistoria* 41-42 (1999-2000), 149-162.

Avram, A. J. Hind & G. Tsetskhladze 2004. The Black Sea Area, in: M.H. Hansen & T.H. Nielsen (eds.), *An Inventory of Archaic and Classical Poleis*. Oxford, 924-973.

Bakirtzis, Ch. 1980. Didymoteichon: un centre de céramique post-byzantine, *Balkan Studies* 21, 147-153.

Barker, G.W. & D. Mattingly (eds.) 1999-2000. *The Archaeology of Mediterranean Landscapes* (5 volumes). Oxford.

Barker, G.W. & J. Lloyd (eds.) 1991. *Roman Landscapes, archaeological survey in the Mediterranean region*. London.

Belov, A.F. & S.N. Ljasko 1991. Ob odnom iz aspektov vozniknovenija kurgannogo pogrebal'nogo obrjada stepej vostočnoi Evropy, *Drevnosti stepnogo Pričernomor'ja i Kryma. Zaporozh'e*, 28-30.

Berkes, F. et al. 1998. Exploring the Basic Ecological Unit: Ecosystem-like Concepts in Traditional Societies, *Ecosystems* 1.5, 409-415.

Bintliff, J.L. 1992. Appearance and reality: understanding the buried landscape through new techniques in field survey, in: M. Bernardi (ed.), *Archeologia del paesaggio*. Firenze, 89-137.

Bintliff, J.L. & A. Snodgrass 1988. Off-site pottery distributions: a regional and interregional perspective, *Current Anthropology* 29, 506-513.

Bintliff, J.L., P. Howard & A. Snodgrass 2007. *Testing the Hinterland: The Work of the Boeotia Survey (1989-1991) in the Southern Approaches to the City of Thespiai* (McDonald Institute Monographs). Cambridge.

Blagovolin, N.S. & A.N. Ščeglov 2002. Archaeological, Paleogeographic, and Geomorphological Researches in the Lake Sasyk (Panskoye) Region, in: Panskoye I 2002, 285-302.

Bokovenko, N.A. 2004. Migrations of the early nomads of the Eurasian steppe in a context of climatic changes, in: Scott, Alekseev & Zaitseva (eds.) 2004, 21-34.

Bolikhovskaja, N.S. et al. 2002. Izmenenija landšaftno-klimatičeskich uslovij Tamanskogo poluostrova na protjaženii poslednich 6 tysjač let, *PIFK* 12, 257-271.

Borisova, V.V. 1958. Gončarnye masterskie Chersonesa (po materialam raskopok 1955-1957 gg.), *SovA* 4, 144-153.

Boltrik, Ju.V. 1990. Suchoputnye kommunikacii Skifii (po materialam novostroečnych issledovanij ot Priazov'ja do Dnepra), *SovA* 4, 30-44.

Boltrik, Ju.V. & E.E. Fialko 1987. K voprosu o lokalizacii gavani Kremny, in: *Skify Severnogo Pričernomor'ja*. Kiev, 40-48.

Bujskich, S.B. 2006. Seraja keramika kak etnopokazatel' grečeskogo naselenija Nižnego Pobuž'ja v VI-I vv. do. n. e., *BI* 9, 29-57.

Buračkov, P.O. 1875. O mestopoloženii drevnego goroda Karkinitesa i monetach, emu prinadležaščich, *ZOOID* 9, 1-133.

Burgers G.-J., P.A.J. Attema & P.M. van Leusen 1998 (published 2004). Walking the Murge: Interim report of the Ostuni field survey (Apulia, Southern Italy), *Studi di Antichità* 11, 257-282.

Buzoianu, L. & M. Bărbulescu 2003. Importuri sinopeene în așezarea greco-indigenă de la Albești (Importations de Synope en Albești – habitat gréco-indigene), in: *Studii istorice dobrogene.* Constanța, 57-67.

Bylkova, V.P. 2005. Novaja nachodka kompleksa amfor na Belozerskom poselenii, *Drevnosti* 3, 256-267.

Carter, J.C. 2006. Towards a Comparative Study of Chorai West and East: Metapontion and Chersonesos, in: Guldager Bilde & Stolba (eds.) 2006, 175-205.

Carter, J.C. et al. 2000. The Chora of Chersonesos in Crimea, *AJA* 104.4, 707-741.

Chang, C. 2000. The Material Culture and Settlement History of Agro-Pastoralism in the Koinotis of Dhidhima: An Ethnoarchaeological Perspective, in: S. Sutton (ed.), *Contingent Countryside: Settlement, Economy, and Land Use in the Southern Argolid since 1700.* Stanford, 125-140.

Chang, C. 2006. The grass is greener on the other side: a study of pastoral mobility on the Eurasian steppe of southeastern Kazakhstan, in: F. Sellet, R. Greaves & P.-L. Yu (eds.), *Archaeology and ethnoarchaeology of mobility*. Gainesville, 184-200.

Chang, C. & P.A. Tourtellotte 1993. Ethnoarchaeological Survey of Pastoral Transhumance Sites in the Grevena Region, Greece, *JFA* 20.3, 249-264.

Chișinău, E.S. 2005. Die spätbronzezeitlichen Aschehügel ("Zol'niki") ein Erklärungsmodell und einige historisch-wirtschaftliche Aspekte, *PrhZ* 2005 no. 1, 65-109.

Chrapunov, I.N. 1995. *Očerki etničeskoj istorii Kryma v rannem železnom veke. Tavry. Skify. Sarmaty*. Simferopol'.

Chtcheglov, A. 1992. *Polis et chora. Cité et territoire dans le Pont-Euxin* (Centre de Recherches d'Histoire ancienne 118: Espaces et paysages 1). Paris.

Clarke, R., 1935. The Flint-knapping Industry at Brandon, *Antiquity* IX, 38-56.

Conrad, S. 2006. Archaeological Survey on the Lower Danube: Results and Perspectives, in: Guldager Bilde & Stolba (eds.) 2006, 309-331.

Cook, R.M. & P. Dupont 1998. *East Greek Pottery*. London-New York.

Cordova, C.E. & P. Lehman 2003. Archaeopalynology of synanthropic vegetation in the *chora* of Chersonesos, Crimea, Ukraine, *JAS* 30, 1483-1501.

Cordova, C.E. & P.H. Lehman 2006. Mediterranean agriculture in southwestern Crimea: paleoenvironments and early adaptations, in: D.L. Peterson, L.M. Popova & A.T. Smith (eds.), *Beyond the Steppe and the Sown*. Proceedings of the 2002 University of Chicago Conference on Eurasian Archaeology (Colloquia Pontica 13). Leiden, 425-447.

Cordova, C.E., A. Rybak & P. Lehman 2001. Vegetation Patterns and Conservation Issues in Southern Crimea, *Post-Soviet Geography and Economics* 42.5, 362-385.

David, B. & J. Thomas 2008. *Handbook of Landscape Archaeology*. Walnut Creek.

Demidoff, M.A. de 1853. *Travels in Southern Russia, and the Crimea; through Hungary, Wallachia, & Modavia, during the Year 1837*. London (electronical form: http://images.google.dk/imgres?imgurl=http://depts.washington.edu/cartah/text_archive/demidoff/d_v2i179.jpg&imgrefurl=http://depts.washington.edu/cartah/text_archive/demidoff/v2c4.shtml&usg=__rDza4W9bBaXr0DZvwM68Vssil9w=&h=459&w=550&sz=104&hl=da&start=23&sig2=6be4eo-WxTAAx21kwFFTHw&um=1&tbnid=c6HdsNN9QBw8cM:&tbnh=111&tbnw=133&prev=/images%3Fq%3Dhorse%2

Bcrimea%2Bsteppe%26ndsp%3D18%26hl%3Dda%26rlz%3D1C1GGLS_daDK324DK325%26sa%3DN%26start%3D18%26um%3D1&ei=0sZMSvvwKtPJ-Qbs2umNBA).

Dmimtrienko, M.V. & A.N. Maslovski 2006. Kompleks 1310-ch godrov iz raskopok Azaka, in: *Historical and Archaeological Investigations in Azov and the Lower Don in 2005*. Azov, 231-257.

Doonan, O. 2004. *Sinop Landscapes: Exploring Connection in the Hinterland of a Black Sea Port*. Philadelphia.

Doonan, O. 2006. Exploring Community in the Hinterland of a Black Sea Port, in: Guldager Bilde & Stolba (eds.) 2006, 47-58.

Dzens-Litovskij, A.I. 1933. Peresypi i kosy krymskich soljanych ozer, *Izvestija gosudarstvennogo geografičeskogo obščestva* 65.6, 585-595.

Embleton, C. (ed.) 1983. The Northern Black Sea Lowlands and Crimea, in: *Geomorphology of Europe*. Basel, 387-392.

Feist, S. 1913. *Kultur und Ausbreitung der Indogermanen*. Berlin.

Foxhall, L. 2004. Small, Rural Farmstead Sites in Ancient Greece: A Material Cultural analysis, in: F. Kolb (ed.), *Chora und Polis* (*Schriften des Historischen Kollegs. Kolloquien 54*). München. 249-270.

Gavriljuk, N.A. 2010. Handmade pottery, in: N.A. Lejpunskaja et al., *The Lower City of Olbia (Sector NGS) in the 6th century BC to the 4th century AD* (BSS 13). Aarhus, 335-354.

Gavrilov, A.V. 2006. Theodosia and its Chora in Antiquity, in: Guldager Bilde & Stolba (eds.) 2006, 249-272.

Gerškovič, J.P. 1999. *Studien zur spätbronzezeitlichen Sabatinovka-Kultur am unteren Dnepr und an der Westküste des Azov'schen Meeres*. Rahden-Westfalen.

Gorlov, Ju.V. et al. 2002. K voprosu ob istoriko-geografičeskoj situacii na Tamanskom poluostrove v period grečeskoj civilizacii, *PIFK* 12, 248-256.

Grace, V.R. 1971. Samian Amphoras, *Hesperia* 40.1, 52-95.

Grammenos, D.V. & E.K. Petropoulos (eds.) 2003. *Ancient Greek Colonies in the Black Sea* (Publications of the Archaeological Institute of Northern Greece, 4). Thessaloniki.

Guldager Bilde, P. et al. 2007. Džarylgačskij issledovatel'nij proekt. Resultaty sesona 2007 g., *Bosporskij Fenomen*. St. Petersburg, 107-118.

Guldager Bilde, P. et al. 2008. Archaeology in the Black Sea Region, *Archaeological Reports for 2007-2008*, 115-173.

Guldager Bilde, J.M. Højte & V.F. Stolba (eds.) 2003. *The Cauldron of Ariantas, Studies Presented to A.N. Sceglov on the Occasion of His 70th Birthday* (Black Sea Studies 1). Aarhus.

Guldager Bilde, P. & V.F. Stolba 2006. Introduction, in: Guldager Bilde & Stolba 2006, 7-12.

Guldager Bilde, P. & V.F. Stolba (eds.) 2006. *Surveying the Greek Chora. The Black Sea Region in a Comparative Perspective* (BSS 4). Aarhus 2006.

Handberg, S. 2011. Amphora Fragments Re-used as Potter's Tools in the Rural Landscape of Panskoe, in: Lawall, M. & J. Lund (eds.). *Pottery in the Archaeological Record: Greece and Beyond. Acts on the International Colloquium held at the Danish and Canadian Institutes in Athens, June 20-22, 2008,* Aarhus, 61-66.

Handberg, S. V.F. Stolba & S.V. Ušakov 2009. Classical and Hellenistic Grey Wares from the Western Crimea, in: V. Bozuanu, P. Dupont & V. Lungu (eds.). Pontica 42 Supplementum I: *Pontic Grey Wares.* Muzeul De Istorie Nationala Si Archeologie Constantza, Romania, 167-185.

Hannestad, L. 2005. The Dating of the Monumental Building U6 at Panskoe I, in: Stolba & Hannestad (eds.) 2005, 179-192.

Hannestad, L. V.F. Stolba & H.B. Hastrup 2002. Black-Glazed, Red-Figure, and Grey Ware Pottery, in: Panskoye I 2002, 127-149.

Harrison, J.E. 1991. *Prolegomena to the study of Greek religion*. Princeton.

Hayes, J. W. 1992. *Excavation at Saraçhane in Istanbul 2. The Pottery*. Princeton.

Holderness, M. 1823. *New Russia: Journey from Riga to the Crimea, by Way of Kiev, with some Account of the Colonization and the Manners and Customs of the Colonists of New Russia to which are added Notes relating to the Crim Tatars*. London.

Jakobson, A.L. 1979. *Keramika i keramičeskoe proizvostvo srednevekovoj Tauriki*. Leningrad.

Jankowski, H. 2006. *A Historical-Etymological Dictionary of Pre-Russian Habitation Names of the Crimea*. Leiden-Boston.

Januševič, Z.V. & A.N. Ščeglov 2002. Palaeoethnobotanical Material, in: Panskoye I 2002, 327-331.

Kac, V.I. 1985. Tipologia i chronologičeskaja klassifikacija chersonesskich magistratskich klejm, *VDI* 1, 83-113.

Kac, V.I. et al. 2002. Tiles and ceramic containers, in: Panskoye I 2002, 101-126.

Karasev, A.N. 1966. Raskopki u sanatorija "Čaika", *AO 1965*, 115-117.

Kasparov, A.K. 2002. Osseous Remains, in: Panskoye I 2002, 332-333.

Kašaev, S.V. 2002. *Commonware*, in: Panskoye I 2002, 150-177.

Klamm, M., G. Viebrok & B. Meyer 1991. Pochvovedčeskie issledovanija skifskogo kurgana Čertomlyk, in: A.Ju. Alexeev, V.Ju. Murzin & R. Rolle, *Čertomlyk. Skifskij tsarskij kurgan 4 veka do n.e.* Kiev, 286-306.

Kovalevskaja, L.A. & I.Ju. Suchanova 2005. Lepnaja keramika iz raskopok sel'skoj usad'by 338 v okruge Chersonesa, *ChSbor* 14, 215-222.

Kolesnikov, A.B. 1984. Antičnye sel'skie usadby u Evpatorijskogo majaka, *VMU Istorija* 4, 76-86.

Kolotuchin, V.A. 1996. *Gornyj Krym v epochu pozdnej bronzy – načale rannego železa.* Kiev.

Kolotuchin, V.A. 1997. Raboty na poselenii epochi bronzy Baj-Kijat u s. Vladimirovka Černomorskogo rajona, *AIK* 1994, 151-153.

Kolotuchin, V.A. 2003. *Pozdnij bronzovij vek Kryma.* Kiev.

Kolotuchin, V.A. 2007. Raskopki poselenija epochi pozdnej bronzy Baj-Kijat u s. Vladimirovka Černomorskogo rajona, *AIK* 1995, 73.

Koltuchov, S.G., A.E. Kislyj & G.N. Toščev 1994. *Kurgannye drevnosti Kryma.* Zaporože.

Korosteleva, P.G. & G.N. Obcinnikova 2002. *Tarchankut. Kratki geografičeskij i istoričeskij ocerk.* Černomorskoe.

Kovnurko, G.M. 2002. Petrophysical investigation of the handmade pottery, in: Panskoye I 2002, 317-326.

Krasil'nikov, K. I. 1976. Gončarnaja masterskaja saltovo-majatskoj kul'tury, *SovA* 3, 267-278.

Krasil'nikov, K. I. 1979. Izdelija iz kosti saltovskoy kul'tury, *SovA* 2, 77-91.

Kravčenko, E.A. 2005. Poselenija kizil-kobinskoj kul'tury Geraklejskogo poluostrova VI–IV vv. do n.e., in: V.M. Zubar' et al. 2005, *Chersones Tavričeskij v tre'ej četverti VI-seredine I vv. do n.e. Očerki istorii i kul'tury.* Char'kov, 547–553.

Kucher, O. 2007. Ukrainian Agriculture and Agri-Environmental Concern, *Diskussionspapier DP-24-2007*, Institut fűr nachhaltige Wirtschaftsentwicklung, Universität fűr Bodenkultur Wien, Department fűr Wirtschafts- und Sozialwissenschaften.

Kurnakov, N.S. et al. 1936. *Soljanye ozera Kryma.* Moskva-Leningrad.

Kutajsov, S.V. 2005. Istorija i archeologija drevnego Kryma. Materialy k archeologičeskoj karte Severo-Zapadnogo Kryma XI-XV vekov, *Istorižeskoe nasledie Kryma žurnal* 29, 1-6.

Kutajsov, V.A. 1996. Cast Money and Coins of Kerkinitis of the Fifth Century B.C., *ACSS* 2.1, 39-60.

Kutajsov, V.A. 2004. *Kerkinitida v antičnuju epochu.* Kiev.

Kutajsov, V.A. 2006. The Chora of Kerkinitis, in: Guldager Bilde & Stolba (eds.) 2006, 141-149.

Kutajsov, V.A. & S.V. Pridnev 2007. Raskopki nekropolja Kalos Limena, *AIK* 1995, 81-86.

Lancov, S.G. 2004. P.N. Šul'c v izučenii antičnych pamjatnikov Severo-Zapdnogo Kryma. Problema opredelenija granic Chersonesskogo gosudarstva v IV-III vv. do n.e., in: *U Ponta Evksinskogo (pamjati Pavla Nikolaevića Šul'ca).* Simferopol', 58-68.

Lantsov, S.B. & V.B. Uzhentzev 2007. Distant chora of Taurian Chersonesos and the city of Kalos Limen, in: D.V. Grammenos & E. K. Petropoulos (eds.), *Ancient Greek Colonies in the Black Sea* 2 (BAR International Series 1675). Oxford, 677-728.

Latyševa, V.A. 1974. Raskopki v Severo-Zapadnom Krymu, *AO* 1973, 298-299.

Latyčeva, V.A. 1978. Raskopki antičnogo poselenija Masliny v Severo-Zapadnom Krymu, *KSIA* 156, 53-60.

Latyčeva, V.A. 1997. O naselenii chersonesskoj chory v Severo-Zapadnom Krymu (po materialam poselenija Masliny), *Drevnosti 1996*, 56-61.

Lawall, M. 1998. Ceramics and positivism revisited: Greek transport amphoras and history, in H. Parkins & C. Smith (eds.), *Trade, traders and the acient city.* London-New York, 75-101.

Lawall, M. et al. 2010. Transport amphoras, in: N.A. Lejpunskaja et al., *The Lower City of Olbia (Sector NGS) in the 6th century BC to the 4th century AD* (BSS 13). Aarhus, 355.405.

Lejpunskaja, N.A. 1981. *Keramičeskaja tara iz Ol'vii.* Kiev.

Lejpunskaja, N.A. 1986. Keramičeskoe proizvodctvo i keramika, in: S.D. Kryžickij et al. (eds.), *Archeologija Ukrainskoj SSR II. Skifo-Sarmatskaja i antičnaja archeologija.* Kiev, 455-487.

Leskov, A.M. 1970. Kirovskoe poselenie, in: A.M. Leskov (ed.), *Drevnosti Vostočnogo Kryma (Predskifskij period i skify).* Kiev, 7-57.

Leskov, A.M. & Merpert , N.Ja. 1967. *Pamjatniki epochi bronzy juga evropejskoj časti SSSR.* Kiev.

Lewarch, D.E. & M.J. O'Brien 1981. The expanding role of surface assemblages in archaeological research, *Advances in Archaeological Method and Theory* 4, 297-342.

Lomtadze, G.A. 1994. Nachodki iz neskol'kich kompleksov Patreja, *Bosporskij Sbornik* 4. 84-106.

Longinov, V.V. 1955. Buchta Jarylgač, *Sbornik rabot instituta okeanologii AN SSSR* 4, 152-166.

Lordkipanidzé, O. & P. Lévêque (eds.) 1996. Sur les traces des Argonautes. Actes du 6e symposium de Vani (Colchide) 22-29 septembre 1990 (Annales Littéraires de l'Université de Besançon 154) Paris.

Lyall, R. 1825. *Travels in Russia, the Krimea, the Caucasus and Georgia.* London (available on-line via Google Books).

MacDonald, A. 1995. All or Nothing at all? Criteria for the Analysis of Pottery from the Surface Survey, in: N. Christie (ed.). *Settlement and Economy in Italy 1500 BC-AD 1500. Papers of the Fifth Conference of Italian Archaeology* (Oxbow Monograph 41). Oxford, 25-29.

Makhortykh, N. 2004. *The Northern Black Sea Steppes in the Cimmerian epoch*, in: Scott, Alekseev & Zaitseva (eds.) 2004, 35-44,

Marčenko, K. & Y. Vinogradov 1989. The Scythian period in the northern Black Sea region (750-250 BC), *Antiquity* 63, 803-813.

Maslennikov, A.A. 1995. *Kamennye Jaciki Vostočnogo Kryma (K istorii sel'skogo naselenija Evropejskogo Bospora v VI-V BB. do n.e.)* (Bosporskij Sbornik 8). Moskva.

Monachov, S.Ju. 1989. *Amfory Chersonesa Tavričeskogo IV-II vv. do. n. e. Opyt sistemnogo analiza.* Saratov.

Monachov, S.Ju. 2003. *Grečeskie amfory v Pričernomor'e. Tipologija amfor veduščich centrov-eksporterov tovarov v keramičeskoj tare.* Moskva-Saratov.

Monachov, S.Ju. & E.Ja. Rogov 1990a. Amfory iz nekropolja Panskoe I, *AMA* 7, 128-153.

Monachov, S.Ju. & E.Ja. Rogov 1990b. Keramičeskie kompleksy nekropolja Panskoe I, *AMA* 8, 122-151.

Mozolevskij, B.N. & S.V. Polin 2005. *Kurgany Skifskogo Herrosa 4 veka do n.e. Babina, Vodjana and Soboleva Mogily.* Kiev.

Murzin, V.Ju. & R. Rolle 1989. Osnovnye itogi sovremennych issledovanij kurgana Čertomlyk. *Archeologia Kiev* 1, 91-101.

Müller, C. et al. 1998. Péninsule de Taman (Russie méridionale), *BCH* 122, 643-654.

Müller, C. et al. 1999. Péninsule de Taman (Russie méridionale), *BCH* 123, 589-598.

Müller, C. et al. 2000. Péninsule de Taman (Russie méridionale), *BCH* 124, 655-657.

Nalivkina, M.A. 1957. Raskopki Kerkinitidy i Kalos Limena (1948-1952), in: P.N. Šul'c (ed.), Istorija i archeologija drevnego Kryma. Kiev, 264-281.

Neustrueva, I.Yu. et al. 2007. Reconstruction of paleobiogeographical and paleoecological conditions of Lake Džarylgač (north-western Crimea) in the Late Holocene based on paleontological data, in: *Paleontology, Paleobiogeography and Paleoecology. Proceedings of the LIII session of the Paleontological Society, 2-6 April 2007.* St Petersburg, 89-90.

Nikolaenko, G. 2001. Excavation at Site 132. The Study of Ancient Territories. Chersonesos and Metaponto. 2001 annual report, Institute of Classical Archaeology, University of Texas, Austin, 57-60.

Nikolaenko, G. M 2003. The Chersonesean Chora in Light of the New Investigations in the Herakleian Peninsula (1991-2003), in: Guldager Bilde, Højte & Stolba (eds.) 2003, 131-144.

Nikolaenko, G.M. 2006. The *Chora* of Tauric Chersonesos and the Cadastre of the 4th-2ndCentury BC, in: Guldager Bilde & Stolba 2006 (eds.), 151-174.

Ol'chovskij, V.S. 1991. *Pogrebal'no-pominal'naja obrjadnost' naselenija stepnoi Skifii (VII-III veka do n.e.).* Moscow.

Osborne, R. 1985. *Demos: The discovery of Classical Attika.* Cambridge.

Pallas, P.S. 2001. *Bemerkungen auf einer Reise in die südlichen Statthalterschaften des Rußischen Reichs in den Jahren 1793 und 1794* vol. 2. Leipzig.

Panskoye I 2002. L. Hannestad, V.F. Stolba & A.N. Ščeglov (eds.) 2002. *Panskoye I.* Vol. 1. *The Monumental Building U6.* Aarhus.

Parzinger, H. 2006. *Die Frühen Völker Eurasiens vom Neolithikum bis zum Mittelalter.* München.

Pavlenkov, V.I. 1986. *Poloveckie kamennye izvajanija iz Evpatorijskogo kraevedčeskogo muzeja. Severnoe Pričernomor'e i Povolž'e vo vznaimootnošenijach Vostoka i Zapada v ×II-XVI vv.* (unpublished thesis). Azov.

Pettegrew, D.K. 2001. Chasing the Classical farmstead: assessing the formation and signature of rural settlement in Greek landscape archaeology, *Journal of Mediterranean Archaeology* 14.2, 189-209.

Podolák, J. 1987. Pastoral Constructions in Slovakia, *Ethnologia Slavica* 19, 45-92 (extensive resume in: http://www.iabsi.com/gen/public/arch3.htm#Podolak, Jan. Pastoral Constructions in Slovakia Ethnologia Slavica XIX 1987 45-92).

Popova, E.A. & S.A. Kovalenko 1996. On the cult of Herakles in the north-western Crimea. Recent finds from the Chaika settlement, in: G.R. Tsetskhladze (ed.), *New studies on the Black Sea littoral* (Colloquia Pontica 1). Oxford,. 63-71.

Poulter, A. 2000. The Roman to Byzantine Tradition in the Balkans: preliminary results on Nikopolis and its hinterland, *JRA* 13, 346-358.

Prummel, W. 1996. Van beesten, bikkels en weekdieren; zes huizen in Hellenistisch Halos (Griekenland), *Paleoaktueel* 7, 70-73.

Rabinowitz, A. 2004. Excavations in the South region of Chersonesos, *The Study of Ancient Territories Chersonesos & South Italy. 2004 Annual Report*, 5-12 (http://www.utexas.edu/research/ica/publications/ICA_publications/ICA_annual_reports/2004AnRpt_SCRN.pdf).

Rădulescu et al. 2000-2001. Albeşti (Departement de Constantza) – site fortifié greco-indigìne, *Talanta*, 189-204.

Randsborg, K. 1994. A Greek Episode, the Early Hellenistic Settlement on the Western Crimea, *Acta Archeologica* 65, 171-196.

Reinders, H.R. & W. Prummel 2003. *Housing in New Halos: A hellenistic town in Thessaly, Greece*. Lisse.

Rice, P.M. 2005. *Pottery analysis. A sourcebook*. Chicago-London.

Robinson, H.S. 1959. *Pottery of the Roman Period: Chronology* (The Athenian Agora V). Princeton.

Rogov, E.Ja. 1989. O chronologii mogil'nika Panskoe I, in: G.A. Košelenko (ed.), *Problemy issledovanija antičnych gorodov. Tezisy*. Moskva, 102.

Rogov, E.Ya. 2002. Ceramic Objects, in: Panskoye I 2002, 272-275.

Rogov, E.Ja. 2011. *Nekropol' Panskoe 1 v Severo-Zapadnom Krymu* (Materialy po archeologii, istorii i etnografii Tavrii Supplementum 10). Simferopol'.

Rogov, E.Ja. & I.V. Tunkina 1998. Raspisnaja i černolakovaja keramika iz necropolja Panskoye I, *Archeologičeskie Vesti* 5, 159-175.

Rotroff, S.I. 1997. *Hellenistic Pottery: The Athenian and imported Wheelmade Tableware* (The Athenian Agora XXIX). Princeton.

Rusanova, I.P. 1998. Kul'tovye zol'niki skifskogo vremeni, *MAIET* VI, 160-172.

Rusjaeva, A.S. 1986. Milet-Didimy-Borisfen-Ol'vija. Problemy kolonizacii Nižnego Pobuž'ja, *VDI* 2, 25-64.

Rybalova, V.D 1974. Poselenie Kamenka v Vostočnom Krimu, *ASGE* 16, 19-49.

Sapelko T.V. & D.A. Subetto 2007. Vegetation changing reconstruction during the Holocene in the North-Western Crimea (on pollen data), in: V.Yanko-Hombach (ed.), Extended Abstracts. IGCP-521-481 Joint Meeting and field trip. Gelendzik-Kerch, 135-136.

Saprykin, S.Ju. 2003. The Chersonesean Farmhouse on Land-plot no. 49 on the Lighthouse Point, in: Guldager Bilde, Højte & Stolba (eds.) 2003, 145-163.

Schmid, E. 1972. *Atlas of Animal Bones for Prehistorians, Archaeologists and Quaternary Geologists*. Amsterdam-London-New York.

Scholl, T. & V. Zinko 1999. *Archaeological Map of Nymphaion (Crimea)*. Warszawa.

Scott, E.M., A.Ju. Alekseev & G. Zaitseva (eds.) 2004. *Impact of the Environment on Human Migration in Eurasia* (Earth and Environmental Sciences). Dordrecht.

Sekerskaja, E.P. 2000. Stock-breeding and hunting in the economy of the Late Sabatinovka culture settlement at the Dremailovka village, *Stratum Plus* 2, 419-422.

Semenov, S.A. & G.F. Korobkova 1983. *Technologia drevnich industrij. Mesolithic-Aeneolithic*. Leningrad.

Senatorov, S.N. 2003. Kizil Kobinskaja keramika konca VI- pervoj poloviny IV vv. do. n.e. iz raskopok v Chersonese, *ChSbor* 12, 14-28.

Shipley, G. 2002. The survey area in the Hellenistic and Roman periods, in: W. Cavanagh et al., *Continuity and change in a Greek rural landscape. The Laconia Survey* I, *Methodology and Interpretation* (Annual of the British School at Athens Supplementary 26). London, 257-337.

Smekalova, T.N. & S.L. Smekalov 2006. Ancient Roads and Land Division in the Chorai of the European Bosporos and Chersonesos on the Evidence of Air Photographs, Mapping and Surface Surveys, in: Guldager & Stolba (eds.) 2006, 207-248.

Smekalova, T.N. 2007. Russkie topografičeskie karty XVIII-XX vekov kak istočnik po izyčeniju archeologičeskich pamjatnikov zapadnogo i vostočnogo Kryma, *BI* XVII, 78-111.

Smit, B.I. 2010. Research Strategies for the study of early prehistoric remains from the Pleistocene soils of the Northern Netherlands, *Groningen Archaeological Studies* 11.

Solov'ev, S.L. & L.G. Šepko 2006. *Archeologičeskie pamjatniki sel'skoj okrugi Akry poselenie Zavetnoe 5* I. St. Petersburg.

Stapert, D., 1976. Some natural surface modifications on flint in the Netherlands, *Palaeohistoria* 18, 7-41.

Stolba, V.F. 1991. Dom IV v. do n.e. na poselenii Panskoe I: raskopki 1987 g., *KSIA* 204, 78-84.

Stolba, V.F. 1996. Ein Münzfund aus Eupatoria von 1917 und der Beginn der Prägung von Kerkinitis, *Hellas und Rom*. Saarbrücken, 225-241.

Stolba, V.F. 2002. Handmade Pottery, in: Panskoye I 2002, 180-200.

Stolba, V.F. 2005a. Hellenistic Chersonesos: Towards Establishing a Local Chronology, in: Stolba & Hannestad (eds.) 2005, 153-177.

Stolba, V.F. 2005b. The Oath of Chersonesos and the Chersonesean Economy in the Early Hellenistic Period, in: Z.H. Archibald, J.K. Davies & V. Gabrielsen (eds.), *Making, Moving and Managing. The New World of Ancient Economies, 323-31 BC*. Oxford, 298-321.

Stolba, V.F. 2005c. Monetary Crises in the Early Hellenistic Poleis of Olbia, Chersonesos and Pantikapaion. A Reassessment, in: C. Alfaro, C. Marcos & P. Otero (eds.), *XIII Congreso Internacional de Numismática (Madrid, 2003). Actas-Proceedings-Actes*. Madrid, 395-403.

Stolba V.F. et al. 2006. Paleolimnologičeskie issledovanija soljanych ozer Zapadnogo Kryma, *AO 2005*, 560-562.

Stolba, V.F. & L. Hannestad (eds.) 2005. *Chronologies in the Black Sea Area in the Period c. 400-100 BC*. Aarhus.

Stolba, V.F., L. Hannestad & A.N. Ščeglov 1995. Raboty rossisko-datskoj Tarchankutskoj ekspedicii v Krymu, *AO 1994*, 335-337.

Subetto D.A., T.V. Sapelko & V.F. Stolba 2007. Issledovanija paleolimnologov v Krymu, *Priroda* 12, 2007, 61-62.

Subetto D.A. et al. 2007a. Stratigraphy and Chronology of the Lake Sediments from Crimea as a Basis for Reconstructing the Black Sea Level Fluctuations, in: A.P. Lisicyn (ed.), Geologija morej i okeanov: Materialy XVII Meždunarodnoj naučnoj konferencii (Školy) po morskoj geologii. Vol. 1. Moscow 2007, 282-284.

Subetto D.A. et al. 2007b. Environmental and Black Sea Level Changes in the Holocene as recorded in Lakes Saki and Dzharylgach, Crimean Peninsula, in: *Extended abstracts IGCP 521-481 Joint Meeting and Field Trip: IGCP 521 "Black Sea-Mediterranean Corridor during the last 30 ky: sea level change and human adaptation" (2005-2009); IGCP 481 "Dating Caspian Sea level change (2003-2007). Gelendzhik-Kerc', September 8-17, 2007*. Moscow, 157-159.

Subetto D.A. et al. 2008. Paleolimnological studies on the Crimean Peninsula, Northern Black Sea, first results // International Geological Congress, Oslo 2008. August 6-14th. CD-Disk. HPQ-02 HPQ-02 Black Sea?Mediterranean Corridor during last 30 ky: Sea level change and human adaptation (www.cprm.gov.br/33IGC/1202633.html).

Suttie, J.M. & S.G. Reynolds (eds.) 2003. *Transhumant grazing systems in temperate Asia*. Rome.

Šul'c, P.N. 1937. O rabotach Evpatorijskoj ekspedicii, *SovA* 3, 252-254.

Šul'c, P.N. 1941. Evpatorijskij rajon: 1933-1934 gg., *Archeologičeskie issledovanija v RSFSR 1934-1936 gg*, 265-277.

Ščeglov, A.N. 1967. Issledovanie sel'skoj okrugi Kalos Limena, *SovA* 3, 234-256.

Ščeglov, A.N. 1968. Issledovanija v Severo-Zapadnom Krymu, *AO 1967*, 213-2

Ščeglov, A.N. 1970. Rannesrednevekovye poselenija na tarchankutskom poluostrove Kryma, *SovA* 1, 254-261.

Ščeglov, A.N. 1976. *Polis i chora (Archeologiceskie pamjatniki Kryma)*. Simferopol'.

Ščeglov, A.N. 1978. *Severo-Zapadnyj Krym v anticnuju epochu*. Leningrad.

Ščeglov, A.N. 1980. Utilisation de la photographie aérienne dans l'étude du cadastre de Chersonésos Taurique (IVe-Ie s. av. n.è.), *DialHistAnc* 6, 59-72.

Ščeglov, A.N. 1985. 25 let rabot Tarchankutskoj ekspedicii: itogi i perspektivy, *KSIA* 182, 3-76.

Ščeglov, A.N. 1986. Process i charakter territorial'noj ekspansii Chersonesa v IV v. do n.e., *Antičnaja grazdanskaja obscina*. Leningrad, 152-176.

Ščeglov, A.N. 1987. Un établissement rural en Crimée: Panskoje I (fouilles de 1969-1985), *DialHistAnc* 13, 239-257.

Ščeglov, A.N. 1988. Bol'šoj Kastel' i svjatilišče II v. do n.e. v uročišče Džangul', *Problemy anticnoj kul'tury. Tezisy dokladov naucnoj konferencii* 3, 273-275.

Ščeglov, A.N. 1989. Izucenie sistem rasselenija v drevnosti s pomosc'ju materialov aerofotos'emki, in: *Novoe v metodike archeologičeskich rabot na novostrojkach*. Moskva, 37-39.

Ščeglov, A.N. 1997. "Staryj" Chersones Strabona. Ukreplenie na peresejke Majacnogo poluostrova: III. Sledy archaiceskogo poselenija na versine Kazac'ej buchty, *Bachcisarajskij istoriko-archeologičeskij sbornik* 1, 42-54.

Ščeglov, A.N. 2002a. Introduction, in: Panskoye I 2002, 11-25.

Ščeglov, A.N. 2002b. Cult sculpture, altars, sacred vessels and votives, in: Panskoye I 2002, 213-227.

Ščeglov, A.N., I.I. Saverkina & V.V. Glazunov 1971. Issledovanija bliz Jarylgacskoj buchty v Severo-Zapadnom Krymu, *AO 1970*, 251-251.

Ščeglov, A.N. & N.B. Selivanova 2002. Petrographic analysis of stamped amphorae, in: Panskoye I 2002, 303-316.

Ščeglov, A.N. et al. 1972. Tarchankutskaja ekspedicija, *AO 1971*, 342-343.

Ščeglov, A.N. et al. 1973. Issledovanija na chore Chersonesa, *AO 1972*, 353-354.

Ščeglov, A.N. et al. 1976. Raboty na Tarchankutskom poluostrove, *AO 1975*, 409-410.

Ščeglov, A.N. et al. 1978. Tarchankutskaja ekspedicija, *AO 1977*, 405-406.

Ščepinskij, A.A. 1973. Severo-Krymskaja ekspedicija, *AO 1972*, 354-355.

Ščepinskij, A.A. & E.N. Čerepanova 1969. *Severnoe Prisivaš'e v V-I tys. do n. e.* Simferopol'.

Šmakov, A.I. 1844. O piramidkach iz gliny, najdennych pri Ak-Mečetskoj buchte, *ZOOID* 1, 630-632.

Terenožkin, A.I. 1955. Skifskij kurgan v g. Melitopole, *KSIA* 5, 59-61.

Tsetskhladze, G.R. 2004. On the Earliest Greek Colonial Architecture in the Pontus, in: C.J. Tuplin (ed.), P*ontus and the Outside World. Studies in Black Sea History, Historiography, and Archaeology* (Colloquia Pontica, 9). Leiden-Boston, 225-278.

Toščev, G.N. 2007. *Krym v epochu bronzy*. Zaparož'e.

Užencev, V.B. 2004. Fortifikacija citadeli Kalos Limena ellinističeskogo perioda, *ChSbor* 13, 250–64.

Užencev, V.B. 2006. *Elliny i varvary Prekrasnoj Gavani*. Simferopol'.

Van Leusen, P.M. 2002. *Pattern to process: methodological investigations into the formation and interpretation of spatial patterns in archaeological landscapes*. Groningen.

Van Leusen, P.M. & P.A.J. Attema 2003. Regional Archaeological Patterns in the Sibaritide, preliminary results of the RPC Field survey campaign 2000, *Palaeohistoria* 42-43 (2001-2002), 397-416.

Van Leusen et al. 2005. Protohistoric to Roman settlement on the Lepine margis near Ninfa (south Lazio, Italy), *Palaeohistoria* 45/46 (2003-2004), 301-346.

Van Zon, H. 2000. *The Political Economy of Independent Ukraine*. London-New York.

Velichko, A.A., & V.P. Nechayev 1984. Late Pleistocene Permafrost in European USSR, in: *Late Quaternary Environments of the Soviet Union*. Minneapolis, 79-86.

Vinogradov, Ju.A. 1989. Olviopolity v Severo-Zapadnoj Tavrike, in: *Drevnee Pričernomor'e. Materialy I Vsesojuznych čtenij pamjati professora Petra Osipoviča Karyškovskogo*. Odessa, 17-19.

Vinogradov, Yu.A., A.M. Butjagin & M.Yu. Vachtina 2003. Myrmekion-Porthmeus. Two "small" towns of Ancient Bosporos, in: V.D. Grammenos & E.K. Petropoulos (eds.), Ancient Greek Colonies in the Black Sea. Thessaloniki, 803-840.

Vlasov, V.P. 1997. O Chronologičeskich ramkach Kizil-Kobinskoj (Tavrskoj) kul'tura, *Drevnosti 1997*, 15-22.

Vnukov, S.Ju. 2001. The North-Western Crimea: an Historical-Archaeological Essay, in: G.R. Tsetskhladze (ed.), *North Pontic Archaeology: Recent Discoveries and Studies* (Colloquia Pontica 6). Leiden, 149-175.

Vroom, J. 2003. *After Antiquity: Ceramics and Society in the Aegean from the 7th to the 20th Century A.C. A Case Study from Boeotia, Central Greece*. Leiden.

Vroom, J. 2005. *Byzantine to Modern Pottery in the Aegean: An Introduction and Field Guide.*

Vuia, R. 1964. *Tipuri de păstorit la români*. Bucuresti.

Vysotskaja, T.N. 1968. Novaja nachodka rel'efa s izobraženiem Gerakla, *AO 1967*, 211-212.

Vysotskaja, T.N. 1969. Drevnegrečeskoe poselenie v pos. Mežvodnoe, *AO 1968*, 295-296.

Wasowicz, A. 1972. Traces de lotissements anciens en Crimée, *Mélanges de l'Ecole française de Rome*. Antiquité, 84, 199-229.

Whitbread, I.K. 1995. *Greek transport amphorae. A petrological and archaeological study* (The British School at Athens Fitch Laboratory Occasional Paper 4). Exeter.

Whitelaw, T.M. 1998. Colonisation and competition in the polis of Koressos: the development of settlements in North-West Keos from the Archaic to the Late Roman periods, in: L.G. Mendoni & A.J. Mazarakis-Ainian (eds.), *Kea-Kythnos: history and archeology.* Paris, 227-257.

Wilkinson, T.J. 1982. The definition of ancient manures zones by means of extensive sherd-sampling techniques, *JFA* 9, 232-333.

Wilkinson, T.J. 1989. Extensive sherd scatters and land-use intensity: some recent results, *JFA* 16, 31-46.

Williams, B.G. 2001. *The Crimean Tatars. The Diaspora Experience and the Forging of a Nation.* Leiden.

Winther-Jacobsen, K. 2010a. *From Pots to People: A ceramic approach to the archaeological interpretation of ploughsoil assemblages in Late Roman Cyprus* (BABESCH Supplement 17). Leuven.

Winther-Jacobsen, K. 2010b. The classical farmstead revisited. Activity differentiation based on a ceramic use-typology, *Annual of the British School at Athens* 105, 269-290.

Zenkovič, V.P. 1960. *Morfologija i dinamika sovetskich beregov Černogo morja* II. Moskva.

Zin'ko, V.N. 2006. The Chora of Nymphaion (6th century BC-6th century AD), in: Guldager Bilde & Stolba 2006 (eds.), 289-308.

Zin'ko, V.N. 2006. The *Chora* of Nymphaion (6th century BC-6th century AD), in: Guldager Bilde & Stolba (eds.) 2006, 289-308.

Zinov'ev, M.K. 1957. Archeologičeskie pamjatniki. Razdol'nenskogo rajona Krymskoj oblasti, IADK, 323-326.

Zographou-Korre, K. 1995. *Ta kerameika tou Hellenikou chorou* (Pottery in the Greek Area). Athens.

Zographou-Korre, K. 2000. *Kerameika tou Tsanak Kale, 1670-1922* (*Pottery from Canakkale: 1670-1922*). Athens.

Zolotarev, M.I. 1986. Novye materially o vznaimootnošenijach Ol'vii I Zapadnogo Kryma v VI-V vv. do n.e., *VDI* 2, 88-93.

Zolotarev, M.I. 1993. *Chersonesskaja Archaika*. Sevastopol'.

Zolotarev, M.I. 1996. Sur la chronologie des Chersonésos à l'époque archaique, in: Lordkipanidzé & Lévêque (eds.) 1996, 311-317.

Zolotarev, M.I. 1998. Rannie etapy gradostroitel'sva v Chersonese, *ChSbor* 9, 26-35.

Zolotarev, M.I. 2005. A Hellenistic Ceramic Deposit from the North-eastern Sector of Chersonesos, in: V.F. Stolba & L. Hannestad (eds.), *Chronologies of the Black Sea Area in the Period C. 400-100 BC*. Aarhus, 193-216.

Zubar', V.M. & E.A. Kravčenko 2003. Interpretation of a Group of Archaeological Sites in the Vicinity of Tauric Chersonesos, in: Guldager Bilde, Højte & Stolba (eds.) 2003, 185-195.

## ABBREVIATIONS

| | |
|---|---|
| *ACSS* | *Ancient Civilizations from Scythia to Siberia.* |
| *AIK* | *Archeologičeskie issledovanija v Krimu. Simferopol'.* |
| *AJA* | *American Journal of Archaeology.* |
| *AMA* | *Antičnyj mir i archeologija,. Saratov.* |
| *AO* | *Archeologičeskie otkrytija. Moskva.* |
| *ASGE* | *Archeologičeskij sbornik Gosudarstvennogo Ermitaža. St. Petersburg.* |
| *BCH* | *BCH Bulletin de Correspondence Hellénique.* |
| *BI* | *Bosporskie issledovanija. Simferopol'-Kerč'.* |
| *BSS* | *Black Sea Studies. Aarhus.* |
| *ChSbor* | *Chersonesskij sbornik. Sevastopol'.* |
| *DialHistAnc* | *Dialogues d'histoire ancienne.* |
| *JAS* | *Journal of Archaeological Science.* |
| *JFA* | *Journal of Field Archaeology.* |
| *JRA* | *Journal of Roman Archaeology.* |
| *KSIA* | *Kratkie soobscenija Instituta Archeologii AN SSSR. St. Petersburg.* |
| *MAIET* | *Materialy po archeologii, istorii i etnografii Tavrii. Simferopol'.* |
| *PIFK* | *Problemy istorii, filologii i kul'tury. Moskva-Magnitogorsk.* |
| *PrhZ* | *Praehistorische Zeitschrift.* |
| *SovA* | *Sovetskaja Archeologija. Moskva.* |
| *VDI* | *Vestnik drevnej istorii. Moskva.* |
| *VMU* | *Vestnik Moskovskogo Universiteta. Moskva.* |
| *ZOOID* | *Zapiski Odesskogo obščestva istorii i drevnostej. Odessa.* |